Nonlinear Functional Analysis and its Applications

II/B: Nonlinear Monotone Operators

David Hilbert (1862–1943)

Eberhard Zeidler

Nonlinear Functional Analysis and its Applications

II/B: Nonlinear Monotone Operators

Translated by the Author and by Leo F. Boron†

With 74 Illustrations

Springer-Verlag
New York Berlin Heidelberg
London Paris Tokyo

Eberhard Zeidler
Sektion Mathematik
Karl-Marx-Platz
7010 Leipzig
German Democratic Republic

Leo F. Boron†
Department of Mathematics
University of Idaho
Moscow, ID 83843
U.S.A.

Mathematics Subject Classification (1980): 46xx

Library of Congress Cataloging in Publication Data
(Revised for vol. 2 pts. A-B)
Zeidler, Eberhard.
Vol. 2, pts. A-B: Translated by the author and
Leo L. Boron.
Vol. 3: Translated by Leo L. Boron.
Vol. 4: Translated by Juergen Quandt.
Includes bibliographies and indexes.
Contents: 1. Fixed point theorems — 2. pt A. Linear
monotone operators. Pt. B Nonlinear operators — [etc.]
— 4. Applications to mathematical physics.
1. Nonlinear functional analysis. I. Title.
QA321.5.Z4513 1985 515.7 83-20455

Printed on acid-free paper
Previous edition, *Vorlesungen über nichtlineare Funktionalanalysis*, Vols. I–III, published by B. G. Teubner Verlagsgesellschaft, 7010 Leipzig, Sternwartenstrasse 8, Deutsche Demokrati Republik.

Typeset by Asco Trade Typesetting Ltd., Hong Kong.
Printed and bound by Edwards Brothers, Inc., Ann Arbor, Michigan
Printed in the United States of America.

9 8 7 6 5 4 3 2 1

ISBN 0-387-97167-X Springer-Verlag New York Berlin Heidelberg
ISBN 3-540-97167-X Springer-Verlag New York Berlin Heidelberg

To the memory of my parents

Preface to Part II/B

The present book is part of a comprehensive exposition of the main principles of nonlinear functional analysis and its numerous applications to the natural sciences and mathematical economics. The presentation is self-contained and accessible to a broader audience of mathematicians, natural scientists, and engineers. The material is organized as follows:

Part I: Fixed-point theorems.
Part II: Monotone operators.
Part III: Variational methods and optimization.
Parts IV/V: Applications to mathematical physics.

Here, Part II is divided into two subvolumes:

Part II/A: Linear monotone operators.
Part II/B: Nonlinear monotone operators.

These two subvolumes form a *unit* equipped with a uniform pagination. The contents of Parts II/A and II/B and the basic strategies of our presentation have been discussed in detail in the Preface to Part II/A. The present volume contains the complete index material for Parts II/A and II/B.

For valuable hints I would like to thank Ina Letzel, Frank Benkert, Werner Berndt, Günther Berger, Hans-Peter Gittel, Matthias Günther, Jürgen Herrler, and Rainer Schumann. I would also like to thank Professor Stefan Hildebrandt for his generous hospitality at the SFB in Bonn during several visits in the last few years. In conclusion, I would like to thank Springer-Verlag for a harmonious collaboration.

Leipzig
Summer 1989

Eberhard Zeidler

Contents (Part II/B)

Contents (Part II/A)

GENERALIZATION TO NONLINEAR STATIONARY PROBLEMS

When the answers to a mathematical problem cannot be found, then the reason is frequently the fact that we have not recognized the general idea, from which the given problem appears only as a single link in a chain of related problems.
David Hilbert (1900)

In the preceding chapters we studied linear monotone problems. In the following chapters we want to generalize these results to nonlinear monotone problems.

(i) In Chapters 25 through 29 we investigate stationary problems, i.e., we study operator equations of the form

$$Au = b, \qquad u \in X,$$

together with applications to quasi-linear elliptic differential equations and to Hammerstein integral equations.

In this connection we consider the following cases:

Lipschitz continuous, strongly monotone operators on H-spaces (Chapter 25);
monotone coercive operators on B-spaces (Chapter 26);
pseudomonotone operators (Chapter 27);
maximal monotone operators (Chapter 32).

For example, strongly continuous perturbations of monotone continuous operators are pseudomonotone. In Part I we considered the two fundamental fixed-point principles of Banach and Schauder. Figure 25.1 shows that these two principles also play a crucial role in the theory of monotone operators.

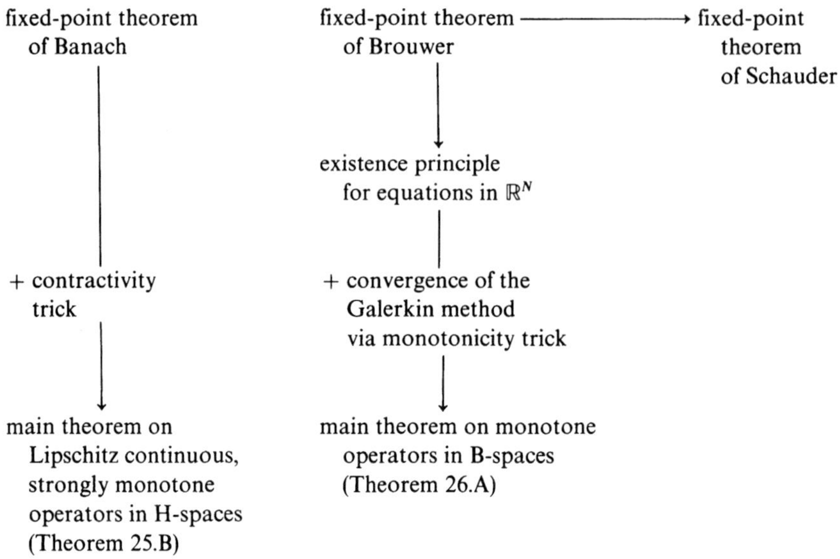

Figure 25.1

The special case of variational methods will be considered in Chapter 25. In this connection, the solutions of the convex minimum problem

$$f(u) = \min!, \qquad u \in X,$$

are solutions of the monotone operator equation

$$f'(u) = b, \qquad u \in X,$$

which generalizes the classical Euler equation.

(ii) In Chapters 30 through 33 we consider nonstationary problems, i.e., we study evolution equations of first and second order

$$u^{(n)} + Au = b, \qquad n = 1, 2,$$

together with applications to quasi-linear parabolic and hyperbolic equations.

In (i) and (ii), the operator A is monotone or, more generally, pseudomonotone. Here, the notion of *maximal monotone operators* plays a fundamental role.

(iii) In Chapters 34 through 36 we investigate a general theory of discretization methods together with applications to Galerkin methods (inner approximation schemes) and difference methods (external approximation schemes).

In this connection, the notion of *A-proper operators* is crucial.

Basic Ideas of the Theory of Monotone Operators

> Riemann has shown us that proofs are better achieved through ideas than through long calculations.
>
> David Hilbert (1897)

> One can understand a mathematical statement, if one
>
> (i) can use it,
> (ii) has completely understood the proof, or
> (iii) one can independently find the proof again at any time.
>
> Only when one reaches the third step can one speak of understanding in a real sense.
>
> Aleksander Ostrowski (1951)

> Male lovers sometimes lack experience before their fortieth year, and later on, there is often a lack of opportunity. In the same way, younger mathematicians often lack the knowledge, and older ones lack the ideas.
>
> Wilhelm Blaschke (1942)

The theory of nonlinear monotone operators is based on only a few tricks. For the convenience of the reader, we summarize these tricks here. This way, we want to make the proofs of the main results as transparent as possible. In the following Chapters 26 through 36, the items (1), (2), etc. below, will be quoted as (25.1), (25.2), etc.

The theory of nonlinear monotone operators generalizes the following elementary result. We consider the real equation

$$\text{(E)} \qquad F(u) = b, \qquad u \in \mathbb{R},$$

and assume that:

(i) The function $F\colon \mathbb{R} \to \mathbb{R}$ is monotone.
(ii) F is continuous.
(iii) $F(u) \to \pm\infty$ as $u \to \pm\infty$.

Then, for each $b \in \mathbb{R}$, equation (E) has a solution. If F is strictly monotone, then the solution is unique (cf. Fig. 25.2).

This classical existence theorem follows from the intermediate value theorem of Bolzano, whereas the uniqueness statement is obvious.

In particular, if $f\colon \mathbb{R} \to \mathbb{R}$ is convex and C^1, then $F = f'$ is monotone, and (E) with $b = 0$ is the Euler equation to the minimum problem

$$\text{(M)} \qquad f(u) = \min!, \qquad u \in \mathbb{R}.$$

This observation is the key to the application of variational methods in the theory of monotone operators. However, note that the general theory of monotone operators concerns operator equations which are not necessarily the Euler equations of extremal problems.

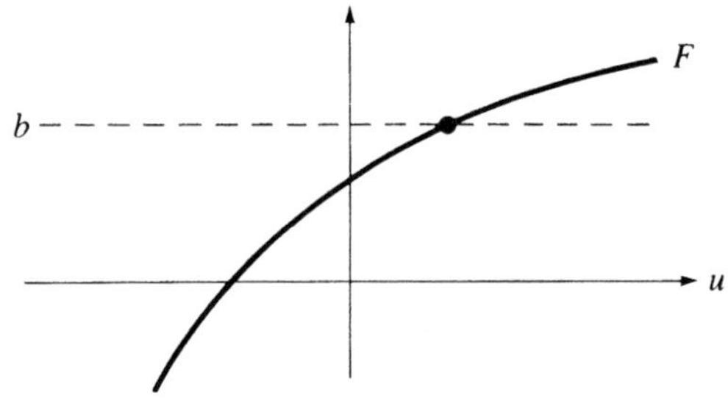

Figure 25.2

We now want to generalize the result above to monotone operator equations of the form

(E*) $$Au = b, \qquad u \in X.$$

Suppose that:

(i*) The operator $A: X \to X^*$ is monotone on the real reflexive B-space X, i.e.,

$$\langle Au - Av, u - v \rangle \geq 0 \qquad \text{for all} \quad u, v \in X.$$

(ii*) A is hemicontinuous, i.e., the map

$$t \mapsto \langle A(u + tv), w \rangle$$

is continuous on $[0, 1]$ for all u, v, $w \in X$.

(iii*) A is coercive, i.e.,

$$\lim_{\|u\| \to \infty} \frac{\langle Au, u \rangle}{\|u\|} = +\infty.$$

Then the main theorem on monotone operators (Theorem 26.A) tells us the following: For each $b \in X^*$, equation (E*) has a solution.

By Theorem 32.H, this fundamental result remains true if we replace (iii*) with the weaker condition that A is weakly coercive, i.e.,

$$\lim_{\|u\| \to \infty} \|Au\| = \infty.$$

If, in addition, A is strictly monotone, i.e.,

$$\langle Au - Av, u - v \rangle > 0 \qquad \text{for all} \quad u, v \in X \quad \text{with } u \neq v,$$

then the solution of (E*) is unique.

The result for equation (E) above is a special case of this theorem. In this connection, set $X = \mathbb{R}$ and

$$F(u) = Au.$$

Note that $X^* = \mathbb{R}$ and

$$\langle Au - Av, u - v \rangle = (F(u) - F(v))(u - v)$$

as well as

$$\langle Au, u\rangle/\|u\| = F(u)u/|u|.$$

Consequently, the assumptions (i), (ii), (iii) above are special cases of (i*), (ii*), (iii*), respectively.

In order to prove the main theorem on monotone operators for (E*) and similar results, we use the *Galerkin method*. Here the existence proof proceeds along the following lines.

Step 1: Solution of the Galerkin equations.
In this connection we obtain equations in $\mathbb{R}^n$. Let u_n be a solution of the nth Galerkin equation.

Step 2: *A priori* estimates.
We show that (u_n) is bounded.

Step 3: Weak convergence.
We show that there is a subsequence $(u_{n'})$ with

$$u_{n'} \rightharpoonup u \qquad \text{as} \quad n \to \infty.$$

Step 4: We show that u is a solution of the original equation $Au = b$, $u \in X$.

We want to discuss this.

(1) *Existence principles for the Galerkin equations.* In Step 1 we can use all the existence results for operator equations considered in Part I. In particular, we will use

(a) the existence principle for equations in $\mathbb{R}^n$ (Section 2.4); and
(b) the antipodal theorem of Borsuk (Section 16.3).

(2) *Coerciveness and a priori estimates.* Step 2 above is based on the following result. Let $A\colon X \to X^*$ be a coercive operator on the real B-space X, i.e.,

$$\text{(C)} \qquad \langle Au, u\rangle/\|u\| \to +\infty \qquad \text{as} \quad \|u\| \to \infty.$$

Then, for fixed $b \in X^*$, the set of the solutions of equation $Au = b$, $u \in X$, is bounded.

Indeed, it follows from (C) that there is a sufficiently large $R > 0$ such that

$$\langle Au, u\rangle \geq (1 + 2\|b\|)\|u\|,$$

for all $u \in X$ with $\|u\| > R$. This implies

$$\begin{aligned}\langle Au, u\rangle - \langle b, u\rangle &\geq (1 + \|b\|)\|u\|\\ &\geq (1 + \|b\|)R,\end{aligned}$$

for all $u \in X$ with $\|u\| > R$. Thus, $Au = b$ implies $\|u\| \leq R$.

(2a) *Noncoercive problems and Fredholm alternatives.* If the operator A: $X \to X^*$ is not coercive, then we need Fredholm alternatives for the equation

$Au = b$, $u \in X$, i.e., this equation has only solutions in the case where the right-hand side b satisfies appropriate conditions. Such nonlinear Fredholm alternatives will be considered in Chapter 29.

In Step 3 above we will use the following result.

(3) *Main theorem on weak convergence.* Each bounded sequence (u_n) in a *reflexive* B-space X has a weakly convergent subsequence (Theorem 21.D).

Step 4 above is based on the following trick which makes essential use of the notion of monotone operators.

(4) *The decisive monotonicity trick.* Let $A: X \to X^*$ be a monotone and hemicontinuous operator on the real reflexive B-space X. Then:

(a) A is *maximal* monotone, i.e.,

$$\langle b - Av, u - v \rangle \geq 0,$$

for all $v \in X$ and fixed $u \in X$, $b \in X^*$, implies $Au = b$.

(b) A satisfies condition (M), i.e., it follows from

$$u_n \rightharpoonup u \quad \text{in } X \qquad \text{as} \quad n \to \infty,$$

$$Au_n \rightharpoonup b \quad \text{in } X^* \qquad \text{as} \quad n \to \infty,$$

and $\langle Au_n, u_n \rangle \to \langle b, u \rangle$ as $n \to \infty$ or, more generally,

$$\overline{\lim_{n \to \infty}} \langle Au_n, u_n \rangle \leq \langle b, u \rangle$$

that $Au = b$.

(c) It follows from either

$$u_n \rightharpoonup u \quad \text{in } X \qquad \text{and} \qquad Au_n \to b \quad \text{in } X^* \qquad \text{as} \quad n \to \infty,$$

or

$$u_n \to u \quad \text{in } X \qquad \text{and} \qquad Au_n \rightharpoonup b \quad \text{in } X^* \qquad \text{as} \quad n \to \infty,$$

that $Au = b$.

PROOF. Ad(a). Let $v = u - tw$, where $t > 0$. Then

$$\langle b - Av, u - v \rangle \geq 0$$

implies $\langle b - A(u - tw), w \rangle \geq 0$. The operator A is hemicontinuous, and hence, letting $t \to 0$, we obtain that

$$\langle b - Au, w \rangle \geq 0 \qquad \text{for all} \quad w \in X.$$

Hence $\langle b - Au, w \rangle = 0$ for all $w \in X$, i.e., $b - Au = 0$.

Ad(b), (c). By the monotonicity of A,

$$\langle Au_n, u_n \rangle - \langle Av, u_n \rangle - \langle Au_n - Av, v \rangle = \langle Au_n - Av, u_n - v \rangle \geq 0$$

for all $v \in X$ and all n. Letting $n \to \infty$, this implies

$$\langle b, u \rangle - \langle Av, u \rangle - \langle b - Av, v \rangle \geq 0,$$

and hence

$$\langle b - Av, u - v \rangle \geq 0 \qquad \text{for all} \quad v \in X.$$

Since A is maximal monotone by (a), we get $Au = b$.

(5) *Monotonicity and uniqueness.*

(5a) *Operator equations.* Let $A: X \to X^*$ be strictly monotone on the real B-space X. Then the equation

$$Au = b, \qquad u \in X,$$

has at most one solution.

Indeed, if $Au = Av$ and $u \neq v$, then $\langle Au - Av, u - v \rangle > 0$ by the strict monotonicity of A. This is a contradiction.

(5b) *Evolution equations.* We consider the initial value problem

$$\text{(P)} \qquad \begin{aligned} u'(t) + Au(t) &= 0 \qquad \text{for} \quad 0 < t < T, \\ u(0) &= u_0. \end{aligned}$$

Let "$V \subseteq H \subseteq V^*$" be an evolution triple. Suppose that $A: V \to V^*$ is monotone. Then, (P) has at most one solution $u \in W_p^1(0, T; V, H)$, where p is a fixed number with $1 < p < \infty$.

To prove this, assume that $u, v \in W_p^1(0, T; V, H)$ are solutions of (P). The formula of integration by parts (Proposition 23.23) yields

$$\begin{aligned} &\tfrac{1}{2}\|u(t) - v(t)\|_H^2 - \tfrac{1}{2}\|u(0) - v(0)\|_H^2 \\ &\qquad = \int_0^t \langle u'(s) - v'(s), u(s) - v(s) \rangle \, ds \\ &\qquad = -\int_0^t \langle Au(s) - Av(s), u(s) - v(s) \rangle \, ds \leq 0 \end{aligned}$$

for almost all $t \in]0, T[$, since A is monotone. Noting $u(0) = v(0) = u_0$, we get

$$\|u(t) - v(t)\|_H \leq 0 \qquad \text{for almost all} \quad t \in]0, T[.$$

Hence $u(t) = v(t)$ for almost all $t \in]0, T[$, i.e., $u = v$ in $W_p^1(0, T; V, H)$.

(6) *Monotonicity and contractivity.* Let the operator $A: X \to X^*$ be Lipschitz continuous and strongly monotone on the real H-space X, i.e.,

$$\langle Au - Av, u - v \rangle \geq c\|u - v\|^2$$

for all $u, v \in X$ and fixed $c > 0$. Then the equation

$$\text{(6a)} \qquad Au = b, \qquad u \in X,$$

is equivalent to the fixed-point problem

(6b) $$u = L_t u, \qquad u \in X,$$

where $L_t u = u - tJ^{-1}(Au - b)$ for fixed $t > 0$. Here, $J: X \to X^*$ is the duality map of X. In Section 25.4, we will show by a simple computation that there is a $t > 0$ such that $L_t: X \to X$ is *k-contractive*. Thus, by the Banach fixed-point theorem, we obtain a unique solution for (6b). Consequently, for each $b \in X^*$, equation (6a) has a unique solution u, and the iterative method

$$u_{n+1} = L_t u_n, \qquad n = 0, 1, 2, \ldots$$

converges to u as $n \to \infty$, for each given $u_0 \in X$.

(7) *Monotonicity and convexity*. Let $f: X \to \mathbb{R}$ be a G-differentiable functional on the real B-space X. In Section 25.5 we will show that:

$$f' \; X \to X^* \text{ is monotone iff } f \text{ is convex.}$$

Moreover, we will show that if f is convex, then the minimum problem

$$f(u) = \min!, \qquad u \in X,$$

is equivalent to the operator equation

$$f'(u) = 0, \qquad u \in X.$$

However, there are monotone operators $A: X \to X^*$ which are *not* potential operators, i.e., A does not allow a representation of the form $A = f'$. In Section 41.3 we will show that an F-differentiable operator $A: X \to X^*$ is a potential operator iff the following *symmetry* condition holds:

$$\langle A'(u)v, w\rangle = \langle A'(u)w, v\rangle \qquad \text{for all} \quad u, v, w \in X.$$

If A is a potential operator, then the potential is given by

$$f(u) = \int_0^1 \langle A(tu), u\rangle \, dt \qquad \text{for all} \quad u \in X.$$

In particular, let $A: X \to X^*$ be linear and continuous on the real B-space X. Then, A is a potential operator iff A is symmetric, i.e., $\langle Av, w\rangle = \langle Aw, v\rangle$ for all $v, w \in X$. The potential is given by

$$f(u) = \tfrac{1}{2}\langle Au, u\rangle \qquad \text{for all} \quad u \in X.$$

(8) *Weak sequential lower semicontinuity and a general minimum principle.* Let M be a nonempty bounded closed convex set in the real reflexive B-space X. Let $f: M \to \mathbb{R}$ be weakly sequentially lower semicontinuous, i.e.,

$$u_n \rightharpoonup u \quad \text{in } M \qquad \text{as} \quad n \to \infty$$

implies

$$f(u) \leq \varliminf_{n \to \infty} f(u_n).$$

Then the minimum problem

(8a) $$f(u) = \min!, \qquad u \in M,$$

has a solution.

To prove this set

$$\alpha = \inf_{u \in M} f(u).$$

By construction of α, there is a sequence (u_n) in M such that

$$\lim_{n\to\infty} f(u_n) = \alpha.$$

Since M is bounded, there exists a subsequence, again denoted by (u_n), such that

$$u_n \rightharpoonup u \qquad \text{as} \quad n \to \infty.$$

Since M is convex and closed, we get $u \in M$ by Proposition 21.23(f). Hence

$$f(u) \leq \varliminf_{n\to\infty} f(u_n) = \alpha,$$

i.e., $f(u) = \alpha$. Consequently, u is a solution of (8a).

In Section 25.5 we will show that each continuous convex functional f: $X \to \mathbb{R}$ is weakly sequentially lower semicontinuous.

(9) *The main theorem on convex minimum problems.* Let $f\colon X \to \mathbb{R}$ be continuous and convex on the real reflexive B-space X. In addition, suppose that f is weakly coercive, i.e.,

(9a) $$\lim_{\|u\|\to\infty} f(u) = +\infty.$$

Then the minimum problem

(9b) $$f(u) = \min!, \qquad u \in X,$$

has a solution.

Indeed, it follows from (9a) that there is an $R > 0$ such that $f(u) > f(0)$ for all $u \in X - M$, where $M = \{u \in X\colon \|u\| \leq R\}$. Consequently, it is sufficient to solve the minimum problem (8a). Now, the assertion follows immediately from (8).

(10) *The idea of regularization.* In order to solve the equation

(10a) $$Au = b, \qquad u \in X,$$

we can consider the modified problem

(10b) $$Au_\varepsilon + \varepsilon B u_\varepsilon = b, \qquad u_\varepsilon \in X,$$

where $\varepsilon > 0$ is a small parameter. Sometimes it is easier to solve (10b) than (10a). For example, it is possible that the operator $A + \varepsilon B$ is coercive for $\varepsilon > 0$ and not coercive for $\varepsilon = 0$. Now the idea of regularization is the following:

(i) We first solve the "regularized" problem (10b) for all small $\varepsilon > 0$.

(ii) We show that the sequence (u_ε) of solutions of (10b) is bounded in the reflexive B-space X for $0 < \varepsilon \leq \varepsilon_0$.
(iii) We choose a weakly convergent subsequence $u_\varepsilon \rightharpoonup u$ as $\varepsilon \to 0$ and show that u is a solution of the original equation (10a).

In Chapter 32 we shall present the theory of *maximal monotone* operators, which forms the *hard core* of the theory of monotone operators. This theory will be based on both the Galerkin method and the method of regularization via duality map.

The following method of the Yosida approximation represents an important regularization method for solving evolution equations.

(11) *The trick of the Yosida approximation.* We want to solve the initial value problem:

$$u'(t) + Au(t) = 0, \qquad 0 < t < \infty, \tag{11a}$$
$$u(0) = u_0,$$

where $u(t)$ lies in the real H-space X. Instead of (11a) we consider the regularized problem:

$$u_\varepsilon'(t) + A(I + \varepsilon A)^{-1}u_\varepsilon(t) = 0, \qquad 0 < t < \infty, \tag{11b}$$
$$u_\varepsilon(0) = u_0.$$

The operator $A(I + \varepsilon A)^{-1}$ is called the Yosida approximation of A. We assume that:

(i) the operator $A\colon D(A) \subseteq X \to X$ is monotone; and
(ii) A is maximal accretive, i.e., for all $\varepsilon > 0$, the operator $(I + \varepsilon A)\colon D(A) \to X$ is bijective and the so-called resolvent

$$(I + \varepsilon A)^{-1}\colon X \to X$$

is nonexpansive.

In Chapter 31 we will show that (ii) implies (i) and that the regularized initial value problem (11b) can be easily solved by using the Picard–Lindelöf theorem (Theorem 3.A), since the Yosida approximation is Lipschitz continuous. In order to solve the original problem (11a), we have to investigate the limiting process

$$u_\varepsilon(t) \to u(t) \qquad \text{as} \quad \varepsilon \to 0,$$

and we have to show that $u(\cdot)$ is a solution of (11a). To this end, we will use the convergence trick of maximal monotonicity (13) below combined with the trick of maximal accretivity (12).

(12) *The trick of maximal accretivity.* Suppose that $A\colon D(A) \subseteq X \to X$ is monotone and maximal accretive on the real H-space X. Then A is *maximal*

monotone, i.e., if

$$(b - Av|u - v) \geq 0 \qquad \text{for all} \quad v \in D(A),$$

and fixed $b, u \in X$, then $u \in D(A)$ and $Au = b$.

To show this, we choose $v = v_t$ where

$$v_t = (I + A)^{-1}(b + u + tz)$$

for fixed $t > 0$. This implies $v_t \in D(A)$. Adding

$$(b - Av_t|u - v_t) \geq 0 \qquad \text{and} \qquad (u - v_t|u - v_t) \geq 0,$$

we get

$$(b + u - (I + A)v_t|u - v_t) \geq 0.$$

This is identical to

$$(z|u - v_t) \leq 0.$$

Letting $t \to 0$, we obtain that

$$(z|u - v_0) = (z|u - (I + A)^{-1}(b + u)) \leq 0 \qquad \text{for all} \quad z \in X.$$

This implies $u - (I + A)^{-1}(b + u) = 0$, and hence

$$u + Au = b + u,$$

i.e., $Au = b$.

(13) *The convergence trick of maximal monotonicity*. Suppose that $A: D(A) \subseteq X \to X$ is maximal monotone on the real H-space X. Then

$$u_n \to u \qquad \text{and} \qquad Au_n \rightharpoonup b \qquad \text{as} \quad n \to \infty$$

implies $Au = b$.

The proof follows easily from

$$(Au_n - Av|u_n - v) \geq 0 \qquad \text{for all} \quad v \in D(A) \quad \text{and all } n.$$

Letting $n \to \infty$, we get

$$(b - Av|u - v) \geq 0 \qquad \text{for all} \quad v \in D(A),$$

and hence $Au = b$, since A is maximal monotone.

This result is closely related to the monotonicity trick (4c) above.

In Chapter 31 we will use (11) through (13) above in order to construct nonexpansive semigroups on H-spaces. In Chapter 57, we will show that maximal accretive operators also play an important role for solving multi-valued evolution equations on B-spaces, and for constructing nonexpansive semigroups on B-spaces. In real H-spaces one has the following:

A is maximal accretive iff A is maximal monotone.

(14) *The convergence of approximation methods*. Let $A: X \to X^*$ be an operator on the real reflexive B-space X. We consider the equation

$$\text{(E)} \qquad Au = b, \qquad u \in X.$$

Suppose that (u_n) is a sequence of approximate solutions in X, and suppose that (u_n) is bounded. By (3), there exists a weakly convergent *subsequence* of (u_n). We want to answer the following two questions:

(i) When is the *total* sequence (u_n) weakly convergent to a solution of (E)?
(ii) When is the *total* sequence (u_n) strongly convergent to a solution of (E)?

(14a) *Uniqueness and weak convergence of the total sequence.* Suppose that the limit of an arbitrary weakly convergent subsequence of (u_n) is a solution of (E), i.e.,

$$u_{n'} \rightharpoonup u \qquad \text{as} \quad n \to \infty$$

implies $Au = b$. Moreover, suppose that equation (E) has at most one solution. Then, equation (E) has a unique solution u and

$$u_n \rightharpoonup u \qquad \text{as} \quad n \to \infty.$$

This follows immediately from the convergence principle in Section 10.5. The same principle yields the following result.

(14b) *Uniqueness and strong convergence of the total sequence.* Suppose that each subsequence of (u_n) has, in turn, a convergent subsequence $u_{n'} \to u$ as $n \to \infty$ and $Au = b$. Moreover, suppose that equation (E) has at most one solution. Then equation (E) has a unique solution u, and

$$u_n \to u \qquad \text{as} \quad n \to \infty.$$

(14c) *Convergence trick for uniformly monotone operators.* Let $A: X \to X^*$ be a uniformly monotone operator on the real reflexive B-space X, i.e., more precisely, there is a $c > 0$ and a $p > 1$ such that

$$c\|u - v\|^p \leq \langle Au - Av, u - v \rangle \qquad \text{for all} \quad u, v \in X.$$

Then the weak convergence

$$u_n \rightharpoonup u \quad \text{in } X \qquad \text{as} \quad n \to \infty,$$

together with $Au_n \rightharpoonup Au$ in X^* and $\langle Au_n, u_n \rangle \to \langle Au, u \rangle$ as $n \to \infty$, implies the strong convergence

$$u_n \to u \qquad \text{as} \quad n \to \infty.$$

This follows from

$$\begin{aligned} c\|u - u_n\|^p &\leq \langle Au - Au_n, u - u_n \rangle \\ &= \langle Au, u \rangle - \langle Au, u_n \rangle - \langle Au_n, u \rangle \\ &\quad + \langle Au_n, u_n \rangle \to 0 \qquad \text{as} \quad n \to \infty. \end{aligned}$$

For the convenience of the reader we have used here a special case of the general definition of uniformly monotone operators, which will be given in (15a) below. It is easily shown that the result above also remains valid for uniformly monotone operators in the general sense.

(14d) *Density trick for weak convergence.* Let (u_n) be a bounded sequence in the B-space X. Suppose that there are a dense set D in X^* and an element u in X such that

$$\langle f, u_n \rangle \to \langle f, u \rangle \quad \text{as} \quad n \to \infty \quad \text{for all} \quad f \in D.$$

Then $u_n \rightharpoonup u$ as $n \to \infty$ (cf. Proposition 21.23).

(14e) *The Toeplitz trick and the convergence of projection–iteration methods.* Let (a_m) be a sequence of real numbers with $a_m \to 0$ as $m \to \infty$, and let $0 \le k < 1$. Then, the classical theorem of Toeplitz tells us that

$$\sum_{m=1}^{n} k^{n-m} a_m \to 0 \quad \text{as} \quad n \to \infty.$$

In Section 25.1, we will use this result in order to prove the convergence of the general iterative method

$$u_{n+1} = A_n u_n, \qquad n = 0, 1, \ldots.$$

In particular, if we set $A_n = P_n A$, where P_n is a projection operator, then we obtain the convergence of the projection–iteration method corresponding to the operator equation $u = Au$.

(15) *Stable operators.* Let X and Y be real B-spaces. The operator $A: X \to Y$ is called stable iff

$$a(\|u - v\|) \le \|Au - Av\| \quad \text{for all} \quad u, v \in X.$$

Here we assume that the function $a: \mathbb{R}_+ \to \mathbb{R}_+$ is strictly monotone increasing and continuous with $a(0) = 0$ and

$$\lim_{t \to +\infty} a(t) = +\infty.$$

For example, we may choose

$$a(t) = ct^{p-1} \quad \text{for all} \quad t \ge 0,$$

and fixed $c > 0$, $p > 1$.

(15a) *Uniformly monotone operators are stable.* If $A: X \to X^*$ is uniformly monotone, i.e.,

$$a(\|u - v\|)\|u - v\| \le \langle Au - Av, u - v \rangle \quad \text{for all} \quad u, v \in X,$$

then A is stable.

Indeed, for all $u, v \in X$, we get

$$a(\|u - v\|)\|u - v\| \le \|Au - Av\| \, \|u - v\|.$$

(15b) *Stable operators and uniqueness.* Let $A: X \to Y$ be stable. Then, for each $b \in Y$, the equation

$$Au = b, \qquad u \in X,$$

has at most one solution.

To prove this, suppose that $Au = Av$. This implies $a(\|u - v\|) = 0$, and hence $u = v$.

(15c) *Stable operators and the continuous dependence of the solutions on the right-hand side.* Let $A: X \to Y$ be stable. Suppose that

$$Au_n = b_n \quad \text{for all } n \quad \text{and} \quad Au = b.$$

Then $b_n \to b$ in Y as $n \to \infty$ implies $u_n \to u$ in X as $n \to \infty$.

This follows from $a(\|u_n - u\|) \le \|Au_n - Au\| \le \|b_n - b\|$.

(15d) *Stable operators and a priori estimates.* Let $A: X \to Y$ be stable. Then, for each $b \in Y$, there is an $R > 0$ such that all the solutions of the equation

$$Au = b$$

satisfy the estimate $\|u\| \le R$.

This follows from $a(\|u\|) \le \|Au - A(0)\| \le \|b\| + \|A(0)\|$ and $a(t) \to +\infty$ as $t \to +\infty$.

(16) *The fundamental properness trick and approximation methods.* Suppose that:

(H1) The operator $A: X \to Y$ is stable.

(H2) If (u_n) is a bounded sequence in X with

$$Au_n \to b \quad \text{as} \quad n \to \infty,$$

then there exists a subsequence

$$u_{n'} \to u \quad \text{as} \quad n \to \infty$$

such that $Au = b$.

Then it follows from

$$Au_n = b_n \quad \text{for all } n$$

and $b_n \to b$ as $n \to \infty$ that the equation

$$Au = b, \qquad u \in X,$$

has a unique solution u, and $u_n \to u$ as $n \to \infty$.

Indeed, the sequence (b_n) is bounded. From

$$a(\|u_n\|) \le \|Au_n - A(0)\| \le \|b_n\| + \|A(0)\| \qquad \text{for all } n$$

we obtain that (u_n) is bounded. By (H2), there exists a subsequence $u_{n'} \to u$ as $n \to \infty$ with $Au = b$. Since A is stable, the equation $Au = b$ has at most one solution. Now the assertion follows from (14b) above.

(16a) *The special case of proper maps.* The map $A: X \to Y$ is called proper iff the compactness of the set M implies the compactness of $A^{-1}(M)$. Obviously, assumption (H2) above is satisfied if $A: X \to Y$ is proper and continuous. This is why (16) is called the properness trick.

(16b) *The fundamental notion of A-proper maps.* In Chapter 20 we have shown that there is a close connection between stability, consistency, existence, and the convergence of approximate methods. Furthermore, in Chapter 21, we have shown that the operator equation $Au = b$, $u \in X$, is approximation solvable for important classes of linear operators A. In Chapters 34 and 35 we will generalize these results to nonlinear operators in the framework of a general theory of discretization methods. In this connection, the notion of A-proper maps will play a fundamental role. The definition of A-proper maps, with respect to inner and external approximation schemes in Chapter 34 and 35, respectively, will be based on a generalization of (H2) above, and the main theorems in Chapters 34 and 35 will be modifications of the properness trick (16), combined with the antipodal theorem of Borsuk.

(17) *General strategies for solving nonlinear partial differential equations.* To prove the existence of solutions of nonlinear partial differential equations, one can use one of the following six basic methods:

(i) compactness and topological methods (Chapter 6);
(ii) variational methods (Chapter 25);
(iii) monotonicity, locally coercive (or semibounded) operators, and the Galerkin method (e.g., Chapters 26, 27, 30, and 83);
(iv) refined iterative methods via interpolation inequalities (Chapters 21 and 83);
(v) semigroups (Chapters 19 and 31);
(vi) compensated compactness (Chapter 62).

The prototype for (i) is the Leray–Schauder principle. This principle tells us that *a priori* estimates and compactness yield existence. In (ii) through (vi) we do not need compactness. However, in (ii) and (iii) we use weak compactness. The basic ideas of the modern method of compensated compactness will be explained in Section 62.12. Furthermore, it is possible to use the theory of A-proper maps from Chapters 34 and 35, in order to obtain constructive existence proofs for nonlinear differential equations.

Summarizing, a modern tendency in the theory of nonlinear differential equations is to weaken the compactness assumptions and to use refined convergence arguments for approximation methods in order to obtain existence proofs.

In this connection, we also mention the important method of geometric measure theory in the calculus of variations (cf. Giusti (1984, M)) and the method of the so-called variational convergence (e.g., G-convergence) of functionals and operators (cf. Attouch (1984, M)).

(18) *The modern strategy of using several spaces.* To prove the convergence of the iterative method or the Galerkin method, it is frequently important to use more than one space. In this connection, we consider the following three basic results in this volume:

(i) the refined Banach fixed-point theorem via interpolation trick (Theorem 21.H);

(ii) the main theorem on locally coercive operators (Theorem 27.B);
(iii) the main theorem on semibounded nonlinear evolution equations (Theorem 30.B).

In (i), we use the fact that the iterative sequence (u_n) is bounded in the "very nice" space X and k-contractive in the "poor" space Z. Then, using the interpolation inequality, we obtain the convergence of (u_n) in the "nice" space Y where

$$X \subseteq Y \subseteq Z.$$

In terms of function spaces, the functions in X are smoother than those in Y and, in turn, the functions in Y are smoother than those in Z.

In (ii) and (iii), we solve the Galerkin equations in the "nice" space Y, and we prove the convergence of the Galerkin sequence in the "poor" space Z where $Y \subseteq Z$.

Recall that in Chapter 5 we have already encountered the hard implicit function theorem (the Moser–Nash theorem), which is based on a modified Newton method working in a family of B-spaces. This method is related to (i).

In Chapter 83, we will continue the study of the strategy (i) through (iii).

(19) *Basic strategies for evolution equations.* To solve evolution equations, one can use one of the following methods:

(i) the Galerkin method (Chapter 30);
(ii) the method of linear semigroups for semilinear and quasi-linear equations (Chapter 19);
(iii) the method of nonlinear semigroups (Chapter 31);
(iv) the method of time-discretization (Chapters 57 and 83).

In connection with (ii) and (iii) we will use several reduction tricks to be discussed in (20) through (22) below.

In (iv), the original equation

$$u'(t) + Au(t) = f(t), \qquad 0 < t < T,$$
$$u(0) = u_0$$

is replaced by the following approximate problem:

$$\frac{u_n - u_{n-1}}{\Delta t} + Au_n = f(n\Delta t), \qquad n = 1, 2, \ldots,$$

where u_n is an approximation for $u(n\Delta t)$. Thus, in each step $n = 1, 2, \ldots$, one has to solve the operator equation

$$u_n + \Delta t \cdot Au_n = \Delta t \cdot f(n\Delta t) + u_{n-1}$$

for u_n. In this connection, the *maximal accretiveness* of A plays a fundamental role. Note that we use backward differences. This is important for proving the convergence of this method.

(20) *The fixed-point trick for solving semilinear evolution equations.* We consider the semilinear initital value problem

(20a)
$$u'(t) + Lu(t) = g(u(t)), \qquad 0 < t < T,$$
$$u(0) = u_0,$$

where L is a linear operator, together with the following linear problem:

(20b)
$$u'(t) + Lu(t) = f(t), \qquad 0 < t < T,$$
$$u(0) = u_0.$$

Let S be the solution operator to equation (20b), i.e.,

$$u = Sf.$$

Then the original problem (20a) is reduced to the fixed-point problem

(20c)
$$u = Sg(u).$$

For example, if we solve (20c) by means of the Banach fixed-point theorem, then the iterative method

$$u_{n+1} = Sg(u_n), \qquad n = 0, 1, 2, \ldots,$$

corresponds to solving the following family of initial value problems:

$$u'_{n+1}(t) + Lu_{n+1}(t) = g(u_n(t)), \qquad 0 < t < T,$$
$$u_{n+1}(0) = u_0,$$

where $n = 0, 1, 2, \ldots$.

Instead of problem (20b), one frequently considers the integral formula

$$u(t) = e^{-tL}u_0 + \int_0^t e^{(s-t)L} f(s)\, ds,$$

which yields a generalized (mild) solution of (20b). Here, e^{-tL} corresponds to the semigroup generated by L. This way the original semilinear problem (20a) can be reduced to the nonlinear integral equation

(20d)
$$u(t) = e^{-tL}u_0 + \int_0^t e^{(s-t)L} g(u(s))\, ds.$$

The solutions of (20d) are called mild solutions of (20a). This method has been used in Chapter 19.

(21) *The fixed-point trick for solving quasi-linear evolution equations.* We consider the quasi-linear initial value problem

(21a)
$$u'(t) + A(u(t))u(t) = f(t), \qquad 0 < t < T,$$
$$u(0) = u_0,$$

where the operator $u \mapsto A(v)u$ is linear for each fixed v. Suppose that, for each

v, we are able to solve the corresponding linear problem

(21b)
$$u'(t) + A(v(t))u(t) = f(t), \qquad 0 < t < T,$$
$$u(0) = u_0.$$

We set

$$u = Sv.$$

Then the original problem (21a) is reduced to the fixed-point problem

(21c)
$$u = Su.$$

For example, if we solve (21c) by using the Banach fixed-point theorem, then the iterative method

$$u_{n+1} = Su_n, \qquad n = 0, 1, 2, \ldots,$$

corresponds to the following family of linear initial value problems:

$$u'_{n+1}(t) + A(u_n(t))u_{n+1}(t) = f(t), \qquad 0 < t < T,$$
$$u_{n+1}(0) = u_0, \qquad n = 0, 1, \ldots.$$

(22) *Two reduction tricks for second-order evolution equations.* We consider the nonlinear initial value problem:

(J)
$$u''(t) + Au'(t) + Bu(t) = f(t), \qquad 0 < t < T,$$
$$u(0) = u_0, \qquad u'(0) = u_1.$$

(22a) *The first reduction trick.* We set

$$v = u'.$$

From (J) we then obtain the first-order system

$$u'(t) - v(t) = 0, \qquad 0 < t < T,$$
$$v'(t) + Av(t) + Bu(t) = f(t),$$

with the initial condition $u(0) = u_0$, $v(0) = u_1$. Letting

$$w = (u, v)$$

and $F = (0, f)$, we obtain the first-order equation:

$$w'(t) + Cw(t) = F(t), \qquad 0 < t < T,$$
$$w(0) = (u_0, u_1).$$

(22b) *The second reduction trick.* We set

$$(Sv)(t) = \int_0^t v(s)\,ds + u_0.$$

Letting $u(t) = (Sv)(t)$, the original problem (J) is reduced to the first-order equation:

$$v'(t) + Av(t) + B(Sv)(t) = f(t), \qquad 0 < t < T,$$
$$v(0) = u_1.$$

The first reduction trick has been used in Chapter 19. The second reduction trick will be used in Chapters 33 and 56.

(23) *The fixed-point trick for semilinear operator equations.* Let $L: X \to Y$ be a linear bijective operator. Then the semilinear operator equation

(23a) $$Lu = g(u), \qquad u \in X,$$

can be reduced to the fixed-point equation

(23b) $$u = L^{-1}g(u), \qquad u \in X.$$

(24) *The reduction of semilinear elliptic boundary value problems to abstract Hammerstein equations.* We consider the semilinear boundary value problem

(24a) $$-\Delta u(x) = g(u(x)) \quad \text{on } G,$$
$$u = 0 \quad \text{on } \partial G,$$

together with the linear problem

(24a*) $$-\Delta u(x) = f(x) \quad \text{on } G,$$
$$u = 0 \quad \text{on } \partial G.$$

Let K be the solution operator to (24a*) in appropriate spaces, i.e., the solution u of (24a*) can be represented in the form

$$u = Kf.$$

This way, the original problem (24a) is reduced to the abstract Hammerstein equation

(24b) $$u = Kg(u).$$

Recall that K is the abstract Green operator introduced in Chapter 22. Equations (24a) and (24b) correspond to (23a) and (23b), respectively.

This method will be used in Chapter 28.

(25) *Application to quasi-linear elliptic differential equations.* Set $Du = (D_1 u, \ldots, D_N u)$, $D_i = \partial/\partial\xi_i$, and

$$|Du|^2 = \sum_{i=1}^{N} |D_i u|^2.$$

In Chapter 25, we study the variational problem

(25a) $$\int_G (\varphi(|Du|^2) - 2fu)\,dx = \min!, \qquad u \in \mathring{W}_2^1(G),$$

in $\mathbb{R}^N$. The corresponding Euler equation reads as follows:

(25b)
$$-\sum_{i=1}^{N} D_i(\varphi'(|Du|^2)D_i u) = f \quad \text{on } G,$$
$$u = 0 \quad \text{on } \partial G.$$

In the special case where $\varphi(t) = t$, we obtain the classical Dirichlet problem. If the real function $t \mapsto \varphi(t^2)$ is convex, then (25a) (resp. (25b)) corresponds to a convex minimum problem (resp. a monotone operator equation) $Au = b$, $u \in \mathring{W}_2^1(G)$.

Problems of type (25b) describe nonlinear stationary conservation laws, which appear frequently in mathematical physics.

(26) *Monotone operators and quasi-linear elliptic equations.* We set

$$Lu = -\sum_{i=1}^{N} D_i(F_i(Du)),$$

and consider the boundary value problem

(26a)
$$Lu + F(u) = f \quad \text{on } G,$$
$$u = 0 \quad \text{on } \partial G.$$

In contrast to (25b), this problem is not always the Euler equation to a variational problem. In Chapters 26 and 27, we reduce (26a) to the operator equation

(26b)
$$Au = b, \qquad u \in X,$$

where $X = \mathring{W}_p^1(G)$, $1 < p < \infty$. Here G is a bounded region in $\mathbb{R}^N$, $N \geq 1$, and $Du = (D_1 u, \ldots, D_N u)$.

In order to guarantee that the operator $A: X \to X^*$ is monotone, coercive, and continuous, we need the following conditions:

(i) *Monotonicity condition for the principal part Lu:*

$$\sum_{i=1}^{N} (F_i(D) - F_i(D'))(D_i - D_i') \geq 0 \qquad \text{for all} \quad D, D' \in \mathbb{R}^N.$$

(ii) *Monotonicity condition for $F(u)$:*

$$(F(u) - F(v))(u - v) \geq 0 \qquad \text{for all} \quad u, v \in \mathbb{R}.$$

(iii) *Coerciveness condition for $Lu + F(u)$:*

$$\sum_{i=1}^{N} F_i(D)D_i + F(u)u \geq c \sum_{i=1}^{N} |D_i|^p - b,$$

for all $D \in \mathbb{R}^N$, $u \in \mathbb{R}$, and fixed $c > 0$, $b \geq 0$.

(iv) *Growth condition:* The functions $F: \mathbb{R} \to \mathbb{R}$ and $F_i: \mathbb{R}^N \to \mathbb{R}$ are continuous for all i, and there is a number $d > 0$ such that

$$|F(u)| \leq d(1 + |u|^{p-1}) \quad \text{for all} \quad u \in \mathbb{R},$$

$$|F_i(D)| \leq d(1 + |D|^{p-1}) \quad \text{for all} \quad D \in \mathbb{R}^N \quad \text{and all } i.$$

For example, these conditions are satisfied for the generalized Laplace operator

$$Lu = -\sum_{i=1}^{N} D_i(|D_i u|^{p-2} D_i u).$$

Let $f \in L_q(G)$ with $p^{-1} + q^{-1} = 1$. As a special case of Proposition 26.12, it follows that the boundary value problem (26a) has a generalized solution $u \in X$ in the case where the assumptions (i) through (iv) above are satisfied.

(27) *Pseudomonotone operators and quasi-linear elliptic differential equations.* We consider again the boundary value problem (26a). But we now assume that the lower-order term $F(u)$ does *not* satisfy the monotonicity condition (ii) in (26) above. In this case, we can use the theory of pseudomonotone operators to be considered in Chapter 27. Then the operator A in (26b) has the form

$$Au = A_1 u + A_2 u,$$

where A_1 and A_2 correspond to the principal part Lu and lower-order part $F(u)$, respectively. More precisely, we have the following typical situation:

(a) the operator $A_1: X \to X^*$ is monotone and continuous;
(b) $A_2: X \to X^*$ is strongly continuous; and
(c) $A = A_1 + A_2$ is pseudomonotone and coercive.

The basic idea of the theory of pseudomonotone operators is that lower-order terms generate strongly continuous perturbations of the monotone principal part of quasi-linear elliptic differential operators.

Compared with the theory of monotone operators, the theory of pseudomonotone operators yields stronger existence results for quasi-linear elliptic equations. For example, it follows from Proposition 27.9 that the existence theorem for (26a) remains valid, if the assumption (ii) drops out in (26) above.

(28) *Noncoercive problems and nonlinear Fredholm alternatives.* As a typical example, we consider the boundary value problem

$$\text{(28a)} \qquad \begin{aligned} -\Delta u - \mu u + h(u) &= f \quad \text{on } G, \\ u &= 0 \quad \text{on } \partial G, \end{aligned}$$

where μ is a given real parameter, and the function $f: G \to \mathbb{R}$ is given. We first consider the corresponding linear problem

$$\text{(28b)} \qquad \begin{aligned} -\Delta u - \mu u &= f \quad \text{on } G, \\ u &= 0 \quad \text{on } \partial G, \end{aligned}$$

together with the eigenvalue problem

(28c)
$$\begin{aligned} -\Delta u - \mu u &= 0 \quad \text{on } G, \\ u &= 0 \quad \text{on } \partial G. \end{aligned}$$

Roughly speaking, we have the following situation:

(i) Problem (28c) has a sequence of eigenvalues (μ_n) with

$$0 < \mu_1 \le \mu_2 \le \dots,$$

and $\mu_n \to +\infty$ as $n \to \infty$.

(ii) If μ is not an eigenvalue of (28c), then problem (28b) has a unique solution for each f.

(iii) If μ is an eigenvalue of (28c), then problem (28b) has only solutions for such f which satisfy a solvability condition.

Our goal is to translate these results to the nonlinear problem (28a). This leads us to nonlinear Fredholm alternatives to be considered in Chapter 29. In the nonlinear case, we have the following situation:

(a) If $\mu < \mu_1$ and $h(u)u \ge 0$ for all $u \in \mathbb{R}$, then (28a) corresponds to a coercive problem of pseudomonotone type. Here we obtain a solution for each f.

(b) If $\mu \ge \mu_1$, then the existence of solutions for (28a) essentially depends on the structure of the nonlinear term h and on the relation between the range of $h: \mathbb{R} \to \mathbb{R}$ and the set of eigenvalues $\{\mu_1, \mu_2, \dots\}$.

(c) In particular, in Chapter 29, we consider an interesting result, due to Landesman and Lazer (1969), where the necessary solvability condition for (28a) is also sufficient.

There exists a very extensive literature on problems of type (28a) in the critical case (b).

(29) *Hammerstein integral equations.* The nonlinear integral equation

(29a)
$$u(x) + \int_G k(x, y) f(u(y))\, dy = 0, \qquad x \in G,$$

can be written in the form of the following operator equation:

(29b)
$$u + KFu = 0, \qquad u \in X,$$

which is called an abstract Hammerstein equation. Here we set $X = L_2(G)$ as well as

$$(Ku)(x) = \int_G k(x, y) u(y)\, dy$$

and $(Fu)(x) = f(u(x))$. In Chapter 28, we will apply the theory of monotone operators to equations (29a) and (29b). In this connection, we need the monotonicity of the function $f\colon \mathbb{R} \to \mathbb{R}$ and the monotonicity of K, i.e.,

$$\langle Ku, u\rangle = \int_{G\times G} k(x,y)u(x)u(y)\,dx\,dy \geq 0 \qquad \text{for all} \quad u \in L_2(G).$$

(30) *The important role of growth conditions in the theory of nonlinear differential and integral equations.* One of the typical difficulties in the theory of nonlinear differential and integral equations is based on the following fact:

(A) The Lebesgue space $L_2(G)$ is *not* a Banach algebra, i.e., $u, v \in L_2(G)$ does *not* imply $uv \in L_2(G)$.

For example, set $G =]-1, 1[$ and consider the functions

$$u(x) = v(x) = |x|^{-1/3}.$$

Then $u, v \in L_2(G)$, but $uv \notin L_2(G)$.

In order to overcome this difficulty, one needs *growth conditions.* To explain this, consider the continuous function $f\colon \mathbb{R} \to \mathbb{R}$, and suppose that there is a $c > 0$ such that the following growth condition holds:

$$|f(u)| \leq c(1 + |u|) \qquad \text{for all} \quad u \in \mathbb{R}. \tag{30a}$$

Let G be a bounded region in $\mathbb{R}^N$, $N \geq 1$. Set

$$(Fu)(x) = f(u(x)) \qquad \text{for all} \quad x \in G.$$

Then

$$u \in L_2(G) \qquad \text{implies} \qquad Fu \in L_2(G),$$

i.e., F is an operator from $L_2(G)$ to $L_2(G)$. Indeed, if $u \in L_2(G)$, then

$$\int_G |(Fu)(x)|^2\,dx \leq 2c^2 \int_G (1 + |u(x)|^2)\,dx < \infty.$$

The following observation is crucial. In order to relax the growth condition (30a), we can use the Sobolev embedding theorems $A_2(45)$. To explain this, suppose that, in contrast to (30a), we have the less restrictive growth condition

$$|f(u)| \leq c(1 + |u|^2) \qquad \text{for all} \quad u \in \mathbb{R}. \tag{30b}$$

Let G be a bounded region in $\mathbb{R}^3$. Then,

$$u \in \mathring{W}_2^1(G) \qquad \text{implies} \qquad Fu \in L_2(G),$$

i.e., F is an operator from $\mathring{W}_2^1(G)$ to $L_2(G)$. Indeed, the embedding

$$\mathring{W}_2^1(G) \subseteq L_4(G)$$

is continuous, i.e., there is a $d > 0$ such that

$$\|u\|_4 \leq d\|u\|_{1,2} \qquad \text{for all} \quad u \in \mathring{W}_2^1(G). \tag{30c}$$

If $u \in \mathring{W}_2^1(G)$, then $u \in L_4(G)$ and hence

$$\int_G |(Fu)(x)|^2\,dx \le 2c^2 \int_G (1 + |u(x)|^4)\,dx < \infty.$$

In order to investigate the properties of nonlinear differential and integral operators, one can use the following tools:

(i) growth conditions;
(ii) the Hölder inequality;
(iii) the Sobolev embedding theorems; and
(iv) the fundamental Gagliardo–Nirenberg interpolation inequalities.

Typical examples for (iv) have been considered in Section 21.23. To explain (ii) and (iii), let us consider the integral

$$J(u) = \int_G |u^2 D_i u|\,dx \qquad \text{for all} \quad u \in \mathring{W}_2^1(G),$$

where G is a bounded region in $\mathbb{R}^3$. We want to show that $J(u) < \infty$. Indeed, the Hölder inequality for three factors yields

$$\begin{aligned}\int_G |uuD_i u|\,dx &\le \left(\int_G u^4\,dx\right)^{1/4}\left(\int_G u^4\,dx\right)^{1/4}\left(\int_G (D_i u)^2\,dx\right)^{1/2}\\ &\le \|u\|_4^2 \|u\|_{1,2}.\end{aligned}$$

Using the Sobolev embedding theorem (30c), we obtain that

$$J(u) \le d^2 \|u\|_{1,2}^3 \qquad \text{for all} \quad u \in \mathring{W}_2^1(G).$$

Observe the following:

(α) In order to relax growth conditions, one can use Orlicz spaces. This will be studied in Chapter 53.
(β) In the case of spaces of smooth functions (e.g., the space $C(\bar{G})$ or $C^{m,\alpha}(\bar{G})$), we do not need any growth conditions.

However, since the spaces in (β) are not reflexive, we cannot apply typical results of the theory of monotone operators to this situation. In Part I we used (β) in order to obtain existence results for nonlinear differential and integral equations via fixed-point theory.

(31) *The difference between K-monotone and monotone operators.* Let $f: \mathbb{R} \to \mathbb{R}$ be a monotone increasing function, i.e.,

$$\text{(31a)} \qquad u \le v \quad \text{implies} \quad f(u) \le f(v),$$

and this is equivalent to

$$\text{(31b)} \qquad (f(u) - f(v))(u - v) \ge 0 \quad \text{for all} \quad u, v \in \mathbb{R}.$$

Let X be a real B-space. Then, (31a) and (31b) lead to the following two different generalizations:

(i) *K-monotone operators.* We first generalize (31a). Let K be an order cone in X. As in Chapter 7, we write $u \leq v$ iff $v - u \in K$. Then the operator $F: X \to X$ is called K-monotone or also monotone increasing iff
$$u \leq v \qquad \text{implies} \qquad F(u) \leq F(v).$$

(ii) *Monotone operators.* We now generalize (31b). The operator $F: X \to X^*$ is called monotone iff
$$\langle F(u) - F(v), u - v \rangle \geq 0 \qquad \text{for all} \qquad u, v \in X.$$

The following two examples should illustrate these two different concepts. Let $f: \mathbb{R} \to \mathbb{R}$ be a continuous monotone increasing function. Set
$$(Fu)(x) = f(u(x)).$$

EXAMPLE 1 (K-Monotone Operator). Let $X = C[a, b]$ with $-\infty < a < b < \infty$, and let $K = C_+[a, b]$, i.e., X consists of all continuous functions $u: [a, b] \to \mathbb{R}$, and K consists of all nonnegative functions in X. Then the operator
$$F: X \to X$$
is K-monotone. Indeed, for all $x \in [a, b]$,
$$u(x) \leq v(x) \qquad \text{implies} \qquad f(u(x)) \leq f(v(x)).$$

EXAMPLE 2 (Monotone Operator). Let $Y = L_2(a, b)$ with $-\infty < a < b < \infty$, and suppose that there is a $c > 0$ such that
$$|f(u)| \leq c(1 + |u|) \qquad \text{for all} \quad u \in \mathbb{R}.$$
Then the operator
$$F: Y \to Y^*$$
is monotone. Indeed, for all $u, v \in L_2(a, b)$, we obtain that
$$\langle Fu - Fv, u - v \rangle = \int_a^b (f(u(x) - f(v(x))(u(x) - v(x))\, dx \geq 0.$$

Roughly speaking, we have the following typical situation:

(α) The theory of K-monotone operators can be applied to semilinear elliptic equations of second order by using the maximum principle (cf. Chapter 7).

(β) The theory of monotone operators can be applied to quasi-linear elliptic equations of order $2m$ with $m = 1, 2, \ldots$. In contrast to (α), we need restrictive growth conditions.

Some relations between K-monotone and monotone operators are considered in Glashoff and Werner (1979).

CHAPTER 25

Lipschitz Continuous, Strongly Monotone Operators, the Projection–Iteration Method, and Monotone Potential Operators

For any continuous function $f: \mathbb{R} \to \mathbb{R}$ growing everywhere strictly faster than the function $g(x) = x$, the equation

$$x = f(x)$$

has a unique solution. This solution can be obtained by the iteration method

$$x_{n+1} = (1 - t)x_n + tf(x_n), \qquad n = 0, 1, 2, \ldots,$$

where t is a sufficiently small positive number.

The purpose of this note is to show that the above theorem, and its corresponding local form, remain valid in Hilbert spaces, provided the notion of a function growing faster than another is adequately extended.

A fundamental tool in our argument is the Banach fixed-point theorem.

Eduardo Zarantonello (1960)

In this chapter we want to study the following topics.

(i) We first consider the operator equation

$$Au = b, \qquad u \in X, \tag{32}$$

where the operator $A: X \to X^*$ is strongly monotone and Lipschitz continuous on the real H-space X. We prove that, for each $b \in X^*$, there exists a unique solution of (32). The idea of proof is to apply the Banach fixed-point theorem to the equivalent operator equation

$$u = u - tJ^{-1}(Au - b), \qquad u \in X, \tag{32a}$$

where $t > 0$ is an appropriate parameter, and $J: X \to X^*$ is the duality map of X. We also prove the convergence of the *iterative method*

$$u_n = u_{n-1} - tJ^{-1}(Au_{n-1} - b), \qquad n = 1, 2, \ldots, \tag{32b}$$

and the convergence of the important *projection–iteration method*

$$u_n = u_{n-1} - tP_nJ^{-1}(Au_{n-1} - b), \qquad u_n \in X_n, \quad n = 1, 2, \ldots, \tag{32c}$$

where $u_0 = 0$, and $X_1 \subseteq X_2 \subseteq \cdots \subseteq X$ with $X = \overline{\bigcup_n X_n}$. Here, $P_n: X \to X_n$ is the orthogonal projection operator from X onto the finite-dimensional linear subspace X_n of X.

(ii) We solve the minimum problem

$$f(u) = \min!, \qquad u \in X,$$

together with the Euler equation

$$f'(u) = 0, \qquad u \in X,$$

where X is a real reflexive B-space. In this connection, we shall prove the following fundamental results:

the main theorem on minimum problems;
the main theorem on convex minimum problems;
the main theorem on monotone potential operators;
the main theorem on pseudomonotone potential operators.

(iii) As an important application of the abstract results to differential equations, we consider the conservation law

$$\begin{aligned} Lu &= f \quad \text{on } G, \\ u &= 0 \quad \text{on } \partial G, \end{aligned} \tag{33}$$

where $Du = (D_1 u, \ldots, D_N u)$, and

$$Lu = -\sum_{i=1}^{N} D_i(\psi(|Du|^2)D_i u).$$

Such problems occur frequently in mathematical physics (e.g., in hydrodynamics, plasticity, rheology, thermodynamics, electrodynamics, etc.). Problem (33) is the Euler equation to the minimum problem

$$\int_G (\varphi(|Du|^2) - 2fu)\,dx = \min!,$$

$$u = 0 \quad \text{on } \partial G.$$

Here, $\psi = \varphi'$. The function φ depends on the constitutive law of the material. Depending on the qualitative behavior of the function φ, we shall prove the following results for (33).

(a) Existence and uniqueness of solutions. In this connection, problem (33) is reduced to an operator equation of the form (32) with $X = \mathring{W}_2^1(G)$. The equivalent operator equation (32a) corresponds to the modified boundary value problem

$$\begin{aligned} -\Delta u &= -\Delta u - t(Lu - f) \quad \text{on } G, \\ u &= 0 \quad \text{on } \partial G, \end{aligned} \tag{33a}$$

where $t > 0$.

(b) Convergence of the iterative method

$$\begin{aligned} -\Delta u_n &= -\Delta u_{n-1} - t(Lu_{n-1} - f) \quad \text{on } G, \\ u_n &= 0 \quad \text{on } \partial G, \qquad n = 1, 2, \ldots, \quad u_0 = 0. \end{aligned} \tag{33b}$$

This corresponds to (32b).

(c) Convergence of the projection–iteration method.
(d) Convergence of the Ritz method.
(e) Duality and two-sided *a posteriori* error estimates for the Ritz method.
(f) Convergence of the Kačanov iterative method

$$\begin{aligned} -\sum_{i=1}^{N} D_i(\psi(|Du_{n-1}|^2)D_i u_n) &= f \quad \text{on } G, \\ u_n &= 0 \quad \text{on } \partial G, \qquad n = 1, 2, \ldots, \end{aligned} \tag{34}$$

where $u_0 = 0$. This is a *linear* boundary value problem for u_n.

The projection–iteration method represents a modification of the iterative method (33b). In contrast to (33b), we only approximately compute the function u_n by using a Ritz method with n basis functions.

The decisive *advantage* of the projection–iteration method for *nonlinear* problems is that, in each step, we have only to solve a finite *linear* system of real equations.

25.1. Sequences of k-Contractive Operators

We consider the convergence of the iterative method

$$u_n = T_n^{m(n)} u_{n-1}, \qquad n = 1, 2, \ldots, \tag{35}$$

for given $u_0 \in M$, where $m(n) \in \mathbb{N}$. In the special case $T_n^{m(n)} = T$, we obtain the iterative method $u_n = Tu_{n-1}$ corresponding to the Banach fixed-point theorem in Section 1.1. We make the following assumptions.

(H1) Let M be a nonempty complete metric space with metric d.
(H2) For each $n \in \mathbb{N}$, the operator $T_n: M \to M$ satisfies the contractivity condition

$$d(T_n u, T_n v) \leq k_n d(u, v), \tag{36}$$

for all $u, v \in M$ and fixed $k_n \in [0, 1[$.

(H3) By the Banach fixed-point theorem (Theorem 1.A), the equation

$$v_n = T_n v_n, \qquad v_n \in M, \tag{37}$$

has a unique solution for each $n \in \mathbb{N}$. Suppose that (v_n) converges to $u \in M$, i.e.,

$$d(v_n, u) \to 0 \qquad \text{as} \quad n \to \infty.$$

Proposition 25.1. *Assume* (H1) *through* (H3). *Then the iterative sequence* (u_n) *constructed in* (35) *converges to* u *in the case where one of the following two conditions is satisfied.*

(i) $m(n) = 1$ *for all* $n \in \mathbb{N}$, *and* $\sup_{n \in \mathbb{N}} k_n < 1$.
(ii) $m(n) \geq c/(1 - k_n)$ *for all* $n \in \mathbb{N}$ *and fixed* $c > 0$, *and* $k_n \to 1$ *as* $n \to \infty$.

This result will be used in the next section in order to prove the convergence of the projection–iteration method.

PROOF. We will use the classical convergence theorem of Toeplitz, which we recall in Problem 25.1.

Ad(i). Let $k = \sup_n k_n$. For $n = 2, 3, \ldots$, we get

$$\begin{aligned} d(u_n, v_n) &= d(T_n u_{n-1}, T_n v_n) \leq k d(u_{n-1}, v_n) \\ &\leq k[d(u_{n-1}, v_{n-1}) + d(v_{n-1}, v_n)]. \end{aligned}$$

Repeated application of this inequality yields the *key* estimate

$$d(u_n, v_n) \leq k^{n-1} d(u_1, v_1) + \sum_{m=1}^{n-1} k^{n-m} d(v_m, v_{m+1}). \tag{38}$$

We set $a_m = d(v_m, v_{m+1})$. By (H3), $a_m \to 0$ as $m \to \infty$. Since $0 \leq k < 1$, the Toeplitz convergence theorem tells us that

$$\sum_{m=1}^{n-1} k^{n-m} a_m \to 0 \qquad \text{as} \quad n \to \infty.$$

Hence $d(u_n, v_n) \to 0$ as $n \to \infty$. By (H3),

$$d(u, u_n) \leq d(u, v_n) + d(v_n, u_n) \to 0 \qquad \text{as} \quad n \to \infty.$$

Thus, $u_n \to u$ as $n \to \infty$.

Ad(ii). We reduce this case to (i). To this end, we set $A_n = T_n^{m(n)}$. By (H2),

$$d(A_n u, A_n v) \leq k_n^{m(n)} d(u, v) \qquad \text{for all} \quad u, v \in M.$$

By (H3), $A_n v_n = v_n$ for all n. Since $k_n \to 1$ as $n \to \infty$, we obtain

$$k_n^{m(n)} \leq (1 - (1 - k_n))^{c/(1-k_n)} \to e^{-c} \qquad \text{as} \quad n \to \infty,$$

and hence $\sup_{n \geq n_0} k_n^{m(n)} < 1$ for sufficiently large n_0. We now apply case (i) to the equation

$$u_n = A_n u_{n-1}, \qquad n = n_0, n_0 + 1, \ldots \qquad \square$$

25.2. The Projection–Iteration Method for k-Contractive Operators

We consider the equation

$$u = Bu, \qquad u \in X, \tag{39}$$

and make the following assumptions.

(A1) The operator $B: X \to X$ is k-contractive on the real separable H-space X, i.e., there is a $k \in [0, 1[$ such that

$$\|Bu - Bv\| \le k\|u - v\| \qquad \text{for all} \quad u, v \in X.$$

(A2) (X_n) is a Galerkin scheme in X, where $X_n = \operatorname{span}\{w_{1n}, \ldots, w_{n'n}\}$.
(A3) $P_n: X \to X_n$ is the orthogonal projection operator from X onto X_n.

We formulate the following three approximation methods.

(i) *Iteration method.* For given $u_0 \in X$ and $n = 1, 2, \ldots$, let

$$u_n = Bu_{n-1}. \tag{40}$$

(ii) *Projection method* (*Galerkin method*). For $\dim X = \infty$ and $n = 1, 2, \ldots$, let

$$v_n = P_n B v_n, \qquad v_n \in X_n. \tag{41}$$

Since P_n is self-adjoint, equation (41) is equivalent to the Galerkin equations

$$(v_n | w_{kn}) = (Bv_n | w_{kn}), \qquad k = 1, \ldots, n',$$

where $v_n = \sum_{k=1}^{n'} c_{kn} w_{kn}$. Here, the real coefficients c_{kn} are unknown.
(iii) *Projection–iteration method.* For $\dim X = \infty$ and $n = 1, 2, \ldots$, let

$$w_n = P_n B w_{n-1}, \qquad w_n \in X_n, \tag{42}$$

where $w_0 = 0$. Equation (42) is equivalent to

$$(w_n | w_{kn}) = (Bw_{n-1} | w_{kn}), \qquad k = 1, \ldots, n',$$

where $w_n = \sum_{k=1}^{n'} d_{kn} w_{kn}$. Thus, we obtain a *linear* system for the unknown real coefficients d_{kn} of w_n.

Theorem 25.A. *Assume* (A1) *through* (A3). *Then*:

(a) *The original equation* (39) *has a unique solution* u.
(b) *The iteration method* (40) *converges, i.e.,* $u_n \to u$ *as* $n \to \infty$, *and we have the error estimates*

$$\|u - u_n\| \le k^n(1 - k)^{-1}\|u_1 - u_0\| \qquad \textit{for} \quad n = 1, 2, \ldots.$$

(c) *The projection method* (41) *converges, i.e., for each* n, *equation* (41) *has a unique solution* v_n, *and* $v_n \to u$ *as* $n \to \infty$. *Moreover, we have the error estimates*

$$\|u - v_n\| \leq (1 - k)^{-1} \operatorname{dist}(u, X_n) \qquad \text{for} \quad n = 1, 2, \ldots.$$

(d) *The projection–iteration method converges, i.e.,* $w_n \to u$ *as* $n \to \infty$.

PROOF. Ad(a). (b). Cf. Theorem 1.A.

Ad(c). From $\|P_n\| = 1$ it follows that

$$\|P_n Bu - P_n Bv\| \leq \|Bu - Bv\| \leq k\|u - v\| \qquad \text{for all} \quad u, v \in X,$$

i.e., $P_n B: X \to X$ is k-contractive. According to the Banach fixed-point theorem (Theorem 1.A), equation (41) has a unique solution v_n, i.e., $v_n = P_n B v_n$. Let $Bu = u$. From

$$\|P_n B P_n u - P_n B v_n\| \leq k\|P_n u - v_n\|$$

and the Schwarz inequality $|(u|v)| \leq \|u\|\,\|v\|$, we obtain the *key* estimate:

$$\begin{aligned}(1 - k)\|v_n - P_n u\|^2 &\leq (v_n - P_n u | v_n - P_n u) + (P_n B P_n u - P_n B v_n | v_n - P_n u)\\ &= (P_n B P_n u - P_n B u | v_n - P_n u) \leq k\|P_n u - u\|\,\|v_n - P_n u\|.\end{aligned}$$

Hence

$$\|v_n - P_n u\| \leq k(1 - k)^{-1}\|P_n u - u\|.$$

This implies

$$\|v_n - u\| \leq \|v_n - P_n u\| + \|P_n u - u\| \leq (1 - k)^{-1}\|P_n u - u\|.$$

Since $P_n u \to u$ as $n \to \infty$, we obtain $v_n \to u$ as $n \to \infty$.

Ad(d). We set $T_n = P_n B$. Since $\|P_n\| = 1$, we obtain

$$\|T_n u - T_n v\| \leq \|Bu - Bv\| \leq k\|u - v\|,$$

for all $u, v \in X$. Now, the assertion follows from Proposition 25.1(i). □

25.3. Monotone Operators

The following definitions are basic.

Definition 25.2. Let X and Y be real B-spaces, and let $A: X \to X^*$ be an operator. Then:

(i) A is called *monotone* iff

$$\langle Au - Av, u - v\rangle \geq 0 \qquad \text{for all} \quad u, v \in X.$$

(ii) A is called *strictly monotone* iff

$$\langle Au - Av, u - v\rangle > 0 \qquad \text{for all} \quad u, v \in X \quad \text{with} \quad u \neq v.$$

(iii) A is called *strongly monotone* iff there is a $c > 0$ such that

$$\langle Au - Av, u - v\rangle \geq c\|u - v\|^2 \qquad \text{for all} \quad u, v \in X.$$

(iv) A is called *uniformly monotone* iff

$$\langle Au - Av, u - v\rangle \geq a(\|u - v\|)\,\|u - v\| \qquad \text{for all} \quad u, v \in X,$$

where the continuous function $a: \mathbb{R}_+ \to \mathbb{R}_+$ is strictly monotone increasing with $a(0) = 0$ and $a(t) \to +\infty$ as $t \to +\infty$.

For example, we may choose $a(t) = c|t|^{p-1}$ with $p > 1$ and $c > 0$. In this case, we obtain

$$\langle Au - Av, u - v\rangle \geq c\|u - v\|^p \qquad \text{for all} \quad u, v \in X.$$

(v) A is called *coercive* iff

$$\lim_{\|u\| \to \infty} \frac{\langle Au, u\rangle}{\|u\|} = +\infty.$$

(vi) A is called *weakly coercive* iff

$$\lim_{\|u\| \to \infty} \|Au\| = \infty.$$

(vii) The operator $A: X \to Y$ is called *stable* iff

$$\|Au - Av\| \geq a(\|u - v\|) \qquad \text{for all} \quad u, v \in X,$$

where the function $a(\cdot)$ is the same as in (iv) above.

(viii) The operator $A: X \to X$ on the H-space X is called *strongly stable* iff there is a $c > 0$ such that

$$|(Au - Av|u - v)| \geq c\|u - v\|^2 \qquad \text{for all} \quad u, v \in X.$$

By the Schwarz inequality, this implies $\|Au - Av\| \geq c\|u - v\|$ for all u, $v \in X$.

Obviously, we have the following implications:

$$\begin{aligned} &A \text{ is strongly monotone} \;\Rightarrow\; A \text{ is uniformly monotone} \;\Rightarrow \\ \Rightarrow\; &A \text{ is strictly monotone} \;\Rightarrow\; A \text{ is monotone.} \end{aligned} \tag{43}$$

Furthermore, we obtain:

$$A \text{ is uniformly monotone} \;\Rightarrow\; A \text{ is coercive and stable.} \tag{44}$$

This follows from

$$\begin{aligned} \langle Au, u\rangle &= \langle Au - A(0), u\rangle + \langle A(0), u\rangle \\ &\geq a(\|u\|)\|u\| - \|A(0)\|\,\|u\| \end{aligned}$$

and

$$\|Au - Av\|\,\|u - v\| \geq \langle Au - Av, u - v\rangle \geq a(\|u - v\|)\|u - v\|.$$

Hence, if $A: X \to X^*$ is uniformly monotone, then

$$\|Au - Av\| \geq a(\|u - v\|) \qquad \text{for all} \quad u, v \in X.$$

EXAMPLE 25.3 (Linear Monotone Operators). Let $A: X \to X^*$ be a linear operator on the real B-space X. Then:

(a) A is monotone iff A is positive, i.e., $\langle Au, u\rangle \geq 0$ for all $u \in X$.
(b) A is strictly monotone iff A is strictly positive, i.e., $\langle Au, u\rangle > 0$ for all $u \in X$ with $u \neq 0$.
(c) A is strongly monotone iff A is strongly positive, i.e., $\langle Au, u\rangle \geq c\|u\|^2$ for all $u \in X$ and fixed $c > 0$.

PROOF. Note that $Au - Av = A(u - v)$. □

EXAMPLE 25.4 (Monotone Real Functions). We consider the function $f: \mathbb{R} \to \mathbb{R}$. We regard f as an operator from X to X^* with $X = \mathbb{R}$. Then,

$$\langle f(u) - f(v), u - v\rangle = (f(u) - f(v))(u - v) \qquad \text{for all} \quad u, v \in \mathbb{R}.$$

This yields the following results.

(a) $f: X \to X^*$ is (strictly) monotone iff $f: \mathbb{R} \to \mathbb{R}$ is (strictly) monotone increasing.
(b) $f: X \to X^*$ is strongly monotone iff

$$\inf_{u \neq v} \frac{f(u) - f(v)}{u - v} > 0.$$

(c) $f: X \to X^*$ is coercive iff

$$\lim_{u \to \pm\infty} f(u) = \pm\infty.$$

(d) If $F: \mathbb{R} \to \mathbb{R}$ is C^2 satisfying

$$F''(u) \geq c \qquad \text{for all} \quad u \in \mathbb{R} \quad \text{and fixed} \quad c > 0,$$

then

$$(F'(u) - F'(v))(u - v) \geq c(u - v)^2 \qquad \text{for all} \quad u, v \in \mathbb{R},$$

i.e., $F': \mathbb{R} \to \mathbb{R}$ is strongly monotone.
(e) If $F: \mathbb{R} \to \mathbb{R}$ is C^1 satisfying

$$F'(u) - F'(v) \geq c(u - v),$$

for all $u, v \in \mathbb{R}$ with $u \geq v$ and fixed $c > 0$, then $F': \mathbb{R} \to \mathbb{R}$ is strongly monotone.

EXAMPLE 25.5. We consider the function $g: \mathbb{R} \to \mathbb{R}$ defined by

$$g(u) = \begin{cases} |u|^{p-2}u & \text{if} \quad u \neq 0, \\ 0 & \text{if} \quad u = 0. \end{cases}$$

Then:

(a) If $p > 1$, then g is strictly monotone.
(b) If $p = 2$, then g is strongly monotone.
(c) If $p \geq 2$, then g is uniformly monotone.

PROOF. This follows immediately from the inequality

$$(|u|^{p-2}u - |v|^{p-2}v)(u - v) \geq c|u - v|^p, \tag{45}$$

for all $u, v \in \mathbb{R}$ and fixed $p \geq 2$, $c > 0$.

To prove (45) we first consider the case $0 \leq v \leq u$. Then,

$$\begin{aligned} u^{p-1} - v^{p-1} &= \int_0^{u-v} (p-1)(t+v)^{p-2}\,dt \\ &\geq \int_0^{u-v} (p-1)t^{p-2}\,dt = (u-v)^{p-1}. \end{aligned}$$

In the case $v \leq 0 \leq u$, we use the inequality A_2(30b) in order to obtain that

$$u^{p-1} + |v|^{p-1} \geq c(u + |v|)^{p-1}. \tag{46}$$

□

Proposition 25.6. *Let $A: X \to X^*$ be an operator on the real* B*-space X. We set*

$$f(t) = \langle A(u + tv), v\rangle \qquad \text{for all} \quad t \in \mathbb{R}.$$

Then the following two statements are equivalent.

(a) *The operator A is monotone.*
(b) *The function $f: [0, 1] \to \mathbb{R}$ is monotone increasing for all $u, v \in X$.*

PROOF. If A is monotone, then for $0 \leq s < t$,

$$f(t) - f(s) = (t - s)^{-1}\langle A(u + tv) - A(u + sv), (t - s)v\rangle \geq 0.$$

Conversely, if $f: [0, 1] \to \mathbb{R}$ is monotone increasing, then for $u, v \in X$,

$$\langle A(u + v) - Au, v\rangle = f(1) - f(0) \geq 0.$$

□

25.4. The Main Theorem on Strongly Monotone Operators, and the Projection–Iteration Method

We consider the operator equation

$$Au = b, \qquad u \in X, \tag{47}$$

and make the following assumptions.

(H1) The operator $A: X \to X^*$ is strongly monotone and Lipschitz continuous on the real H-space X, i.e., there are numbers $c > 0$ and $L > 0$ such that, for all $u, v \in X$,

$$\langle Au - Av, u - v\rangle \geq c\|u - v\|^2$$

and

$$\|Au - Av\| \leq L\|u - v\|.$$

(H2) Let $\dim X = \infty$, and let (X_n) be a Galerkin scheme in the separable H-space X, where $X_n = \operatorname{span}\{w_{1n}, \ldots, w_{n'n}\}$. Moreover, let $P_n\colon X \to X_n$ be the orthogonal projection operator from X onto X_n.

The idea of our existence proof is to replace (47) by the equivalent operator equation

$$u = Bu, \qquad u \in X, \tag{48}$$

where

$$Bu = u - tJ^{-1}(Au - b),$$

for fixed $t > 0$. Here, $J\colon X \to X^*$ is the duality map of X. We shall show that, for $0 < t < 2c/L^2$, the operator $B\colon X \to X$ is k-contractive with

$$k^2 = 1 - 2ct + t^2L^2 < 1.$$

Therefore, we can apply Theorem 25.A to equation (48). In particular, we can solve equation (48) by means of the following approximation methods.

(i) *Iteration method.* For given $u_0 \in X$ and $n = 1, 2, \ldots$, let

$$u_n = u_{n-1} - tJ^{-1}(Au_{n-1} - b). \tag{49a}$$

(ii) *Projection method (Galerkin method).* For $n = 1, 2, \ldots$, let

$$v_n = P_n(v_n - tJ^{-1}(Av_n - b)), \qquad v_n \in X_n,$$

i.e.,

$$P_nJ^{-1}(Av_n - b) = 0, \qquad v_n \in X_n. \tag{49b}$$

(iii) *Projection–iteration method.* For $w_0 = 0$ and $n = 1, 2, \ldots$, let

$$w_n = P_n(w_{n-1} - tJ^{-1}(Aw_{n-1} - b)), \qquad w_n \in X_n. \tag{49c}$$

This equation is equivalent to

$$(w_n|w_{kn}) = (w_{n-1}|w_{kn}) - t\langle Aw_{n-1} - b, w_{kn}\rangle,$$

where $k = 1, \ldots, n'$ and

$$w_n = \sum_{k=1}^{n'} d_{kn}w_{kn}$$

with the unknown real coefficients d_{kn} of w_n. In this connection, note that $(J^{-1}b|w) = \langle b, w\rangle$ for all $b \in X^*$, $w \in X$.

The following theorem shows that the original problem (47) is *well-posed* and uniquely approximation-solvable.

Theorem 25.B (Zarantonello (1960)). *Assume* (H1). *Then, for each $b \in X^*$, the operator equation $Au = b$, $u \in X$, has a unique solution.*

The solution u depends continuously on b. More precisely, it follows from $Au_j = b_j$, $j = 1, 2$, that

$$\|u_1 - u_2\| \le c^{-1}\|b_1 - b_2\|,$$

and the inverse operator $A^{-1}: X^* \to X$ *is Lipschitz continuous with Lipschitz constant* c^{-1}.

Corollary 25.7. *Assume* (H1) *and* (H2). *Choose a fixed* t *with* $0 < t < 2c/L^2$. *Then, as* $n \to \infty$, *the approximation methods* (49a) *through* (49c) *converge to the unique solution of the equation* $Au = b, u \in X$.

In particular, for each $n \in \mathbb{N}$, *the Galerkin equation* (49b) *has a unique solution* v_n, *and we get the error estimate*

(E) $$\|v_n - u\| \le c^{-1} L \operatorname{dist}(u, X_n),$$

where $\operatorname{dist}(u, X_n) \to 0$ *as* $n \to \infty$.

Similarly as in Section 22.22e, relation (E) is very useful for the finite element method. In fact, (E) allows us to estimate the rapidity of convergence of the finite element method.

Proof. The case $X = \{0\}$ is trivial. Assume that $X \neq \{0\}$. Set $C = J^{-1}A$. For all $u, v \in X$, we obtain that

$$(Cu - Cv|u - v) = \langle Au - Av, u - v\rangle \ge c\|u - v\|^2 \tag{50}$$

and

$$\|Cu - Cv\| \le \|Au - Av\| \le L\|u - v\|,$$

by (H1). Note that $\|J^{-1}\| = 1$. This implies the *key* inequality:

$$\begin{aligned}\|Bu - Bv\|^2 &= \|u - v\|^2 - 2t(Cu - Cv|u - v) + t^2\|Cu - Cv\|^2 \\ &\le k^2\|u - v\|^2 \qquad \text{for all} \quad u, v \in X,\end{aligned}$$

where

$$k^2 = 1 - 2tc + t^2L^2.$$

By (50), $\|Cu - Cv\| \ge c\|u - v\|$ for all $u, v \in X$. Hence $0 < c \le L$. Now, an elementary discussion shows that $0 \le k < 1$ if $0 < t < 2c/L^2$.

Consequently, the operator $B: X \to X$ is k-contractive. By the Banach fixed-point theorem (Theorem 1.A), the equation $u = Bu, u \in X$, has a unique solution. Hence the equivalent equation $Au = b$, $u \in X$, also has a unique solution.

Now let $Au_j = b_j$. Then

$$\begin{aligned}c\|u_1 - u_2\|^2 &\le \langle Au_1 - Au_2, u_1 - u_2\rangle \\ &\le \|Au_1 - Au_2\|\,\|u_1 - u_2\|,\end{aligned}$$

and hence $\|u_1 - u_2\| \le c^{-1}\|Au_1 - Au_2\|$. This implies $\|A^{-1}b_1 - A^{-1}b_2\| \le c^{-1}\|b_1 - b_2\|$ for all $b_1, b_2 \in X^*$. The proof of Theorem 25.B is complete.

Corollary 25.7 follows from Theorem 25.A. In order to prove (E), we first note that the Galerkin equation (49b) is equivalent to

$$(J^{-1}(Av_n - b)|v) = 0 \qquad \text{for all} \quad v \in X_n.$$

By (21.30), this is equivalent to the equation

$$\langle Av_n - b, v\rangle = 0 \qquad \text{for all} \quad v \in X_n,$$

where we are looking for $v_n \in X_n$. This implies the *key* relation

$$\langle Av_n - b, v - v_n\rangle = 0 \qquad \text{for all} \quad v \in X_n.$$

Noting that $Au = b$, we get

$$\langle Av_n - Au, v_n - u\rangle + \langle Av_n - Au, u - v\rangle = 0,$$

for all $v \in X_n$. Hence

$$\begin{aligned} c\|v_n - u\|^2 &\le \langle Av_n - Au, v_n - u\rangle \\ &\le \|Av_n - Au\|\,\|u - v\| \\ &\le L\|v_n - u\|\,\|u - v\|, \end{aligned}$$

for all $v \in X_n$. This implies (E). □

25.5. Monotone and Pseudomonotone Operators, and the Calculus of Variations

In the following it is our goal to study the important connection between variational problems and the theory of monotone operators.

In the preceding section we considered the monotone operator equation

$$Au = b, \qquad u \in X. \tag{51}$$

We now investigate the special case that there exists a functional $f: X \to \mathbb{R}$ with $f'(u) = Au - b$ on X, i.e., we consider the operator equation

$$f'(u) = 0, \qquad u \in X, \tag{52}$$

together with the corresponding minimum problem

$$f(u) = \min!, \qquad u \in X. \tag{53}$$

An operator $B: X \to X^*$ on the B-space X is called a *potential operator* iff there exists a G-differentiable functional $f: X \to \mathbb{R}$ such that $B = f'$. Our plan is the following:

(i) We discuss the connection between the convexity of f and the monotonicity of f'.

(ii) We show that the solutions of (53) are also solutions of (52). This way, we obtain an important method in order to solve operator equations of the form (52) by solving minimum problems. Equation (52) is the abstract Euler equation to the variational problem (53).

(iii) We prove existence theorems for convex minimum problems and, more generally, for minimum problems with weakly sequentially lower semicontinuous functionals.
(iv) This leads very simply to existence theorems for monotone and pseudomonotone potential operators (Theorems 25.F and 25.G).
(v) In the following two chapters we shall show that the existence theorems in (iv) can be generalized to monotone and pseudomonotone operators which are *not* potential operators.
(vi) In Section 25.9 we shall consider applications to nonlinear conservation laws which allow many applications in various fields of physics.
(vii) Parallel to the minimum problem (53) on the entire space X we also consider minimum problems of the form

$$f(u) = \min!, \qquad u \in C,$$

where C is a convex subset of X. Then the solutions of this minimum problem are also solutions of the *variational inequality*

$$\langle f'(u), v - u \rangle \geq 0 \qquad \text{for all} \quad v \in C.$$

(viii) In Part III we shall study potential operators in greater detail. In particular, we shall consider eigenvalue problems for potential operators and a duality theory which leads, for example, to two-sided error estimates for the Ritz method.

Recall that a set C in a linear space (e.g., a B-space) is called *convex* iff

$$u, v \in C \qquad \text{and} \qquad t \in [0, 1] \quad \text{imply} \quad (1 - t)u + tv \in C,$$

i.e., if the points u and v belong to C, then the segment joining them also belongs to C (Fig. 25.3). A functional

$$f: C \to \mathbb{R}$$

on a convex set C is called *convex* iff

$$f((1 - t)u + tv) \leq (1 - t)f(u) + tf(v),$$

for all $t \in [0, 1]$ and all u, $v \in C$, i.e., the chords always lie above the curve corresponding to f (Figs. 25.4(a), (b)).

The functional f is called *strictly convex* iff

$$f((1 - t)u + tv) < (1 - t)f(u) + tf(v),$$

for all $t \in]0, 1[$ and all $u, v \in C$ with $u \neq v$, i.e., the interior points of the chords lie properly above the curve (Fig. 25.4(a)). The function f in Figure 25.4(b) is convex, but not strictly convex.

Moreover, f is called *concave* (resp. strictly concave) iff $-f$ is convex (resp. strictly convex). As we will see in Part III, convexity is one of the most important notions in mathematics and physics. For example, in many cases *energy* is convex and *entropy* is concave.

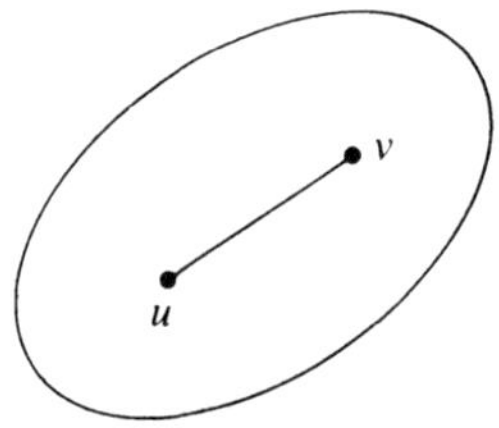

Figure 25.3

Classical Prototype 25.8. *Let C be a convex set in $\mathbb{R}$, i.e., C is an interval, and let $f: C \to \mathbb{R}$ be a differentiable function. Then the following facts are well-known:*

(i) *f is* convex *on C iff f' is* monotone *on C (Fig. 25.5).*
(ii) *f is strictly convex on C iff f' is strictly monotone on C.*
(iii) *If f is strictly convex, then f has at most one minimal point on C.*
(iv) *If $f: \mathbb{R} \to \mathbb{R}$ is convex and $f(u) \to +\infty$ as $|u| \to \infty$ holds, then f has a minimum on $\mathbb{R}$ (Fig. 25.5).*

The proof can be found, for example, in Fichtenholz (1972, M), Section 143.

In the following we want to generalize those results to B-spaces. In this connection, we will obtain very simple proofs by reducing the general case to

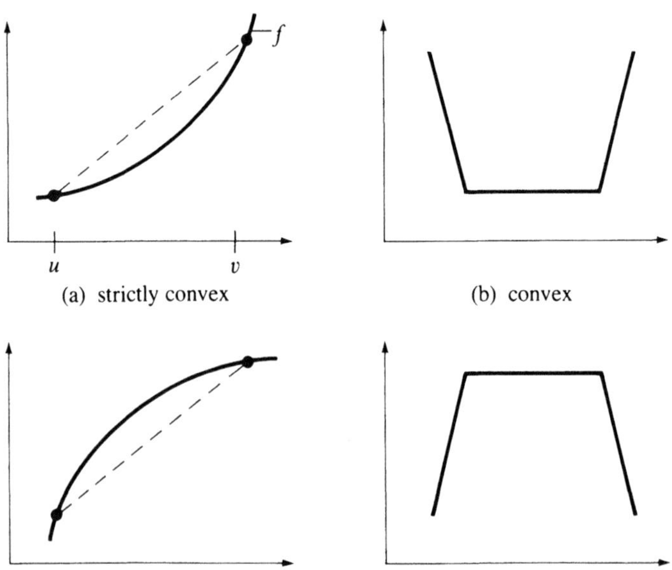

(a) strictly convex (b) convex

(c) strictly concave (d) concave

Figure 25.4

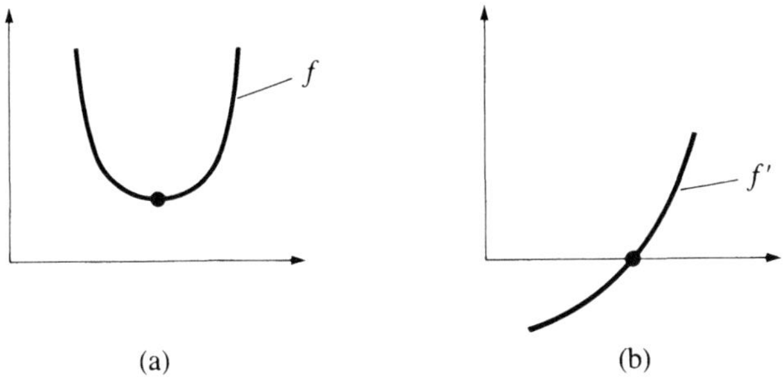

Figure 25.5

the Classical Prototype 25.8. To this end, we introduce the function

$$\varphi(t) = f((1 - t)u + tv)$$

for fixed u and v, where f is a functional on a linear space. Then φ is a real function. Since the convexity of f depends only on the values of f on segments, we immediately obtain the following result.

Lemma 25.9. *Let C be a convex set in a linear space (e.g., a* B*-space). Then the following two conditions are equivalent.*

(i) *The functional $f: C \to \mathbb{R}$ is convex (resp. strictly convex).*
(ii) *The real function $\varphi: [0, 1] \to \mathbb{R}$ is convex for all $u, v \in C$ (resp. φ is strictly convex on $]0, 1[$ for all $u, v \in C$ with $u \neq v$).*

The following result shows that the theory of monotone operators is a natural generalization of the calculus of variations for convex functionals.

Proposition 25.10. *Let $f: C \subseteq X \to \mathbb{R}$ be a G-differentiable functional on the convex set C in the real* B*-space X. Then the following two conditions are equivalent:*

(i) *f is convex (resp. strictly convex).*
(ii) *$f': C \to X^*$ is monotone (resp. strictly monotone).*

Convention. Recall the following from Section 4.2. First let C be an *open* subset of the B-space X (e.g., $C = X$). Then the functional $f: C \subseteq X \to \mathbb{R}$ is called G-*differentiable* iff it is G-differentiable at each point $u \in C$. That is, for each $u \in C$, there exists a functional $a \in X^*$ such that

$$\lim_{t \to 0} \frac{f(u + th) - f(u)}{t} = \langle a, h \rangle \qquad \text{for all} \quad h \in X.$$

We set $f'(u) = a$.

If C is an *arbitrary* subset of the B-space X, then $f: C \subseteq X \to \mathbb{R}$ is called G-*differentiable* iff f is defined on an open neighborhood of C and f is G-differentiable at each point $u \in C$.

PROOF. We fix $u, v \in C$ and set

$$\varphi(t) = f(u + t(v - u)), \qquad 0 \le t \le 1. \tag{54}$$

Differentiation yields $\varphi'(t) = f'(u + t(v - u))(v - u)$. Since $f'(w) \in X^*$, we can also write

$$\varphi'(t) = \langle f'(u + t(v - u)), v - u \rangle. \tag{55}$$

(I) Let $f: C \to \mathbb{R}$ be convex. Then $\varphi: [0, 1] \to \mathbb{R}$ is convex and φ' is monotone. From $\varphi'(1) \ge \varphi'(0)$ we obtain that

$$\langle f'(v) - f'(u), v - u \rangle \ge 0 \qquad \text{for all} \quad u, v \in C,$$

i.e., f' is monotone.

(II) Let $f': C \to X^*$ be monotone. If $s < t$, then

$$\varphi'(t) - \varphi'(s) = \langle f'(u + t(v - u)) - f'(u + s(v - u)), v - u \rangle \ge 0,$$

since f' is monotone. Hence φ' is monotone. Thus, φ is convex and hence f is convex. □

The following proposition describes the connection between the minimum problem (53) and the operator equation (52) above. At the same time we will show how variational inequalities arise from variational problems.

Proposition 25.11 (Abstract Euler Equation). *Let $f: C \subseteq X \to \mathbb{R}$ be G-differentiable on the subset C of the* B-*space X. Suppose that u is a solution of the minimum problem*

$$f(u) = \min!, \qquad u \in C.$$

Then:

(a) *If C is open, then u is a solution of the operator equation*

$$f'(u) = 0.$$

(b) *If C is convex, then u is a solution of the variational inequality*

$$\langle f'(u), v - u \rangle \ge 0 \qquad \textit{for all} \quad v \in C.$$

PROOF. We set $\varphi(t) = f(u + t(v - u))$ for fixed $u, v \in C$.

(I) Let C be convex. If $f: C \to \mathbb{R}$ has a minimum at u, then $\varphi: [0, 1] \to \mathbb{R}$ has a minimum at $t = 0$, i.e.,

$$\varphi'(0) \ge 0.$$

This implies

$$\langle f'(u), v - u \rangle \ge 0 \qquad \text{for all} \quad v \in C.$$

(II) If C is open, then this inequality holds for all v in a neighborhood of u, i.e., $f'(u) = 0$. □

25.5a. A General Minimum Principle

Our next goal is to obtain existence theorems for minimum problems on B-spaces.

Classical Prototype 25.12 (Theorem of Weierstrass). *Let* $-\infty < a < b < \infty$. *If* $f: [a,b] \to \mathbb{R}$ *is continuous, then* f *has a minimum and a maximum on* $[a,b]$ (*Fig.* 25.6(a)).

Let $f: M \subseteq X \to \mathbb{R}$ be a continuous function on a nonempty bounded closed set M of a B-space X with $\dim X < \infty$. Then f has a minimum and a maximum on M.

This is a classical generalization of Example 25.12. Note that X can be identified with $\mathbb{R}^N$ for some N. Unfortunately, this result does *not* remain true in the case where $\dim X = \infty$. Many of the troubles in the history of the calculus of variations came from this fact. The deeper reason for this negative result is based on the theorem that the closed unit ball in X is compact iff $\dim X < \infty$. Thus, we need a notion which is weaker than continuity. This notion is the weak sequential lower semicontinuity of functionals. The following definition is the *most important definition* in the calculus of variations.

Definition 25.13. Let $f: M \subseteq X \to \mathbb{R}$ be a functional on the subset M of the B-space X. Then f is called *weakly sequentially lower semicontinuous* on M iff,

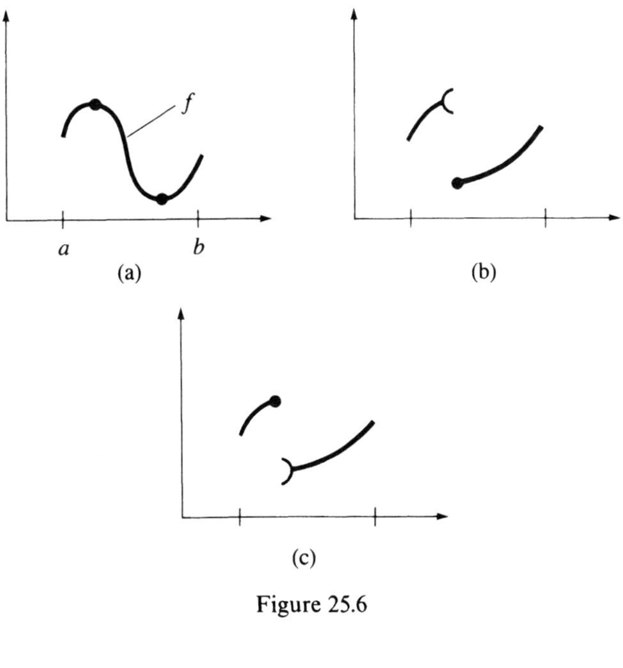

Figure 25.6

for each $u \in M$ and each sequence (u_n) in M,

$$u_n \rightharpoonup u \quad \text{as } n \to \infty \qquad \text{implies} \qquad f(u) \leq \varliminf_{n\to\infty} f(u_n).$$

EXAMPLE 25.14. Let $f: [a, b] \to \mathbb{R}$ be a function with $-\infty < a < b < \infty$.

(i) If f is continuous, then f is also weakly sequentially lower semicontinuous.
(ii) The function f in Figure 25.6(b) is weakly sequentially lower semicontinuous, but f in Figure 25.6(c) does not have this property.

PROOF. We prove (i). Note that in a finite-dimensional B-space weak convergence and strong convergence coincide. Thus, if $u_n \rightharpoonup u$ as $n \to \infty$, then $u_n \to u$ and hence $f(u) = \lim_{n\to\infty} f(u_n)$. □

The following theorem is the *most important theorem* in the calculus of variations. It is motivated by Figure 25.6(b).

Theorem 25.C (Main Theorem on Minimum Problems). *Suppose that the functional $f: M \subseteq X \to \mathbb{R}$ has the following two properties.*

(i) *M is a nonempty bounded closed convex set in the reflexive* B-*space X.*
(ii) *f is weakly sequentially lower semicontinuous on M.*

Then f has a minimum.

In Proposition 25.20 below we will show that large classes of functionals possess the property (ii). For example, each continuous convex functional $f: M \to \mathbb{R}$ on a closed convex set M in a B-space is weakly sequentially lower semicontinuous. Thus, we can replace condition (ii) by the following condition: f is continuous and convex on M.

Corollary 25.15 (Uniqueness). *A strictly convex functional $f: M \to \mathbb{R}$ on a convex set M has at most one minimal point.*

PROOF. We set $\alpha = \inf_{u\in M} f(u)$. Then there is a sequence (u_n) in M such that $f(u_n) \to \alpha$ as $n \to \infty$. Since M is bounded and X is reflexive, there is a subsequence, again denoted by (u_n), so that $u_n \rightharpoonup u$ as $n \to \infty$. The set M is closed and convex. Hence $u \in M$. By (ii),

$$f(u) \leq \varliminf_{n\to\infty} f(u_n) = \alpha.$$

Therefore, $f(u) = \alpha$. □

PROOF OF COROLLARY 25.15. Suppose that $f(u) = f(v) = \alpha$ and $u \neq v$. Then

$$f(2^{-1}(u + v)) < 2^{-1}(f(u) + f(v)) = \alpha.$$

This is a contradiction to the construction of α above. □

25.5b. The Main Theorem on Weakly Coercive Functionals

If the functional f in Theorem 25.C attains its minimum at an inner point u of M, then $f'(u) = 0$ by Proposition 25.11. The following coerciveness properties of f guarantee this.

Definition 25.16. Let $f: M \subseteq X \to \mathbb{R}$ be a functional on the subset M of the B-space X.

(i) f is called *coercive* iff $f(u)/\|u\| \to +\infty$ as $\|u\| \to \infty$ on M.
(ii) f is called *weakly coercive* iff $f(u) \to +\infty$ as $\|u\| \to \infty$ on M.

EXAMPLE 25.17. Let $a: X \times X \to \mathbb{R}$ be a strongly positive bilinear functional on the B-space X. Then a is coercive.

PROOF. For all $u \in X$ and fixed $c > 0$, $a(u, u) \geq c\|u\|^2$. Hence $a(u, u)/\|u\| \to +\infty$ as $\|u\| \to \infty$. □

In contrast to Theorem 25.C, the set M can be unbounded in the following theorem.

Theorem 25.D (Main Theorem on Weakly Coercive Functionals). *Suppose that the functional $f: M \subseteq X \to \mathbb{R}$ has the following three properties:*

(i) *M is a nonempty closed convex set in the reflexive* B-*space X (e.g., $M = X$).*
(ii) *f is weakly sequentially lower semicontinuous on M.*
(iii) *f is weakly coercive.*

Then f has a minimum on M.

By Proposition 25.20 below, assumption (ii) is satisfied if, for example, f is convex and continuous on M.

Corollary 25.18. *If, in addition, $M = X$ and f is* G-*differentiable on X, then the operator equation $f'(u) = 0$ has a solution on X.*

PROOF. Let u_0 be a fixed element in M. Since $f(u) \to +\infty$ as $\|u\| \to \infty$ we find a closed ball $B = \{u \in X: \|u\| \leq R\}$ so that $u_0 \in B \cap M$ and

$$f(u) \geq f(u_0) \qquad \text{outside} \qquad B \cap M.$$

Hence it is sufficient to consider the minimum problem for f on $B \cap M$. Then Theorem 25.C yields the assertion. □

Corollary 25.18 follows from Proposition 25.11.

25.5c. Criteria for the Weak Sequential Lower Semicontinuity of Functionals

The following three standard examples show that large classes of functionals are weakly sequentially lower semicontinuous.

Proposition 25.19. *Let $f: M \subseteq X \to \mathbb{R}$ be a functional on the* closed *set M of the* B*-space X with* $\dim X < \infty$. *Then f is weakly sequentially lower semicontinuous if one of the following two conditions is satisfied*:

(i) *f is continuous.*
(ii) *f is lower semicontinuous.*

PROOF. We set

$$M_r = \{u \in M: f(u) \le r\}. \tag{56}$$

The point is that it follows from (i) or (ii) that the set M_r is *closed* for each r (cf. Definition 9.11).

If f is not weakly sequentially lower semicontinuous on M, then there exists a sequence (u_n) in M such that $u_n \rightharpoonup u$ as $n \to \infty$ and

$$f(u) > \lim_{n\to\infty} f(u_n).$$

Hence there is an r so that $r < f(u)$ and $u_n \in M_r$ for all $n \ge n_0$. Since $\dim X < \infty$, we obtain $u_n \to u$ as $n \to \infty$; therefore, $u \in M_r$. This contradicts $r < f(u)$. □

Proposition 25.20. *Let $f: M \subseteq X \to \mathbb{R}$ be a functional on the* convex closed *set M of the* B*-space X. Then f is weakly sequentially lower semicontinuous if one of the following three conditions is satisfied*:

(i) *f is continuous and convex.*
(ii) *f is lower semicontinuous and convex.*
(iii) *f is* G*-differentiable on M and f' is monotone on M.*

PROOF. Ad(i), (ii). The set M_r in (56) is *closed and convex* for all r. If the assertion is not true, then there is a sequence (u_n) in M with $u_n \rightharpoonup u$ as $n \to \infty$ and

$$f(u) > \lim_{n\to\infty} f(u_n).$$

Consequently, there is an r so that $r < f(u)$ and $u_n \in M_r$ for all $n \ge n_0$. Since M_r is convex and closed, $u \in M_r$. This is a contradiction.

Ad(iii). We set $\varphi(t) = f(u + t(v - u))$. Then φ: $[0, 1] \to \mathbb{R}$ is convex and φ' is monotone. By the classical mean value theorem,

$$\varphi(1) - \varphi(0) = \varphi'(\vartheta) \ge \varphi'(0), \qquad 0 < \vartheta < 1,$$

i.e.,

$$f(v) \ge f(u) + \langle f'(u), v - u\rangle \qquad \text{for all} \quad u, v \in M. \tag{57}$$

If $u_n \rightharpoonup u$, then $\langle f'(u), u_n - u \rangle \to 0$ as $n \to \infty$. Hence

$$\underline{\lim}_{n \to \infty} f(u_n) \geq f(u). \qquad \square$$

In Chapter 27 we shall show that important classes of quasi-linear elliptic differential operators correspond to pseudomonotone operators which generalize monotone operators. In the following, we want to show that this class of operators also plays an important role in the calculus of variations. By definition, an operator

$$A: M \subseteq X \to X^*$$

on the subset M of the real B-space X is called *pseudomonotone* iff, for each $u \in M$ and each sequence (u_n) in M,

$$u_n \rightharpoonup u \quad \text{as } n \to \infty \qquad \text{and} \qquad \overline{\lim}_{n \to \infty} \langle Au_n, u_n - u \rangle \leq 0$$

imply

$$\langle Au, u - w \rangle \leq \underline{\lim}_{n \to \infty} \langle Au_n, u_n - w \rangle \qquad \text{for all} \quad w \in X.$$

This definition seems to be obscure. However, in Section 27.2 we shall show that pseudomonotone operators are quite natural objects. The *prototype* of a pseudomonotone operator $A: X \to X^*$ on a real reflexive B-space X is

$$A = B + C,$$

where $B: X \to X^*$ is monotone and continuous and $C: X \to X^*$ is strongly continuous, i.e., $u_n \rightharpoonup u$ implies $Cu_n \to Cu$ as $n \to \infty$. Thus, sufficiently regular perturbations of continuous monotone operators are pseudomonotone. In particular, if $f: X \to \mathbb{R}$ is a C^1-functional on the real reflexive B-space X, then f' is pseudomonotone if f is a sufficiently regular perturbation of a convex C^1-functional.

Proposition 25.21. *Let $f: M \subseteq X \to \mathbb{R}$ be a C^1-functional on the open convex set M of the real* B*-space X, and let f' be pseudomonotone and bounded.*

Then, f is weakly sequentially lower semicontinuous on M.

PROOF.

(I) Let $v_n \rightharpoonup v$ on M as $n \to \infty$. The pseudomonotonicity of f' implies

$$\underline{\lim}_{n \to \infty} \langle f'(v_n), v_n - v \rangle \geq 0.$$

Otherwise there would exist a subsequence, again denoted by (v_n), such that

$$\lim_{n \to \infty} \langle f'(v_n), v_n - v \rangle < 0.$$

Since f' is pseudomonotone, we obtain $\underline{\lim}_{n \to \infty} \langle f'(v_n), v_n - v \rangle \geq 0$. This is a contradiction.

(II) We set $\varphi(t) = f(u + t(v - u))$. From

$$\varphi(1) - \varphi(0) = \int_0^1 \varphi'(t)\,dt$$

we obtain the *key* relation

$$f(v) = f(u) + \int_0^1 \langle f'(u + t(v - u)), v - u\rangle\,dt \qquad \text{for all} \quad u, v \in M. \tag{58}$$

(III) Let $u_n \rightharpoonup u$ on M as $n \to \infty$. We set

$$v_n(t) = u + t(u_n - u) \qquad \text{for} \quad t \in [0, 1].$$

Then $v_n(t) \rightharpoonup u$ on M as $n \to \infty$. Let $t > 0$. By (I),

$$t^{-1} \varliminf_{n\to\infty} \langle f'(v_n(t)), v_n(t) - u\rangle = \varliminf_{n\to\infty} \langle f'(v_n(t)), u_n - u\rangle \geq 0. \tag{59}$$

From (58) we obtain that

$$f(u_n) - f(u) \geq \int_0^1 [\langle f'(v_n(t)), u_n - u\rangle]_-\,dt,$$

where we set $[a]_- = \min(a, 0)$. Since f' is continuous, the integrand is continuous with respect to t. Because f' is bounded, the integrand is uniformly bounded. Finally, by (59), the integrand goes to zero as $n \to \infty$. Hence

$$\varliminf_{n\to\infty} f(u_n) \geq f(u). \qquad \square$$

Further sharp criteria for the weak sequential lower semicontinuity of functionals can be found in Section 41.4.

25.5d. The Main Theorem on Convex Minimum Problems

Theorem 25.E. *Let $f: X \to \mathbb{R}$ be a convex, continuous, and weakly coercive functional on the reflexive* B-*space X. Then, f has a minimum on X.*

The minimal point is unique if f is strictly convex.

Proof. By Proposition 25.20, the existence and uniqueness of a minimal point follows from Theorem 25.D and Corollary 25.15, respectively. $\square$

25.6. The Main Theorem on Monotone Potential Operators

Theorem 25.F. *Let $f: X \to \mathbb{R}$ be a* G-*differentiable functional on the real reflexive* B-*space X with the following two properties*:

(i) *f' is monotone on X.*
(ii) *f is weakly coercive.*

Then:

(a) *The minimum problem*

$$f(u) = \min!, \qquad u \in X, \tag{60}$$

and the operator equation

$$f'(u) = 0, \qquad u \in X, \tag{60*}$$

are equivalent.
(b) *Both problems have a solution.*
(c) *If f' is strictly monotone on X, then the solutions of* (60) *and* (60*) *are unique.*

Proof. Ad(a). If u is a solution of (60), then u is also a solution of (60*) by Proposition 25.11.

Conversely, let u be a solution of (60*). From (57) it follows that

$$f(u) + \langle f'(u), v - u\rangle \leq f(v) \qquad \text{for all} \quad v \in X,$$

i.e., u is a solution of (60).

Ad(b). By Proposition 25.20, f is weakly sequentially lower semicontinuous. Theorem 25.D yields the assertion.

Ad(c). If $f'(u) = f'(v) = 0$, then $\langle f'(v) - f'(u), v - u\rangle = 0$. The strict monotonicity of f' yields $u = v$. □

The following proposition is an interesting special case of Theorem 25.F. The key condition is

$$\delta^2 f(u; h) \geq c(\|h\|)\|h\| \qquad \text{for all} \quad u, h \in X. \tag{61}$$

Proposition 25.22. *Suppose that*:

(i) *The functional $f: X \to \mathbb{R}$ is* G*-differentiable on the real reflexive* B*-space X, and the second variation $\delta^2 f(u; h)$ exists for all $u, h \in X$.*
(ii) *There is a function $c: \mathbb{R}_+ \to \mathbb{R}_+$ with $c(t) \to +\infty$ as $t \to +\infty$ so that* (61) *holds.*

Then:

(a) *The minimum problem* (60) *and the operator equation* (60*) *are equivalent and both problems have a solution.*
(b) *For all $u, v \in X$,*

$$\langle f'(v) - f'(u), v - u\rangle \geq c(\|v - u\|)\|v - u\|.$$

(c) *If $c(t) > 0$ for all $t > 0$, then f' is strictly monotone on X and the solutions of* (60) *and* (60*) *are unique.*

Corollary 25.23. *If, in addition, C is a nonempty closed convex subset of X, then the minimum problem*

$$f(u) = \min!, \qquad u \in C, \tag{62}$$

and the variational inequality

$$\langle f'(u), v - u \rangle \geq 0 \qquad \text{for all} \quad v \in C \quad \text{and fixed} \quad u \in C \tag{62*}$$

are equivalent and both problems have a solution.

If $c(t) > 0$ for all $t > 0$, then the solutions of (62) *and* (62*) *are unique.*

PROOF. We set $\varphi(t) = f(u + t(v - u))$ for all $t \in [0, 1]$ and fixed $u, v \in X$. Then

$$\varphi'(t) = \langle f'(u + t(v - u)), v - u \rangle,$$

$$\varphi''(t) = \delta^2 f(u + t(v - u); v - u).$$

(I) For $s < t$, the mean value theorem yields

$$\varphi'(t) - \varphi'(s) = \varphi''(\vartheta)(t - s) \geq 0, \qquad 0 < \vartheta < 1.$$

Therefore, φ' is monotone and hence φ is convex. Consequently, f is convex and f' is monotone.

(II) It follows from $\varphi'(1) - \varphi'(0) = \varphi''(\vartheta)$, $0 < \vartheta < 1$, that

$$\langle f'(v) - f'(u), v - u \rangle \geq c(\|v - u\|)\|v - u\|.$$

(III) The Taylor formula yields $\varphi(1) = \varphi(0) + \varphi'(0) + \frac{1}{2}\varphi''(\vartheta)$, i.e., for $u = 0$,

$$f(v) \geq f(0) + \langle f'(0), v \rangle + \tfrac{1}{2}c(\|v\|)\|v\|.$$

Since $|\langle f'(0), v \rangle| \leq \|f'(0)\| \|v\|$, we obtain that $f(v) \to +\infty$ as $\|v\| \to \infty$.

Theorem 25.F yields the assertion. □

Corollary 25.23 follows from Proposition 25.11 and Theorem 25.D.

Remark 25.24. The proof shows that Corollary 25.23 also holds if we replace (61) with the condition

$$\varphi''(t) \geq c(\|v - u\|)\|v - u\|$$

for all $t \in [0, 1]$ and all $u, v \in C$.

25.7. The Main Theorem on Pseudomonotone Potential Operators

Theorem 25.G. *Let $f: X \to \mathbb{R}$ be a C^1-functional on the real reflexive* B*-space X with the following two properties:*

(i) *f' is pseudomonotone and bounded on X.*
(ii) *f is weakly coercive.*

Then the minimum problem

$$f(u) = \min!, \qquad u \in X,$$

has a solution which is also a solution of the operator equation

$$f'(u) = 0, \qquad u \in X.$$

PROOF. By Proposition 25.21, f is weakly sequentially lower semicontinuous. Theorem 25.D yields the assertion. □

25.8. Application to the Main Theorem on Quadratic Variational Inequalities

We consider the quadratic variational inequality

$$a(u, v - u) \geq b(v - u) \qquad \text{for all} \quad v \in C \quad \text{and fixed} \quad u \in C, \tag{63}$$

and we make the following assumptions:

(H1) C is a nonempty closed convex set in the real H-space X (e.g., $C = X$).
(H2) $a: X \times X \to \mathbb{R}$ is bilinear, bounded, and strongly positive, i.e., $a(u, u) \geq c\|u\|^2$ for all $u \in X$ and fixed $c > 0$.
(H3) $b: X \to \mathbb{R}$ is linear and continuous.

In the special case $C = X$, problem (63) is equivalent to the equation

$$a(u, v) = b(v) \qquad \text{for all} \quad v \in X \quad \text{and fixed} \quad u \in X. \tag{64}$$

We first consider the easier case that $a(\cdot, \cdot)$ is symmetric. To this end, we study the minimum problem

$$2^{-1}a(u, u) - b(u) = \min!, \qquad u \in C. \tag{65}$$

Proposition 25.25. *Suppose that* (H1) *through* (H3) *hold and* $a(\cdot, \cdot)$ *is symmetric. Then the variational inequality* (63) *and the minimum problem* (65) *are equivalent and both problems have a unique solution.*

The functional in (65) *is convex and weakly coercive on* X.

PROOF. We set $f(u) = 2^{-1}a(u, u) - b(u)$ and

$$\varphi(t) = f(u + t(v - u)),$$

for all $t \in \mathbb{R}$ and fixed $u, v \in X$. A simple computation yields

$$\varphi''(t) = a(u - v, u - v),$$

since

$$f(u + th) = 2^{-1}a(u, u) + ta(u, h) + 2^{-1}t^2 a(h, h) - b(u) - tb(h).$$

Hence

$$\langle f'(u), v - u\rangle = \varphi'(0) = a(u, v - u) - b(v - u),$$

$$\delta^2 f(u; v - u) = \varphi''(0) = a(u - v, u - v) \geq c\|u - v\|^2.$$

Corollary 25.23 yields the existence and uniqueness result.

Finally, it follows from $\varphi''(t) \geq 0$ that φ is convex, i.e., f is convex on X. Since

$$f(u) \geq 2^{-1}c\|u\|^2 - \|b\|\,\|u\|,$$

f is weakly coercive on X. □

This proposition shows that the main theorem on monotone potential operators (Theorem 25.F) *generalizes* the main theorem on quadratic variational problems (Theorem 18.A).

We now consider the general case, i.e., $a(\cdot,\cdot)$ need *not* be symmetric. The point is that in this case we *cannot* apply variational methods. However, we can use the same trick as in the proof of Theorem 25.B, i.e., instead of the original problem (63) we consider the following *equivalent* problem:

$$(u|v - u) \geq (z|v - u) - t[a(z, v - u) - b(v - u)] \qquad \text{for all} \quad v \in C, \tag{66a}$$

$$z = u, \qquad u \in C, \tag{66b}$$

where $t > 0$ is fixed. As in the proof of Theorem 25.B, we will apply the Banach fixed-point theorem.

Theorem 25.H. *If* (H1) *through* (H3) *hold, then the variational inequality* (63) *has a unique solution.*

Proof.

(I) Uniqueness. Let u and $\bar{u}$ be solutions of (63). Then, for all $v \in C$,

$$a(u, v - u) \geq b(v - u),$$

$$a(\bar{u}, v - \bar{u}) \geq b(v - \bar{u}).$$

Choosing $v = \bar{u}$ and $v = u$ in the first and second equation, respectively, we obtain

$$-a(u - \bar{u}, u - \bar{u}) \geq 0,$$

i.e., $u = \bar{u}$.

(II) Existence. By Proposition 25.25, for each $z \in C$, the variational inequality (66a) has a solution $u \in C$. We set

$$u = Sz.$$

We shall show that the operator $S: C \to C$ is k-contractive. Then S has a fixed-point u by the Banach fixed-point theorem (Theorem 1.A). From $u = Su$ we obtain that u is a solution of (66), i.e., u is a solution of (63).

(II-1) Since $a(\cdot,\cdot)$ is bounded, there exists a linear continuous operator A:

$X \to X$ with

$$a(u, v) = (Au|v) \qquad \text{for all} \quad u, v \in X.$$

Let $u = Sz$ and $\bar{u} = S\bar{z}$. Then (66a) yields

$$(u|v - u) \geq (z|v - u) - t[(Az|v - u) - b(v - u)],$$

$$(\bar{u}|v - \bar{u}) \geq (\bar{z}|v - \bar{u}) - t[(A\bar{z}|v - \bar{u}) - b(v - \bar{u})].$$

Choosing $v = \bar{u}$ and $v = u$ in the first and second equation, respectively, we obtain

$$-\|u - \bar{u}\|^2 \geq -((I - tA)(z - \bar{z})|u - \bar{u}).$$

Since $|(x|y)| \leq \|x\|\,\|y\|$, it follows that

$$\|u - \bar{u}\| \leq \|(I - tA)(z - \bar{z})\|. \tag{67}$$

(II-2) We set $w = z - \bar{z}$. From

$$\|(I - tA)w\|^2 = \|w\|^2 + t^2\|Aw\|^2 - 2t(Aw|w)$$

we obtain

$$\|(I - tA)w\|^2 \leq k^2\|w\|^2,$$

where $k^2 = 1 + t^2\|A\|^2 - 2tc$. By (67), $\|u - \bar{u}\| \leq k\|z - \bar{z}\|$, i.e.,

$$\|Sz - S\bar{z}\| \leq k\|z - \bar{z}\| \qquad \text{for all} \quad z, \bar{z} \in C.$$

From $c\|u\|^2 \leq (Au|u) \leq \|Au\|\,\|u\|$, we obtain $c \leq \|A\|$. Thus, if we choose t such that $0 < t < 2c/\|A\|^2$, than $k < 1$, and hence the operator $S: C \to C$ is k-contractive. □

In Part V we shall show that the famous problem of the filtration of water through a dam leads to the variational inequality (63). Moreover, in Parts III through V we shall show that many problems in elasticity, plasticity, gas dynamics, heat conduction, control theory, etc., can be treated by using variational inequalities which are generalizations of (63).

25.9. Application to Nonlinear Stationary Conservation Laws

We want to study the fundamental nonlinear stationary conservation law

$$\begin{aligned} \operatorname{div} j &= f \quad \text{on } G, \\ u &= g \quad \text{on } \partial_1 G, \\ jn &= h \quad \text{on } \partial_2 G \end{aligned} \tag{68}$$

with the current density vector

$$j = -\alpha(|\operatorname{grad} u|^2)\operatorname{grad} u. \tag{69}$$

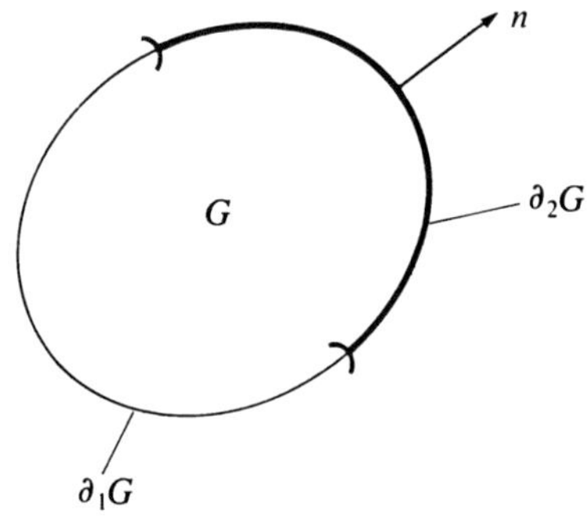

Figure 25.7

We assume:

G is a bounded region in $\mathbb{R}^N$, $N \geq 2$, with $\partial G \in C^{0,1}$. Moreover, $\partial_1 G$ and $\partial_2 G$ are disjoint open subsets of ∂G such that

$$\partial G = \overline{\partial_1 G} \cup \overline{\partial_2 G}. \tag{70}$$

Let n denote the outer unit normal vector on ∂G (Fig. 25.7).

Parallel to (68) we study the variational problem

$$\begin{gathered}\int_G (\beta(|\operatorname{grad} u|) - fu)\,dx + \int_{\partial_2 G} hu\,dO = \min!,\\ u = g \quad \text{on } \partial_1 G,\end{gathered} \tag{71}$$

where

$$\beta(s) = \frac{1}{2}\int_0^{s^2} \alpha(t)\,dt. \tag{72}$$

In the special case $\alpha \equiv 1$, we have $\beta(s) = s^2/2$ and the original problem (68) passes to the mixed boundary value problem for the Poisson equation

$$\begin{aligned} -\Delta u &= f \quad \text{on } G,\\ u &= g \quad \text{on } \partial_1 G,\\ -\frac{\partial u}{\partial n} &= h \quad \text{on } \partial_2 G.\end{aligned} \tag{73}$$

If $\partial_2 G = \varnothing$ and $\partial_1 G = \varnothing$, then we obtain the first and second boundary value problem, respectively. Of course, in case $\partial_2 G = \varnothing$ the boundary integral in (71) drops out.

Consequently, from the mathematical point of view, problem (68) is the simplest *nonlinear generalization* of the classical Dirichlet problem. In Part V we shall show that (68) comprehends many completely different physical problems in hydrodynamics and gas dynamics (subsonic and supersonic flow), electrostatics, magnetostatics, heat conduction, elasticity, and plasticity (e.g., the plastic torsion of rods), etc.

In order to have an *intuitive* picture at hand, let $N = 3$ and regard $u(x)$ as the temperature of a body G at the point x. Then j in (69) is the current density vector of the stationary heat flow in G. The function f describes outer heat sources. The boundary conditions prescribe the temperature u on the boundary part $\partial_1 G$ and the heat flow through the complementary boundary part $\partial_2 G$. Note that jn is the normal component of j. A detailed discussion of the precise physical meaning of (68) can be found in Sections 69.1 and 69.2. Equation (69) represents a *constitutive law* which depends on the specific properties of the material. In the case where $\alpha =$ constant, the positive number α is called the heat conductivity, and (69) is called the Fourier law of heat conductivity. The general case (69) corresponds to a nonlinear constitutive law, where the heat conductivity depends on the gradient of temperature. The negative sign in (69) reflects the fact that heat flows from points with higher temperature to points with lower temperature. The larger the gradient of temperature is the larger is the heat flow.

Our plan is the following:

(i) We prove existence and uniqueness results which only depend on the qualitative behavior of the material function $\alpha(\cdot)$. This allows an abundance of applications to physical problems. Furthermore, this way, we obtain a unified approach to many special results in the mathematical and physical literature.

(ii) We show that Theorem 25.B on Lipschitz continuous, strongly monotone operators can be applied to (68). Hence, for example, we obtain effective projection–iteration methods for solving (68).

(iii) We show that Theorem 25.C (the general minimum principle) can be applied to (68).

(iv) For solving equation (68) by a rapidly convergent iteration method, engineers invented the so-called Kačanov method. We prove the convergence of this method.

(v) We develop a nice duality theory for (68) which yields, for example, two-sided error estimates for the Ritz method. In this connection, the dual problem arises in a quite natural way, namely, it corresponds to a *dual* constitutive law.

(vi) We show that the *convexity* and monotonicity of the function β in (72) leads to a convex variational problem (71) and the corresponding Euler equation (68) is elliptic.

The latter result gives us considerable insight into the structure of (68). In Part V we shall show that subsonic flow corresponds to (68), where β is convex. The passage from subsonic flow to supersonic flow causes serious mathematical difficulties since β loses its convexity and equation (68) changes the type from ellipticity to hyperbolicity. In nature, there occur shock waves (e.g., sonic booms caused by supersonic aircraft). In Section 25.5 we have seen that convex minimum problems are relatively easy to handle. In contrast to this, there arise great difficulties if the functional loses its convexity. Exactly this situation occurs in the case of transonic flow. In Part V we shall prove existence theorems for transonic flow via variational methods by using the recent method of compensated compactness which overcomes the lack of convexity (resp. compactness). In this connection, an additional so-called entropy condition plays a fundamental role which ensures that, roughly speaking, the second law of thermodynamics is not violated. We postpone this until Part V, since in that volume we shall be able to understand completely the physical background.

The following proofs are based on the abstract results which we have proved in the preceding sections. Only the duality results (v) form an exceptional case. In this connection, we will use the abstract duality theory for monotone potential operators which will be developed in Part III. This should help the reader to obtain a complete picture with respect to the fundamental equation (68).

The key to our approach is the following result. We consider the function $\gamma\colon \mathbb{R}^N \to \mathbb{R}$ with

$$\gamma(x) = \beta(|x|),$$

where $\beta\colon \mathbb{R} \to \mathbb{R}$ is given by

$$\beta(s) = \frac{1}{2}\int_0^{s^2} \alpha(t)\,dt.$$

Lemma 25.26. *Let the function $\alpha\colon \mathbb{R}_+ \to \mathbb{R}_+$ be continuous. Then*

$$\langle \gamma'(x)|h\rangle = \alpha(|x|^2)\langle x|h\rangle \qquad \text{for all} \quad x, h \in \mathbb{R}^N.$$

(a) *If β is (strictly) convex, then so is γ.*

(b) *If β' is strongly monotone, i.e., there is a $d > 0$ so that*

$$\beta'(t) - \beta'(s) \geq d(t - s) \qquad \text{for all real} \quad t \geq s,$$

then γ' is strongly monotone, i.e.,

$$\langle \gamma'(x) - \gamma'(y)|x - y\rangle \geq d|x - y|^2 \qquad \text{for all} \quad x, y \in \mathbb{R}^N.$$

For example, this condition is satisfied if α is C^1 and $\beta''(s) \geq d > 0$ on $\mathbb{R}_+$.

(c) *Let α be C^1. If β is convex on $]a, b[$, then the quadratic form*

$$\gamma''(x)h^2 = 2\alpha'(|x|^2)\langle x|h\rangle^2 + \alpha(|x|^2)|h|^2 \tag{74}$$

is positive on $\mathbb{R}^N$ for all x with $a < |x| < b$.

(d) *If β' is Lipschitz continuous, i.e., there is an $L > 0$ so that*

$$|\beta'(t) - \beta'(s)| \leq L|t - s| \qquad \text{for all} \quad t, s \in \mathbb{R}_+,$$

then γ' is Lipschitz continuous, i.e.,

$$|\gamma'(x) - \gamma'(y)| \leq 3L|x - y| \qquad \text{for all} \quad x, y \in \mathbb{R}^N.$$

We give the proof in Problem 25.2. Note the following. If the function $\beta\colon \mathbb{R}_+ \to \mathbb{R}_+$ is convex and monotone increasing, then the function

$$x \mapsto \beta(|x|)$$

is convex on $\mathbb{R}^N$, $N \geq 1$. In fact, for all $x, y \in \mathbb{R}^N$ and $t \in [0,1]$, we get

$$\begin{aligned}\beta(|tx + (1-t)y|) &\leq \beta(t|x| + (1-t)|y|)\\ &\leq t\beta(|x|) + (1-t)\beta(|y|).\end{aligned}$$

Lemma 25.26 is closely connected with this simple observation.

25.9a. Convexity and Ellipticity

We use Cartesian coordinates $x = (\xi_1, \ldots, \xi_N)$ and set $D_i = \partial/\partial\xi_i$ as well as

$$Du = (D_1 u, \ldots, D_N u), \qquad DuDv = \sum_{i=1}^N D_i u D_i v,$$

$$|Du|^2 = \sum_{i=1}^N |D_i u|^2.$$

For brevity, we write Du instead of $\operatorname{grad} u$. Then the original equation (68) becomes

$$\begin{aligned}-\sum_{i=1}^N D_i(\alpha(|Du|^2)D_i u) &= f \quad \text{on } G,\\ u &= g \quad \text{on } \partial_1 G,\\ -\alpha(|Du|^2)\frac{\partial u}{\partial n} &= h \quad \text{on } \partial_2 G.\end{aligned} \tag{75}$$

The variational problem (71) becomes

$$\begin{gathered}\int_G (\beta(|Du|) - fu)\,dx + \int_{\partial_2 G} hu\,dO = \min!,\\ u = g \quad \text{on } \partial_1 G.\end{gathered} \tag{76}$$

Using (74), equation (75) can be written in the form

$$-(\gamma''(Du)D^2)u = f \quad \text{on } G. \tag{77}$$

In fact, it follows from (74) that

$$\begin{aligned}\gamma''(Du)D^2 = {}& 2\alpha'(|Du|^2)\sum_{i,j=1}^N D_i u D_j u D_i D_j\\ &+ \alpha(|Du|^2)\sum_{i=1}^N D_i^2.\end{aligned}$$

Equation (77) is called *elliptic* (resp. weakly elliptic) at the point x iff the corresponding quadratic form

$$h \mapsto \gamma''(Du(x))h^2$$

is *strictly positive* (resp. positive) on $\mathbb{R}^N$. Hence from Lemma 25.26 we immediately obtain the following result.

Proposition 25.27. *Let $\alpha\colon \mathbb{R}_+ \to \mathbb{R}_+$ be C^1 and let u be a C^2-solution of equation* (75).

If β is convex in a neighborhood of the point $|Du(x)|$, then equation (75) *is weakly elliptic at the point x with respect to u.*

25.9b. The Classical Variational Problem

Proposition 25.28. *Let G be a bounded region in $\mathbb{R}^N$ such that* (70) *holds. Suppose that the functions β, f, g, and h are sufficiently smooth. Then each sufficiently smooth solution u of the variational problem* (76) *is also a solution of the boundary value problem* (75).

PROOF. We write (76) in the form

$$F(u) = \min!,$$

$$u = g \quad \text{on } \partial_1 G,$$

and we set $\varphi(t) = F(u + tv)$, where v is a sufficiently smooth function on G with

$$v = 0 \quad \text{on } \partial_1 G.$$

If u is a solution of (76), then $\varphi'(0) = 0$, i.e.,

$$\int_G (\alpha(|Du|^2)DuDv - fv)\,dx + \int_{\partial_2 G} hv\,dO = 0.$$

Note that $\beta'(s) = \alpha(s^2)s$. Integration by parts yields

$$\int_G Av\,dx + \int_{\partial_2 G} Bv\,dO = 0, \tag{78}$$

where

$$A = -\sum_i D_i(\alpha(|Du|^2)D_i u) - f,$$

$$B = h + \alpha(|Du|^2)\frac{\partial u}{\partial n}.$$

In particular, equation (78) holds for all $v \in C_0^\infty(G)$. Hence $A = 0$. Variation of v then also yields $B = 0$. $\square$

25.9c. The Classical Variational Inequality

Let C be a convex set of sufficiently smooth functions. Instead of (76) we consider now the minimum problem

$$\begin{gathered}\int_G (\beta(|Du|) - fu)\,dx + \int_{\partial_2 G} hu\,dO = \min!,\\ u = g \quad \text{on } \partial_1 G, \qquad u \in C.\end{gathered} \tag{79}$$

For example, we may set

$$C = \{u \in C^1(\bar{G}) \colon |Du| \le c\},$$

where c is a constant.

Proposition 25.29. *Let G be a bounded region in $\mathbb{R}^N$ such that* (70) *holds. Suppose that the functions β, f, g, and h are sufficiently smooth and C is a convex set of sufficiently smooth functions.*

If u is a sufficiently smooth solution of the variational problem (79), *then u is a solution of the variational inequality*

$$\int_G \alpha(|Du|^2)Du(Dv - Du) - f(v - u)\,dx + \int_{\partial_2 G} h(v - u)\,dO \ge 0 \tag{80}$$

for all sufficiently smooth functions v on $\bar{G}$ with

$$v = g \quad \text{on } \partial_1 G \qquad \text{and} \qquad v \in C.$$

PROOF. We write (79) in the form

$$\begin{gathered}F(u) = \min!,\\ u = g \quad \text{on } \partial_1 G, \qquad u \in C,\end{gathered}$$

and we set $\varphi(t) = F(u + t(v - u))$, where $v = g$ on $\partial_1 G$ and $v \in C$. If u is a solution of (79), then $\varphi\colon [0, 1] \to \mathbb{R}$ has a minimum at $t = 0$, i.e.,

$$\varphi'(0) \ge 0.$$

This yields (80). □

25.10. Projection–Iteration Method for Conservation Laws

We set

$$Lu = -\sum_{i=1}^{N} D_i(\alpha(|Du|^2)D_i u)$$

and consider the boundary value problem

$$\begin{aligned} Lu &= f \quad \text{on } G, \\ u &= g \quad \text{on } \partial_1 G, \\ -\alpha(|Du|^2)\frac{\partial u}{\partial n} &= h \quad \text{on } \partial_2 G. \end{aligned} \tag{81}$$

Recall that

$$\beta(s) = \frac{1}{2}\int_0^{s^2} \alpha(t)\,dt.$$

The special case $\alpha \equiv 1$ corresponds to the mixed boundary value problem for the Poisson equation.

We make the following assumptions:

(H1) G is a bounded region in $\mathbb{R}^N$ such that (70) holds and $\partial_1 G \neq \emptyset$.

(H2) The function $\alpha\colon \mathbb{R}_+ \to \mathbb{R}_+$ is continuous and β' is strongly monotone and Lipschitz continuous, i.e., there are numbers $d > 0$ and $L > 0$ so that

$$\begin{aligned} \beta'(t) - \beta'(s) &\geq d(t - s) \qquad \text{for all real} \quad t \geq s, \\ |\beta'(t) - \beta'(s)| &\leq L|t - s| \qquad \text{for all} \quad t, s \geq 0. \end{aligned}$$

(H3) We set

$$X = \{w \in W_2^1(G)\colon w = 0 \text{ on } \partial_1 G\},$$

and we equip X with the scalar product

$$(u|v) = \int_G DuDv\,dx.$$

By A_2(53c), X becomes an H-space with respect to $(\cdot|\cdot)$. More precisely, the two norms $\|u\|_{1,2}$ and $\|u\|_X = (u|u)^{1/2}$ are equivalent on X.

(H4) We are given

$$f \in X^*, \qquad g \in W_2^{1/2}(\partial_1 G), \qquad h \in W_2^{1/2}(\partial_2 G)^*.$$

These conditions are fulfilled if, for example, we have

$$f \in L_2(G), \qquad g \in W_2^1(G), \qquad h \in L_2(\partial_2 G). \tag{82}$$

(H5) Let (Y_m) be a Galerkin scheme in X with

$$Y_m = \operatorname{span}\{w_{1m}, \ldots, w_{m'm}\}.$$

Assumption (H4) will be discussed in Remark 25.32 below. We shall show that, in some sense, condition (H4) is the most general condition which guarantees that the right-hand side b of the generalized problem (84) below belongs to the dual space X^*, and hence (84) has a solution.

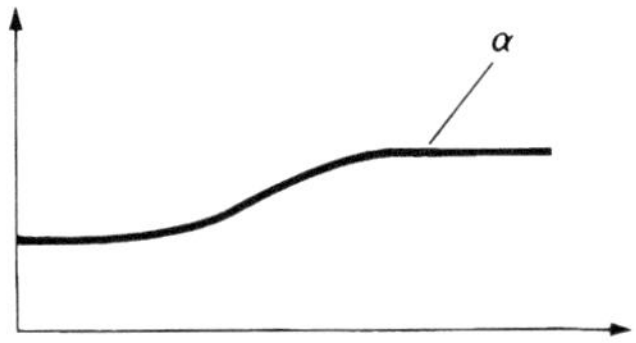

Figure 25.8

EXAMPLE 25.30. Condition (H2) is satisfied if the function $\alpha: \mathbb{R}_+ \to \mathbb{R}_+$ is C^1 and there are numbers $d > 0$ and $L > 0$ so that

$$0 < d \le \beta''(s) \le L \qquad \text{for all} \quad s \ge 0. \tag{83}$$

Since $\beta''(s) = 2\alpha'(s^2)s^2 + \alpha(s^2)$, condition (H2) is satisfied if, for example, the function $\alpha(\cdot)$ has the qualitative behavior of Figure 25.8, i.e., α is monotone and bounded, and

$$\alpha(0) > 0, \qquad \lim_{s \to +\infty} \alpha'(s^2)s^2 < \infty.$$

If we regard $\alpha(|\operatorname{grad} u|^2)$ as heat conductivity and u as temperature, then from the physical point of view, such a behavior of α is reasonable.

Definition 25.31. The *generalized problem* for (81) reads as follows: We seek a function $u \in W_2^1(G)$ such that

$$\begin{aligned} a(u, v) &= b(v) \qquad \text{for all} \quad v \in X, \\ u &= g \quad \text{on } \partial_1 G, \end{aligned} \tag{84}$$

where we set

$$a(u, v) = \int_G \alpha(|Du|^2) Du Dv \, dx,$$

$$b(v) = \int_G fv \, dx - \int_{\partial_2 G} hv \, dO.$$

The precise meaning of the boundary condition "$u = g$ on $\partial_1 G$" will be discussed in Remark 25.32 below.

If all the functions are sufficiently smooth, then we obtain (84) by multiplying (81) with $v \in X$ and subsequent integration by parts.

Let $\bar{g} \in W_2^1(G)$ be a *fixed* extension of the given function $g: \partial_1 G \to \mathbb{R}$ in the sense of Remark 25.32 below, i.e., $\bar{g} = g$ on $\partial_1 G$.

The *projection–iteration method* for the generalized problem (84) reads as follows. For $m = 1, 2, \ldots,$ we seek a function

$$u_m \in Y_m + \bar{g}$$

such that $u_0 = \bar{g}$ on G and

$$(u_m|w_{km}) = (u_{m-1}|w_{km}) - t[a(u_{m-1}, w_{km}) - b(w_{km})], \qquad k = 1, \ldots, m'. \quad (85)$$

Since $u_m = \bar{g} + \sum_k c_{km} w_{km}$, this is a *linear* system for the unknown real numbers $c_{1m}, \ldots, c_{m'm}$. If all the functions are sufficiently smooth, then integration by parts shows that (85) is exactly the Galerkin method for the following linear differential equation with respect to u_m:

$$\begin{aligned} -\Delta u_m &= -\Delta u_{m-1} - t(Lu_{m-1} - f) \quad \text{on } G, \\ u_m &= g \quad \text{on } \partial_1 G, \\ \frac{\partial u_m}{\partial n} &= \frac{\partial u_{m-1}}{\partial n} - t(\alpha(|Du_{m-1}|^2)\frac{\partial u_{m-1}}{\partial n} + h) \quad \text{on } \partial_2 G. \end{aligned} \qquad (85^*)$$

Theorem 25.I (The Mixed Boundary Value Problem for Conservation Laws). *Assume* (H1) *through* (H5). *Then*:

(i) Existence and uniqueness. *The generalized problem* (84) *has a unique solution u.*
(ii) Projection–iteration method. *If t is a fixed number with* $0 < t < 2d/9L^2$, *then the sequence* (u_m), *constructed by the projection–iteration method* (85) *above, converges to u in* $W_2^1(G)$ *as* $m \to \infty$.
(iii) Linear problem. *In the special case* $\alpha \equiv 1$, *we obtain the mixed boundary value problem for the Poisson equation. Here, the unique solution u of* (84) *satisfies the estimate*

$$\|u\|_W \le \text{const}(\|f\|_{X^*} + \|g\|_Y + \|h\|_{Z^*})$$

for all given $f \in X^*$, $g \in Y$, $h \in Z^*$, *where*

$$W = W_2^1(G), \qquad X = \{w \in W: w = 0 \text{ on } \partial_1 G\},$$
$$Y = W_2^{1/2}(\partial_1 G), \qquad Z = W_2^{1/2}(\partial_2 G).$$

In particular, for all given $f \in L_2(G)$, $g \in W_2^{1/2}(\partial_1 G)$, $h \in L_2(\partial_2 G)$, *we have* $f \in X^*$, $g \in Y$, $h \in Z^*$, *and the unique solution u of* (84) *satisfies the estimate*:

$$\|u\|_W \le \text{const}(\|f\|_{L_2(G)} + \|g\|_{W_2^{1/2}(\partial_1 G)} + \|h\|_{L_2(\partial_2 G)}).$$

Remark 25.32 (Interpretation of the Generalized Problem (84)).

(a) We discuss the assumption $g \in W_2^{1/2}(\partial_1 G)$. By $A_2(49)$ and $A_2(51)$, each function $g \in W_2^1(G)$ has generalized boundary values $g \in W_2^{1/2}(\partial G)$, and we have

$$\|g\|_{W_2^{1/2}(\partial G)} \le \text{const}\|g\|_{W_2^1(G)} \qquad \text{for all} \quad g \in W_2^1(G).$$

Conversely, each function $g \in W_2^{1/2}(\partial G)$ can be extended to a function

$\bar{g} \in W_2^1(G)$ such that

$$\bar{g} = g \quad \text{on } \partial G,$$

and $\|\bar{g}\|_{W_2^1(G)} \leq \text{const}\|g\|_{W_2^{1/2}(\partial G)}$ for all $g \in W_2^{1/2}(\partial G)$.

By definition, a function $g\colon \partial_1 G \to \mathbb{R}$ belongs to the space $W_2^{1/2}(\partial_1 G)$ iff it can be extended to a function $\bar{g} \in W_2^{1/2}(\partial G)$. We set

$$\|g\|_{W_2^{1/2}(\partial_1 G)} = \inf_{\bar{g}} \|\bar{g}\|_{W_2^{1/2}(\partial G)},$$

where the infimum is taken over all possible extensions of g. Recall that $\partial_1 G \subseteq \partial G$.

This way, $W_2^{1/2}(\partial_1 G)$ becomes a real normed space if we identify two functions whose values differ on a subset of $\partial_1 G$ of surface measure zero.

The space $W_2^{1/2}(\partial_2 G)$ is defined analogously.

(b) We show that the embeddings

$$W_2^{1/2}(\partial_j G) \subseteq L_2(\partial_j G), \qquad j = 1, 2,$$

are continuous. For example, let $j = 1$. By $A_2(51)$,

$$\left(\int_{\partial G} g^2\, dO\right)^{1/2} \leq \text{const}\|g\|_{W_2^{1/2}(\partial G)} \qquad \text{for all} \qquad g \in W_2^{1/2}(\partial G).$$

Hence

$$\left(\int_{\partial_1 G} g^2\, dO\right)^{1/2} \leq \text{const}\|g\|_{W_2^{1/2}(\partial_1 G)} \qquad \text{for all} \quad g \in W_2^{1/2}(\partial_1 G).$$

(c) We discuss the assumption $f \in X^*$. We first consider the special case $f \in L_2(G)$. The Hölder inequality yields

$$\left|\int_G fv\, dx\right| \leq \|f\|_2 \|v\|_{1,2} \leq \text{const}\|f\|_2 \|v\|_X,$$

for all $v \in X$, by (H3). Thus, f generates the linear continuous functional

$$v \mapsto \int_G fv\, dx$$

on the B-space X. In this sense we write $f \in X^*$. Note that

$$\|f\|_{X^*} \leq \text{const}\|f\|_2.$$

In the general case, $f \in X^*$, the integral $\int_G fv\, dx$ does not always make sense. However, we agree that $\int_G fv\, dx$ denotes the value of the functional f at the point v, i.e., we write

$$\langle f, v\rangle_X = \int_G fv\, dx \qquad \text{for all} \quad f \in X^*.$$

(d) We discuss the assumption $h \in Z^*$, where $Z = W_2^{1/2}(\partial_2 G)$. We first consider the special case $h \in L_2(\partial_2 G)$. Since the embeddings

$$X \subseteq W_2^1(G) \subseteq W_2^{1/2}(\partial G) \subseteq Z \subseteq L_2(\partial_2 G)$$

are continuous, we obtain that

$$\left| \int_{\partial_2 G} hv\, dO \right| \leq \left(\int_{\partial_2 G} h^2\, dO \right)^{1/2} \left(\int_{\partial_2 G} v^2\, dO \right)^{1/2}$$
$$\leq \text{const}\, \|h\|_{L_2(\partial_2 G)} \|v\|_Z \qquad \text{for all} \quad v \in Z.$$

Thus, h generates the linear continuous functional

$$v \mapsto \int_{\partial_2 G} hv\, dO$$

on Z. In this sense we write $h \in Z^*$. Note that

$$\|h\|_{Z^*} \leq \text{const}\, \|h\|_{L_2(\partial_2 G)} \qquad \text{for all} \quad h \in Z.$$

In the general case, $h \in Z^*$, we agree that $\int_{\partial_2 G} hv\, dO$ denotes the value of the functional h at the point v, i.e., we write

$$\langle h, v \rangle_Z = \int_{\partial_2 G} hv\, dO \qquad \text{for all} \quad v \in Z.$$

Since the embedding $X \subseteq Z$ is continuous by (a), it follows from $h \in Z^*$ that, for all $v \in X$,

$$|\langle h, v \rangle_Z| \leq \|h\|_{Z^*} \|v\|_Z \leq \text{const}\, \|h\|_{Z^*} \|v\|_X.$$

This implies $h \in X^*$ and

$$\|h\|_{X^*} \leq \text{const}\, \|h\|_{Z^*} \qquad \text{for all} \quad h \in Z^*.$$

(e) We show that $b \in X^*$. In fact, let $f \in X^*$ and $h \in Z^*$. By Definition 25.31,

$$b(v) = \int_G fv\, dx - \int_{\partial_2 G} hv\, dO.$$

Hence, for all $v \in X$,

$$|b(v)| = |\langle f, v \rangle_X - \langle h, v \rangle_Z| \leq \text{const}(\|f\|_{X^*} + \|h\|_{Z^*}) \|v\|_X.$$

Thus, we obtain $b \in X^*$ and

$$\|b\|_{X^*} \leq \text{const}(\|f\|_{X^*} + \|h\|_{Z^*}). \tag{86}$$

PROOF OF THEOREM 25.I IN THE LINEAR CASE. We shall use the main theorem on quadratic minimum problems (Theorem 22.A). Let $\alpha \equiv 1$.

(I) Equivalent norm on X. Recall that $X \subseteq W$, where

$$W = W_2^1(G) \qquad \text{and} \qquad X = \{w \in W \colon w = 0 \text{ on } \partial_1 G\}.$$

We equip X with the scalar product

$$(u|v) = \int_G DuDv\,dx \qquad \text{for all} \quad u, v \in X,$$

and we set

$$\|u\|_X = (u|u)^{1/2} = \left(\int_G |Du|^2\,dx\right)^{1/2},$$

$$\|u\|_W = \left(\int_G |u|^2 + |Du|^2\,dx\right)^{1/2}.$$

From A_2(53c) we obtain the important fact that the two norms $\|\cdot\|_W$ and $\|\cdot\|_X$ are equivalent on the subspace X of W. Thus, X becomes an H-space with respect to the scalar product $(\cdot|\cdot)$.

(II) The equivalent problem. The generalized problem (84) reads as follows:

$$\text{(P)} \qquad \begin{aligned} a(u, v) &= b(v) \qquad \text{for all} \quad v \in X, \\ u &= g \quad \text{on } \partial_1 G, \qquad u \in W, \end{aligned}$$

where $g \in Y = W_2^{1/2}(\partial_1 G)$ is given. According to Remark 25.32(a), we choose a *fixed* function $\bar{g} \in W$ such that

$$\bar{g} = g \quad \text{on } \partial_1 G \qquad \text{and} \qquad \|\bar{g}\|_W \le \text{const}\,\|g\|_Y.$$

Then, problem (P) is equivalent to:

$$a(u, v) = b(v) \qquad \text{for all} \quad v \in X, \quad u \in \bar{g} + X.$$

Finally, letting $w = u - \bar{g}$, we obtain that the original generalized problem (P) is equivalent to:

$$a(w, v) = b_1(v) \qquad \text{for all} \quad v \in X \quad \text{and fixed} \quad w \in X, \tag{87}$$

where

$$b_1(v) = b(v) - a(\bar{g}, v) \qquad \text{for all} \quad v \in X.$$

(III) We show that $b_1 \in X^*$. By (86), $b \in X^*$. For all $v \in X$,

$$|a(\bar{g}, v)| = \left|\int_G D\bar{g}Dv\,dx\right| \le \|\bar{g}\|_W \|v\|_X.$$

Therefore, we obtain from (86) that

$$\|b_1\|_{X^*} \le \text{const}(\|f\|_{X^*} + \|h\|_{Z^*} + \|\bar{g}\|_W).$$

(IV) Existence and uniqueness. Note that $a(w, v) = (w|v)$ for all $w, v \in X$ in the case where $\alpha \equiv 1$. Thus, it follows from Theorem 22.A that equation (87) has a unique solution w and

$$\|w\|_X \le \text{const}\,\|b_1\|_{X^*} \qquad \text{for all} \quad b_1 \in X^*.$$

Consequently, the original problem (P) has a unique solution u.

(V) Estimates. By (I),

$$\|w\|_W \le \text{const}\,\|b_1\|_{X^*} \qquad \text{for all} \quad b_1 \in X^*.$$

Noting that $u = w + \bar{g}$, we obtain

$$\|u\|_W \le \text{const}(\|b_1\|_{X^*} + \|\bar{g}\|_W).$$

Since $\|\bar{g}\|_W \le \text{const}\,\|g\|_Y$, this implies

$$\|u\|_W \le \text{const}(\|f\|_{X^*} + \|h\|_{Z^*} + \|g\|_Y). \qquad \square$$

PROOF OF THEOREM 25.I IN THE GENERAL CASE. We shall use Theorem 25.B.

(I) The equivalent problem. According to the preceding proof, the generalized problem (84) is equivalent to

$$\bar{a}(w, v) = b(v) \qquad \text{for all} \quad v \in X \quad \text{and fixed} \quad w \in X, \tag{87*}$$

where

$$\bar{a}(w, v) = \int_G \alpha(|D\bar{w}|^2) D\bar{w} Dv\, dx,$$

and $b \in X^*$, $\bar{g} \in W$ as well as

$$\bar{v} = v + \bar{g} \qquad \text{and} \qquad \bar{w} = w + \bar{g}.$$

(II) The equivalent operator equation. By (81), $\beta'(s) = \alpha(s^2)s$ and hence $\beta'(0) = 0$. Thus, it follows from assumption (H2) that $|\alpha(s^2)| \le L$ for all $s \in \mathbb{R}$ and hence

$$|\bar{a}(w, v)| \le L\|\bar{w}\|_X \|v\|_X \qquad \text{for all} \quad w, v \in X.$$

From (22.1a), we obtain that there exists an operator $A: X \to X^*$ such that

$$\langle Aw, v\rangle = \bar{a}(w, v) \qquad \text{for all} \quad w, v \in X.$$

Consequently, problem (87*) is equivalent to the operator equation

$$Aw = b, \qquad w \in X. \tag{87**}$$

(III) Properties of A. According to Lemma 25.26 we obtain that, for all v, w, $z \in X$,

$$\bar{a}(w, v) = \int_G \langle \gamma'(D\bar{w}) | Dv\rangle\, dx$$

as well as

$$\begin{aligned}\bar{a}(v, v - w) - \bar{a}(w, v - w) &= \int_G \langle \gamma'(D\bar{v}) - \gamma'(D\bar{w}) | D\bar{v} - D\bar{w}\rangle\, dx \\ &\ge d \int_G |D\bar{v} - D\bar{w}|^2\, dx = d\|v - w\|_X^2\end{aligned}$$

and

$$|\bar{a}(v,z) - \bar{a}(w,z)| = \left| \int_G \langle \gamma'(D\bar{v}) - \gamma'(D\bar{w}) | Dz \rangle \, dx \right|$$

$$\leq 3L \int_G |D\bar{v} - D\bar{w}||Dz| \, dx \leq 3L \|v - w\|_X \|z\|_X.$$

This implies that, for all $v, w \in X$,

$$\langle Av - Aw, v - w \rangle \geq d \|v - w\|_X^2,$$

$$\|Av - Aw\|_{X^*} \leq 3L \|v - w\|_X.$$

Thus, $A: X \to X^*$ is strongly monotone and Lipschitz continuous.

(IV) By Theorem 25.B, the operator equation (87**) has a unique solution and the corresponding projection–iteration method (85) converges. In connection with (85), note that $u = w + \bar{g}$, $w \in X$. □

25.11. The Main Theorem on Nonlinear Stationary Conservation Laws

In Sections 25.9b and 25.9c we have studied classical variational problems and variational inequalities with respect to conservation laws. We now want to prove the existence of generalized solutions for those problems. To this end, we set

$$F(u) = \int_G (\beta(|Du|) - fu) \, dx + \int_{\partial_2 G} hu \, dO$$

and consider the variational problem

$$F(u) = \min!, \qquad u \in W_2^1(G),$$
$$u = g \quad \text{on } \partial_1 G. \tag{88}$$

More generally, we consider

$$F(u) = \min!, \qquad u \in C,$$
$$u = g \quad \text{on } \partial_1 G, \tag{89}$$

where C is a convex subset of $W_2^1(G)$.

Problem (88) corresponds to the variational problem in Section 25.9b. If C is a proper subset of $W_2^1(G)$, then (89) corresponds to the variational inequality in Section 25.9c. Recall that

$$\beta(s) = \frac{1}{2} \int_0^{s^2} \alpha(t) \, dt.$$

We make the following assumptions:

(H1) G is a bounded region in $\mathbb{R}^N$ such that (70) holds and $\partial_1 G \neq \emptyset$. We set $W = W_2^1(G)$ and equip W with the equivalent scalar product

$$(u|v) = \int_G DuDv\,dx + \int_{\partial_1 G} uv\,dO$$

(cf. A_2(53c)), and we set $\|u\| = (u|u)^{1/2}$.

(H2) We are given

$$f \in W^*, \qquad g \in W_2^{1/2}(\partial_1 G), \qquad h \in W_2^{1/2}(\partial_2 G)^*. \tag{90}$$

For example, this assumption is fulfilled if

$$f \in L_2(G), \qquad g \in W_2^1(G), \qquad h \in L_2(\partial_2 G).$$

In the general case (90), the integrals $\int_G fu\,dx$ and $\int_{\partial_2 G} hu\,dO$ are to be understood in the sense of Remark 25.32.

(H3) We set

$$M = \{u \in W: u = g \text{ on } \partial_1 G\}.$$

Since the embedding $W \subseteq W_2^{1/2}(\partial_1 G)$ is continuous, the set M is closed and convex in W. Let C be a closed convex set in W with $M \cap C \neq \emptyset$ (e.g., $C = W$).

(H4) The function $\alpha: \mathbb{R}_+ \to \mathbb{R}_+$ is continuous and $\beta: \mathbb{R} \to \mathbb{R}$ is convex. There is a number $c > 0$ so that $\alpha(s) \leq 2c$ on $\mathbb{R}_+$, i.e.,

$$0 \leq \beta(s) \leq cs^2 \quad \text{on } \mathbb{R}.$$

If C is unbounded (e.g., $C = W$), then there is a number $a > 0$ so that $\alpha(s) \geq 2a$ on $\mathbb{R}_+$, i.e.,

$$as^2 \leq \beta(s) \quad \text{on } \mathbb{R}.$$

Figure 25.9 shows the typical behavior of α and β.

Theorem 25.J. *Under the assumptions* (H1) *through* (H4), *the variational problem* (89) *has a solution.*

If β is strictly convex on $\mathbb{R}$, then the solution is unique.

PROOF. We set

$$b(u) = \int_G fu\,dx - \int_{\partial_2 G} hu\,dO.$$

By Remark 25.32, $b \in W^*$.

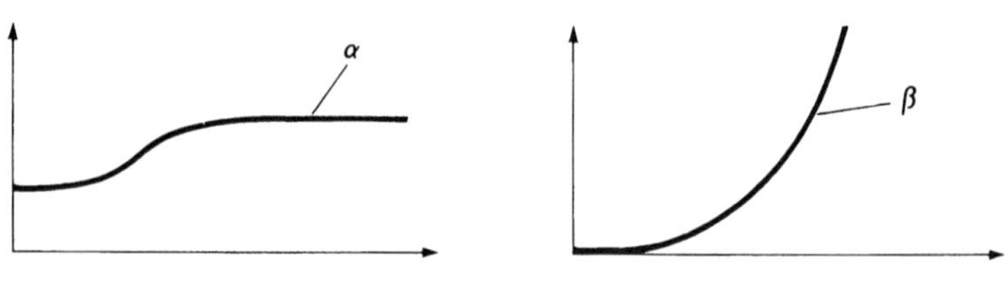

Figure 25.9

(I) The functional $F: W \to \mathbb{R}$ is continuous. In fact, we have the growth condition

$$0 \le \beta(|Du|) \le c|Du|^2. \tag{91}$$

Hence the integral $F(u)$ exists for all $u \in W$. If

$$u_n \to u \quad \text{in } W \qquad \text{as} \quad n \to \infty,$$

then $|Du_n| \to |Du|$ in $L_2(G)$ as $n \to \infty$. It follows from (91) and the continuity of the Nemyckii operator (Proposition 26.6) that

$$\beta(|Du_n|) \to \beta(|Du|) \quad \text{in } L_1(G) \qquad \text{as} \quad n \to \infty,$$

i.e., $F(u_n) \to F(u)$.

(II) F is convex on W, since $x \mapsto \beta(|x|)$ is convex on $\mathbb{R}^N$ by Lemma 25.26.

Moreover, if β is strictly convex on $\mathbb{R}$, then $x \mapsto \beta(|x|)$ is strictly convex on $\mathbb{R}^N$ and hence F is strictly convex on W.

(III) If C is bounded, then the assertion follows from Theorem 25.C.

(IV) Let C be unbounded. Then $\beta(s) \ge as^2$ on $\mathbb{R}$. For $u \in M$, this implies

$$\begin{aligned}\int_G \beta(|Du|)\,dx &\ge a \int_G |Du|^2\,dx \\ &\ge a\|u\|^2 - a \int_{\partial_1 G} g^2\,dO\end{aligned}$$

and

$$F(u) \ge a\|u\|^2 - \|b\|\,\|u\| + \text{const}.$$

Hence $F(u) \to +\infty$ as $\|u\| \to \infty$ on M, i.e., F is weakly coercive on M.

The assertion follows now from Theorem 25.D for $C \cap M$. □

25.12. Duality Theory for Conservation Laws and Two-Sided *a posteriori* Error Estimates for the Ritz Method

We consider the original problem

$$\begin{aligned} F(u) = \min!, \qquad & u \in W_2^1(G), \\ u = g \quad & \text{on } \partial_1 G, \end{aligned} \tag{92}$$

together with the so-called dual problem

$$F_*(p) = \max!, \qquad p \in L_2(G)^N, \tag{92a*}$$

$$\int_G pDu\,dx = b(u) \qquad \text{for all} \quad u \in W_2^1(G) \quad \text{with} \quad u = 0 \quad \text{on } \partial_1 G. \tag{92b*}$$

Here, we set $p = (p_1, \ldots, p_N)$, $pDu = \sum_{i=1}^N p_i D_i u$, and

$$b(u) = \int_G fu\,dx - \int_{\partial_2 G} hu\,dO,$$

$$F(u) = \int_G \beta(|Du|)\,dx - b(u),$$

$$F_*(p) = -\int_G (\beta^*(|p|) - pD\bar{g})\,dx - b(\bar{g}),$$

where $g = \bar{g}$ on $\partial_1 G$ and

$$\beta^*(s) = \int_0^s \beta'^{-1}(t)\,dt, \qquad \beta(s) = \frac{1}{2}\int_0^{s^2} \alpha(t)\,dt.$$

Here, β'^{-1} denotes the inverse function to $\beta'(s) = \alpha(s^2)s$. The function β^* is called the *conjugate* function to β. This notion coincides with the general definition of conjugate functionals in Section 51.1.

Condition (92b*) is the generalized formulation of the classical equation

$$\begin{aligned} -\sum_{i=1}^N D_i p_i &= f \quad \text{on } G, \\ -np &= h \quad \text{on } \partial_2 G, \end{aligned} \tag{93}$$

where n denotes the outer unit normal vector of ∂G. In fact, choose $u \in C^1(\bar{G})$ with $u = 0$ on $\partial_1 G$. Then multiplication of the first equation in (93) with u and subsequent integration by parts yield (92b*).

Recall that the classical Euler equation to the original problem (92) reads as follows:

$$\begin{aligned} -\sum_{i=1}^N D_i(\alpha(|Du|^2)D_i u) &= f \quad \text{on } G, \\ u &= g \quad \text{on } \partial_1 G, \\ -\alpha(|Du|^2)\frac{\partial u}{\partial n} &= h \quad \text{on } \partial_2 G. \end{aligned} \tag{94}$$

We shall show that the unique solution u of the original problem (92) and the unique solution p of the dual problem (92*) are related by the equation

$$p = \alpha(|Du|^2)Du. \tag{95}$$

Regarding (93) and (94), this is a very natural result.

The classical formulation of the original problem (92) reads as follows:

$$\begin{aligned} \int_G (\beta(|Du|) - fu)\,dx + \int_{\partial_2 G} hu\,dO &= \min!, \\ u &= g \quad \text{on } \partial_1 G. \end{aligned} \tag{96}$$

The classical formulation of the dual problem (92*) reads as follows:

$$-\int_G \beta^*(|Du|)\,dx = \max!,$$

$$-\sum_{i=1}^{N} D_i(\alpha(|Du|^2)D_i u) = f \quad \text{on } G, \tag{96*}$$

$$-\alpha(|Du|^2)\frac{\partial u}{\partial n} = h \quad \text{on } \partial_2 G.$$

This is motivated by (93), (95) and by the relation

$$\int_G pD\bar{g}\,dx - b(\bar{g}) = \int_G (pD\bar{g} - f\bar{g})\,dx + \int_{\partial_2 G} h\bar{g}\,dO = 0$$

if we suppose that the given function $g\colon \partial_1 G \to \mathbb{R}$ can be extended to a sufficiently smooth function $\bar{g}\colon \bar{G} \to \mathbb{R}$ such that $\bar{g} = 0$ on $\partial_2 G$.

The classical formulation (96) and (96*) shows clearly a remarkable duality which was first discovered for the Laplace equation by Trefftz (1927). Namely, the side condition of the original problem (96) is given by the boundary condition, whereas the side condition of the dual problem (96*) is given by the Euler equation and the natural boundary condition of the original problem (96).

EXAMPLE 25.33. Let $\alpha = \text{const} > 0$. Then

$$\beta(s) = \alpha s^2/2, \qquad \beta^*(s) = s^2/2\alpha.$$

In this case the original problem (96) corresponds to the mixed boundary value problem for the Poisson equation.

We make the following assumptions:

(H1) G is a bounded region in $\mathbb{R}^N$ with (70) and $\partial_1 G \neq \varnothing$. We equip $W_2^1(G)$ with the equivalent scalar product

$$(u|v) = \int_G DuDv\,dx + \int_{\partial_1 G} uv\,dO.$$

We are given

$$f \in W_2^1(G)^*, \qquad g \in W_2^{1/2}(\partial_1 G), \qquad h \in W_2^{1/2}(\partial_2 G)^*.$$

Let $\bar{g} \in W_2^1(G)$ be a *fixed* extension of g to G. For example, by Remark 25.32, we can choose

$$f \in L_2(G), \qquad g \in W_2^1(G), \qquad h \in L_2(\partial_2 G)$$

with $\bar{g} = g$.

(H2) The function $\alpha\colon \mathbb{R}_+ \to \mathbb{R}_+$ is continuous and β is strictly convex on $\mathbb{R}$.

There are numbers a, $c > 0$ so that $2a \le \alpha(s) \le 2c$ on $\mathbb{R}_+$, i.e.,

$$as^2 \le \beta(s) \le cs^2 \quad \text{on } \mathbb{R}.$$

(H2*) The function $\alpha: \mathbb{R}_+ \to \mathbb{R}_+$ is C^1 and there is a number $d > 0$ so that $\beta''(s) \ge d$ on $\mathbb{R}$.

Theorem 25.K (Main Theorem of the Duality Theory for Conservation Laws).

(a) *If* (H1), (H2) *hold, then the original problem* (92) *and the dual problem* (92*) *have a unique solution* $\bar{u}$ *and* $\bar{p}$, *respectively, and*

$$\bar{p} = \alpha(|D\bar{u}|^2)D\bar{u} \qquad \textit{almost everywhere} \quad \textit{on } G. \tag{97}$$

Moreover, we have

$$F_*(p) \le F_*(\bar{p}) = F(\bar{u}) \le F(u), \tag{98}$$

for all $u \in W_2^1(G)$ *with* $u = g$ *on* $\partial_1 G$ *and all* $p \in L_2(G)^N$ *with* (92b*).

(b) *If, in addition,* (H2*) *holds, then there exists a number* $d_0 > 0$ *so that*

$$\begin{aligned} d_0 \int_G |u - \bar{u}|^2\,dx &\le \frac{d}{2}\int_G |Du - D\bar{u}|^2\,dx \\ &\le F(u) - F_*(p), \end{aligned} \tag{99}$$

for all u, p *with the same properties as in* (98).

Remark 25.34 (Error Estimates). Relation (98) yields two-sided error estimates for the minimal value $F(\bar{u})$ of the original problem. For example, we can compute u and p by a Ritz method for the original problem (92) and the dual problem (92*), respectively. The Ritz method for (92*) is called the *Trefftz method*.

Furthermore, relation (99) yields error estimates for the solution $\bar{u}$ of the original problem. The existence of $d_0 > 0$ in (99) results from the fact that $(\cdot|\cdot)$ in (H1) is an equivalent scalar product on $W_2^1(G)$ and we have $u = \bar{u}$ on $\partial_1 G$.

PROOF. Theorem 25.J yields existence and uniqueness of a solution for the original problem (96).

For the dual problem, we need the duality theory for potential operators in Chapter 51. In particular, we will use Theorem 51.B. To this end, we set

$$X = \{w \in W_2^1(G): w = 0 \text{ on } \partial_1 G\},$$
$$Y = L_2(G)^N,$$

and we equip Y with the scalar product

$$(p|q)_Y = \int_G pq\,dx.$$

For brevity, we write pq instead of $\langle p|q\rangle = \sum_{i=1}^N p_i q_i$. We identify Y^* with Y.

(I) Function β^*. By Definition 51.1,

$$\beta^*(s^*) = \sup_{s \in \mathbb{R}} (s^* s - \beta(s)). \tag{100}$$

By assumption (H2), the function $\beta'\colon \mathbb{R} \to \mathbb{R}$ is strictly monotone and surjective. Hence β' is bijective. By Proposition 51.5, $\beta^{*\prime} = \beta'^{-1}$, i.e.,

$$\beta^*(s) = \int_0^s \beta'^{-1}(t)\,dt.$$

It follows from (100) that

$$\beta^*(s^*) = s^* s - \beta(s), \qquad s^* = \beta'(s)$$

and hence

$$\beta^*(s^*) = \alpha(s^2)s^2 - \beta(s), \qquad s = \beta'^{-1}(s^*). \tag{100*}$$

(II) Function γ^*. We set

$$\gamma(p) = \beta(|p|) \qquad \text{for all} \quad p \in \mathbb{R}^N.$$

By Definition 51.1,

$$\gamma^*(p^*) = \sup_{p \in \mathbb{R}^N} (p^* p - \gamma(p)). \tag{101}$$

By Lemma 25.26, γ is strictly convex. By (H2), γ is coercive and C^1. Hence, the function $p \mapsto \gamma(p) - p^* p$ is strictly convex and weakly coercive. Consequently, for each $p^* \in \mathbb{R}^N$, the maximum problem (101) has a unique solution p and $p^* - \gamma'(p) = 0$, i.e., $p^* = \alpha(|p|^2)p$, and hence

$$|p^*| = \alpha(p^2)|p| = \beta'(|p|).$$

By (101), $\gamma^*(p^*) = p^* p - \gamma(p)$, i.e.,

$$\gamma^*(p^*) = \alpha(p^2)p^2 - \beta(|p|), \qquad |p| = \beta'^{-1}(|p^*|).$$

Thus we obtain from (100*) the *key* formula

$$\gamma^*(p^*) = \beta^*(|p^*|) \qquad \text{for all} \quad p^* \in \mathbb{R}^N.$$

(III) Function δ^*. We set $\delta(p) = \gamma(p + c)$ for all $p \in \mathbb{R}^N$ and fixed $c \in \mathbb{R}^N$. Then

$$\begin{aligned}\delta^*(p^*) &= \sup_{p \in \mathbb{R}^N} (p^* p - \delta(p)) \\ &= \sup_{q \in \mathbb{R}^N} (p^* q - \gamma(q)) - p^* c \\ &= \gamma^*(p^*) - p^* c.\end{aligned}$$

(IV) The dual problem. Let

$$H(p) = \int_G \gamma(p + D\bar{g})\,dx - b(\bar{g})$$

and let $w = u - \bar{g}$. Then the original problem (92) reads as follows:

(P) $$H(Dw) - b(w) = \min!, \qquad w \in X.$$

By Theorem 51.B, the dual problem reads as follows:

$$(\mathrm{P}^*) \qquad \begin{aligned} -H^*(p) &= \max!, \qquad p \in Y, \\ \langle p, Dw \rangle_Y &= b(w) \qquad \text{for all} \quad w \in X. \end{aligned}$$

(V) We compute H^*. Let $\delta(p) = \gamma(p + D\bar{g})$. By Problem 51.7, the formation of the conjugate functional and integration can be interchanged. Hence

$$\begin{aligned} H^*(p) &= \left(\int_G \delta(p)\, dx \right)^* + b(\bar{g}) = \int_G \delta^*(p)\, dx + b(\bar{g}) \\ &= \int_G (\gamma^*(p) - pD\bar{g})\, dx + b(\bar{g}) \\ &= \int_G (\beta^*(|p|) - pD\bar{g})\, dx + b(\bar{g}), \end{aligned}$$

i.e., (P*) coincides with the dual problem (92*).

(VI) We compute H'. For all $p, q \in Y$,

$$\langle H'(p), q \rangle_Y = \int_G \gamma'(p + D\bar{g}) q\, dx,$$

i.e.,

$$H'(p) = \gamma'(p + D\bar{g}) = \alpha(|p + D\bar{g}|^2)(p + D\bar{g}).$$

By Lemma 25.26,

$$\langle H'(p) - H'(q), p - q \rangle_Y \geq d \int_G |p - q|^2\, dx.$$

Now, Theorem 51.B yields the assertions. □

25.13. The Kačanov Method for Stationary Conservation Laws

We consider again the conservation law

$$\begin{aligned} -\operatorname{div}(\alpha(|Du|^2)Du) &= f \quad \text{on } G, \\ u &= g \quad \text{on } \partial_1 G, \\ -\alpha(|Du|^2)\frac{\partial u}{\partial n} &= h \quad \text{on } \partial_2 G. \end{aligned} \tag{102}$$

Around 1950, engineers began to apply the following iteration method:

$$\begin{aligned} -\operatorname{div}[\alpha(|Du_k|^2)Du_{k+1}] &= f \quad \text{on } G, \\ u_{k+1} &= g \quad \text{on } \partial_1 G, \\ -\alpha(|Du_k|^2)\frac{\partial u_{k+1}}{\partial n} &= h \quad \text{on } \partial_2 G, \end{aligned} \tag{103}$$

where $k = 0, 1, \ldots$, and u_0 is given with $u_0 = g$ on $\partial_1 G$. The advantage of this method is that the unknown function u_{k+1} is determined by a *linear* equation and, in practice, the method converges rapidly.

This method is called the secant modulus method or the *Kačanov method.* In order to obtain a simple physical interpretation of this method, we regard u as temperature. Then

$$j = -\alpha(|Du|^2)Du$$

is the current density vector of the heat flow and α is the heat conductivity. By (103), the unknown approximation u_{k+1} is determined by the heat conductivity $\alpha(|Du_k|^2)$ which corresponds to the known approximation u_k.

Equation (102) corresponds to the variational problem

$$\begin{gathered}\int_G (\beta(|Du|) - fu)\,dx + \int_{\partial_2 G} hu\,dO = \min!,\\ u = g \quad \text{on } \partial_1 G,\end{gathered} \tag{104*}$$

and the iteration method (103) corresponds to the quadratic variational problem for u_{k+1}, where $u_{k+1} = g$ on $\partial_1 G$ and

$$\int_G (2^{-1}\alpha(|Du_k|^2)|Du_{k+1}|^2 - fu_{k+1})\,dx + \int_{\partial_2 G} hu_{k+1}\,dO = \min!. \tag{105*}$$

Our goal is to prove the *convergence* of this method.

We set $Y = W_2^1(G)$ and

$$C = \{u \in Y : u = g \text{ on } \partial_1 G\}.$$

Then (104*) and (105*) correspond to

$$F(u) - b(u) = \min!, \qquad u \in C, \tag{104}$$

and

$$2^{-1}B(u_k; u_{k+1}, u_{k+1}) - b(u_{k+1}) = \min!, \qquad u_{k+1} \in C, \tag{105}$$

respectively. Here, we set

$$b(u) = \int_G fu\,dx - \int_{\partial_2 G} hu\,dO, \qquad \beta(s) = 2^{-1} \int_0^{s^2} \alpha(t)\,dt,$$

$$F(u) = \int_G \beta(|Du|)\,dx,$$

$$B(u; v, w) = \int_G \alpha(|Du|^2)DvDw\,dx.$$

We make the following assumptions:

(H1) G is a bounded region in $\mathbb{R}^N$ such that (70) holds and $\partial_1 G \neq \varnothing$. We are given

$$f \in W_2^1(G)^*, \qquad g \in W_2^{1/2}(\partial_1 G), \qquad h \in W_2^{1/2}(\partial_2 G)^*.$$

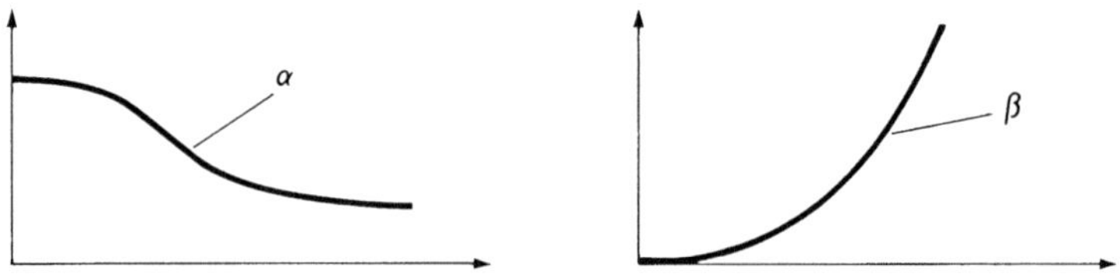

Figure 25.10

For example, we can choose

$$f \in L_2(G), \qquad g \in W_2^1(G), \qquad h \in L_2(\partial_2 G).$$

(H2) The function $\alpha\colon \mathbb{R}_+ \to \mathbb{R}_+$ is C^1 and there are positive numbers a, c, d so that

$$a \le \alpha(s) \le c, \qquad \alpha'(s) \le 0, \qquad \beta''(s) = \alpha'(s^2)s^2 + \alpha(s^2) \ge d,$$

for all $s \ge 0$ (Fig. 25.10).

Proposition 25.35. *Assume* (H1), (H2). *Then the original variational problem* (104) *has a unique solution* u.

Let $u_0 \in C$ *be given. Then, for* $k = 0, 1, \dots$, *the quadratic variational problem* (105) *has a unique solution* u_{k+1} *and the Kačanov method converges, i.e.,*

$$u_k \to u \quad \text{in } W_2^1(G) \qquad \text{as} \quad k \to \infty.$$

Proof. This proposition is a special case of Theorem 25.L in the next section. Using our considerations about conservation laws in the previous sections, it is easy to check the fulfillment of the assertions of Theorem 25.L. We want to discuss this.

Recall that $Y = W_2^1(G)$ and

$$C = \{w \in Y\colon w = g \text{ on } \partial_1 G\}, \qquad X = \{w \in Y\colon w = 0 \text{ on } \partial_1 G\}.$$

As in Section 25.10 we set

$$\|u\|_X = \left(\int_G |Du|^2\,dx\right)^{1/2} \qquad \text{and} \qquad \|u\|_Y = \left(\int_G |u|^2 + |Du|^2\,dx\right)^{1/2}.$$

It is important for our proof that $\|\cdot\|_X$ and $\|\cdot\|_Y$ are equivalent norms on X. In the following note that $v, w \in C$ implies $v - w \in X$.

(I) It follows as in Remark 25.32 that $b \in Y^*$.

(II) The functional $F\colon Y \to \mathbb{R}$. Let $\psi(s) = 2^{-1}\int_0^s \alpha(t)\,dt$. Then

$$F(u) = \int_G \psi(|Du|^2)\,dx \qquad \text{for all} \quad u \in Y.$$

Since $a \le \alpha(s) \le c$ for all $s \ge 0$, it follows from Problem 25.4 that $F\colon Y \to$

$\mathbb{R}$ is C^1 and

$$\begin{aligned}\langle F'(u), v\rangle &= \int_G 2\psi'(|Du|^2)DuDv\,dx \\ &= \int_G \alpha(|Du|^2)DuDv\,dx = B(u; u, v) \qquad \text{for all} \quad u, v \in Y.\end{aligned}$$

By Lemma 25.26(b), for all $v, w \in C$ and fixed $\rho_0 > 0$, we get

$$\langle F'(v) - F'(w), v - w\rangle \geq \int_G d|Dv - Dw|^2\,dx \geq \rho_0\|v - w\|_Y^2,$$

since $\|\cdot\|_X$ and $\|\cdot\|_Y$ are equivalent norms on X.

(III) The functional B: $Y \times Y \times Y \to \mathbb{R}$. From $a \leq \alpha(s) \leq c$ for all $s \geq 0$ it follows that, for all $v, w \in C$ and fixed $\rho > 0$,

$$B(u; v - w, v - w) \geq \int_G a|Dv - Dw|^2\,dx \geq \rho\|v - w\|_Y^2,$$

and, for all $u, v, w \in Y$, we get

$$|B(u; v, w)| \leq \int_G c|DvDw|\,dx \leq c\|v\|_Y\|w\|_Y,$$

by the Hölder inequality.

(IV) The key condition. From $\alpha'(s) \leq 0$ for all $s \geq 0$ it follows that

$$\alpha(t)(t - s) \leq \int_s^t \alpha(\tau)\,d\tau \leq \alpha(s)(t - s),$$

for all $0 \leq s \leq t < \infty$. This implies

$$\beta(t) - \beta(s) \leq 2^{-1}(\alpha(s^2)t^2 - \alpha(s^2)s^2),$$

for all $s, t \in \mathbb{R}$. Hence we obtain the *key* relation

$$F(v) - F(u) \leq 2^{-1}(B(u; v, v) - B(u; u, u)) \qquad \text{for all} \quad u, v \in X.$$

(V) Theorem 25.L yields the assertions. □

25.14. The Abstract Kačanov Method for Variational Inequalities

We make the following assumptions:

(H1) Let C be a nonempty closed convex set in the real H-space Y (e.g., $C = Y$). Let $b \in Y^*$ and $u_0 \in C$ be given.

(H2) The functional F: $C \to \mathbb{R}$ is G-differentiable. There exists a functional B: $C \times Y \times Y \to \mathbb{R}$ such that, for all $u, v, w \in C$, we have the representation formula

$$\langle F'(u), v - w\rangle = B(u; u, v - w)$$

and the *key* inequality

$$F(v) - F(u) \le 2^{-1}(B(u; v, v) - B(u; u, u)). \tag{106}$$

(H3) For each $u \in C$, the map

$$(v, w) \mapsto B(u; v, w)$$

is bilinear, bounded, and symmetric from $Y \times Y$ to $\mathbb{R}$. There are numbers $\rho > 0$ and $\delta > 0$ such that

$$|B(u; v, w)| \le \delta \|v\| \|w\| \quad \text{for all} \quad u \in C, \quad v, w \in Y,$$

and

$$B(u; w - v, w - v) \ge \rho \|w - v\|^2 \quad \text{for all} \quad u, v, w \in C.$$

(H3*) The operator $F'\colon C \to Y^*$ is continuous and strongly monotone, i.e., there is a number $\rho_0 > 0$ such that

$$\langle F'(u) - F'(v), u - v\rangle \ge \rho_0 \|u - v\|^2 \quad \text{for all} \quad u, v \in C.$$

Theorem 25.L. *Assume* (H1), (H2), (H3). *Then*:

(a) *For* $k = 0, 1, \ldots$, *the quadratic variational problem*

(Q) $$2^{-1} B(u_k; u_{k+1}, u_{k+1}) - b(u_{k+1}) = \min!, \qquad u_{k+1} \in C,$$

has a unique solution u_{k+1}, *and*

$$B(u_k; u_{k+1}, v - u_{k+1}) \ge b(v - u_{k+1}) \quad \textit{for all} \quad v \in C. \tag{107}$$

(b) *If, in addition,* (H3*) *holds, then the original variational problem*

(V) $$F(u) - b(u) = \min!, \qquad u \in C,$$

has a unique solution u. *Moreover,* u *is the unique solution of the variational inequality*

$$\langle F'(u), v - u\rangle \ge b(v - u) \quad \textit{for all} \quad v \in C \quad \textit{and fixed} \quad u \in C, \tag{108}$$

and the Kačanov method converges, that is

$$u_k \to u \quad \text{in } Y \qquad \textit{as} \quad k \to \infty.$$

Corollary 25.36. *Assume* (H1), (H2), (H3). *Suppose that the set* C *is compact and that the map*

$$u \mapsto B(u; v, w)$$

is equicontinuous on C *with respect to all* $v, w \in Y$.

Then the sequence (u_k) *has at least one cluster point and each such cluster point* u *is a solution of the variational inequality* (108).

In Part V we shall show that Corollary 25.36 can be used to compute transonic flow. In this connection, the compactness of C results from a so-called entropy condition and the method of compensated compactness.

PROOF OF THEOREM 25.L. Ad(a). For fixed $u \in C$, we set

$$H(v) = 2^{-1} B(u; v, v) - b(v) \qquad \text{for all} \quad v \in Y.$$

By (H3),

$$\langle H'(v), w \rangle = B(u; v, w) - b(w) \qquad \text{for all} \quad v, w \in Y,$$

and hence, for all $v, w \in C$,

$$\langle H'(v) - H'(w), v - w \rangle = B(u; v - w, v - w) \geq \rho \|v - w\|^2.$$

It follows from Problem 25.3 that the functional $H: C \to \mathbb{R}$ is weakly coercive, strictly convex, and continuous. Therefore, the minimum problem

$$H(w) = \min!, \qquad w \in C,$$

has a unique solution w, and $\langle H'(w), v - w \rangle \geq 0$ for all $v \in C$.

Ad(b). We set $G(u) = F(u) - b(u)$. By (H3*) and Problem 25.3 the functional $G: C \to \mathbb{R}$ is weakly coercive, strictly convex, and continuous. Therefore, the minimum problem

$$G(u) = \min!, \qquad u \in C,$$

has a unique solution u, and $\langle G'(u), v - u \rangle \geq 0$ for all $v \in C$, i.e., (108) holds.

Since F' is strongly monotone on C, the solution u of the variational inequality (108) is unique.

(I) We set

$$g(u) = 2^{-1}(B(u_k; u, u) - B(u_k; u_k, u_k)) + F(u_k) - b(u).$$

By (106), we obtain the *key condition*

$$G(u_{k+1}) \leq g(u_{k+1}). \tag{109}$$

From the quadratic minimum problem (Q) we obtain $\min_{u \in C} g(u) = g(u_{k+1})$. Hence

$$g(u_{k+1}) \leq g(u_k). \tag{110}$$

This yields

$$G(u_{k+1}) \leq g(u_{k+1}) \leq g(u_k) = G(u_k).$$

Thus, the sequence $(G(u_k))$ is monotone decreasing. Since G has a minimum on C, this sequence is also bounded below and hence it is convergent. This yields

$$\lim_{k \to \infty} G(u_{k+1}) - G(u_k) = 0.$$

(II) The quadratic variational inequality (107) implies

$$\begin{aligned} G(u_k) - G(u_{k+1}) &\geq G(u_k) - g(u_{k+1}) \\ &= b(u_{k+1} - u_k) - 2^{-1} B(u_k; u_{k+1}, u_{k+1}) + 2^{-1} B(u_k; u_k, u_k) \\ &\geq 2^{-1} B(u_k; u_{k+1} - u_k, u_{k+1} - u_k) \\ &\geq 2^{-1} \rho \|u_{k+1} - u_k\|^2. \end{aligned}$$

Hence

$$\|u_{k+1} - u_k\| \to 0 \quad \text{as} \quad k \to \infty. \tag{111}$$

(III) By the original variational inequality (108),

$$\begin{aligned}\rho_0\|u_k - u\|^2 &\le \langle F'(u_k) - F'(u), u_k - u\rangle \\ &\le \langle F'(u_k), u_k - u\rangle + b(u - u_k). \\ &= B(u_k; u_k, u_k - u) + b(u - u_k).\end{aligned}$$

From the quadratic variational inequality (107) we get

$$\begin{aligned}B(u_k; u_k, u_k - u) + b(u - u_k) \le B(u_k; u_k &- u_{k+1}, u_k - u + u_{k+1}) \\ &+ b(u_{k+1} - u_k).\end{aligned}$$

This implies

$$\rho_0\|u_k - u\|^2 \le \|u_{k+1} - u_k\|(\delta\|u_k - u + u_{k+1}\| + \|b\|).$$

Hence $\|u_k - u\| \to 0$ as $k \to \infty$. □

PROOF OF COROLLARY 25.36. The sequence (u_k) lies in the compact set C. Hence there is a subsequence $(u_{k'})$ with

$$u_{k'} \to u \quad \text{as} \quad k' \to \infty,$$

and $u \in C$. By (111),

$$u_{k'+1} \to u \quad \text{as} \quad k' \to \infty.$$

Letting $k' \to \infty$, it follows from the quadratic variational inequality

$$B(u_k; u_{k+1}, v - u_{k+1}) \ge b(v - u_{k+1}) \quad \text{for all} \quad v \in C$$

that

$$B(u; u, v - u) \ge b(v - u) \quad \text{for all} \quad v \in C.$$

This is the original variational inequality (108).

In this connection, use the equicontinuity of $u \mapsto B(u; v, w)$. This implies

$$\begin{aligned}B(u; v, w) - B(\bar{u}; \bar{v}, \bar{w}) &= B(u; v, w) - B(\bar{u}; v, w) \\ &\quad + B(\bar{u}; v - \bar{v}, w) + B(\bar{u}; \bar{v}, w - \bar{w}) \to 0\end{aligned}$$

as $u \to \bar{u}$, $v \to \bar{v}$, and $w \to \bar{w}$. □

PROBLEMS

25.1. *Theorem of Toeplitz for real sequences.* Let $T = (t_{ij})$ be an infinite matrix of real numbers t_{ij} which has the following three properties:

(i) T is a triangular matrix, i.e., $t_{ij} = 0$ for $j > i$.

(ii) $t_{ij} \to 0$ as $i \to \infty$ for each j.

(iii) $\sup_i \sum_{j=1}^{i} |t_{ij}| < \infty$.

Let (a_n) be a given sequence of real numbers. We set

$$b_n = t_{n1}a_1 + t_{n2}a_2 + \cdots + t_{nn}a_n.$$

Show: If $a_n \to 0$ as $n \to \infty$, then also $b_n \to 0$ as $n \to \infty$.
Hint: Cf. Fichtenholz (1972, M), Vol. 2, Section 391.

25.2. *Proof of Lemma* 25.26.
Solution: We will make essential use of the Schwarz inequality

$$-|x||y| \leq \pm\langle x|y\rangle \leq |x||y|,$$

for all $x, y \in \mathbb{R}^N$. We set

$$\varphi(t) = \gamma(x + th).$$

Then

$$\varphi'(0) = \langle \gamma'(x)|h\rangle = \alpha(|x|^2)\langle x|h\rangle,$$
$$\varphi''(0) = \gamma''(x)h^2.$$

Note that $\beta'(s) = \alpha(s^2)s$ and $\alpha(t) \geq 0$ for all $t \geq 0$.

(I) Proof of (a). Let β be (strictly) convex. Then β' is (strictly) monotone. The Schwarz inequality yields

$$\langle \gamma'(x) - \gamma'(y)|x - y\rangle = \alpha(|x|^2)\langle x|x - y\rangle - \alpha(|y|^2)\langle y|x - y\rangle$$
$$\geq [\alpha(|x|^2)|x| - \alpha(|y|^2)|y|](|x| - |y|) \geq 0,$$

for all $x, y \in \mathbb{R}$. Hence γ' is (strictly) monotone; therefore, γ is (strictly) convex, by Proposition 25.10.

(II) Proof of (b). Let

$$\beta'(s) - \beta'(t) \geq d(s - t) \qquad \text{for all} \quad s \geq t \quad \text{and fixed} \quad d > 0.$$

Then the function $s \mapsto \beta'(s) - ds$ is monotone and hence

$$s \mapsto \beta(s) - 2^{-1}ds^2$$

is convex. Application of (I) to the corresponding function

$$x \mapsto \gamma(x) - 2^{-1}d|x|^2$$

yields

$$\langle (\gamma'(x) - dx) - (\gamma'(y) - dy)|x - y\rangle \geq 0$$

and hence

$$\langle \gamma'(x) - \gamma'(y)|x - y\rangle \geq d|x - y|^2.$$

(III) Proof of (c). Let β be (strictly) convex on $]a, b[$. Fix a point x with $a < |x| < b$. The same argument as in (I) shows that γ is (strictly) convex in a neighborhood of x. Consequently, for $h \neq 0$, the function φ is (strictly) convex in a neighborhood of $t = 0$, i.e., $\varphi''(0) \geq 0$.

(IV) Proof of (d). The inequality

$$|\beta'(t) - \beta'(s)| \leq L|t - s| \qquad \text{for all} \quad t, s \in \mathbb{R}_+$$

means

$$|\alpha(t^2)t - \alpha(s^2)s| \leq L|t - s| \qquad \text{for all} \quad t, s \in \mathbb{R}_+.$$

This implies

$$|\alpha(t^2)| \le L \qquad \text{for all } t. \tag{112}$$

We use the decomposition

$$\begin{aligned}\langle \gamma'(x) - \gamma'(y)|z\rangle &= \alpha(|x|^2)\langle x|z\rangle - \alpha(|y|^2)\langle y|z\rangle \\ &= \alpha(|x|^2)\langle x - y|z\rangle + \delta,\end{aligned} \tag{113}$$

where

$$\delta = [\alpha(|x|^2) - \alpha(|y|^2)]\langle y|z\rangle.$$

The Schwarz inequality yields

$$\begin{aligned}|\delta| &\le |\alpha(|x|^2) - \alpha(|y|^2)||y||z| \\ &= |\alpha(|x|^2)(|y| - |x|) + \alpha(|x|^2)|x| - \alpha(|y^2|)|y||z|.\end{aligned}$$

By (112),

$$|\delta| \le 2L||x| - |y|||z| \le 2L|x - y||z|.$$

By (113),

$$|\langle \gamma'(x) - \gamma'(y)|z\rangle| \le L|x - y||z| + |\delta| \le 3L|x - y||z|.$$

Hence $|\gamma'(x) - \gamma'(y)| \le 3L|x - y|$ for all $x, y \in \mathbb{R}^N$.

25.3. *Coerciveness of functionals.* Let C be a convex subset of a real H-space X. Let $F: C \to \mathbb{R}$ be G-differentiable and suppose that $F': C \to X^*$ is continuous and strongly monotone, i.e.,

$$\langle F'(u) - F'(v), u - v\rangle \ge c\|u - v\|^2$$

for all $u, v \in C$ and fixed $c > 0$. Set $G(u) = F(u) - b(u)$, where $b \in X^*$. By Proposition 25.10, G is strictly convex on C.

Show that G is weakly coercive and continuous.

Solution: Set $\varphi(t) = F(u + t(v - u))$ for fixed $u, v \in C$. From

$$\varphi(1) - \varphi(0) = \int_0^1 \varphi'(t)\,dt$$

we obtain that

$$\begin{aligned}G(v) - G(u) &= \int_0^1 \langle F'(u + t(v - u)) - F'(u), v - u\rangle\,dt \\ &\quad + \langle F'(u), v - u\rangle - b(v - u) \\ &\ge \frac{c}{2}\|u - v\|^2 - \|F'(u)\|\,\|v - u\| - \|b\|\,\|v - u\|.\end{aligned} \tag{114}$$

Hence $G(v) \to +\infty$ as $\|v\| \to \infty$ on C.

The continuity of G follows from (114) by using the continuity of F' on C and $|\langle a, b\rangle| \le \|a\|\,\|b\|$.

25.4. *Typical properties of variational integrals.* Let G be a bounded region in $\mathbb{R}^N$, $N \ge 1$, and let $Du = (D_1 u, \ldots, D_N u)$, where $x = (\xi_1, \ldots, \xi_N)$ and $D_i = \partial/\partial\xi_i$.

25.4a. *Special case.* Let

$$F(u) = \int_G \psi(|Du|^2)\,dx, \tag{115}$$

and let $X = W_2^1(G)$. Use the continuity of the Nemyckii operator (Proposition 26.6) in order to show the following:

(i) If $\psi\colon \mathbb{R}_+ \to \mathbb{R}$ is continuous and we have the growth condition

$$|\psi(t)| \le \text{const}(1 + t) \qquad \text{for all} \quad t \ge 0,$$

then the functional $F\colon X \to \mathbb{R}$ is continuous.

(ii) If, in addition, $\psi\colon \mathbb{R}_+ \to \mathbb{R}$ is C^1 and

$$|\psi'(t)| \le \text{const} \qquad \text{for all} \quad t \ge 0,$$

then $F\colon X \to \mathbb{R}$ is C^1 and

$$\langle F'(u), v\rangle = \int_G 2\psi'(|Du|^2)\,DuDv\,dx \qquad \text{for all} \quad u, v \in X. \tag{116}$$

Solution: Ad(i). If $u_n \to u$ in X as $n \to \infty$, then

$$D_i u_n \to D_i u \quad \text{in } L_2(G) \qquad \text{as} \quad n \to \infty \qquad \text{for all } i.$$

Since $\psi(|Du|^2) \le \text{const}(1 + |Du|^2)$ it follows from Proposition 26.6 that

$$\psi(|Du_n|^2) \to \psi(|Du|^2) \quad \text{in } L_1(G) \qquad \text{as} \quad n \to \infty,$$

and hence $F(u_n) \to F(u)$ as $n \to \infty$.

Ad(ii). For fixed $u, v \in X$, set

$$\varphi(t) = \int_G \psi(|Du + tDv|^2)\,dx \qquad \text{for all} \quad t \in [-1, 1].$$

Formal differentiation yields

$$\varphi'(t) = \int_G 2\psi'(|Du + tDv|^2)(Du + tDv)\,Dv\,dx \qquad \text{for all} \quad t \in [-1, 1].$$

To justify this formal argument by means of the majorant criterion A_2(25b), note that, for all $t \in [-1, 1]$,

$$|\psi'(|Du + tDv|^2)(Du + tDv)\,Dv| \le \text{const}(|Du| + |Dv|)|Dv|,$$

where the majorant function $(|Du| + |Dv|)|Dv|$ belongs to $L_1(G)$, according to the Hölder inequality and $u, v \in X$.

To prove the continuity of $F'\colon X \to X^*$, let $h_i(u) = 2\psi'(|Du|^2)D_i u$. Since $|\psi'(t)| \le \text{const}$ for all $t \ge 0$, it follows from Proposition 26.6 that

$$u_n \to u \quad \text{in } X \qquad \text{as} \quad n \to \infty$$

implies that

$$h_i(u_n) \to h_i(u) \quad \text{in } L_2(G) \qquad \text{as} \quad n \to \infty \qquad \text{for all } i,$$

and hence

$$|\langle F'(u_n) - F'(u), v\rangle| \le \sum_{i=1}^{N} \|h_i(u_n) - h_i(u)\|_2 \|v\|_X,$$

for all $v \in X$, according to the Hölder inequality. Therefore,

$$\|F'(u_n) - F'(u)\|_{X^*} \le \sum_{i=1}^{N} \|h_i(u_n) - h_i(u)\|_2 \to 0 \qquad \text{as} \quad n \to \infty.$$

25.4b. *Generalization.* We now consider the following more general variational integral

$$F(u) = \int_G L(u, Du, x)\, dx.$$

Let $X = W_p^1(G)$, $1 < p < \infty$. Show the following.

(i) If $L: \mathbb{R}^{N+1} \times G \to \mathbb{R}$ is continuous and, for all $u \in \mathbb{R}$, $D \in \mathbb{R}^N$, $x \in G$, we have the *growth condition*

$$|L(u, D, x)| \le \text{const}(1 + |u|^p + |D|^p),$$

then the functional $F: X \to \mathbb{R}$ is continuous.

(ii) If, in addition, the function L is C^1 and, for all $u \in \mathbb{R}$, $D \in \mathbb{R}^N$, $x \in G$, we have, for the partial derivatives of L, the growth conditions

$$|L_u(u, D, x)|, |L_{D_i u}(u, D, x)| \le \text{const}(1 + |u|^{p-1} + |D|^{p-1}), \qquad i = 1, \ldots, N,$$

then $F: X \to \mathbb{R}$ is C^1 and, for all $u, v \in X$,

$$\langle F'(u), v\rangle = \int_G L_u(u, Du, x)v + \sum_{i=1}^{N} L_{D_i u}(u, Du, x) D_i v\, dx.$$

Hint: Use the same arguments as in Problem 25.4a. In particular, let $p^{-1} + q^{-1} = 1$ and use Proposition 26.6 with $p/q = p - 1$ (continuity of the Nemyckii operator from $L_p(G)$ to $L_q(G)$), in connection with the Hölder inequality.

References to the Literature

Classical work on Lipschitz continuous monotone operators: Zarantonello (1960).

Monotone potential operators: Vainberg (1956, M), (1972, M), Langenbach (1959), (1966), (1976, M), Browder (1966), (1970a), Ekeland and Temam (1974, M), Gajewski, Gröger, and Zacharias (1974, M), Kluge (1979, M).

Projection–iteration methods: Kurpel (1976, M) (comprehensive representation), Gajewski and Kluge (1970, S), Gajewski, Gröger, and Zacharias (1974, M), Langenbach (1976, M), Scheurle (1977) (bifurcation theory).

Iteration methods: Browder and Petryshyn (1967, S), Krasnoselskii (1973, M), Kluge (1979, M) (general Toeplitz technique) (cf. also the References to the Literature for Chapter 1).

Variational inequalities: Lions (1968, M), (1969, M), Kinderlehrer and Stampacchia (1980, M), Friedman (1982, M).

Conservation laws and rheology: Langenbach (1959), (1965), (1976, M), Gajewski (1970), (1970a), (1972), Gajewski, Gröger, and Zacharias (1974, M).

W_p^1-estimates for mixed boundary value problems on nonsmooth domains: Gröger (1989).

Kačanov's method: Fučik, Kratochvil, and Nečas (1973), Nečas and Hlaváček (1981, M) (applications in elasticity).

Kačanov's method for variational inequalities and transonic flow in gas dynamics: Feistauer, Mandel, and Nečas (1985), Gittel (1987).

Monotone operators and parameter identification: Kluge (1985, M).

CHAPTER 26

Monotone Operators and Quasi-Linear Elliptic Differential Equations

Without involving ourselves in the controversy over the fatherhood of monotone operators, we would merely like to point out that this concept, among others, appeared in the papers of Golomb (1935), Kačurovskii (1960) (influenced by the papers of Vainberg), and Zarantonello (1960).[1]

Haïm Brézis (1973)

The fundamental step in the theory of monotone operators was made by George J. Minty (1962).[2]

Mark Aleksandrovič Krasnoselskii and Pjotr Pjotrovič Zabreiko (1975)

Even before the creation of a general theory of monotone operators, Višik (1961) used certain monotonicity conditions in order to give new existence proofs for quasi-linear partial differential equations of strongly elliptic type....

The modern theory of monotone operators was developed in fundamental papers by Minty (1962), (1963), Browder (1963), and Leray and Lions (1965), and in the works of a whole series of other authors during the following years.

Morduchaj Moiseevič Vainberg (1972)

The recently developed theory of monotone nonlinear operators from a Banach space to its dual space can be considered most naturally as an extension to nonvariational problems of the basic ideas of the direct method of the calculus of variations. First of all, in practice, its simplest and most basic application is to a class of nonlinear elliptic boundary value problems which is an extension of the class of Euler–Lagrange equations of multiple integral problems of general order, paralleling the application of Hilbert space methods for general linear elliptic problems as an extension of the variational method for self-adjoint problems.

Felix E. Browder (1966)

The truly new ideas are extremely rare in mathematics.

Folclore

[1] Golomb (1935) and Vainberg (1956) worked on Hammerstein integral equations; Kačurovskii (1960) investigated derivatives of convex functionals, and Zarantonello (1960) proved existence theorems for strongly monotone, Lipschitz continuous operators on H-spaces (see Chapter 25).

[2] George J. Minty of Indiana University (U.S.A.) died in 1986, at the age of 56.

In this chapter, we study the operator equation

$$\text{(E)} \qquad Au = b, \qquad u \in X,$$

where $A: X \to X^*$ is a monotone operator on the real B-space X. In order to treat more general quasi-linear elliptic partial differential equations than those in Chapter 25, we weaken the assumptions on the operator A, in contrast to Theorem 25.B. In particular, we free ourselves of the assumption that A is Lipschitz continuous or that (E) is related to a variational problem, i.e., A is a potential operator. The main result of this chapter is the following:

If the operator $A: X \to X^*$ is monotone, coercive, and hemicontinuous on the real reflexive B-*space X, then A is surjective.*

This means that for each $b \in X^*$, equation (E) has a solution. If A is strictly monotone, then this solution is unique. The proof of this important existence result in Section 26.2 is based on the following two ideas:

(i) We use a Galerkin method and solve the Galerkin equations by means of the following existence principle: If $f: \mathbb{R}^N \to \mathbb{R}^N$ is continuous and

$$\langle f(x)|x\rangle \geq 0 \quad \text{on } \partial B,$$

where $B = \{x \in \mathbb{R}^N: \|x\| < R\}$, then the equation

$$f(x) = 0, \qquad x \in \bar{B},$$

has a solution.

(ii) The convergence of the Galerkin method is proved by using the fundamental monotonicity trick (25.4).

Let $N = 1$. Then the existence principle in (i) follows immediately from $f(R) \geq 0$, $f(-R) \leq 0$ and the intermediate value theorem of Bolzano. In the case where $N \geq 2$ we proved this existence principle in Section 2.4 as an easy consequence of the Brouwer fixed-point theorem.

26.1. Hemicontinuity and Demicontinuity

Definition 26.1. Let $A: X \to X^*$ be an operator on the real B-space X.

(a) A is said to be *demicontinuous* iff

$$u_n \to u \qquad \text{as} \quad n \to \infty$$

implies $Au_n \rightharpoonup Au$ as $n \to \infty$.

(b) A is said to be *hemicontinuous* iff the real function

$$t \mapsto \langle A(u + tv), w\rangle$$

is continuous on $[0, 1]$ for all $u, v, w \in X$.

(c) A is said to be *strongly continuous* iff

$$u_n \rightharpoonup u \qquad \text{as} \quad n \to \infty$$

implies $Au_n \to Au$ as $n \to \infty$.

(d) A is said to be *bounded* iff A maps bounded sets into bounded sets.

These definitions will be used frequently. Demicontinuity, strong continuity, and boundedness can be defined entirely analogously for operators A: $X \to Y$, where X and Y are B-spaces. For a more convenient formulation of the following results, we make the following general assumption:

(H) Let $A: X \to Y$ be an operator where X and Y are real reflexive B-spaces.

We first of all investigate the relation between strong continuity and compactness.

Proposition 26.2. *Under the assumption* (H), *the following two assertions are valid*:

(a) *A is strongly continuous implies A is compact.*
(b) *A is linear and compact implies A is strongly continuous.*

In (b) *we do not need that the* B-*spaces are reflexive.*

We treat the proof in Problem 26.1. We next consider demicontinuous operators.

Proposition 26.3. *Under the assumption* (H), *the following two assertions are valid*:

(a) *A is demicontinuous implies A is locally bounded.*
(b) *A is demicontinuous iff A is continuous as an operator from X into Y, in the case where X is equipped with the norm topology and Y is equipped with the weak topology.*

The reader will find the proof in Problem 26.2. Finally, we study the important continuity properties of monotone operators.

Proposition 26.4. *Let $A: X \to X^*$ be an operator on the real* B-*space X. Then*:

(a) *If A is monotone, then A is locally bounded.*
(b) *If A is linear and monotone, then A is continuous.*
(c) *If A is monotone and hemicontinuous on the real reflexive* B-*space X, then A is demicontinuous.*

The local boundedness of monotone operators plays an important role in the theory of monotone operators.

Proof Ad(a). Local boundedness of A means that for each $u \in X$ there exists a neighborhood $U(u)$ such that $A(U(u))$ is bounded.

Let A be monotone. Assume, to the contrary, that A is not locally bounded. Then there exist a point $u \in X$ and a sequence (u_n) with

$$u_n \to u \quad \text{and} \quad \|Au_n\| \to \infty \quad \text{as} \quad n \to \infty.$$

Without loss of generality, let $u = 0$. We set

$$a_n = (1 + \|Au_n\| \, \|u_n\|)^{-1}.$$

It follows from the monotonicity of the operator A that

$$\begin{aligned} \pm a_n \langle Au_n, v\rangle &\le a_n(\langle Au_n, u_n\rangle - \langle A(\pm v), u_n \mp v\rangle) \\ &\le a_n(\|Au_n\| \, \|u_n\| + \|A(\pm v)\| \, \|u_n \mp v\|). \end{aligned}$$

Therefore,

$$\sup_n |\langle a_n Au_n, v\rangle| < \infty \quad \text{for all} \quad v \in X.$$

According to the *Banach–Steinhaus theorem* A_1(35a), there exists a number N such that

$$\sup_n \|a_n Au_n\| \le N.$$

We set $b_n = \|Au_n\|$. Then

$$b_n \le a_n^{-1} N = (1 + b_n \|u_n\|)N \quad \text{for all } n.$$

Since $\|u_n\| \to 0$ as $n \to \infty$, the sequence (b_n) is bounded. This contradicts $\|Au_n\| \to \infty$ as $n \to \infty$.

Ad(b). Consider statement (a) and note the fact that a linear locally bounded operator A is bounded on a neighborhood of zero and thus also on the closed unit ball. This means that $\|Au\| \le \text{const}\,\|u\|$ for all $u \in X$, i.e., A is continuous.

Ad(c). Let $u_n \to u$ as $n \to \infty$. Since (u_n) is bounded, the sequence (Au_n) is also bounded, by (a).

Let $Au_{n'} \rightharpoonup b$ as $n \to \infty$ for a subsequence $(u_{n'})$ of (u_n). Then

$$\langle Au_{n'}, u_{n'}\rangle \to \langle b, u\rangle \quad \text{as} \quad n \to \infty.$$

The monotonicity trick (25.4) gives $Au = b$. Hence, the convergence principle (Proposition 21.23(i)) yields

$$Au_n \rightharpoonup Au \quad \text{as} \quad n \to \infty. \qquad \square$$

26.2. The Main Theorem on Monotone Operators

We consider the operator equation

$$Au = b, \qquad u \in X, \tag{1}$$

together with the corresponding Galerkin equations

$$a(u_n, w_k) = \langle b, w_k \rangle, \qquad k = 1, \ldots, n. \tag{2}$$

In this connection, let

$$a(u, v) = \langle Au, v \rangle, \qquad X_n = \operatorname{span}\{w_1, \ldots, w_n\}.$$

We seek $u_n \in X_n$, i.e.,

$$u_n = \sum_{k=1}^{n} c_{kn} w_k,$$

where the coefficients c_{kn} are unknown real numbers.

Theorem 26.A (Browder (1963), Minty (1963)). *Let $A: X \to X^*$ be a monotone, coercive, and hemicontinuous operator on the real, separable, reflexive* B-*space X. Assume $\{w_1, w_2, \ldots\}$ is a basis in X. Then the following assertions hold:*

(a) Solution set. *For each $b \in X^*$, equation* (1) *has a solution. The solution set of* (1) *is bounded, convex, and closed.*

(b) Galerkin method. *If $\dim X = \infty$, then for each $n \in \mathbb{N}$, the Galerkin equation* (2) *has a solution $u_n \in X_n$ and the sequence (u_n) has a weakly convergent subsequence*

$$u_{n'} \rightharpoonup u \quad \text{in } X \qquad \text{as} \quad n \to \infty,$$

where u is a solution of the original equation (1).

(c) Uniqueness. *If the operator A is strictly monotone, then equation* (1) *(resp. equation* (2)*) is uniquely solvable in X (resp. X_n).*

(d) Inverse operator. *If A is strictly monotone, then the inverse operator A^{-1}: $X^* \to X$ exists. This operator is strictly monotone, demicontinuous, and bounded.*

If A is uniformly monotone, then A^{-1} is continuous.

If A is strongly monotone, then A^{-1} is Lipschitz continuous.

(e) Strong convergence of the Galerkin method. *Let $\dim X = \infty$. If the operator A is strictly monotone, then the sequence of Galerkin solutions (u_n) converges weakly in X to the unique solution u of equation* (1).

If A is uniformly monotone, then (u_n) converges strongly in X to the unique solution u of (1).

(f) Nonseparable spaces. *If X is not separable, then the assertions* (a), (c), *and (d) remain true.*

Proof. The basic idea of our proof is the following:

(i) We solve the Galerkin equations (2) by means of Proposition 2.8 which follows from the Brouwer fixed-point theorem.

(ii) The convergence of the Galerkin method is based on the monotonicity trick (25.4).

The monotonicity of the operator A is essentially used in (ii). We need the coerciveness of A to get a *priori* estimates for the Galerkin solutions.

Proof of (b).

Step 1: Solution of the Galerkin equations.
We set

$$g(u) = \langle Au - b, u \rangle, \qquad g_k(u) = \langle Au - b, w_k \rangle.$$

The operator A is coercive, i.e.,

$$g(u)/\|u\| \to +\infty \qquad \text{as} \quad \|u\| \to \infty.$$

Consequently, there exists a number $R > 0$ such that

$$g(u) > 0 \qquad \text{for all} \quad u\colon \|u\| \geq R. \tag{3}$$

This is the key condition.

The Galerkin equations read as follows:

$$\text{(G)} \qquad g_k(u_n) = 0, \qquad u_n \in X_n, \qquad k = 1, 2, \ldots, n,$$

that is

$$u_n = \sum_{k=1}^{n} c_{kn} w_k.$$

Thus, (G) is a nonlinear system of real equations with respect to the numbers $c_{1n}, \ldots, c_{nn}$. By Proposition 26.4(c), the operator A is demicontinuous. Hence the functions

$$u \mapsto g_k(u)$$

are continuous on X.

In particular, the functions g_k in (G) are continuous with respect to $(c_{1n}, \ldots, c_{nn})$. From (3) it follows that, for all $u_n \in X_n$ with $\|u_n\| = R$,

$$\sum_{k=1}^{n} g_k(u_n) c_{kn} = g(u_n) > 0.$$

By Proposition 2.8, the Galerkin equation (G) has a solution.

Step 2: *A priori* estimates.
If u_n is a solution of the Galerkin equation (G), then

$$g(u_n) = 0.$$

From (3) it follows that

$$\|u_n\| \leq R \qquad \text{for all } n. \tag{3*}$$

If u is a solution of equation (1), then $g(u) = 0$, and (3) implies

$$\|u\| \leq R.$$

Step 3: Boundedness of (Au_n).
By Proposition 26.4(a), the operator A is locally bounded, i.e., there exist positive numbers r and δ such that

$$\|v\| \leq r \qquad \text{implies} \qquad \|Av\| \leq \delta.$$

The operator A is monotone; therefore,

$$\langle Au_n - Av, u_n - v\rangle \geq 0.$$

By the Galerkin equations (2),

$$\langle Au_n, u_n\rangle = \langle b, u_n\rangle \qquad \text{for all } n.$$

Hence

$$|\langle Au_n, u_n\rangle| \leq \|b\|\,\|u_n\| \leq \|b\|R \qquad \text{for all } n,$$

by (3*). The definition of the norm in X^* yields

$$\begin{aligned}\|Au_n\| &= \sup_{\|v\|=r} r^{-1}\langle Au_n, v\rangle \\ &\leq \sup_{\|v\|=r} r^{-1}(\langle Av, v\rangle + \langle Au_n, u_n\rangle - \langle Av, u_n\rangle) \\ &\leq r^{-1}(\delta r + \|b\|R + \delta R).\end{aligned}$$

Step 4: Convergence of the Galerkin method.

The B-space X is reflexive. Thus, the bounded sequence (u_n) has a weakly convergent subsequence, again denoted by (u_n), i.e.,

$$u_n \rightharpoonup u \quad \text{in } X \qquad \text{as} \quad n \to \infty.$$

From the Galerkin equations (2) it follows that

$$\lim_{n\to\infty} \langle Au_n, w\rangle = \langle b, w\rangle \qquad \text{for all} \quad w \in \bigcup_{n=1}^{\infty} X_n. \tag{4}$$

Since $\bigcup_n X_n$ is dense in X and (Au_n) is bounded in X^*, relation (4) holds for all $w \in X$, i.e.,

$$Au_n \rightharpoonup b \quad \text{in } X^* \qquad \text{as} \quad n \to \infty,$$

by Proposition 21.26(c), (f). Furthermore, by the Galerkin equations, we obtain

$$\lim_{n\to\infty} \langle Au_n, u_n\rangle = \lim_{n\to\infty} \langle b, u_n\rangle = \langle b, u\rangle. \tag{5}$$

By the monotonicity trick (25.4), it follows from

$$u_n \rightharpoonup u \quad \text{in } X \qquad \text{as} \quad n \to \infty,$$

$$Au_n \rightharpoonup b \quad \text{in } X^* \qquad \text{as} \quad n \to \infty,$$

and from (5) that $Au = b$, i.e., u is a solution of the original operator equation (1).

Proof of (a). Let S be the solution set of equation (1) for fixed $b \in X^*$. This set is nonempty by Step 4 above.

(I) S is bounded by Step 2 above.

(II) S is convex. To prove this, let $u_1, u_2 \in S$, i.e.,

$$Au_i = b, \qquad i = 1, 2.$$

For $u = t_1 u_1 + t_2 u_2$ and $0 \le t_1, t_2 \le 1$, $t_1 + t_2 = 1$, there follows the inequality

$$\langle b - Av, u - v\rangle = \sum_{i=1}^{2} t_i \langle Au_i - Av, u_i - v\rangle \ge 0,$$

for all $v \in X$. The monotonicity trick (25.4a) yields $Au = b$, i.e., $u \in S$.

(III) S is closed. Indeed, from $Av_n = b$ for all n and $v_n \to u$ as $n \to \infty$ it follows that

$$\langle b - Av, u - v\rangle = \lim_{n\to\infty} \langle Av_n - Av, v_n - v\rangle \ge 0,$$

for all $v \in X$. The monotonicity trick (25.4a) yields $Au = b$.

Proof of (c). Let the operator A be strictly monotone. By the uniqueness trick (25.5), the equation $Au = b$ has at most one solution u.

The same argument yields the unique solvability of the Galerkin equations.

Proof of (d). Let the operator $A: X \to X^*$ be strictly monotone. Then A is injective by statement (c). By (a), the operator A is surjective. Hence A is bijective, i.e., the inverse operator $A^{-1}: X^* \to X$ exists.

(I) A^{-1} is strictly monotone. Indeed, we get

$$\langle b - \bar{b}, A^{-1}b - A^{-1}\bar{b}\rangle = \langle Au - A\bar{u}, u - \bar{u}\rangle > 0 \text{ for } u \ne \bar{u}.$$

(II) A^{-1} is bounded. This follows from the coerciveness of A as in Step 1 above.

(III) A^{-1} is demicontinuous. For, let $v_n = A^{-1}b_n$ and let

$$b_n \to b \qquad \text{as} \quad n \to \infty.$$

From the boundedness of A^{-1} there follows the boundedness of (v_n). Moreover, from $v_{n'} \rightharpoonup v$ as $n \to \infty$ it follows that

$$\langle b - Aw, v - w\rangle = \lim_{n\to\infty} \langle b_{n'} - Aw, v_{n'} - w\rangle \ge 0,$$

for all $w \in X$. The monotonicity trick (25.4a) yields $Av = b$, i.e., $v = A^{-1}b$. Hence the convergence principle (Proposition 21.23(i)) ensures that

$$v_n \rightharpoonup v \qquad \text{as} \quad n \to \infty.$$

(IV) Let A be uniformly monotone. By (25.15), A is stable, i.e.,

$$\|Au - Av\| \ge a(\|u - v\|) \qquad \text{for all} \quad u, v \in X,$$

where the function $a: \mathbb{R}_+ \to \mathbb{R}_+$ is strictly monotone and $a(0) = 0$. This implies the continuity of $A^{-1}: X^* \to X$. If A is strongly monotone, then $a(\|u - v\|) = c\|u - v\|$, $c > 0$. Therefore, A^{-1} is Lipschitz continuous, i.e.,

$$\|A^{-1}b - A^{-1}d\| \le c^{-1}\|b - d\| \qquad \text{for all} \quad b, d \in X^*.$$

Proof of (e). This follows immediately from the convergence tricks in (25.14).

Proof of (f). Suppose that X is not separable. Then we prove the convergence of the Galerkin method by means of M–S sequences, which we introduced in the Appendix of Part I.

To be precise let $\Sigma = \{Y\}$ be the system of all the finite-dimensional subspaces Y of X. We define an order relation on Σ by means of

$$Y \le Z \qquad \text{iff} \qquad Y \subseteq Z.$$

Then Σ is a directed set in the sense of $A_1(16)$. In fact, if $Y, Z \in \Sigma$, then $Y, Z \le \operatorname{span}(Y \cup Z)$.

For each $Y \in \Sigma$, we now consider the corresponding Galerkin equation and as above we obtain a solution u_Y instead of u_n. Then (u_Y) is an M–S sequence which is bounded in the reflexive B-space X. Since each closed ball in X is weakly compact, there exists an M–S subsequence $(u_{Y'})$ with

$$u_{Y'} \rightharpoonup u$$

according to $A_1(17f)$. We now proceed as in the proof of (b) above. □

26.3. The Nemyckii Operator

In order to be able to apply Theorem 26.A to differential and integral equations, we require the properties of the so-called *Nemyckii operator* F defined by

$$(Fu)(x) = f(x, u_1(x), \dots, u_n(x)) \tag{6}$$

with $u = (u_1, \dots, u_n)$. Thus, F results when one replaces all the variables u_j by $u_j(x)$ in $f(x, u_1, \dots, u_n)$. We formulate the following conditions:

(H1) *Carathéodory condition.* Let $f\colon G \times \mathbb{R}^n \to \mathbb{R}$ be a given function, where G is a nonempty measurable set in $\mathbb{R}^N$ and $n, N \ge 1$. Moreover, the following hold:

$$x \mapsto f(x, u) \text{ is measurable on } G \text{ for all } u \in \mathbb{R}^n;$$

$$u \mapsto f(x, u) \text{ is continuous on } \mathbb{R}^n \text{ for almost all } x \in G.$$

(H2) *Growth condition.* For all $(x, u) \in G \times \mathbb{R}^n$,

$$|f(x, u)| \le a(x) + b \sum_{i=1}^{n} |u_i|^{p_i/q}.$$

Here, b is a fixed positive number, the function $a \in L_q(G)$ is nonnegative, and $1 \le q, p_i < \infty$ for all i.

EXAMPLE 26.5. Condition (H1) is satisfied in the case where $f\colon G \times \mathbb{R}^n \to \mathbb{R}$ is continuous.

Proposition 26.6. *Under the assumptions* (H1) *and* (H2), *the Nemyckii operator*

$$F\colon \prod_{i=1}^{n} L_{p_i}(G) \to L_q(G)$$

is continuous and bounded with

$$\|Fu\|_q \leq \text{const}(\|a\|_q + \sum_{i=1}^{n} \|u_i\|_{p_i}^{p_i/q}) \tag{7}$$

for all $u \in \prod_{i=1}^{n} L_{p_i}(G)$.

Before proving this, we consider the special case $n = 1$, i.e., we consider a function $f\colon G \times \mathbb{R} \to \mathbb{R}$. Then the Nemyckii operator F is given by

$$(Fu)(x) = f(x, u(x)).$$

In particular, we are interested in the case $F\colon X \to X^*$, where $X = L_p(G)$. This corresponds to (H1) and (H2) with $n = 1$, $p = p_1$ and $p^{-1} + q^{-1} = 1$, i.e., $p/q = p - 1$. More precisely, we make the following assumptions:

(A1) *Carathéodory condition.* Let $f\colon G \times \mathbb{R} \to \mathbb{R}$ be a given function, where G is a nonempty measurable set in $\mathbb{R}^N$, $N \geq 1$. Suppose that:

$$x \mapsto f(x, u) \text{ is measurable on } G \text{ for all } u \in \mathbb{R};$$

$$u \mapsto f(x, u) \text{ is continuous on } \mathbb{R} \text{ for almost all } x \in G.$$

(A2) *Growth condition.* Let $1 < p, q < \infty$ and $p^{-1} + q^{-1} = 1$. Suppose that there exist a nonnegative function $a \in L_q(G)$ and a fixed positive number b such that

$$|f(x, u)| \leq a(x) + b|u|^{p-1} \quad \text{for all} \quad (x, u) \in G \times \mathbb{R}.$$

(A3) *Monotonicity of f.* The function f is monotone with respect to u, i.e.,

$$f(x, u) \leq f(x, v)$$

holds for all $u, v \in \mathbb{R}$ with $u \leq v$ and for all $x \in G$.

(A3*) *Strict monotonicity of f.* The function f is strictly monotone with respect to u, i.e.,

$$f(x, u) < f(x, v)$$

holds for all $u, v \in \mathbb{R}$ with $u < v$ and for all $x \in G$.

(A4) *Coerciveness of f.* For a fixed number $d > 0$ and a fixed function $g \in L_1(G)$,

$$f(x, u)u \geq d|u|^p + g(x) \quad \text{for all} \quad (x, u) \in G \times \mathbb{R}.$$

(A5) *Positivity of f.* For all $(x, u) \in G \times \mathbb{R}$,

$$f(x, u)u \geq 0.$$

(A6) *Asymptotic positivity of f.* There exists a number $R > 0$ such that

$$f(x, u)u \geq 0$$

holds for all $(x, u) \in G \times \mathbb{R}$ with $|u| \geq R$, and meas $G < \infty$.

Let $X = L_p(G)$. Then $X^* = L_q(G)$.

Proposition 26.7. *Under the assumptions* (A1), (A2), *the following are valid:*

(a) *The Nemyckii operator* $F: X \to X^*$ *is continuous and bounded with*

$$\|Fu\|_q \leq \text{const}(\|a\|_q + \|u\|_p^{p-1}) \qquad \text{for all} \quad u \in X$$

and

$$\langle Fu, u\rangle_X = \int_G f(x, u(x))u(x)\,dx \qquad \textit{for all} \quad u \in X.$$

(b) (A3) *implies F is monotone.*
(c) (A3*) *implies F is strictly monotone.*
(d) (A4) *implies F is coercive and*

$$\langle Fu, u\rangle_X \geq d\|u\|^p + \int_G g(x)\,dx \qquad \textit{for all} \quad u \in X.$$

(e) (A3*) *and* (A4) *imply F satisfies the condition* $(S)_+$, *i.e., it follows from* $u_n \rightharpoonup u$ *in X as* $n \to \infty$ *and*

$$\overline{\lim_{n\to\infty}}\ \langle Fu_n - Fu, u_n - u\rangle \leq 0$$

that $u_n \to u$ *in X as* $n \to \infty$.
(f) (A5) *implies*

$$\langle Fu, u\rangle_X \geq 0 \qquad \textit{for all} \quad u \in X.$$

(g) (A6) *implies*

$$\langle Fu, u\rangle_X \geq -c \qquad \textit{for all} \quad u \in X,$$

where c is a positive constant.

In Propositions 32.44 and 41.10 we will prove the following important additional results. If (A1), (A2), and (A3) are satisfied and the set G is bounded, then the Nemyckii operator $F: X \to X^*$ is a monotone continuous potential operator, i.e., there exists a convex C^1-functional $f: X \to \mathbb{R}$ such that

$$F = f'.$$

Moreover, F is maximal monotone, cyclic monotone, and hence 3*-monotone.

If (A1) and (A2) are satisfied and G is bounded, then the Nemyckii operator $F: X \to X^*$ is a continuous potential operator.

Proof of Proposition 26.6. We set $n = 1$, $u = u_1$, $p = p_1$. The deliberations run analogously in the general case.

(I) Measurability. Since $u \in L_p(G)$, the function $x \mapsto u(x)$ is measurable on G, and, by (H1) and $A_2(12)$, the function

$$x \mapsto f(x, u(x))$$

is measurable on G.

(II) F is bounded. Indeed, the estimate (7) follows from (H2) and $A_2(31)$.

(III) F is continuous from $L_p(G)$ into $L_q(G)$. To prove this, let

$$u_n \to u \quad \text{in } L_p(G) \qquad \text{as} \quad n \to \infty.$$

By the convergence principle in Problem 26.4, there exists a subsequence $(u_{n'})$ and a function $v \in L_p(G)$ with

$$u_{n'}(x) \to u(x) \qquad \text{as} \quad n \to \infty \qquad \text{for almost all} \quad x \in G$$

and the majorant condition

$$|u_{n'}(x)| \le v(x) \qquad \text{for all } n' \quad \text{and almost all} \quad x \in G.$$

Therefore, by the inequalities $A_2(30b)$ and (H2),

$$\begin{aligned} \|Fu_{n'} - Fu\|_q^q &= \int_G |f(x, u_{n'}(x)) - f(x, u(x))|^q \, dx \\ &\le \text{const} \int_G (|f(x, u_{n'}(x))|^q + |f(x, u(x))|^q) \, dx \\ &\le \text{const} \int_G (|a(x)|^q + |v(x)|^p + |u(x)|^p) \, dx. \end{aligned}$$

By (H1),

$$f(x, u_{n'}(x)) - f(x, u(x)) \to 0 \qquad \text{as} \quad n \to \infty \qquad \text{for almost all} \quad x \in G.$$

Majorized convergence $A_2(19)$ gives $\|Fu_{n'} - Fu\|_q \to 0$, i.e.,

$$Fu_{n'} \to Fu \quad \text{in } L_q(G) \qquad \text{as} \quad n \to \infty.$$

The convergence principle (Proposition 10.13(1)) tells us that the entire sequence converges, i.e.,

$$Fu_n \to Fu \quad \text{in } L_q(G) \qquad \text{as} \quad n \to \infty. \qquad \square$$

The continuity of the Nemyckii operator F can also be proved in a simple way by using the generalized principle of majorized convergence $A_2(19a)$ instead of the convergence principle in (III) above.

The proof of Proposition 26.7 will be given in Problem 26.3.

26.4. Generalized Gradient Method for the Solution of the Galerkin Equations

The Galerkin equations in Theorem 26.A represent a nonlinear system of equations of the form

$$g_i(x) = 0, \qquad x \in \mathbb{R}^n, \qquad i = 1, \ldots, n. \tag{8}$$

Let $x = (\xi_1, \ldots, \xi_n)$ and $y = (\eta_1, \ldots, \eta_n)$. We investigate when (8) can be solved with the aid of the iteration method

$$\xi_i^{(k+1)} = \xi_i^{(k)} - tg_i(x^{(k)}), \qquad i = 1, \dots, n \tag{9}$$

for $k = 0, 1, 2, \dots$ with $x^{(0)} = 0$.

Proposition 26.8. *Assume that the functions $g_i\colon \mathbb{R}^n \to \mathbb{R}$ are continuously differentiable or at least locally Lipschitz continuous for all i. Suppose there exists a number $c > 0$ such that*

$$\sum_{i=1}^{n} (g_i(x) - g_i(y))(\xi_i - \eta_i) \geq c \sum_{i=1}^{n} (\xi_i - \eta_i)^2 \qquad \text{for all} \quad x, y \in \mathbb{R}^n.$$

Then the system (8) has exactly one solution $x \in \mathbb{R}^n$, and the sequence $(x^{(k)})$ constructed in (9) converges to x as $k \to \infty$ for sufficiently small $t > 0$.

This proposition is a special case of Theorem 26.B below. In connection with this theorem, we proceed from the operator equation

$$Gx = 0, \qquad x \in X, \tag{10}$$

with the iteration method

$$x_{k+1} = x_k - tLGx_k, \qquad k = 0, 1, \dots, \tag{11}$$

where $x_0 = 0$. Our assumptions are the following:

(H1) $L, G\colon X \to X$ are operators on the real H-space X.

(H2) G is strongly monotone and locally Lipschitz continuous. To be precise, there exist positive numbers a and $M(r)$ such that

$$(Gx - Gy|x - y) \geq a\|x - y\|^2 \qquad \text{for all} \quad x, y \in X$$

and

$$\|Gx - Gy\| \leq M(r)\|x - y\|,$$

for all $x, y \in X$ with $\|x\|, \|y\| \leq r$ and all $r > 0$.

(H3) L is linear, continuous, self-adjoint, and strongly monotone, i.e.,

$$(Lx|x) \geq b\|x\|^2 \qquad \text{for all} \quad x \in X \quad \text{and fixed} \quad b > 0.$$

(H4) We assume the nontrivial case $G(0) \neq 0$ and choose the numbers

$$c = \|L\|, \qquad r_0 = a^{-1}(1 + \sqrt{c/b})\|G(0)\|, \qquad t_0 = 2a/cM(r_0)^2.$$

Theorem 26.B (Generalized Gradient Method). *Under the assumptions* (H1) *through* (H4), *the original equation* (10) *has exactly one solution $x \in X$.*

If one chooses a fixed number $t \in \,]0, t_0[$, then the iterative sequence (x_k) in (11) *converges in the* H-*space X to the solution x as $k \to \infty$. For all $k = 0, 1, 2, \dots$, we have the error estimate*

$$\|x_k - x\| \leq a^{-1}\|Gx_k\|.$$

Gradient methods will be considered in detail in Part III.

PROOF. We will use a majorant method. The basic idea of the proof is contained in (14) below.

(I) Existence and uniqueness. We set $X = X^*$. It follows from Theorem 26.A that equation (10) has a unique solution.

(II) Convergence of (x_k). Since the operator L is self-adjoint it follows from

$$b\|y\|^2 \le (Ly|y) \le c\|y\|^2 \qquad \text{for all} \quad y \in X,$$

that $bI \le L \le cI$. This implies $c^{-1}I \le L^{-1} \le b^{-1}I$, i.e.,

$$c^{-1}\|y\|^2 \le (L^{-1}y|y) \le b^{-1}\|y\|^2 \qquad \text{for all} \quad y \in X$$

(cf. Problem 19.5g).

Let x be the solution of (10), i.e., $Gx = 0$. Our proof of convergence is based on the investigation of the functional

$$h(y) \stackrel{\text{def}}{=} (L^{-1}(y-x)|y-x) \qquad \text{for all} \quad y \in X.$$

(II-1) By (11), for $k = 0, 1, \ldots$, we have

$$\begin{aligned} h(x_k) - h(x_{k+1}) &= 2t(Gx_k - Gx|x_k - x) - t^2(LGx_k|Gx_k) \\ &\ge 2ta\|x_k - x\|^2 - t^2c\|Gx_k\|^2. \end{aligned} \tag{12}$$

From (H2) and $Gx = 0$ it follows that $a\|x\|^2 \le \|Gx - G(0)\|\,\|x\|$, therefore, we obtain the *a priori* estimate:

$$\|x\| \le a^{-1}\|G(0)\| \le r_0.$$

(II-2) Assume we know that

$$\|x_k\| \le r_0.$$

Then, from (H2) we obtain the estimate

$$\|Gx_k\| = \|Gx_k - Gx\| \le M(r_0)\|x_k - x\|. \tag{13}$$

(II-3) Let $0 < t < t_0$. We set

$$d = t(2a - tcM(r_0)^2),$$

i.e., $d > 0$. We want to show that, for all $k = 1, 2, \ldots$, we have the following two *key estimates*:

$$\begin{aligned} d\|x_{k-1} - x\|^2 &\le h(x_{k-1}) - h(x_k), \\ \|x_{k-1}\| &\le r_0. \end{aligned} \tag{14}$$

We prove this by induction. By (12) and (13), we have that (14) holds for $k = 1$.

Suppose that (14) is valid up to a fixed $k > 1$. Then

$$\begin{aligned} c^{-1}\|x_k - x\|^2 &\le h(x_k) \le h(x_{k-1}) \le \cdots \le h(x_0) = h(0) \\ &\le b^{-1}\|x\|^2 \le b^{-1}a^{-2}\|G(0)\|^2 \end{aligned}$$

and hence

$$\|x_k\| \le \|x_k - x\| + \|x\| \le r_0.$$

Moreover, it follows from (12) and (13) that

$$d\|x_k - x\|^2 \le h(x_k) - h(x_{k+1}).$$

(II-4) The key relation (14) tells us that the sequence $(h(x_k))$ is *monotone increasing* and not negative, therefore convergent. Moreover, (14) shows that $x_k \to x$ as $k \to \infty$.

(III) The error estimate follows from $Gx = 0$ and (H2). This yields

$$a\|x_k - x\|^2 \le \|Gx_k - Gx\| \, \|x_k - x\|. \qquad \square$$

26.5. Application to Quasi-Linear Elliptic Equations of Order $2m$

As a first model equation we consider the boundary value problem

$$\begin{gathered} -\sum_{i=1}^{N} D_i(|D_i u|^{p-2} D_i u) + su = f \quad \text{on } G, \\ u = 0 \quad \text{on } \partial G. \end{gathered} \tag{15}$$

Let G be a bounded region in $\mathbb{R}^N$ with $N \ge 1$. Furthermore, let $2 \le p < \infty$, $p^{-1} + q^{-1} = 1$, and let s be a nonnegative real number. We set $x = (\xi_1, \ldots, \xi_N)$ and $D_i = \partial/\partial \xi_i$.

Definition 26.9. Let $X = \mathring{W}_p^1(G)$, and let

$$a(u, v) = \int_G \left(\sum_{i=1}^{N} |D_i u|^{p-2} D_i u D_i v + suv\right) dx,$$

$$b(v) = \int_G fv \, dx.$$

The *generalized problem* corresponding to (15) reads as follows: For given $f \in L_q(G)$, we seek $u \in X$ such that

$$a(u, v) = b(v) \qquad \text{for all} \quad v \in X. \tag{16}$$

Formally, one obtains (16) by multiplying (15) by $v \in C_0^\infty(G)$ and subsequent integration by parts. The Galerkin equations corresponding to (16) read as follows:

$$a(u_n, w_k) - b(w_k) = 0, \qquad k = 1, \ldots, n, \tag{17}$$

with

$$u_n = \sum_{k=1}^{n} c_{kn} w_k.$$

In this connection, let $\{w_1, w_2, \ldots\}$ be a basis in X. We set $X_n = \operatorname{span}\{w_1, \ldots, w_n\}$.

Proposition 26.10 (Galerkin Method).

(a) *The generalized problem* (16) *corresponding to* (15) *has exactly one solution* $u \in X$. *The Galerkin equation* (17) *has exactly one solution* $u_n \in X_n$ *for each* n, *and* (u_n) *converges in* X *to* u *as* $n \to \infty$.

(b) *For* $s > 0$ *and fixed* n, *the solution* u_n *of the Galerkin equation* (17) *can be computed by the iteration method of Proposition* 26.8 *in the case where the functions* w_k *and* $D_i w_k$ *are bounded on* G *for all* i *and* k.

(c) *The generalized problem* (16) *is equivalent to the operator equation*

$$Au = b, \qquad u \in X,$$

where $\langle Au, v\rangle = a(u, v)$ *for all* u, $v \in X$, *and the operator* $A: X \to X^*$ *is continuous, uniformly monotone, coercive, and bounded.*

PROOF. Ad(a), (c). It first follows from $u \in \mathring{W}_p^1(G)$ that $D_i u \in L_p(G)$. Because $(p-1)q = p$, we have

$$|D_i u|^{p-2} D_i u \in L_q(G).$$

By $A_2(45)$, the embedding $\mathring{W}_p^1(G) \subseteq L_2(G)$ is continuous, i.e.,

$$\|u\|_2 \le \text{const}\|u\|_X \qquad \text{for all} \quad u \in X.$$

According to $A_2(53b)$, we introduce the *equivalent norm*

$$\|u\| = \left(\int_G \sum_{i=1}^N |D_i u|^p \, dx\right)^{1/p}$$

on the Sobolev space $X = \mathring{W}_p^1(G)$. This is an important step of our proof.

Let $\|\cdot\|_*$ denote the corresponding norm on X^*.

(I) First key inequality. The Hölder inequality along with $(p-1)q = p$ yields

$$\begin{aligned} |a(u, v)| &\le \sum_{i=1}^N \left(\int_G |D_i u|^p \, dx\right)^{1/q} \left(\int_G |D_i v|^p \, dx\right)^{1/p} + s\|u\|_2\|v\|_2 \\ &\le \text{const}(\|u\|^{p/q} + \|u\|)\|v\| \qquad \text{for all} \quad u, v \in X. \end{aligned}$$

(II) Second key inequality. By (25.45),

$$(|\lambda|^{p-2}\lambda - |\mu|^{p-2}\mu)(\lambda - \mu) \ge c|\lambda - \mu|^p,$$

for all λ, $\mu \in \mathbb{R}$ and fixed $c > 0$. This implies

$$a(u, u - v) - a(v, u - v) \ge c\|u - v\|^p + s\int_G (u - v)^2 \, dx \text{ for all } u, v \in X.$$

(III) Equivalent operator equation. By (I) and (22.1a), there exists an operator

$$A: X \to X^*$$

with

$$\langle Au, v\rangle = a(u, v) \qquad \text{for all} \quad u, v \in X.$$

By (22.1b), we obtain $b \in X^*$. Thus the generalized problem (16) is equivalent to the operator equation

$$Au = b, \qquad u \in X. \tag{18}$$

We want to apply Theorem 26.A to this equation.

(IV) Properties of A. By (I),

$$\|Au\|_* \leq \text{const}(\|u\|^{p/q} + \|u\|) \qquad \text{for all} \quad u \in X,$$

i.e., A is bounded. By (II),

$$\langle Au - Av, u - v\rangle \geq c\|u - v\|^p \qquad \text{for all} \quad u, v \in X,$$

i.e., A is uniformly monotone, and hence A is coercive.

(V) Continuity of A via continuity of the Nemyckii operator. In order to prove the continuity of $A: X \to X^*$ let

$$u_n \to u \quad \text{in } X \qquad \text{as} \quad n \to \infty.$$

This implies

$$D_i u_n \to D_i u \quad \text{in } L_p(G) \qquad \text{as} \quad n \to \infty,$$

according to the definition of the convergence in $X = \mathring{W}_p^1(G)$. Set

$$F(u) = |u|^{p-2}u \qquad \text{for all} \quad u \in \mathbb{R}.$$

It follows from the growth condition

$$|F(u)| \leq |u|^{p-1} \qquad \text{for all} \quad u \in \mathbb{R}$$

and from Proposition 26.7 that the Nemyckii operator

$$F: L_p(G) \to L_q(G)$$

is *continuous*. Therefore, it follows from $D_i u_n \to D_i u$ in $L_p(G)$ as $n \to \infty$ that

$$F(D_i u_n) \to F(D_i u) \quad \text{in } L_q(G) \qquad \text{as} \quad n \to \infty.$$

By the Hölder inequality, for all $v \in X$, we obtain that

$$\begin{aligned}|\langle Au_n - Au, v\rangle| &= \left|\int_G \sum_i (F(D_i u_n) - F(D_i u))D_i v + s(u_n - u)v\,dx\right| \\ &\leq \sum_i \|F(D_i u_n) - F(D_i u)\|_q \|D_i v\|_p + s\|u_n - u\|_2 \|v\|_2.\end{aligned}$$

From the continuity of the embedding $X \subseteq L_2(G)$ it follows that

$$|\langle Au_n - Au, v\rangle| \leq \text{const}\left(\sum_i \|F(D_i u_n) - F(D_i u)\|_q + \|u_n - u\|\right)\|v\|,$$

for all $v \in X$. This implies the *key estimate*

$$\|Au_n - Au\|_* \leq \text{const}\left(\sum_i \|F(D_i u_n) - F(D_i u)\|_q + \|u_n - u\|\right),$$

and hence $\|Au_n - Au\|_* \to 0$ as $n \to \infty$, i.e., the operator $A: X \to X^*$ is continuous.

(VI) Now it follows from Theorem 26.A that, for each $b \in X$, the operator equation (18) has a unique solution $u \in X$. Moreover, we obtain the convergence of the Galerkin method.

Ad(b). We want to apply Proposition 26.8. To this end, we need the following inequality

$$||\lambda|^{p-2}\lambda - |\mu|^{p-2}\mu| \leq (p-1)K^{p-2}|\lambda - \mu|, \tag{19}$$

for all λ, $\mu \in \mathbb{R}$ with $|\lambda|$, $|\mu| \leq K$. Recall that $p \geq 2$. In fact, in the case where λ, $\mu \geq 0$, inequality (19) results from the mean value theorem of differential calculus. In the case where $\mu \leq 0 \leq \lambda$ we use $|\lambda|^{p-1} \leq (p-1)K^{p-2}|\lambda|$.

We now set

$$g_k(x) = a(u, w_k) - b(w_k)$$

with $x = (\xi_1, \ldots, \xi_n)$ and

$$u = \sum_{k=1}^{n} \xi_k w_k,$$

where $\xi_1, \ldots, \xi_n$ are real numbers.

For all x, $y \in \mathbb{R}^n$, we obtain that

$$\begin{aligned}\sum_{k=1}^{n} (g_k(x) - g_k(y))(\xi_k - \eta_k) &\geq s \int_G \sum_{k=1}^{n} ((\xi_k - \eta_k)w_k)^2\,dx \\ &\geq C \sum_{k=1}^{n} (\xi_k - \eta_k)^2,\end{aligned}$$

where C denotes a positive constant. Moreover, the functions $g_i: \mathbb{R}^n \to \mathbb{R}$ are locally Lipschitz continuous. This follows from inequality (19) and the boundedness of the functions w_k and $D_i w_k$ on G for all i, k.

The assertion (b) now follows from Proposition 26.8. □

We now want to generalize Proposition 26.10 to quasi-linear elliptic equations of order $2m$. To this end, we set

$$(Lu)(x) = \sum_{|\alpha| \leq m} (-1)^{|\alpha|} D^\alpha A_\alpha(x, Du(x)) \tag{20}$$

with $m \geq 1$. We consider the boundary value problem

$$\begin{aligned}Lu &= f \quad \text{on } G, \\ D^\beta u &= 0 \quad \text{on } \partial G \qquad \text{for all} \quad \beta \quad \text{with} \quad |\beta| \leq m-1,\end{aligned} \tag{21}$$

where G is a bounded region in $\mathbb{R}^N$ with $N \geq 1$. We use the notation of Definition 21.1. In (20), the summation is over all

$$\alpha = (\alpha_1, \ldots, \alpha_N) \qquad \text{with} \quad |\alpha| \leq m,$$

where $|\alpha| = \alpha_1 + \cdots + \alpha_N$, i.e., the summation is over all partial derivatives D^α

up to order m including u. In this connection, note our convention $D^0 u = u$. We set

$$Du = (D^\gamma u)_{|\gamma| \le m},$$

i.e., Du denotes the tuple of all partial derivatives up to order m including u. The boundary condition in (21) means that all partial derivatives up to order $m - 1$ should vanish on ∂G.

We now formulate several conditions on L. In this connection, we think of A_α as a real function of the variables $x \in G$ and $D \in \mathbb{R}^M$, where

$$D = (D^\gamma)_{|\gamma| \le m}.$$

(H1) *Carathéodory condition.* For all α with $|\alpha| \le m$, the function A_α: $G \times \mathbb{R}^M \to \mathbb{R}$ has the following two properties:

$$x \mapsto A_\alpha(x, D) \text{ is measurable on } G \text{ for all } D \in \mathbb{R}^M;$$

$$D \mapsto A_\alpha(x, D) \text{ is continuous on } \mathbb{R}^M \text{ for almost all } x \in G.$$

For example, this condition is satisfied if A_α is continuous.

In the following conditions, let

$$1 < p < \infty, \qquad p^{-1} + q^{-1} = 1, \qquad g \in L_q(G), \qquad h \in L_1(G).$$

Moreover, let c, d, and C denote positive numbers and suppose that the conditions are valid for all D, $D' \in \mathbb{R}^M$, $|\alpha| \le m$, and $x \in G$.

(H2) *Growth condition*

$$|A_\alpha(x, D)| \le C\left(g(x) + \sum_{|\gamma| \le m} |D^\gamma|^{p-1}\right).$$

(H3) *Monotonicity condition*

$$\sum_{|\alpha| \le m} (A_\alpha(x, D) - A_\alpha(x, D'))(D^\alpha - D'^\alpha) \ge 0.$$

(H4) *Coerciveness condition*

$$\sum_{|\alpha| \le m} A_\alpha(x, D) D^\alpha \ge c \sum_{|\gamma| = m} |D^\gamma|^p - h(x).$$

(H5*) *Uniform monotonicity*

$$\sum_{|\alpha| \le m} (A_\alpha(x, D) - A_\alpha(x, D'))(D^\alpha - D'^\alpha) \ge d \sum_{|\gamma| = m} |D^\gamma - D'^\gamma|^p.$$

(H6*) *Degeneracy.* All the functions A_α depend only on derivatives of u up to order $m - 1$.

A differential operator L in (20) is called a *monotone coercive quasi-linear elliptic differential operator* iff the conditions (H1) through (H4) are satisfied.

Definition 26.11. Let $X = \mathring{W}_p^1(G)$ with $1 < p < \infty$ and $p^{-1} + q^{-1} = 1$. The *generalized problem* corresponding to the boundary value problem (21) reads as follows. Let $f \in L_q(G)$ be given. We seek a function $u \in X$ such that

$$a(u, v) = b(v) \qquad \text{for all} \quad v \in X. \tag{22}$$

Here, for all u, $v \in X$, we set

$$a(u, v) = \int_G \sum_{|\alpha| \leq m} A_\alpha(x, Du(x)) D^\alpha v(x)\, dx,$$

$$b(v) = \int_G fv\, dx.$$

Formally, one obtains (22) by multiplying (21) by $v \in C_0^\infty(G)$ and subsequent integration by parts. The choice of the space X is meaningful for, it follows from $u \in \mathring{W}_p^m(G)$ that $D^\beta u = 0$ holds on ∂G, in the generalized sense, for all β with $|\beta| \leq m - 1$ (cf. $A_2(48)$). Therefore, the *boundary condition* in (21) is valid in the *generalized* sense.

Proposition 26.12 (Generalized Solutions of (21)). *Assume* (H1), (H2), (H3), *and* (H4). *Then there exists exactly one operator $A: X \to X^*$ such that*

$$\langle Au, v \rangle = a(u, v) \qquad \textit{for all} \quad u, v \in X.$$

The generalized problem (22) *corresponding to* (21) *is equivalent to the operator equation*

$$Au = b, \qquad u \in X. \tag{23}$$

The operator $A: X \to X^*$ is monotone, coercive, continuous, and bounded. Thus, all the assertions of the main theorem on monotone operators (*Theorem* 26.A) *are valid for* (23) *and hence for* (22).

In particular, for each $f \in L_q(G)$, the generalized problem (22) *has a solution.*

Corollary 26.13. *If, in addition, condition* (H5*) *is valid, then the operator $A: X \to X^*$ is uniformly monotone. In this case, the solution $u \in X$ of* (22) *is unique.*

Corollary 26.14. *Assume* (H1), (H2), *and* (H6*). *Then there exists exactly one operator $A: X \to X^*$ such that*

$$\langle Au, v \rangle = a(u, v) \qquad \textit{for all} \quad u, v \in X.$$

The operator A is strongly continuous.

Corollary 26.14 contains the observation, important for the next chapter, that terms of *lower order* lead to *strongly continuous* operators.

PROOF. We use analogous arguments as in the proof of Proposition 26.10. For brevity, the norm $\|\cdot\|_{m,p}$ on the Sobolev space $X = \mathring{W}_p^m(G)$ is denoted by $\|\cdot\|$, i.e.,

$$\|u\| = \left(\int_G \sum_{|\gamma| \leq m} |D^\gamma u(x)|^p \, dx\right)^{1/p}.$$

The corresponding norm on X^* is denoted by $\|\cdot\|_*$. Recall that

$$\|u\|_{m,p,0} = \left(\int_G \sum_{|\gamma| = m} |D^\gamma u(x)|^p \, dx\right)^{1/p}.$$

By A_2(53b), the norm $\|\cdot\|_{m,p,0}$ is equivalent to $\|\cdot\|$ on X, i.e., there are positive constants c_1 and c_2 such that

$$c_1\|u\| \leq \|u\|_{m,p,0} \leq c_2\|u\| \qquad \text{for all} \quad u \in X.$$

This relation will be used critically below in order to prove the coerciveness and the uniform monotonicity of the operator A.

(I) The Nemyckii operator. We set

$$(F_\alpha u)(x) = A_\alpha(x, Du(x)) \qquad \text{for all} \quad x \in G.$$

Obviously, we have

$$D^\gamma u \in L_p(G) \qquad \text{if} \quad u \in \mathring{W}_p^m(G) \qquad \text{and} \qquad |\gamma| \leq m. \tag{24}$$

Thus, it follows from (H1), (H2), and from Proposition 26.6 on the Nemyckii operator that the operator

$$F_\alpha\colon X \to L_q(G)$$

is continuous and

$$\|F_\alpha u\|_q \leq \text{const}(\|g\|_q + \|u\|^{p/q}) \qquad \text{for all} \quad u \in X.$$

(II) The key inequality. By the Hölder inequality,

$$\begin{aligned}|a(u,v)| &\leq \sum_{|\alpha| \leq m} \|F_\alpha u\|_q \|D^\alpha v\|_p \\ &\leq \text{const}(\|g\|_q + \|u\|^{p/q})\|v\| \qquad \text{for all} \quad u, v \in X.\end{aligned}$$

(III) Equivalent operator equation. By (II) and (22.1a), there exists an operator $A\colon X \to X^*$ such that

$$\langle Au, v\rangle = a(u,v) \qquad \text{for all} \quad u, v \in X$$

and

$$\|Au\|_* \leq \text{const}(\|g\|_q + \|u\|^{p/q}) \qquad \text{for all} \quad u \in X.$$

Hence, A is bounded. Let $f \in L_q(G)$. By the Hölder inequality,

$$|b(v)| \leq \|f\|_q \|v\| \qquad \text{for all} \quad v \in X.$$

Hence $b \in X^*$.

Consequently, problem (22) is equivalent to the operator equation $Au = b$, $u \in X$. This is (23).

(IV) Properties of A. By (H3), for all $u, v \in X$,

$$\langle Au - Av, u - v\rangle = a(u, u - v) - a(v, u - v) \geq 0,$$

i.e., A is monotone.

By (H4), for all $u \in X$,

$$\langle Au, u\rangle = a(u, u) \geq c\|u\|_{m,p,0}^p - \int_G h(x)\,dx$$

$$\geq cc_1^p\|u\|^p - \text{const}.$$

This implies $\langle Au, u\rangle/\|u\| \to +\infty$ as $\|u\| \to \infty$, i.e., A is coercive. Note that $p > 1$.

(V) Continuity of A. Let

$$u_n \to u \quad \text{in } X \qquad \text{as} \quad n \to \infty.$$

Since $F_\alpha: X \to L_q(G)$ is continuous,

$$F_\alpha u_n \to F_\alpha u \quad \text{in } L_q(G) \qquad \text{as} \quad n \to \infty.$$

By the Hölder inequality,

$$|a(u_n, v) - a(u, v)| \leq \sum_{|\alpha \leq m} \|F_\alpha u_n - F_\alpha u\|_q \|v\|,$$

for all $v \in X$. This implies

$$\|Au_n - Au\|_* \leq \sum_{|\alpha| \leq m} \|F_\alpha u_n - F_\alpha u\|_q,$$

and hence $\|Au_n - Au\|_* \to 0$ as $n \to \infty$. Thus, the operator $A: X \to X^*$ is continuous. □

Proof of Corollary 26.13. By (H5*), for all $u, v \in X$, we obtain that

$$\langle Au - Av, u - v\rangle = a(u, u - v) - a(v, u - v) \geq d\|u - v\|_{m,p,0}^p \geq dc_1^p\|u - v\|^p,$$

i.e., $A: X \to X^*$ is uniformly monotone. □

Proof of Corollary 26.14. By $A_2(45)$, the embedding operator corresponding to

$$\mathring{W}_p^m(G) \subseteq \mathring{W}_p^{m-1}(G)$$

is linear and compact, therefore strongly continuous. Let

$$u_n \rightharpoonup u \quad \text{in } X \qquad \text{as} \quad n \to \infty.$$

Since $X = \mathring{W}_p^m(G)$, this implies

$$u_n \to u \quad \text{in } \mathring{W}_p^{m-1}(G).$$

By assumption (H6*), the function A_α depends only on the partial derivatives of u up to order $m - 1$. Thus, as in the proof of Proposition 26.12 above,

we obtain

$$\|F_\alpha u_n - F_\alpha u\|_q \to 0 \qquad \text{as} \quad n \to \infty$$

and hence $\|Au_n - Au\|_* \to 0$ as $n \to \infty$, i.e.,

$$Au_n \to Au \quad \text{in } X^* \qquad \text{as} \quad n \to \infty.$$

Consequently, the operator $A: X \to X^*$ is strongly continuous. □

Using the Sobolev embedding theorems, we want to relax the growth condition.

(H2*) *Relaxed growth condition.* For all $x \in G$, $D \in \mathbb{R}^M$, $|\alpha| \le m$, and fixed $C > 0$, we suppose that

$$|A_\alpha(x, D)| \le C\left(g(x) + \sum_{|\gamma| \le m} |D^\gamma|^{p_\gamma/q_\alpha}\right),$$

where $g \in L_{q_\alpha}(G)$ is given. Here, we choose the numbers $p_\alpha, q_\alpha \in]1, \infty[$ in such a way that

$$\frac{1}{p} - \frac{m - |\alpha|}{N} \le \frac{1}{p_\alpha} \tag{25}$$

and $p_\alpha^{-1} + q_\alpha^{-1} = 1$. Recall that m denotes the order of the differential operator and N denotes the dimension of the bounded region G.

(H2**) Assume (H2*), where "$\le$" in (25) is replaced by "$<$".

Proposition 26.15. *All the assertions of Proposition* 26.12 *and Corollary* 26.13 *remain valid if we replace the growth condition* (H2) *by* (H2*).

Corollary 26.14 *remains valid if we replace* (H2) *by* (H2**).

Applications of this result will be considered in Section 27.4.

PROOF. By A_2(45), it follows from (25) that the embedding

$$\mathring{W}_p^m(G) \subseteq \mathring{W}_{p_\alpha}^{|\alpha|}(G) \tag{26}$$

is continuous. Hence we can replace (24) above by the following more general relation:

$$D^\alpha u \in L_{p_\alpha}(G) \qquad \text{if} \quad u \in \mathring{W}_p^m(G) \quad \text{and} \quad |\alpha| \le m.$$

In the case (H2**) the embedding (26) is compact for $|\alpha| < m$.

Now, the proof proceeds as the corresponding proofs of Proposition 26.12 and Corollaries 26.13 and 26.14 above. In this connection, we make essential use of the continuity of the Nemyckii operator described in Proposition 26.6. □

Using more sophisticated arguments (e.g., the Hölder inequality for several factors together with the Gagliardo–Nirenberg interpolation inequalities), it is also possible to prove the continuity or the strong continuity of the operator $A: X \to X^*$ under much weaker assumptions than those made above. In Proposition 27.11 we will consider a typical example in this direction, which is closely related to our existence proofs for the Navier–Stokes equations (viscous flow) and the von Kármán plate equations in Part IV.

26.6. Proper Monotone Operators and Proper Quasi-Linear Elliptic Differential Operators

A map is called *proper* iff the preimages of compact sets are again compact. As we have seen in Section 4.14, the notion of proper maps plays a fundamental role in the study of the *global* properties of the solution set of the operator equation

$$Au = b, \qquad u \in X.$$

Further important results in this direction will be considered in Section 29.10. The following proposition shows that important classes of monotone-like operators are indeed proper.

Proposition 26.16. *Let $A_1, A_2: X \to X^*$ be operators on the real reflexive* B-*space X. Set $A = A_1 + A_2$, and suppose that*:

(i) *A_1 is uniformly monotone and continuous.*
(ii) *A_2 is compact (e.g., strongly continuous).*
(iii) *A is coercive.*

Then, the operator $A: X \to X^*$ is proper. Moreover, if A_2 is strongly continuous, then A is pseudomonotone.

PROOF. By Theorem 26.A, the inverse operator $A_1^{-1}: X^* \to X$ is continuous. Hence A_1 is proper by Section 4.14.

We want to prove that A is proper. To this end, let C be a compact subset of X^*. We have to show that the set $A^{-1}(C)$ is compact. In fact, let (u_n) be a sequence in $A^{-1}(C)$. We set

$$b_n = A_1 u_n + A_2 u_n.$$

Since C is compact, there exists a subsequence, again denoted by (b_n), such that

$$b_n \to b \quad \text{as} \quad n \to \infty.$$

Since C is bounded and A is coercive, the sequence (u_n) is bounded. From the compactness of A_2 it follows that there exists a subsequence, again denoted by (u_n), such that

$$A_2 u_n \to c \quad \text{as} \quad n \to \infty.$$

Hence $u_n = A_1^{-1}(b_n - A_2 u_n)$ converges, i.e.,

$$u_n \to u \quad \text{as} \quad n \to \infty, \qquad \text{where} \quad u = A_1^{-1}(b - c).$$

This implies $A_1 u + A_2 u = b$, since A_1 and A_2 are continuous. Hence $u \in A^{-1}(C)$, i.e., $A^{-1}(C)$ is compact.

The pseudomonotonicity of A follows from Proposition 27.6 below. □

As in Section 26.5, we consider the differential operator (20), i.e., we set

$$(Lu)(x) = \sum_{|\alpha| \le m} (-1)^{|\alpha|} D^\alpha A_\alpha(x, Du(x)),$$

and we assume the following.

(A1) The Carathéodory condition (H1), the relaxed growth condition (H2*), and the coerciveness condition (H4) from Section 26.5 are satisfied.

(A2) The highest-order terms of L satisfy the uniform monotonicity condition (H5*), i.e., there is a number $d > 0$ such that

$$\sum_{|\alpha| = m} (A_\alpha(x, D) - A_\alpha(x, D'))(D^\alpha - D'^\alpha) \ge d \sum_{|\alpha| = m} |D^\alpha - D'^\alpha|^p,$$

for all $x \in G$ and $D, D' \in \mathbb{R}^M$.

(A3) All the terms A_α with $|\alpha| < m$ depend only on derivatives of u up to order $m - 1$.

We set $X = \mathring{W}_p^m(G)$ and consider the operator $A: X \to X^*$ which corresponds to L, according to Proposition 26.12.

Proposition 26.17. *Assume* (A1) *through* (A3). *Then the operator $A: X \to X^*$ is proper and pseudomonotone.*

Proof. According to Propositions 26.12 and 26.15, there exist operators A_1, $A_2: X \to X^*$ such that

$$\langle A_1 u, v \rangle = \int_G \sum_{|\alpha| = m} A_\alpha(x, Du(x)) D^\alpha v(x)\, dx,$$

$$\langle A_2 u, v \rangle = \int_G \sum_{|\alpha| < m} A_\alpha(x, Du(x)) D^\alpha v(x)\, dx,$$

for all $u, v \in X$, where $A = A_1 + A_2$, and A_1 and A_2 satisfy the assumptions of Proposition 26.16.

Thus, Proposition 26.17 is a consequence of Proposition 26.16. □

Problems

26.1. *Proof of Proposition* 26.2.

Solution: Let $A: X \to Y$ be strongly continuous, where X and Y are B-spaces and X is reflexive. We want to show that A is compact. To this end, let (u_n) be a bounded sequence in X. Since X is reflexive, there exists a weakly convergent subsequence $u_{n'} \rightharpoonup u$ as $n \to \infty$. Hence $Au_{n'} \to Au$ as $n \to \infty$. Thus, A is compact.

Proposition 26.2(b) follows from Proposition 21.29.

26.2. *Proof of Proposition* 26.3. Ad(a). Let $A: X \to Y$ be demicontinuous. If A is not locally bounded, then there exists a point $u \in X$ and a sequence $u_n \to u$ as $n \to \infty$ with $\|Au_n\| \to \infty$ as $n \to \infty$. Since A is demicontinuous, $Au_n \rightharpoonup Au$ as $n \to \infty$. Therefore, the sequence (Au_n) is bounded in contradiction to $\|Au_n\| \to \infty$ as $n \to \infty$.

Ad(b). Let $A: X \to Y$ be demicontinuous. If A is not continuous from X (norm topology) to Y (weak topology), then there exists a point $u \in X$, a neighborhood $V(Au)$ of Au in the weak topology of Y, and a sequence $u_n \to u$ as $n \to \infty$ such that $Au_n \notin V(Au)$ for all n. This is in contradiction to $Au_n \rightharpoonup Au$ as $n \to \infty$.

Conversely, if A is continuous from X (norm topology) to Y (weak topology), then $A: X \to Y$ is demicontinuous, since continuity implies sequential continuity, by A_1(17e).

26.3. *Proof of Proposition* 26.7.

Solution: Let $X = L_p(G)$ with $1 < p < \infty$ and $p^{-1} + q^{-1} = 1$. Then $X^* = L_q(G)$. We set

$$(Fu)(x) = f(x, u(x)).$$

By Proposition 26.6, the Nemyckii operator $F: X \to X^*$ is continuous and bounded. If $u, v \in X$, then $Fu \in X^*$ and hence

$$\langle Fu, v\rangle_X = \int_G f(x, u(x))v(x)\,dx.$$

Let f be monotone with respect to u. Then

$$[f(x,u) - f(x,v)](u - v) \geq 0 \qquad \text{for all} \quad u, v \in \mathbb{R}, \quad x \in G.$$

This implies

$$\langle Fu - Fv, u - v\rangle_X = \int_G [f(x,u(x)) - f(x,v(x))](u(x) - v(x))\,dx \geq 0,$$

for all $u, v \in X$, i.e., the operator $F: X \to X^*$ is monotone.

Let f be strictly monotone with respect to u. Then

$$\langle Fu - Fv, u - v\rangle_X = 0$$

implies

$$[f(x,u(x)) - f(x,v(x))](u(x) - v(x)) = 0$$

for almost all $x \in G$ and hence $u(x) = v(x)$ for almost all $x \in G$. This means $u = v$ in X. Consequently, F is strictly monotone.

From $f(x,u)u \geq d|u|^p + g(x)$ it follows that

$$\langle Fu, u\rangle_X \geq d\int |u|^p\,dx + \int g\,dx \geq d\|u\|_X^p + \int g\,dx.$$

This implies $\langle Fu, u\rangle/\|u\| \to +\infty$ as $\|u\| \to \infty$, i.e., F is coercive.

Suppose that there exists an $R > 0$ such that

$$f(x,u)u \geq 0$$

for all $(u, x) \in G \times \mathbb{R}$ with $|u| \geq R$. Let

$$A = \{x \in \mathbb{R}^N : |u(x)| \geq R\}.$$

Then the growth condition $|f(x, u)| \leq C(a + b|u|^{p-1})$ yields

$$\begin{aligned}\langle Fu, u \rangle_X &= \int_{G-A} f(x, u)u\,dx + \int_A f(x, u)u\,dx \\ &\geq \int_{G-A} f(x, u)u\,dx \geq -\int_G C(aR + bR^p)\,dx = -c,\end{aligned}$$

for all $u \in X$. If meas $G < \infty$, then $c < \infty$.

The more difficult proof of Proposition 26.7(e) concerning the condition $(S)_+$ can be found in Browder (1971), p. 431.

26.4. *A convergence theorem.* Let $1 \leq p < \infty$, and let G be a nonempty measurable set in $\mathbb{R}^N$, $N \geq 1$. Show that if

$$u_n \to u \quad \text{in } L_p(G) \qquad \text{as} \quad n \to \infty,$$

then there exists a subsequence $(u_{n'})$ and a function $v \in L_p(G)$ with

$$u_{n'}(x) \to u(x) \qquad \text{as} \quad n \to \infty \qquad \text{for almost all} \quad x \in G$$

and

$$|u_{n'}(x)| \leq v(x) \qquad \text{for almost all} \quad x \in G \quad \text{and all } n.$$

Hint: This assertion results as a by-product in the proof of the completeness of $L_p(G)$. Cf. Kufner, John, and Fučik (1977, M), p. 74.

References to the Literature

Classical works: Višik (1961), (1963) (partial differential equations), Minty (1962), (1963) and Browder (1963), (1963a) (functional analytic theory).

Monographs on the theory of monotone operators: Browder (1968/76), Lions (1969) (application to partial differential equations), Vainberg (1972) (connection with variational methods and Hammerstein integral equations), Brézis (1973) (maximal monotone operators and time-dependent problems), Skrypnik (1973), (1986) (mapping degree and elliptic differential equations), Gajewski, Gröger and Zacharias (1974), Langenbach (1976) (monotone potential operators), Barbu (1976, M) (time-dependent problems), Pascali and Sburlan (1978), Kluge (1979), Deimling (1985).

Application to quasi-linear elliptic differential equations: Lions (1969, M), Browder (1970, S), Dubinskii (1976, S), Fučik and Kufner (1980, M), Petryshyn (1980a), (1981), Nečas (1983, M), Skrypnik (1986, M). Cf. also the References to the Literature for Chapter 27.

Nemyckii operator: Krasnoselskii (1956, M), Vainberg (1956, M), Browder (1971), Pascali and Sburlan (1978, M), Appell (1987, S, B).

Numerical methods: Cf. the References to the Literature for Chapter 35.

History of the theory of monotone operators: Petryshyn (1970, S), Vainberg (1972, M), Brézis (1973, M), Dubinskii (1976, S).

Recent trends: Browder (1986, P).

CHAPTER 27

Pseudomonotone Operators and Quasi-Linear Elliptic Differential Equations

The classical development of nonlinear functional analysis arose contemporaneously with the beginnings of linear functional analysis at about the beginning of the twentieth century in the work of such men as Picard, S. Bernstein, Ljapunov, E. Schmidt, and Lichtenstein, and was motivated by the desire to study the existence and properties of boundary value problems for nonlinear partial differential equations. Its most classical tool was the Picard contraction principle (put in its sharpest form by Banach in his thesis of 1920—the Banach fixed-point theorem).

Beyond the early development of bifurcation theory by Ljapunov and E. Schmidt around 1905, the second, and even more fruitful, branch of the classical methods in nonlinear functional analysis was developed in the theory of compact nonlinear mappings in Banach spaces in the late 1920's and early 1930's. These included Schauder's well-known fixed-point theorem and the extension of the Brouwer topological degree by Leray and Schauder in 1934 to mappings in Banach spaces of the form $I + C$ with C compact (as well as interesting related results of Caccioppoli on nonlinear Fredholm mappings).

The central role of *compact* mappings in this phase of the development of nonlinear functional analysis was due in part to the nature of the technical apparatus being developed, but also in part to a *not* always fruitful tendency to see the theory of integral equations as the predestined domain of application of the theory to be developed. Since, however, the more *significant* analytical problems lie in the somewhat different domain of boundary value problems for partial differential equations, and since the efforts to apply the theory of compact operators (and in particular the Leray–Schauder theory) to the latter problems have given rise to demands for ever more inaccessible (and sometimes, invalid) *a priori* estimates in these problems; the hope of applying nonlinear functional analysis to problems of this type centers on a general program of creating new theories for significant classes of *noncompact* nonlinear operators. The focus of this study is then to find such classes of operators which have the opposed characteristics of being narrow enough to have a significant structure of results while also being wide enough to have a significant variety of applications....

From the point of view of applications to partial differential equations, the most important class is that of monotone-like operators.

Felix E. Browder (1968)

The first substantial results concerning monotone operators were obtained by G. Minty (1962), (1963) and F. E. Browder (1963). Then the properties of monotone operators were studied systematically by F. E. Browder in order to obtain existence theorems for quasi-linear elliptic and parabolic partial differential equations. The existence theorems of F. E. Browder were generalized to more general classes of quasi-linear elliptic differential equations by J. Leray and J. L. Lions (1965), and P. Hartman and G. Stampacchia (1966).

In this paper we introduce two vast classes of operators, namely, operators of type (M) and pseudomonotone operators. These classes of operators contain many of those monotone-like operators which were used by the authors mentioned above.

Haïm Brézis (1968)

Many problems in analysis reduce to solving an equation of the form

$$Au = b, \qquad u \in D(A),$$

where A is an operator on a space into another space. In this paper we assume that A is a map on a subset $D(A)$ of a Banach space X into another Banach space Y^+, where $\{Y, Y^+\}$ is a dual pair of Banach spaces. In an ideal situation, our equation will have a solution u for every $b \in Y^+$. There is a large literature on the "surjectivity" of this kind, including those related to monotone operators and their generalizations.

In the present paper we want to generalize the problem and seek sufficient conditions for our equation to have a solution u for all "sufficiently small b." Our theorem, together with its companion for evolution equations,[1] have been found useful in applications to many nonlinear partial differential equations.

Tosio Kato (1984)

In this chapter our goal is to treat, in contrast to Section 26.5, more general quasi-linear elliptic equations having terms of *lower order* which satisfy *no* monotonicity condition. We introduce pseudomonotone operators for this purpose, i.e., we study the solvability of operator equations of the form

$$\text{(E)} \qquad A_1 u + A_2 u = b, \qquad u \in X,$$

where $A_1 + A_2: X \to X^*$ is a pseudomonotone and coercive operator on the real reflexive B-space X. The prototype of a pseudomonotone operator is the sum operator $A_1 + A_2$, where

(i) the operator $A_1: X \to X^*$ is monotone and hemicontinuous; and
(ii) the operator $A_2: X \to X^*$ is strongly continuous.

Hence we obtain:

The theory of pseudomonotone operators unifies both monotonicity arguments and compactness arguments.

[1] See Theorems 27.B and 30.B.

In connection with quasi-linear differential operators we use the following important principle:

Lower order terms correspond frequently to strongly continuous operators.

As we will show in Section 27.4, this is a consequence of the Sobolev embedding theorems (compact embeddings). In order to transform quasi-linear elliptic boundary value problems to the operator equation (E) above and to be able to apply our general existence theorem for pseudomonotone operators (Theorem 27.A), we need the following structure of the differential operator:

(a) The principal part of the differential operator is a monotone quasi-linear elliptic differential operator in the sense of Section 26.5 (e.g., a linear strongly elliptic operator).
(b) There are nonlinear lower order terms which need not be monotone.
(c) The differential operator is coercive.

This way we obtain the operator equation (E), where the monotone and hemicontinuous operator A_1 corresponds to (a) and the strongly continuous operator A_2 corresponds to (b). Condition (c) ensures that the sum $A_1 + A_2$ is coercive. This restricts the structure of the lower order terms.

In Part IV we will consider the following applications of the theory of pseudomonotone operators to interesting problems in mathematical physics:

(α) The nonlinear von Kármán plate equations in elasticity (Chapter 65).
(β) The Navier–Stokes equations for viscous fluids (Chapter 72).

The proof of the main theorem on monotone operators in Section 26.2 was based on the Galerkin method and the monotonicity trick. An inspection of this proof shows that exactly the same proof works for more general classes of operators than monotone operators. This is the *basic idea* of this chapter. To this end, we introduce the conditions (M) and (S) as well as pseudomonotone operators. The condition (S) will also play an important role in Part III in connection with variational problems and eigenvalue problems. Pseudomonotone operators have the following nice property:

The strongly continuous perturbation of a pseudomonotone operator is again a pseudomonotone operator.

The precise formulation may be found in Proposition 27.7. In Figure 27.1 in Section 27.5 we present graphically a series of important interrelations between operator properties. We recommend a study of this diagram.

The main results of this chapter are the following:

(A) The main theorem on pseudomonotone operators of Brézis (1968) (Theorem 27.A).
(B) The main theorem for locally coercive operators of Kato (1984) (Theorem 27.B).

This latter theorem is closely related to the main theorem for the semi-coercive evolution equations of Kato and Lai (1984) (Theorem 30.B). Both Theorems 27.B and 30.B represent an important recent technique in order to solve nonlinear partial differential equations of time-independent as well as of time-dependent type.

27.1. The Conditions (*M*) and (*S*), and the Convergence of the Galerkin Method

Definition 27.1. Let X be a real reflexive B-space. The operator $A: X \to X^*$ satisfies the conditions (M), $(S)_+$, (S), $(S)_0$ and $(S)_1$ iff, as $n \to \infty$, the following hold.

(i) *Condition* (M):

$$u_n \rightharpoonup u, \qquad Au_n \rightharpoonup b, \qquad \overline{\lim_{n\to\infty}} \langle Au_n, u_n\rangle \leq \langle b, u\rangle$$

$$\text{implies} \qquad Au = b.$$

(ii) *Condition* $(S)_+$:

$$u_n \rightharpoonup u, \qquad \overline{\lim_{n\to\infty}} \langle Au_n - Au, u_n - u\rangle \leq 0 \qquad \text{implies} \qquad u_n \to u.$$

(iii) *Condition* (S):

$$u_n \rightharpoonup u, \qquad \lim_{n\to\infty} \langle Au_n - Au, u_n - u\rangle = 0 \qquad \text{implies} \qquad u_n \to u.$$

(iv) *Condition* $(S)_0$:

$$u_n \rightharpoonup u, \qquad Au_n \rightharpoonup b, \qquad \lim_{n\to\infty} \langle Au_n, u_n\rangle = \langle b, u\rangle$$

$$\text{implies} \qquad u_n \to u.$$

(v) *Condition* $(S)_1$:

$$u_n \rightharpoonup u, \qquad Au_n \to b \qquad \text{implies} \qquad u_n \to u.$$

Obviously, the following holds:

$$(S)_+ \quad \Rightarrow \quad (S) \quad \Rightarrow \quad (S)_0 \quad \Rightarrow \quad (S)_1, \tag{1}$$

i.e., if the operator A satisfies the condition $(S)_+$, then A also satisfies condition (S), etc.

We now introduce the *prototypes* for (M) and $(S)_+$.

Example 27.2. The following hold for $A: X \to X^*$ on the real reflexive B-space X:

(a) The operator A is monotone and hemicontinuous implies that A satisfies (M).
(b) The operator A is uniformly monotone implies that A satisfies $(S)_+$.

PROOF. Ad(a). This is the monotonicity trick (25.4).

Ad(b). It follows from

$$0 \leq \overline{\lim_{n\to\infty}}\, a(\|u_n - u\|)\|u_n - u\| \leq \overline{\lim_{n\to\infty}}\, \langle Au_n - Au, u_n - u\rangle \leq 0$$

that

$$\lim_{n\to\infty} a(\|u_n - u\|)\|u_n - u\| = 0;$$

therefore also $\lim_{n\to\infty} \|u_n - u\| = 0$ because of the properties of $a(\cdot)$ in Definition 25.2. □

The importance of (M) and $(S)_+$ consists in the fact that, roughly speaking, these conditions are invariant with respect to compact perturbations. We make this precise in the following example.

EXAMPLE 27.3. Let $A, B: X \to X^*$ be operators on the real reflexive B-space X. Then the following hold:

(a) The operator A satisfies $(S)_+$ and B is strongly continuous or, more generally, B is compact implies that $A + B$ satisfies $(S)_+$.
(b) The operator A satisfies (S) and B is strongly continuous implies that $A + B$ satisfies (S).
(c) The operator A satisfies (M) and B is strongly continuous implies that $A + B$ satisfies (M).

We deal with a simple proof in Problem 27.1. We now study the significance of (M) and $(S)_0$ for the convergence of the Galerkin method

$$\langle Au_n - b, w_k\rangle = 0, \qquad u_n \in X_n, \qquad k = 1, \ldots, n \tag{2}$$

with $X_n = \operatorname{span}\{w_1, \ldots, w_n\}$ for the operator equation

$$Au = b, \qquad u \in X. \tag{3}$$

Proposition 27.4 (Convergence of the Galerkin Method). *Assume*:

(i) *Let $A: X \to X^*$ be a bounded operator satisfying condition (M) on the real, separable, reflexive, and infinite-dimensional* B-*space X. Let $b \in X^*$.*
(ii) *Let $\{w_1, w_2, \ldots\}$ be a basis in X.*
(iii) *There exist an $R > 0$ and an n_0 such that, for each $n \geq n_0$, the Galerkin equation* (2) *has a solution u_n with $\|u_n\| \leq R$.*

Then:

(a) *There exists a subsequence $(u_{n'})$ with $u_{n'} \rightharpoonup u$ as $n \to \infty$ such that u is a solution of* (3).

(b) *If equation* (3) *has a unique solution u, then* $u_n \rightharpoonup u$ *as* $n \to \infty$.
(c) *If A satisfies the condition* $(S)_0$ *instead of* (M), *and A is demicontinuous, then one can replace weak convergence in* (a), (b) *by strong convergence.*

PROOF. Ad(a). It follows from (2) that

$$\langle Au_n - b, v \rangle \to 0 \qquad \text{as} \quad n \to \infty \tag{4}$$

for all $v \in \operatorname{span}\{w_1, w_2, \ldots\}$. The operator A is bounded. Therefore, together with (u_n), the sequence (Au_n) is also bounded. Consequently,

$$Au_n \rightharpoonup b \quad \text{in } X^* \qquad \text{as} \quad n \to \infty, \tag{5}$$

by Proposition 21.26(c), (f).

The sequence (u_n) is bounded in the reflexive B-space X. Thus, there exists a subsequence $(u_{n'})$ with

$$u_{n'} \rightharpoonup u \qquad \text{as} \quad n \to \infty.$$

It follows from (2) that

$$\langle Au_{n'}, u_{n'} \rangle = \langle b, u_{n'} \rangle \to \langle b, u \rangle \qquad \text{as} \quad n \to \infty.$$

Now condition (M) yields $Au = b$.

Ad(b). Use the convergence trick (25.14).

Ad(c). In the proof of (a) we found a subsequence $(u_{n'})$ with

$$u_{n'} \rightharpoonup u, \qquad Au_{n'} \rightharpoonup b, \qquad \langle Au_{n'}, u_{n'} \rangle \to \langle b, u \rangle \qquad \text{as} \quad n \to \infty.$$

It follows from $(S)_0$ that

$$u_{n'} \to u \qquad \text{as} \quad n \to \infty.$$

The demicontinuity of A yields

$$Au_{n'} \rightharpoonup Au \qquad \text{as} \quad n \to \infty$$

and hence $Au = b$.

If the equation $Au = b$ has exactly one solution $u \in X$, then it follows that the total sequence converges, i.e.,

$$u_n \to u \qquad \text{as} \quad n \to \infty,$$

by the convergence trick (25.14). □

27.2. Pseudomonotone Operators

In order to be able to apply Proposition 27.4 to a comprehensive class of operators, we make available the following definition.

Definition 27.5. Let $A: X \to X^*$ be an operator on the real reflexive B-space X.

(i) The operator A is called *pseudomonotone* iff $u_n \rightharpoonup u$ as $n \to \infty$ and

$$\overline{\lim_{n\to\infty}} \langle Au_n, u_n - u\rangle \le 0$$

implies

$$\langle Au, u - w\rangle \le \underline{\lim_{n\to\infty}} \langle Au_n, u_n - w\rangle \qquad \text{for all} \quad w \in X.$$

(ii) The operator A satisfies the *condition* (P) iff $u_n \rightharpoonup u$ as $n \to \infty$ implies

$$\overline{\lim_{n\to\infty}} \langle Au_n, u_n - u\rangle \ge 0.$$

We first give some prototypes.

Proposition 27.6. *Let A, B: $X \to X^*$ be operators on the real reflexive* B-*space X. Then*:

(a) *If A is monotone and hemicontinuous, then A is pseudomonotone.*
(b) *If A is strongly continuous, then A is pseudomonotone.*
(c) *If A is demicontinuous with $(S)_+$, then A is pseudomonotone.*
(d) *If A is continuous and* $\dim X < \infty$, *then A is pseudomonotone.*
(e) Additivity. *If A and B are pseudomonotone, then $A + B$ is pseudomonotone.*
(f) *If A is monotone and hemicontinuous and B is strongly continuous, then $A + B$ is pseudomonotone.*
(g) *If A is monotone and B is strongly continuous, then $A + B$ satisfies (P).*

Statement (f) plays a fundamental role in applications of the theory of pseudomonotone operators to quasi-linear elliptic differential equations. Obviously, property (f) is an immediate consequence of the additivity principle (e) along with (a), (b).

PROOF. Ad(a). Let $u_n \rightharpoonup u$ as $n \to \infty$ and

$$\overline{\lim_{n\to\infty}} \langle Au_n, u_n - u\rangle \le 0.$$

Since the operator A is monotone, we obtain that

$$\langle Au_n, u_n - u\rangle \ge \langle Au, u_n - u\rangle \to 0 \qquad \text{as} \quad n \to \infty.$$

This implies

$$\langle Au_n, u_n - u\rangle \to 0 \qquad \text{as} \quad n \to \infty,$$

since for this sequence, the upper and the lower limits are equal to zero.

Let $z = u + t(w - u)$ with $t > 0$. The monotonicity of A yields

$$\langle Au_n - Az, u_n - z\rangle \ge 0$$

and hence

$$t\langle Au_n, u - w\rangle \ge \langle -Au_n, u_n - u\rangle + \langle Az, u_n - u\rangle + t\langle Az, u - w\rangle.$$

This implies

$$\langle Az, u - w\rangle \leq \underline{\lim}_{n\to\infty} \langle Au_n, u_n - w\rangle \qquad \text{for all} \quad w \in X.$$

The operator A is hemicontinuous. Thus, letting $t \to 0$, we obtain

$$\langle Au, u - w\rangle \leq \underline{\lim}_{n\to\infty} \langle Au_n, u_n - w\rangle \qquad \text{for all} \quad w \in X,$$

i.e., A is pseudomonotone.

Ad(b). If $u_n \rightharpoonup u$ as $n \to \infty$, then $Au_n \to Au$, and hence

$$\langle Au, u - w\rangle = \lim_{n\to\infty} \langle Au_n, u_n - w\rangle \qquad \text{for all} \quad w \in X.$$

Ad(c). Let $u_n \rightharpoonup u$ and $\overline{\lim}\langle Au_n, u_n - u\rangle \leq 0$ as $n \to \infty$. This implies

$$\overline{\lim}_{n\to\infty} \langle Au_n - Au, u_n - u\rangle \leq 0.$$

The operator A satisfies $(S)_+$, and hence $u_n \to u$ as $n \to \infty$. Since A is demicontinuous, we obtain $Au_n \rightharpoonup Au$ and hence

$$\langle Au, u - w\rangle = \lim_{n\to\infty} \langle Au_n, u_n - w\rangle \qquad \text{for all} \quad w \in X.$$

Ad(d). This follows from (b). Note that weak convergence and strong convergence coincide on finite-dimensional B-spaces.

Ad(e). Let $u_n \rightharpoonup u$ as $n \to \infty$ and

$$\overline{\lim}_{n\to\infty} \langle Au_n + Bu_n, u_n - u\rangle \leq 0.$$

This implies

$$\overline{\lim}_{n\to\infty} \langle Au_n, u_n - u\rangle \leq 0 \qquad \text{and} \qquad \overline{\lim}_{n\to\infty} \langle Bu_n, u_n - u\rangle \leq 0. \tag{6}$$

Otherwise, we may assume that there exists a subsequence, again denoted by (u_n), such that

$$\lim_{n\to\infty} \langle Au_n, u_n - u\rangle = a > 0,$$

and hence $\overline{\lim}_{n\to\infty} \langle Bu_n, u_n - u\rangle \leq -a$. Since B is pseudomonotone, this implies

$$\langle Bu, u - w\rangle \leq \underline{\lim}_{n\to\infty} \langle Bu_n, u_n - w\rangle \qquad \text{for all} \quad w \in X.$$

Letting $w = u$, we obtain the contradiction $0 \leq -a$.

Since A and B are pseudomonotone, it follows from (6) that, for all $w \in X$,

$$\langle Au, u - w\rangle \leq \underline{\lim}_{n\to\infty} \langle Au_n, u_n - w\rangle,$$

$$\langle Bu, u - w\rangle \leq \underline{\lim}_{n\to\infty} \langle Bu_n, u_n - w\rangle,$$

and hence

$$\langle Au + Bu, u - w\rangle \leq \underline{\lim}_{n\to\infty} \langle Au_n + Bu_n, u_n - w\rangle \qquad \text{for all} \quad w \in X,$$

i.e., $A + B$ is pseudomonotone.

Ad(f). This follows from (a), (b), and (e).

Ad(g). The proof will be given in Problem 27.2. □

Proposition 27.7 (Properties of Pseudomonotone Operators). *Let $A, B: X \to X^*$ be operators on the real reflexive* B-*space* X. *Then*:

(a) *If A is pseudomonotone, then A satisfies the two conditions* (P) *and* (M).
(b) *If A is pseudomonotone and locally bounded, then A is demicontinuous.*
(c) *If A is pseudomonotone and B is monotone and hemicontinuous, then $A + B$ is pseudomonotone.*
(d) *If A is pseudomonotone and B is strongly continuous, then $A + B$ is pseudomonotone.*

Statements (c) and (d) represent perturbation principles for pseudomonotone operators, which follow immediately from Proposition 27.6(a), (b), (e).

PROOF. Ad(a). Let A be pseudomonotone. If A does not satisfy condition (P), then there is a sequence (u_n) with

$$u_n \rightharpoonup u \quad \text{as} \quad n \to \infty \qquad \text{and} \qquad \overline{\lim_{n\to\infty}}\, \langle Au_n, u_n - u\rangle < 0.$$

From the pseudomonotonicity of A we obtain the contraction

$$0 = \langle Au, u - u\rangle \leq \underline{\lim_{n\to\infty}}\, \langle Au_n, u_n - u\rangle.$$

We shall now show that A satisfies (M). Let $u_n \rightharpoonup u$, $Au_n \rightharpoonup b$ as $n \to \infty$ and

$$\overline{\lim_{n\to\infty}}\, \langle Au_n, u_n\rangle \leq \langle b, u\rangle.$$

Then $\overline{\lim}_{n\to\infty} \langle Au_n, u_n - u\rangle \leq 0$. The pseudomonotonicity of A yields

$$\langle Au, u - w\rangle \leq \underline{\lim_{n\to\infty}}\, \langle Au_n, u_n - w\rangle \qquad \text{for all} \quad w \in X;$$

therefore

$$\langle Au, u - w\rangle \leq \langle b, u\rangle - \langle b, w\rangle = \langle b, u - w\rangle \qquad \text{for all} \quad w \in X.$$

Hence $Au = b$.

Ad(b). Let (u_n) be a sequence with

$$u_n \to u \qquad \text{as} \quad n \to \infty.$$

Since A is locally bounded, the sequence (Au_n) is bounded. Let

$$Au_{n'} \rightharpoonup b \qquad \text{as} \quad n \to \infty.$$

Then $\lim_{n\to\infty} \langle Au_{n'}, u_{n'} - u\rangle = 0$. The operator A is pseudomonotone; therefore,

$$\langle Au, u - w\rangle \leq \underline{\lim_{n\to\infty}}\, \langle Au_{n'}, u_{n'} - w\rangle \qquad \text{for all} \quad w \in X.$$

From this it follows that

$$\langle Au, u - w\rangle \le \langle b, u - w\rangle \qquad \text{for all} \quad w \in X,$$

and hence $Au = b$. By the convergence principle (Proposition 21.23(i)), the total sequence (Au_n) is weakly convergent, i.e.,

$$Au_n \rightharpoonup b \qquad \text{as} \quad n \to \infty.$$

Thus, A is demicontinuous. □

27.3. The Main Theorem on Pseudomonotone Operators

We consider the operator equation

$$Au = b, \qquad u \in X, \tag{7}$$

along with the Galerkin method

$$\langle Au_n - b, w_k\rangle = 0, \qquad u_n \in X_n, \qquad k = 1, \ldots, n, \tag{8}$$

where $X_n = \operatorname{span}\{w_1, \ldots, w_n\}$.

Theorem 27.A (Brézis (1968)). *Assume*:

(i) *The operator* $A\colon X \to X^*$ *is pseudomonotone, bounded, and coercive on the real, separable, and reflexive* B*-space* X *with* $\dim X = \infty$.
(ii) *Let* $\{w_1, w_2, \ldots\}$ *be a basis in* X.

Then the following hold:

(a) Existence. *For each* $b \in X^*$, *the original equation* (7) *has a solution.*
(b) Galerkin method. *For fixed* $b \in X^*$ *and for each* $n \in \mathbb{N}$, *the Galerkin equation* (8) *has a solution* u_n. *There exists a subsequence* $(u_{n'})$ *which converges weakly to a solution of the original equation* (7).

If the operator A *satisfies* $(S)_+$, *then* $(u_{n'})$ *converges strongly to a solution of equation* (7).

If equation (7) *has a unique solution* u, *then the total sequence* (u_n) *converges to* u.

Proof. By Proposition 27.7, the operator A is demicontinuous and satisfies (M).

As in the proof of Theorem 26.A, it follows that the Galerkin equation (8) has a solution u_n, where $\|u_n\| \le R$ for all n and fixed $R > 0$. Now Proposition 27.4 yields the assertion. □

27.4. Application to Quasi-Linear Elliptic Differential Equations

We consider two simple examples which nonetheless illustrate two typical situations. In both cases, we have a monotone principal part. The lower order terms yield strongly continuous perturbations. In particular, we shall show how we can use the Sobolev embedding theorems in order to verify the strong continuity.

We begin with the boundary value problem

$$\begin{gathered} -\sum_{i=1}^{N} D_i(|D_i u|^{p-2} D_i u) + g(u) = f \qquad \text{on } G, \\ u = 0 \quad \text{on } \partial G. \end{gathered} \tag{9}$$

Let G be a bounded region in $\mathbb{R}^N$ with $N \geq 1$. Let $x \in \mathbb{R}^N$. We set $x = (\xi_1, \ldots, \xi_N)$, $D_i = \partial/\partial\xi_i$, and we place the following conditions on the nonlinearity g.

(H1) *Coerciveness condition for g.* The function $g: \mathbb{R} \to \mathbb{R}$ is continuous and $\inf_{u \in \mathbb{R}} g(u)u > -\infty$.

(H2) *Growth condition for g.* For all $u \in \mathbb{R}$,

$$|g(u)| \leq \text{const}(1 + |u|^{r-1}),$$

where $1 < p, q, r < \infty$, $p^{-1} + q^{-1} = 1$, and $p^{-1} - N^{-1} < r^{-1}$.

In particular, we can choose $r = p$. In the special case $N = 1, 2$ and $p \geq 2$, the number r can be chosen arbitrarily in the interval $]1, \infty[$, since $p^{-1} - N^{-1} \leq 0$.

Instead of (9) we also consider the more general boundary value problem

$$\begin{gathered} -\sum_{i=1}^{N} D_i(F_i(Du)) + g(u) = f \quad \text{on } G, \\ u = 0 \quad \text{on } \partial G, \end{gathered} \tag{9*}$$

where $Du = (D_1 u, \ldots, D_N u)$. In this connection, we make the following additional assumptions.

(H3) *Monotonicity condition for the principal part.* For all $D, D' \in \mathbb{R}^N$,

$$\sum_{i=1}^{N} (F_i(D) - F_i(D'))(D_i - D_i') \geq 0.$$

(H4) *Coerciveness condition for the principal part.* There is a number $c > 0$ such that

$$\sum_{i=1}^{N} F_i(D) D_i \geq c \sum_{i=1}^{N} |D_i|^p \qquad \text{for all} \quad D \in \mathbb{R}^N.$$

(H5) *Growth condition for the principal part.* The functions $F_i: \mathbb{R}^N \to \mathbb{R}$ are continuous for all i, and

$$|F_i(D)| \le \text{const}(1 + |D|^{p-1}) \qquad \text{for all} \quad D \in \mathbb{R}^N \quad \text{and all } i.$$

Obviously, problem (9) is a special case of (9*) with

$$F_i(D) = \begin{cases} |D_i|^{p-2} D_i & \text{if} \quad D_i \ne 0, \\ 0 & \text{if} \quad D_i = 0. \end{cases}$$

Definition 27.8. Let $X = \mathring{W}_p^1(G)$. The *generalized problem* for (9*) reads as follows. The function $f \in L_q(G)$ is given. We seek $u \in X$ with

$$a_1(u, v) + a_2(u, v) = b(v) \qquad \text{for all} \quad v \in X. \tag{10}$$

Here, for all $u, v \in X$, we set

$$a_1(u, v) = \int_G \sum_{i=1}^N F_i(Du) D_i v \, dx,$$

$$a_2(u, v) = \int_G g(u) v \, dx,$$

$$b(v) = \int_G f v \, dx.$$

Formally, problem (10) results from (9*) upon multiplication by $v \in C_0^\infty(G)$ and subsequent integration by parts.

Proposition 27.9. *Under assumptions* (H1) *through* (H5), *the generalized problem* (10) *corresponding to* (9*) *has a solution* $u \in X$.

In particular, under assumptions (H1) *and* (H2), *the generalized problem for* (9) *has a solution* $u \in X$.

Proof. We will use Theorem 27.A and the results from Section 26.5.

(I) According to Proposition 26.12, there exists an operator $A_1: X \to X^*$ with

$$a_1(u, v) = \langle A_1 u, v \rangle \qquad \text{for all} \quad u, v \in X.$$

The operator A_1 is monotone, coercive, continuous, and bounded.

(II) According to Corollary 26.14 in the case where $r = p$, and according to Proposition 26.15 in the general case, there exists an operator $A_2: X \to X^*$ with

$$a_2(u, v) = \langle A_2 u, v \rangle \qquad \text{for all} \quad u, v \in X.$$

The operator A_2 is strongly continuous.

(III) By (22.1b), $b \in X^*$. Thus, the generalized problem (10) is equivalent to the operator equation

$$A_1 u + A_2 u = b, \qquad u \in X. \tag{11}$$

We set $A = A_1 + A_2$.

The operator $A: X \to X^*$ is continuous. By Proposition 27.6(f), the operator A is pseudomonotone as a strongly continuous perturbation of the continuous monotone operator A_1.

The operator A_2 is bounded, since it is strongly continuous, and hence $A = A_1 + A_2$ is also bounded.

It follows from assumption (H1) that

$$\langle A_2 u, u \rangle = \int_G g(u) u \, dx \geq \text{const} \qquad \text{for all} \quad u \in X.$$

Hence $A = A_1 + A_2$ is coercive, since A_1 is coercive.

(IV) The main theorem on pseudomonotone operators (Theorem 27.A) yields the assertion. □

In the special case of problem (9), the operator $A_1: X \to X^*$ is uniformly monotone, by Proposition 26.10. Thus, it follows from Examples 27.2(b) and 27.3(a) that the operator $A_1 + A_2: X \to X^*$ satisfies condition $(S)_+$.

Recall from the proof of Proposition 26.15 that the growth condition $|g(u)| \leq \text{const}(1 + |u|^{r-1})$ is related to the compactness of the embedding $\mathring{W}_p^1(G) \subseteq L_r(G)$.

Using our results from Section 26.5, it is possible to generalize Proposition 27.9 to quasi-linear elliptic differential equations of order $2m$. A general result about such equations will be considered in Problem 27.6. There we will use the notion of *semimonotone* operators, which are special cases of pseudomonotone operators.

As a second simple but typical example, we consider in $\mathbb{R}^N$ with $N = 1, 2, 3$, parallel to (9), the boundary value problem

$$\begin{gathered} -\Delta u + \alpha \sum_{i=1}^{N} (\sin u) D_i u = f \quad \text{on } G, \\ u = 0 \quad \text{on } \partial G, \end{gathered} \tag{12}$$

where α is a real number. The appearance of the first-order derivatives makes it more difficult to prove that the lower order terms correspond to a strongly continuous operator. To show this, we will use critically the *compactness* of the embedding

$$\mathring{W}_2^1(G) \subseteq L_4(G)$$

in $\mathbb{R}^N$ with $N = 1, 2, 3$, and the Hölder inequality for three factors. In Part IV we will use the same technique in order to investigate the nonlinear von Kármán plate equations in elasticity and the Navier–Stokes equations for viscous fluids. The coerciveness of (12) can be guaranteed if $|\alpha|$ is sufficiently small. In this connection, we will use the Poincaré–Friedrichs inequality.

Definition 27.10. Let G be a bounded region in $\mathbb{R}^N$ with $N = 1, 2, 3$, and let $X = \mathring{W}_2^1(G)$. The *generalized problem* for (12) reads as follows: The function

$f \in L_2(G)$ is given. We seek $u \in X$ such that

$$a_1(u, v) + a_2(u, v) = b(v) \qquad \text{for all} \quad v \in X. \tag{13}$$

Here, we set

$$a_1(u, v) = \int_G \sum_{i=1}^{N} D_i u D_i v \, dx,$$

$$a_2(u, v) = c(u, u, v), \qquad c(u, w, v) = \alpha \int_G \sum_{i=1}^{N} (\sin u)(D_i w) v \, dx,$$

$$b(v) = \int_G f v \, dx.$$

Proposition 27.11. *There is a number $\alpha_0 > 0$ such that the generalized problem* (13) *corresponding to* (12) *has a solution for all α: $|\alpha| \leq \alpha_0$ and all $f \in L_2(G)$.*

PROOF.

Step 1: Operator A_1.

According to the proof of Proposition 26.10, there exists an operator $A_1: X \to X^*$ with

$$\langle A_1 u, v \rangle = a_1(u, v) \qquad \text{for all} \quad u, v \in X.$$

The operator A_1 is uniformly monotone (and hence coercive), continuous, and bounded. More precisely, the operator A_1 is linear, continuous, and strongly monotone.

Step 2: Operator A_2.

The Hölder inequality yields

$$\left| \int (\sin u)(D_i u) v \, dx \right| \leq \left(\int (D_i u)^2 \, dx \right)^{1/2} \left(\int v^2 \, dx \right)^{1/2}$$

and hence

$$|a_2(u, v)| \leq \text{const} \, \|u\|_X \|v\|_X \qquad \text{for all} \quad u, v \in X.$$

By (22.1a), there exists an operator $A_2: X \to X^*$ such that

$$\langle A_2 u, v \rangle = a_2(u, v) \qquad \text{for all} \quad u, v \in X.$$

Thus, the generalized problem (13) is equivalent to the operator equation

$$A_1 u + A_2 u = b, \qquad u \in X,$$

where $b \in X^*$.

Step 3: Strong continuity of A_2.

(I) By $A_2(45)$, the embedding $X \subseteq L_4(G)$ is compact.

(II) Let $u_n \rightharpoonup u$ in X as $n \to \infty$. Then the sequence (u_n) is bounded in X. By (I),

$$u_n \to u \quad \text{in } L_4(G) \qquad \text{as} \quad n \to \infty.$$

We must verify

$$A_2u_n \to A_2u \quad \text{in } X^* \qquad \text{as} \quad n \to \infty,$$

i.e., as $n \to \infty$,

$$\|A_2u_n - A_2u\| = \sup_{\|v\| \leq 1} |\langle A_2u_n - A_2u, v\rangle| \to 0.$$

Otherwise, there would exist an $\varepsilon_0 > 0$ and a sequence $(v_{n'})$, which we denote briefly by (v_n), such that

$$\|v_n\|_X \leq 1 \qquad \text{for all} \quad n$$

with

$$\langle A_2u_n - A_2u, v_n\rangle \geq \varepsilon_0 \qquad \text{for all } n. \tag{14}$$

Passing to a subsequence, if necessary, we can assume that $v_n \rightharpoonup v$ in X and hence

$$v_n \to v \quad \text{in } L_4(G) \qquad \text{as} \quad n \to \infty.$$

In order to obtain a contradiction, we use the *decomposition*:

$$\begin{aligned}(\sin u_n)(D_iu_n)v_n - (\sin u)(D_iu)v_n = {} & (\sin u_n - \sin u)(D_iu_n)v_n \\ & + (\sin u)(D_iu_n)(v_n - v) \\ & + (\sin u)(D_iu_n - D_iu)v \\ & + (\sin u)(D_iu)(v - v_n).\end{aligned} \tag{14*}$$

The decisive trick of our argument will be to use the weak convergence $u_n \rightharpoonup u$ in X with respect to the term $(\sin u)(D_iu_n - D_iu)v$. From

$$|\sin u_n - \sin u| \leq |u_n - u|$$

and the Hölder inequality for three factors, we obtain that

$$\begin{aligned}&\left|\int (\sin u_n - \sin u)(D_iu_n)v_n\,dx\right| \\ &\qquad \leq \left(\int |u_n - u|^4\,dx\right)^{1/4}\left(\int |D_iu_n|^2\,dx\right)^{1/2}\left(\int |v_n|^4\,dx\right)^{1/4} \\ &\qquad \leq \|u_n - u\|_4\|u_n\|_X\|v_n\|_X.\end{aligned}$$

By (14*), the same argument yields

$$\begin{aligned}|\langle A_2u_n - A_2u, v_n\rangle| \leq {} & \|u_n - u\|_4\|u_n\|_X\|v_n\|_X + \|u_n\|_X\|v_n - v\|_4 \\ & + c(u, u_n - u, v) + \|u\|_X\|v_n - v\|_4 \to 0 \qquad \text{as} \quad n \to \infty.\end{aligned} \tag{14**}$$

In fact, we have $u_n \to u$ and $v_n \to v$ in $L_4(G)$ as $n \to \infty$, i.e., $\|u_n - u\|_4 \to 0$ and $\|v_n - v\|_4 \to 0$. Moreover, the sequences (u_n) and (v_n) are bounded in X. Finally, we have

$$c(u, u_n - u, v) \to 0 \qquad \text{as} \quad n \to \infty.$$

This follows from $u_n \rightharpoonup u$ in X and

$$|c(u, w, v)| \le \text{const}\, \|w\|_X \|v\|_X \qquad \text{for all} \quad u, v, w \in X,$$

according to the Hölder inequality, and hence the linear functional $w \mapsto c(u, w, v)$ is continuous on X.

Relation (14**) contradicts (14).

Step 4: Coerciveness of $A_1 + A_2$.

For all $u \in X$,

$$\begin{aligned}|a_2(u,u)| &\le |\alpha| \sum_i \left| \int (\sin u)(D_i u) u \, dx \right| \\ &\le |\alpha| \sum_i \left(\int (D_i u)^2 \, dx \right)^{1/2} \left(\int u^2 \, dx \right)^{1/2} \\ &\le |\alpha| \, \|u\|_X^2.\end{aligned}$$

By the Poincaré–Friedrichs inequality, there exists a $c > 0$ such that

$$a_1(u, u) \ge c\|u\|_X^2 \qquad \text{for all} \quad u \in X.$$

This implies

$$\begin{aligned}\langle A_1 u + A_2 u, u \rangle &= a_1(u, u) + a_2(u, u) \\ &\ge (c - |\alpha|)\|u\|_X^2 \qquad \text{for all} \quad u \in X,\end{aligned}$$

i.e., $A_1 + A_2$ is coercive if $|\alpha| < c$.

Step 5: Application of Theorem 27.A.

As in the proof of Proposition 27.9, we obtain that the operator $A_1 + A_2$ is pseudomonotone, continuous, and bounded. Now the assertion follows from Theorem 27.A.

Moreover, $A_1 + A_2$ satisfies $(S)_+$. □

27.5. Relations Between Important Properties of Nonlinear Operators

Proposition 27.12. *If X and Y are real reflexive* B-*spaces, then the implications in Figure* 27.1 *hold for operators $A: X \to Y$.*

In this connection, let $Y = X^*$ in all the assertions that refer to monotonicity, pseudomonotonicity, hemicontinuity, coerciveness, (M), $(S)_+$, (S), $(S)_0$, $(S)_1$, and (P).

Explanation 27.13. In Figure 27.1 we only consider operators between real reflexive B-spaces. The domain of definition is the total B-space. The arrows are to be understood in the sense of an implication. The plus sign means the operator sum, i.e., for example, $(S)_+$ + (compact) is the sum of an operator

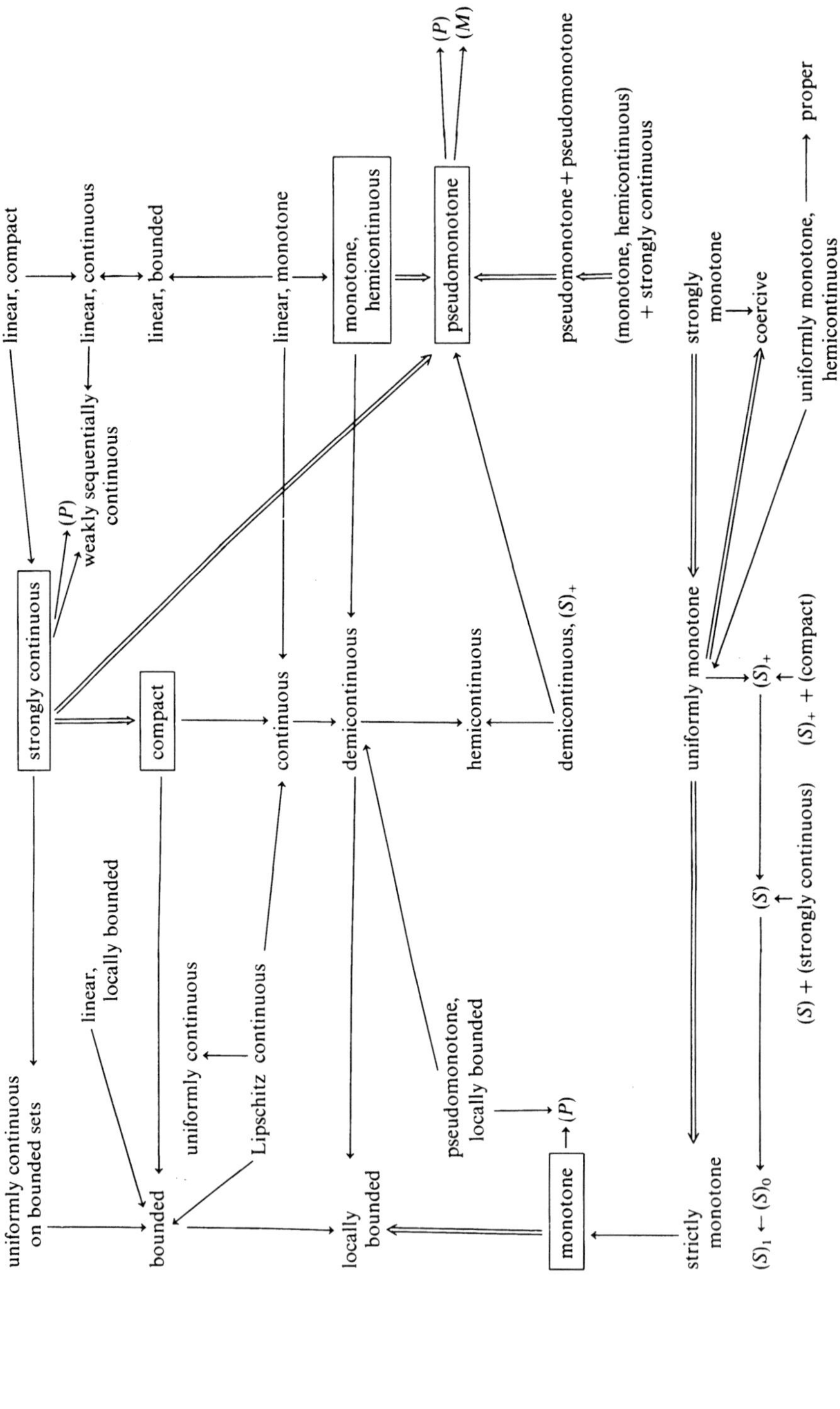

Figure 27.1 (See Explanation 27.13)

with $(S)_+$ and a compact operator. In particular, one infers from Figure 27.1 that every compact operator is bounded, and that monotone hemicontinuous operators are pseudomonotone, etc.

Most of the assertions in Figure 27.1 have already been proved. We prove the remaining assertions in Problem 27.3. In the following we summarize once more all the important definitions concerning continuity and boundedness properties of operators.

Definition 27.14. Let X, Y be B-spaces over $\mathbb{K} = \mathbb{R}, \mathbb{C}$. We define the following properties for an operator $A: X \to Y$:

A is *continuous* iff ($u_n \to u$ implies $Au_n \to Au$).

A is *strongly continuous* iff ($u_n \rightharpoonup u$ implies $Au_n \to Au$).

A is *demicontinuous* iff ($u_n \to u$ implies $Au_n \rightharpoonup Au$).

A is *weakly sequentially continuous* iff ($u_n \rightharpoonup u$ implies $Au_n \rightharpoonup Au$).

A is *compact* iff A is continuous and maps bounded sets into relatively compact sets.

A is *image compact* (i-compact) iff A is continuous and the range of A is relatively compact.

A is *proper* iff the preimage of compact sets is again compact.

A is *Lipschitz continuous* on $M \subseteq X$ iff

$$\|Au - Av\| \leq L\|u - v\| \qquad \text{for all} \quad u, v \in M \quad \text{and fixed } L.$$

A is *k-contractive* iff A is Lipschitz continuous with $L = k$ and $0 \leq k < 1$.

A is *nonexpansive* iff A is Lipschitz continuous with $L = 1$.

A is *locally Lipschitz continuous* iff each $u \in X$ has a neighborhood $V(u)$ on which A is Lipschitz continuous.

A is *uniformly continuous* on $M \subseteq X$ iff for each $\varepsilon > 0$ there exists a $\delta(\varepsilon) > 0$ such that, for $u, v \in M$,

$$\|u - v\| < \delta(\varepsilon) \qquad \text{implies} \qquad \|Au - Av\| < \varepsilon.$$

A is *bounded* iff A maps bounded sets into bounded sets.

A is *locally bounded* iff each $u \in X$ has a neighborhood $V(u)$ such that $A(V(u))$ is bounded.

A is *hemicontinuous* in case $Y = X^*$ iff $t \mapsto \langle A(u + tv), w\rangle$ is continuous on $[0, 1]$ for all $u, v, w \in X$.

Note that $u_n \to u$ stands for $u_n \to u$ as $n \to \infty$, where $n \in \mathbb{N}$.

For mappings $A: D(A) \subseteq X \to Y$ these definitions are to be modified in an obvious way, in that one considers only elements from $D(A)$. One sometimes uses the term completely continuous for compact. Note, however, that some authors understand a compact operator to be a continuous operator whose range is relatively compact. We call such operators image compact (i-compact).

27.6. Dual Pairs of B-Spaces

In the preceding sections we considered operators of the form

$$A: X \to X^*$$

from a B-space X to its dual space X^*. In order to be able to apply our abstract results to more general partial differential equations, we now study operators of the form

$$A: X \to X^+,$$

where $\{X, X^+\}$ is a dual pair of B-spaces.

Definition 27.15. Let X and X^+ be B-spaces over $\mathbb{K} = \mathbb{R}, \mathbb{C}$. Then $\{X, X^+\}$ is called a *dual pair* iff the following hold:

(i) There exists a bilinear, bounded map

$$(v, u) \to \langle v, u\rangle_X$$

from $X^+ \times X$ to $\mathbb{K}$.

(ii) If $\langle v, u\rangle_X = 0$ for all $u \in X$, then $v = 0$.

(iii) If $\langle v, u\rangle_X = 0$ for all $v \in X^+$, then $u = 0$.

Standard Example 27.16. Let X be a B-space X and let X^* be its dual space. Then $\{X, X^*\}$ is a dual pair if we set, in the usual way,

$$\langle v, u\rangle_X = v(u) \qquad \text{for all} \quad v \in X^*, \quad u \in X.$$

Example 27.17. Let $X = C(\bar{G})$, where G denotes a bounded region in $\mathbb{R}^N$, $N \geq 1$. If we set

$$\langle v, u\rangle_X = \int_G uv\,dx, \tag{15}$$

then $\{X, X\}$ is a dual pair.

Moreover, if we set $X = W_2^m(G)$, $m = 0, 1, \ldots,$ and $X^+ = L_2(G)$, then $\{X, X^+\}$ forms a dual pair with respect to (15).

In contrast to the Standard Example 27.16, in this example we do not have $X^+ = X^*$. In fact, dual pairs are a useful generalization of the usual duality for B-spaces.

27.7. The Main Theorem on Locally Coercive Operators

We consider the operator equation

$$Au = b, \qquad u \in D(A), \tag{16}$$

with respect to the operator

$$A: D(A) \subseteq X \to Y^+, \tag{17}$$

where

$$(K \cap Y) \subseteq D(A).$$

We are given $b \in X^+$ with $X^+ \subseteq Y^+$. Our goal is Theorem 27.B below. This theorem allows interesting applications to nonlinear partial differential equations. In Section 27.8 we will consider such an application. In this connection, the local coerciveness condition (H3) below is essential. Furthermore, it is a typical *pecularity* of our situation that the operator A is not defined on the total B-space X but only on a subset $D(A)$ of X. In order to compensate for this fact, we use *two* B-spaces X and Y with

$$Y \subseteq X.$$

Roughly speaking, the space Y is "better" than the space X. For example, in applications to differential equations, the space Y contains smoother functions than the space X. The *basic idea* of our existence proof is to use a Galerkin method in the "better" space Y which converges only in the "worse" space X.

This idea is related to the basic idea of the hard implicit function theorem in Chapter 5. There we constructed an iteration sequence (u_n) with $u_n \in X_n$ for all n and

$$X_1 \subseteq X_2 \subseteq \cdots \subseteq X,$$

where (u_n) converges in the "bad" space X.

The use of different spaces combined with convergence in "bad" spaces is important for the modern theory of nonlinear partial differential equations. In Chapter 30 we will apply this idea to evolution equations (Theorem 30.B).

In what follows the reader should think of the standard situation that X and Y are B-spaces with the continuous and dense embedding

$$Y \subseteq X$$

(e.g., $Y = X$). This implies the continuous embedding

$$X^* \subseteq Y^*$$

in the sense that a linear continuous functional $b: X \to \mathbb{K}$ is also a linear continuous functional of the form $b: Y \to \mathbb{K}$, by restricting b to Y. If we set

$$X^+ = X^*, \qquad Y^+ = Y^*,$$

then we obtain dual pairs $\{X, X^+\}$ and $\{Y, Y^+\}$. In this situation, the set K can be chosen as a closed ball in X.

We make the following assumptions:

(H1) *Dual pairs.* Let the dual pairs $\{X, X^+\}$ and $\{Y, Y^+\}$ be given, where X, X^+, Y, Y^+ are real B-spaces with the corresponding bilinear forms $\langle \cdot, \cdot \rangle_X$ and $\langle \cdot, \cdot \rangle_Y$ and the continuous embeddings

$$Y \subseteq X \quad \text{and} \quad X^+ \subseteq Y^+.$$

The two dual pairs are compatible, i.e.,

$$\langle b, u\rangle_X = \langle b, u\rangle_Y \qquad \text{for all} \quad b \in X^+, \quad u \in Y.$$

Moreover, the B-spaces X and Y are separable and X is reflexive.

(H2) *Operator A.* Let the operator $A: D(A) \subseteq X \to Y^+$ be given, and let K be a bounded closed convex set in X containing the zero point as an interior point and $K \cap Y \subseteq D(A)$.

(H3) *Local coerciveness.* There exists a number $\alpha \geq 0$ such that

$$\langle Av, v\rangle_Y \geq \alpha \qquad \text{for all} \quad v \in Y \cap \partial K,$$

where ∂K denotes the boundary of K in the B-space X.

(H4) *Continuity.* For each finite-dimensional subspace Y_0 of the B-space Y, the mapping

$$w \mapsto \langle Aw, v\rangle_Y$$

is continuous on $K \cap Y_0$ for all $v \in Y_0$.

(H5) *Generalized condition (M).* Let (u_n) be a sequence in $Y \cap K$ and let $b \in X^+$. Then, from

$$u_n \rightharpoonup u \quad \text{in } X \qquad \text{as} \quad n \to \infty \tag{18}$$

and from

$$\langle Au_n, v\rangle_Y \to \langle b, v\rangle_Y \qquad \text{as} \quad n \to \infty \qquad \text{for all} \quad v \in Y,$$

and

$$\overline{\lim_{n\to\infty}}\, \langle Au_n, u_n\rangle_Y \leq \langle b, u\rangle_X,$$

it follows that $Au = b$.

(H6) *Quasi-boundedness.* Let (u_n) be a sequence in $Y \cap K$. Then, from (18) and

$$\langle Au_n, u_n\rangle_Y \leq \text{const}\, \|u_n\|_X \qquad \text{for all} \quad n \in \mathbb{N}$$

it follows that the sequence (Au_n) is bounded in Y^+.

Note that the conditions (H4), (H5) and (H6) are satisfied if, for example, $D(A) = K$ and the operator $A: K \subseteq X \to Y^+$ is weakly sequentially continuous.

(H7) *Local strict monotonicity.* There exists a dual pair $\{Z, Z^+\}$, where Z and Z^+ are real B-spaces with the continuous embeddings

$$X \subseteq Z \qquad \text{and} \qquad Y^+ \subseteq Z^+,$$

such that

$$\langle Au - Av, u - v\rangle_Z > 0 \qquad \text{for all} \quad u, v \in D(A) \quad \text{with} \quad u \neq v.$$

(H8) *Local strong monotonicity.* In addition to (H7), there exists a number $d > 0$ such that

$$\langle Au - Av, u - v\rangle_Z \geq d\|u - v\|_Z^2 \qquad \text{for all} \quad u, v \in D(A).$$

Theorem 27.B (Hess (1973), Kato (1984)). *Suppose that the assumptions* (H1) *through* (H6) *are satisfied. Then the following hold*:

(a) Existence. *For each* $b \in X^+$ *with*

$$\langle b, v\rangle_X \le \alpha \qquad \text{for all} \quad v \in K \cap Y,$$

equation (16) *has a solution* u, *i.e.*, $b \in R(A)$.

(b) Uniqueness. *This solution is unique if, in addition, condition* (H7) *holds.*

(c) Continuous dependence of the solution on the data. *If, in addition, condition* (H8) *holds, then it follows from* $Au_i = b_i$ *and* $b_i \in X^+$ *for* $i = 1, 2$ *that*

$$\|u_1 - u_2\|_Z \le \text{const}\,\|b_1 - b_2\|_{Z^+}.$$

This is a generalization of a related result in Kato (1984). Our generalization is useful with respect to the applications to strongly nonlinear differential equations in Section 27.8.

Obviously, Theorem 27.B is a generalization of the main theorem on pseudomonotone operators (Theorem 27.A). In fact, Theorem 27.B is a powerful tool for handling nonlinear partial differential equations.

Corollary 27.18 (The Special Case of Balls). *Suppose that conditions* (H1) *through* (H6) *hold. Let* K^0 *denote the polar set of* K, *i.e.*,

$$K^0 = \{b \in X^+ \colon \langle b, v\rangle_X \le 1 \text{ for all } v \in K\}.$$

Then $\alpha K^0 \subseteq R(A)$, *i.e.*, *the original equation* (16) *has a solution* u *for each* $b \in \alpha K^0$.

In particular, if K *is a ball of radius* $R > 0$, *i.e.*,

$$K = \{v \in X \colon \|v\|_X \le R\}$$

and if

$$\langle b, v\rangle_X \le c\|b\|_{X^+}\|v\|_X \qquad \text{for all} \quad b \in X^+, \quad v \in X,$$

and fixed $c > 0$, *then*

$$\{b \in X^+ \colon \|b\|_{X^+} \le 1/cR\} \subseteq K^0,$$

i.e., *equation* (16) *has a solution* u *for each* $b \in X^+$ *with* $\|b\|_{X^+} \le \alpha/cR$.

This is an immediate consequence of Theorem 27.B.

Corollary 27.19 (Global Coerciveness). *Suppose that the conditions* (H1) *through* (H6) *are valid for all balls* K *in* X, *and suppose that the global coerciveness condition*

$$\langle Av, v\rangle_Y/\|v\|_X \to +\infty \qquad \text{as} \quad \|v\|_X \to \infty, \quad v \in Y,$$

is satisfied. Then $X^+ \subseteq R(A)$, *i.e.*, *equation* (16) *has a solution* u *for each* $b \in X^+$.

PROOF. This follows from Corollary 27.18. Note that we can choose $\alpha = \beta(R)R$ with $\beta(R) \to +\infty$ as $R \to +\infty$. Hence $\alpha/cR \to +\infty$ as $R \to \infty$. $\square$

PROOF OF THEOREM 27.B(a). We use a similar argument as in the proof of Theorem 27.A. More precisely, we consider a Galerkin method in the space Y which converges in the space X.

If $Y = \{0\}$, then $Y^+ = \{0\}$, and the statement of Theorem 27.B(a) is trivial. Therefore, let us suppose that $Y \neq \{0\}$ and $X \neq \{0\}$.

Since Y is separable, there exists a sequence

$$Y_1 \subseteq Y_2 \subseteq \cdots \subseteq Y$$

of nontrivial subspaces Y_n of Y such that the set $\bigcup_n Y_n$ is dense in Y. We set

$$K_n = K \cap Y_n.$$

The embedding $Y \subseteq X$ is continuous. This implies that each open (resp. closed) set in X is also open (resp. closed) in Y. Consequently, K_n is a closed convex subset of Y_n containing the zero point as an interior point according to (H2). Moreover, $\partial K_n \subseteq \partial K$.

For each nonzero $v \in Y_n$, let

$$\|v\| = \inf\{\lambda > 0 : \lambda^{-1} v \in \partial K_n\},$$

i.e., $\|\cdot\|$ is the Minkowski functional of K_n. Therefore, $\|\cdot\|$ is a norm on Y_n. Since $\dim Y_n < \infty$, the mapping $u \mapsto \|u\|$ is continuous on Y_n. Thus, we obtain that

$$\operatorname{int} K_n = \{v \in Y_n : \|v\| < 1\}$$

and $\partial K_n = \{v \in Y_n : \|v\| = 1\}$.

Step 1: The Galerkin equations in Y.

We consider the Galerkin equation

$$\langle Au_n, v\rangle_Y = \langle b, v\rangle_Y \qquad \text{for all} \quad v \in Y_n. \tag{19}$$

We are given $b \in X^+$ with $\langle b, v\rangle_Y \le \alpha$ for all $v \in K_n$, and we are looking for

$$u_n \in K_n.$$

By the local coerciveness condition (H3), we obtain the *key condition*

$$\langle Av, v\rangle_Y - \langle b, v\rangle_Y \ge 0 \qquad \text{for all} \quad v \in \partial K_n. \tag{20}$$

Step 2: Solution of the Galerkin equations by using the existence principle in Section 2.4.

Let $\{w_1, \ldots, w_N\}$ be a basis in Y_n. We set

$$x = \sum_{i=1}^{N} \xi_i w_i$$

and identify $\mathbb{R}^N$ with Y_n. Moreover, let

$$g_i(x) = \langle Ax, w_i \rangle_Y - \langle b, w_i \rangle_Y.$$

Then the function $g\colon K_n \subseteq \mathbb{R}^N \to \mathbb{R}^N$ is continuous, by (H4). From (20) it follows that

$$\sum_{i=1}^N g_i(x)\xi_i \geq 0 \qquad \text{for all} \quad x \in \partial K_n,$$

that is, for all $x \in \mathbb{R}^N$ with $\|x\| = 1$. By Section 2.4, the system

$$g_i(x) = 0, \qquad x \in K_n, \qquad i = 1, \ldots, N,$$

has a solution, i.e., the Galerkin equation (19) has a solution $u_n \in K_n$.

Step 3: Convergence of the Galerkin method in X.

By (H2), the set K is bounded in X. Thus, the sequence (u_n) with $u_n \in K$ for all n is bounded in the reflexive B-space X. Consequently, there exists a subsequence, again denoted by (u_n), such that

$$u_n \rightharpoonup u \quad \text{in } X \qquad \text{as} \quad n \to \infty. \tag{21}$$

From the Galerkin equation (19) and (H1) it follows that

$$\langle Au_n, u_n \rangle_Y = \langle b, u_n \rangle_Y = \langle b, u_n \rangle_X$$

and hence

$$\langle Au_n, u_n \rangle_Y \leq \text{const}\, \|b\|_{X^+} \|u_n\|_X \qquad \text{for all } n.$$

By (H6), the sequence (Au_n) is bounded in Y^+.

Let $v \in \bigcup_n Y_n$, i.e., $v \in Y_k$ for a fixed k. By (19),

$$\langle Au_n, v \rangle_Y \to \langle b, v \rangle_Y \qquad \text{as} \quad n \to \infty. \tag{22}$$

Since the set $\bigcup_n Y_n$ is dense in Y and the sequence (Au_n) is bounded in Y^+, the relation (22) also holds true for all $v \in Y$. This follows from a simple approximation argument, by using the fact that the bilinear form $(w, v) \mapsto \langle w, v \rangle_Y$ is bounded from $Y^+ \times Y$ to $\mathbb{R}$.

From (19) and (21) we obtain

$$\lim_{n \to \infty} \langle Au_n, u_n \rangle_Y = \lim_{n \to \infty} \langle b, u_n \rangle_X = \langle b, u \rangle_X.$$

By assumption (H5), $Au = b$.

Proof of Theorem 27.B(b). From $Au_1 = Au_2$ we obtain $u_1 = u_2$ by (H7).

Proof of Theorem 27.B(c). Condition (H8) implies

$$d\|u_1 - u_2\|_Z^2 \leq \text{const}\, \|u_1 - u_2\|_Z \|Au_1 - Au_2\|_{Z^+}$$

and hence

$$\|u_1 - u_2\|_Z \leq \text{const}\, \|Au_1 - Au_2\|_{Z^+}. \qquad \square$$

27.8. Application to Strongly Nonlinear Differential Equations

We consider the boundary value problem

$$\begin{aligned} -\Delta u + g(u) &= f \quad \text{on } G, \\ u &= 0 \quad \text{on } \partial G. \end{aligned} \tag{23}$$

In contrast to Section 27.4, there are no growth conditions for g. Therefore, we speak of a "strongly nonlinear" problem. For example, think of $g(u) = e^u$.

We make the following assumptions:

(A1) The function $g\colon \mathbb{R} \to \mathbb{R}$ is continuous with $(g(u) - a)u \geq 0$ for all $u \in \mathbb{R}$ and fixed real a.

For example, this condition is satisfied if $g\colon \mathbb{R} \to \mathbb{R}$ is continuous and monotone increasing (e.g., $g(u) = e^u$). In this case, we choose $a = g(0)$.

(A2) G is a bounded region in $\mathbb{R}^N$ with piecewise smooth boundary, i.e., $\partial G \in C^{0,1}$.

The basic idea of the following investigation of problem (23) is to reduce (23) to an operator equation

$$Au = b, \qquad u \in D(A),$$

where

$$D(A) = \{u \in X\colon h \in L_1(G)\}$$

and $h(x) = (g(u(x)) - a)u(x)$, i.e., the operator A is not defined on the total B-space

$$X = \mathring{W}_2^1(G).$$

Moreover, we set

$$Y = W_2^k(G) \cap X, \qquad k > N/2,$$

and $\|u\|_Y = \|u\|_{k,2}$. Since $k > N/2$, the embedding

$$Y \subseteq C(\bar{G})$$

is continuous by the Sobolev embedding theorems. Note that

$$|g(u)| \leq (g(u) - a)u + \text{const} \qquad \text{for all} \quad u \in \mathbb{R},$$

since

$$|g(u) - a| = |u^{-1}(g(u) - a)u| \leq (g(u) - a)u \qquad \text{for} \quad |u| \geq 1.$$

Therefore, from $u \in D(A)$, it follows that $x \mapsto g(u(x))$ belongs to $L_1(G)$.

Definition 27.20. The *generalized problem* corresponding to (23) reads as follows. Let $f \in L_2(G)$ be given. We seek $u \in D(A)$ such that

$$a_1(u, v) + a_2(u, v) = b(v) \qquad \text{for all} \quad v \in Y, \tag{24}$$

where

$$a_1(u, v) = \int_G \sum_{i=1}^N D_i u D_i v \, dx,$$

$$a_2(u, v) = \int_G g(u(x)) v(x) \, dx,$$

$$b(v) = \int_G fv \, dx.$$

Formally, we obtain (24) by multiplication of (23) with $v \in C_0^\infty(G)$ and subsequent integration by parts. The *idea* behind the construction of the set $D(A)$ is to guarantee the existence of $a_2(u, v)$ and $a_2(u, u)$ for all $u \in D(A)$ and $v \in Y$.

Proposition 27.21. *Suppose that* (A1), (A2) *hold. Then, for each* $f \in L_2(G)$, *the boundary value problem* (23) *has a generalized solution* $u \in D(A)$.

PROOF. Passing from (23) to

$$-\Delta u + (g(u) - a) = f - a,$$

we can assume that $a = 0$.

We apply Theorem 27.B to a sufficiently large ball K in the B-space X. Moreover, we set

$$X^* = X^+, \qquad Y^* = Y^+.$$

The key to our proof will be (V) below.

The Hölder inequality yields

$$|b(v)| \le \text{const} \|f\|_2 \|v\|_X \qquad \text{for all} \quad v \in X.$$

Hence $b \in X^*$.

(I) Operator A_1. By the Hölder inequality, we have

$$|a_1(u, v)| \le \text{const} \|u\|_X \|v\|_Y \qquad \text{for all} \quad u \in X, \quad v \in Y.$$

Hence there exists exactly one linear continuous operator $A_1: X \to Y^*$ with

$$\langle A_1 u, v \rangle_Y = a_1(u, v) \qquad \text{for all} \quad u \in X, \quad v \in Y.$$

(II) Operator A_2. Let $u \in D(A)$. Then

$$\begin{aligned} |a_2(u, v)| &\le \int_G |g(u)| \, dx \|v\|_{C(\bar{G})} \\ &\le \text{const} \|v\|_Y \qquad \text{for all} \quad v \in Y. \end{aligned}$$

Thus, there exists a uniquely determined operator $A_2: D(A) \subseteq X \to Y^*$ with

$$\langle A_2 u, v\rangle_Y = a_2(u, v) \qquad \text{for all} \quad u \in D(A), \quad v \in Y.$$

(III) Operator A. We define $A: D(A) \subseteq X \to Y^*$ by

$$A = A_1 + A_2.$$

Then, the generalized problem (24) corresponds to the operator equation

$$Au = b, \qquad u \in D(A). \tag{25}$$

Note that $Y \subseteq C(\bar{G}) \subseteq D(A) \subseteq X$.

(IV) Global coerciveness of A. Noting $g(u)u \geq 0$ on $\mathbb{R}$ and the inequality of Poincaré–Friedrichs, we obtain that there exists a constant $d > 0$ such that

$$\begin{aligned}\langle Av, v\rangle_Y &= a_1(v, v) + a_2(v, v) \geq a_1(v, v)\\ &\geq d\|v\|_X^2 \qquad \text{for all} \quad v \in Y.\end{aligned}$$

(V) Generalized condition (M). Let $b \in X^*$ and let (u_n) be a sequence in Y with

$$u_n \rightharpoonup u \quad \text{in } X \qquad \text{as} \quad n \to \infty$$

and

$$\langle Au_n, v\rangle_Y \to b(v) \qquad \text{as} \quad n \to \infty \qquad \text{for all} \quad v \in Y, \tag{26}$$

$$\overline{\lim_{n \to \infty}}\, \langle Au_n, u_n\rangle_Y \leq b(u). \tag{27}$$

We want to show that this implies $Au = b$.

In the following, c denotes an arbitrary constant. The operator $A_1: X \to Y^*$ is linear and continuous, therefore weakly sequentially continuous, i.e.,

$$\langle A_1 u_n, v\rangle_Y \to \langle A_1 u, v\rangle_Y \qquad \text{as} \quad n \to \infty \qquad \text{for all} \quad v \in Y.$$

Because of (26) it is sufficient to prove that $u \in D(A)$ and that

$$\langle A_2 u_n, v\rangle_Y \to \langle A_2 u, v\rangle_Y \qquad \text{as} \quad n \to \infty \qquad \text{for all} \quad v \in Y.$$

Noting $Y \subseteq C(\bar{G})$ and $\langle A_2 u, v\rangle_Y = \int_G g(u)v\,dx$ along with the estimate (II), it is sufficient to show that

$$g(u_n(x)) \to g(u(x)) \quad \text{in } L_1(G) \qquad \text{as} \quad n \to \infty. \tag{28}$$

Moreover, note that it is sufficient to prove (28) for a subsequence of (u_n).

The following proof of the conditions $u \in D(A)$ and (28) is based on the Lemma of Fatou and the Vitali convergence theorem.

(V-1) We show that $g(u(\cdot))u(\cdot) \in L_1(G)$, i.e., $u \in D(A)$. The embedding $X \subseteq L_2(G)$ is compact. This implies

$$u_n \to u \quad \text{in } L_2(G) \qquad \text{as} \quad n \to \infty.$$

Thus, there exists a subsequence, again denoted by (u_n), such that

$$u_n(x) \to u(x) \quad \text{as} \quad n \to \infty \quad \text{for almost all} \quad x \in G.$$

Moreover, from $u_n \rightharpoonup u$ in X it follows that

$$\sup_n \|u_n\|_X < \infty$$

and hence

$$\langle A_1 u_n, u_n \rangle_Y \le c\|u_n\|_X^2 \le \text{const.}$$

Relation (27) yields

$$\overline{\lim_{n\to\infty}} \langle A_2 u_n, u_n \rangle_Y = \overline{\lim_{n\to\infty}} \int_G g(u_n)u_n \, dx \le c, \tag{29}$$

and the continuity of g implies

$$g(u_n(x))u_n(x) \to g(u(x))u(x) \quad \text{as} \quad n \to \infty \quad \text{for almost all} \quad x \in G.$$

Therefore, the *Lemma of Fatou* $A_2(19c)$ tells us that

$$g(u(\cdot))u(\cdot) \in L_1(G).$$

(V-2) We prove (28). Let $d > 0$ be fixed. For each $x \in G$, we have either

$$|u_n(x)| \le d$$

or

$$|g(u_n(x))| \le d^{-1} g(u_n(x))u_n(x).$$

Note that $g(u) = u^{-1}g(u)u$ if $u \neq 0$. Moreover, we get

$$|g(u)| \le c(d) \quad \text{for all} \quad u\colon |u| \le d.$$

Let H be a measurable subset of G. Then,

$$\begin{aligned} \int_H |g(u_n)| \, dx &\le c(d) \operatorname{meas} H + d^{-1} \int_H g(u_n)u_n \, dx \\ &\le c(d) \operatorname{meas} H + d^{-1} c_1. \end{aligned}$$

By (29), the constant c_1 is independent of n, d and H. Hence

$$\int_H |g(u_n)| \, dx < \varepsilon/2 \quad \text{for all } n$$

if d is sufficiently large and meas H is sufficiently small. Noting the absolute continuity $A_2(20)$ of the integral, we obtain the following. For each $\varepsilon > 0$, there exists a $\delta(\varepsilon) > 0$ such that

$$\int_H |g(u_n) - g(u)| \, dx \le \int_H (|g(u_n)| + |g(u)|) \, dx < \varepsilon,$$

for all n and for all subsets H of G with meas $H < \delta(\varepsilon)$.

Therefore, the *Vitali convergence theorem* $A_2(21)$ tells us that (28) holds true.

(VI) Quasi-boundedness of the operator A. Let (u_n) be a sequence in the space Y with

$$u_n \rightharpoonup u \quad \text{in } X \quad \text{as} \quad n \to \infty$$

and suppose that

$$\langle Au_n, u_n \rangle_Y \leq \text{const} \|u_n\|_X \quad \text{for all} \quad n.$$

We want to show that the sequence (Au_n) is bounded in Y^*. The boundedness of (u_n) in X implies

$$\overline{\lim_{n\to\infty}} \langle Au_n, u_n \rangle_Y \leq \text{const}.$$

In order to obtain a contradiction, suppose that (Au_n) is unbounded in Y^*. Then there exists a subsequence, again denoted by (u_n), such that

$$\|Au_n\|_{Y^*} \to \infty \quad \text{as} \quad n \to \infty. \tag{30}$$

By the same argument as in (V) above, we obtain that

$$\langle Au_n, v \rangle_Y \to \langle Au, v \rangle_Y \quad \text{as} \quad n \to \infty \quad \text{for all} \quad v \in Y,$$

by passing to a subsequence, if necessary. The *uniform boundedness principle* $A_1(35)$ tells us that the sequence (Au_n) is bounded. This contradicts (30).

Now it follows from Corollary 27.19 that the equation $Au = b$, $u \in D(A)$, has a solution if $b \in X^*$. □

Problems

27.1. *Proof of Example* 27.3.

Solution: Ad(a). Let

$$u_n \rightharpoonup u, \quad \overline{\lim} \langle Au_n + Bu_n - Au - Bu, u_n - u \rangle \leq 0 \quad \text{as} \quad n \to \infty.$$

Since (u_n) is bounded in the reflexive B-space X and the operator B is compact, there exists a subsequence $(u_{n'})$ such that $Bu_{n'} \to b$ as $n \to \infty$ and hence

$$\overline{\lim_{n\to\infty}} \langle Au_{n'} - Au, u_{n'} - u \rangle \leq 0.$$

The operator A satisfies $(S)_+$; therefore $u_{n'} \to u$ as $n \to \infty$.

By the convergence principle (Proposition 10.13(1)), the total sequence (u_n) converges, i.e., $u_n \to u$ as $n \to \infty$.

Ad(b). Let

$$u_n \rightharpoonup u, \quad \langle Au_n + Bu_n - Au - Bu, u_n - u \rangle \to 0 \quad \text{as} \quad n \to \infty.$$

The operator B is strongly continuous; therefore, $Bu_n \to Bu$ and hence

$$\langle Au_n - Au, u_n - u \rangle \to 0 \quad \text{as} \quad n \to \infty.$$

The operator A satisfies (S); consequently $u_n \to u$ as $n \to \infty$.

Ad(c). Let $C = A + B$, and let

$$u_n \rightharpoonup u, \qquad Cu_n \rightharpoonup b, \qquad \overline{\lim}\langle Cu_n, u_n\rangle \le \langle b, u\rangle \qquad \text{as} \quad n \to \infty.$$

The operator B is strongly continuous, i.e., $Bu_n \to Bu$ and hence

$$\overline{\lim_{n\to\infty}}\ \langle Au_n, u_n\rangle \le \langle b - Bu, u\rangle.$$

The operator A satisfies (M). Thus, from $Au_n \rightharpoonup b - Bu$ as $n \to \infty$ it follows that $Au = b - Bu$, i.e., $Cu = b$.

27.2. *Proof of Proposition* 27.6(g). Let $u_n \rightharpoonup u$ as $n \to \infty$. The operator B is strongly continuous; consequently

$$\lim_{n\to\infty} \langle Bu_n, u_n - u\rangle = 0.$$

The operator A is monotone, i.e.,

$$\langle Au_n, u_n - u\rangle \ge \langle Au, u_n - u\rangle \to 0 \qquad \text{as} \quad n \to \infty.$$

Hence $\overline{\lim}_{n\to\infty} \langle Au_n + Bu_n, u_n - u\rangle \ge 0$.

27.3. *Proof of the assertions in Figure* 27.1. Prove the assertions that have not yet been proved in Chapters 26 and 27.

Let $A\colon X \to Y$ be an operator, where X and Y are B-spaces over $\mathbb{K} = \mathbb{R}, \mathbb{C}$.

(I) Show that if the operator A is strongly continuous, then A is uniformly continuous on bounded sets in the case where X is reflexive.

Otherwise, there exist an $\varepsilon_0 > 0$ and bounded sequences (u_n), (v_n) with $\|u_n - v_n\| \le 1/n$ and

$$\|Au_n - Av_n\| \ge \varepsilon_0 \qquad \text{for all} \quad n \in \mathbb{N}.$$

Because of the reflexivity of X there then exist subsequences $u_{n'} \rightharpoonup u$ and $v_{n'} \rightharpoonup v$ as $n \to \infty$; therefore, $u = v$ and $Au_{n'} - Av_{n'} \to 0$ as $n \to \infty$. This is a contradiction.

(II) Show that if the operator A is uniformly continuous on bounded sets, then A is bounded.

Let $M = \{u \in X\colon \|u\| \le r\}$. The operator A is uniformly continuous on M. Therefore, there is a $\delta > 0$ such that it follows from $u, v \in M$ and $\|u - v\| < \delta$ that $\|Au - Av\| < 1$.

Then there is an integer n that is independent of $u \in M$ such that one can partition the segment from 0 to u by subdivision points $u_0 = 0$, $u_1, \ldots, u_n = u$ with $\|u_i - u_{i-1}\| < \delta$ for all i; therefore,

$$\|Au - A(0)\| \le \sum_{i=1}^{n} \|Au_i - Au_{i-1}\| < n \qquad \text{for all} \quad u \in M.$$

(III) Show that if the operator A is demicontinuous, then A is locally bounded.

Otherwise, there is a $u \in X$ and a sequence $u_n \to u$ with $\|Au_n\| \to \infty$ as $n \to \infty$. This is a contradiction to the boundedness of (Au_n) that follows from $Au_n \rightharpoonup Au$ as $n \to \infty$.

27.4. *Semimonotone operators are pseudomonotone.* Let X be a real separable reflexive B-space, and let the operator $B: X \times X \to X^*$ be given. We set

$$Au = B(u, u) \qquad \text{for all} \quad u \in X.$$

The operator $A: X \to X^*$ is called semimonotone iff the following hold.

(i) For all $u, v \in X$,

$$\langle B(u,u) - B(u,v), u - v\rangle \geq 0.$$

(ii) For each $u \in X$, the operator

$$v \mapsto B(u, v)$$

is hemicontinuous and bounded from X to X^* and, for each $v \in X$, the operator

$$u \mapsto B(u, v)$$

is hemicontinuous and bounded from X to X^*.

(iii) If $u_n \rightharpoonup u$ in X as $n \to \infty$ and

$$\langle B(u_n, u_n\rangle - B(u_n, u), u_n - u\rangle \to 0 \qquad \text{as} \quad n \to \infty,$$

then $B(u_n, v) \rightharpoonup B(u, v)$ in X^* as $n \to \infty$ for all $v \in X$.

(iv) Let $v \in X$. If $u_n \rightharpoonup u$ in X as $n \to \infty$ and

$$B(u_n, v) \rightharpoonup w \quad \text{in } X^* \qquad \text{as} \quad n \to \infty,$$

then $\langle B(u_n, v), u_n\rangle \to \langle w, u\rangle$ as $n \to \infty$.

(v) $A: X \to X^*$ is bounded.

Show that $A: X \to X^*$ is pseudomonotone.

Hint: Use similar arguments as in the proof of Proposition 27.6. Cf. Lions (1969, M), Section 2.5.

27.5. *Existence theorem for semimonotone operators.* Let $A: X \to X^*$ be semimonotone and coercive (cf. Problem 27.4). Then, for each $b \in X^*$, the equation

$$Au = b, \qquad u \in X,$$

has a solution.

This is an immediate consequence of Problem 27.4 and Theorem 27.B.

27.6. *A general existence theorem for quasi-linear elliptic differential equations of order* 2*m*. We want to generalize Proposition 26.12. As in (26.21), we consider the boundary value problem

$$\text{(E)} \qquad \begin{aligned} &\sum_{|\alpha| \leq m} (-1)^{|\alpha|} D^\alpha A_\alpha(x, Du(x)) = f(x) \quad \text{on } G, \\ &D^\beta u = 0 \quad \text{on } \partial G \qquad \text{for all} \quad \beta: |\beta| \leq m - 1, \end{aligned}$$

where G is a bounded region in $\mathbb{R}^N$, $N, m \geq 1$, and $Du = (D^\alpha u)_{|\alpha| \leq m}$. We set $Du = (du, D_m u)$, where

$$du = (D^\beta u)_{|\beta| \leq m-1} \qquad \text{and} \qquad D_m u = (D^\beta u)_{|\beta| = m}.$$

Let $1 < p < \infty$, $p^{-1} + q^{-1} = 1$. We assume:

(A1) The differential operator of order $2m$ in (E) satisfies the Carathéodory condition (H1), the growth condition (H2), and the coerciveness condition (H4) from Proposition 26.12.

(A2) The highest order terms are *strictly monotone* with respect to the highest order derivatives, i.e.,

$$\sum_{|\alpha|=m} (A_\alpha(x, d, D_m) - A_\alpha(x, d, D'_m))(D^\alpha - D'^\alpha) > 0$$

for all $D_m \neq D'_m$, all d, and almost all $x \in G$.

(A3) The highest order terms are *coercive* with respect to the highest order derivatives, i.e.,

$$\lim_{|D_m| \to \infty} \sup_{d \in \Omega} \sum_{|\alpha|=m} \frac{A_\alpha(x, d, D_m) D^\alpha}{|D_m| + |D_m|^{p-1}} = +\infty$$

for almost all $x \in G$ and all bounded sets Ω.

Show that, for each $f \in L_q(G)$, equation (E) has a generalized solution $u \in \mathring{W}_p^m(G)$ in the sense of Definition 26.11.

Hint: Use the main theorem on semimonotone operators (Problems 27.4 and 27.5) and use similar arguments as in the proof of Proposition 26.12. Cf. Lions (1969, M), Section 2.6.

This theorem dates back to Višik (1961) and Leray and Lions (1965). Further general results can be found in Browder (1970, S). Compare also Chapter 42, where the relation between (E) and variational problems is studied.

27.7. *Quasi-linear elliptic differential equations in unbounded regions.* In this connection, study the work by Hess (1978) and Pascali and Sburlan (1978, M), pp. 299–310. Such problems are characterized by a lack of compactness. For example, the embedding $W_2^1(G) \subseteq L_2(G)$ is *not* compact if $G = \mathbb{R}^N$. This causes the typical difficulties for elliptic differential equations in unbounded domains.

27.8. *Application of the main theorem on locally coercive operators to periodic solutions of general first-order partial differential equations.*

27.8a. *Torus and periodicity.* Let $x = (\xi_1, \ldots, \xi_N)$, $y = (\eta_1, \ldots, \eta_N)$ and $p = (p_1, \ldots, p_N)$, where $0 < p_i < \infty$ for all i. Furthermore, let

$$C = \{x \in \mathbb{R}^N \colon 0 \leq \xi_i \leq p_i \text{ for all } i\}$$

be an N-dimensional cuboid. A function $f\colon \mathbb{R}^N \to \mathbb{R}$ is called *p-periodic* iff

$$f(x + p) = f(x) \qquad \text{for all} \quad x \in \mathbb{R}^N.$$

If we identify the corresponding points on opposite sides of C, then we obtain an N-dimensional torus T^N (cf. Fig. 27.2 for $N = 2$). Each function

$$f\colon T^N \to \mathbb{R}$$

is equivalent to a p-periodic function

$$f\colon \mathbb{R}^N \to \mathbb{R}.$$

By definition, the space $C^k(T^N)$ consists of all p-periodic C^k-functions $f\colon \mathbb{R}^N \to \mathbb{R}$ with the norm

$$\|f\| = \max_{x \in C} |f(x)|.$$

The Sobolev spaces $W_p^k(T^N)$ are defined similarly. For brevity, we set $H^s = W_2^s(T^N)$.

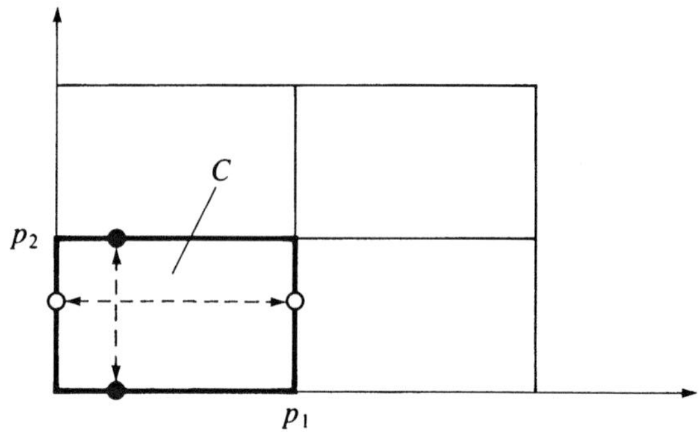

Figure 27.2

27.8b.* *Existence theorem.* We seek a p-periodic solution $u: \mathbb{R}^N \to \mathbb{R}$ of the general first-order partial differential equation

$$a(x, u(x), Du(x)) = f(x) \quad \text{on } \mathbb{R}^N, \tag{31}$$

where $D_i = \partial/\partial\xi_i$ and $D = (D_1, \ldots, D_N)$. We make the following assumptions:
(H1) The function $x \mapsto a(x, u, D)$ is p-periodic on $\mathbb{R}^N$ for all $(u, D) \in \mathbb{R}^{N+1}$ and

$$a \in C^{s+1}(T^N \times \mathbb{R}^{N+1}), \qquad s \geq [N/2] + 3.$$

(H2) $a(x, 0) \equiv 0$. There is a positive constant c such that, for all $x \in C$, $y \in \mathbb{R}^N$,

$$A(x) \stackrel{\text{def}}{=} \frac{\partial a(x,0)}{\partial u} - \frac{1}{2}\sum_{j=1}^{N} \frac{\partial^2 a(x,0)}{\partial x_j \partial D_j} \geq c$$

and

$$A(x)|y|^2 + s \sum_{j,k=1}^{N} \frac{\partial^2 a(x,0)}{\partial x_j \partial D_k} \eta_j \eta_k \geq c|y|^2.$$

(H3) The function $f: \mathbb{R}^N \to \mathbb{R}$ is p-periodic and belongs to H^s.

Show that (31) is *locally uniquely solvable*. To be precise, there are positive numbers R and r so that, for each $f \in H^s$ with $\|f\|_s \leq R$, there exists a solution $u \in H^s$ of (31) with $\|u\|_s \leq r$. Moreover, the solution u depends on f Lipschitz-continuously in H^{s-1}-norm.

Hint: Use Theorem 27.B with $Y \subseteq X = X^+ \subseteq Y^+$ and

$$Y = H^{s+1}, \qquad X = H^s, \qquad Y^+ = H^{s-1}.$$

Let $L = (-\Delta)^{1/2}$. Moreover, we set $(f|g) = \int_C fg\,dx$ and

$$\langle w, v \rangle_Y = (L^{s-1}w | L^{s+1}v) + \lambda^2(w|v),$$

$$\langle w, v \rangle_X = (L^s w | L^s v) + \lambda^2(w|v),$$

where λ is an appropriate real number. Finally, let

$$\|f\|_s^2 = \langle f, f \rangle_X.$$

The space X together with the scalar product $\langle \cdot, \cdot \rangle_X$ becomes an H-space. Note that the smooth functions in H^s can be represented by Fourier series.

This way, it is easy to prove many properties of the spaces H^s via Fourier series, e.g., embedding theorems (cf. Triebel (1972, M), Chapter 6).

We now write the original equation (31) in the form

$$Au = f, \qquad u \in X.$$

Show that the operators $A: X \to Y^+$ and $A: Y \to X$ are weakly sequentially continuous. Moreover, show that, for sufficiently large λ, we obtain the *key* estimate:

$$\langle Av, v \rangle_Y \geq \frac{c}{2} \|v\|_X^2 - d\|v\|_X^3, \tag{32}$$

for all $v \in Y$ with $\|v\|_X \leq 1$. Set $K = \{v \in X: \|v\|_X \leq r\}$. The point is that, for sufficiently small r, relation (32) implies the *local coerciveness* of A, i.e.,

$$\langle Av, v \rangle_Y \geq \frac{c}{2} r^2 - dr^3 \geq \alpha > 0,$$

for all $v \in Y \cap \partial K$.

Cf. Kato (1984). A variant of this theorem was first proved by Moser (1966) via the technique of the hard implicit function theorem.

27.8c. *Generalization.* The preceding result can be generalized to first-order partial differential equations of the form (31) on finite-dimensional, compact, proper Riemannian manifolds. It is also possible to show that the solution u of (31) depends continuously on f in the H^s-norm. This can be found in Günther (1988a) (cf. also Chapter 85).

References to the Literature

Classical papers: Višik (1963), Leray and Lions (1965), Brézis (1968).

Pseudomonotone operators and their generalizations: Browder and Hess (1972), Pascali and Sburlan (1978, M), Kluge (1979, M).

Applications to quasi-linear elliptic differential equations: Morrey (1966, M), Lions (1969, M), Browder (1970, S, H), (1986, P), Skrypnik (1973, M), (1986, M), Dubinskii (1976, S, B), Berger (1977, M), Ivanov (1982, S).

Weighted Sobolev spaces and quasi-linear elliptic differential equations: Kufner and Sändig (1987, M).

Sobolev spaces of infinite order and partial differential equations: Dubinskii (1984, M).

Regularity of solutions of nonlinear elliptic equations: Ladyženskaja and Uralceva (1964, M), Morrey (1966, M), Skrypnik (1973, M), (1976, S), (1986, M), Giaquinta (1981, M), Nečas (1983, M), Koshelev (1985, L), (1986, M), Giaquinta and Hildebrandt (1989, M).

Singularities of the solutions of quasi-linear elliptic differential equations: Serrin (1964), (1965), Ni and Serrin (1985), Kichenassamy (1987).

Linear elliptic equations in nonsmooth domains: Kondratjev (1967), Kondratjev and Oleinik (1983, S), Grisvard (1985, M), Kufner and Sändig (1987, M), Maslennikova (1988).

Quasi-linear elliptic equations in nonsmooth domains and asymptotic expansions near conical points: Miersemann (1982), (1988), (1988a) (applications to the capillary surface problem).

Fully nonlinear equations and the Monge–Ampère equation: Gilbarg and Trudinger (1983, M) (recommended as an introduction), Aubin (1982, M), Evans (1982), (1983), Lions (1983a, S), (1985), Caffarelli, Nirenberg, and Spruck (1984/88), Lieberman and Trudinger (1986).

Continuity properties of operators: Vainberg (1956, M), (1972, M).

Main theorem on locally coercive operators: Hess (1973), Kato (1984).

Difference method and finite element method for quasi-linear elliptic differential equations: cf. the References to the Literature for Chapter 35.

CHAPTER 28

Monotone Operators and Hammerstein Integral Equations

Around 1900, Fredholm was the first to study systematically linear integral equations, by solving linear systems of equations and by passing to the limit. In this paper, we investigate nonlinear integral equations by solving nonlinear systems of equations via a variational method and by passing to the limit.

Adolf Hammerstein (1930)

The first application of the concept of monotone operator equations to Hammerstein integral equations was made implicitly by Golomb (1935) and explicitly by Vainberg (1956).... Our method consists of splitting the linear kernel operator K via a Hilbert space H and reducing the original equation

$$u + KFu = 0, \qquad u \in X^*,$$

with the splitting $K = S^*CS$, to the equivalent equation

$$C^{-1}w + SFS^*w = 0, \qquad w \in H,$$

with $w = (S^*)^{-1}u$, which is then solved by using the results of Browder (1963) and Minty (1963) for monotone operator equations.

Felix E. Browder and Chaitan Gupta (1969)

In this chapter we investigate the abstract Hammerstein equation

$$u + KFu = 0, \qquad u \in X^*, \tag{1}$$

with the nonlinear operator $F: X^* \to X$ and the linear operator $K: X \to X^*$, with the aid of the theory of monotone operators and the fixed-point index. The applications relate to Hammerstein integral equations and boundary value problems for semilinear elliptic partial differential equations. Whereas, in Chapter 7, we made use of B-spaces of smooth functions and monotone increasing operators in ordered B-spaces, we now work in $L_p(G)$-spaces and apply the theory of monotone operators.

A *Hammerstein integral equation* has the form

$$u(x) + \int_G k(x, y) f(y, u(y))\, dy = 0. \tag{2}$$

Equation (1) follows from this in the case where we define the so-called *kernel operator* K by

$$(Kw)(x) = \int_G k(x, y) w(y)\, dy, \tag{2*}$$

and where F is the *Nemyckii operator* for f, i.e.,

$$(Fu)(y) = f(y, u(y)).$$

In this connection, the kernel $k(\cdot, \cdot)$ can be regular, weakly singular, or singular. On application of $L_q(G)$-spaces, the function $u \mapsto f(y, u)$ must not increase more rapidly than certain polynomials. If one wants to allow stronger growth, then one must work with either smooth kernels and B-spaces of smooth functions (cf. Chapter 7), or one must employ Orlicz spaces for nonsmooth kernels (cf. Chapter 53). Nonhomogeneous Hammerstein integral equations

$$v(x) + \int_G k(x, y) g(y, v(y))\, dy = b(x)$$

can be reduced to the homogeneous equation (2) if one sets

$$u(x) = v(x) - b(x) \qquad \text{and} \qquad f(y, u(y)) = g(y, u(y) + b(y)).$$

Analogously, one can reduce the nonhomogeneous operator equation $v + KGv = b$ to the homogeneous equation (1). Hence it is sufficient to consider homogeneous problems.

Boundary value problems for *semilinear elliptic equations* also lead to equations of type (1). For example, we consider the boundary value problem

$$\begin{aligned} -\Delta u(x) &= -f(x, u(x)) \quad \text{on } G, \\ u &= 0 \quad \text{on } \partial G. \end{aligned} \tag{3}$$

If we define the operator K by $u = Kg$, where u is a solution of the corresponding linear boundary value problem

$$\begin{aligned} -\Delta u(x) &= g(x) \quad \text{on } G, \\ u &= 0 \quad \text{on } \partial G, \end{aligned} \tag{3*}$$

and if F is the Nemyckii operator for f, then there results from (3) the operator equation

$$u = -KFu,$$

which coincides with our original equation (1). Here K is called the solution operator for (3*).

If the boundary ∂G and the function g are sufficiently smooth, then one can represent K by means of the classical Green function $k(\cdot, \cdot)$ in the form (2*) above. Then the Hammerstein integral equation (2) results from (3).

The approach used in this chapter, concerning the *abstract* solution operator K that also comprises generalized solutions of (3*), has the advantage that it can be applied to problems with *nonsmooth* boundary ∂G and *nonsmooth* right member f. In this connection, we do *not* need any sophisticated investigations concerning the properties of the Green function in the nonsmooth case.

For the original operator equation (1) we investigate the following special cases:

(a) K is monotone and angle-bounded, F is monotone and hemicontinuous (Theorem 28.A).
(b) K is monotone and compact, F is demicontinuous and bounded (Theorem 28.B).
(c) K is monotone, F is pseudomonotone, bounded, and coercive (Theorem 32.B).
(d) K is symmetric (variational method in Chapter 41).

Roughly speaking, the following holds:

The requirements on the Nemyckii operator F are weaker, to the extent that the assumptions on the kernel operator K are stronger, and conversely.

In the concrete cases (2) and (3) above, the monotonicity of F means that the real function

$$u \mapsto f(y, u)$$

is monotone increasing for each fixed $y \in G$. The monotonicity of K corresponds to a certain definiteness property of the kernel $k(\cdot, \cdot)$. Moreover, we consider for equation (1) the convergence of the Galerkin method. In Section 7.13, we have already investigated the convergence of the iteration method for equation (1).

The *basic idea* of all proofs of existence for (1) consists in that, in contrast to (1), one studies the equivalent equations (4), (5), and (6) below. We sketch briefly the arguments used in this connection.

First equivalence principle. If we denote by K^{-1} the, in general, multivalued inverse operator to K, then equation (1) is equivalent to $-Fu \in K^{-1}u$, therefore also equivalent to

$$0 \in K^{-1}u + Fu. \tag{4}$$

We make use of this modified equation in (b) and (c). In case (b), the operator KF is compact. Therefore, we can use the fixed-point index. In fact, let U be an open ball around the origin. Using (4), we give conditions in Theorem 28.B guaranteeing that

$$-tKFu \neq u \qquad \text{for all} \quad (u, t) \in \partial U \times [0, 1].$$

Then, on the basis of the homotopy $H(u,t) = -tKFu$, we obtain for the fixed-point index:

$$i(-KF, U) = i(H(\cdot,0), U) = 1.$$

Therefore, the operator $-KF$ has a fixed point on U, i.e., equation (1) is solvable.

In case (c), the operator K^{-1} is maximal monotone. The solvability of equation (4), and thus of (1), then results from the main theorem on maximal monotone operators (Theorem 32.A).

Second equivalence principle. We consider the case (a). If, first of all, the linear operator K is monotone and *symmetric* on the real separable H-space X, then K is continuous and there exists a symmetric continuous *square root*

$$S = K^{1/2}$$

on X, i.e., $K = S^2$ and $S = S^{*\prime}$. If K^{-1} exists as a single-valued operator, then S^{-1} also exists. Consequently, equation (1), with $X = X^*$, is equivalent to

$$w + SFSw = 0, \qquad w \in X, \tag{5}$$

where $w = S^{-1}u$. Let $A = I + SFS$. If F is hemicontinuous and monotone, then $(SFSw|v) = (FSw|Sv)$ and, for all $w, z \in X$,

$$(Aw - Az|w - z) = (w - z|w - z) + (FSw - FSz|Sw - Sz) \geq (w - z|w - z).$$

Consequently, A is hemicontinuous and strongly monotone on X. The main theorem on monotone operators (Theorem 26.A) yields a solution of the equation $Aw = 0$, $w \in X$, i.e., (5) has a solution w. Thus, $u = Sw$ is a solution of the original equation (1).

More generally, if $K: X \to X^*$ is an *angle-bounded* operator on the real B-space X, then we make use of the decomposition

$$K = S^*CS$$

that is more general in contrast to $K = S^2$ with appropriate operators S and C (cf. the factorization theorem in Section 28.1). Now, equation (1) is equivalent to

$$C^{-1}w + SFS^*w = 0, \qquad w \in H, \tag{6}$$

with $w = (S^*)^{-1}u$. Here H is an appropriate H-space. Similarly, as in (5), one can apply to (6) the main theorem on monotone operators. The modified equation (6) also results in the case where K^{-1} does *not* exist as a single-valued operator.

Angle-bounded operators generalize symmetric operators. If K is symmetric, then $C = I$. The significance of angle-bounded operators consists in that the solution operators of certain strongly elliptic differential equations are angle-bounded. Therefore, boundary value problems for semilinear equations of type (3), with a strongly elliptic linear differential operator Lu instead of $-\Delta u$, can be reduced to the original operator equation (1) with angle-bounded K.

In Chapter 41 (resp. Chapter 44) of Part III we deal with the equation

$u + KFu = 0$ (resp. the eigenvalue problem $u + \lambda KFu = 0$) for symmetric K with variational methods.

28.1. A Factorization Theorem for Angle-Bounded Operators

In order to prepare for Theorem 28.A we construct for the linear operator K a decomposition of the form

$$K = S^*CS \tag{7}$$

with the following commutative diagram:

$$\begin{array}{ccc} X & \xrightarrow{K} & X^* \\ {\scriptstyle S}\downarrow & & \uparrow{\scriptstyle S^*}\ \text{injective} \\ H & \underset{\text{bijective}}{\xrightarrow{C}} & H = H^* \end{array} \tag{8}$$

Let I denote the identity operator on the H-space H. We make the following assumptions:

(H1) The operator $K: X \to X^*$ is linear and monotone on the real B-space X.
(H2) The operator K is angle-bounded, i.e., there is a number $a \geq 0$ such that

$$|\langle Ku, v\rangle - \langle Kv, u\rangle|^2 \leq 4a^2\langle Ku, u\rangle\langle Kv, v\rangle \qquad \text{for all} \quad u, v \in X. \tag{9}$$

Proposition 28.1 (Browder and Gupta (1969)). *Suppose that* (H1) *and* (H2) *hold. Then*:

(a) Decomposition. *There exist a real* H-*space* H *with scalar product* $(\cdot|\cdot)$ *and operators* S, C *such that* (7) *holds and the diagram* (8) *is commutative. To be precise, the following hold*:

$$S: X \to H \text{ is linear and continuous with } \|S\|^2 \leq \|K\|. \text{ The range } S(X) \text{ is dense in } H. \tag{10}$$

$$C: H \to H \text{ is linear, continuous, and bijective.} \tag{11}$$

$$C - I \text{ is skew-adjoint, i.e., } (C - I)^{*\prime} = -(C - I) \text{ and } \|C - I\| \leq a. \tag{12}$$

C and C^{-1} are strongly monotone. To be precise, for all $h \in H$,

$$(C^{-1}h|h) \geq (1 + a^2)^{-1}(h|h), \qquad (Ch|h) = (h|h). \tag{13}$$

$$(Su|Sv) = 2^{-1}(\langle Ku, v\rangle + \langle Kv, u\rangle) \text{ for all } u, v \in X. \tag{14}$$

$$((C - I)Su|Sv) = 2^{-1}(\langle Ku, v\rangle - \langle Kv, u\rangle) \text{ for all } u, v \in X. \tag{15}$$

$$(h|Su) = \langle S^*h, u\rangle_X \text{ for all } u \in X, h \in H. \tag{16}$$

(b) *If* $K \neq 0$, *then*

$$\langle Ku, u\rangle \geq (1 + a^2)^{-1} \|K\|^{-1} \|Ku\|_{X^*}^2 \qquad \text{for all} \quad u \in X.$$

(c) *If* $K: X \to X^*$ *is strongly continuous, then* $S: X \to H$ *is strongly continuous.* *If* $K: X \to X^*$ *is compact on the reflexive* B-*space* X, *then* $S: X \to H$ *is compact.*

(d) *If* K *is symmetric, then* $C = I$.

The proof will be given in Problem 28.1. The simple idea of proof consists in constructing the H-space H with the aid of the bilinear form

$$[u, v]_+ = 2^{-1}(\langle Ku, v\rangle + \langle Kv, u\rangle) \qquad \text{for all} \quad u, v \in X.$$

The idea for the construction of the operators S and C is contained in (14) and (15).

28.2. Abstract Hammerstein Equations with Angle-Bounded Kernel Operators

We study the operator equation

$$u + KFu = 0, \qquad u \in X^*, \tag{17}$$

with the operators $F: X^* \to X$ and $K: X \to X^*$. To construct the approximate solution u_n we use the Galerkin method

$$\langle u_n + KFu_n, w_i\rangle_X = 0, \qquad i = 1, \ldots, n, \tag{18}$$

with

$$u_n = \sum_{j=1}^{n} c_j K^* w_j.$$

Letting $c = (c_1, \ldots, c_n)$ and $g_i(c) = \langle u_n + KFu_n, w_i\rangle_X$, the Galerkin equation (18) is equivalent to the nonlinear real system

$$g_i(c) = 0, \qquad c \in \mathbb{R}^n, \qquad i = 1, \ldots, n.$$

For an approximate determination of a solution c we use the iteration method

$$c^{(k+1)} = c^{(k)} - tg(c^{(k)}), \qquad k = 0, 1, \ldots, \tag{19}$$

with $c^{(0)} = 0$. Here t is a sufficiently small positive parameter. We make the following assumptions:

(H1) The operator $K: X \to X^*$ is linear, monotone, and angle-bounded on the real separable B-space X.

(H2) The operator $F: X^* \to X$ is monotone and hemicontinuous. In this connection, X is identified with a subset of X^{**}.

(H3) Let $\{w_1, w_2, \ldots\}$ be a basis in X and set $X_n = \text{span}\{w_1, \ldots, w_n\}$.

Theorem 28.A (Amann (1969a)). *Suppose that* (H1) *through* (H3) *hold. Then*:

(a) Existence and uniqueness. *The abstract Hammerstein equation* (17) *has exactly one solution.*
(b) Galerkin method. *If* $\dim X = \infty$, *then the Galerkin equation* (18) *has exactly one solution* $u_n \in X_n$ *for each* $n \in \mathbb{N}$. *As* $n \to \infty$, *the sequence* (u_n) *converges in the norm topology of* X^* *to the solution* u *of equation* (17).
(c) Iteration method. *Suppose that the real functions* $g_1, \ldots, g_n$ *are continuously differentiable on* $\mathbb{R}^n$ *or at least locally Lipschitz continuous and*

$$\det(\langle Kw_i, w_j \rangle_X) \neq 0, \tag{20}$$

where $i, j = 1, \ldots, n$. *Then, for sufficiently small and fixed parameter* $t > 0$, *the sequence* $(c^{(k)})$ *in* (19) *converges as* $k \to \infty$ *to the solution of the Galerkin equation* (18).

Note that, according to Theorem 26.B, the value of t can be given explicitly.

Remark 28.2 (Interpretation of $X \subseteq X^{**}$). In order for the monotonicity of $F\colon X^* \to X$ to be meaningfully defined, one must think of F as an operator of the form

$$F\colon X^* \to X^{**}.$$

We achieve that by $X \subseteq X^{**}$. As usual, this inclusion is to be understood as follows. To each element $x \in X$, one can assign a linear continuous functional $\bar{x}$ on X^* by means of

$$\bar{x}(x^*) = x^*(x) \qquad \text{for all} \quad x^* \in X^*,$$

i.e., $\bar{x} \in X^{**}$. The mapping $\varphi\colon X \to X^{**}$ with $\varphi(x) = \bar{x}$ is injective and norm preserving. We identify x with $\bar{x}$. In this sense, $X \subseteq X^{**}$ and

$$\langle x, x^* \rangle_{X^*} = \langle x^*, x \rangle_X \qquad \text{for all} \quad x \in X, \quad x^* \in X^*. \tag{21}$$

In the following proof of Theorem 28.A, we use Proposition 28.1 (splitting of K) and Theorem 26.A (main theorem on monotone operators).

The reader should always have in mind the splitting

$$K = S^*CS$$

and the basic diagram (8). This diagram contains the operators $S\colon X \to H$, $C\colon H \to H^*$, $S^*\colon H^* \to X^*$, and $K\colon X \to X^*$. Hence the corresponding dual operators have the form $C^*\colon H^{**} \to H^*$,

$$S^{**}\colon X^{**} \to H^{**} \qquad \text{and} \qquad K^*\colon X^{**} \to X^* \qquad \text{where} \quad K^* = S^*C^*S^{**}.$$

We identify the H-space H with its dual space H^*, i.e., we set

$$H^* = H, \qquad H^{**} = H.$$

Then the dual operator $C^*\colon H^{**} \to H^*$ coincides with the adjoint operator

$C^{*'}: H \to H$, i.e., we set $C^* = C^{*'}$ in Proposition 28.1. Let $(\cdot|\cdot)$ denote the scalar product on H.

From (16) and (21) we immediately obtain the *first* important formula

$$(Sv|h) = \langle v, S^*h \rangle_{X^*} \qquad \text{for all} \quad v \in X, \quad h \in H. \tag{22}$$

By $H = H^{**}$ and $X \subseteq X^{**}$,

$$(S^{**}v|h) = \langle v, S^*h \rangle_{X^*} \qquad \text{for all} \quad v \in X, \quad h \in H.$$

From (22) it follows that

$$(S^{**}v|h) = (Sv|h) \qquad \text{for all} \quad v \in X, \quad h \in H,$$

that is,

$$S^{**}v = Sv \qquad \text{for all} \quad v \in X.$$

Since $K^* = S^*C^*S^{**}$, we obtain the *second* important formula

$$K^*v = S^*C^*Sv \qquad \text{for all} \quad v \in X. \tag{23}$$

PROOF OF THEOREM 28.A(a).

(I) Equivalent equation. By Proposition 28.1, $K = S^*CS$. Consequently, the original equation (17), i.e., $u + KFu = 0$, $u \in X^*$, is equivalent to the equation

$$C^{-1}w + SFS^*w = 0, \qquad w \in H, \tag{24}$$

with $w = (S^*)^{-1}u$. We set

$$Aw = C^{-1}w + SFS^*w.$$

(II) We show that the operator $A: H \to H$ is hemicontinuous. For all h, $w \in H$, it follows from (22) that

$$(Aw|h) = (C^{-1}w|h) + \langle FS^*w, S^*h \rangle_{X^*}.$$

Let $w = u + tv$ with $u, v \in H$. Then the function $t \mapsto (Aw|h)$ is continuous on $[0, 1]$, since by assumption the operator F is hemicontinuous. Hence A is hemicontinuous.

(III) We show that the operator $A: H \to H$ is strongly monotone. Let $w, z \in H$. By (22),

$$(Aw - Az|w - z) = (C^{-1}(w - z)|w - z) + \langle FS^*w - FS^*z, S^*w - S^*z \rangle_{X^*}.$$

The monotonicity of F and (13) imply

$$(Aw - Az|w - z) \geq (C^{-1}(w - z)|w - z) \geq (1 + a^2)^{-1} \|w - z\|_H^2,$$

i.e., A is strongly monotone.

(IV) The operator $A: H \to H$ is coercive, since it is strongly monotone.

Theorem 26.A in Section 26.2 yields that the operator $A: H \to H$ is bijective. In particular, there exists exactly one $w \in H$ with $Aw = 0$, i.e., equation (24) has exactly one solution. Consequently, the original equation (17) has exactly one solution. □

PROOF OF THEOREM 28.A(b).

(I) We construct a basis in H. Let

$$H_n = S^{*-1}K^*(X_n).$$

From $X_n \subseteq X_{n+1}$ it follows that $H_n \subseteq H_{n+1} \subseteq H$. We shall show that

$$(u|h) = 0 \qquad \text{for all} \quad h \in \bigcup_n H_n \qquad \text{implies} \qquad u = 0. \tag{25}$$

Consequently, the set $\bigcup_n H_n$ is dense in H. If one chooses suitable basis elements in H_n, then there arises a basis for the H-space H.

In order to prove (25), assume that $(u|h) = 0$ for all $h \in \bigcup_n H_n$. Let $v \in \bigcup_n X_n$. Then

$$(u|S^{*-1}K^*v) = 0.$$

By (23),

$$(Cu|Sv) = (u|C^*Sv) = (u|S^{*-1}K^*v) = 0.$$

Since, by assumption, the set $\bigcup_n X_n$ is dense in X and the operator $S: X \to H$ is continuous, we obtain

$$(Cu|Sv) = 0 \qquad \text{for all} \quad v \in X.$$

Moreover, by Proposition 28.1, the range $S(X)$ is dense in H, therefore, $Cu = 0$, i.e., $u = 0$.

(II) Galerkin equation for the operator A. We consider equation (24), i.e., $Aw = 0$, $w \in H$. Then the corresponding Galerkin equation is given by

$$(Aw_n|h) = 0 \qquad \text{for all} \quad h \in H_n. \tag{26}$$

We seek $w_n \in H_n$. According to Theorem 26.A, equation (26) has exactly one solution $w_n \in H_n$ and

$$w_n \to w \quad \text{in } H \qquad \text{as} \quad n \to \infty,$$

where $Aw = 0$. We set

$$u_n = S^*w_n$$

and $u = S^*w$. Then we obtain $u + KFu = 0$, $u \in X^*$, and the continuity of the operator S^* implies

$$u_n \to u \quad \text{in } X^* \qquad \text{as} \quad n \to \infty.$$

(III) Galerkin equation of the original problem. In order to finish the convergence proof for the Galerkin method, we show that u_n is a solution of the original Galerkin equation (18), i.e.,

$$\langle u_n + KFu_n, v\rangle_X = 0 \qquad \text{for all} \quad v \in X_n. \tag{26*}$$

Here we seek $u_n \in K^*(X_n)$. More precisely, we show the equivalence of (26) and (26*).

In fact, by (22), we have

$$(Aw_n|h) = (C^{-1}w_n|h) + \langle FS^*w_n, S^*h\rangle_{X^*} \qquad \text{for all} \quad h \in H_n.$$

By (23),

$$K^*v = S^*C^*Sv \qquad \text{for all} \quad v \in X_n.$$

Let $v \in X_n$ and consider

$$h = S^{*-1}K^*v,$$

i.e., $h = C^*Sv$. From (22) it follows that

$$\begin{aligned}(C^{-1}w_n|h) &= (C^{-1}S^{*-1}u_n|h) = (C^{-1}S^{*-1}u_n|C^*Sv) = (S^{*-1}u_n|Sv)\\ &= \langle v, u_n\rangle_{X^*} = \langle u_n, v\rangle_X\end{aligned}$$

and

$$\begin{aligned}\langle FS^*w_n, S^*h\rangle_{X^*} &= \langle Fu_n, S^*C^*Sv\rangle_{X^*}\\ &= (SFu_n|C^*Sv) = (CSFu_n|Sv)\\ &= \langle v, S^*CSFu_n\rangle_{X^*} = \langle S^*CSFu_n, v\rangle_X = \langle KFu_n, v\rangle_X.\end{aligned}$$

In this connection, note that $X_n \subseteq X \subseteq X^{**}$ and use (21). This implies

$$(Aw_n|h) = \langle u_n + KFu_n, v\rangle_X.$$

From $H_n = S^{*-1}K^*(X_n)$ and $u_n = S^*w_n$ with $w_n \in H_n$ it follows that

$$u_n \in K^*(X_n).$$

Hence (26) is equivalent to (26*). □

Proof of Theorem 28.A(c). We set

$$v = \sum_{j=1}^{n} c_j w_j, \qquad u = K^*v,$$

and

$$g_j(c) = \langle u + KFu, w_j\rangle_X.$$

We want to apply Proposition 26.8 to the Galerkin equation

$$g(c) = 0, \qquad c \in \mathbb{R}^n.$$

To this end, we need the *key* condition

$$\sum_{j=1}^{n} (g_j(c) - g_j(\bar{c}))(c_j - \bar{c}_j) \geq \gamma \|c - \bar{c}\|^2$$

for all $c, \bar{c} \in \mathbb{R}^n$ and fixed $\gamma > 0$. To this end, we set

$$a(c, \bar{c}) = \langle u, \bar{v}\rangle_X.$$

By (21), $\langle Ku, v\rangle_X = \langle v, Ku\rangle_{X^*}$ for all $u, v \in X$. From this it follows that

$$a(c,\bar{c}) = \langle K^*v, \bar{v}\rangle_X = \langle v, K\bar{v}\rangle_{X^*}$$

$$= \langle K\bar{v}, v\rangle_X = \sum_{i,j=1}^{n} \langle Kw_i, w_j\rangle_X \bar{c}_i c_j,$$

and

$$a(c,c) = \langle Kv, v\rangle_X \geq 0,$$

since K is monotone.

Let $a(c,c) = 0$. Then $\langle Kv, v\rangle = 0$. By Problem 28.2b, this implies $Kv = 0$, i.e.,

$$\langle Kv, w_j\rangle = \sum_{i=1}^{n} \langle Kw_i, w_j\rangle_X c_i = 0$$

for $j = 1, \ldots, n$. Hence we obtain $c = 0$ since $\det(\langle Kw_i, w_j\rangle_X) \neq 0$. Consequently, the bilinear functional $a(\cdot,\cdot)$ is positive definite, i.e., there is a $\gamma > 0$ such that

$$a(c,c) \geq \gamma \|c\|^2 \qquad \text{for all} \quad c \in \mathbb{R}^n.$$

Now the key condition above follows from the monotonicity of F. In fact, we obtain that

$$\sum_{j=1}^{n} (g_j(c) - g_j(\bar{c}))(c_j - \bar{c}_j) = \langle u - \bar{u}, v - \bar{v}\rangle_X + \langle KFu - KF\bar{u}, v - \bar{v}\rangle_X$$

$$= \langle u - \bar{u}, v - \bar{v}\rangle_X + \langle Fu - F\bar{u}, u - \bar{u}\rangle_{X^*}$$

$$\geq \langle u - \bar{u}, v - \bar{v}\rangle_X$$

$$= a(c - \bar{c}, c - \bar{c})$$

$$\geq \gamma \|c - \bar{c}\|^2.$$

Now Proposition 26.8 yields the convergence of the iteration method $c^{(k+1)} = c^{(k)} - tg(c^{(k)})$ for the equation $g(c) = 0$. □

28.3. Abstract Hammerstein Equations with Compact Kernel Operators

We consider again the abstract Hammerstein equation

$$u + KFu = 0, \qquad u \in X^*, \tag{27}$$

with the operators $F: X^* \to X$ and $K: X \to X^*$. In contrast to Theorem 28.A, we now require the compactness of K. However, we can weaken the assumptions on F, in particular, we do *not* require F to be monotone. In this connection, we formulate the following two conditions:

(C1) K is monotone, i.e.,

$$\langle Kw, w\rangle_X \geq 0 \qquad \text{for all} \quad w \in X,$$

and there exists a number $R > 0$ such that

$$\langle u, Fu\rangle_X > 0, \qquad \text{for all} \quad u \in X^* \quad \text{with} \quad \|u\|_{X^*} = R.$$

(C2) There exists a number $\gamma > 0$ such that

$$\langle Kw, w\rangle_X \geq \gamma \|Kw\|_{X^*}^2 \qquad \text{for all} \quad w \in X.$$

Moreover, there exists a function $\varphi\colon \mathbb{R}_+ \to \mathbb{R}$ and a number $R > 0$ such that $\varphi(R)/R^2 < \gamma$ and

$$\langle u, Fu\rangle_X \geq -\varphi(\|u\|) \qquad \text{for all} \quad u \in X^*.$$

By Proposition 28.1(b), the requirement on K in (C2) is fulfilled in the case where K is angle-bounded.

Theorem 28.B (Hess (1971) and Amann (1972)). *The abstract Hammerstein equation* (27) *has a solution in the case where the following three assumptions are fulfilled*:

(i) *The operator* $K\colon X \to X^*$ *is linear and compact on the* B-*space* X.
(ii) *The operator* $F\colon X^* \to X$ *is demicontinuous and bounded.*
(iii) *Either* (C1) *or* (C2) *is valid.*

PROOF. We will use the properties (A1), (A2), and (A4) of the fixed-point index in Section 12.3. We set

$$U = \{u \in X^*\colon \|u\|_{X^*} < R\}$$

for fixed $R > 0$. Below we shall show that the operator $KF\colon X^* \to X^*$ is *compact* and that

$$-tKFu \neq u \qquad \text{for all} \quad (u, t) \in \partial U \times [0, 1]. \tag{28}$$

Then the assertion follows from standard arguments. In fact, letting $H(u, t) = -tKFu$, the homotopy invariance (A4) of the fixed-point index yields

$$i(H(\cdot, 0), U) = i(H(\cdot, 1), U).$$

Since $H(u, 0) \equiv 0$ and $0 \in U$, we obtain from (A1) that

$$i(H(\cdot, 0), U) = 1,$$

and $H(u, 1) = -KFu$ implies

$$i(-KF, U) = 1.$$

Now the existence principle (A2) yields the existence of a fixed point $u \in U$ of the operator $-KF$. Therefore, the equation $KFu + u = 0$, $u \in U$, has a solution.

(I) We show that the operator $KF: X^* \to X^*$ is continuous. From

$$u_n \to u \quad \text{in } X^* \qquad \text{as} \quad n \to \infty,$$

it follows that

$$Fu_n \rightharpoonup Fu \quad \text{in } X \qquad \text{as} \quad n \to \infty,$$

since F is demicontinuous. The operator $K: X \to X^*$ is linear and compact and hence strongly continuous. This implies

$$KFu_n \to KFu \quad \text{in } X^* \qquad \text{as} \quad n \to \infty.$$

(II) We show that $KF: X^* \to X^*$ is compact. Let (u_n) be a bounded sequence in X^*; then (Fu_n) is bounded in X, since F is bounded. Because of the compactness of K there exists a subsequence $(u_{n'})$ such that $(KFu_{n'})$ converges in X^*.

(III) We prove (28). If (28) does not hold, then there exists a point $u \in \partial U$ and a $t \in [0, 1]$ with $-tKFu = u$. Because $u \neq 0$, we have $t > 0$ and there exists a $w \in K^{-1}(u)$ with

$$w + tFu = 0. \tag{29}$$

In the case (C1), from $\langle Kw, w\rangle_X \geq 0$ and $Kw = u$, it follows that $\langle u, w\rangle_X \geq 0$ and hence

$$\langle u, w\rangle_X + t\langle u, Fu\rangle_X > 0.$$

This contradicts (29).

In the case (C2), we obtain that

$$\begin{aligned}\langle u, w\rangle_X + t\langle u, Fu\rangle_X &\geq \gamma\|u\|^2 - \varphi(\|u\|)\\ &= R^2(\gamma - \varphi(R)/R^2) > 0.\end{aligned}$$

This also contradicts (29). □

28.4. Application to Hammerstein Integral Equations

We consider the Hammerstein integral equation

$$u(x) + \int_G k(x, y)f(y, u(y))\,dy = 0, \qquad u \in L_p(G). \tag{30}$$

Our goal is to obtain an overview of a series of different conditions on the kernel $k(x, y)$ and the nonlinearity $f(y, u)$ that guarantee the existence of solutions for (30). We set

$$X = L_q(G), \qquad 1 < q < \infty.$$

We then have

$$X^* = L_p(G), \qquad p^{-1} + q^{-1} = 1.$$

In order to be able to apply our abstract results, we write (30) in the form

$$u + KFu = 0, \qquad u \in X^*. \tag{31}$$

In this connection, the linear operator $K: X \to X^*$ is generated by the kernel $k(\cdot, \cdot)$, i.e.,

$$(Kv)(x) = \int_G k(x, y)v(y)\,dy.$$

The operator F is the Nemyckii operator for f, i.e.,

$$(Fu)(y) = f(y, u(y)).$$

We formulate the following four basic assumptions:

(H1) G is a bounded region in $\mathbb{R}^N$ with $N \geq 1$, and $1 < p, q < \infty, p^{-1} + q^{-1} = 1$.

(H2) *Linearity and monotonicity of* K. The kernel $k: G \times G \to \mathbb{R}$ is so constituted that the operator $K: X \to X^*$ is linear, i.e., $Kv \in X^*$ for all $v \in X$. This implies

$$\langle Kv, w\rangle_X = \int_G \left[\int_G k(x, y)v(y)\,dy\right] w(x)\,dx \qquad \text{for all} \quad v, w \in X.$$

Moreover, let K be monotone, i.e.,

$$\langle Kv, v\rangle_X \geq 0 \qquad \text{for all} \quad v \in X.$$

(H3) *Growth condition for* f. The function $f: G \times \mathbb{R} \to \mathbb{R}$ satisfies the Carathéodory condition (e.g., f is continuous) and there is a nonnegative function $a \in L_q(G)$ and a number $b \geq 0$ such that the growth condition

$$|f(y, u)| \leq a(y) + b|u|^{p-1}$$

holds for all $(y, u) \in G \times \mathbb{R}$.

(H4) The functions $\{w_1, w_2, \ldots\}$ form a basis in $X = L_q(G)$.

In addition, we will make use of one of the following two conditions:

(H5) *Coerciveness of* f. There is a number $d > 0$ such that

$$f(y, u)u \geq d|u|^p \qquad \text{for all} \quad (y, u) \in G \times \mathbb{R}.$$

(H5*) *Asymptotic positivity of* f. There is a number $R > 0$ such that

$$f(y, u)u \geq 0$$

holds for all $(y, u) \in G \times \mathbb{R}$ with $|u| \geq R$.

We say that f is *monotone* (resp. strictly monotone) with respect to u iff the function

$$u \mapsto f(y, u)$$

is monotone increasing (resp. strictly monotone increasing) on $\mathbb{R}$ for all $y \in G$.

Recall that $K: X \to X^*$ is *strictly monotone* iff

$$\langle Kv, v\rangle_X > 0 \qquad \text{for all nonzero} \quad v \in X.$$

Moreover, the linear monotone operator K is *angle-bounded* iff there exists a number $a \geq 0$ such that

$$|\langle Kv, w\rangle_X - \langle Kw, v\rangle_X| \leq 4a^2\langle Kv, v\rangle_X\langle Kw, w\rangle_X \qquad \text{for all} \quad v, w \in X.$$

In particular, this condition is satisfied in the case where K is symmetric, i.e.,

$$\langle Kv, w\rangle_X = \langle Kw, v\rangle_X \qquad \text{for all} \quad v, w \in X. \tag{32}$$

Standard Example 28.3 (Operator K). Suppose that (H1) holds and suppose that the kernel

$$k\colon \overline{G \times G} \to \mathbb{R}$$

is continuous. Let $X = L_q(G)$ and $X^* = L_p(G)$. Then:

(i) The operator $K\colon X \to X^*$ is linear and compact.
(ii) The dual operator $K^*\colon X \to X^*$ is given by

$$(K^*v)(x) = \int_G k(y,x)v(y)\,dy \qquad \text{for all} \quad v \in X.$$

(iii) Let $k(x,y) = k(y,x)$ for all $x, y \in G$. Then $K\colon X \to X^*$ is symmetric. If K is monotone, then K is also angle-bounded.

Let $p = 2$, i.e., $X = X^* = L_2(G)$. Then the compact symmetric operator $K\colon X \to X$ has a complete orthonormal system of eigenvectors on the H-space X. If all the eigenvalues of K are nonnegative (resp. positive), then K is monotone (resp. strictly monotone).

In Corollaries 28.5 and 28.6 below we will generalize this well-known result to regular kernels $k \in L_p(G \times G)$ and to weakly singular kernels.

In order to compute approximate solutions u_n of the Hammerstein integral equation (30), we apply the Galerkin method. The Galerkin equation (18) corresponding to the operator equation (31) reads as follows:

$$\int_G \left[u_n(x) + \int_G k(x,y)f(y, u_n(y))\,dy\right] w_i(x)\,dx = 0 \tag{33}$$

for $i = 1, \ldots, n$ with

$$u_n(x) = \sum_{j=1}^{n} c_j(K^*w_j)(x).$$

This is a nonlinear real system of n equations for the n unknown real numbers $c_1, \ldots, c_n$.

Proposition 28.4 (Solution of the Hammerstein Equation (30) for a Monotone Kernel Operator K). *Under the assumptions* (H1), (H2), (H3), *and* (H4) *the following assertions are valid*:

(a) Convergence of the Galerkin method. *Suppose that K is angle-bounded (e.g., symmetric) and f is monotone with respect to u.*

Then equation (30) *has exactly one solution u and, for each n, the Galerkin equation* (33) *has exactly one solution u_n where the sequence (u_n) converges in $L_p(G)$ to u as $n \to \infty$.*

(b) Existence. *Equation* (30) *has a solution in the case where one of the following three conditions is valid*:

(i) *f is monotone with respect to u and satisfies the coerciveness condition* (H5). *If $p = q = 2$, then* (H5) *drops out.*

(ii) *K is compact and f satisfies* (H5).

(iii) *K is compact and angle-bounded, and f satisfies the asymptotic positivity condition* (H5*).

(c) Existence and uniqueness. *Equation* (30) *has exactly one solution in the case where one of the following two conditions is valid*:

(i) *f is strictly monotone with respect to u and satisfies the coerciveness condition* (H5). *If $p = q = 2$, then* (H5) *drops out.*

(ii) *K is strictly monotone, and f is monotone with respect to u and satisfies* (H5).

Stronger results for symmetric kernel operators K can be found in Section 41.6. There we will use a variational approach.

PROOF. We use, in an essential way, the structure theorems on the Nemyckii operator in Section 26.3. By the growth condition (H2), the Nemyckii operator $F: X^* \to X$ is continuous and bounded and

$$\langle u, Fu \rangle_X = \int_G u(y) f(y, u(y))\, dy \qquad \text{for all} \quad u \in X^*.$$

If f is monotone (resp. strictly monotone) with respect to u, then $F: X^* \to X$ is monotone (resp. strictly monotone).

The coerciveness condition (H5), i.e.,

$$uf(y,u) \geq d|u|^p \qquad \text{for all} \quad (y,u) \in G \times \mathbb{R}$$

implies

$$\langle u, Fu \rangle_X \geq d \int_G |u|^p\, dy = d\|u\|_{X^*}^p \qquad \text{for all} \quad u \in X^*,$$

i.e., F is coercive.

The asymptotic positivity condition (H5*) implies

$$\langle u, Fu \rangle_X \geq -c \qquad \text{for all} \quad u \in X^*,$$

where c is a positive constant. This follows from Proposition 26.7.

Now, the assertions of Proposition 28.4 follow from our results on abstract Hammerstein equations.

Ad(a). Cf. Theorem 28.A in Section 28.2.

Ad(b), (i). Cf. Theorem 32.B in Section 32.5 and Theorem 32.O in Section 32.23.

Ad(b), (ii). Cf. Theorem 28.B in Section 28.3 with condition (C1).
Ad(b), (iii). Cf. Theorem 28.B with condition (C2).
Ad(c). Cf. Theorem 32.B and Theorem 32.O. □

In the following three corollaries to Proposition 28.4 we summarize several well-known results concerning linear integral operators of the form

$$(Kv)(x) = \int_G k(x, y)v(y)\,dy.$$

In this connection, we generalize Standard Example 28.3 and we obtain results which are useful for applying Proposition 28.4 to broad classes of Hammerstein integral equations.

(H) Let G be a bounded region in $\mathbb{R}^N$ with $N \geq 1$ and let $1 < p, q < \infty$ with $p^{-1} + q^{-1} = 1$. We set $X = L_q(G)$. Then $X^* = L_p(G)$.

Recall that $L_p(G \times G)$ consists of all measurable functions $k: G \times G \to \mathbb{R}$ with

$$\int_{G \times G} |k(x, y)|^p\,dx\,dy < \infty.$$

Moreover, by definition, $k \in L_\infty(G \times G)$ iff $k: G \times G \to \mathbb{R}$ is measurable and bounded (e.g., continuous and bounded).

Corollary 28.5 (Regular Kernels and Compact K). *Suppose that* (H) *holds and that the kernel k is regular, i.e., $k \in L_p(G \times G)$.*

Then the operator $K: X \to X^*$ is linear and compact and the dual operator $K^*: X \to X^*$ is given by the relation

$$(K^*v)(x) = \int_G k(y, x)v(y)\,dy,$$

for almost all $x \in G$ and all $v \in X^*$. In particular, if

$$k(x, y) = k(y, x) \qquad \text{for almost all} \quad x, y \in G,$$

then the operator $K: X \to X^*$ is symmetric.

PROOF. For example, see Kantorovič and Akilov (1964, M), Section 2. □

Corollary 28.6 (Weakly Singular Kernels and Compact K). *Suppose that* (H) *holds and that the kernel k is weakly singular, i.e.,*

$$k(x, y) = \frac{m(x, y)}{|x - y|^\alpha}, \qquad 0 < \alpha < N,$$

for all $x, y \in G$ with $x \neq y$. Here, $m \in L_\infty(G \times G)$ and N denotes the dimension of G. Let $X = L_q(G)$, $1 < q < \infty$.

Then the operator $K: X \to X$ is linear and compact.

If $2 \le q < \infty$, then $1 < p \le 2$ and therefore, the embedding $X \subseteq X^*$ is continuous. Hence the linear operator $K: X \to X^*$ is compact. The dual operator $K^*: X \to X^*$ has the same properties as in Corollary 28.5.

PROOF. For example, see Triebel (1972, M), p. 122. □

We now consider the limit case $\alpha = N$, i.e., we consider the *singular kernel*

$$k(x, y) = \frac{c(x, s)}{|x - y|^N},$$

where $s = (x - y)/|x - y|$. The function $c(\cdot, \cdot)$ is called the characteristic of the singular kernel. We have $s \in S$, where

$$U(x, \varepsilon) = \{y \in \mathbb{R}^N: |y - x| < \varepsilon\}, \qquad S = \{x \in \mathbb{R}^N: |x| = 1\}.$$

For almost all $x \in G$, we define the integral operator K by the limit

$$(Kv)(x) = \lim_{\varepsilon \to 0} \int_{G - U(x, \varepsilon)} k(x, y) v(y)\, dy,$$

i.e., the integral is to be understood in the sense of the Cauchy principal value.

Corollary 28.7 (Singular Kernels and Continuous K). *Suppose that* (H) *holds with $N \ge 2$ and that the kernel k is singular, i.e., c is measurable and*

$$\int_S c(x, s)\, dO(s) = 0 \qquad \text{for all} \quad x \in G,$$

$$\sup_{x \in G} \int_S |c(x, s)|^p\, dO(s) < \infty.$$

Let $X = L_q(G)$, $1 < q < \infty$.

Then the linear operator $K: X \to X$ is continuous.

If $2 \le q < \infty$, then the embedding $X \subseteq X^*$ is continuous. Hence the linear operator $K: X \to X^*$ is continuous.

This is the famous theorem of Calderon and Zygmund (1956) which plays a fundamental role in the modern theory of linear elliptic partial differential operators. Corollary 28.7 remains valid for $G = \mathbb{R}^N$.

28.5. Application to Semilinear Elliptic Differential Equations

We consider the boundary value problem

$$\begin{aligned} -\Delta u(x) &= -f(x, u(x)) \quad \text{on } G, \\ u &= 0 \quad \text{on } \partial G, \end{aligned} \tag{34}$$

or, more generally,

$$\begin{aligned} Lu - \mu u &= -f(x,u) \quad \text{on } G, \\ D^\gamma u &= 0 \quad \text{on } \partial G \qquad \text{for all} \quad \gamma\colon |\gamma| \le m-1, \end{aligned} \tag{35}$$

with the fixed real number μ and the $2m$th order linear differential operator

$$Lu = \sum_{|\alpha|,|\beta| \le m} (-1)^{|\alpha|} D^\alpha (a_{\alpha\beta} D^\beta u).$$

We make the following assumptions:

(H1) G is a bounded region in $\mathbb{R}^N$, $N \ge 1$.
(H2) All the functions $a_{\alpha\beta}\colon G \to \mathbb{R}$ are measurable and bounded, and the ellipticity condition (E) below is satisfied.
(H3) The function $f\colon G \times \mathbb{R} \to \mathbb{R}$ satisfies the Carathéodory condition (e.g., f is continuous), and there exists a nonnegative function $a \in L_2(G)$ and a positive number b such that the following growth condition holds:

$$|f(x,u)| \le a(x) + b|u| \qquad \text{for all} \quad (x,u) \in G \times \mathbb{R}.$$

Definition 28.8. Let $X = \mathring{W}_2^m(G)$. Then the *generalized problem* for (35) reads as follows: We seek a function $u \in X$ *such that*

$$a(u,v) = -\int_G f(x,u(x))v(x)\,dx \qquad \text{for all} \quad v \in X, \tag{36}$$

where

$$a(u,v) = \int_G \sum_{|\alpha|,|\beta| \le m} (a_{\alpha\beta} D^\alpha v D^\beta u - \mu uv)\,dx.$$

As usual, the generalized problem (36) results from (35) if one multiplies (35) by $v \in C_0^\infty(G)$ and then integrates by parts. In particular, the special case (34) corresponds to $m = 1$ and

$$a(u,v) = \int_G \sum_{i=1}^N D_i u D_i v\,dx, \tag{37}$$

where $x = (\xi_1, \ldots, \xi_N)$ and $D_i = \partial/\partial\xi_i$. As a further important assumption we formulate the following *ellipticity condition*:

(E) There exists a number $c > 0$ such that

$$a(u,u) \ge c\|u\|_X^2 \qquad \text{for all} \quad u \in X.$$

In the special case (37), condition (E) is valid because of the Poincaré–Friedrichs inequality.If the linear differential operator L is strongly elliptic, then it follows from the Gårding inequality in Section 22.15 that

$$a(u,u) \ge c\|u\|_X^2 - (C + \mu)\|u\|_2^2$$

for all $u \in X$ and fixed $c > 0$, $C \in \mathbb{R}$. Consequently, condition (E) is valid for all μ with $\mu \le -C$.

If, in addition, L is symmetric, then condition (E) is valid for all μ with $\mu < \mu_{\min}$, where $\mu_{\min}$ is the smallest eigenvalue of the eigenvalue problem (35) with $f \equiv 0$.

Proposition 28.9. *Under the assumptions* (H1)–(H3), *the following hold*:

(a) Existence and uniqueness. *Suppose that the function* $u \mapsto f(x, u)$ *is monotone increasing on* $\mathbb{R}$ *for all* $x \in G$. *Then the generalized problem for* (34) (*resp.* (35)) *has exactly one solution.*
(b) Existence. *Suppose that there exists a number* $R > 0$ *such that*

$$f(x,u)u \geq 0 \qquad \textit{for all} \quad (x,u) \in G \times \mathbb{R} \quad \textit{with} \quad |u| \geq R. \tag{38}$$

Then the generalized problem for (34) (*resp.* (35)) *has a solution.*

PROOF. Ad(a). Let $Y = L_2(G)$. Then $Y^* = L_2(G)$. By Section 26.3, the Nemyckii operator $F\colon Y^* \to Y$ corresponding to f is continuous, bounded, and monotone.

We consider the linear boundary value problem

$$\begin{aligned} Lu - \mu u &= g \quad \text{on } G, \\ D^\gamma u &= 0 \quad \text{on } \partial G \qquad \text{for all} \quad \gamma\colon |\gamma| \leq m - 1, \end{aligned}$$

i.e., we set $-f(x, u(x)) \equiv g(x)$. By (E) and Theorem 22.C the corresponding generalized problem (36) has exactly one solution u. We set

$$u = Kg.$$

By Corollary 22.20, the solution operator $K\colon Y \to Y^*$ is linear, monotone, compact, and angle-bounded.

Consequently, the generalized problem for (35) is equivalent to the operator equation

$$u = -KFu, \qquad u \in Y^*. \tag{39}$$

Now Theorem 28.A yields the assertion.

Ad(b). By Proposition 26.7, it follows from (38) that

$$\langle u, Fu \rangle_Y \geq -c \qquad \text{for all} \quad u \in Y^*,$$

where c is a positive constant. Theorem 28.B with condition (C2) yields the assertion. □

PROBLEMS

28.1. *Proof of Proposition* 28.1.

Solution:

(I) Construction of the H-space H with the scalar product $(\cdot|\cdot)$. As a linear monotone operator, K is continuous. We construct the two bilinear functionals

$$[u, v]_\pm = 2^{-1}(\langle Ku, v\rangle \pm \langle Kv, u\rangle), \qquad \text{for all} \quad u, v \in X.$$

Note that $[\cdot,\cdot]_+$ is symmetric and $[\cdot,\cdot]_-$ is skew-symmetric. It follows from the monotonicity of K that

$$[u,u]_+ = \langle Ku, u\rangle \geq 0 \qquad \text{for all} \quad u \in X.$$

According to the Schwarz inequality for positive symmetric bilinear functionals, we obtain

$$|[u,v]_+|^2 \leq [u,u]_+[v,v]_+ \qquad \text{for all} \quad u, v \in X \tag{40}$$

(cf. Problem 21.16). The operator K is angle-bounded. This means

$$|[u,v]_-|^2 \leq a^2[u,u]_+[v,v]_+ \qquad \text{for all} \quad u, v \in X. \tag{41}$$

We now consider the zero set

$$Z = \{u \in X : [u,u]_+ = 0\}.$$

By (40) and (41), $[u,v]_\pm = 0$ for $u \in Z$ or $v \in Z$. This implies

$$[u + w, v + z]_\pm = [u,v]_\pm \qquad \text{for all} \quad u, v \in X, \quad w, z \in Z. \tag{42}$$

The *key* to our proof is the equivalence relation

$$u \sim v \qquad \text{iff} \qquad u - v \in Z.$$

Let X/Z denote the set of equivalence classes arising this way. We equip the factor space X/Z with the scalar product

$$(U|V) \overset{\text{def}}{=} [u,v]_+, \tag{43}$$

where $u \in U$ and $v \in V$. By (42), this definition is independent of the choice of the representatives $u \in U$ and $v \in V$. Moreover, $(\cdot|\cdot)$ is indeed a scalar product on X/Z, for it follows from $(U|U) = 0$ that $[u,u]_+ = 0$, i.e., $u \in Z$, therefore $U = 0$.

According to the completion principle, we can extend X/Z to an H-space H such that

$$X/Z \subseteq H$$

and X/Z is dense in H (see Problem 18.4).

(II) Construction of the canonical operator $S: X \to H$. We define

$$Su = U, \qquad u \in U,$$

i.e., the operator S assigns to each $u \in X$ the corresponding equivalence class U. Therefore,

$$(Su|Sv) = [u,v]_+ = 2^{-1}(\langle Ku, v\rangle + \langle Kv, u\rangle).$$

The operator S is linear and continuous, for

$$\|Su\|_H^2 = (Su|Su) = \langle Ku, u\rangle \leq \|Ku\|\,\|u\| \leq \|K\|\,\|u\|^2,$$

and hence $\|S\|^2 \leq \|K\|$.

(III) Construction of the operator $B: H \to H$. For all $U, V \in X/Z$, we define

$$b(U,V) = [u,v]_-, \qquad \text{where} \quad u \in U, \qquad v \in V.$$

By (42), this definition is independent of the choice of the representatives

$u \in U$ and $v \in V$. From (41) it follows that

$$|b(U, V)| \leq a\|U\|_H \|V\|_H \quad \text{for all} \quad U, V \in X/Z. \tag{44}$$

Since X/Z is dense in H, we can extend $b(\cdot, \cdot)$ to $H \times H$ such that (44) holds for all $U, V \in H$. Consequently, there exists a linear continuous operator $B: H \to H$ with

$$b(U, V) = (BU|V) \quad \text{for all} \quad U, V \in H. \tag{45}$$

By (44), $\|B\| \leq a$. Because $b(U, V) = -b(V, U)$, we have

$$B^{*'} = -B,$$

i.e., B is skew-symmetric. It follows from (45) that

$$(BSu|Bv) = b(Su, Sv) = [u, v]_- \quad \text{for all} \quad u, v \in X. \tag{46}$$

(IV) Properties of the operator $C = I + B$.

(IV-1) C is strongly monotone, for $(BU|U) = b(U, U) = 0$ and thus

$$(CU|U) = ((I + B)U|U) = \|U\|_H^2 \quad \text{for all} \quad U \in H. \tag{47}$$

(IV-2) C is bijective. In fact, because $\|B\| \leq a$,

$$\begin{aligned} \|CV\|^2 &= ((I + B)V|(I + B)V) \\ &= \|V\|^2 + \|BV\|^2 \\ &\leq (1 + a^2)\|V\|^2 \quad \text{for all} \quad V \in H. \end{aligned} \tag{48}$$

From $\|CV\| \geq \|V\|$ for all $V \in H$ it follows that C is injective and $C(H)$ is closed. We show that $C(H) = H$. If we had $C(H) \neq H$, then there would exist $V \neq 0$ with $(CU|V) = 0$ for all $U \in H$. This means that $(CV|V) = 0$, therefore $V = 0$ by (47). This contradicts $V \neq 0$. Thus $C(H) = H$.

(IV-3) C^{-1} is strongly monotone on H. For, by (47) and (48),

$$\begin{aligned} (U|C^{-1}U) &= (C(C^{-1}U)|C^{-1}U) = \|C^{-1}U\|^2 \\ &\geq (1 + a^2)^{-1}\|U\|^2 \quad \text{for all} \quad U \in H. \end{aligned}$$

(V) Properties of the dual operator $S^*: H^* \to X^*$.

(V-1) By the definition of the dual operator and $H^* = H$, we obtain that

$$(h|Su) = \langle S^*h, u \rangle_X \quad \text{for all} \quad u \in X, \quad h \in H. \tag{49}$$

(V-2) S^* is injective. For, by (49), it follows from $S^*h = 0$ that $(h|Su) = \langle S^*h, u \rangle = 0$ for all $u \in X$. Since $S(X) = X/Z$ is dense in H, we have $h = 0$.

(V-3) The norm of a linear continuous operator is always equal to the norm of the corresponding dual operator. Hence $\|S\| = \|S^*\|$.

(VI) The decomposition $K = S^*(I + B)S$ follows from

$$\begin{aligned} \langle Ku, v \rangle = [u, v]_+ + [u, v]_- &= (Su|Sv) + (BSu|Sv) \\ &= ((I + B)Su|Sv) \\ &= \langle S^*(I + B)Su, v \rangle \quad \text{for all} \quad u, v \in X. \end{aligned}$$

Equation (48) together with $\|S^*\| = \|S\|$ yields

$$\|Ku\|^2 = \|S^*CSu\|^2 \leq \|S\|^2\|CSu\|^2$$
$$\leq \|K\|(1 + a^2)\|Su\|^2 = \|K\|(1 + a^2)\langle Ku, u\rangle.$$

This is Proposition 28.1(b).

Let K be strongly continuous. Then, from $u_n \rightharpoonup u$ as $n \to \infty$, it follows that $Ku_n \to Ku$ and hence

$$\|Su_n - Su\|^2 = \langle Ku_n - Ku, u_n - u\rangle \to 0 \quad \text{as} \quad n \to \infty,$$

i.e., $Su_n \to Su$. Therefore, S is strongly continuous.

Let $K: X \to X^*$ be compact on the reflexive B-space X. Then $S: X \to H$ is also compact, since a linear operator on a reflexive B-space is compact iff it is strongly continuous.

If K is symmetric, then $[u, v]_- = 0$ for all $u, v \in X$, i.e., $B = 0$.

The proof of Proposition 28.1 is complete.

28.2. *Linear monotone angle-bounded operators.*

28.2a. Let $K: X \to X^*$ be a linear and monotone operator on the B-space X. Show the equivalence of the following two assertions:

(i) K is angle-bounded, i.e., for all $u, v \in X$ and fixed $a \geq 0$,

$$|\langle Ku, v\rangle - \langle Kv, u\rangle|^2 \leq 4a^2\langle Ku, u\rangle\langle Kv, v\rangle.$$

(ii) For all $u, v, z \in X$,

$$\langle Ku - Kz, z - v\rangle \leq 4^{-1}(1 + a^2)\langle Ku - Kv, u - v\rangle.$$

Hint: Cf. Pascali and Sburlan (1978, M), p. 191.

28.2b. Let $K: X \to X^*$ be a linear monotone angle-bounded operator on the B-space X. Show that $\langle Kw, w\rangle = 0$ implies $Kw = 0$.

Solution: We set $z = v + ty$, $t > 0$ and $u = w + v$. By (ii) above,

$$\langle Ku - Kz, z - v\rangle \leq 0.$$

Letting $t \to 0$, we obtain $\langle Kw, y\rangle \leq 0$ for all $y \in X$, i.e., $Kw = 0$.

References to the Literature

Classical work: Hammerstein (1930).

General expositions on Hammerstein integral equations: Krasnoselskii (1956, M), (1966, M), Zabreiko and Krasnoselskii (1975, M), Vainberg (1956, M), (1972, M), Pogorzelski (1966, M), Pascali and Sburlan (1978, M), Guseinov (1980, M).

Angle-bounded monotone operators and Hammerstein integral equations: Amann (1969), (1969a), (1972), Browder and Gupta (1969), Pascali and Sburlan (1978, M).

Monotone operators and Hammerstein integral equations: Browder (1971, S, B), Hess (1971), Vainberg (1972, M), Brézis and Browder (1975), (1978), Brézis and Haraux (1976), Pascali and Sburlan (1978, M).

Monotone operators, singular integral equations, and nonlinear boundary value problems in complex function theory: v. Wolfersdorf (1982), (1983), (1987), Wegert (1987) (topological methods).

Galerkin method: Amann (1969a), Vainberg (1972, M).

Numerical methods: Amann (1972a), Delves and Walsh (1974, M) (numerous physical applications), Fenyö and Stolle (1982, M), Vol. 4, Hackbusch (1985, M) (multigrid methods), Reinhardt (1985, M).

Subroutine package QUADPACK for automatic integration: Piessens (1983).

Monotone operators and nonlinear integral equations of Levin–Nohel type: Barbu (1976, M).

Handbook on linear and nonlinear integral equations: Zabreiko and Krasnoselskii (1975).

(Cf. also the References to the Literature for Chapter 1 (integral equations), Chapter 7 (semilinear equations), and Chapter 53 (integral equations in Orlicz spaces)).

CHAPTER 29

Noncoercive Equations, Nonlinear Fredholm Alternatives, Locally Monotone Operators, Stability, and Bifurcation

It came as a complete surprise, when, in a short note published in 1900, Fredholm showed that the general theory of all *integral equations* considered prior to him was, in fact, extremely simple.

Ivar Fredholm (1866–1927) was a student of Mittag-Leffler in Stockholm in 1888–1890; he published only a few papers during his lifetime, mostly concerned with partial differential equations. After a visit to Paris, where he had been in contact with all the French analysts, and had become familiar with the recent papers of Poincaré, he communicated, in August 1899, his first results on integral equations to his former teacher; they were published in 1900 and completed two years later in a paper published in *Acta Mathematica.*

Jean Dieudonné (1981)

The purpose of this note is to introduce a *nonlinear* version of *Fredholm operators,* and to prove that in this context Sard's theorem (1942) holds if zero measure is replaced by first category.

Steve Smale (1965)

These notes formed the basis of a course entitled "Nonlinear Problems" given at the Department of Mathematical Analysis, Charles University, Prague, during the years 1976 and 1977.

The problems of solvability of *noncoercive* nonlinear equations and of nonlinear Fredholm alternatives have been very popular recently. Therefore, it was the author's desire to give a survey of results concerning these problems which are known at present, as well as to show the methods which have been used to obtain them and some of their consequences for concrete problems.

Svatopluk Fučik (1977)

Svatopluk Fučik died prematurely on May 18, 1979. He was 34 years old and knew since 1973 that his time was severely limited. 1973 is also the year when Fučik wrote the first of twenty-one papers devoted to nonlinear noncoercive problems.... This domain of analysis owes so much to his remarkable ingenuity and formidable energy.

Jean Mawhin (1981)

We consider the operator equation

$$Au = b, \qquad u \in X.$$

The main theorems on monotone and pseudomonotone operators in Chapters 26 and 27, respectively, ensure the existence of a solution u for *each* $b \in X^*$ if the operator $A\colon X \to X^*$ is *coercive*. In this chapter we want to study the case that A is *not* coercive. Here we need *solvability conditions* for the right member b. In particular, we will study the asymptotically linear operator equation

$$Bu + Nu = b, \qquad u \in X, \tag{1}$$

i.e., the operator B is linear and continuous, and the nonlinear operator N satisfies the condition

$$\lim_{\|u\| \to \infty} \frac{\|Nu\|}{\|u\|} = 0.$$

Roughly speaking, we obtain the following results:

(i) If the linearized equation $Bu = 0$ has only the trivial solution $u = 0$, then equation (1) has a solution for each $b \in X^*$.
(ii) If $Bu = 0$ has a nontrivial solution $u \neq 0$, then (1) has a solution if b satisfies an appropriate solvability condition.

Statement (i) can be formulated like this: If the linearized equation $Bu = b$ has at most one solution u, then the nonlinear equation $Bu + Nu = b$ possesses a solution u. Roughly speaking, that means:

Uniqueness implies existence.

This principle is of general interest because, as a rule, it is much easier to prove the uniqueness of solutions than the existence.

Concerning the nonlinear operator N, we consider the following two cases:

(a) $R(N) \subseteq R(B)$ (Theorems 29.A and 29.C);
(b) The operator $B + N$ has so-called weak asymptotes (Theorem 29.D).

The applications of (a) and (b) concern integral equations and boundary value problems for semilinear elliptic equations, respectively. In connection with (b), we consider theorems of the so-called Landesman–Lazer type in Section 29.9. Here, we obtain the surprising fact that the *necessary* solvability conditions are also *sufficient*.

The general idea of proof is the following:

(α) We replace equation (1) by an equivalent problem, where we do *not* assume that the single-valued inverse operator B^{-1} exists. In Section 29.1, we will study several different possibilities for formulating such important equivalent problems.
(β) We apply the mapping degree to the equivalent problem corresponding to (1). In particular, we will use the Borsuk antipodal theorem and the Leray–Schauder principle from Part I.

Further important results on the range of sum operators can be found in Section 32.22.

In Section 29.10, we study in detail the operator equation

$$\text{(E)} \qquad Au = b, \qquad u \in S,$$

where A is a proper *nonlinear Fredholm operator* and S is a subset of a B-space. The class of Fredholm operators plays a fundamental role with respect to integral equations, differential equations, and the calculus of variations. We want to show what the abstract operator equation (E) behaves like in the special case of reasonable real functions $A: \mathbb{R} \to \mathbb{R}$. This is a crucial observation. In particular, we will show that, in most cases, equation (E) has at most a *finite* number of solutions u, and the number of solutions is *locally constant* with respect to the right member b. Roughly speaking, the fundamental Sard–Smale theorem tells us that:

Pathological behavior of equation (E) *is rare with respect to the right member b.*

In Section 29.11, we introduce the notion of *locally monotone operators.* This class of operators allows important applications to the calculus of variations and to elasticity. In Sections 29.12 through 29.22, we investigate the following topics:

(i) sufficient conditions for strict local minima (locally regularly monotone operators, the abstract Gårding inequality, the Hestenes theorem, and eigenvalue criteria);
(ii) nonlinear Fredholm operators and an abstract stability theory;
(iii) a general bifurcation theorem for operators of variational type;
(iv) local and global multiplicity theorems for nonlinear Fredholm operators.

We want to emphasize the following general principle:

Loss of stability can lead to bifurcation.

In connection with applications to the calculus of variations, it is important in (iii) that the so-called operators of "variational type" *generalize* potential operators. This way, it is possible to prove a general bifurcation theorem for Euler–Lagrange equations in spaces of smooth functions. In contrast to the corresponding theory in Sobolev spaces, we do *not* need any restrictive growth conditions for the differential equations.

Bifurcation theory studies equations of the form

$$\text{(B)} \qquad G(u, b) = 0, \qquad u \in X, \qquad b \in P,$$

where the "parameter" b lives in the parameter space P. Obviously, equation (B) generalizes the equation $Au = b$, which we have considered above. Let

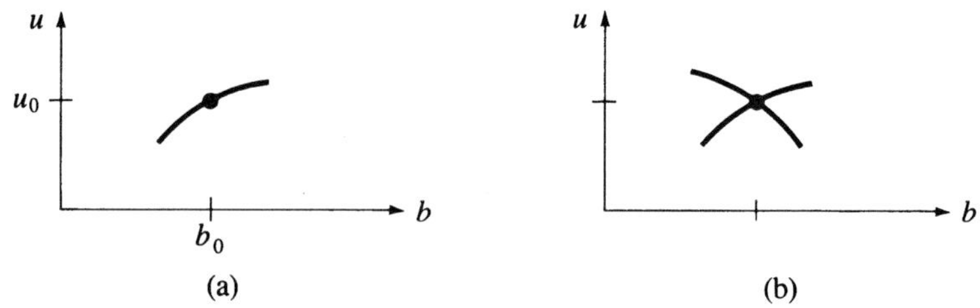

Figure 29.1

$G(u_0, b_0) = 0$, and let the operator

$$G\colon U(u_0, b_0) \subseteq X \times P \to Y$$

be C^1 in a neighborhood of the point (u_0, b_0), where X, P, and Y are real B-spaces. Suppose that the linearization

$$G_u(u_0, b_0)\colon X \to Y$$

is a Fredholm operator of index zero. Then, according to Sections 8.1 and 8.4, we have the following two different situations:

(α) *Regular case.* If $G_u(u_0, b_0)h = 0$ implies $h = 0$, then $G_u(u_0, b_0)$ is bijective and the solution set of (B) in a neighborhood of (u_0, b_0) is given by a C^1-curve $u = u(b)$ through the point (u_0, b_0). This follows from the implicit function theorem (Fig. 29.1(a)).

(β) *Singular case.* If the linearized equation $G_u(u_0, b_0)h = 0$ has a nontrivial solution $h \neq 0$, then the solution set of (B) in a neighborhood of (u_0, b_0) may have a complex structure. In particular, it is possible that this solution set bifurcates at (u_0, b_0) (Fig. 29.1(b)).

In terms of elasticity, we have:

$$u = \text{displacement of the elastic body};$$

$$b = \text{outer forces}.$$

Roughly speaking, the regular case (α) means that the outer forces cause uniquely determined displacements of the body, whereas bifurcation in (β) corresponds, for example, to the buckling of rods, beams, plates, and shells.

In Sections 29.13 and 29.19, we consider:

(a) applications of bifurcation and stability theory to the buckling of beams; and

(b) applications of the general theory for minimum problems to the calculus of variations.

In particular, we study the following in (b):

the Legendre–Hadamard condition and strongly elliptic systems, strongly stable solutions of the Euler–Lagrange equation, and sufficient conditions for local strict minima;

the Jacobi equation and sufficient eigenvalue criteria for local strict minima;
the continuation method and an approximation method for solving the Euler–Lagrange equations in the domain of stability;
loss of stability and a general bifurcation theorem for Euler–Lagrange equations;
stability of the bifurcation branches.

Applications of these results to nonlinear elasticity will be considered in Chapter 61.

Our abstract results represent the final functional analytic formulation of classical results due to Euler, Lagrange, Legendre, Jacobi, and Fredholm, mentioned in Section 18.7. Roughly speaking, we obtain the following:

Euler–Lagrange equation	$\Rightarrow$	*necessary conditions for local minima of functionals*;
Legendre condition	$\Rightarrow$	*locally regularly monotone operators and sufficient conditions for minima of functionals*;
Jacobi equation	$\Rightarrow$	*sufficient eigenvalue criteria for local minima of functionals*;
Fredholm's theory of integral equations	$\Rightarrow$	*nonlinear Fredholm operators.*

29.1. Pseudoresolvent, Equivalent Coincidence Problems, and the Coincidence Degree

We consider the semilinear operator equation

$$Bu = Mu, \qquad u \in D(B) \cap D(M), \tag{2}$$

where $B\colon D(B) \subseteq X \to Y$ is a linear operator and $M\colon D(M) \subseteq X \to Y$ is a nonlinear operator. This equation is called a *coincidence equation* because one wants to find a point u for which the images under B and M coincide. If the inverse operator $B^{-1}\colon Y \to X$ exists, then equation (2) is equivalent to $u = B^{-1}Mu$. However, in this section, we consider operators B which are *not* necessarily invertible and we want to obtain a number of different problems which are equivalent to (2). Those equivalent problems are very useful in order to prove existence theorems for (2). Our goal is to show the connections between apparently completely different approaches in the literature. To this end, we shall use the *pseudoresolvent* as a simple tool. Before considering the general case, we begin with an important special case.

We make the following assumptions:

(H1) X and Y are B-spaces over $\mathbb{K} = \mathbb{R}, \mathbb{C}$ and $M\colon X \to Y$ is a given operator.
(H2) The linear continuous operator $B\colon X \to Y$ is Fredholm of *index zero* with

$\dim N(B) > 0$, i.e., the range $R(B)$ is closed and

$$\dim N(B) = \operatorname{codim} R(B) < \infty.$$

We studied Fredholm operators in Section 8.4. Recall that $\dim N(B^*) = \dim N(B)$. The dual operator B^* maps Y^* into X^*.

(H3) Let $\{u_1, \ldots, u_n\}$ and $\{u_1^*, \ldots, u_n^*\}$ be a basis in $N(B)$ and $N(B^*)$, respectively. Note that $u_i \in X$ and $u_i^* \in Y^*$ for all i.

(H4) By $A_1(51)$, we choose elements $v_i \in Y$ and $v_i^* \in X^*$ such that

$$\langle u_i^*, v_j \rangle = \langle v_i^*, u_j \rangle = \delta_{ij}, \qquad i, j = 1, \ldots, n, \tag{3}$$

and we set

$$Qu = \sum_{i=1}^{n} \langle v_i^*, u \rangle u_i,$$

$$Pb = b - \sum_{i=1}^{n} \langle u_i^*, b \rangle v_i.$$

Recall that the equation

$$Bu = b, \qquad u \in X, \tag{4}$$

has a solution iff

$$\langle u^*, b \rangle = 0 \qquad \text{for all} \quad u^* \in N(B^*), \tag{5}$$

i.e., $Pb = b$. Consequently, the operators $Q: X \to N(B)$ and $P: Y \to R(B)$ are projection operators onto the null space $N(B)$ and the range $R(B)$ of B, respectively.

Now we are ready to give our *key definition*:

$$Su = Bu + \sum_{i=1}^{n} \langle v_i^*, u \rangle v_i. \tag{6}$$

We shall show below that $S: X \to Y$ is bijective. Hence the inverse operator $S^{-1}: Y \to X$ exists. This operator is called the *pseudoresolvent* of B. Moreover, we shall show below that the following problems are mutually equivalent:

(A) *Original equation*

$$Bu = Mu, \qquad u \in X. \tag{7}$$

(B) *Fixed-point problem of E. Schmidt* (1908)

$$u = S^{-1}Mu + \sum_{i=1}^{n} \langle v_i^*, u \rangle u_i, \qquad u \in X. \tag{8}$$

(C) *Branching equations of E. Schmidt* (1908)

$$u = S^{-1}Mu + \sum_{i=1}^{n} c_i u_i, \tag{9a}$$

$$c_i = \langle v_i^*, u \rangle, \qquad i = 1, \ldots, n. \tag{9b}$$

We seek $u \in X$ and $c_1, \ldots, c_n \in \mathbb{K}$.

(D) *Branching equations of Ljapunov* (1906)

$$w = S^{-1}PM(v + w), \tag{10a}$$

$$0 = (I - P)M(v + w). \tag{10b}$$

We seek $v \in N(B)$, $w \in N(B)^{\perp}$ and set $u = v + w$. Here, $N(B)^{\perp} = (I - Q)(X)$.

(E) *Alternative problem of Cesari* (1964)

$$u = S^{-1}Mu + \sum_{i=1}^{n} c_i u_i,$$

$$c_i = c_i - t\langle u_i^*, Mu\rangle, \qquad i = 1, \ldots, n.$$

In this connection, let $t \in \mathbb{K}$ be a fixed nonzero number. We seek $u \in X$ and $c_1, \ldots, c_n \in \mathbb{K}$.

Obviously, this problem is equivalent to the following problem:

$$\begin{aligned} u &= S^{-1}Mu + \sum_{i=1}^{n} c_i u_i, \\ c_i &= c_i - t\left\langle u_i^*, M\left(S^{-1}Mu + \sum_{i=1}^{n} c_i u_i\right)\right\rangle. \end{aligned} \tag{11}$$

We seek $u \in X$ and $c_1, \ldots, c_n \in \mathbb{K}$.

(F) *Alternative problem of Mawhin* (1972)

$$u = S^{-1}\left(Mu - \sum_{i=1}^{n} \langle u_i^*, Mu\rangle v_i\right) + \sum_{i=1}^{n} \langle u_i^*, Mu\rangle u_i + \sum_{i=1}^{n} \langle v_i^*, u\rangle u_i. \tag{12}$$

Proposition 29.1. *Assume* (H1) *through* (H4). *Then*:

(a) *The operator $S: X \to Y$ is a linear homeomorphism.*
(b) *If $b \in R(B)$, then $BS^{-1}b = b$.*
(c) *Problems* (A) *through* (F) *are mutually equivalent.*

Before giving the proof, we want to discuss this result. In (C) one first solves equation (9a). Then the solution $u = u(c_1, \ldots, c_n)$ has to be substituted into (9b). This yields the branching equations of E. Schmidt

$$c_i = \langle v_i^*, u(c_1, \ldots, c_n)\rangle, \qquad i = 1, \ldots, n.$$

Similarly, in (D) one first solves equation (10a). Then substitution of the corresponding solution $w = w(v)$ into (10b) yields the branching equation of Ljapunov

$$(I - P)M(v + w(v)) = 0, \qquad v \in N(B).$$

Sometimes it is better to solve (10b) first and to substitute the solution $v = v(w)$ into (10a).

In order to solve the equations (B)–(F) above and the more general equations (B*)–(F*) below, one can use fixed-point theorems, mapping degree, the theory of monotone operators, variational methods, etc. Many papers in the literature are based on this approach, in order to prove a steadily increasing number of existence theorems for nonlinear ordinary and partial differential equations. Our approach should help the reader to understand the *substance* of the turbulently accumulating literature.

Remark 29.2 (Coincidence Degree). If one wants to apply the Leray–Schauder mapping degree from Part I to the original problem (A) above, then one can proceed as follows. We consider the equivalent fixed-point problem (B) of E. Schmidt and we set

$$Cu = S^{-1}Mu + \sum_{i=1}^{n} \langle v_i^*, u\rangle u_i.$$

Then the equation

$$Bu = Mu, \qquad u \in G,$$

is equivalent to the equation

$$Cu = u, \qquad u \in G.$$

Suppose that $Bu \neq Mu$ on ∂G. This implies $Cu \neq u$ on ∂G. Moreover, suppose that $C\colon X \to X$ is compact (e.g., M is compact). Then the Leray–Schauder degree $\deg(I - C, G)$ is well defined and we can define the so-called *coincidence degree* by means of

$$\deg(B, M; G) = \deg(I - C, G). \tag{13}$$

One can show that the following hold:

(i) If we fix an orientation in both $N(B)$ and $N(B^*)$, then $\deg(B, M; G)$ does not depend on the choice of the positively oriented basis $\{u_1, \ldots, u_n\}$ and $\{u_1^*, \ldots, u_n^*\}$ in $N(B)$ and $N(B^*)$, respectively.
(ii) If we change the orientation in either $N(B)$ or $N(B^*)$, then $\deg(B, M; G)$ changes its sign.

Using (B^*) and (F^*) below, it is possible to generalize the coincidence degree in a simple way to more general situations. This will be studied in detail in Problems 29.1 through 29.4.

The coincidence degree was introduced by Mawhin (1972) in order to prove systematically existence theorems for differential equations (e.g., the existence of periodic solutions).

Proof of Proposition 29.1. Ad(a). The operator $B\colon X \to Y$ is a Fredholm operator of index zero, and $S\colon X \to Y$ is a compact perturbation of B. Hence S is a Fredholm operator of index zero. Thus, in order to prove the bijectivity of S, it is sufficient to show that $Su = 0$ implies $u = 0$.

Let $Su = 0$. Then

$$Bu = -\sum_{i=1}^{n} \langle v_i^*, u\rangle v_i.$$

From $\langle u_k^*, Bu\rangle = \langle B^*u_k^*, u\rangle = 0$ for all k it follows from (3) that $\langle v_k^*, u\rangle = 0$ for all k and hence $Bu = 0$. This implies $u = \sum_i c_i u_i$. Finally, we obtain from (3) that

$$c_k = \langle v_k^*, u\rangle = 0 \qquad \text{for all } k,$$

i.e., $u = 0$.

Thus, the operator $S\colon X \to Y$ is bijective and $S^{-1}\colon Y \to X$ is continuous by the open mapping theorem $A_1(36)$. By definition of S, we obtain $Su_i = v_i$, i.e.,

$$S^{-1}v_i = u_i \qquad \text{for all } i.$$

Ad(b). Let $b \in R(B)$. Then $\langle u_k^*, b\rangle = 0$ and $\langle u_k^*, Bu\rangle = 0$ for all $u \in X$ and all k.

Let $u = S^{-1}b$. Then $Su = b$ and hence

$$Bu = b - \sum_{i=1}^{n} \langle v_i^*, u\rangle v_i.$$

Applying u_k^* to this equation, we obtain from $\langle u_k^*, v_i\rangle = \delta_{ki}$ that $\langle v_k^*, u\rangle = 0$ for all k, i.e., $Bu = b$. Thus, $BS^{-1}b = b$.

Ad(c).

(I) (A)$\Leftrightarrow$(B). By definition of S, equation (A) is equivalent to

$$Su = Mu + \sum_{i=1}^{n} \langle v_i^*, u\rangle v_i.$$

This is equivalent to (B).

(II) (B)$\Leftrightarrow$(C)$\Leftrightarrow$(E)$\Leftrightarrow$(F). This is obvious. For example, if u is a solution of (E), then $\langle u_i^*, Mu\rangle = 0$ for all i, since $t \neq 0$. Hence $Mu \in R(B)$. Applying the operator B to the first equation of (E), we get $Bu = Mu$, since $Bu_i = 0$ and $BS^{-1}Mu = Mu$, according to Proposition 29.1(b).

(III) (A)$\Leftrightarrow$(D). We set $u = v + w$ with $v = Qu$ and $w = (I - Q)u$. Then (A) is equivalent to the system

$$B(v + w) = PM(v + w), \qquad (I - P)M(v + w) = 0,$$

and, in turn, this is equivalent to

$$Sw = PM(v + w), \qquad (I - P)M(v + w) = 0.$$

Note that $Bv = 0$, and that $Qw = 0$ implies $\langle v_i^*, w\rangle = 0$ for all i. □

The approach above is convenient for treating concrete problems because we have explicit constructions at hand. However, in order to get more insight, we now consider a geometrical approach for a more general situation. The basic idea is contained in Figure 29.2.

Note that the following considerations contain the situation above as a special case, if we set

$$Ju_i = v_i, \qquad i = 1, \ldots, n.$$

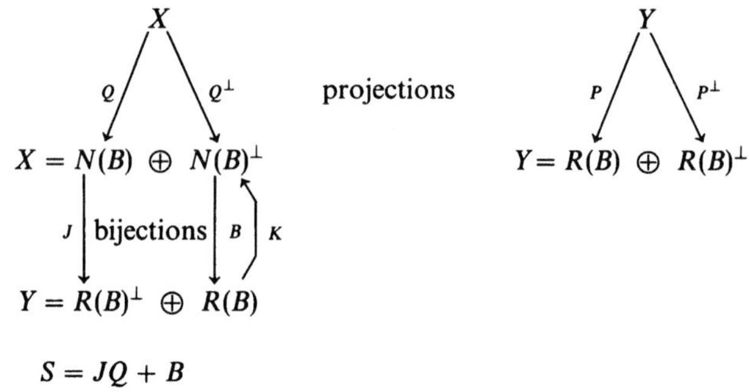

Figure 29.2

We make the following assumptions:

(H1*) X and Y are B-spaces over $\mathbb{K} = \mathbb{R}, \mathbb{C}$, and the operator $M: D(M) \subseteq X \to Y$ is given.

(H2*) The linear operator $B: D(B) \subseteq X \to Y$ is Fredholm of index zero, i.e., the range $R(B)$ is closed and

$$\dim N(B) = \operatorname{codim} R(B) < \infty.$$

(H3*) Let

$$Q: X \to N(B), \qquad P: Y \to R(B),$$

be projection operators onto $N(B)$ and $R(B)$, respectively.

Such operators always exist because $\dim N(B) < \infty$ and $\operatorname{codim} R(B) < \infty$. We set $Q^\perp = I - Q$ and $P^\perp = I - P$ as well as

$$N(B)^\perp = Q^\perp(X), \qquad R(B)^\perp = P^\perp(Y).$$

This yields the topological direct sums

$$X = N(B) \oplus N(B)^\perp, \qquad Y = R(B) \oplus R(B)^\perp.$$

Recall that

$$Qu = u \qquad \text{iff} \qquad u \in N(B)$$

and $Qu = 0$ iff $u \in N(B)^\perp$. Hence $Q^2 = Q$. It follows that the operator

$$B: N(B)^\perp \cap D(B) \to R(B)$$

is bijective. In fact, $Bu = 0$ and $u \in N(B)^\perp$ imply $u \in N(B)$, i.e., $u = 0$. Let

$$K: R(B) \to N(B)^\perp \cap D(B)$$

be the corresponding inverse operator, i.e., $BKb = b$ on $R(B)$.

(H4*) Let $J: N(B) \to R(B)^\perp$ be a bijective linear operator. Such an operator always exists because (H2*) implies that $\dim N(B) = \dim R(B)^\perp < \infty$.

Now we are ready to give our *key definition*:

$$Su = Bu + JQu \qquad \text{for all} \quad u \in D(B).$$

We shall show below that the operator $S\colon D(B) \to Y$ is bijective and

$$Su = Bu \qquad \text{for all} \quad u \in N(B)^\perp \cap D(B).$$

Hence

$$S^{-1}b = Kb \qquad \text{for all} \quad b \in R(B).$$

Moreover, we shall show below that the following problems are mutually equivalent:

(A*) *Original equation*

$$Bu = Mu, \qquad u \in D(B) \cap D(M).$$

(B*) *Fixed-point problem of E. Schmidt*

$$u = S^{-1}Mu + Qu, \qquad u \in D(M).$$

(C*) *Branching equation of E. Schmidt*

$$u = S^{-1}Mu + v, \qquad v = Qu.$$

We seek $u \in D(M)$ and $v \in N(B)$.

(D*) *Branching equation of Ljapunov*

$$S^{-1}PM(v + w) = w, \qquad P^\perp M(v + w) = 0.$$

We seek $v \in N(B)$, $w \in N(B)^\perp$ and set $u = v + w$.

Since $S^{-1}b = Kb$ on $R(B)$, this problem is identical with

$$KPM(v + w) = w, \qquad P^\perp M(v + w) = 0.$$

(E*) *Alternative problem of Cesari*

$$u = v + S^{-1}Mu, \qquad v = v - tJ^{-1}P^\perp Mu.$$

Here let $t \in \mathbb{K}$ be a fixed nonzero number. We seek $u \in D(M)$ and $v \in N(B)$.

Obviously, this problem is equivalent to

$$u = v + S^{-1}Mu, \qquad v = v - tJ^{-1}P^\perp M(v + S^{-1}Mu).$$

(F*) *Alternative problem of Mawhin*

$$u = S^{-1}PMu + S^{-1}P^\perp Mu + Qu.$$

As we shall show, this problem is the same as

$$u = KPMu + J^{-1}P^\perp Mu + Qu, \qquad u \in D(M).$$

Proposition 29.3. *Suppose that* (H1*) *through* (H4*) *hold. Then the operator* $S\colon D(B) \to Y$ *is bijective and*

$$Su = Bu \quad \text{on } N(B)^\perp \cap D(B), \qquad Su = Ju \quad \text{on } N(B).$$

Problems (A*) *through* (F*) *are mutually equivalent.*

PROOF.

(I) S is injective. In fact, if $Su = 0$, then it follows from $Su = Bu + JQu$ and $Bu \in R(B)$, $JQu \in R(B)^\perp$, that

$$Bu = 0, \qquad JQu = 0.$$

This implies $u \in N(B)$ and $Qu = 0$, i.e., $u = 0$.

(II) S is surjective. To show this let $y \in Y$. Then there exists a decomposition

$$y = b + c, \qquad b \in R(B), \qquad c \in R(B)^\perp.$$

Let $u = Kb + J^{-1}c$. Then, by the definition of S and Figure 29.2,

$$Su = SKb + SJ^{-1}c = b + c = y.$$

(III) Note that $Q^2 = Q$. From $SQ = JQ$ it follows that

$$S^{-1}JQ = Q.$$

Furthermore, from $J^{-1} = QJ^{-1}$ and hence $SJ^{-1}P^\perp = JQJ^{-1}P^\perp = P^\perp$ we obtain

$$S^{-1}P^\perp = J^{-1}P^\perp.$$

(IV) (A*)$\Leftrightarrow$(B*). Equation (A*) is equivalent to

$$Su = Mu + JQu,$$

and, in turn, this is equivalent to

$$u = S^{-1}Mu + S^{-1}JQu = S^{-1}Mu + Qu.$$

(B*)$\Leftrightarrow$(C*)$\Leftrightarrow$(E*). This is obvious.

(B*)$\Leftrightarrow$(F*). This follows from $S^{-1}b = Kb$ on $R(B)$ and $S^{-1}P^\perp = J^{-1}P^\perp$.

(A*)$\Leftrightarrow$(D*). This follows as in the proof of (A)$\Leftrightarrow$(D) above. □

29.2. Fredholm Alternatives for Asymptotically Linear, Compact Perturbations of the Identity

We consider the nonlinear operator equation

$$Bu + Nu = b, \qquad u \in X, \tag{14}$$

together with the linearized problem

$$Bu = b, \qquad u \in X. \tag{15}$$

We set $B = I + L$.

Theorem 29.A (Kačurovskii (1970)). *Suppose that:*

(i) *The operators* $L, N: X \to X$ *are compact on the* B-*space* X.
(ii) L *is linear, and* $L + N$ *is asymptotically linear, i.e.,* $\|Nu\|/\|u\| \to 0$ *as* $\|u\| \to \infty$.

Then:

(a) *If* $Bu = 0$ *implies* $u = 0$, *then equation* (14) *has a solution for each* $b \in X$.
(b) *If* $R(N) \subseteq R(B)$, *then equation* (14) *has a solution* u iff $b \in R(B)$.

By (i), the operator $B: X \to X$ is Fredholm of index zero. Thus, $R(N) \subseteq R(B)$ iff

$$\langle u^*, Nv \rangle = 0 \qquad \text{for all} \quad u^* \in N(B^*), \quad v \in X. \tag{16}$$

Moreover, $b \in R(B)$ iff

$$\langle u^*, b \rangle = 0 \qquad \text{for all} \quad u^* \in N(B^*). \tag{17}$$

Proof. We will use the pseudoresolvent S^{-1} corresponding to B and the fixed-point index introduced in Chapter 12.

Ad(a). If $N(B) = \{0\}$, then $R(B) = X$. Consequently, (a) is a special case of (b).

Ad(b).

(I) Suppose that (14) has the solution u. Then $b \in Nu + R(B)$, i.e., $b \in R(B)$.

(II) Conversely, let $b \in R(B)$. We want to show that equation (14) has a solution.

(II-1) Modified equation. Suppose that u is a solution of the equation

$$Su + Nu - b = 0. \tag{18}$$

Since $R(N) \subseteq R(B)$ and $b \in R(B)$, we get $Su \in R(B)$. By Proposition 29.1(b), $Su = Bu$. Hence $Bu + Nu - b = 0$, i.e., u is a solution of (14).

(II-2) Solution of (18) via homotopy. We may suppose that $X \neq \{0\}$. Define $H(u,t)$ by the relation

$$u - H(u,t) = Su + t(Nu - b),$$

where $0 \leq t \leq 1$, and the operator S is defined by (6) if $N(B) \neq \{0\}$. In the special case $N(B) = \{0\}$ let $S = B$. Since $S = B +$ compact, we obtain that the map

$$H: X \times [0,1] \to X$$

is compact. By Proposition 29.1, the inverse $S^{-1}: X \to X$ is linear and continuous.

Let $U = \{u \in X: \|u\| < R\}$. For sufficiently large R we have the *key* relation

$$H(u,t) \neq u \qquad \text{for all} \quad (u,t) \in \partial U \times [0,1]. \tag{19}$$

Indeed, this follows from $\|u\| = \|S^{-1}Su\| \leq \|S^{-1}\| \, \|Su\|$ and hence

$$\|u - H(u,t)\| \geq \|S^{-1}\|^{-1}\|u\| - \|Nu\| - \|b\|,$$

noting that $\|Nu\|/\|u\| \to 0$ as $\|u\| \to \infty$. By the homotopy invariance (A4) of the fixed-point index in Section 12.3, it follows from (19) that

$$i(H(\cdot,0), U) = i(H(\cdot,1), U).$$

Since $u \mapsto H(u,0)$ is odd, we obtain $i(H(\cdot,0), U) \neq 0$, by the antipodal theorem in Section 16.3. Hence

$$i(H(\cdot,1), U) \neq 0.$$

According to the existence principle (A2) in Section 12.3, the equation $u = H(u,1)$, $u \in U$, has a solution, i.e., $Su + Nu - b = 0$. $\square$

29.3. Application to Nonlinear Systems of Real Equations

We consider the nonlinear system

$$\sum_{j=1}^{n} b_{ij}\xi_j + N_i(\xi_1,\ldots,\xi_n) = b_i, \qquad i = 1,\ldots,n. \tag{20}$$

For given $b = (b_1,\ldots,b_n)$ in $\mathbb{R}^n$, we seek $x = (\xi_1,\ldots,\xi_n)$ in $\mathbb{R}^n$. We make the following assumptions on the nonlinearities N_i:

(H1) Growth condition. The functions $N_i\colon \mathbb{R}^n \to \mathbb{R}$ are continuous for all i. There are numbers $c > 0$ and $0 \le \alpha < 1$ such that

$$|N_i(x)| \le c|x|^\alpha \qquad \text{for all} \quad x \in \mathbb{R}^n \text{ and all } i.$$

(H2) Orthogonality condition. We have

$$\sum_{i=1}^{n} \xi_i^* N_i(x) = 0$$

for all $x \in \mathbb{R}^n$ and all solutions $(\xi_1^*,\ldots,\xi_n^*)$ of the *dual* linear system

$$\sum_{i=1}^{n} \xi_i^* b_{ij} = 0, \qquad j = 1,\ldots,n. \tag{20*}$$

Proposition 29.4. *Let the real $(n \times n)$-matrix (b_{ij}) be given.*

(a) *Assume* (H1) *and* $\det(b_{ij}) \neq 0$. *Then, for each $b \in \mathbb{R}^n$, equation* (20) *has a solution.*

(b) *Assume* (H1) *and* (H2). *Then, equation* (20) *has a solution $x \in \mathbb{R}^n$ iff*

$$\sum_{i=1}^{n} \xi_i^* b_i = 0$$

for all solutions $(\xi_1^*,\ldots,\xi_n^*)$ of (20*).

PROOF. By (H1), $|N(x)|/|x| \le \text{const}|x|^{\alpha-1} \to 0$ as $|x| \to \infty$. Now, the assertion follows from Theorem 29.A with $X = \mathbb{R}^n$. $\square$

29.4. Application to Integral Equations

Let $-\infty < a < b < \infty$. We consider the Hammerstein integral equation

$$u(t) - \int_a^b k(t,s)[u(s) + f(s,u(s))]\,ds = g(t), \qquad a < t < b, \tag{21}$$

together with the linearized equation

$$u(t) - \int_a^b k(t,s)u(s)\,ds = 0, \qquad a < t < b. \tag{22}$$

Proposition 29.5. *Suppose that:*

(i) *The functions $k: [a,b] \times [a,b] \to \mathbb{R}$ and $f: [a,b] \times \mathbb{R} \to \mathbb{R}$ are continuous, and there are numbers $c > 0$ and $0 \le \alpha < 1$ such that*

$$|f(s,u)| \le c|u|^\alpha \qquad \text{for all} \quad (s,u) \in [a,b] \times \mathbb{R}.$$

(ii) *The linearized equation (22) has only the trivial solution $u \equiv 0$ in $C[a,b]$.*

Then, for each $g \in C[a,b]$, equation (21) has a solution u in $C[a,b]$.

PROOF. Set $X = C[a,b]$ and

$$(Nu)(t) = -\int_a^b k(t,s)f(s,u(s))\,ds.$$

By (i),

$$\|Nu\| = \max_{a \le t \le b} |(Nu)(t)| \le c(b-a) \cdot \max_{a \le t,s \le b} |k(t,s)|\,\|u\|^\alpha.$$

Hence $\|Nu\|/\|u\| \to 0$ as $\|u\| \to \infty$. Now use Theorem 29.A(a). $\square$

29.5. Application to Differential Equations

Let $-\infty < a < b < \infty$. We consider the semilinear boundary value problem

$$\begin{gathered} -u''(t) + q(t)u(t) + f(t,u(t)) = h(t), \qquad a < t < b, \\ u(a) = u(b) = 0, \end{gathered} \tag{23}$$

together with the linearized problem

$$\begin{gathered} -u''(t) + q(t)u(t) = 0, \qquad a < t < b, \\ u(a) = u(b) = 0. \end{gathered} \tag{24}$$

Proposition 29.6. *Suppose that:*

(i) *The functions* $q: [a,b] \to \mathbb{R}$ *and* $f: [a,b] \times \mathbb{R} \to \mathbb{R}$ *are continuous, and there are numbers* $c > 0$ *and* $0 \leq \alpha < 1$ *such that*

$$|f(t,u)| \leq c|u|^\alpha \qquad \text{for all} \quad (t,u) \in [a,b] \times \mathbb{R}.$$

(ii) *The linearized problem* (24) *has only the trivial solution* $u \equiv 0$ *in* $C^2[a,b]$.

Then, for each $h \in C[a,b]$, *the original problem* (23) *has a solution* $u \in C^2[a,b]$.

PROOF. Let $k(\cdot,\cdot)$ be the Green function corresponding to the differential operator $Lu = u''$ with the boundary conditions $u(a) = u(b) = 0$. Then problem (23) is equivalent to the integral equation

$$u(t) = \int_a^b k(t,s)(q(s)u(s) + f(s,u(s)) - h(s))\,ds.$$

Note that $k: [a,b] \times [a,b] \to \mathbb{R}$ is continuous. Now, the proof proceeds analogously to the proof of Proposition 29.5. □

29.6. The Generalized Antipodal Theorem

In order to generalize Theorem 29.A to *noncompact* operators, we need a generalization of the antipodal theorem. To this end, we consider the operator equation

$$Bu + Nu = b, \qquad u \in X, \tag{25}$$

together with the Galerkin equations

$$\langle Bu_n + Nu_n, w_i \rangle = \langle b, w_i \rangle, \qquad u_n \in X_n, \qquad i = 1, \ldots, n, \tag{26}$$

where $X_n = \text{span}\{w_1, \ldots, w_n\}$, i.e.,

$$u_n = \sum_{i=1}^n c_{in} w_i.$$

Here we seek the real coefficients c_{in}. We make the following assumptions:

(H1) The operators $B, N: X \to X^*$ are bounded and demicontinuous on the real separable reflexive B-space X with $\dim X = \infty$. Let $\{w_1, w_2, \ldots\}$ be a basis in X.

(H2) B is odd, i.e., $B(-u) = -B(u)$ for all $u \in X$.

(H3) For fixed $b \in X^*$ and each $t \in [0,1]$, the operator A_t defined by

$$A_t u = Bu + t(Nu - b)$$

satisfies condition $(S)_0$ on X.

(H4) The decisive homotopy condition. There is an $R > 0$ such that

$$A_t u \neq 0 \qquad \text{for all} \quad u \in X \quad \text{with} \quad \|u\| = R \quad \text{and all} \quad t \in [0,1].$$

Theorem 29.B (Generalized Antipodal Theorem). *Assume* (H1) *through* (H4). *Then*:

(a) Existence. *Equation* (25) *has a solution* u *with* $\|u\| < R$.
(b) Galerkin method. *There is an* n_0 *such that, for each* $n \geq n_0$, *the Galerkin equation* (26) *has a solution* $u_n \in X_n$ *with* $\|u_n\| < R$. *The sequence* (u_n) *possesses a subsequence which converges strongly in* X *to* u.

If the original equation (25) *has a unique solution* u *with* $\|u\| < R$, *then the entire sequence* (u_n) *converges to* u.

Proof. According to Proposition 27.4 it suffices to prove that there is an n_0 such that, for each $n \geq n_0$, the Galerkin equation (26) has a solution u_n with $\|u_n\| < R$. To this end, we will use the classical antipodal theorem in finite-dimensional spaces.

We set $U_n = \{u \in X_n : \|u\| < R\}$ and

$$H_n(u, t) = u - \sum_{i=1}^{n} \langle A_t u, w_i \rangle w_i.$$

Our goal is to solve the equation

$$H_n(u_n, 1) = u_n, \qquad u_n \in U_n. \tag{27}$$

That means $\langle A_1 u_n, w_i \rangle = 0$ for $i = 1, \ldots, n$; therefore, u_n is a solution of the Galerkin equation (26).

(I) The decisive homotopy property. We will show below that there is an n_0 such that

$$H_n(u, t) \neq u \tag{28}$$

for all $(u, t) \in \partial U_n \times [0, 1]$ and all $n \geq n_0$.

(II) Solution of (27). From (28) and the homotopy invariance (A4) of the fixed-point index in Section 12.3 it follows that

$$i(H_n(\cdot, 0), U_n) = i(H_n(\cdot, 1), U_n).$$

By (H2), the map $u \mapsto H_n(u, 0)$ is odd. The antipodal theorem in Section 16.3 tells us that $i(H_n(\cdot, 0), U_n) \neq 0$ and hence

$$i(H_n(\cdot, 1), U_n) \neq 0.$$

By the existence principle (A2) in Section 12.3, equation (27) has a solution.

(III) Proof of (28). Let $E_n: X_n \to X$ be the embedding operator corresponding to $X_n \subseteq X$. Then, the dual operator $E_n^*: X^* \to X_n^*$ satisfies

$$\langle E_n^* u, v \rangle_{X_n} = \langle u, E_n v \rangle_X = \langle u, v \rangle_X \tag{29}$$

for all $u \in X^*$, $v \in X_n$.

(III-1) We first prove the following. For each $t \in [0, 1]$, there is a $\delta(t) > 0$ and an n_0 such that

$$\|E_n^* A_t u\| \geq \delta(t) \qquad \text{for all} \quad u \in \partial U_n, \quad n \geq n_0. \tag{30}$$

Otherwise, there is a t and a sequence $(u_{n'})$ with

$$\|E_{n'}^* A_t u_{n'}\| \to 0 \qquad \text{as} \quad n' \to \infty,$$

and $\|u_{n'}\| = R$ for all n'. For brevity, we write u_n instead of $u_{n'}$. By (29),

$$\begin{aligned} |\langle A_t u_n, u_n \rangle| &= |\langle E_n^* A_t u_n, u_n \rangle| \\ &\leq \|E_n^* A_t u_n\| R \to 0 \qquad \text{as} \quad n \to \infty. \end{aligned} \tag{30a}$$

Similarly, for all $w \in \operatorname{span}\{w_1, w_2, \ldots\}$, we obtain

$$\langle A_t u_n, w \rangle \to 0 \qquad \text{as} \quad n \to \infty. \tag{30b}$$

Since (u_n) is bounded, there is a subsequence, again denoted by (u_n), such that

$$u_n \rightharpoonup u \qquad \text{as} \quad n \to \infty.$$

Since B and N are bounded, the sequence $(A_t u_n)$ is bounded. By (30b),

$$A_t u_n \rightharpoonup 0 \qquad \text{as} \quad n \to \infty$$

(cf. Proposition 21.26(c), (f)). The operator A_t satisfies condition $(S)_0$. Thus, it follows from (30a) that

$$u_n \to u \qquad \text{as} \quad n \to \infty.$$

This implies

$$A_t u = 0.$$

In this connection, note that A_t is demicontinuous, since B and N have this property. Furthermore, $\|u\| = R$. This contradicts (H4).

(III-2) We show that δ and n_0 in (III-1) can be chosen independently of t. Indeed, note that $\|E_n\| = 1$ and hence $\|E_n^*\| = \|E_n\| = 1$. Moreover, note that B and N are bounded. Thus, for each $t_0 \in [0, 1]$, there exists an open neighborhood $U(t_0)$ of t_0 such that $\delta(t)$ and $n_0(t)$ can be chosen as constants on $U(t_0)$. Now the assertion follows from the fact that the compact set $[0, 1]$ can be covered by finitely many open sets $U(t_0)$, $U(t_1), \ldots$.

(III-3) The norm of the functional

$$w \mapsto \langle A_t u, w \rangle_X \quad \text{on } X_n$$

is equal to $\|E_n^* A_t u\|$. Indeed, it follows from (29) that

$$\|E_n^* A_t u\| = \sup_{\|w\|_{X_n} \leq 1} \langle E_n^* A_t u, w \rangle_{X_n} = \sup_{\|w\|_{X_n} \leq 1} \langle A_t u, w \rangle_X.$$

By (III-2), there is an n_0 such that

$$E_n^* A_t u \neq 0$$

for all $u \in \partial U_n$, $t \in [0, 1]$, $n \geq n_0$. Thus, we obtain

$$\langle A_t u, w_i \rangle \neq 0 \qquad \text{for some } i.$$

This implies (28). □

29.7. Fredholm Alternatives for Asymptotically Linear (S)-Operators

We consider the operator equation

$$Bu + Nu = b, \qquad u \in X, \tag{31}$$

and make the following assumptions:

(H1) The operator $B: X \to X^*$ is linear and continuous on the real separable reflexive B-space X.
(H2) The operator $N: X \to X^*$ is demicontinuous and bounded. Moreover, $B + N$ is asymptotically linear, i.e., $\|Nu\|/\|u\| \to 0$ as $\|u\| \to \infty$.
(H3) For each $b \in X^*$ and each $t \in [0, 1]$, the operator $u \mapsto Bu + t(Nu - b)$ satisfies condition (S) on X.

Theorem 29.C (Hess (1972)). *Assume* (H1) *through* (H3). *Then*:

(a) *Suppose that* $Bu = 0$ *implies* $u = 0$. *Then, for each* $b \in X^*$, *equation* (31) *has a solution* u.
(b) *Suppose that* $R(N) \subseteq R(B)$. *Then, equation* (31) *has a solution* u *iff* $b \in R(B)$.
(c) *The operator* B *is Fredholm of index zero.*

PROOF. Assertion (c) will be proved in Problem 29.5. Now the proof proceeds analogously to the proof of Theorem 29.A by replacing the classical antipodal theorem with the generalized antipodal theorem (Theorem 29.B). In this connection, note that the operator

$$u \mapsto Su + t(Nu - b) \tag{32}$$

is a strongly continuous perturbation of $u \mapsto Bu + t(Nu - B)$, according to (6). By (H3), the latter operator satisfies condition (S). Thus, it follows from Figure 27.1 that the operator (32) also satisfies condition (S) and hence condition $(S)_0$. □

Since B is a Fredholm operator of index zero, we have $R(N) \subseteq R(B)$ iff

$$\langle u^*, Nv \rangle = 0 \qquad \text{for all} \quad u^* \in N(B^*), \quad v \in X,$$

and $b \in R(B)$ iff

$$\langle u^*, b \rangle = 0 \qquad \text{for all} \quad u^* \in N(B^*).$$

29.8. Weak Asymptotes and Fredholm Alternatives

We again study the operator equation

$$Bu + Nu = b, \qquad u \in X. \tag{33}$$

We want to relax the condition $R(N) \subseteq R(B)$, which we employed in the preceding sections. Roughly speaking, we will use the asymptotic condition

$$(v|N(v+w)) = a(v) + o(\|v\|) \qquad \text{as} \quad \|v\| \to \infty,$$

for all $v \in N(B)$ and all $w \in X$, in order to obtain the necessary and sufficient *solvability condition*:

$$a(v) > (v|b) \qquad \text{for all} \quad v \in N(B) - \{0\}. \tag{34}$$

More precisely, we make the following assumptions:

(H1) The operator $B: X \to X$ is linear, continuous, and self-adjoint on the real H-space X. The range $R(B)$ is closed, and $0 < \dim N(B) < \infty$.
(H2) The operator $N: X \to X$ is compact and $\sup\{\|Nu\|: u \in X\} < \infty$.
(H3) The operator N has weak asymptotes on the null space $N(B)$ of B, i.e., for all $v \in N(B)$ and $w \in X$, we have the decomposition

$$(v|N(v+w)) = a(v) + r(v, w), \tag{35}$$

where

$$\lim_{\|v\| \to \infty} \sup_{w \in K} \frac{|r(v, w)|}{\|v\|} = 0,$$

for all bounded sets K in X, and

$$a(tv) = ta(v) \qquad \text{for all} \quad t > 0, \quad v \in N(B).$$

Theorem 29.D (Nečas (1973)). *Assume* (H1) *through* (H3). *Let* $b \in X$ *be given. Then*:

(a) *If condition* (34) *is satisfied, then the original equation* (33) *has a solution* u.
(b) *Conversely, if* (33) *has a solution* u *and if* $r(v, w) < 0$, *i.e.*,

$$a(v) > (v|N(v+w)) \qquad \textit{for all} \quad v \in N(B) - \{0\}, \quad w \in X, \tag{36}$$

then condition (34) *is satisfied.*

Corollary 29.7. *The same is true if we replace* "$>$" *by* "$<$" *in* (34) *and* (36).

Example 29.8. Suppose that (H1) holds with $\dim N(B) = 1$ and let e be a unit vector in $N(B)$. Moreover, set

$$Nv = h((v|e))e \qquad \text{for all} \quad v \in X,$$

and suppose that the function $h: \mathbb{R} \to \mathbb{R}$ is continuous and the finite limits

$$h(\pm\infty) = \lim_{s \to \pm\infty} h(s)$$

exist (Fig. 29.3). Then, the conditions (H2) and (H3) are satisfied if we set

$$a(v) = \begin{cases} h(+\infty)(v|e) & \text{if} \quad (v|e) > 0, \\ h(-\infty)(v|e) & \text{if} \quad (v|e) < 0, \\ 0 & \text{if} \quad v = 0. \end{cases}$$

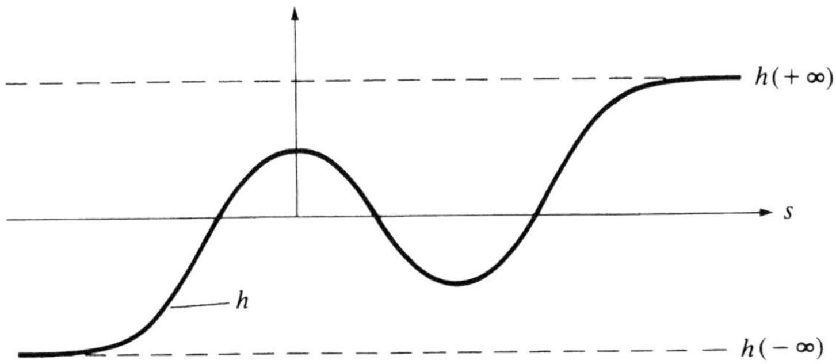

Figure 29.3

By Theorem 29.D, the equation $Bu + Nu = b$, $u \in X$, has a solution if (34) holds, i.e.,

$$h(-\infty) < (b|e) < h(+\infty)$$

in the case where $h(-\infty) < h(+\infty)$ (resp.

$$h(+\infty) < (b|e) < h(-\infty)$$

in the case where $h(+\infty) < h(-\infty)$). This sufficient solvability condition is also necessary if

$$h(-\infty) < h(s) < h(+\infty) \qquad \text{for all} \quad s \in \mathbb{R}$$

(resp. $h(+\infty) < h(s) < h(-\infty)$ for all $s \in \mathbb{R}$).

Proof of Theorem 29.D. We combine the alternative method of Cesari in Section 29.1 with the Leray–Schauder principle in Chapter 6. The necessary *a priori* estimates follow from the existence of weak asymptotes and from the solvability condition (34).

Ad(a).

(I) The equivalent fixed-point problem. Using the Identification Principle 21.18, we set $X = X^*$. Then, $\langle u^*, v\rangle = (u^*|v)$ and $B^* = B$. Let $S^{-1}: X \to X$ be the pseudoresolvent to B introduced in Section 29.1. Set

$$g(u) = S^{-1}(b - Nu).$$

We now use the alternative method of Cesari. By (11) and Proposition 29.1, the original problem (33) is equivalent to the fixed-point problem

$$\begin{aligned} u &= \lambda(v + g(u)), \\ c_i &= \lambda[c_i - t(u_i|N(v + g(u)) - b)], \qquad i = 1, \ldots, n, \end{aligned} \tag{37}$$

with $\lambda = 1$ and fixed $t > 0$. Here $\{u_1, \ldots, u_n\}$ is an orthonormal basis in the null space $N(B)$, and

$$v = \sum_{i=1}^{n} c_i u_i.$$

Equation (37) is equivalent to

$$\begin{aligned} u &= \lambda(v + g(u)), \qquad u \in X, \quad v \in N(B), \\ v &= \lambda\left(v - t\sum_{i=1}^{n} \alpha_i u_i\right), \end{aligned} \tag{37*}$$

where

$$\alpha_i = (u_i | N(v + g(u)) - b).$$

We write (37*) in the form

$$(u, v) = \lambda C(u, v), \qquad (u, v) \in X \times N(B). \tag{37**}$$

(II) Solution of (37**) via the Leray–Schauder principle (continuation with respect to λ). We will show below that the following two conditions are satisfied:

(i) The operator $C: X \times N(B) \to X \times N(B)$ is compact.
(ii) There is a number $k > 0$ such that

$$\|u\| + \|v\| \le k$$

for all solutions (u, v) of (37**) with $0 < \lambda < 1$.

Then the Leray–Schauder principle (Theorem 6.A) tells us that equation (37**) has a solution for $\lambda = 1$, i.e., the original equation (33) has a solution.

(II-1) We prove (i). The operator $N: X \to X$ is compact and $S^{-1}: X \to X$ is linear and continuous. This yields (i). Note that $\dim N(B) < \infty$, and hence continuous bounded operators on $N(B)$ are compact.

(II-2) We prove (ii). Since $\dim N(B) < \infty$, the set $Y = \{v \in N(B): \|v\| = 1\}$ is compact. By (34),

$$\mu \stackrel{\text{def}}{=} \min_{v \in Y} [a(v) - (v|b)] > 0.$$

By (H2),

$$\sup\{\|g(u)\|: u \in X\} < \infty. \tag{38}$$

Thus it follows from (H3) that

$$\lim_{\|v\| \to \infty} \sup_{u \in X} \frac{|r(v, g(u))|}{\|v\|} = 0.$$

Again by (H3),

$$\begin{aligned} \beta &\stackrel{\text{def}}{=} (v|N(v + g(u)) - b) = a(v) - (v|b) + r(v, g(u)) \\ &\ge \mu\|v\| + r(v, g(u)) \qquad \text{for all} \quad v \in N(B), \quad u \in X. \end{aligned} \tag{39}$$

Now let (u, v) be a solution of (37*). By the second equation of (37*),

$$\begin{aligned} \|v\|^2 &= \lambda^2\left(\|v\|^2 - 2t\beta + t^2 \sum_{i=1}^{n} \alpha_i^2\right) \\ &\le \lambda^2(\|v\|^2 - 2t\mu(\|v\| + o(\|v\|) + t^2 O(1)), \qquad \|v\| \to \infty. \end{aligned} \tag{40}$$

Note that $\sup_i |\alpha_i| < \infty$. By (40), there is a number k_1 such that

$$\|v\| \leq k_1$$

for all solutions (u, v) of (37*) with $0 < \lambda < 1$. Otherwise, there is a sequence (u_n, v_n) of solutions of (37*) with $\|v_n\| \to \infty$ as $n \to \infty$. Letting $v = v_n$ in (40) and dividing by $\|v_n\|^2$, we obtain $1 \leq \lambda^2$ as $n \to \infty$. This is a contradiction.

By the first equation of (37*), there is a number k_2 such that $\|u\| \leq k_2$ for all solutions (u, v) of (37*) with $0 < \lambda < 1$, by (38).

Ad(b). Let $Bu + Nu = b$. Suppose that

$$(v|N(v + w)) < a(v) \qquad \text{for all} \quad v \in N(B) - \{0\}, \quad w \in X.$$

The operator B is self-adjoint. From $(v|Bu) = (Bu|v) = 0$ for all $v \in N(B)$ it follows that

$$(v|Nu) = (b|v).$$

Hence $(b|v) < a(v)$ for all $v \in N(B) - \{0\}$. □

Replacing t by $-t$, we obtain Corollary 29.7.

29.9. Application to Semilinear Elliptic Differential Equations of the Landesman–Lazer Type

We want to apply Theorem 29.D to the boundary value problem

$$\begin{aligned} -\Delta u - \mu u + h(u) &= f \quad \text{on } G, \\ u &= 0 \quad \text{on } \partial G, \end{aligned} \tag{41}$$

where μ is a fixed real number. The corresponding linearized problem reads as follows:

$$\begin{aligned} -\Delta u - \mu u &= 0 \quad \text{on } G, \\ u &= 0 \quad \text{on } \partial G. \end{aligned} \tag{42}$$

If μ is *not* an eigenvalue of (42) (resp. if μ is an eigenvalue of (42)), then we speak of the *nonresonance* case (resp. resonance case). We assume:

(H1) G is a bounded region in $\mathbb{R}^n$, $n \geq 1$.
(H2) The function $h: \mathbb{R} \to \mathbb{R}$ is continuous and bounded.
(H3) There exist the finite limits

$$h(\pm\infty) = \lim_{s \to \pm\infty} h(s)$$

and $h(-\infty) < h(s) < h(+\infty)$ for all $s \in \mathbb{R}$ (cf. Fig. 29.3 above).

Let $u \in \mathring{W}_2^m(G)$, $m \geq 1$. We set

$$G_{\pm} = \{x \in G: u(x) \gtrless 0\}$$

and

$$a_{\pm}(u) = \int_{G_+} h(\pm\infty)u(x)\,dx + \int_{G_-} h(\mp\infty)u(x)\,dx$$

for $u \neq 0$. In the case where $u = 0$ let $a(0) = 0$. As usual, we set $\int_M \ldots dx = 0$ if $M = \emptyset$. Note that $a_{\pm}$ is well defined. Indeed, if we change u on a set of measure zero, then $G_{\pm}$ may change, but $a_{\pm}(u)$ remains unchanged.

We first formulate a simple necessary solvability condition for (41) in the resonance case. To this end, let (μ, u_0) be a classical eigensolution of (42) and suppose that the original problem (41) has a classical solution. Using integration by parts, it follows from (41) that

$$\int_G f u_0\,dx = \int_G h(u_0)u_0\,dx.$$

By (H3), we obtain

$$a_-(u_0) < \int_G u_0 f\,dx < a_+(u_0). \tag{43}$$

In Proposition 29.10 below we will obtain the *surprising fact* that the simple necessary solvability condition (43) for problem (41) is also sufficient in the case where the eigenvalue μ is simple.

Instead of (41) we consider the more general problem

$$\begin{aligned} Lu - \mu u + h(u) &= f \quad \text{on } G, \\ D^\gamma u &= 0 \quad \text{on } \partial G \qquad \text{for all} \quad \gamma\colon |\gamma| \leq m-1, \end{aligned} \tag{44}$$

together with the linearized problem

$$\begin{aligned} Lu - \mu u &= 0 \quad \text{on } G, \\ D^\gamma u &= 0 \quad \text{on } \partial G \qquad \text{for all} \quad \gamma\colon |\gamma| \leq m-1, \end{aligned} \tag{45}$$

where $m \geq 1$ and

$$Lu(x) = \sum_{|\alpha|, |\beta| \leq m} (-1)^{|\alpha|} D^\alpha(a_{\alpha\beta}(x) D^\beta u(x)).$$

(H4) Let the functions $a_{\alpha\beta}\colon G \to \mathbb{R}$ be measurable and bounded with $a_{\alpha\beta} = a_{\beta\alpha}$ for all α, β. Suppose that L is regularly strongly elliptic or strongly elliptic in the sense of Definition 22.42.

Note that (41) is a special case of (44) with $m = 1$.

Definition 29.9. Let $X = \mathring{W}_2^m(G)$. The *generalized problem* for (44) reads as follows. For given $f \in L_2(G)$, we seek $u \in X$ such that

$$c(u, v) + d(u, v) = b(v) \qquad \text{for all} \quad v \in X. \tag{46}$$

Here we set

$$c(u, v) = \int_G \left(\sum_{|\alpha|, |\beta| \le m} a_{\alpha\beta} D^\beta u D^\alpha v - \mu uv \right) dx,$$

$$d(u, v) = \int_G h(u) v \, dx, \qquad b(v) = \int_G fv \, dx.$$

Formally, we obtain (46) by multiplying (44) with $v \in C_0^\infty(G)$ and using subsequent integration by parts.

In the special case (41), we obtain

$$c(u, v) = \int_G \left(\sum_{i=1}^n D_i u D_i v - \mu uv \right) dx,$$

where $x = (\xi_1, \ldots, \xi_n)$ and $D_i = \partial / \partial \xi_i$.

Proposition 29.10 (Landesman and Lazer (1969)). *Assume* (H1) *through* (H4). *Then*:

(a) The nonresonance case. *If μ is not an eigenvalue of the generalized problem for* (45), *then for each $f \in L_2(G)$, the generalized problem for* (44) *has a solution.*
(b) The resonance case. *If μ is a simple eigenvalue of the generalized problem for* (45) *with the eigenfunction u_0, then the generalized problem for* (44) *has a solution iff condition* (43) *is satisfied.*

Corollary 29.11. *Statement* (a) *remains true if L is not symmetric and assumption* (H3) *drops out.*

Proof. Ad(a). We apply Theorem 29.C. Using the Identification Principle 21.18, we set $X = X^*$. We define the operators $B, N: X \to X$ through

$$(Bu|v)_X = c(u, v), \qquad (Nu|v)_X = d(u, v) \qquad \text{for all} \quad u, v \in X.$$

By (22.1b), $b \in X^*$. Hence the generalized problem (46) is equivalent to the operator equation

$$Bu + Nu = b, \qquad u \in X. \tag{47}$$

It follows from (H4) and Proposition 22.45 that $c(\cdot, \cdot)$ is a regular Gårding form. By Proposition 21.31 and Lemma 22.38, B is the sum of a linear strongly monotone operator and a linear compact operator. Hence it follows from Theorem 21.F that B is Fredholm of index zero. By Figure 27.1, B satisfies condition (S).

By Corollary 26.14, the operator $N: X \to X$ is strongly continuous. Since h is bounded,

$$|(Nu|v)| = \left| \int_G h(u) v \, dx \right| \le \text{const} \left(\int_G v^2 \, dx \right)^{1/2}$$

$$\le \text{const} \, \|v\|_X \qquad \text{for all} \quad u, v \in X,$$

i.e., $\|Nu\| \le \text{const}$ for all $u \in X$.

For each $t \in [0, 1]$, the operator $u \mapsto Bu + t(Nu - b)$ is a strongly continuous perturbation of the (S)-operator B. Thus, this operator also satisfies condition (S), by Figure 27.1.

Since μ is not an eigenvalue of (45), $Bu = 0$ implies $u = 0$. Now, statement (a) and Corollary 29.11 follow from Theorem 29.C.

Ad(b). We apply Theorem 29.D. By assumption, $\dim N(B) = 1$ and

$$N(B) = \{\pm tu_0 : t \geq 0\}.$$

Since $a_{\alpha\beta} = a_{\beta\alpha}$, we obtain $c(u, v) = c(v, u)$ for all $u, v \in X$, i.e., B is self-adjoint. The operator B is Fredholm. Hence $R(B)$ is closed.

(I) We show that N has weak asymptotes on $N(B)$. By (H2), the principle of majorized convergence $A_2(19)$ yields the *key* relation:

$$\lim_{t \to +\infty} (N(tu_0 + w)|u_0) = \lim_{t \to +\infty} \int_G h(tu_0 + w)u_0 \, dx$$

$$= \int_{G_+} h(+\infty)u_0 \, dx + \int_{G_-} h(-\infty)u_0 \, dx = a_+(u_0). \tag{48}$$

This convergence is uniform with respect to w on each bounded subset of X. Otherwise, there are a number $\varepsilon > 0$, a bounded sequence (w_n) in X, and a real sequence (t_n) with $t_n \to +\infty$ as $n \to \infty$ such that

$$|(N(t_n u_0 + w_n)|u_0) - a_+(u_0)| \geq \varepsilon \qquad \text{for all} \quad n \in \mathbb{N}. \tag{49}$$

Since the embedding $\mathring{W}_2^m(G) \subseteq L_2(G)$ is compact, we may assume that $w_n \to w$ in $L_2(G)$ as $n \to \infty$. This implies

$$w_{n'}(x) \to w(x) \qquad \text{as} \quad n \to \infty \qquad \text{for almost all} \quad x \in G.$$

Using majorized convergence, it follows from (49) that $|a_+(u_0) - a_+(u_0)| \geq \varepsilon$. This is a contradiction.

We set

$$a(tu_0) = \begin{cases} ta_\pm(u_0) & \text{if} \quad t \gtrless 0, \\ 0 & \text{if} \quad t = 0. \end{cases}$$

Relation (48) and a similar relation as $t \to -\infty$ imply

$$(N(tu_0 + w)|tu_0) = a(tu_0) + r(tu_0, w),$$

for all $t \in \mathbb{R}$, where

$$\lim_{t \to \pm\infty} \sup_{w \in K} \frac{|r(tu_0, w)|}{|t|} = 0,$$

for each bounded subset K of X. This is relation (35).

(II) We show that

$$(N(v + w)|v) < a(v) \qquad \text{for all} \quad v \in N(B) - \{0\}, \quad w \in X.$$

This follows from

$$(N(tu_0 + w)|tu_0) = \int_G h(tu_0 + w)tu_0\,dx < ta_\pm(u_0) \qquad \text{if} \quad t \gtrless 0,$$

since $h(-\infty) < h(s) < h(+\infty)$ for all $s \in \mathbb{R}$.

(III) By Theorem 29.D, the condition

$$b(v) < a(v) \qquad \text{for all} \quad v \in N(B) - \{0\} \tag{50}$$

is necessary and sufficient for the solvability of the equation $Bu + Nu = b$, $u \in X$. Condition (50) means

$$\int_G tu_0 f\,dx < ta_\pm(u_0) \qquad \text{if} \quad t \gtrless 0.$$

This is the solvability condition (43). □

29.10. The Main Theorem on Nonlinear Proper Fredholm Operators

Let $n_X(b)$ denote the *number of solutions* of the equation

$$Au = b, \qquad u \in X. \tag{51}$$

We want to study the properties of the function $n_X(\cdot)$ on the image space Y of the operator A. We set

$$D_{\text{sing}} = \{u \in X: A'(u)h = 0 \text{ has a solution } h \neq 0\},$$

i.e., D_{sing} is the set of singular points of A.

Theorem 29.E (Main Theorem). *Let X and Y* be B-*spaces over* $\mathbb{K} = \mathbb{R}, \mathbb{C}$, *and let $A: X \to Y$ be a proper C^k-Fredholm operator of index zero with $1 \le k \le \infty$. Then*:

(a) *For each $b \in Y - A(D_{\text{sing}})$, the number $n_X(b) \ge 0$ is finite.*
(b) *The function $n_X(\cdot)$ is constant on each connected subset of the open and dense subset $Y - A(D_{\text{sing}})$ of Y.*
(c) *For each u in the open set $X - D_{\text{sing}}$, the operator A is locally invertible, i.e., A is a local C^k-diffeomorphism at u.*
(d) *If the set D_{sing} is empty, then $A: X \to Y$ is a C^k-diffeomorphism.*
(e) *The set $A(D_{\text{sing}})$ of singular values of A is closed and nowhere dense in Y.*

The proof of Theorem 29.E will be given in Section 29.10e after a detailed discussion of Theorem 29.E.

Recall that (c) is equivalent to the following *important fact*. Let $u \in X - D_{\text{sing}}$ be a solution of the original equation (51). Then there are neighborhoods

U and V of u and b, respectively, such that for each $\bar{b} \in V$ the equation

$$A\bar{u} = \bar{b}, \qquad \bar{u} \in U,$$

has a unique solution $\bar{u}$. If we set $\bar{u} = B(\bar{b})$, then the solution operator B is C^k on V.

The fundamental Theorem 29.E tells us that proper Fredholm mappings of index zero behave like "reasonable" functions $A: \mathbb{R} \to \mathbb{R}$ (Fig. 29.4(a), (b)). Roughly speaking, we obtain that, for "almost all" right members b, equation (51) has at most a finite number of solutions, and this number may *bifurcate* only at the singular values $b \in A(D_{\text{sing}})$. This critical set $A(D_{\text{sing}})$ is nowhere dense in Y, i.e., it is *meagre*.

The more general equation $Au = b$, $u \in S$, where S is an arbitrary subset of the B-space X, will be considered in Theorem 29.F below.

EXAMPLE 29.12. The C^1-function

$$A: \mathbb{R} \to \mathbb{R} \tag{52}$$

satisfies the assumptions of Theorem 29.E iff

$$|Au| \to \infty \qquad \text{as} \quad |u| \to \infty. \tag{53}$$

Moreover, we have

$$u \in D_{\text{sing}} \qquad \text{iff} \qquad A'(u) = 0.$$

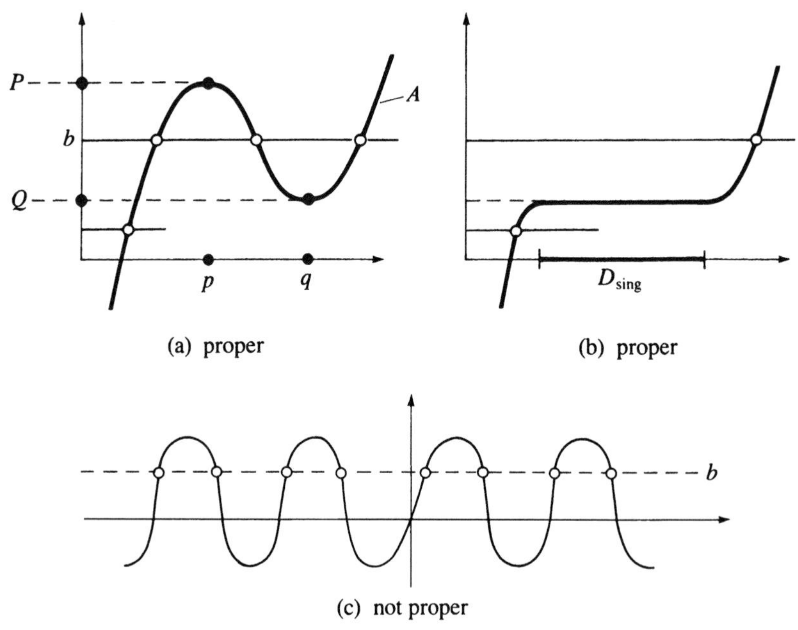

(a) proper (b) proper

(c) not proper

Figure 29.4

The function A in (52) is always Fredholm of index zero. The condition (53) is equivalent to the properness of A.

In Figure 29.4 (a) the set D_{sing} consists of the two points p and q, and the set $A(D_{\text{sing}})$ of the singular values of A consists of the two points P and Q. The function $n_{\mathbb{R}}(\cdot)$ bifurcates at P and Q. Figure 29.4(a) allows an intuitive interpretation of Theorem 29.E.

Figure 29.4(b) shows that, in contrast to $A(D_{\text{sing}})$, the set D_{sing} is not always meagre.

If A is *not* proper. i.e., the growth condition (53) is violated, then Theorem 29.E is not always true. Look for an example in Figure 29.4(c) which shows $Au = \sin u$. Thus, the assumption of *properness* is essential in Theorem 29.E.

PROOF. This follows easily from the definition of the notions "proper" and "Fredholm operator" given in Part I, which we recall in what follows. □

29.10a. Basic Definitions

Recall from Section 4.16 that a mapping is called *proper* iff the preimages of compact sets are again compact. If the operator $A: D(A) \subseteq X \to Y$ is continuous, where X and Y are B-spaces, then A is proper iff:

$Au_n \to a$ as $n \to \infty$ implies the existence of a subsequence $u_{n'} \to u$ as $n \to \infty$ such that $u \in D(A)$.

Recall from Section 8.4 the following. Let X and Y be B-spaces over $\mathbb{K} = \mathbb{R}$, $\mathbb{C}$, and let $L(X, Y)$ denote the set of all linear continuous operators $C: X \to Y$. Then, $L(X, Y)$ is a B-space over $\mathbb{K}$ with respect to the operator norm. The *linear* operator

$$C: X \to Y$$

is called *Fredholm* iff C is continuous and

$$\dim N(C) < \infty \qquad \text{and} \qquad \operatorname{codim} R(C) < \infty,$$

where $N(C) = \{u \in X: Cu = 0\}$ and $R(C) = C(X)$. This implies that the range $R(C)$ is closed. The *index* of C is defined through

$$\operatorname{ind} C = \dim N(C) - \operatorname{codim} R(C).$$

The number $\dim R(C)$ is called the *rank* of C.

The *nonlinear* operator

$$A: D(A) \subseteq X \to Y \tag{54}$$

on the open set $D(A)$ is called *Fredholm* iff A is C^1 and the linearization, i.e., the F-derivative

$$A'(u): X \to Y$$

is Fredholm for all $u \in D(A)$. The number $\dim R(A'(u))$ is called the *rank* of A

at u. Furthermore, the number $\operatorname{ind} A'(u)$ is called the *index* of A at u. If $D(A)$ is connected, then the index of A is constant on $D(A)$, by (56) below. If A in (54) is a Fredholm operator with $\operatorname{ind} A'(u) = \text{const.} = m$ on $D(A)$, then A is called a Fredholm operator of index m, and we write $\operatorname{ind} A = m$. For example, the C^1-operator A in (54) is Fredholm of index zero iff

$$\dim N(A'(u)) = \operatorname{codim} R(A'(u)) < \infty \qquad \text{for all} \quad u \in D(A). \tag{55}$$

Let $A: D(A) \subseteq X \to Y$ be Fredholm on the open set $D(A)$. The point u is called a *singular point* of A iff $A'(u): X \to Y$ is *not* surjective. Otherwise, u is called a *regular point* of A. Furthermore, the point $b \in Y$ is called a *singular value* of A iff there is a singular point u of A such that $Au = b$. Otherwise, b is called a *regular value* of A. In particular, if A is Fredholm of index zero, then u is a singular point of A iff the equation $A'(u)h = 0$ has a solution $h \neq 0$.

Singular points (resp. singular values) are also called critical points (resp. critical values).

Proper Fredholm operators play a fundamental role for the following two reasons:

(i) *Localization.* Fredholm mappings reflect the fact that the linearizations of large classes of nonlinear integral equations and differential equations have the following two important properties:

The linearized homogeneous problem has only a finite *number of linearly independent solutions* ($\dim N(A'(u)) < \infty$).

The linearized inhomogeneous problem has only a finite number of linearly independent solvability conditions ($\operatorname{codim} R(A'(u)) < \infty$).

(ii) *Globalization.* In order to globalize local results one uses proper maps.

29.10b. Typical Properties of Linear Fredholm Operators

Summary 29.13. *Let $C, D \in L(X, Y)$, where X and Y are* B-*spaces over $\mathbb{K} = \mathbb{R}$, $\mathbb{C}$. The following properties are fundamental:*

(a) *If C is Fredholm and $\|D\|$ is sufficiently small, then the perturbation $C + D$ is also Fredholm and we have:*

$$\operatorname{ind}(C + D) = \operatorname{ind} C \tag{56}$$

and

$$\operatorname{rank} C \leq \operatorname{rank}(C + D),$$

$$\operatorname{codim} R(C) \geq \operatorname{codim} R(C + D),$$

$$\dim N(C) \geq \dim N(C + D).$$

That is, the set $\mathscr{F}$ of linear Fredholm operators from X to Y is open in $L(X, Y)$, and the index is locally constant and hence continuous on $\mathscr{F}$.

Moreover, the rank is lower semicontinuous on $\mathscr{F}$. *Finally, the codimension of the range and the dimension of the null space are upper semicontinuous on* $\mathscr{F}$.

(b) *If* C *is Fredholm and* D *is compact, then* $C + D$ *is also Fredholm and* $\operatorname{ind}(C + D) = \operatorname{ind} C$.

(c) *Let* C *be Fredholm of index zero. Then* $C: X \to Y$ *is a* C^∞*-diffeomorphism iff* $Ch = 0$ *implies* $h = 0$.

(d) *Let* C *be Fredholm of negative index. Then* $C: X \to R(C)$ is a C^∞*-diffeomorphism iff* $Ch = 0$ *implies* $h = 0$.

(e) *A Fredholm operator* $C: X \to Y$ *of negative (resp. positive) index can never be surjective (resp. injective).*

(f) *If* $\dim X < \infty$ *and* $\dim Y < \infty$, *then each linear operator* $C: X \to Y$ *is Fredholm and* $\operatorname{ind} C = \dim X - \dim Y$.

Statement (c) underlines the importance of Fredholm operators of index zero. From (f) it follows that Fredholm operators between B-spaces are a natural generalization of linear operators between finite-dimensional linear spaces.

Further important properties of linear Fredholm operators and references to the relevant literature can be found in Section 8.4.

29.10c. Typical Examples

We consider two typical situations.

EXAMPLE 29.14 (Finite-Dimensional Spaces). Let X and Y be B-spaces over $\mathbb{K} = \mathbb{R}, \mathbb{C}$ with

$$\dim X = \dim Y < \infty$$

(e.g., $X = Y = \mathbb{R}^N, N \geq 1$). Let $A: D(A) \subseteq X \to Y$ be a C^1-map on the open set $D(A)$. Then:

(a) A is Fredholm of index zero.

(b) Let $D(A) = X$. Then, A is proper iff A is weakly coercive, i.e.,

$$\|Au\| \to \infty \qquad \textit{as} \quad \|u\| \to \infty. \tag{57}$$

(c) Let $D(A)$ be bounded. Then A is proper iff

$$\|Au\| \to \infty \qquad \text{as} \quad \operatorname{dist}(u, \partial D(A)) \to 0. \tag{58}$$

(d) A is proper on each compact subset of $D(A)$.

PROOF. Cf. Example 4.42. □

In the case where $X = Y = \mathbb{R}$, Figure 29.5 shows examples for proper maps, whereas the functions pictured in Figure 29.6 are *not* proper.

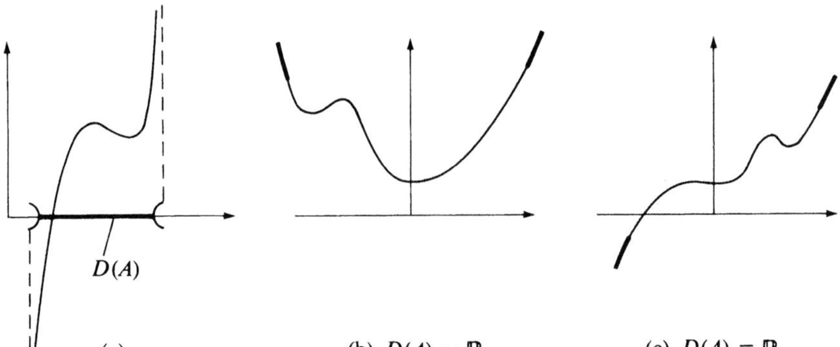

(a) (b) $D(A) = \mathbb{R}$ (c) $D(A) = \mathbb{R}$

Figure 29.5

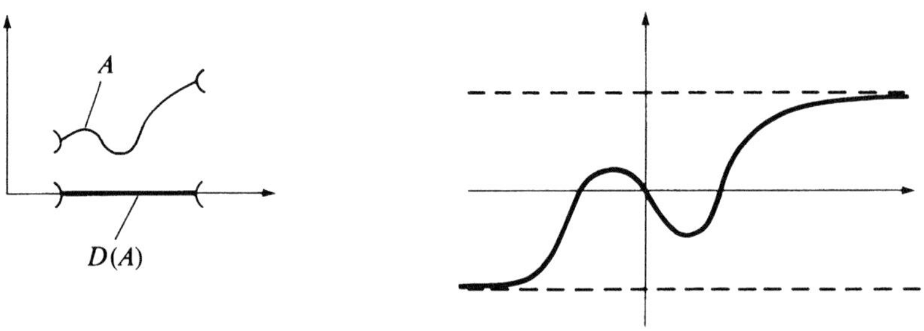

(a) (b)

Figure 29.6

STANDARD EXAMPLE 29.15 (Compact Perturbations of Diffeomorphisms). Let X and Y be B-spaces over $\mathbb{K} = \mathbb{R}, \mathbb{C}$. Suppose that:

(i) The map $B: X \to Y$ is a C^1-diffeomorphism.
(ii) The map $C: X \to Y$ is compact and C^1.
(iii) The map $A = B + C$ is weakly coercive, i.e., $\|Au\| \to \infty$ as $\|u\| \to \infty$.

Then, $A: X \to Y$ is a proper C^1-Fredholm operator of index zero, i.e., Theorem 29.E can be applied to the operator A.

PROOF. This is a special case of the following Example 29.16. □

EXAMPLE 29.16 (Compact Perturbations of Local Diffeomorphisms). Let X and Y be B-spaces over $\mathbb{K} = \mathbb{R}, \mathbb{C}$. Consider the map

$$A: D(A) \subseteq X \to Y \qquad \text{with} \quad A = B + C$$

on the open set $D(A)$. Suppose that:

(i) The C^1-map $B: D(A) \subseteq X \to Y$ is a local diffeomorphism at each point of

$D(A)$. By the inverse mapping theorem, this is equivalent to the fact that $B'(u)\colon X \to Y$ is bijective for all $u \in D(A)$.

(ii) The C^1-map $C\colon D(A) \subseteq X \to Y$ is compact.

Then:

(a) A is Fredholm of index zero.

(b) Let B be proper (e.g., B is bijective) and suppose that, for each bounded set M in Y, the set $A^{-1}(M)$ is bounded and if, in addition, $D(A) \neq X$, then $\operatorname{dist}(A^{-1}(M), \partial D(A)) > 0$. Then, A is proper.

(c) Suppose that B is injective and the set $B(N)$ is closed for each closed bounded subset N of $D(A)$ with the additional property that $\operatorname{dist}(N, \partial D(A)) > 0$ if $D(A) \neq X$. Suppose that the operator A has the same properties as in (b). Then, A is proper.

(d) A is proper on each compact subset of $D(A)$.

PROOF. Ad(a). By (i), $B'(u)$ is bijective and hence $B'(u)$ is Fredholm of index zero. By Proposition 7.33, the compactness of C implies the compactness of $C'(u)$. Thus $A'(u) = B'(u) + C'(u)$ is Fredholm of index zero, by Summary 29.13(b).

Ad(b). Let M be compact, and let (u_n) be a sequence in $A^{-1}(M)$. We have to show that there exists a subsequence $u_{n'} \to u$ as $n \to \infty$ such that $u \in A^{-1}(M)$. Since A is continuous and M is compact it suffices to show that $u \in D(A)$.

Since (u_n) is bounded and the operator C is compact, there exists a subsequence, again denoted by (u_n), such that $Cu_n \to c$ *as* $n \to \infty$ for some c. Since the set M is compact, there is a subsequence, again denoted by (u_n), such that $Au_n \to a$ as $n \to \infty$ for some a. Hence

$$Bu_n = Au_n - Cu_n \to a - c \qquad \text{as} \quad n \to \infty.$$

Since B is proper, the set $\{u_n : n \in \mathbb{N}\}$ is relatively compact. Thus there is a subsequence, again denoted by (u_n), such that $u_n \to u$ *as* $n \to \infty$ for some u. From $\operatorname{dist}(A^{-1}(M), \partial D(A)) > 0$ we obtain that $u \in D(A)$.

Ad(c). As in the proof of (b) we obtain that $Bu_n \to a - c$ as $n \to \infty$. Let N be the closure of the bounded set $\{u_n : n \in \mathbb{N}\}$. Then $\operatorname{dist}(N, \partial D(A)) > 0$, and hence $B(N)$ is closed. Consequently, $a - c \in R(B)$. Since B is locally invertible and injective, we get $u_n \to u$ as $n \to \infty$, where $u = B^{-1}(a - c)$. Hence $u \in D(A)$.

Ad(d). We consider the operator $A\colon N \to Y$, where N is a compact subset of $D(A)$. Let M be a compact subset of Y. Since A is continuous, the set $A^{-1}(M)$ is a closed subset of the compact set N, and hence $A^{-1}(M)$ is compact (cf. $A_1(11)$). □

29.10d. Typical Properties of Nonlinear Fredholm Operators

The following proposition collects fundamental properties of nonlinear Fredholm operators.

Proposition 29.17. *Let X and Y be B-spaces over $\mathbb{K} = \mathbb{R}, \mathbb{C}$, and let*

$$A\colon D(A) \subseteq X \to Y$$

be a C^k-Fredholm operator on the open set $D(A)$, where $1 \le k \le \infty$. Then:

(a) Compact perturbations. *If $C\colon D(A) \subseteq X \to Y$ is compact and C^1, then $A + C$ is a Fredholm operator with the same index as A at each point of $D(A)$.*

(b) Stability of the index. *Let $\mathscr{C}$ be a connected subset of $D(A)$. Then,*

$$\operatorname{ind} A'(v) = \text{const} \qquad \text{for all} \quad v \in \mathscr{C}.$$

(c) Lower semicontinuity of the rank. *For each $u \in D(A)$, there exists a neighborhood U of u such that, for all $v \in U$:*

$$\operatorname{rank} A'(u) \le \operatorname{rank} A'(v),$$

$$\operatorname{codim} R(A'(u)) \ge \operatorname{codim} R(A'(v)),$$

$$\dim N(A'(u)) \ge \dim N(A'(v)).$$

Recall that $\operatorname{rank} A'(u) = \dim R(A'(u))$.

(d) Local closedness. *The map A is locally closed, i.e., for each $u \in D(A)$, there exists a neighborhood U such that A maps every closed subset of U onto a closed set.*

(e) Regular points. *The set of regular points of A is open. If u is a regular point of A, then the index of A at u is nonnegative.*

(f) Regular values (Sard–Smale theorem). *Let*

$$k > \max(\operatorname{ind} A'(u), 0) \qquad \textit{for all} \quad u \in D(A), \tag{59}$$

and let X and Y be separable. Then the set of singular values of A is nowhere dense and hence of first Baire category in Y (i.e., this set is meagre), whereas the set of regular values of A is dense and of second Baire category in Y (i.e., this set is "big").

(g) Regular values of proper maps. *Suppose that* (59) *holds and that A is proper. Then the set of singular values of A is nowhere dense and hence of first Baire category in Y, whereas the set of regular values of A is open and dense and hence of second Baire category in Y.*

(h) Preimage theorem. *If b is a regular value of A, then the solution set M of the equation*

$$Au = b, \qquad u \in D(A), \tag{60}$$

is a C^k-submanifold of X.

The dimension of M at each point u is equal to $\dim N(A'(u))$. *If* $\dim X < \infty$ *and* $\dim Y < \infty$, *then*

$$\dim M = \dim X - \dim Y.$$

(i) Local diffeomorphism. *Suppose that A has index zero at u. Then A is a local C^k-diffeomorphism at u iff $A'(u)h = 0$ implies $h = 0$.*

(j) Well-posedness of equation (60). *If A has index zero and b_0 is a regular value of A, then $b_0 \in \operatorname{int} R(A)$, i.e., the equation* (60) *has a solution for all b in a neighborhood of b_0.*

(k) Ill-posedness of equation (60). *Suppose that* (59) *holds and that X and Y are separable. If A has a negative index at each point of $D(A)$, then* $\operatorname{int} R(A) = \varnothing$, *i.e., if equation* (60) *has a solution for $b = b_0$, then in each neighborhood of b_0, there is a b such that* (60) *has no solution.*

The reader should check that all the statements above can be easily verified in the case where the operator $A: X \to Y$ is linear with $X = \mathbb{R}^n$ and $Y = \mathbb{R}^m$. Here, $\operatorname{ind} A = n - m$. Statement (i) above underlines the importance of Fredholm operators of index zero.

The deep results of Proposition 29.17 are contained in (f) and (g). These statements are special cases of the Sard–Smale theorem on Banach manifolds which will be proved in Section 78.9 of Part IV. There we will also give the simple proof for (d). In the proof of Theorem 29.E below, we only need (e), (g), and (i).

PROOF. Ad(a). Since C is compact, so is $C'(u)$. The theory of linear Fredholm operators tells us that the compact perturbation $A'(u) + C'(u)$ of the Fredholm operator $A'(u)$ is again a Fredholm operator with the same index as $A'(u)$.

Ad(b). Set $f(u) = \operatorname{ind} A'(u)$. By (56), the function $f: D(A) \to \mathbb{R}$ is locally constant and hence continuous. Thus, $f(\mathscr{C})$ is connected, since $\mathscr{C}$ is connected. However, f is integer-valued and hence f is constant on $\mathscr{C}$.

Ad(c). This follows from (56).

Ad(d). Cf. Section 78.9.

Ad(e). The point u is regular iff

$$\operatorname{codim} R(A'(u)) = 0.$$

This implies $\operatorname{ind} A'(u) = \dim N(A'(u)) \geq 0$. By (c), there is a neighborhood U of u such that

$$\operatorname{codim} R(A'(v)) \leq \operatorname{codim} R(A'(u)) = 0$$

for all $v \in U$, i.e., all the points v in U are regular.

Ad(f), (g). Cf. Section 78.9.

Ad(h). Theorem 4. J tells us that the solution set of (60) is a C^k-Banach manifold. The fact that this set is indeed a submanifold of X will be proved in Section 73.11.

Ad(i). Since $A'(u)$ has index zero, this operator is bijective iff $A'(u)h = 0$ implies $h = 0$. Thus, the assertion follows from the inverse mapping theorem (Theorem 4.F).

Ad(j). This follows from (i).

Ad(k). Suppose that $\operatorname{int} R(A) \neq \varnothing$. By the Sard–Smale theorem (f), there is a regular value b in $R(A)$, i.e., there is a u such that $A'(u)$ is surjective. But this is impossible, since the index of $A'(u)$ is negative. □

29.10e. Proof of Theorem 29.E

(I) By Proposition 29.17(e), the set $X - D_{\text{sing}}$ of the regular points of A is open. If $u \in X - D_{\text{sing}}$, then A is a local diffeomorphism at u, according to Proposition 29.17(i).

(II) Let $b \in Y - A(D_{\text{sing}})$. Since A is proper, the set $A^{-1}(b)$ is compact. By (I), the points of $A^{-1}(b)$ are isolated. Hence the set $A^{-1}(b)$ is finite.

(III) We show that $n_X(\cdot)$ is constant on some neighborhood of b.

Since D_{sing} is closed and A is proper, the set $A(D_{\text{sing}})$ is closed, by Proposition 4.44.

(III-1) Let $A^{-1}(b) = \emptyset$. Hence $n_X(b) = 0$. If the assertion (III) is not true, then there exists a sequence (v_n) such that

$$Av_n \to b \quad \text{as} \quad n \to \infty,$$

and hence $n_X(Av_n) > 0$. Since A is proper, there exists a subsequence, again denoted by (v_n), such that

$$v_n \to v \quad \text{as} \quad n \to \infty.$$

Hence $Av = b$. This contradicts $A^{-1}(b) = \emptyset$.

(III-2) Let $A^{-1}(b) = \{u_1, \ldots, u_k\}$. Since $u_i \in X - D_{\text{sing}}$ for all i and the set $X - D_{\text{sing}}$ is open, there exist pairwise disjoint sufficiently small *open* neighborhoods $U_1, \ldots, U_k$ of $u_1, \ldots, u_k$, respectively, such that

$$A: U_i \to A(U_i)$$

is a C^k-diffeomorphism, where $A(U_i)$ is a neighborhood of b for each i. Set

$$U = \bigcup_{i=1}^{k} U_i.$$

If $n_X(\cdot)$ is not constant on some neighborhood of b, then there exists a sequence (b_n) with

$$b_n \to b \quad \text{as} \quad n \to \infty$$

and $n_X(b_n) > k$ for all n. Consequently, there is a sequence (v_n) with

$$Av_n = b_n \quad \text{and} \quad v_n \in X - U \quad \text{for all } n.$$

Since A is proper, there exists a subsequence, again denoted by (v_n), such that $v_n \to v$ as $n \to \infty$ and hence

$$Av = b, \qquad v \in X - U,$$

i.e., $v \neq u_i$ for all i. This is a contradiction.

(IV) Let $\mathscr{C}$ be a connected subset of $Y - A(D_{\text{sing}})$. We show that $n_X(\cdot)$ is constant on $\mathscr{C}$. In fact, it follows from (III) that the function $n_X: \mathscr{C} \to \mathbb{R}$ is continuous. Hence $n_X(\mathscr{C})$ is connected. Since n_X is integer-valued, this function is constant on $\mathscr{C}$.

(V) By the Sard–Smale theorem (Proposition 29.17(g)), the set $Y - A(D_{\text{sing}})$ is open and dense, and $A(D_{\text{sing}})$ is nowhere dense in Y.

(VI) Let $D_{\text{sing}} = \varnothing$. Since the proper map $A: X \to Y$ is a local C^k-diffeomorphism at each point of X, it follows from the global inverse mapping theorem (Theorem 4.G) that A is a C^k-diffeomorphism.

The proof of Theorem 29.E is complete. □

29.10f. Generalization to Arbitrary Sets

We now want to study the *number of solutions* $n_S(b)$ of the equation

$$Au = b, \qquad u \in S, \tag{61}$$

where S is an *arbitrary* set in a B-space. To this end, we set

$$D_{\text{sing}} = \{u \in \operatorname{int} S: A'(u)h = 0 \text{ has a solution } h \neq 0\}.$$

We define the set Y_{crit} of the *critical* right members b of equation (61) through

$$Y_{\text{crit}} = A(D_{\text{sing}}) \cup A(\partial S \cap S),$$

and we set $Y_{\text{reg}} = Y - Y_{\text{crit}}$.

The following theorem is fundamental:

Theorem 29.F (Solution set of (61)). *Let X and Y be* B*-spaces over* $\mathbb{K} = \mathbb{R}, \mathbb{C}$. *Suppose that*:

(i) *The map $A: S \subseteq X \to Y$ is proper and continuous.*
(ii) *The map $A: \operatorname{int} S \to Y$ is C^k and Fredholm of index zero,* $1 \leq k \leq \infty$.

Then:

(a) *For each point u in the open set* $\operatorname{int} S - D_{\text{sing}}$, *the map A is a local C^k-diffeomorphism at u.*
(b) *Let $b \in Y_{\text{reg}}$. Then the number $n_S(b) \geq 0$ is finite. Moreover, the map A is a local C^k-diffeomorphism at each point u of the solution set of equation* (61).
(c) *The function $n_S(\cdot)$ is constant on each connected subset of the open set Y_{reg}.*
(d) *If A is proper on* $\operatorname{int} S$, *then the set*

$$Y_{\text{reg}} \cup A(\partial S \cap S)$$

is open and dense in Y and hence of second Baire category in Y, and the set $A(D_{\text{sing}})$ is nowhere dense in Y.

(e) *If X and Y are separable, then the set $Y_{\text{reg}} \cup A(\partial S \cap S)$ is dense and of second Baire category in Y, and the set $A(D_{\text{sing}})$ is nowhere dense in Y.*
(f) *If A is injective on* $\operatorname{int} S - D_{\text{sing}}$, *then the map*

$$A: \operatorname{int} S - D_{\text{sing}} \to A(\operatorname{int} S - D_{\text{sing}})$$

is a C^k-diffeomorphism.

(g) *If $S = X$ and $D_{\text{sing}} = \varnothing$, then $A: X \to Y$ is a C^k-diffeomorphism.*

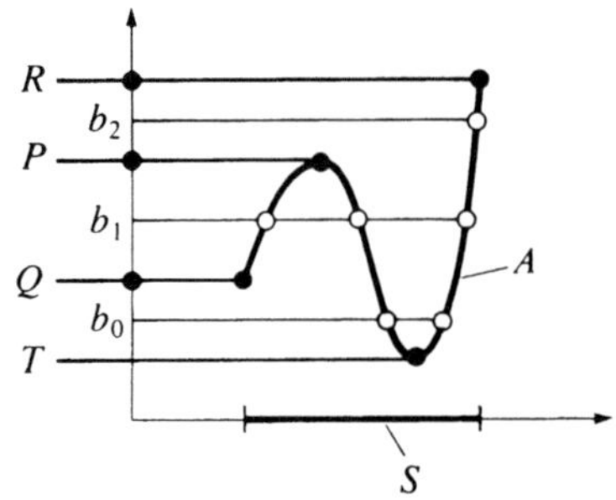

(a) $n_S(b_0) = 2$, $n_S(b_1) = 3$, $n_S(b_2) = 1$

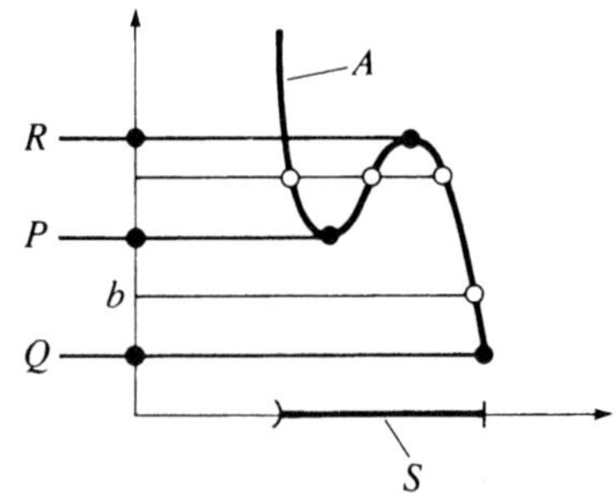

(b)

Figure 29.7

This theorem tells us that the function $n_S(\cdot)$ may only bifurcate at the points of the critical set Y_{crit}. Up to possible pathological situations, the set $A(\partial S \cap S)$ is "thin" and hence the set Y_{crit} is also "thin" and the set Y_{reg} is "big", according to (d) and (e). In connection with Example 29.19 below, simple counter-examples for $X = \mathbb{R}$ show that Theorem 29.F is "*sharp.*"

In Section 29.10c we have proved the properness of maps $A: D(A) \subseteq X \to Y$, where $D(A)$ is open (e.g., $D(A) = X$). The following corollary shows how to use this information in order to prove the properness of A on closed sets S.

Corollary 29.18 (Properness). *Let X and Y be* B-*spaces over* $\mathbb{K} = \mathbb{R}, \mathbb{C}$. *Then*:

(a) *If $A: D(A) \subseteq X \to Y$ is proper, then $A: S \subseteq X \to Y$ is proper on each closed subset S of $D(A)$.*
(b) *If S is compact, then each continuous map $A: S \subseteq X \to Y$ is proper.*

This follows from the fact that closed subsets of compact sets are compact.

Example 29.19. Figure 29.7 illustrates the intuitive meaning of Theorem 29.F. The number $n_S(b)$ of solutions of the equation $Au = b$, $b \in S$, may only bifurcate at the points P, Q, R, T. Here, we have

$$Y_{\text{crit}} = \{P, Q, R, T\} \qquad \text{and} \qquad Y_{\text{reg}} = \mathbb{R} - Y_{\text{crit}}.$$

In Figure 29.7(a), we obtain that

$$A(\partial S \cap S) = \{Q, R\} \qquad \text{and} \qquad A(D_{\text{sing}}) = \{P, T\},$$

and in Figure 29.7(b) we have $A(\partial S \cap S) = \{Q\}$ and $A(D_{\text{sing}}) = \{P, R\}$.

Proof of Theorem 29.F. Ad(a). This follows from Proposition 29.17(e), (i).

Ad(b). Let $b \in Y_{\text{reg}}$. Then $b \notin A(\partial S \cap S) \cup A(D_{\text{sing}})$ and hence

$$A^{-1}(b) \subseteq \operatorname{int} S - D_{\text{sing}}. \tag{62}$$

Since the map A is proper, the set $A^{-1}(b)$ is compact. By (a), all the points of $A^{-1}(b)$ are isolated. Hence $A^{-1}(b)$ is finite.

Ad(c). We first prove that the set Y_{reg} is open. In fact, by (a) the set $S_0 = \operatorname{int} S - D_{\text{sing}}$ is open in X. Thus, the set S_0 is also open with respect to the induced topology on S, and hence the set

$$(\partial S \cap S) \cup D_{\text{sing}} = S - S_0$$

is closed with respect to the induced topology on S (cf. $A_1(8)$). Since $A: S \to Y$ is continuous and proper with respect to the topology on X, this map has the same properties with respect to the induced topology on S. Therefore, by Proposition 4.44, the set $Y_{\text{crit}} = A(S - S_0)$ is closed in Y, and hence $Y_{\text{reg}} = Y - Y_{\text{crit}}$ is open in Y.

The following simple observation is crucial. Let

$$Av_n = b_n, \qquad v_n \in S \qquad \text{for all } n$$

and $b_n \to b$ as $n \to \infty$. Since A is proper, the sequence (v_n) lies in a compact subset of S. Thus, there exists a subsequence, again denoted by (v_n), such that $v_n \to v$ as $n \to \infty$ and $v \in S$. Since A is continuous,

$$Av = b.$$

Using this observation, the proof of (c) now proceeds as the corresponding proof of Theorem 29.E in Section 29.10e. In this connection, observe that $b \in Y_{\text{reg}}$ implies (62) above.

Ad(d), (e). The set of regular values of the map $A: \operatorname{int} S \to Y$ is equal to

$$Y - A(D_{\text{sing}}) = Y_{\text{reg}} \cup A(\partial S \cap S).$$

Now the assertion follows from the Sard–Smale theorem (Proposition 29.17(f), (g)).

Ad(f). This follows from (a).

Ad(g). This follows from Theorem 29.E. $\square$

In the following sections we will consider applications of Theorem 29.F.

29.11. Locally Strictly Monotone Operators

Definition 29.20. Let X be a real B-space. The operator $A: D(A) \subseteq X \to X^*$ is called *locally strictly monotone* at the point u iff $u \in \operatorname{int} D(A)$ and the F-derivative $A'(u): X \to X^*$ is strictly monotone, i.e.,

$$\langle A'(u)h, h\rangle > 0 \qquad \text{for all} \quad h \in X \quad \text{with} \quad h \neq 0. \tag{63}$$

The operator A is called locally strictly monotone on the set S iff it has this property at each point of S.

The operator A is called *locally strongly monotone* at u iff $u \in \operatorname{int} D(A)$ and the F-derivative $A'(u)$ is strongly monotone, i.e., there is a $c > 0$ such that

$$\langle A'(u)h, h\rangle \geq c\|h\|^2 \qquad \text{for all} \quad h \in X. \tag{64}$$

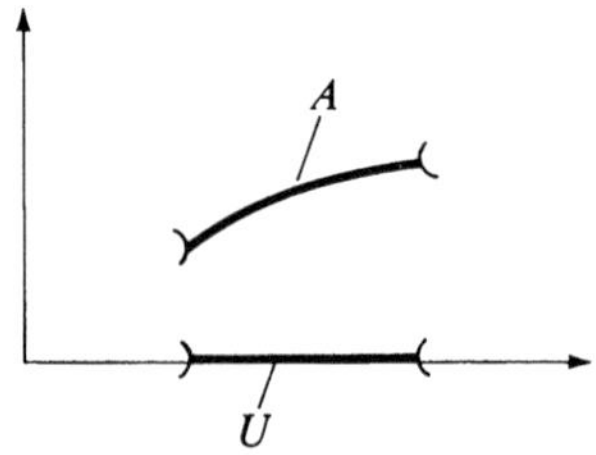

Figure 29.8

Analogously, the operator A is called *locally monotone* at u iff $u \in \operatorname{int} D(A)$ and the F-derivative $A'(u)$ is monotone.

In this connection, we assume tacitly that the F-derivative $A'(u)$ exists.

In Section 29.12, we will also introduce the notion of locally regularly monotone operators which play a fundamental role in the calculus of variations for obtaining sufficient conditions for strict local minima. If the B-space X is finite dimensional, then the following three notions coincide: locally strictly monotone, locally strongly monotone, and locally regularly monotone.

EXAMPLE 29.21. The function $A: D(A) \subseteq \mathbb{R} \to \mathbb{R}$ is locally strictly monotone at the point u iff $u \in \operatorname{int} D(A)$ and

$$A'(u) > 0.$$

Thus, if A is locally strictly monotone on the open interval $U =]a, b[$, then the real function $A(\cdot)$ is strictly monotone increasing on U (Fig. 29.8). The following results generalize the behavior of such functions.

We first investigate the connection between locally monotone operators and monotone operators.

Proposition 29.22. *Let $A: U \subseteq X \to \mathbb{R}$ be C^1 on the open subset U of the real B-space X, and let C be a convex subset of U. Then:*

(a) *If A is locally strictly monotone on C, then A is strictly monotone on C.*
(b) *If A is locally monotone on C, then A is monotone on C.*

PROOF. Ad(a). Let $u, v \in C$ with $u \neq v$. Set $h = v - u$. By Taylor's theorem (Theorem 4.A),

$$Av - Au = \int_0^1 A'(u + th)h \, dt$$

and hence

$$\langle Av - Au, v - u \rangle = \int_0^1 \langle A'(u + th)h, h \rangle \, dt > 0.$$

Note that the integrand is positive for each $t \in [0, 1]$, since A is locally strictly monotone on the convex set C.

Ad(b). Use the same argument. $\square$

Proposition 29.23 (Local Diffeomorphisms). *Let* $A: U \subseteq X \to X^*$ *be a* C^k-*Fredholm operator of index zero on the open subset* U *of the real* B-*space* X, *and let* A *be locally strictly monotone on* U, *where* $1 \leq k \leq \infty$. *Then*:

(a) *The map* A *is a local* C^k-*diffeomorphism at each point of* U.
(b) *If* U *is convex, then* $A: U \to A(U)$ *is a strictly monotone* C^k-*diffeomorphism.*

PROOF. Ad(a). If $A'(u)h = 0$, then $h = 0$, by (63). Thus, the assertion follows from Proposition 29.17(i).

Ad(b). By (a), it is sufficient to prove that A is injective. But this follows from the strict monotonicity of A, according to Proposition 29.22. □

We now study the *number of solutions* $n_S(b)$ of the equation

$$Au = b, \qquad u \in S. \tag{65}$$

Theorem 29.G (Benkert (1985)). *Let* S *be a subset of the real* B-space X. *Suppose that*:

(i) *The map* $A: S \subseteq X \to X^*$ *is proper and continuous.*
(ii) *The map* A: int $S \to X^*$ *is a* C^k-*Fredholm operator of index zero and locally strictly monotone on* int S, *where* $1 \leq k \leq \infty$.

Then:

(a) *For each* $b \in X^* - A(\partial S \cap S)$, *the number* $n_S(b) \geq 0$ is finite.
(b) *The function* $n_S(\cdot)$ *is constant on each connected subset of the open set* $X^* - A(\partial S \cap S)$.
(c) *In the case where* $S = X$, *the map* $A: X \to X^*$ *is a* C^k-*diffeomorphism.*

This result is related to the main theorem on monotone operators (Theorem 26.A).

PROOF. This is a special case of Theorem 29.F. Note that $A'(u)h = 0$, $u \in \text{int}\, S$, implies $h = 0$, and hence $D_{\text{sing}} = \varnothing$. □

29.12. Locally Regularly Monotone Operators, Minima, and Stability

We want to study the local minimum problem

$$F(u) - b(u) = \min!, \qquad u \in X, \tag{66}$$

together with the Euler equation

$$F'(u) = b, \qquad u \in X. \tag{67}$$

Let u be a solution of (67). Our goal is to prove important criteria which guarantee that u is a strict local minimal point of (66). In this connection, we will use *locally strongly monotone* operators and *locally regularly monotone* operators. We assume:

(H1) The functional $F: U(u) \subseteq X \to \mathbb{R}$ is defined on a neighborhood $U(u)$ of the point u in the real B-space X.
(H2) The functional $b: X \to \mathbb{R}$ is linear and continuous.

Recall that u is called a *local minimal point* of (66) iff

$$F(v) - b(v) \geq F(u) - b(u)$$

for all v in some neighborhood of u. Moreover, u is called a *strict* local minimal point of (66) iff

$$F(v) - b(v) > F(u) - b(u)$$

for all v in some neighborhood of u with $v \neq u$.

In nonlinear elasticity, problem (66) allows the following physical interpretation:

$$\begin{aligned} u &= \text{displacement of the elastic body,} \\ F &= \text{elastic potential energy,} \\ b &= \text{outer force,} \\ b(u) &= \text{work of the outer force corresponding to the displacement } u, \\ E_{\text{pot}}(u) &= F(u) - b(u) \text{ (potential energy).} \end{aligned} \tag{68}$$

Then problem (66) corresponds to the principle of minimal potential energy. This will be studied in Part IV.

Definition 29.24. Assume (H1), (H2). Set $E_{\text{pot}} = F - b$.

(i) The point u is called *stable* with respect to E_{pot} iff u is a strict local minimal point of E_{pot}.
(ii) The point u is called *weakly stable* iff the second variation satisfies the following condition:

$$\delta^2 F(u; h) \geq 0 \qquad \text{for all} \quad h \in X.$$

(iii) The point u is called *unstable* iff u is not weakly stable, i.e., there is an h such that $\delta^2 F(u; h) < 0$.
(iv) The point u is called *critically stable* iff u is weakly stable and there is an $h \neq 0$ such that $\delta^2 F(u; h) = 0$.
(v) The *boundary of stability* consists of the boundary of the set of all stable points.

Below, we shall show the following:

(a) If u is a local minimal point of E_{pot}, then u is weakly stable provided $\delta^2 F$ exists at u.

(b) Let $\dim X < \infty$. Then u is stable if

$$\text{(F)}\quad \delta F(u;h) = 0 \quad \text{and} \quad \delta^2 F(u;h) > 0 \quad \text{for all} \quad h \in X - \{0\}.$$

(c) Let $\dim X = \infty$. Then u is stable if condition (F) holds, and the second variation $\delta^2 F$ is *regular* at u.

In Section 18.7 we mentioned the classical error of Legendre (1786) in the calculus of variations. In modern language, Legendre's error was to think that (b) above also holds for infinite-dimensional B-spaces X. The reason for this failure will be explained in Remark 29.30 below.

Let F be C^2 in a neighborhood of u. Then, in terms of locally monotone operators, we shall show the following.

(a) The point u is weakly stable iff u is locally monotone at u.
(b) Let $\dim X < \infty$. Then u is stable if

$$E'_{\text{pot}}(u) \equiv F'(u) - b = 0$$

and F' is locally *strictly* monotone at u.
(c) Let $\dim X = \infty$. Then u is stable if $E'_{\text{pot}}(u) = 0$ and F' is locally *regularly* monotone at u.

The notion of regular second variations and locally regularly monotone operators is closely related to the Gårding inequality (cf. Definition 29.38 below).

In Part IV we shall discover the following remarkable fact:

The elastic potential energy F of an elastic body is not a convex functional of the displacement u in realistic models in elasticity.

This reflects the possible appearance of rupture and plasticity. Thus, the theory of monotone operators is *not* applicable to realistic models in elasticity, but only to approximation models. However, locally strictly monotone (more precisely, locally regularly monotone) operators are applicable, as we will discuss in Section 29.19.

29.12a. Basic Ideas

In order to explain the basic ideas for equations (66) and (67) above, we first consider the simple finite-dimensional case. The following assertions are special cases of our results below for arbitrary B-spaces. Recall that the original problem (66) is identical to the minimum problem

$$\text{(M)} \qquad F(u) - b(u) = \min!, \qquad u \in X,$$

and (67) is identical to the Euler equation

$$\text{(E)} \qquad F'(u) = b, \qquad u \in X.$$

Proposition 29.25 (Finite-Dimensional Minimum Problems). *Assume* (H1) *and* (H2) *with* $\dim X < \infty$ (*e.g.., $X = \mathbb{R}^N$*). *Let F be C^k on a neighborhood of u with $k \geq 1$. Then*:

(A) Necessary condition for a local minimum. *If u is a local minimal point of the original problem* (66), *then*:
  (i) *$F'(u) = b$, i.e., u is a solution of the Euler equation.*
  (ii) *F' is locally monotone at u in the case where $k \geq 2$, i.e.,*

$$\langle F''(u)h, h\rangle \geq 0 \qquad \textit{for all} \quad h \in X.$$

(B) Sufficient condition for a strict local minimum. *Let $F'(u) = b$. Then u is a strict local minimal point of* (66) *if one of the following two conditions is satisfied*:
  (i) *F' is strictly monotone on a neighborhood of u.*
  (ii) *F' is locally strictly monotone at u in the case where $k \geq 2$, i.e.,*

$$\langle F''(u)h, h\rangle > 0 \qquad \textit{for all} \quad h \in X - \{0\}. \tag{69}$$

(C) Sufficient condition for a local minimum. *Let $F'(u) = b$. Then u is a local minimal point of* (66) *if one of the following two conditions is satisfied*:
  (i) *F' is monotone on a neighborhood of u.*
  (ii) *F' is locally monotone on a neighborhood V of u in case $k \geq 2$, i.e.,*

$$\langle F''(v)h, h\rangle \geq 0 \qquad \textit{for all} \quad v \in V, \quad h \in X. \tag{70}$$

From (i) *or* (ii) *it follows that F is convex on a neighborhood of u.*

(D) Sufficient condition for a unique global minimal point. *Let $F: X \to X^*$ be C^k with $k \geq 1$, and let $F'(u) = b$. Then u is the unique global minimal point of* (66) *if one of the following two conditions is satisfied*:
  (i) *F' is strictly monotone on X.*
  (ii) *F' is locally strictly monotone on X in case $k \geq 2$, i.e.,*

$$\langle F''(v)h, h\rangle > 0 \qquad \textit{for all} \quad v \in X, \quad h \in X - \{0\}. \tag{71}$$

From (i) *or* (ii) *it follows that F is strictly convex on X.*

(E) Sufficient condition for a global minimum. *Let $F: X \to X^*$ be C^k with $k \geq 1$, and let $F'(u) = b$. Then u is a global minimal point of* (66) *if one of the following two conditions is satisfied*:
  (i) *F' is monotone on X.*
  (ii) *F' is locally monotone on X in case $k \geq 2$, i.e.,*

$$\langle F''(v)h, h\rangle \geq 0 \qquad \text{for all} \quad v, h \in X.$$

From (i) *or* (ii) *it follows that F is convex on X.*

(F) Eigenvalue criterion and stability. *Let $k \geq 2$. Along with the Euler equation*

$$F'(u) = b, \qquad u \in X,$$

we consider the linearized *eigenvalue equation of Jacobi*:

$$F''(u)h = \mu h, \qquad h \in X, \quad \mu \in \mathbb{R}. \tag{72}$$

Suppose that u is a solution of the Euler equation. Let $\mu_{\min}$ *denote the smallest eigenvalue of* (72), *i.e.*,

$$\mu_{\min} = \min_{h \in X - \{0\}} \frac{\delta^2 F(u; h)}{|h|^2},$$

where $\delta^2 F(u; h) = \langle F''(u)h, h \rangle$, *and* $|h|$ *denotes the Euclidean norm of* h. *Then*:

(i) *If* $\mu_{\min} > 0$, *then u is a strict local minimal point of the original minimum problem* (66), *i.e., u is* stable.
(ii) *If* $\mu_{\min} < 0$, *then u is not a local minimal point of* (66), *and u is* unstable.
(iii) *The point is* weakly stable *iff* $\mu_{\min} \geq 0$.
(iv) *The point u is* critically stable *iff* $\mu_{\min} = 0$. *In this case, the behavior of* $E_{\text{pot}} = F - b$ *depends sensitively on the structure of the higher-order derivatives* $F^{(r)}(u)$ *with* $r > 2$.

Moreover, the minimal value of the so-called accessory quadratic minimum problem

$$\delta^2 F(u; h) = \min!, \qquad |h| = 1, \quad h \in X,$$

is equal to $\mu_{\min}$.

(G) Local diffeomorphism. *Let* $k \geq 2$. *The operator* F' *is a local* C^k*-diffeomorphism at the point u iff*

$$F''(u)h = 0 \qquad \textit{implies} \qquad h = 0,$$

i.e., $\mu = 0$ *is not an eigenvalue of the Jacobi equation* (72).

(H) Bifurcation and the Jacobi equation. *We now consider the problem*

$$F(u, \lambda) - b(u) = \min!, \qquad u \in X, \tag{73}$$

where F depends on the real parameter λ. *The corresponding Euler equation*[1] *reads as follows*:

$$F_u(u, \lambda) = b, \qquad u \in X, \quad \lambda \in \mathbb{R}, \quad b \in X. \tag{74}$$

Let $F: U(u_0, \lambda_0) \subseteq X \times \mathbb{R} \to \mathbb{R}$ *be* C^k *on a neighborhood of the point* (u_0, λ_0) *with* $k \geq 2$, *and suppose that* (u_0, λ_0, b_0) *is a solution of* (74). *The corresponding Jacobi equation is given by*

$$F_{uu}(u_0, \lambda_0)h = \mu h, \qquad h \in X, \quad \mu \in \mathbb{R}. \tag{75}$$

Then:

(i) *If* $\mu = 0$ *is* not *an eigenvalue of the Jacobi equation* (75) (*e.g.*, $\delta^2 F(u, \lambda_0; h) \neq 0$ *for all* $h \neq 0$), *then the Euler equation* (74) *has a* C^{k-1}*-solution* $u = u(\lambda, b)$ *through the point* (u_0, λ_0, b_0), *which is unique in a sufficiently small neighborhood of this point* (*Fig.* 29.9(a)).
(ii) *If* (u_0, λ_0, b_0) *is a bifurcation point of the Euler equation* (74) *with respect to the parameter* (λ, b), *then* $\mu = 0$ *is an eigenvalue of the Jacobi*

[1] In Section 29.13 we consider the buckling of beams. In this case, the parameter λ corresponds to the outer boundary force.

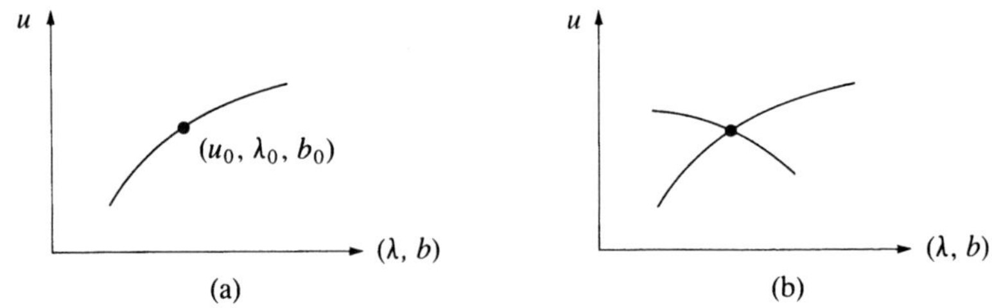

Figure 29.9

equation (75), *i.e., there is an* $h \neq 0$ *such that* $\delta^2 F(u_0, \lambda_0; h) = 0$ (*Fig.* 29.9(b)).

(iii) *Conversely, if* F *has the structure*

$$F(u, \lambda) = (\lambda - \lambda_0)a(u - u_0, u - u_0) + r(u, \lambda),$$

then (u_0, λ_0) *is a bifurcation point of the Euler equation* (74) *with* $b = 0$ *in the case where* $a: X \times X \to \mathbb{R}$ *is a bilinear, symmetric, strictly positive functional, and*

$$\lim_{u \to u_0} \frac{\|r(u, \lambda)\|}{\|u - u_0\|^2} = 0,$$

uniformly for all λ *in a sufficiently small neighborhood of* λ_0.

This proposition shows clearly the importance of the theory of monotone operators and of spectral theory for minimum problems and stability theory. The crucial connection between stability and bifurcation contained in Proposition 29.25(H) will be discussed in Remark 29.29 below. Assertions (i) and (ii) in (H) above follow from the implicit function theorem (Theorem 4.B) and Proposition 8.2, respectively. The proof of (iii) in (H) will be discussed in Problem 29.7. This proof is extremely simple in the case where the remainder $r(\cdot)$ does not depend on the parameter λ.

EXAMPLE 29.26. Let $X = \mathbb{R}$, and let $F: U(u) \subseteq X \to \mathbb{R}$ be C^2 on a neighborhood of u. Then

$$\langle F''(u)h, h\rangle = F''(u)h^2 \qquad \text{for all} \quad h \in \mathbb{R},$$

and hence we have that:

(i) F' is locally strictly monotone at u iff $F''(u) > 0$.
(ii) F' is locally monotone at u iff $F''(u) \geq 0$.
(iii) F' is (strictly) monotone on the open interval U iff F' is (strictly) monotonically increasing on U.

Consequently, Proposition 29.25 and the following results below generalize well-known properties of real functions. Figure 29.10 corresponds to the case $b = 0$.

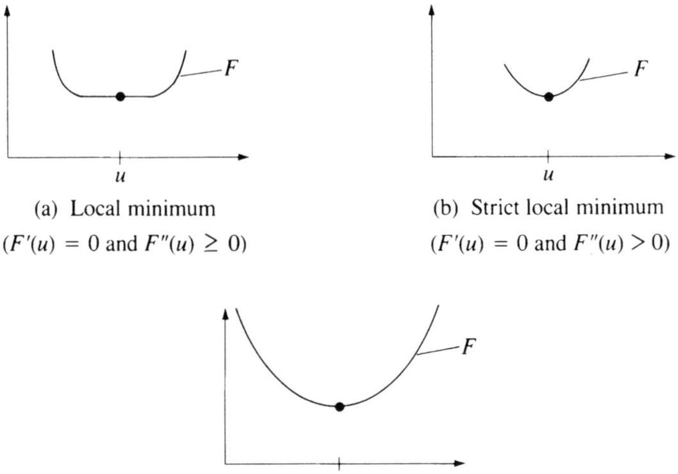

(a) Local minimum
($F'(u) = 0$ and $F''(u) \geq 0$)

(b) Strict local minimum
($F'(u) = 0$ and $F''(u) > 0$)

(c) Global minimum
($F''(v) \geq 0$ for all v)

Figure 29.10

EXAMPLE 29.27. Let $X = \mathbb{R}^N$, and let $F\colon U(u) \subseteq X \to \mathbb{R}$ be C^2 on a neighborhood of u. Then

$$F'(u)h = \sum_{i=1}^{N} D_i F(u) h_i,$$

$$\langle F''(u)h, h\rangle = \sum_{i,j=1}^{N} D_i D_j F(u) h_i h_j \qquad \text{for all} \quad h \in X,$$

where $h = (h_1, \ldots, h_N)$, $u = (u_1, \ldots, u_N)$, $b = (b_1, \ldots, b_N)$, and $D_i = \partial/\partial u_i$. The Euler equation $F'(u) = b$ is equivalent to the system

$$D_i F(u) = b_i, \qquad i = 1, \ldots, N.$$

The Jacobi equation $F''(u)h = \mu h$ corresponds to the eigenvalue problem

$$\sum_{j=1}^{N} D_i D_j F(u) h_j = \mu h_i, \qquad i = 1, \ldots, N,$$

i.e., the eigenvalues $\mu_1, \mu_2, \ldots$ of the Jacobi equation are identical to the eigenvalues of the symmetric matrix $(D_i D_j F(u))_{i,j=1,\ldots,N}$. Then:

(i) F' is locally strictly monotone at u iff $\mu_i > 0$ for all i.
(ii) $F'(u)$ is locally monotone at u iff $\mu_i \geq 0$ for all i.

Remark 29.28 (The Importance of Jacobi's Eigenvalue Criterion). It is quite remarkable that the eigenvalue criterion in Proposition 29.25(F) can be generalized to large classes of problems in infinite-dimensional B-spaces and hence it can be generalized to variational problems. This will be shown below. In order to explain the basic idea of this *fundamental* technique let us consider the variational problem

$$\text{(V)}\qquad \begin{aligned} &\int_G \tfrac{1}{2}(u_\xi^2 + u_\eta^2) + (f(u) - bu)\,dx = \min!,\\ &u = 0 \quad \text{on } \partial G, \end{aligned}$$

where G is a bounded region in $\mathbb{R}^2$. If u is a solution of (V), then u satisfies the Euler equation

$$\text{(E)}\qquad \begin{aligned} -\Delta u + f'(u) &= b \quad \text{on } G,\\ u &= 0 \quad \text{on } \partial G, \end{aligned}$$

which corresponds to $F'(u) = b$. The linearized Euler equation $F''(u)h = b$ is obtained from the Euler equation by replacing u with $u + th$ and differentiating with respect to t at $t = 0$. This yields

$$(\text{E}_{\text{lin}})\qquad \begin{aligned} -\Delta h + f''(u)h &= b \quad \text{on } G,\\ h &= 0 \quad \text{on } \partial G. \end{aligned}$$

Consequently, the Jacobi equation $F''(u)h = \mu h$ corresponds to

$$\text{(J)}\qquad \begin{aligned} -\Delta h + f''(u)h &= \mu h \quad \text{on } G,\\ h &= 0 \quad \text{on } \partial G. \end{aligned}$$

If the smallest eigenvalue $\mu_{\min}$ of (J) is positive, then we expect that the function u corresponds to a strict local minimum of the original problem (V), i.e., u is stable. It follows from our general results in Section 29.19 that this criterion is true in the case where the functions u and f are sufficiently smooth.

Remark 29.29 (Loss of Stability and Bifurcation). We want to show that Proposition 29.25 (H) represents the simplest model for the following two important principles in physics.

(P1) Bifurcation cannot occur at stable states.
(P2) Loss of stability of a state with respect to an outer parameter can lead to bifurcation.

To explain this, we will use the language of elasticity introduced in (68) above. Principle (P1) is an immediate consequence of Proposition 29.25(H).

We now discuss (P2). In Proposition 29.25(H) (ii), the elastic potential energy has the form

$$F(u, \lambda) = (\lambda - \lambda_0)a(u - u_0, u - u_0) + o(\|u - u_0\|^2), \qquad u \to u_0,$$

where λ is an outer parameter. Since the symmetric bilinear functional $a(\cdot, \cdot)$ is strictly positive, we have:

u_0 is stable if $\lambda > \lambda_0$;

u_0 is unstable if $\lambda < \lambda_0$.

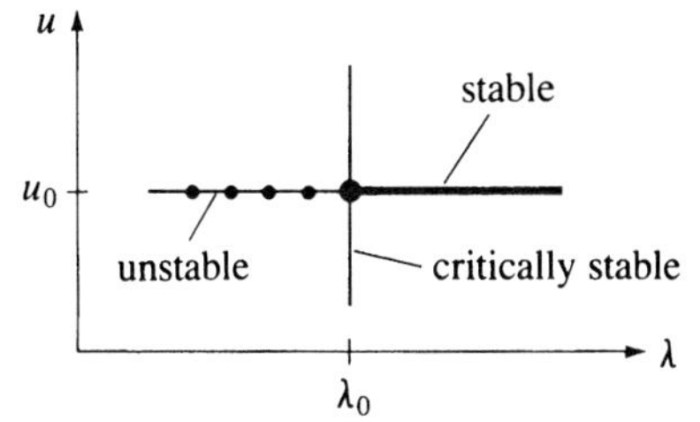

(a)

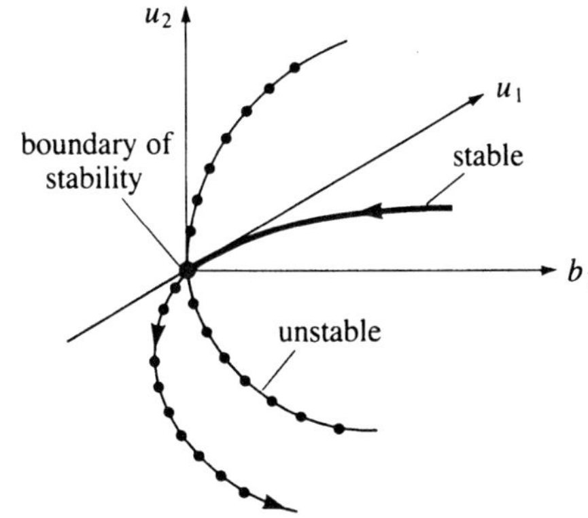

(b)

Figure 29.11

Roughly speaking, this change of stability of u_0 is responsible for the bifurcation of the Euler equation $F'(u, \lambda) = 0$ at (u_0, λ_0). To understand this, we consider the simplest possible case, namely,

$$F(u, \lambda) = \tfrac{1}{2}(\lambda - \lambda_0)(u - u_0)^2,$$

where u and λ are real numbers. In fact, the Euler equation

$$F_u(u, \lambda) = (\lambda - \lambda_0)(u - u_0) = 0$$

has the bifurcation point (u_0, λ_0) (cf. Fig. 29.11(a)).

As a second example for (P2), we consider the minimum problem

$$F(u) - b(u) = \min!, \qquad u \in \mathbb{R}^2,$$

with $F(u) = u_1 u_2^2 + u_1^3/3$ and $b(u) = b_1 u_1 + b_2 u_2$, where we regard b as an outer parameter. The Euler equation $F'(u) = b$ has the form

(E) $$u_1^2 + u_2^2 = b_1, \qquad 2u_1 u_2 = b_2,$$

and the Jacobi equation $F''(u)h = \mu h$ reads as follows:

(J) $$\begin{pmatrix} 2u_1 & 2u_2 \\ 2u_2 & 2u_1 \end{pmatrix}\begin{pmatrix} h_1 \\ h_2 \end{pmatrix} = \mu \begin{pmatrix} h_1 \\ h_2 \end{pmatrix}.$$

Obviously, the point $u_1 = 0, u_2 = 0$, is *critically stable*, since $\mu_{\min} = 0$. Furthermore, this point lies on the *boundary of stability*. In fact, in each neighborhood of $(0, 0)$, there are points (u_1, u_2) with $\mu_{\min} > 0$.

Figure 29.11(b) shows the solution set of equation (E) and its stability properties for $b_2 = 0$ and $b_1 \in \mathbb{R}$. In this case, b_1 plays the role of a bifurcation parameter. In fact, the point $u = 0, b = 0$, is a bifurcation point of the solution set of (E) with respect to the parameter $b = (b_1, 0)$. For $b_2 = 0$, the two solution branches of (E) are given by

$$u_2^2 = b_1, \qquad u_1 = 0 \qquad \text{and} \qquad u_1^2 = b_1, \qquad u_2 = 0.$$

In order to avoid misunderstandings, note the following. If the functional $F: D(F) \subseteq X \to \mathbb{R}$ is C^2 on the open set $D(F)$ of the real B-space X, then the set of all the solutions (u, b) of the corresponding Euler equation $F'(u) - b = 0$ forms a C^1-manifold M in the product space $X \times X^*$, according to the preimage theorem (Theorem 4.J). In contrast to this fact, the intersection of M with a linear subspace of $X \times X^*$ need *not* be a manifold. This can lead to bifurcation with respect to appropriate parameters b living in a subspace of X^*. For example, the set of all the solutions (u, b) of equation (E) above forms a C^∞-manifold M in $\mathbb{R}^4$. The solution set of Figure 29.11(b) corresponds to the intersection between M and the hyperplane $b_2 = 0$. From a physical point of view, the restriction of the parameters is meaningful. For example, in Section 65.7 we shall study the buckling of clamped plates. In this case, the outer boundary forces b have no vertical component.

Remark 29.30 (Typical Difficulties in the Case where $\dim X = \infty$). In the case where $\dim X < \infty$, the local *strict* monotonicity of F' at u, i.e.,

$$\langle F''(u)h, h\rangle > 0 \qquad \text{for all} \quad h \in X - \{0\} \tag{76}$$

is equivalent to the local *strong* monotonicity of F' at u, i.e.,

$$\langle F''(u)h, h\rangle \geq c\|h\|^2 \qquad \text{for all} \quad h \in X \quad \text{and fixed} \quad c > 0. \tag{76*}$$

In fact, if we set $a(h, k) = \langle F''(u)h, k\rangle$, then the bilinear functional $a: X \times X \to \mathbb{R}$ is symmetric, and it follows from (76) that $a(\cdot, \cdot)$ is strictly positive. Since $\dim X < \infty$, the functional $a(\cdot, \cdot)$ is also strongly positive.

Observe that this argument fails in the case where $\dim X = \infty$. Roughly speaking, this follows from the fact that the positive eigenvalues of $a(\cdot, \cdot)$ may converge to zero. This causes the typical difficulties in the calculus of variations. We have:

If $F'(u) = 0$, $u \in X$, and $\dim X = \infty$, then the local strict monotonicity of F' at u does not guarantee that F has a local minimum at u.

We will show below that one has to replace "locally strictly monotone at u" by "locally strongly monotone at u" or "locally regularly monotone at u." The latter condition plays a fundamental role in the calculus of variations as we shall see below.

Remark 29.31 (Importance of Variations). Assume (H1). In Proposition 29.25 we supposed that F is C^2. However, this assumption is too strong with respect to many applications in infinite-dimensional B-spaces X. Therefore, we work with variations. Recall the following from Chapter 4. Let

$$\varphi(t) = F(u + th)$$

for all real numbers in some neighborhood of $t = 0$ and fixed u and h in X. By definition, the nth *variation* of F at the point u in the direction of h is given by

$$\delta^n F(u; h) = \varphi^{(n)}(0), \qquad n = 1, 2, \ldots \tag{77}$$

We say that the nth variation $\delta^n F$ exists at the point u iff $\delta^n F(u; h)$ exists for all $h \in X$. If F is C^n in a neighborhood of u, then $\delta^n F$ exists for all v in a neighborhood of u and

$$\delta^n F(v; h) = F^{(n)}(v)h^n \qquad \text{for all} \quad h \in X. \tag{78}$$

For $n = 1, 2$, this can be written in the form:

$$\begin{aligned} \delta F(v; h) &= F'(v)h, \\ \delta^2 F(v; h) &= \langle F''(v)h, h\rangle \qquad \text{for all} \quad h \in X. \end{aligned} \tag{79}$$

In the following, a condition like "$\delta^2 F(u, h) \geq 0$" includes tacitly the existence of $\delta^2 F(u; h)$.

Remark 29.32 (First Reduction Trick). Assume (H1) and (H2). We set

$$\psi(t) = F(u + th) - b(u + th) \tag{80}$$

for all real numbers t in some neighborhood of $t = 0$ and fixed u and h in X. Note that $b(u + th) = b(u) + tb(h)$. Obviously, we have

$$\begin{aligned} \psi'(t) &= \delta F(u + th; h) - b(h), \\ \psi''(t) &= \delta^2 F(u + th; h), \end{aligned} \tag{81}$$

in the case where the corresponding variations exist. Using the *real function* ψ, the proofs below will follow *very easily* from the corresponding well-known properties of real functions.

Remark 29.33 (Second Reduction Trick). Boundary value problems lead frequently to problems of the following form:

$$F(u) - b(u) = \min!, \qquad u \in g + X, \tag{82}$$

where we assume that "$g + X$" makes sense, i.e., we assume that there is a linear space L such that $X \subseteq L$ and $g \in L$. In elasticity, for example, g corresponds to the displacement of the boundary of the elastic body. Suppose that $b: L \to \mathbb{R}$ is linear. Now, if we set $v = u - g$, then problem (82) is equivalent to the following problem:

$$G(v) - b(v) = \min!, \qquad v \in X, \tag{82*}$$

where $G(v) = F(g + v)$. Using this simple method, all the following results can also be applied to problem (82). In this connection, note that

$$\delta^n F(u; h) = \delta^n G(v; h) \qquad \text{for all} \quad h \in X,$$

where $u \in g + X$ and $u = g + v$.

29.12b. Necessary Conditions for Minima via Locally Monotone Operators

Proposition 29.34. *Assume* (H1) *and* (H2) *above. Suppose that u is a local minimal point of the problem*

$$F(u) - b(u) = \min!, \qquad u \in X. \tag{83}$$

Then:

(i) *If the first variation δF exists at u, then*

$$\delta F(u; h) = b(h) \qquad \text{for all} \quad h \in X. \tag{84}$$

(ii) *If the second variation $\delta^2 F$ exists at u, then*

$$\delta^2 F(u; h) \geq 0 \qquad \text{for all} \quad h \in X. \tag{85}$$

Corollary 29.35. *Assume* (H1), (H2), *and assume that F is C^2 in a neighborhood of u. Suppose that u is a local minimal point of* (83). *Then*

$$F'(u) = b, \tag{84*}$$

$$\langle F''(u)h, h\rangle \geq 0 \qquad \text{for all} \quad h \in X, \tag{85*}$$

i.e., F' is locally monotone at u.

Proof. For fixed $h \in X$, we set

$$\psi(t) = F(u + th) - b(u + th),$$

where t is a real number in some neighborhood of $t = 0$. If u is a local minimal point of (83), then $t = 0$ is a local minimal point of the real function ψ. Hence

$$\psi'(0) = 0 \qquad \text{and} \qquad \psi''(0) \geq 0.$$

This is (84) and (85).

The corollary follows from (79). □

29.12c. Sufficient Condition for a Strict Local Minimum via Locally Strongly Monotone Operators

The *key* condition reads as follows:

$$\delta^2 F(u; h) \geq c\|h\|^2 \qquad \text{for all} \quad h \in X \quad \text{and fixed} \quad c > 0. \tag{86}$$

Proposition 29.36. *Assume* (H1), (H2), *and assume that*:

(i) $\delta F(u; h) = b(h)$ *for fixed u and all $h \in X$, and condition* (86) *holds true.*
(ii) *The second variation $\delta^2 F$ exists on a neighborhood of u, and for each*

$\varepsilon > 0$, *there is an* $\eta(\varepsilon) > 0$ *such that*

$$|\delta^2 F(v; h) - \delta^2 F(u; h)| \leq \varepsilon \|h\|^2,$$

for all $h, v \in X$ *with* $\|v - u\| < \eta(\varepsilon)$.

Then u *is a strict local minimal point of* (83).

Corollary 29.37. *Assume* (H1), (H2). *Let* F *be* C^2 *in a neighborhood of* u. *Suppose that* $F'(u) = b$ *and that* F' *is locally strongly monotone at* u, *i.e.*,

$$\langle F''(u)h, h\rangle \geq c\|h\|^2 \qquad \text{for all} \quad h \in X \quad \text{and fixed} \qquad c > 0.$$

Then u *is a strict local minimal point of* (83).

PROOF. For fixed $h \in X$, we set

$$\psi(t) = F(u + th) - b(u + th),$$

where the real number t lies in some neighborhood $U(0)$ of $t = 0$. By (ii), the second derivative $\psi''(t)$ exists for all $t \in U(0)$. The classical Taylor theorem tells us that

$$\psi(t) = \psi(0) + t\psi'(0) + \frac{t^2}{2}\psi''(\vartheta t)$$

for all $t \in U(0)$ and $0 < \vartheta < 1$, where ϑ depends on t. By (i), $\psi'(0) = 0$ and

$$\psi''(0) \geq c\|h\|^2.$$

Let $|t| \leq 1$. It follows from (ii) that

$$|\psi''(\vartheta t) - \psi''(0)| = |\delta^2 F(u + \vartheta th; h) - \delta^2 F(u; h)| \leq \varepsilon \|h\|^2$$

for all $h \in X$ with $\|h\| < \eta(\varepsilon)$. Set $\varepsilon = c/2$. Then, for all $h \in X$ with $\|h\| < \eta$, we obtain

$$\psi(1) \geq \psi(0) + \frac{c}{4}\|h\|^2 \qquad \text{for all} \quad h \in X, \tag{87}$$

i.e., u is a strict local minimal point of (83).

The corollary follows from (79). $\square$

If the space X consists of smooth functions, then the key condition (86) cannot be verified for variational integrals. Our goal is to *weaken* the condition (86) in order to obtain a fundamental sufficient criterium for local minima in the calculus of variations. The simple idea is to show that the inequality (87) is valid

$$\text{when } \|h\|^2 \text{ is replaced by } \|h\|_Y^2,$$

where $\|\cdot\|_Y$ is a weaker norm on X, i.e., $\|h\|_Y \leq \text{const}\,\|h\|$ for all $h \in X$. When applying the following results to variational problems, one uses typically the following spaces:

$$X = \{u \in C^m(\bar{G}) \colon D^\alpha u = 0 \text{ on } \partial G \text{ for } |\alpha| \leq m - 1\},$$

$$Y = \mathring{W}_2^m(G), \qquad Z = L_2(G),$$

where G is a bounded region in $\mathbb{R}^N$, $N \geq 1$, and $m \geq 1$.

29.12d. Locally Regularly Monotone Operators

The *key* condition is the following Gårding inequality:

$$a(h, h) \geq c\|h\|_Y^2 - C\|h\|_Z^2 \tag{88}$$

for all $h \in Y$ with fixed $c > 0$ and $C \geq 0$, where

$$\delta^2 F(u; h) = a(h, h) \qquad \text{for all} \quad h \in X, \tag{89}$$

and $X \subseteq Y \subseteq Z$, i.e., $a(\cdot, \cdot)$ is an extension of the second variation from X to the larger space Y.

Definition 29.38. Let $F\colon U(u) \subseteq X \to \mathbb{R}$ be a functional on a neighborhood of the point u in the real B-space X. The second variation $\delta^2 F$ is called *regular* at the point u iff the following conditions are satisfied.

(i) $\delta^2 F$ exists on a neighborhood of u.

(ii) There exist real H-spaces Y and Z such that the embeddings $X \subseteq Y \subseteq Z$ are continuous and the embedding $Y \subseteq Z$ is *compact*.

(iii) There exists a bilinear, symmetric, bounded functional $a\colon Y \times Y \to \mathbb{R}$ such that (88) and (89) hold, i.e., $a(\cdot, \cdot)$ is a *regular Gårding form.*

(iv) The second variation is *uniformly continuous* with respect to the larger space Y, i.e., for each $\varepsilon > 0$, there is an $\eta(\varepsilon) > 0$ such that

$$|\delta^2 F(v; h) - \delta^2 F(u; h)| \leq \varepsilon \|h\|_Y^2,$$

for all $v, h \in X$ with $\|v - u\|_X < \eta(\varepsilon)$.

If condition (88) holds with $C = 0$, then the space Z drops out.

This concept corresponds to the general strategy in analysis of using *several* spaces in order to obtain sophisticated results.

Definition 29.39. Let $F\colon U(u) \subseteq X \to \mathbb{R}$ be C^2 on a neighborhood of the point u in the real B-space X. Then, F' is called *locally regularly monotone* at the point u iff F' is locally strictly monotone at u, and the second variation $\delta^2 F$ is regular at u.

Recall that $\delta^2 F(v; h) = \langle F''(v)h, h\rangle$.

29.12e. Sufficient Condition for a Strict Local Minimum via Locally Regularly Monotone Operators

The *key* condition reads as follows:

$$\begin{aligned} \delta F(u;h) &= b(h) \qquad \text{for all} \quad h \in X, \\ a(h,h) &> 0 \qquad \text{for all} \quad h \in Y - \{0\}, \end{aligned} \tag{90}$$

where $\delta^2 F(u;h) = a(h,h)$ for all $h \in X$ and $X \subseteq Y \subseteq Z$. We assume:

(H) Let $F\colon U(u) \subseteq X \to \mathbb{R}$ be a functional on a neighborhood of the point u in the real B-space X, and let $b \in X^*$ be given. Suppose that the second variation $\delta^2 F$ is *regular* at u in the sense of Definition 29.38 above. Let $Y \neq \{0\}$.

Theorem 29.H. *Assume* (H) *and assume that condition* (90) *is satisfied. Then u is a strict local minimal point of the problem*

$$F(u) - b(u) = \min!, \qquad u \in X. \tag{91}$$

Moreover, there are numbers $d > 0$ and $\varepsilon > 0$ such that

$$E_{\text{pot}}(v) - E_{\text{pot}}(u) \geq d\|v - u\|_Y^2 \tag{92}$$

for all $v \in X$ with $\|v - u\|_Y < \varepsilon$, where we set $E_{\text{pot}}(v) = F(v) - b(v)$.

Corollary 29.40. *Let $F\colon U(u) \subseteq X \to \mathbb{R}$ be C^2 on a neighborhood of the point u in the real* B-*space X, and let $b \in X^*$ be given. Suppose that*

$$F'(u) = b,$$

and suppose that F' is locally regularly monotone at the point u. Then u is a strict local minimal point of (91).

In terms of elasticity, condition (92) says that small changes of the potential energy E_{pot} correspond to small changes of the displacement u with respect to the norm $\|\cdot\|_Y$. Thus, inequality (92) describes the *stability* of the state u of the elastic body. The norm $\|\cdot\|_Y$ is called the energetic norm.

The following proof is essentially based on the Hestenes theorem (Proposition 22.39).

PROOF. It follows from (90) that $a\colon Y \times Y \to \mathbb{R}$ is strictly positive. By Proposition 22.39, the Gårding form $a(\cdot, \cdot)$ is *strongly* positive, i.e., there is a $d > 0$ such that

$$a(h,h) \geq d\|h\|_Y^2 \qquad \text{for all} \quad h \in Y.$$

We set $\psi(t) = F(u + th) - b(u + th)$ for fixed $h \in X$ and real t in a neighborhood of $t = 0$. As in the proof of Proposition 29.36, we obtain that

$$\psi(1) \geq \psi(0) + \frac{c}{4}\|h\|_Y^2, \tag{93}$$

for all $h \in X$ with $\|h\|_Y < \eta$. Since the embedding $X \subseteq Y$ is continuous, we have $\|h\|_Y \leq \text{const}\,\|h\|_X$ for all $h \in X$. Thus, there is an $\eta_0 > 0$ such that (93) holds for all $h \in X$ with $\|h\|_X < \eta_0$. □

The corollary is an immediate consequence of Theorem 29.H by Definition 29.39.

29.12f. Sufficient Condition for a Strict Local Minimum via Linear Eigenvalue Problems

Along with the original minimum problem

(P) $$F(u) - b(u) = \min!, \qquad u \in X,$$

we consider the so-called *accessory quadratic minimum problem*

(A) $$a(h, h) = \min! \qquad h \in Y, \qquad (h|h)_Z = 1$$

and the *eigenvalue problem of Jacobi*

(E) $$a(h, k) = \mu(h|k)_Z,$$

for fixed $h \in Y - \{0\}$, $\mu \in \mathbb{R}$, and all $k \in Y$.

Recall that $a(h, h) = \delta^2 F(u; h)$ for all $h \in X$. Moreover, the embeddings $X \subseteq Y \subseteq Z$ are continuous, and the embedding $Y \subseteq Z$ of the H-spaces Y and Z is *compact*.

Theorem 29.I. *Assume* (H) *from Theorem* 29.H, *and assume that*

$$\delta F(u; h) = b(h) \qquad \textit{for all} \quad h \in X.$$

Then u is a strict local minimal point of the original problem (P) *in the case where one of the following four mutually equivalent conditions is satisfied*:

(i) Strict Legendre condition: $a(h, h) > 0$ *for all* $h \in Y - \{0\}$.
(ii) Strong Legendre condition: $a(h, h) \geq d\|h\|_Y^2$ *for all* $h \in Y$ *and fixed* $d > 0$.
(iii) Accessory problem. *The infimum of problem* (A) *is positive.*
(iv) Eigenvalue criterion: *The smallest eigenvalue μ of the Jacobi equation* (E) *is positive.*

Corollary 29.41. *Assume* (H). *Then the accessory problem* (A) *has a solution, and the minimal value $\mu_{\min}$ of* (A) *is the smallest eigenvalue of* (E). *Let*

$\delta F(u;h) = b(h)$ *for all* $h \in X$. *Then*:

(a) *If* $\mu_{\min} > 0$, *then* u *is a strict local minimal point of* (P) *and* u *is stable.*
(b) *If* $\mu_{\min} < 0$ *and* X *is dense in* Y, *then* u *is not a local minimal point of* (P) *and* u *is unstable.*

Proof of Theorem 29.I. (i) $\Leftrightarrow$ (ii). This follows from the proof of Theorem 29.H.

(ii) $\Rightarrow$ (iii). Since the embedding $Y \subseteq Z$ is continuous, we have $\|h\|_Z \leq \text{const}\,\|h\|_Y$ for all $h \in Y$.

(iii) $\Rightarrow$ (i). This is obvious.

(iii) $\Leftrightarrow$ (iv). This follows from Theorem 22.G. Note that, in the case of the smallest eigenvalue, the proof of Theorem 22.G does not depend on the separability of the H-space.

If (i) holds, then u is a strict local minimal point of (P), by Theorem 29.H. $\square$

Proof of Corollary 29.41. The first statement follows from Theorem 22.G.

Ad(a). This is a consequence of Theorem 29.I.

Ad(b). If $\mu_{\min} < 0$, then there exists an $h \in Y$ such that $a(h,h) < 0$. Since $a\colon Y \times Y \to \mathbb{R}$ is continuous and X is dense in Y, there is an $\bar{h} \in X$ such that $a(\bar{h},\bar{h}) < 0$, and hence $\delta^2 F(u;\bar{h}) < 0$. $\square$

Theorem 29.I is an abstract variant of the classical theory of Jacobi mentioned in Section 18.7.

We now want to study the behavior of the Euler equation

$$F'(u) = b, \qquad u \in X, \tag{94}$$

in a neighborhood of a fixed solution u. To this end, we will use the strict Legendre condition in the weak form

$$\delta^2 F(u;h) > 0 \qquad \text{for all} \quad h \in X - \{0\}. \tag{94*}$$

Proposition 29.42. *Let* $F\colon U(u) \subseteq X \to \mathbb{R}$ *be* C^k *on an open neighborhood of the point* u *in the real* B-*space* X, *where* $2 \leq k \leq \infty$. *Let* u *be a solution of the Euler equation* (94). *Suppose that* (94*) *holds and that* $F''(u)\colon X \to X^*$ *is Fredholm of index zero.*

Then the operator $F'\colon U(u) \subseteq X \to X^*$ *is a local* C^k-*diffeomorphism at the point* u.

This proposition tells us the following important fact. There are neighborhoods V and W of u and b, respectively, such that the equation

$$F'(\bar{u}) = \bar{b}, \qquad \bar{u} \in V,$$

has a unique solution $\bar{u}$ for each given $\bar{b} \in W$. Moreover, if we set $\bar{u} = G(\bar{b})$, then the solution operator G is C^k on W. Figure 29.12 illustrates the intuitive meaning of this result.

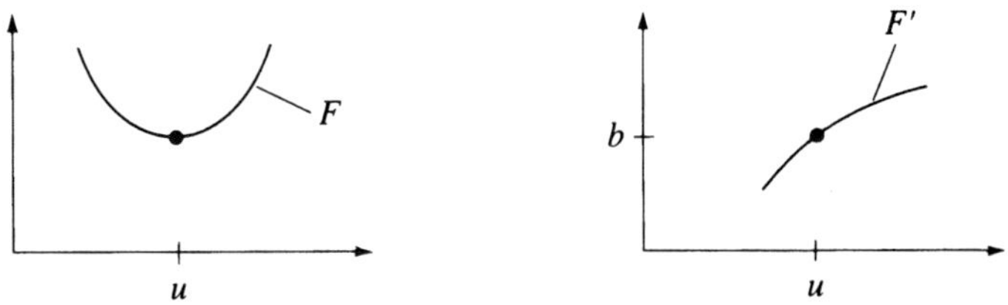

Figure 29.12

PROOF. If $F''(u)h = 0$, then $\langle F''(u)h, h\rangle = 0$ and hence $h = 0$, by (94*). Thus, the assertion follows from Proposition 29.17(i). □

29.12g. Sufficient Condition for a Global Minimum via Convexity

We want to study the minimum problem

$$F(u) - b(u) = \min!, \qquad u \in C, \tag{95}$$

where F is convex on the convex set C (e.g., $C = X$). Let $b \in X^*$.

Proposition 29.43. *Let $F: C \subseteq X \to \mathbb{R}$ be convex on the convex subset C of the real* B*-space X. Then each local minimal point of* (95) *is also a global minimal point.*

If, in addition, F is strictly convex on C, then the global minimal point u is unique

PROOF. Set $G(v) = F(v) - b(v)$. Let u be a local minimal point of (95) and suppose that there is a $v \in C$ such that $G(v) < G(u)$. Since G is convex, we obtain

$$G((1-t)u + tv) \leq (1-t)G(u) + tG(v) < G(u) \qquad \text{for all} \quad t \in\,]0, 1].$$

For small t, we obtain that u is not a local minimal point. This is a contradiction. The uniqueness of the minimal point follows from Corollary 25.15. □

Proposition 29.44. *Let $F: C \subseteq X \to \mathbb{R}$ be a functional on the open convex neighborhood C of the point u in the real* B*-space X.* Let

$$\delta F(u; h) = 0 \qquad \textit{for all} \quad h \in X.$$

Suppose that $\delta^2 F$ exists at each point of the set C and suppose that

$$\delta^2 F(v; h) \geq 0 \quad (\textit{resp.}\ \delta^2 F(v; h) > 0)$$

for all $v \in C$ and all $h \in X$ (resp. all $h \in X - \{0\}$).

Then u is a global minimal point of (95) (*resp. u is the unique global minimal point of* (95)). *Moreover, F is convex (resp. strictly convex) on C.*

Recall that

$$\delta^2 F(v; h) = \langle F''(v)h, h\rangle \qquad \text{for all} \quad v \in C, \quad h \in X,$$

in the case where F is C^2 on the set C.

Corollary 29.45. *Let $F\colon C \subseteq X \to \mathbb{R}$ be C^1 on the open convex neighborhood C of the point u in the real* B*-space X. Let $F'(u) = b$. Suppose that F' is monotone (resp. strictly monotone) on C.*

Then the assertions of Proposition 29.44 *hold true.*

PROOF. We set $\psi(t) = G(u + th)$, where $G = F - b$, and u, $u + h \in C$. For all $t \in [0, 1]$,

$$\psi'(t) = \delta F(u + th; h) - b(h), \qquad \psi''(t) = \delta^2 F(u + th; h).$$

If F is C^1 on the set C, then $\psi'(t) = \langle F'(u + th) - b, h\rangle$.

The functional G is (strictly) convex on C iff it is (strictly) convex on each line in C, i.e., the real function ψ is (strictly) convex on $[0, 1]$ for each $h \in X$ with $u + h \in C$.

Now use the following well-known property of real functions. It follows from either

$$\psi''(t) \geq 0 \quad (\text{resp. } \psi''(t) > 0) \qquad \text{for all} \quad t \in [0, 1], \tag{96}$$

or

$$\psi' \text{ is monotone} \quad (\text{resp. strictly monotone}) \qquad \text{on } [0, 1], \tag{96*}$$

that ψ is convex (resp. strictly convex) on $[0, 1]$.

Finally, note that the assumptions of Proposition 29.44 and Corollary 29.45 ensure (96) and (96*), respectively. □

29.13. Application to the Buckling of Beams

We investigate the buckling of a clamped beam of length l under the influence of the boundary force $P > 0$ as pictured in Figure 29.13. According to Euler, we use the following classical variational problem:

$$\begin{gathered} E_{\text{pot}}(u) \stackrel{\text{def}}{=} \int_0^l \left(\tfrac{1}{2}A\varphi'^2 + P(\cos\varphi - 1)\right) d\tau = \min!, \\ \varphi(0) = \varphi(l) = 0, \end{gathered} \tag{97}$$

which corresponds to the principle of minimal potential energy. The corresponding Euler equation is given by

$$\begin{gathered} \varphi'' + \frac{P}{A}\sin\varphi = 0, \\ \varphi(0) = \varphi(l) = 0. \end{gathered} \tag{98}$$

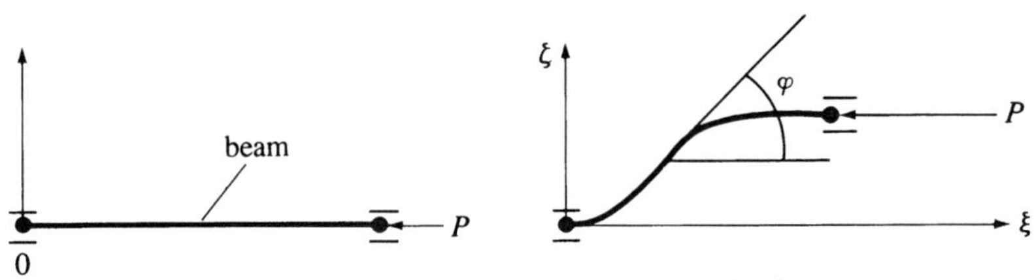

(a) $P < P_{\text{crit}}$ (b) $P > P_{\text{crit}}$

Figure 29.13

A detailed physical motivation for (97) will be given in Section 64.7 of Part IV. Here, the parameter τ denotes *arclength* counted from the origin O, and φ denotes the angle between the tangent at a point of the beam and the ξ-axis. Note that φ' corresponds to the curvature of the beam. The positive parameter A depends on the material of the beam. The precise meaning of A will be discussed in Section 64.7. Set

$$\|\varphi\| = \max_{0 \le \tau \le l} |\varphi(\tau)|.$$

From the physical point of view, we expect a behavior of the beam as pictured in Figure 29.13:

(i) If the force is small enough, i.e., $P < P_{\text{crit}}$, then there is *no* buckling.
(ii) If the force becomes supercritical, i.e., $P > P_{\text{crit}}$, then buckling occurs.

The decisive problem of Euler was to calculate P_{crit}. From Proposition 29.48 below we will obtain for the first buckling that:

$$P = P_{\text{crit}}(1 + \tfrac{1}{8}s^2 + O(s^3)), \qquad P_{\text{crit}} = \frac{\pi^2 A}{l^2},$$

$$\varphi(\tau) = s \sin \frac{\pi}{l}\tau + O(s^3), \qquad s \to 0,$$

where s is a small real parameter. This solution corresponds to Figure 29.13(b).

Mathematically, we have the following situation. The Euler equation has the trivial solution $\varphi \equiv 0$ for all P. However, this solution looses its stability at P_{crit}, and at this point, a new stable solution appears (Fig. 29.14(a)). To understand the bifurcation mechanism, we first study the following simple model in $\mathbb{R}^1$:

$$F(u) \stackrel{\text{def}}{=} \tfrac{1}{2}u^2 + P(\cos u - 1) = \min!, \qquad u \in \mathbb{R}^1, \tag{99}$$

where $P > 0$.

EXAMPLE 29.46. The Euler equation to (99) is given by

$$F'(u) \equiv u - P \sin u = 0$$

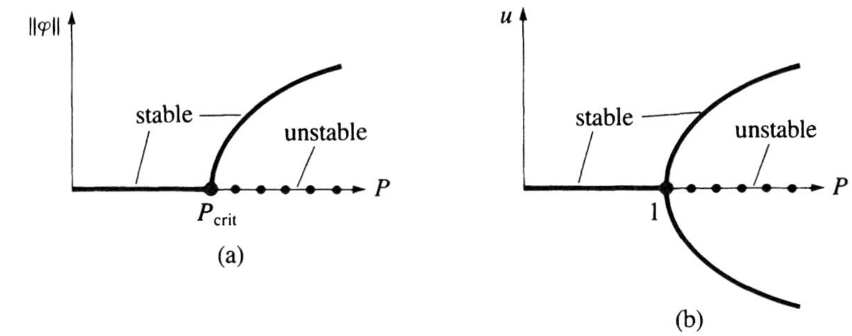

Figure 29.14

with the following two solutions:

(i) $u = 0$, $P =$ arbitrary,

(ii) $P = P(u) = 1 + \dfrac{u^2}{6} + O(u^3)$, $u \to 0$.

In addition, we have:

$$F''(0) = 1 - P \quad \text{for (i)},$$

$$F''(P(u)) = \frac{u^2}{3} + O(u^3), \qquad u \to 0 \quad \text{for (ii)}.$$

Thus, the trivial solution $u = 0$ is a strict local minimal point of (99) for $P < 1$, and a strict local maximal point for $P > 1$. Figure 29.14(b) shows the solutions of (99) in a neighborhood of the line $u = 0$ and their stability.

This analogy is remarkable. Already, in Chapter 7, we have encountered the important fact that the behavior of the solutions of semilinear elliptic partial differential equations can be understood by considering simple analogous models in $\mathbb{R}^1$.

Remark (More General Models for The Beam). Problem (97) corresponds to the simplest nonlinear model for a beam. The general form of the solutions of the Euler equation $\varphi'' + PA^{-1} \sin \varphi = 0$, independent of boundary conditions, will be considered in Proposition 29.49 by means of elliptic functions. The boundary condition $\varphi(0) = \varphi(l) = 0$ in (98) postulates that the beam is horizontal at the two end points.

If we represent the shape of the beam by the curve

$$\xi = \alpha + a(\alpha), \qquad \zeta = c(\alpha), \qquad 0 \le \alpha \le l,$$

where ξ, ζ are Cartesian coordinates (Fig. 29.13) and $u = (a, c)$ is the displacement vector, then it is possible to formulate more general variational problems than (97) and to consider more general boundary conditions (e.g., $a(0) = c(0) = 0$ and $a(l) = 0$). This will be studied in Problem 29.12.

In what follows we will investigate problem (97) in detail.

29.13a. Heuristic Arguments

In order to explain the basic idea as clearly as possible, we start with heuristic considerations based on Remark 29.28 above. Let $\varphi = \varphi(\tau)$ be a solution of the Euler equation (98). As in Remark 29.28, linearization of (98) at φ yields the Jacobi equation:

$$-h'' - \frac{P}{A} h \cos \varphi = \mu h, \qquad h(0) = h(l) = 0. \tag{100}$$

The sign in (100) has been chosen in such a way that the eigenvalues μ of (100) are bounded below. Let $\mu_{\min}$ be the smallest eigenvalue of (100). Similarly, as in the finite-dimensional case, we expect the following:

(i) If $\mu_{\min} > 0$, then φ is a strict local minimal point of the variational problem (97), i.e., φ is *stable*.
(ii) if $\mu_{\min} < 0$, then φ is *not* a strict local minimal point of (97), i.e., φ is *unstable*.

Case 1: Let $\varphi \equiv 0$. Then equation (100) has the eigensolutions

$$h(\tau) = \sin \frac{\pi n}{l} \tau, \qquad \mu = \frac{\pi^2 n^2}{l^2} - \frac{P}{A}, \qquad n = 1, 2, \ldots,$$

and hence

$$\mu_{\min} = \frac{\pi^2}{l^2} - \frac{P}{A}.$$

This yields the famous *formula of Euler*:

$$P_{\text{crit}} = \frac{\pi^2 A}{l^2},$$

i.e., the trivial solution $\varphi \equiv 0$ is stable for $P < P_{\text{crit}}$ and unstable for $P > P_{\text{crit}}$.

Case 2: If $\varphi \not\equiv 0$ is a solution of the Euler equation (98), then we have to check the sign of $\mu_{\min}$.

In Chapter 8 we studied in detail the bifurcation at simple eigenvalues. In the following, we want to show that our general functional analytic results from Chapter 8 allow us to construct very easily solutions bifurcating at $P = P_{\text{crit}}$ and to investigate their stability. Roughly speaking, it follows from Section 8.17 that the bifurcating solutions are stable in the case where the bifurcation is supercritical, as pictured in Figure 29.14, and as expected for the beam.

29.13b. A General Result

In order to see that our method does not depend on the special form of the variational problem (97), we consider the more general problem

$$F(\varphi) \stackrel{\text{def}}{=} \int_0^l \tfrac{1}{2}\varphi'^2 - Pf(\varphi)\,d\tau = \min!,$$
$$\varphi(0) = \alpha, \qquad \varphi(l) = \beta, \tag{101}$$

where α, β, P, and l are fixed real numbers. We also consider the corresponding Euler equation

$$-\varphi'' - Pf'(\varphi) = 0,$$
$$\varphi(0) = \alpha, \qquad \varphi(l) = \beta, \tag{102}$$

and the Jacobi equation

$$-h'' - Pf''(\varphi)h = \mu h, \qquad \mu \in \mathbb{R},$$
$$\varphi(0) = \varphi(l) = 0. \tag{103}$$

Here, P and μ are real numbers. Let $\mu_{\min}$ denote the smallest eigenvalue of (103). Let $\rho \in C^1[0, l]$ with $\rho(0) = \alpha$, $\rho(l) = \beta$. We need the following spaces:

$$X = \{\varphi \in C^1[0, l]: \varphi(0) = \varphi(l) = 0\},$$
$$Y = \mathring{W}_2^1(0, l), \qquad Z = L_2(0, l)$$

as well as $V = C^2[0, l] \cap X$. Then problem (101) can be written in the following form:

$$F(\varphi) = \min!, \qquad \varphi \in \rho + X. \tag{101*}$$

Proposition 29.47. *Let $f\colon \mathbb{R} \to \mathbb{R}$ be C^2, and let φ be a C^2-solution of the Euler equation* (102). *Then*:

(a) *If $\mu_{\min} > 0$, then φ is a strict local minimal point of* (101*), *and φ is stable.*
(b) *If $\mu_{\min} < 0$, then φ is not a minimal point of* (101*), *and φ is unstable.*

Proof. According to Remark 29.33, it is sufficient to consider the homogeneous case with $\alpha = \beta = 0$, i.e., $\rho = 0$.

(I) We show that the first variation vanishes. For $h \in X$, we set $\psi(t) = F(\varphi + th)$. Then

$$\psi'(0) = \delta F(\varphi; h) = \int_0^l h'\varphi' - Pf'(\varphi)h\,d\tau,$$

$$\psi''(0) = \delta^2 F(\varphi; h) = \int_0^l h'^2 - Pf''(\varphi)h^2\,d\tau.$$

For all $h \in X$, it follows from (102) that

$$\delta F(\varphi; h) = \int_0^l (-\varphi'' - Pf'(\varphi))h\,d\tau = 0.$$

(II) We show that the second variation is *regular* at φ. By the inequality of

Poincaré–Friedrichs, there is a $c > 0$ such that

$$\int_0^l h'^2\,d\tau \geq c\int_0^l (h'^2 + h^2)\,d\tau \qquad \text{for all} \quad h \in Y.$$

We set

$$a(h, g) = \int_0^l h'g' - Pf''(\varphi)hg\,d\tau.$$

Hence

$$\delta^2 F(\varphi; h) = a(h, h) \qquad \text{for all} \quad h \in X.$$

Moreover, for all $h \in Y$,

$$a(h, h) \geq c\|h\|_Y^2 - \text{const}\,\|h\|_Z^2.$$

Thus, $\delta^2 F$ is regular at φ.

(III) The eigenvalue problem

$$a(h, g) = \mu(h|g)_Z \qquad \text{for fixed} \quad h \in Y \quad \text{and all} \quad g \in Y \tag{104}$$

is obtained from the Jacobi equation (103) by using integration by parts. Hence, the generalized problem corresponding to (103) is identical to (104). According to the regularity theory for linear strongly elliptic equations, each solution of (104) is a classical solution of (103) and vice versa. Thus, the smallest eigenvalue of (104) is identical to $\mu_{\min}$.

Now, the assertions follow from Corollary 29.41. □

Proposition 29.48 (Bifurcation). *Let $f\colon \mathbb{R} \to \mathbb{R}$ be analytic with $f'(0) = 0$ and $f''(0) > 0$. Let $\alpha = \beta = 0$. We set $P_{\text{crit}} = \pi^2/l^2 f''(0)$. Then*:

(a) *The point $(\varphi, P) = (0, P_{\text{crit}})$ is a bifurcation point of the Euler equation* (102) *in the space $V \times \mathbb{R}$.*

(b) *In a sufficiently small neighborhood of $(0, P_{\text{crit}})$ in $X \times \mathbb{R}$, there exists a unique curve through the point $(0, P_{\text{crit}})$ which consists of nontrivial solutions (φ, P) of the Euler equation* (102). *This curve depends analytically on a small real parameter s.*

(c) *Let $f'''(0) = 0$. Then the solutions have the following form*:

$$\varphi(\tau) = s\sin\frac{\pi}{l}\tau + O(s^3), \qquad s \to 0,$$

$$P = P_{\text{crit}} + \varepsilon_2 s^2 + O(s^3),$$

$$\mu_{\min}(s) = 2f''(0)\varepsilon_2 s^2 + O(s^3),$$

where $\varepsilon_2 = -f^{(4)}(0)\pi^2/8l^2 f''(0)^2$.

(d) *The trivial solution $\varphi \equiv 0$ is stable for $P < P_{\text{crit}}$ and unstable for $P > P_{\text{crit}}$.*

(e) *The nontrivial solution φ in* (c) *is stable for $\varepsilon_2 > 0$ (supercritical bifurcation) and unstable for $\varepsilon_2 < 0$ (subcritical bifurcation).*

In the case of the beam, we have $f(\varphi) = A^{-1}(1 - \cos\varphi)$, and hence $\varepsilon_2 > 0$.

PROOF. We will apply Theorems 8.A and 8.E. The following proof proceeds completely analogously to Example 8.31. We set

$$P = P_{\text{crit}} + \varepsilon.$$

Let

$$V = \{\varphi \in C^2[0,l]: \varphi(0) = \varphi(l) = 0\}, \qquad W = C[0,l].$$

(I) Operator equation. We write the *Euler equation* (102) in the form

$$H(\varphi,\varepsilon) = 0, \qquad \varphi \in V, \qquad \varepsilon \in \mathbb{R}. \tag{105}$$

The operator $H: V \times \mathbb{R} \to W$ is analytic and $H(0,\varepsilon) \equiv 0$. The *Jacobi equation* (103) can be written in the form

$$H_\varphi(\varphi,\varepsilon)h = \mu h, \qquad h \in V, \quad \mu \in \mathbb{R}. \tag{106}$$

(II) The inhomogeneous linearized problem. The linearized equation

$$H_\varphi(0,0)h = b, \qquad h \in V, \quad b \in W, \tag{107*}$$

corresponds to the boundary value problem

$$-h'' - P_{\text{crit}} f''(0)h = b, \qquad h(0) = h(l) = 0. \tag{107}$$

We have:

(i) For $b = 0$, problem (107) has the simple eigensolution $h_1(\tau) = \sin(\pi\tau/l)$.
(ii) For $b \in W$, problem (107) has a solution iff

$$(b|h_1) = 0,$$

where $(v|w) = \int_0^l vw\,d\tau$.
Hence the operator $H_\varphi(0,0): V \to W$ is Fredholm of index zero.

(III) The generic branching condition. Letting $b = -f''(0)h_1$, we obtain $(b|h_1) \neq 0$, i.e., problem (107) has no solution. This implies the branching condition

$$R(H_{\varphi\varepsilon}(0,0)h_1) \notin R(H_\varphi(0,0)).$$

Observe that $H(\varphi,\varepsilon) = -\varphi'' - (P_{\text{crit}} + \varepsilon)f'(\varphi)$, and hence $H_{\varphi\varepsilon}(0,0)h_1 = -f''(0)h_1$.

Now, assertions (a) and (b) above follow immediately from Theorem 8.A.

(IV) Computation of the nontrivial solution. According to Theorem 8.A., the nontrivial solution branch has the form

$$\varphi = sh_1 + s^2h_2 + \ldots, \qquad P = P_{\text{crit}} + \varepsilon_1 s + \varepsilon_2 s^2 + \ldots, \tag{108}$$

with $(h_1|h_k) = 0$ for all $k \geq 2$. Substituting this into the Euler equation (102), a comparison of the terms with s^2 yields:

$$-h_2'' - P_{\text{crit}} f''(0)h_2 = b, \qquad h_2(0) = h_2(l) = 0,$$

where $b = \varepsilon_1 f''(0)h_1$. From $(b|h_1) = 0$ it follows that $\varepsilon_1 = 0$. Furthermore, since $(h_1|h_2) = 0$, we get $h_2 = 0$.

A comparison of the terms with s^3 yields:

$$-h_3'' - P_{\text{crit}} f''(0)h_3 = b, \qquad h_3(0) = h_3(l) = 0,$$

where $b = \varepsilon_2 f''(0)h_1 + P_{\text{crit}} f^{(4)}(0)h_1^3/6$. From $(b|h_1) = 0$ we get ε_2.

(V) Stability of the trivial solution. Letting $\varphi = 0$ in the Jacobi equation (103) with $P = P_{\text{crit}} + \varepsilon$, we obtain

$$\mu_{\min} = -f''(0)\varepsilon, \qquad f''(0) > 0.$$

Thus, the trivial solution $\varphi = 0$ is stable for $\varepsilon < 0$ and unstable for $\varepsilon > 0$.

(VI) Stability of the nontrivial solution. By Theorem 8.E, the smallest eigenvalue $\mu_{\min}(s)$ of equation (106) along the nontrivial solution $(\varphi(s), \varepsilon(s))$ in $V \times \mathbb{R}$ has the form

$$\mu_{\min}(s) = 2\varepsilon_2 s^2 + O(s^3), \qquad s \to 0.$$

Note that (106) corresponds to the classical Jacobi equation (103). Since $\mu_{\min}(s) \gtrless 0$, for small $s \neq 0$ if $\varepsilon_2 \gtrless 0$, we get the assertion (e). □

29.13c. Explicit Solution of the Beam Problem

Proposition 29.49. *For every $a \in \mathbb{R}$, the initial value problem*

$$\begin{gathered} \varphi'' + \frac{P}{A}\sin\varphi = 0, \\ \varphi(0) = 0, \qquad \varphi'(0) = a, \end{gathered} \tag{109}$$

has a unique global solution on $\mathbb{R}$.

For $a = 2k\sqrt{P/A}$ with $0 \le k < 1$, the solution of (109) *is given by*

$$\varphi(\tau) = 2\arcsin\left(k\,\mathrm{sn}\left(\sqrt{\frac{P}{A}}\tau, k\right)\right) \quad \text{for all} \quad \tau \in \mathbb{R}. \tag{110}$$

The function $\varphi = \varphi(\tau)$ has the same zeros τ_n as the elliptic function $\tau \mapsto \mathrm{sn}(\sqrt{P/A}\,\tau, k)$, and hence

$$\tau_n = 2n\sqrt{\frac{A}{P}}K(k), \qquad n = 0, \pm 1, \pm 2, \ldots,$$

where

$$K(k) = \int_0^{\pi/2} \frac{d\psi}{\sqrt{1 - k^2\sin^2\psi}}.$$

Example 29.50 (The Beam). By Proposition 29.49 each solution of the Euler equation (98) for the beam is obtained from (110) by letting $\varphi(l) = 0$. Thus,

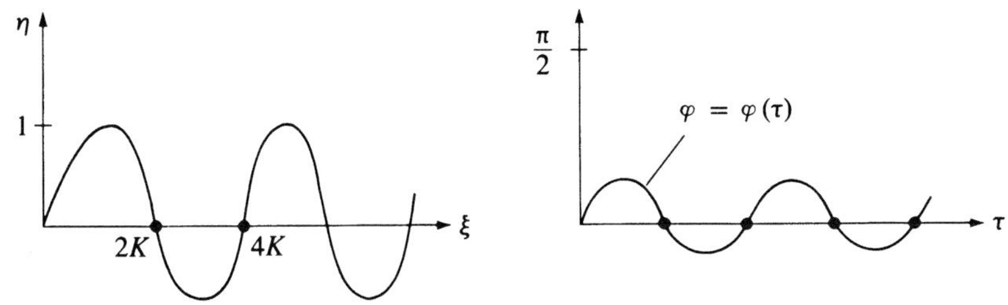

(a) $\eta = sn(\xi, k)$ (b) $\varphi(0) = \varphi(l) = 0$

Figure 29.15

$l = \tau_n$, and hence we obtain for the corresponding outer force

$$P = \frac{4n^2 A}{l^2} K(k)^2.$$

Note that the function $k \mapsto K(k)$ is monotonically increasing and $K(0) = \pi/2$. More precisely, we obtain

$$P = P_n(1 + 8k^2 + O(k^4)), \qquad \text{as} \quad k \to 0,$$

where $P_n = n^2\pi^2 A/l^2$. By (110), bifurcation occurs for $k = 0$ and hence for the critical forces $P = P_n$, $n = 1, 2, \ldots$. Note that $P_{\text{crit}} = P_1$.

The functions sn(·) and $\varphi(\cdot)$ are pictured in Figure 29.15. Both functions are periodic. The behavior of $\varphi = \varphi(\tau)$ can be understood best if one observes that the beam equation (109) describes the motion of a pendulum as well (Fig. 29.16(a)). In the case of a pendulum, we have: $\varphi =$ angle, $\tau =$ time, $m =$ mass, $L =$ length, $g =$ gravitational acceleration, and in (109) we have to set $P/A = g/L$. The larger a is in (109), the larger is the initial velocity of the pendulum. If a is sufficiently large, i.e., if $k > 1$, then the pendulum completely surrounds the circle in Figure 29.16(a). Such solutions lead to strange forms of the beam, like those pictured in Figure 29.16(b), which may correspond to a metallic rod.

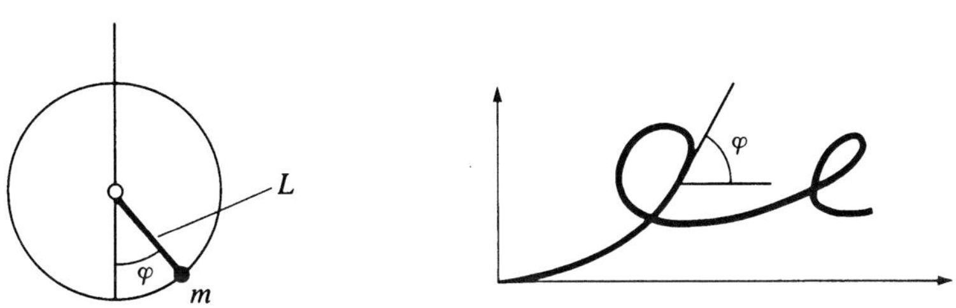

(a) (b)

Figure 29.16

PROOF OF PROPOSITION 29.49. If a solution $\varphi = \varphi(\tau)$ of (109) exists on an open interval, then φ is bounded, since $\sin(\cdot)$ is bounded. Thus, the existence of a unique global solution follows from the continuation principle in Section 3.3.

We want to show that φ in (110) solves equation (109). For brevity we set $P/A = 1$. It follows from (110) that

$$\sin \varphi/2 = k\sigma, \tag{111}$$

where $\sigma(\tau) = \mathrm{sn}(\tau, k)$. Hence $\cos \varphi = 1 - 2k^2\sigma^2$. Differentiation of (111) yields

$$\begin{aligned}\frac{\varphi'^2}{4}\cos^2\frac{\varphi}{2} &= k^2\sigma'^2 = k^2(1 - k^2\sigma^2)(1 - \sigma^2)\\ &= k^2\left(\cos^2\frac{\varphi}{2}\right)(1 - \sigma^2).\end{aligned} \tag{111*}$$

Therefore, $\varphi'^2/2 = \cos \varphi + \text{const}$, and hence

$$(\varphi'' + \sin \varphi)\varphi' = 0.$$

Moreover, from (111*), we get $\varphi'(0)/2 = k\,\mathrm{cn}(0, k) = k$. □

29.14. Stationary Points of Functionals

By definition, the problem

$$F(u) = \text{stationary!}, \qquad u \in S,$$

corresponds to the determination of all the stationary points of the functional $F\colon S \to \mathbb{R}$ in the sense of Definition 29.51 below. For example, if S is an open interval in $\mathbb{R}$, then u is called a stationary point of F iff $F'(u) = 0$, i.e., the tangent of the graph of F is horizontal (Fig. 29.17(a)). The following definition generalizes this situation.

Definition 29.51. Let $F\colon S \to \mathbb{R}$ be a functional on a subset S of a real B-space, and let $u_0 \in S$. We set

$$\psi(t) = F(u(t)),$$

where t is a real number. By an admissible curve through the point u_0, we understand a map $u\colon U(0) \subseteq \mathbb{R} \to S$ such that $u(0) = u_0$ and the derivative $u'(0)$ exists (Fig. 29.18). The point u_0 is called a *stationary point* of F iff

$$\psi'(0) = 0$$

for all admissible curves through u_0.

EXAMPLE 29.52. Let $S = \{(\xi, \eta) \in \mathbb{R}^2\colon \xi^2 + \eta^2 = 1\}$, and let $F(\xi, \eta) = \eta$. Then the function $F\colon S \to \mathbb{R}$ has the two stationary points $(0, \pm 1)$ (Fig. 29.17(b)).

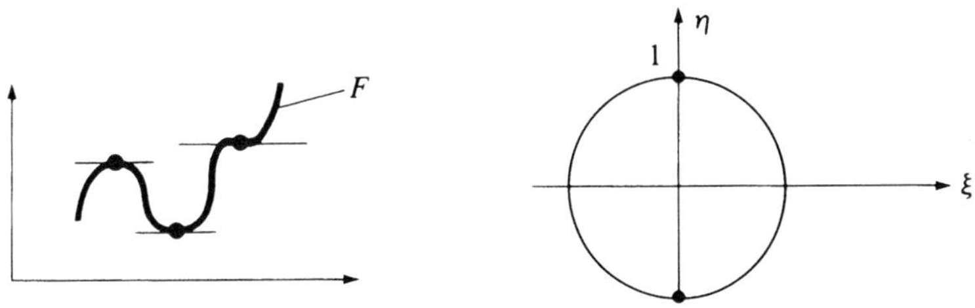

(a) (b)

Figure 29.17

Proof. We set $\psi(t) = F(\xi(t), \eta(t))$. If $\xi = \xi(t)$, $\eta = \eta(t)$ is a differentiable curve on S with $\xi(0) = 0$, $\eta(0) = \pm 1$, then $\psi'(0) = \eta'(0) = 0$. $\square$

Note that $(0, \pm 1)$ is *not* a stationary point of $F: \mathbb{R}^2 \to \mathbb{R}$, where again $F(\xi, \eta) = \eta$.

Proposition 29.53. *Let $F: U(u_0) \subseteq X \to \mathbb{R}$ be a functional on a neighborhood of the point u_0 in the real* B*-space X. Then:*

(a) *If u_0 is a stationary point of F, then*

$$\delta F(u_0; h) = 0 \qquad \text{for all} \quad h \in X.$$

(b) *Let F be* F*-differentiable at u_0. Then u_0 is a stationary point of F iff $F'(u_0) = 0$.*

Stationary points of functionals are frequently also called *critical points*. By Proposition 29.53(b), this convention is meaningful with respect to the more special definition of critical points for maps given in Section 29.10.

Proof. Ad(a). Set $\psi(t) = F(u_0 + th)$, where $h \in X$. If u_0 is a stationary point of F, then $\psi'(0) = \delta F(u_0; h) = 0$.

Ad(b). Let $F'(u_0) = 0$, and let $u = u(t)$ be an admissible curve through u_0. Letting $\psi(t) = F(u(t))$, we obtain that $\psi'(0) = F'(u(0))u'(0) = 0$, since $u(0) = u_0$. Thus, u_0 is a stationary point of F.

Conversely, if u_0 is a stationary point of F, then it follows from (a) that $\delta F(u_0, h) = F'(u_0)h = 0$ for all $h \in X$ and hence $F'(u_0) = 0$. $\square$

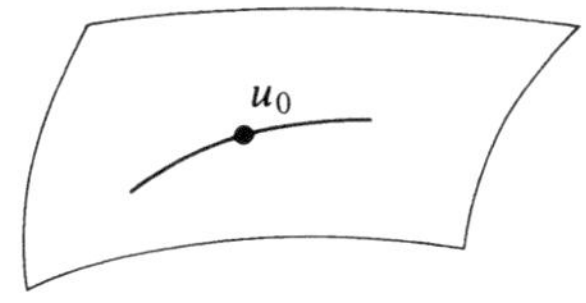

Figure 29.18

Corollary 29.54 (Stationary Points on Planes S). *Let* $F\colon U(u_0) \subseteq X \to \mathbb{R}$ *be* F-*differentiable at* u_0. *Let* $S = a + Y$, *where* Y *is a linear subspace of* X *and* $a \in X$. *Let* $u_0 \in S$. *Then* u_0 *is a stationary point of* $F\colon S \to \mathbb{R}$ *iff*

$$\delta F(u_0; h) = 0 \qquad \textit{for all} \quad h \in Y. \tag{112}$$

PROOF. We set $\psi(t) = F(u(t))$. Let u_0 be a stationary point of $F\colon S \to \mathbb{R}$. Let $h \in Y$. Then $u(t) = u_0 + th$ is an admissible curve through u_0, since $u(t) \in S$ for all $t \in \mathbb{R}$. From $\psi'(0) = 0$ we obtain (112).

Conversely, suppose that (112) holds. Let $u\colon U(0) \subseteq \mathbb{R} \to S$ be an admissible curve through u_0. Then $u'(0) \in Y$. By (112), $\psi'(0) = F'(u_0)u'(0) = \delta F(u_0; u'(0)) = 0$. $\square$

29.15 Application to the Principle of Stationary Action

> The integral, considered in the principle of least action, can never have a maximum, as Lagrange mistakenly believed; in no way, however, will it always have a minimum.
>
> Carl Gustav Jacob Jacobi (1837)

We consider the motion $q = q(t)$ of a point of mass m on the real line, i.e., t and $q(t)$ are real numbers. Let $U\colon \mathbb{R} \to \mathbb{R}$ be a C^1-function which we regard as the potential energy of the mass point. The principle of stationary action reads as follows:

$$F(q) \overset{\text{def}}{=} \int_0^T \left(\frac{m}{2} q'^2 - U(q)\right) dt = \text{stationary!}, \qquad q \in S, \tag{113}$$

where

$$S = \{q \in C^1[0, T]\colon q(0) = a, q(T) = b\}.$$

Here, a, b, and T are fixed real numbers.

Proposition 29.55. *The* C^2-*function* $q(\cdot)$ *is a solution of* (113) *iff*

$$mq'' = -U'(q), \qquad q \in S. \tag{114}$$

Equation (114) is called Newton's equation of motion. Here, $-U'(q)$ corresponds to the force acting on the mass point. This will be considered in greater detail in Chapter 58.

PROOF. We set $X = C^1[0, T]$ and $Y = \{q \in X\colon q(0) = q(T) = 0\}$. For all $h \in Y \cap C^2[0, T]$, integration by parts yields

$$\delta F(q; h) = \int_0^T (mq'h' - U'(q)h)\,dt$$
$$= -\int_0^T (mq'' + U'(q))h\,dt.$$

By Corollary 29.54, q is a solution of (113) iff $\delta F(q; h) = 0$ for all $h \in Y$. Since the set $Y \cap C^2[0, T]$ is dense in Y, this is equivalent to

$$\delta F(q; h) = 0 \qquad \text{for all} \quad h \in Y \cap C^2[0, T]$$

and, in turn, this is equivalent to the Euler equation (114). □

Corollary 29.56. *A solution q of the Euler equation* (114) *can never be a local maximal point of* (113).

PROOF. Otherwise, q is a local minimal point of $-F(q) = \min!$, $q \in S$. By the Legendre condition in Section 18.17b,

$$L_{q'q'} \leq 0,$$

where $L = \frac{m}{2}q'^2 - U(q)$; but this is impossible. □

The following example shows that *not* all the solutions of the Euler equation (114) are local minimal points of (113).

EXAMPLE 29.57. We set $U(q) = \omega^2 q^2/2$ for fixed $\omega > 0$. Then the Euler equation (114) describes the motion of a harmonic oscillator. The corresponding Jacobi equation

$$-h'' - \omega^2 h = \mu h, \qquad h(0) = h(T) = 0,$$

has the smallest eigenvalue $\mu_{\min} = (\pi^2/T^2) - \omega^2$. From Proposition 29.47, we obtain that:

(a) If $T > \pi/\omega$, then *no* solution of the Euler equation (114) is a local minimal point of (113).
(b) However, if T is sufficiently small, i.e., if $T < \pi/\omega$, then every solution of (114) is a strict local minimal point of (113).

29.16. Abstract Statical Stability Theory

The following abstract model allows, for example, important applications in nonlinear elasticity. Let $n_S(b)$ denote the *number of solutions* of the equation

$$Au = b, \qquad u \in S. \tag{115}$$

Our goal is Theorem 29.J below. We assume:

(H1) Let X and Y be B-spaces over $\mathbb{K} = \mathbb{R}, \mathbb{C}$. The C^k-operator $A: D \subseteq X \to Y$ is Fredholm of index zero on the open set D, where $1 \leq k \leq \infty$.

(H2) Let D_{stab} be a subset of D such that

$$A'(u)h = 0, \qquad u \in D_{\text{stab}}, \quad h \in X,$$

implies $h = 0$.

(H3) Let S be a subset of D_{stab} such that $A: S \to Y$ is proper.

In elasticity, the set D_{stab} corresponds to (strongly) stable states u of the elastic body, and b corresponds to the outer forces and to the displacements of the boundary of the elastic body. In Section 29.19 we consider general variational problems. There, we will set:

$$D = \{u\text{: the strong Legendre–Hadamard condition holds at } u\},$$

$$D_{\text{stab}} = \{u \in D\text{: the second variation is strictly positive at } u\}.$$

EXAMPLE 29.58. Let $A: D \subseteq X \to X^*$ be a C^k-Fredholm operator of index zero on the open subset D of the real B-space X, where $1 \leq k \leq \infty$. We set

$$D_{\text{stab}} = \{u \in D\text{: } A \text{ is locally strictly monotone at } u\}.$$

Then assumptions (H1) and (H2) are statisfied with $Y = X^*$.

This follows directly from Definition 29.20.

Recall that the operator $A: S \to X^*$ is *injective* on each convex subset S of D_{stab}, by Proposition 29.22.

EXAMPLE 29.59. We consider the local minimum problem

$$F(u) - b(u) = \min!, \qquad u \in D. \tag{116}$$

Let $F: D \subseteq X \to \mathbb{R}$ be a C^{k+1}-functional on the open subset D of the real B-space X, where $1 \leq k \leq \infty$. Suppose that $F': D \subseteq X \to X^*$ is Fredholm of index zero. Let

$$D_{\text{stab}} = \{u \in D\text{: } F' \text{ is locally regularly monotone at } u\}.$$

If we set $A = F'$, then the assumptions (H1) and (H2) above are satisfied with $Y = X^*$. Let S be a subset of D_{stab}. Then, each solution of the Euler equation (115) is a strict local minimal point of (116), by Corollary 29.40.

If S is convex, then A is injective on S.

Theorem 29.J. *Assume* (H1) *through* (H3). *Let* $b \in Y - A(\partial S \cap S)$ *be given. Then*:

(a) *The number* $n_S(b) \geq 0$ *is finite.*

(b) *The function* $n_S(\cdot)$ *is constant on each connected subset of the open set* $Y - A(\partial S \cap S)$.

Corollary 29.60. *Assume* (H1) *and* (H2). *Let S be a subset of D_{stab}. Then*:

(a) *For each $u \in D_{\text{stab}}$, the operator A is a local C^k-diffeomorphism at u.*
(b) *If A is injective on S, then $A: S \to A(S)$ is a C^k-diffeomorphism.*

PROOF. Theorem 29.J follows immediately from Theorem 29.F, since the subset D_{sing} of D_{stab} is empty, according to (H2).

Corollary 29.60 follows from Proposition 29.17(i). □

The following corollary is related to the important fact that the apparently special assumption (H3) for the operator $A: D \subseteq X \to Y$ can be satisfied for a large class of subsets S of D_{stab}.

Corollary 29.61 (Properness). *Assume* (H1). *Then*:

(a) *The operator A is proper on each compact subset S of D_{stab}.*
(b) *If C is a compact subset of D_{stab}, then there exists an open neighborhood U of C such that A is proper on $S = \bar{U}$.*
(c) *Let the* B-*space X be separable. If C is a bounded closed subset of D_{stab}, then for each $\varepsilon > 0$, there exists an open neighborhood U of C such that*

$$\operatorname{dist}(\partial U, \partial C) < \varepsilon,$$

and A is proper on $S = \bar{U}$ (Fig. 29.19).
(d) *If A is proper on a set D_0, then A is also proper on each closed subset S of D_0.*
(e) *If A is proper on a finite number of subsets of D, then A is also proper on the union of these subsets.*

Statements (b) and (c) are especially important in connection with Theorem 29.J.

PROOF. Ad(a), (d), (e). Note that closed subsets of compact sets are compact, and the union of a finite number of compact sets is compact.

Ad(b). Let $u \in D_{\text{stab}}$. By Corollary 29, 60, A is a local diffeomorphism at u. Moreover, by Proposition 29.17(d), A is locally closed. Thus, there exists a closed neighborhood V of u such that $A: V \to A(V)$ is a homeomorphism, and the set $A(V)$ is closed in Y. Hence A is proper on V.

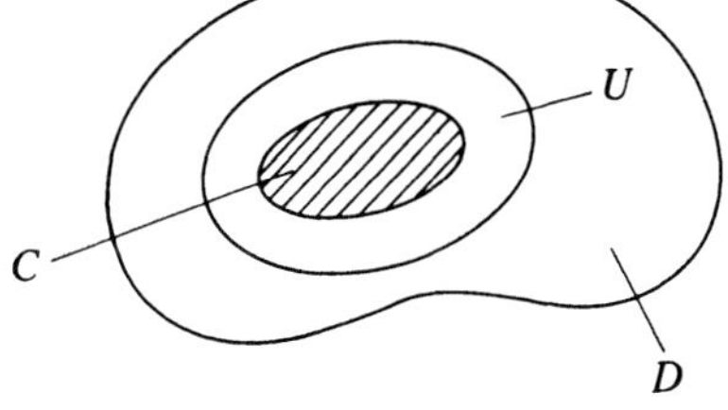

Figure 29.19

Now note that the compact set C can be covered by a finite number of such sets V.

Ad(c). Let $\{u_1, u_2, \ldots\}$ be a dense subset of C. Set $C_n = C \cap \operatorname{span}\{u_1, \ldots, u_n\}$, and apply the argument (b) to C_n for sufficiently large n. □

29.17 The Continuation Method

Let u_0 be a point in the domain of stability D_{stab} and let $b = b(t)$ be a C^1-curve with $Au_0 = b(0)$, where t is a real parameter. Since A is a local C^1-diffeomorphism at u_0, there exists a C^1-curve $u = u(t)$, for real t in a neighborhood of $t = 0$, such that $u(0) = u_0$ and

$$A(u(t)) = b(t). \tag{117}$$

This solution curve can be continued as long as the curve remains in the domain of stability D_{stab}. In elasticity, the curve $u = u(t)$ corresponds to a deformation process. Bifurcation may occur if the curve reaches the boundary of stability ∂D_{stab} (cf. Fig. 29.11(b)). Differentiation of (117) yields $A'(u(t))u'(t) = b'(t)$, and hence

$$u'(t) = A'(u(t))^{-1} b(t). \tag{118}$$

Discretization of the time derivative $u'(t)$ yields the approximation method

$$v_{n+1} = v_n + \Delta t A'(v_n)^{-1} b'(n\Delta t), \qquad n = 0, 1, \ldots, \tag{119}$$

where v_n corresponds to $u(n\Delta t)$ and $v_0 = u_0$. The convergence of this method will be proved in Section 61.6 in connection with nonlinear elasticity. This proof proceeds completely analogously to the proof of the classical Peano existence theorem for ordinary differential equations. The necessary compactness is obtained through embedding theorems.

29.18. The Main Theorem of Bifurcation Theory for Fredholm Operators of Variational Type

We want to prove that (0, 0) is a *bifurcation point* of the equation

$$H(u, \lambda) = 0, \qquad u \in X, \quad \lambda \in \Lambda, \tag{120}$$

where the parameter λ lives in the parameter space Λ. More precisely, we want to show how to reduce problem (120) to a *finite-dimensional* variational problem (Proposition 29.64 below), and we will use this information in order to prove an important bifurcation theorem for (120).

A *pecularity* of our approach is that we do *not* restrict ourselves to the case of H-spaces as this is usually done in the literature. This allows us in Section 29.20 to prove bifurcation theorems for variational problems *without* any growth conditions. In this connection, we will use the B-spaces of *smooth* functions

$$X = \{v \in C^{2,\alpha}(\bar{G}) : v = 0 \text{ on } \partial G\}, \qquad Y = C^{\alpha}(\bar{G}), \quad 0 < \alpha < 1,$$

and we will set

$$(v|w) = \int_G vw\,dx \qquad \text{for all} \quad v, w \in Y.$$

Note that Y is *not* an H-space. The H-space approach corresponds to the Sobolev spaces

$$X = Y = \mathring{W}_2^1(G)$$

with the usual scalar product $(\cdot|\cdot)$ on X. In this case, one needs growth conditions.

In (H3) below, we introduce a new class of operators which we call operators of *variational type*. These operators can be regarded as *generalized* potential operators. Roughly speaking, operators of variational type are "potential operators" with respect to a generalized scalar product $(\cdot|\cdot)$ on a B-space Y. In the special case of an H-space X with the usual scalar product $(\cdot|\cdot)$, we set $Y = X$. Then, operators of variational type and potential operators coincide. This will be discussed below.

We assume:

(H1) Let X, Y, Λ be real B-spaces such that the embedding $X \subseteq Y$ is continuous. Let the operator

$$H: U(0,0) \subseteq X \times \Lambda \to Y$$

be C^k on a neighborhood of the point $(0,0)$, where $1 \le k \le \infty$.

(H2) Linearization. Suppose that $H(0,\lambda) \equiv 0$ and that the linearized equation

$$H_u(0,0)h = 0, \qquad h \in X,$$

has exactly n linearly independent solutions with $n \ge 1$. Suppose that $H_u(0,0)\colon X \to Y$ is Fredholm of index zero.

(H3) Variational structure. Assume that there is a C^{k+1}-functional f: $U(0,0) \to \mathbb{R}$ such that the following *key condition* holds:

$$\delta f(u,\lambda;h) = (H(u,\lambda)|h) \qquad \text{for all} \quad h \in X, \quad (u,\lambda) \in U(0,0).$$

Here, δf denotes the first variation of the functional f with respect to the first variable u, that is,

$$\delta f(u,\lambda;h) = \lim_{t\to 0} \frac{f(u+th,\lambda) - f(u,\lambda)}{t}$$

and $(u,v) \mapsto (u|v)$ denotes a given bilinear, bounded, symmetric, strictly positive functional from $Y \times Y$ to $\mathbb{R}$.

We say that the operator $u \mapsto H(u,\lambda)$ is of *variational type* iff the condition (H3) is satisfied.

(H4) Generic branching condition. Suppose that the F-derivative $H_{u\lambda}(0,0)$ exists and that there is a special parameter λ_1 such that

$$(v|H_{u\lambda}(0,0)v\lambda_1) \neq 0$$

for all $v \in N(H_u(0,0))$ with $v \neq 0$.

(H5) Remainder. Let

$$R(u,\lambda) = H(u,\lambda) - H_u(0,0)u - H_{u\lambda}(0,0)u\lambda$$

and suppose that

$$\|R(u,\lambda)\| \leq \text{const}\,\|u\|^2$$

for all (u,λ) in a sufficiently small neighborhood of $(0,0)$.

The following assumptions will be used in Corollary 29.63 below, but not in Theorem 29.K.

(H5*) Special remainder. Define $R(\cdot)$ as in (H5). Suppose that

$$R(u,\varepsilon\lambda_1) = Su + \varepsilon Tu$$

for all (u,ε) in a neighborhood of $(0,0)$ in $X \times \mathbb{R}$, where the operators S, T: $U(0) \subseteq X \to Y$ are C^k and $S(0) = T(0) = 0$, $S'(0) = T'(0) = 0$.

(H6*) The operator $v \mapsto H_{u\lambda}(0,0)v\lambda_1$ transforms the null space $N(H_u(0,0))$ into itself.

EXAMPLE 29.62. Let λ be a real parameter, and suppose that

$$H(u,\lambda) = Lu + \lambda Au + Su + \lambda Tu, \tag{121}$$

where L, A: $X \to Y$ are linear continuous operators, and S, T: $U(0) \subseteq X \to Y$ are C^k, $1 \leq k \leq \infty$ with $S(0) = T(0) = 0$ and $S'(0) = T'(0) = 0$. Then

$$H_u(0,0) = L, \qquad H_{u\lambda}(0,0)u\lambda = \lambda Au,$$

and conditions (H5), (H5*) are satisfied. Moreover, in the important special case

$$Au = u \qquad \text{for all} \quad u \in X,$$

the generic branching condition (H4) and hypothesis (H6*) are satisfied trivially with $\lambda_1 = 1$.

If X is an H-space with scalar product $(\cdot|\cdot)$, then we set $Y = X$. In this case, condition (H3) is satisfied iff

$$(H_u(u,\lambda)h|k) = (H_u(u,\lambda)k|h) \qquad \text{for all} \quad h,k \in X,$$

and all (u,λ) in the convex open neighborhood $U(0.0)$. The functional f: $U(0,0) \subseteq X \times \Lambda \to \mathbb{R}$ is then given by

$$f(u,\lambda) = \int_0^1 (H(tu,\lambda)|u)\,dt.$$

This follows from Proposition 41.5. In this special case, operators of variational type and potential operators coincide.

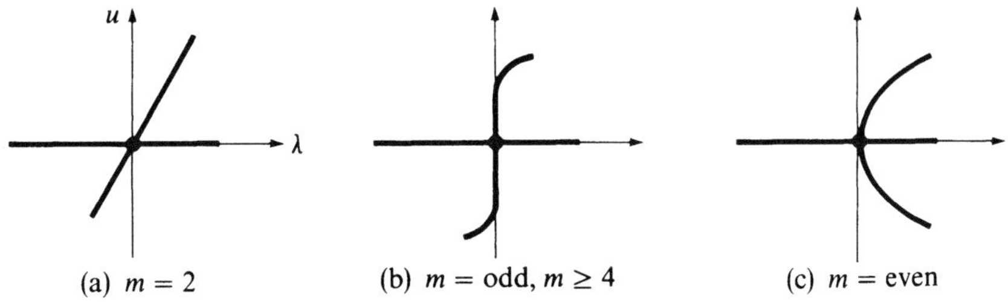

Figure 29.20

Theorem 29.K. *If* (H1)–(H5) *hold with* $k \geq 2$, *then* $(0, 0)$ *is a bifurcation point of the equation* $H(u, \lambda) = 0$, $u \in X$, $\lambda \in \Lambda$.

More precisely, the bifurcation takes place in the direction of the parameter λ_1, *i.e., the point* $u = 0$, $\varepsilon = 0$ *is a bifurcation point of the equation* $H(u, \varepsilon\lambda_1) = 0$, $u \in X$, $\varepsilon \in \mathbb{R}$.

The *prototype* for this theorem is given by the real equation

$$-\lambda u + u^m = 0, \qquad u, \lambda \in \mathbb{R}, \quad m \geq 2,$$

with the trivial solution $u = 0$, $\lambda =$ arbitrary, and the nontrivial solution $\lambda = u^{m-1}$ (Fig. 29.20).

Recall that n is the number of the linearly independent solutions of the linearized equation $H_u(0, 0)h = 0$, $h \in X$.

Corollary 29.63. *Assume* (H1)–(H4) *and* (H5*), (H6*), *and assume that*

$$H(-u, \lambda) = -H(u, \lambda) \qquad \textit{for all} \quad (u, \lambda) \in U(0, 0).$$

Then there exist n "solution branches" in a neighborhood of $(0, 0)$ *in the direction of the parameter* λ_1.

More precisely, there exists a family of $(n - 1)$*-dimensional* C^k*-manifolds* M_r

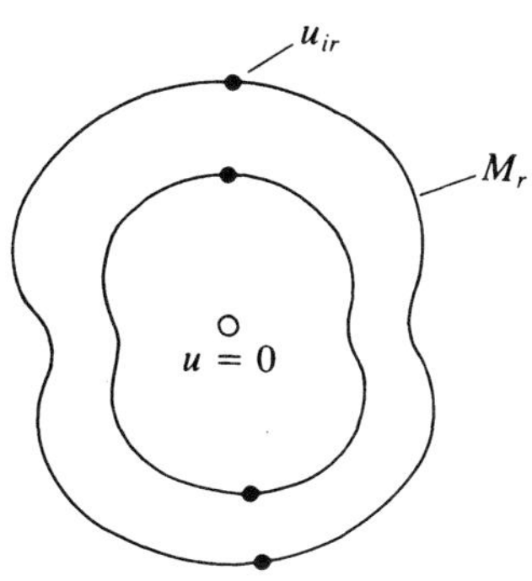

Figure 29.21

in the null space $N(H_u(0,0))$ *such that, for each* $r \in]0, r_0[$, *the following hold* (*Fig.* 29.21):

(a) M_r *is* C^k-*diffeomorphic to a sphere of radius* r *in* $\mathbb{R}^n$, $0 \notin M_r$, *and*

$$\max\{\|u\|: u \in M_r\} \to 0 \qquad as \quad r \to 0.$$

(b) M_r *contains* n *pairs* $\{u_{ir}, -u_{ir}\}$ *and there are real numbers* ε_{ir} *such that*

$$H(\pm u_{ir}, \varepsilon_{ir}\lambda_1) = 0, \qquad i = 1, \ldots, n,$$

and $(\pm u_{ir}, \varepsilon_{ir}\lambda_1) \to (0,0)$ *as* $r \to 0$ *for all* i.

The proof will be based on Proposition 29.64 below. Since $\dim N(H_u(0,0)) = n$, there exist elements $v_1, \ldots, v_n$ of X such that $H_u(0,0)v_i = 0$ and

$$(v_i|v_j) = \delta_{ij} \qquad \text{for} \quad i, j = 1, \ldots, n.$$

For all $u \in Y$, define

$$Pu = \sum_{i=1}^{n} (v_i|u)v_i.$$

Then the operators $P\colon Y \to N(H_u(0,0))$ and $P\colon X \to N(H_u(0,0))$ are linear continuous projection operators onto the null space $N(H_u(0,0))$, since the embedding $X \subseteq Y$ is continuous. We set

$$P^\perp = I - P.$$

Observe the important fact that, with respect to $(\cdot|\cdot)$, the operator P has the same properties as an orthogonal projection operator on an H-space, i.e., we have $P^2 = P$ and, for all $h, k \in Y$,

$$(Ph|k) = (h|Pk), \qquad (Ph|P^\perp k) = 0. \tag{122}$$

We now use the method of the *branching equation of Ljapunov* which we have carefully studied in Section 8.6. Letting $u = v + w$, the original equation

$$H(u, \lambda) = 0, \qquad u \in X, \quad \lambda \in \Lambda, \tag{123}$$

is equivalent to the system:

$$P^\perp H(v + w, \lambda) = 0, \qquad v \in PX, \quad w \in P^\perp X, \tag{124}$$

$$PH(v + w, \lambda) = 0. \tag{124*}$$

The following proposition tells us the crucial fact that the solution of the branching equation corresponding to (124*) can be reduced to the solution of a finite-dimensional variational problem. This is the key to our proof below.

Proposition 29.64. (Reduction Trick). *Assume* (H1), (H2), (H3). *Then*:

(a) *There exists a* C^k-*function* $w = w(v, \lambda)$ *on some neighborhood of* $(0,0)$ *in* $P^\perp X \times \Lambda$ *such that all the solutions of* (124) *have the form* $(v + w(v, \lambda), \lambda)$ *in a sufficiently small neighborhood of* $(0,0)$ *in* $X \times \Lambda$.

(b) $w(0, \lambda) \equiv 0$ *and* $w_v(0,0) = 0$.

(c) *Set*

$$\varphi(v,\lambda) = f(v + w(v,\lambda),\lambda), \qquad v \in PX, \quad \lambda \in \Lambda.$$

There exists a neighborhood V of $(0,0)$ in $X \times \Lambda$ such that the point $(u,\lambda) \in V$ is a solution of the original equation (123) *iff $u = v + w(v,\lambda)$ and v is a critical point of φ, i.e.,*

$$\varphi_v(v,\lambda) = 0, \qquad v \in PV, \quad \lambda \in \Lambda.$$

(d) *If, in addition, condition* (H5) *holds, then the functional φ has the form*

$$\varphi(v,\lambda) = \tfrac{1}{2}(v|H_{u\lambda}(0,0)v\lambda) + r(v,\lambda),$$

where $r(v,\lambda) = o(\|v\|^2)$, $v \to 0$, uniformly with respect to all λ in a sufficiently small neighborhood of $\lambda = 0$ in Λ.

Proof. Ad(a). We proceed analogously to Section 8.6.

(I) The range of $H_u(0,0)$. Let $b \in Y$. We show that the equation

$$H_u(0,0)v = b, \qquad v \in X, \tag{125}$$

has a solution iff $(b|v_i) = 0$ for all i, i.e., $Pb = 0$. Hence $R(H_u(0,0)) = P^\perp Y$. Consequently, the operator $P^\perp H_u(0,0)$ is a *linear homeomorphism* from $P^\perp X$ onto $P^\perp Y$.

Indeed, it follows from (H3) that $\delta^2 f(0,0;h,k) = (H_u(0,0)h|k)$, and hence

$$(H_u(0,0)h|k) = (H_u(0,0)k|h) \qquad \text{for all} \quad h, k \in X,$$

since $\delta^2 f$ is symmetric. We now construct $f_i \in Y^*$ through $\langle f_i, v\rangle = (v_i|v)$ for all $v \in Y$. Recall that $H_u(0,0)v_i = 0$. Thus,

$$\langle f_i, H_u(0,0)v\rangle = 0 \qquad \text{for all} \quad v \in X,$$

and hence $f_i \in N(H_u(0,0)^*)$. Since $H_u(0,0)$ is Fredholm of index zero, $\dim N(H_u(0,0)^*) = \dim N(H_u(0,0)) = n$ (cf. Section 8.4). Consequently, $N(H_u(0,0)^*) = \operatorname{span}\{f_1,\ldots,f_n\}$, and hence equation (125) has a solution iff $\langle f_i, b\rangle = 0$ for all i.

(II) Solution of (124). Letting $F(u,\lambda) = P^\perp H(u,\lambda)$, we obtain $F_u(0,0) = P^\perp H_u(0,0)$. By (I), this operator is surjective from X onto $P^\perp Y$. Thus, assertion (a) follows from the surjective implicit function theorem (Theorem 4.H).

Ad(b). From $H(0,\lambda) \equiv 0$ and (a) we get $w(0,\lambda) \equiv 0$. Letting $w = w(v,\lambda)$, differentiation of (124) with respect to v yields

$$P^\perp H_u(0,0)(h + w_v(0,0)h) = 0 \qquad \text{for all} \quad h \in PX.$$

Hence

$$P^\perp H_u(0,0)w_v(0,0)h = 0 \qquad \text{for all} \quad h \in PX,$$

where $w_v(0,0)h \in P^\perp X$. By (I), $w_v(0,0)h = 0$.

Ad(c). Substituting $w = w(v, \lambda)$ into (124), we obtain

$$P^{\perp} H(v + w(v, \lambda), \lambda) = 0. \tag{126}$$

Thus, it is sufficient to prove that the equation $\varphi_v(v, \lambda) = 0$ is *equivalent* to the *branching equation*

$$PH(v + w(v, \lambda), \lambda) = 0. \tag{126*}$$

In fact, noting (H3) and (126), this follows from

$$\begin{aligned} \varphi_v(v, \lambda)h &= (H(v + w(v, \lambda), \lambda)|h + w_v(v, \lambda)h) \\ &= (PH(v + w(v, \lambda), \lambda)|h + w_v(v, \lambda)h) \\ &= (PH(v + w(v, \lambda), \lambda)|h) \qquad \text{for all} \quad h \in PX. \end{aligned}$$

In this connection, observe $Pw_v(v, \lambda) = 0$ and (122).

Ad(d). From (b), we obtain

$$w(v, \lambda) = w(v, \lambda) - w(0, \lambda) = \int_0^1 w_v(tv, \lambda)v \, dt,$$

and hence

$$w(v, \lambda) = o(\|v\|), \qquad v \to 0,$$

uniformly with respect to small $\|\lambda\|$. By (H3),

$$f(u, \lambda) = \int_0^1 (H(tu, \lambda)|u) \, dt.$$

The assertion now follows from (H5). In this connection, note that $H_u(0, 0)v = 0$ for all $v \in PX$ and

$$\begin{aligned} |(H_{u\lambda}(0, 0)tw(v, \lambda)\lambda|v + w(v, \lambda))| &\leq \|H_{u\lambda}(0, 0)\| \, \|tw(v, \lambda)\| \, \|\lambda\|(\|v\| + \|w(v, \lambda)\|) \\ &= o(\|v\|^2), \qquad v \to 0. \end{aligned}$$

□

Proof of Theorem 29.K. We set

$$a(h, k) = (H_{u\lambda}(0, 0)h\lambda_1|k) \qquad \text{for all} \quad h, k \in X.$$

Let $(0, \lambda) \in U(0, 0)$. The symmetry of $\delta^2 f$ yields

$$(H_u(0, \lambda)h|k) = (H_u(0, \lambda)k|h) \qquad \text{for all} \quad h, k \in X.$$

Differentiation with respect to λ implies the symmetry of $a(\cdot, \cdot)$. Moreover, $a(\cdot, \cdot)$ is definite on $PX \times PX$ according to (H4). Letting

$$\lambda = \varepsilon\lambda_1 \qquad \text{for real } \varepsilon,$$

the assertion follows immediately from Propositions 29.25(H) (iii) and 29.64.

□

The proof of Corollary 29.63 will be based on the following simple "multiplier trick."

Lemma 29.65. *Suppose that we are given operators A, B: $D \subseteq X \to Y$, points $u \in X, v \in PX, z(h) \in P^{\perp}X$, and real numbers $\varepsilon(v)$, μ such that the following hold:*

$$P^{\perp}(Au + \varepsilon(v)Bu) = 0, \tag{127}$$

$$(Au|v) + \varepsilon(v)(Bu|v) = 0, \tag{128}$$

$$(Au|h + z(h)) + \mu(Bu|h + z(h)) = 0 \qquad \textit{for all} \quad h \in PX, \tag{129}$$

and $(Bu|v + z(v)) \neq 0$. Then u is a solution of the equation

$$Au + \varepsilon(v)Bu = 0. \tag{130}$$

PROOF. From (127) and (128), we obtain

$$(Au|v + z(v)) + \varepsilon(v)(Bu|v + z(v)) = 0.$$

By (129), $\mu = \varepsilon(v)$. Hence, by (127) and (129),

$$(Au|h) + \varepsilon(v)(Bu|h) = 0 \qquad \text{for all} \quad h \in PX,$$

i.e., $PAu + \varepsilon(v)PBu = 0$. From (127) we get (130). □

PROOF OF COROLLARY 29.63. By (H5*),

$$H(u, \varepsilon\lambda_1) = Au + \varepsilon Bu,$$

where

$$Au = H_u(0,0)u + Su \qquad \text{and} \qquad Bu = H_{u\lambda}(0,0)u\lambda_1 + Tu,$$

and Su, $Tu = o(\|u\|)$, $\|u\| \to 0$. By (H3),

$$f(u, \varepsilon\lambda_1) = a(u) + \varepsilon b(u),$$

where

$$a'(u)h = (Au|h) \qquad \text{and} \qquad b'(u)h = (Bu|h) \qquad \text{for all} \quad h \in X.$$

Finally, we set

$$u = v + w(v, \varepsilon(v)\lambda_1)$$

and

$$\alpha(v) = a(u), \qquad \beta(v) = b(u).$$

(I) The basic ideas. The following considerations will be justified below.

Ad(127). By (124) and Proposition 29.64, equation (127) is valid for each small real number $\varepsilon(v)$.

Ad(128). We fix $\varepsilon(v)$ in such a way that (128) holds.

Ad(129). We consider the problem

$$\alpha(v) = \text{stationary!}, \qquad \beta(v) = r^2, \quad v \in V, \tag{131}$$

where V is a fixed sufficiently small neighborhood of $v = 0$ in PX, and the real number $r > 0$ is sufficiently small.

Let v be a solution of (131). By the Lagrange multiplier rule (cf. Section 43.10), there is a real number μ such that $\alpha'(v) = -\mu\beta'(v)$, i.e.,

$$\alpha'(v)h + \mu\beta'(v)h = 0 \qquad \text{for all} \quad h \in PX.$$

This is equation (129), since

$$\begin{aligned} \alpha'(v)h &= (Au|h + w_v h + w_\lambda \varepsilon'(v)h), \\ \beta'(v)h &= (Bu|h + w_v h + w_\lambda \varepsilon'(v)h) \qquad \text{for all} \quad h \in PX, \end{aligned} \tag{132}$$

where $w = w(v, \varepsilon(v)\lambda_1)$.

Equation (130) is our original equation $H(u, \varepsilon\lambda_1) = 0$. Therefore, we are done if we can justify the following:

(i) We can solve equation (128) for ε if $\|v\|$ is small.
(ii) The finite-dimensional variational problem (131) possesses n pairs of solutions $(v, -v)$.
(iii) The solutions of (131) have the property $(v, \varepsilon(v)) \to (0, 0)$ as $r \to 0$.
(iv) We can use the Lagrange multiplier rule from Section 43.10, i.e., we have $\beta'(v)v \neq 0$ for $\beta(v) = r^2$.
(v) The set $\{v \in V\colon \beta(v) = r^2\}$ is C^k diffeomorphic to a sphere for small $r > 0$.

In the following, "$o(\|v\|^n)$, $v \to 0$," always means "uniformly for small $|\varepsilon|$."

(II) Justification.

Ad(i). Differentiation of (124), $P^\perp H(v + w(v, \varepsilon\lambda_1), \varepsilon\lambda_1) = 0$, with respect to ε yields

$$\text{(D)} \qquad (P^\perp H_u(0,0) + C)w_\lambda = -P^\perp H_{u\lambda}(0,0)w\lambda_1 - P^\perp T(v + w),$$

where

$$C = \varepsilon P^\perp H_{u\lambda}(0,0)\lambda_1 + P^\perp S'(u) + \varepsilon P^\perp T'(u)$$

and $u = v + w(v, \varepsilon\lambda_1)$. In this connection, observe the important fact that

$$P^\perp H_{u\lambda}(0,0)v\lambda_1 = 0$$

by (H6*). Recall that

$$w(v, \varepsilon\lambda_1) = o(\|v\|), \qquad v \to 0,$$

and Su, $Tu = o(\|u\|)$, $u \to 0$. Since $P^\perp H_u(0,0)$ is a linear homeomorphism from $P^\perp X$ onto $P^\perp Y$ and $\|C\|$ is small for small $\|v\|$, we obtain from (D) that

$$\|w_\lambda\| \leq \text{const}\,\|(P^\perp H_u(0,0) + C)^{-1}\|(\|w\| + o(\|v + w\|)).$$

Hence we get the *key relation*:

$$w_\lambda(v, \varepsilon\lambda_1) = o(\|v\|), \qquad v \to 0.$$

Equation (128) has the form

$$\text{(E)} \qquad \varepsilon a(v, v) = -(Su|v) - \varepsilon(Tu|v), \qquad v \in PX, \quad \varepsilon \in \mathbb{R},$$

with $u = v + w(v, \varepsilon\lambda_1)$ and

$$a(v, v) = (H_{u\lambda}(0,0)v\lambda_1 | v).$$

Now note the decisive fact that $a(\cdot, \cdot)$ is definite, according to the *generic branching* condition (H4). Without any loss of generality, we may assume that $a(\cdot, \cdot)$ is *positive* definite on $PX \times PX$.

In Problem 29.8 we prove a simple variant of the *implicit function theorem*. This result shows that there is a neighborhood W of the origin in PX such that equation (E) has a C^k-solution $\varepsilon = \varepsilon(v)$ in $W - \{0\}$ and

$$\varepsilon(v) \to 0 \qquad \text{as} \quad v \to 0.$$

In this connection, note that $Su = o(\|u\|)$ and $Tu = o(\|u\|)$, $u \to 0$, and

$$w_\lambda(v, \varepsilon\lambda_1) = o(\|v\|) \qquad \text{as} \quad v \to 0.$$

Ad(ii). Differentiation of (E) with respect to v yields

$$\varepsilon'(v)v = o(1), \qquad v \to 0.$$

By (132),

$$\begin{aligned} \beta'(v)v &= a(v, v) + o(\|v\|^2), \qquad v \to 0, \\ \beta(v) &= \int_0^1 (B(tu)|u)\,dt = \tfrac{1}{2}a(v, v) + o(\|v\|^2). \end{aligned} \tag{133}$$

Recall that $u = v + w(v, \varepsilon\lambda_1)$.

We now come to the *point* of our proof. The finite-dimensional variational problem (131) has always at least one solution (e.g., a minimum). This way we again obtain *Theorem* 29.*K*. However, if H is odd, then f is even and $w = w(v, \lambda)$ is odd with respect to v. Thus, $\alpha(\cdot)$ and $\beta(\cdot)$ are *even*. In this case, the *Ljusternik–Schnirelman theory* tells us the deep result that problem (131) has n solution pairs $(v, -v)$, since $\dim PX = n$. This result together with important generalizations to B-spaces will be proved in Chapter 44 (cf. Theorem 44.B).

Ad(iii), (iv). This follows from (133).

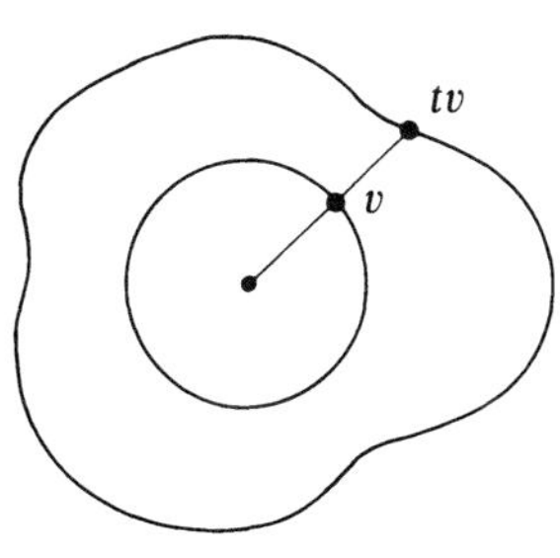

Figure 29.22

Ad(v). Passing to an equivalent scalar product on PX, if necessary, we can assume that $a(v, w) = (v|w)$ for all v, $w \in PX$. Let $v \in PX$ be given with $2^{-1}(v|v) = r^2$ for small $r > 0$. We now consider the equation

$$\beta(tv) = r^2, \qquad t \in \mathbb{R}. \tag{133*}$$

Letting $t = 1 + \varepsilon$ for small ε and using (133), it follows from the implicit function theorem (Problem 29.8) that equation (133*) has a locally unique solution $t(v)$ near $t = 1$. The C^k-map $v \mapsto t(v)v$ yields the desired C^k-diffeomorphism (Fig. 29.22). □

29.19. Application to the Calculus of Variations

Let us study the variational problem

$$\begin{gathered} \int_G L(x, u(x), u'(x))\, dx - \int_G K(x)u(x)\, dx = \min!, \\ u = g \quad \text{on } \partial G. \end{gathered} \tag{134}$$

Our main goal is to obtain *sufficient* conditions for local minima. We write problem (134) in the form

$$f(u) = \min!, \qquad u \in g + X, \tag{134*}$$

and assume the following:

(H1) Let G be a bounded region in $\mathbb{R}^N$ with $\partial G \in C^{0,1}$, and let

$$u = (u_1, \dots, u_M), \qquad x = (\xi_1, \dots, \xi_N),$$

where N, $M \geq 1$. We set $D_i = \partial/\partial\xi_i$. Then, the F-derivative $u'(x)$ corresponds to the matrix $(D_i u_j(x))$ with $i = 1, \dots, N$ and $j = 1, \dots, M$.

(H2) Let the function $L: \bar{G} \times \mathbb{R}^M \times \mathbb{R}^{MN} \to \mathbb{R}$ be C^∞, and let

$$K \in L_2(G)^M, \qquad g \in C^2(\bar{G})^M.$$

As usual, we set $Ku = \sum_{j=1}^M K_j u_j$.

We now introduce the following spaces:

$$X = \{u \in C^1(\bar{G})^M : u = 0 \text{ on } \partial G\},$$

$$Y = \mathring{W}_2^1(G)^M, \qquad Z = L_2(G)^M.$$

Recall that, by definition, the product space W^M consists exactly of all the tupels $w = (w_1, \dots, w_M)$ with $w_j \in W$ for all j. The norm on W^M is given by

$$\|w\|_{W^M} = \sum_{j=1}^M \|w_j\|_W.$$

For example, in elasticity, we have $N = M = 3$, and the quantities allow the following physical interpretation:

u = displacement of the elastic body,

L = density of the elastic potential energy,

K = density of the outer volume forces,

g = displacement of the boundary of the body.

In the following, we *sum* over two equal indices, where $i, j = 1, \ldots, N$ and k, $m = 1, \ldots, M$.

29.19a. The Euler Equation and the Legendre–Hadamard Condition

Proposition 29.66 (Necessary Conditions for a Local Minimum). *Assume* (H1), (H2). *Let $u \in C^2(\bar{G})^M$ be a local minimal point of the original problem* (134*). *Then u satisfies the* Euler equation

$$\begin{aligned} -\operatorname{div} L_{u'}(x, u(x), u'(x)) + L_u(x, u(x), u'(x)) &= K(x) \quad \textit{on } G, \\ u &= g \quad \textit{on } \partial G. \end{aligned} \tag{135}$$

Moreover, the second variation $\delta^2 f$ is positive, *i.e.*,

$$\delta^2 f(u; h) \geq 0 \qquad \textit{for all} \quad h \in C_0^\infty(\bar{G})^M, \tag{136}$$

where $q = (x, u(x), u'(x))$ and

$$\delta^2 f(u; h) = \int_G L_{u'u'}(q)h'^2 + 2L_{u'u}(q)h'h + L_{uu}(q)h^2\, dx,$$

and the Legendre–Hadamard condition *is valid, i.e.*,

$$L_{u'u'}(x, u(x), u'(x))(d \circ v)^2 \geq 0 \tag{137}$$

for all $x \in \bar{G}$, $d \in \mathbb{R}^N$, $v \in \mathbb{R}^M$.

Corollary 29.67. *If $u \in C^2(\bar{G})^M$ is a solution of the modified problem*

$$\int_G L(x, u(x), u'(x))\, dx - \int_G K(x)u(x)\, dx = \text{stationary!},$$

$$u = g \quad \textit{on } \partial G,$$

then u is also a solution of the Euler equation (135).

Explicitly, the *Euler equation* (135) reads as follows:

$$\begin{aligned} -D_i\left(\frac{\partial L}{\partial D_i u_k}\right) + \frac{\partial L}{\partial u_k} &= K_k \quad \text{on } G, \\ u_k = g_k \quad \text{on } \partial G, &\qquad k = 1, \ldots, M. \end{aligned} \tag{135*}$$

Furthermore, if we set

$$a_{ikjm}(x) = \frac{\partial^2 L(x, u(x), u'(x))}{\partial D_i u_k \, \partial D_j u_m},$$

then $L_{u'u'}h'^2 = a_{ikjm}D_ih_kD_jh_m$, and the *Legendre–Hadamard condition* (137) is identical to

$$a_{ikjm}(x)d_iv_kd_jv_m \geq 0 \qquad \text{for all} \quad x \in \bar{G}, \quad d \in \mathbb{R}^N, \quad v \in \mathbb{R}^M.$$

PROOF. Let $h \in X$. We set

$$\varphi(t) = f(u + th), \qquad t \in \mathbb{R}.$$

Then the real function φ has a local minimum at $t = 0$, i.e.,

$$\varphi'(0) = 0 \qquad \text{and} \qquad \varphi''(0) \geq 0.$$

Integration by parts yields

$$0 = \varphi'(0) = \int_G (L_{u'}h' + L_u h - Kh)\,dx$$

$$= \int_G (-\operatorname{div} L_{u'} + L_u - K)h\,dx \qquad \text{for all} \quad h \in X.$$

This implies the Euler equation (135).

Since $\delta^2 f(u; h) = \varphi''(0)$, we obtain (136) from $\varphi''(0) \geq 0$.

Finally, the Legendre–Hadamard condition (137) follows from Section 18.17b.

Corollary 29.67 follows from $\varphi'(0) = 0$. □

29.19b. Sufficient Conditions for a Strict Local Minimum and Strongly Stable Solutions

Definition 29.68. Let $u \in C^2(\bar{G})^M$. Then u is called *strongly stable* iff

$$\delta^2 f(u; h) > 0 \qquad \text{for all} \quad h \in Y - \{0\}, \tag{138}$$

and the so-called *strong* Legendre–Hadamard condition is valid, i.e.,

$$L_{u'u'}(x, u(x), u'(x))(d \circ v)^2 > 0, \tag{139}$$

for all $x \in \bar{G}$ and all nonzero $d \in \mathbb{R}^N$, $v \in \mathbb{R}^M$.

Theorem 29.L. *Assume* (H1), (H2). *Let $u \in C^2(\bar{G})^M$ be a strongly stable solution of the Euler equation* (135). *Then u is a strict local minimal point of the original problem* (134*).

PROOF. Let c_0 be the minimum of the functional

$$\psi(x, d, v) = L_{u'u'}(x, u(x), u'(x))(d \circ v)^2$$

on the compact set $\{(x, d, v) \in \bar{G} \times \mathbb{R}^{N+M}: |d| = |v| = 1\}$. Then $c_0 > 0$ and we obtain that

$$L_{u'u'}(x, u(x), u'(x))(d \circ v)^2 \geq c_0 |d|^2 |v|^2, \tag{140}$$

for all $x \in \bar{G}$, $d \in \mathbb{R}^N$, $v \in \mathbb{R}^M$. Explicitly, this means

$$a_{ikjm}(x) d_i v_k d_j v_m \geq c_0 |d|^2 |v|^2 \qquad \text{for all} \quad x \in \bar{G}, \quad d \in \mathbb{R}^N, \quad v \in \mathbb{R}^M. \tag{140*}$$

By Problem 22.7b, the strong Legendre–Hadamard condition (140) implies the decisive *Gårding inequality*:

$$\delta^2 f(u; h) \geq c\|h\|_Y^2 - C\|h\|_Z^2, \tag{141}$$

for all $h \in Y$ and fixed numbers $c > 0$, $C \geq 0$. Hence the second variation $\delta^2 f$ is *regular* at u.

In fact, the assumptions of Problem 22.7b are satisfied since integration by parts yields

$$\delta^2 f(u; h) = \int_G hJ(u)h \, dx \qquad \text{for all} \quad h \in C_0^\infty(G)^M,$$

where $J(u)$ denotes the so-called Jacobi differential operator. Our discussion after (142) below shows that the strong Legendre–Hadamard condition implies the strong ellipticity of $J(u)$ which we need in Problem 22.7b.

From (138) and (141) it follows, by the Hestenes theorem (Proposition 22.39), that

$$\delta^2 f(u; h) \geq c_1 \|h\|_Y^2 \qquad \text{for all} \quad h \in Y \quad \text{and fixed} \quad c_1 > 0. \tag{141*}$$

The assertion now follows from Theorem 29.I and the reduction trick in Remark 29.33. □

29.19c. The Jacobi Equation and the Eigenvalue Criterion

We set

$$\mu_{\min} = \inf_{h \in S} \delta^2 f(u; h),$$

where $S = \{h \in Y: (h|h)_Z = 1\}$. Moreover, we consider the so-called *Jacobi equation*

$$\begin{aligned} J(u)h &= \mu h \quad \text{on } G, \\ h &= 0 \quad \text{on } \partial G, \end{aligned} \tag{142}$$

where $q = (x, u(x), u'(x))$ and

$$J(u)h = -\operatorname{div}(L_{u'u'}(q)h' + L_{u'u}(q)h)) + L_{uu'}(q)h' + L_{uu}(q)h.$$

Explicitly, equation (142) means

$$-D_j(a_{ikjm} D_i h_k) - D_j(L_{u_k D_j u_m} h_k) + L_{u_m D_i u_k} D_i h_k + L_{u_k u_m} h_k = \mu h_m \quad \text{on } G,$$

$$h_m = 0 \quad \text{on } \partial G, \qquad m = 1, \ldots, M. \tag{142*}$$

The principal part of this equation is given by

$$-D_j(a_{ikjm}D_ih_k) = -a_{ikjm}D_iD_jh_k + \cdots,$$

and (142) represents a linear strongly elliptic system in the case where the strong Legendre–Hadamard condition (140*) holds.

Observe that the Jacobi operator $J(u)$ corresponds to the *linearization* of the Euler equation (135) at u.

Proposition 29.69. *Assume* (H1), (H2). *Let $u \in C^2(\bar{G})M$ be a solution of the Euler equation* (135), *which satisfies the strong Legendre–Hadamard condition* (139). *Then*:

(a) *If $\mu_{\min} > 0$, then u is a strict local minimal point of the original problem* (134*).

(b) *If $\mu_{\min} < 0$, then u is not a local minimal point of* (134*).

Corollary 29.70. *In addition, let $u \in C^{2,\alpha}(\bar{G})^M$ and $\partial G \in C^{2,\alpha}$ for fixed α with $0 < \alpha < 1$. Then the number $\mu_{\min}$ is equal to the smallest eigenvalue of the Jacobi equation* (142).

Furthermore, for all $h, k \in C_0^\infty(G)^M$, we have

$$\delta^2 f(u; h, k) = \int_G kJ(u)h\,dx. \tag{143}$$

Relation (143) shows the natural connection between the Jacobi differential operator $J(u)$ and the second variation $\delta^2 f$.

PROOF. We set

$$a(h,k) = \delta^2 f(u; h, k) \qquad \text{for all} \quad h, k \in Y,$$

i.e.,

$$a(h,k) = \int_G L_{u'u'}h'k' + L_{u'u}h'k + L_{uu'}hk' + L_{uu}hk\,dx,$$

where the argument of L is $(x, u(x), u'(x))$.

The assertions now follow from Corollary 29.41. □

PROOF OF COROLLARY 29.70. Integration by parts yields (143).

By Corollary 29.41, the number $\mu_{\min}$ is equal to the smallest eigenvalue of the equation

$$a(h,k) = \mu(h|k)_Z \tag{144}$$

for fixed $h \in Y$ and all $k \in Y$, where $(h|k)_Z = \int_G hk\,dx$. Noting (143), we obtain (144) from (142) in the case where the functions are sufficiently smooth. Thus, equation (144) is nothing other than the *generalized* problem to the classical Jacobi problem (142). Now, the regularity theory for strongly elliptic systems tells us that each generalized solution of (142) is also a classical solution (see

Section 61.11). In this connection, observe that $u \in C^{2,\alpha}(\bar{G})^M$ implies that the coefficients of the Jacobi equation (142) belong to the space $C^\alpha(\bar{G})$. $\square$

29.19d. The Euler operator in Spaces of Smooth Functions

We write the *Euler equation* (135) in the form of the operator equation

$$Au = (K, g), \qquad u \in \mathscr{X}, \quad (K, g) \in \mathscr{Y}, \tag{145}$$

where we set

$$\mathscr{X} = C^{2,\alpha}(\bar{G})^M, \qquad \mathscr{Y} = C^\alpha(\bar{G})^M \times C^{2,\alpha}(\partial G),$$

and

$$D = \{u \in \mathscr{X} : u \text{ satisfies the strong Legendre–Hadamard condition (139)}\},$$

$$D_{\text{stable}} = \{u \in \mathscr{X} : u \text{ is strongly stable}\}.$$

It follows from (140) and (141*) that D and D_{stable} are open subsets of the B-space $\mathscr{X}$.

For the following, it is very important that $0 < \alpha < 1$. We assume:

(A1) Let G be a bounded region in $\mathbb{R}^N$ with $\partial G \in C^{2,\alpha}$, where $N \geq 1$ and $0 < \alpha < 1$.
(A2) Let $L: \bar{G} \times \mathbb{R}^N \times \mathbb{R}^{NM} \to \mathbb{R}$ be C^∞.

The following deep result (b) is crucial.

Proposition 29.71. *Assume* (A1), (A2). *Then*:

(a) *For each* $(K, g) \in \mathscr{Y}$ *and* $u \in \mathscr{X}$, *the linearized equation*

$$A'(u)h = (K, g), \qquad h \in \mathscr{X}, \tag{146}$$

corresponds to the Jacobi equation

$$\begin{aligned} J(u)h &= K \quad \text{on } G, \\ h &= g \quad \text{on } \partial G. \end{aligned} \tag{146*}$$

(b) *The operator*

$$A: D \subseteq \mathscr{X} \to \mathscr{Y}$$

is C^∞ *and Fredholm of index zero.*
(c) *Let* $u \in D_{\text{stable}}$. *If* $A'(u)h = 0$, $h \in \mathscr{X}$, *then* $h = 0$.

PROOF. Ad(a). From the second-order Euler equation (135) it follows that $A(\mathscr{X}) \subseteq \mathscr{Y}$. Linearization of (135) yields (146*).

Ad(b). Recall that the Jacobi equation (146*) represents a linear strongly elliptic system if $u \in D$. A deep theorem on such systems tells us that $A'(u): \mathscr{X} \to \mathscr{Y}$ is Fredholm of index zero (cf. Proposition 61.17 in Part IV).

Ad(c). If $h \in \mathscr{X}$ is a solution of (146*) with $K = g = 0$, then $h = 0$ on ∂G and integration by parts yields

$$\delta^2 f(u; h) = \int_G hJ(u)h\,dx = 0.$$

Hence $h = 0$ by (138). □

29.19e. The Number of Strongly Stable Solutions of the Euler Equation

We write the Euler equation (135) in the form

$$Au = (K, g), \qquad u \in S, \tag{147}$$

where S is a subset of D_{stable}, and where $(K, g) \in \mathscr{Y}$ is given. Let $n_S(K, g)$ be the *number of solutions* of (147).

Proposition 29.72. *Assume* (A1), (A2). *Then*:

(a) *For each* $u \in D_{\text{stable}}$, *the operator*

$$A: D \subseteq \mathscr{X} \to \mathscr{Y}$$

is a local C^∞*-diffeomorphism at* u.

(b) *Let* S *be a subset of* D_{stable} *such that* A *is proper on* S. *Then, for each*

$$(K, g) \in \mathscr{Y} - A(\partial S \cap S),$$

the number $n_S(K, g)$ *is finite.*

Furthermore, $n_S(\cdot)$ *is constant on each connected subset of the open set* $\mathscr{Y} - A(\partial S \cap S)$.

Corollary 29.73. *If* C *is a compact subset of* D_{stable}, *then there exists an open set* U *in* $\mathscr{X}$ *such that*

$$C \subseteq \bar{U} \subseteq D_{\text{stable}},$$

and A *is proper on* $S = \bar{U}$.

If the operator A *is proper on the subsets* $S_1, \ldots, S_n$ *of* D_{stable}, *then* A *is also proper on* $S = \bigcup_i S_i$.

PROOF. This follows from Proposition 29.71 and Section 29.16. □

These results show that the Euler equation (135) has nice properties with respect to smooth strongly stable solutions.

Remark 29.74 (Bifurcation). Proposition 29.72 tells us the important fact that the local *one-to-one correspondence* between the right-hand sides $(K, g) \in \mathscr{Y}$ and the solution $u \in \mathscr{X}$ of the Euler equation (135) can be *violated* merely at points

$$u \in \mathscr{X} - D_{\text{stable}},$$

i.e., u is *not* strongly stable. In this case, bifurcation may occur. Since the set D_{stable} is open in $\mathscr{X}$, bifurcation may occur at the points

$$u \in \partial D_{\text{stable}},$$

i.e., u belongs to the boundary of stability. In this connection, a typical example has been considered in Remark 29.29 and Figure 29.11. We thus obtain the following important general principle:

Loss of strong stability can lead to bifurcation.

29.19f. The Continuation Method

Let $u_0 \in D_{\text{stable}}$ be a given solution of the Euler equation (135), i.e.,

$$Au_0 = (K_0, g_0).$$

In order to continue this known solution u_0, we consider the equation

$$Au_t = (1 - t)(K_0, g_0) + t(K, g), \quad u_t \in \mathscr{X}, \tag{148}$$

with the real parameter $t \in [0, 1]$. By Proposition 29.72(a), we obtain the following results:

(i) Let $(K, g) \in \mathscr{Y}$. Then, for each t in a sufficiently small neighborhood of $t = 0$, equation (148) has a unique solution $u_t \in D_{\text{stable}}$ in some neighborhood of u_0.
(ii) The map $t \mapsto u_t$ is C^∞.
(iii) The solution u_t can be uniquely continued as long as u_t remains in D_{stable}, i.e., u_t remains strongly stable.
(iv) If u_t reaches the boundary of stability $\partial D_{\text{stable}}$, then bifurcation of $t \mapsto u_t$ may occur.

Explicitly, equation (148) means:

$$\begin{aligned} -\operatorname{div} L_{u'}(q_t) + L_u(q_t) &= (1 - t)K_0 + tK \quad \text{on } G, \\ u_t &= (1 - t)g_0 + tg \quad \text{on } \partial G, \end{aligned} \tag{148*}$$

where $q_t = (x, u_t(x), u_t'(x))$.

29.19g. An Approximation Method

Differentiation of equation (148) with respect to the real parameter t yields

$$A'(u_t)\frac{du_t}{dt} = (K, g) - (K_0, g_0),$$

and discretization of the derivative du_t/dt leads to the following approximation method:

$$A'(u_{n\Delta t})(u_{(n+1)\Delta t} - u_{n\Delta t}) = \Delta t((K,g) - (K_0,g_0)) \tag{149}$$

for fixed $\Delta t > 0$ and $n = 0, 1, \ldots$. Explicitly, equation (149) means

$$\begin{aligned} J(u_{n\Delta t})u_{(n+1)\Delta t} &= J(u_{n\Delta t})u_{n\Delta t} + \Delta t(K - K_0) \quad \text{on } G, \\ u_{(n+1)\Delta t} &= u_{n\Delta t} + \Delta t(g - g_0) \quad \text{on } \partial G, \end{aligned} \tag{149*}$$

where J denotes the Jacobi differential operator (142).

Therefore, in each step $n = 0, 1, 2, \ldots$, we have to solve a *linear* strongly elliptic system in order to find $u_{(n+1)\Delta t}$. The convergence of this approximation method as $\Delta t \to 0$, and the simple physical interpretation of this method, in terms of elasticity, will be considered in Section 61.16.

29.20. A General Bifurcation Theorem for the Euler Equations and Stability

Along with the variational problem

$$\begin{gathered} \int_G L(x, u(x), u'(x))\,dx - P\int_G R(u(x))\,dx = \text{stationary!}, \\ u = 0 \quad \text{on } \partial G, \end{gathered} \tag{150}$$

where P is a real parameter, we consider the corresponding *Euler equation*

$$\begin{aligned} -\operatorname{div} L_{u'}(q) + L_u(q) - PR'(u) = 0 &\quad \text{on } G, \\ u = 0 &\quad \text{on } \partial G, \end{aligned} \tag{151}$$

where $q = (x, u(x), u'(x))$. Moreover, we also consider the corresponding Jacobi equation at $u = 0$:

$$\begin{aligned} J(0, P_{\text{crit}})h = 0 &\quad \text{on } G, \\ h = 0 &\quad \text{on } \partial G, \end{aligned} \tag{152}$$

where we set

$$J(u,P) = -\operatorname{div}(L_{u'u'}(q)h' + L_{u'u}(q)h) + L_{uu'}(q)h' + L_{uu}(q)h - PR''(0)h.$$

Observe that (152) follows from (151) by linearization with respect to u at $u = 0$. The explicit form of equation (151) and (152) follows from (135*) and (142*), respectively.

We make the following assumptions.

(H1) G is a bounded region in $\mathbb{R}^N$ with $\partial G \in C^{2,\alpha}$ where $N \geq 1$ and $0 < \alpha < 1$. Let $u = (u_1, \ldots, u_M)$, $M \geq 1$.

(H2) The functions $L\colon \bar{G} \times \mathbb{R}^N \times \mathbb{R}^{NM} \to \mathbb{R}$ and $R\colon \mathbb{R}^M \to \mathbb{R}$ are C^∞. Suppose that

$$L_{u'}(x,0,0) \equiv 0, \qquad L_u(x,0,0) \equiv 0, \qquad R'(0) = 0,$$

i.e., $u \equiv 0$ is a *trivial* solution of the Euler equation (151), and suppose that

$$R''(0)h^2 > 0 \qquad \text{for all} \quad h \in \mathbb{R}^M - \{0\}.$$

(H3) The *strong Legendre–Hadamard condition* is satisfied at $u = 0$, i.e.,

$$L_{u'u'}(x,0,0)(d \circ v)^2 > 0$$

for all $x \in \bar{G}$ and all nonzero $d \in \mathbb{R}^N$, $v \in \mathbb{R}^M$.

We set

$$X = \{u \in C^{2,\alpha}(\bar{G})^M\colon u = 0 \text{ on } \partial G\},$$
$$Y = C^\alpha(\bar{G})^M,$$

and, on the B-space Y, we introduce the scalar product

$$(f|g) = \int_G fg\,dx \qquad \text{for all} \quad f, g \in Y.$$

In terms of elasticity, the function u describes the displacement of an elastic body, and P describes an outer force. The variational problem (150) corresponds to the principle of stationary potential energy, where we have:

$$\int_G L\,dx = \text{elastic potential energy},$$

$$P\int_G R(u)\,dx = \text{work of the outer force}.$$

We are looking for nontrivial solutions $u \neq 0$ of the Euler equation (151). This corresponds to the buckling of beams, rods, and plates for the critical outer force P_{crit}. The following propositions generalize our results about the buckling of beams in Section 29.13.

Theorem 29.M (Bifurcation). *Assume* (H1)–(H3), *and assume that, for fixed real* P_{crit}, *the linearized equation* (152) *has exactly n linearly independent solutions* $h \neq 0$, *where* $n \geq 1$.

Then $(u, P) = (0, P_{\text{crit}})$ *is a bifurcation point of the Euler equation* (151) *in the space* $X \times \mathbb{R}$.

If, in addition, the functions L and R are even with respect to u and u', then there exist "n branches" of nontrivial solutions of the Euler equation (151) *in a neighborhood of the point* $(0, P_{\text{crit}})$ in the space $X \times \mathbb{R}$. *This is to be understood in the sense of Corollary* 29.63.

A solution u of the Euler equation (151) is called *stable* if it corresponds to a strict local minimum of the original variational problem (150) in the space X.

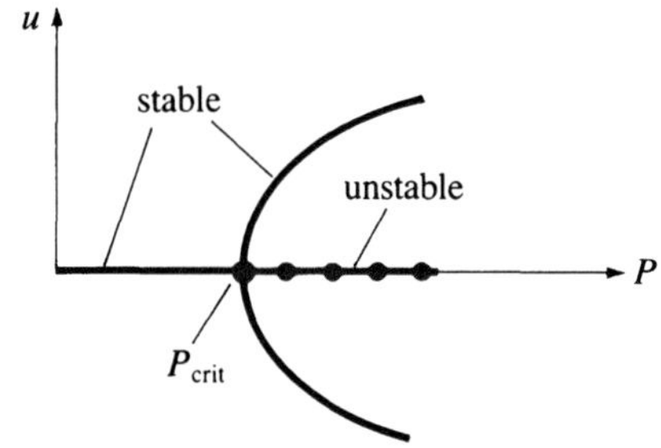

Figure 29.23

Corollary 29.75 (Simple Eigenvalue P_{crit} and Stability). *Assume* (H1)–(H3), *and assume that the functions L and R are analytic with respect to the variables u and u'. Moreover, assume that the linearized equation* (152) *has exactly one linearly independent solution denoted by* h_1, *where* $h_1 \neq 0$, *i.e.,* P_{crit} *is simple.*

Then there exists a neighborhood V of the point $(u, P) = (0, P_{\text{crit}})$ *in the space* $X \times \mathbb{R}$ *such that all the nontrivial solutions of the Euler equation* (151) *in V are given by the analytic curve* $s \mapsto (u, P)$, *where*

$$\begin{aligned} P &= P_{\text{crit}} + \varepsilon_1 s + \varepsilon_2 s^2 + O(s^3), \qquad s \to 0, \\ u &= sh_1 + s^2 u_2 + O(s^3), \end{aligned} \tag{153}$$

and s is a small real parameter in a neighborhood of $s = 0$. *The trivial solutions of* (151) *are given by* $u = 0$, $P =$ *arbitrary. If the bifurcation is supercritical* (*Fig.* 29.23), *i.e., for fixed* $k = 1, 2, \ldots$ *and* $\varepsilon_{2k} > 0$, *we have*

$$P = P_{\text{crit}} + \varepsilon_{2k} s^{2k} + O(s^{2k+1}), \qquad s \to 0,$$

and if P_{crit} *is the smallest eigenvalue of* (152), *then there exist positive numbers* s_0 *and* ε_0 *such that the following hold.*

(a) *The bifurcation solution* (153) *is stable for* $0 < |s| \le s_0$.
(b) *The trivial solution* $u = 0$ *is stable for* $P_{\text{crit}} - \varepsilon_0 < P < P_{\text{crit}}$ *and unstable for* $P_{\text{crit}} < P < P_{\text{crit}} + \varepsilon_0$.

Proof of Theorem 29.M. Let $P = P_{\text{crit}} + \varepsilon$. We write problems (150) and (151) in the form

$$f(u) = \text{stationary!}, \qquad u \in X,$$

and

$$H(u, \varepsilon) = 0, \qquad u \in X, \quad \varepsilon \in \mathbb{R},$$

respectively. Then the linearized equation (152) corresponds to the equation

$$H_u(0, 0)h = 0, \qquad h \in X.$$

The assertion now follows from Theorem 29.K, Corollary 29.63, and Proposition 29.71 (properties of the Euler operator).

In this connection, observe that we may assume that $R''(0) = I$ after a linear transformation of the coordinates $(u_1, \ldots, u_M)$, if necessary. This immediately implies the generic branching condition (H4) from Theorem 29.K, since $H_{u\varepsilon}(0,0)h\varepsilon = -\varepsilon R''(0)h = -\varepsilon h$, and hence

$$(h|H_{u\varepsilon}(0,0)h) = -(h|h) \neq 0 \qquad \text{if} \quad h \neq 0.$$

□

PROOF OF COROLLARY 29.75. The proof proceeds completely analogously to the corresponding proof for the buckling of beams (Proposition 29.48) by using Theorems 8.A and 8.E and the eigenvalue criterion from Proposition 29.69.

In this connection, note that the eigenvalue equation

$$H_u(u, \varepsilon)h = \mu h, \qquad h \in X, \quad \mu \in \mathbb{R},$$

from Theorem 8.E corresponds to the Jacobi equation

$$J(u, P)h = \mu h, \qquad h \in X, \quad \mu \in \mathbb{R},$$

where $P = P_{\text{crit}} + \varepsilon$. Theorem 8.E tells us the sign of the smallest eigenvalue $\mu_{\min}$ along the two solution branches (153) and $u = 0$, $\varepsilon =$ small. This implies assertions (a) and (b). Note that P_{crit} is the smallest eigenvalue of (152) and, as above, we may assume that $R''(0)h = h$ for all $h \in X$. □

29.21. A Local Multiplicity Theorem

We want to study the equation

$$Au = b, \qquad u \in U(u_0), \tag{154}$$

where $U(u_0)$ denotes an open neighborhood of the point u_0 in the B-space X. Our goal is to generalize the situation pictured in Figure 29.24(a), where we have the following: If $b = b_0$, $b > b_0$, and $b < b_0$, then the number of solutions of equation (154) is equal to one, two, and zero, respectively.

Definition 29.76. Let C be a subset of a B-space Y over $\mathbb{K} = \mathbb{R}, \mathbb{C}$, and let $r \geq 1$. Then C is called a *C^r-manifold of codimension one* in Y iff, for each point $b_0 \in C$,

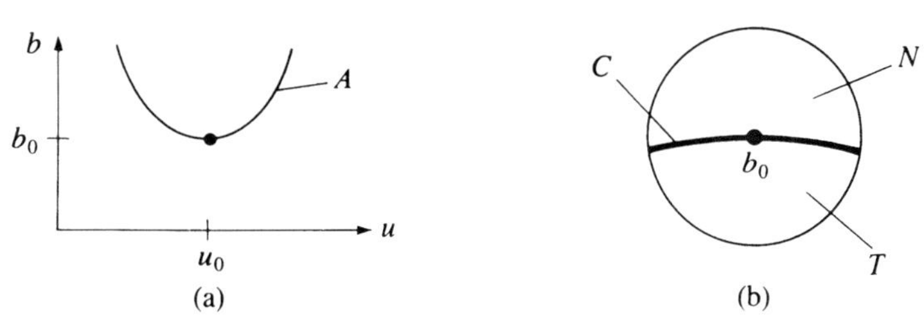

Figure 29.24

there exists a C^r-functional $f\colon U(b_0) \subseteq Y \to \mathbb{K}$ such that $f'(b_0) \neq 0$ and there is some neighborhood V of b_0 such that

$$C \cap V = \{b \in V\colon f(b) = 0\},$$

i.e., C can be described locally by the equation $f(b) = 0$ (see Fig. 29.24(b) for $Y = \mathbb{R}^2$).

We assume:

(H1) The operator $A\colon U(u_0) \subseteq X \to Y$ is C^k and Fredholm of index zero, where X and Y are real B-spaces and $2 \le k \le \infty$.

(H2) Let $\dim N(A'(u_0)) = 1$ and suppose that the crucial generic branching condition

$$A''(u_0)h^2 \notin R(A'(u_0)) \tag{155}$$

is valid for some $h \in N(A'(u_0))$. Finally, let D_{sing} denote the set of singular points of the operator A.

In Figure 29.24(a), we have $A'(u_0) = 0$, and condition (155) means $A''(u_0) \neq 0$.

Proposition 29.77. *Assume* (H1), (H2). *Then there exist a neighborhood* $V(u_0)$ *of* u_0 *and a neighborhood* $W(Au_0)$ *of* Au_0 *such that the following hold*:

(a) *The local set of singular values of the operator* A,

$$C = A(D_{\text{sing}} \cap V(u_0)),$$

is a C^{k-1}*-manifold of codimension one in* Y.

(b) *There exist nonempty open connected sets* T *and* N *with*

$$W(Au_0) = C \cup T \cup N,$$

such that the original equation (154) *has*

(i) *exactly one solution for* $b \in C$,

(ii) *exactly two solutions for* $b \in T$, *and*

(iii) *no solution for* $b \in N$ (*Fig.* 29.24(b)).

PROOF. Letting

$$F(\varepsilon, x) \stackrel{\text{def}}{=} A(u_0 + x) - (b_0 + \varepsilon) = 0, \qquad x \in X, \quad \varepsilon \in Y,$$

the assertion is a special case of Theorem 8.F. Condition (155) corresponds to the generic branching condition of Theorem 8.F:

$$\alpha \stackrel{\text{def}}{=} \langle x_1^*, F_{xx}(0,0)x_1^2\rangle \neq 0,$$

where $h = x_1$. Furthermore, observe that the proof of Theorem 8.F shows that, for all $\varepsilon \in Y$,

$$\gamma'(0)\varepsilon = \alpha B_\varepsilon(0,0)\varepsilon = \alpha\langle x_1^*, F_\varepsilon(0,0)\varepsilon\rangle.$$

In our present case, we have $F_\varepsilon(0,0)\varepsilon = -\varepsilon$. Hence $\gamma'(0) = -\alpha x_1^*$. This implies

$$\gamma'(0) \neq 0.$$

Therefore, the equation $\gamma(\varepsilon) = 0$ in Theorem 8.F describes a C^{k-1}-manifold of codimension one in a neighborhood of $\varepsilon = 0$ in Y. □

29.22. A Global Multiplicity Theorem

We now want to globalize the preceding result. To this end, we consider the operator equation

$$Au = b, \qquad u \in X. \tag{156}$$

(H1) Let $A: X \to Y$ be a *proper* C^k-Fredholm operator of index zero, where X and Y are real B-spaces and $2 \leq k \leq \infty$.
(H2) The set D_{sing} of the singular points of the operator A is nonempty, closed, and connected. Furthermore, A is injective on D_{sing}.
(H3) If $u \in D_{\text{sing}}$ then $\dim N(A'(u)) = 1$ and

$$A''(u)h^2 \notin R(A'(u)) \qquad \text{for some} \quad h \in N(A'(u)).$$

In the special case $X = Y = \mathbb{R}$, condition (H3) means that $A'(u) = 0$ implies $A''(u) \neq 0$.

Theorem 29.N (Ambrosetti and Prodi (1972)). *Assume* (H1)–(H3). *Then the set $C = A(D_{\text{sing}})$ of the singular values of the operator A is a closed connected C^{k-1}-manifold of codimension one in Y.*

There exist two nonempty open connected subsets T and N of Y with

$$Y = C \cup T \cup N,$$

such that the original equation (156) *has*

(i) *exactly one solution for $b \in C$,*
(ii) *exactly two solutions for $b \in T$, and*
(iii) *no solution for $b \in N$.*

Proof. According to Theorem 29.E and Proposition 29.77, it is sufficient to show that the set $Y - C$ consists precisely of two components.

Since the set D_{sing} is connected, so is C. Moreover, since D_{sing} is closed and the operator A is proper, the set C is also closed. Let B be a *component* of the set $Y - C$, i.e., B is a maximally connected open subset of $Y - C$ with respect to the induced topology on $Y - C$. Since $Y - C$ is open in Y, so is B. The set ∂B is not empty. Otherwise, we would have $B = Y$.

Let $b \in \partial B$. Then $b \in C$. The set ∂B is closed. By Proposition 29.77, there exists a neighborhood $W(b)$ of the point b such that $W(b) \cap C \subseteq \partial B$. Hence the set ∂B is *open and closed* in C. Since C is connected, we obtain the *key result*

$$\partial B = C.$$

Again by Proposition 29.77, the set $Y - C$ has locally two components. Consequently, $Y - C$ has globally two components. □

Applications of this theorem to semilinear elliptic differential equations can be found in Problem 29.9.

PROBLEMS

29.1. *Coincidence degree in* H-*spaces*. In order to explain the simple basic idea of the general coincidence degree in Problem 29.2, we first consider a special case. We investigate the equation

$$Bu = Mu, \qquad u \in G, \tag{157}$$

where G is a nonempty bounded open set in the H-space X. We make the following assumptions:

(i) The linear continuous operator $B: X \to X$ is Fredholm of index zero, i.e., the range $R(B)$ is closed and

$$\dim N(B) = \dim R(B)^\perp < \infty.$$

Here, $R(B)^\perp$ denotes the orthogonal complement to $R(B)$.

(ii) The operator $M: X \to X$ is compact.

(iii) We fix an orientation in both $N(B)$ and $R(B)^\perp$ and choose a linear bijective orientation-preserving operator

$$J: N(B) \to R(B)^\perp.$$

Let $Q: X \to N(B)$ be the orthogonal projector onto the null space $N(B)$ of B. Now, our key definition reads as follows:

$$Su = Bu + JQu \qquad \text{for all} \quad u \in X.$$

29.1a. Show that $S: X \to X$ is bijective.
Solution: See the proof of Proposition 29.3.

29.1b. Show that the original equation (157) is equivalent to the equation

$$u = S^{-1}Mu + S^{-1}JQu. \tag{157*}$$

Solution: Equation (157) is equivalent to $Su = Mu + JQu$.

29.1c. Show that the operator

$$C = S^{-1}M + S^{-1}JQ$$

is compact on X.

Solution: The operator $S^{-1}: X \to X$ is continuous according to the open mapping theorem $A_1(36)$. The operator $M: X \to X$ is compact by hypothesis. Finally, the operator Q is compact since $\dim Q(X) = \dim N(B) < \infty$. Hence C is compact.

29.1d. Definition of the coincidence degree. Suppose that

$$Bu \neq Mu \quad \text{on } \partial G.$$

Then $Cu \neq u$ on ∂G. Thus, the Leray–Schauder degree $\deg(I - C, G)$ is well defined, and we set

$$\deg(B, M; G) = \deg(I - C, G).$$

29.1e.* Show that $\deg(B, M; G)$ does not depend on the choice of J, i.e., $\deg(B, M; G)$ depends only on the orientation of $N(B)$ and $R(B)^\perp$. Moreover, show that if we change the orientation of either $N(B)$ or $R(B)^\perp$, then $\deg(B, M; G)$ changes its sign.

Hint: This is a special case of Problem 29.2e.

29.2. *The general coincidence degree of Mawhin* (1972). We now study the equation

$$Bu = Mu, \qquad u \in G \cap D(B). \tag{158}$$

We make the following assumptions:

(H1) X and Y are B-spaces over $\mathbb{K} = \mathbb{R}, \mathbb{C}$.

(H2) G is a nonempty bounded open set in X.

(H3) The linear operator $B\colon D(B) \subseteq X \to Y$ is Fredholm of index zero, i.e., $R(B)$ is closed and

$$\dim N(B) = \operatorname{codim} R(B) < \infty. \tag{159}$$

(H4) The operator $M\colon \bar{G} \subseteq X \to Y$ is B-compact. By definition, this means that the operator $M\colon \bar{G} \to Y$ is continuous and bounded and the operator $KPM\colon \bar{G} \to X$ below is compact.

As we shall see below, this notion depends only on M and B. For example, M is B-compact if one of the following two conditions is valid:

(i) M is compact.

(ii) M is continuous and bounded and $K\colon R(B) \to X$ is compact.

We call the pair $\{B, M\}$ *admissible* on $\bar{G}$ iff (H1) through (H4) hold.

29.2a. Projections. Let

$$Q\colon X \to N(B), \qquad P\colon Y \to R(B),$$

be projection operators onto $N(B)$ and $R(B)$, respectively.

We set $Q^\perp = I - Q$ and $P^\perp = I - P$ as well as

$$N(B)^\perp = Q^\perp(X), \qquad R(B)^\perp = P^\perp(Y).$$

Then we obtain the topological direct sums

$$X = N(B) \oplus N(B)^\perp, \qquad Y = R(B) \oplus R(B)^\perp.$$

The operator

$$B\colon N(B)^\perp \cap D(B) \to R(B)$$

is bijective. Let

$$K\colon R(B) \to N(B)^\perp \cap D(B)$$

be the corresponding inverse operator, i.e., $BKb = b$ on $R(B)$.

We fix an orientation in both $N(B)$ and the factor space $Y/R(B)$. The canonical mapping

$$j(u) = u + R(B)$$

is a linear bijective map from $R(B)^\perp$ onto $Y/R(B)$. Let

$$J\colon N(B) \to R(B)^\perp$$

be a linear bijective map so that $j \circ J: N(B) \to Y/R(B)$ is orientation-preserving.

29.2b. Show that the B-compactness of M in (H4) depends only on M and B.
Hint: One has to show that the compactness of $KPM: \bar{G} \to X$ does not depend on the choice of the projection operators P and Q. Cf. Gaines and Mawhin (1977), p. 13.

29.2c. Our *key* definition reads as follows:

$$Su = Bu + JQu \quad \text{for all} \quad u \in D(B).$$

As in the proof of Proposition 29.3 it follows that $S: D(B) \to Y$ is bijective and

$$S^{-1}b = Kb \quad \text{on } R(B). \tag{160}$$

The original equation $Bu = Mu$, $u \in G \cap D(B)$, is equivalent to the *basic equation*

(E) $$u = S^{-1}Mu + S^{-1}JQu.$$

Show that the operator

$$C = S^{-1}M + S^{-1}JQ$$

is compact on $\bar{G}$.
Solution: It follows from (160) that

$$C = KPM + S^{-1}P^{\perp}M + S^{-1}JQ.$$

By (H4), the operator KPM is compact on $\bar{G}$. The projection operators $P^{\perp}$ and Q have finite-dimensional range, and the linear operator S^{-1} is continuous on finite-dimensional spaces.

29.2d. *Definition of the coincidence degree.* Let $\{B, M\}$ be admissible on $\bar{G}$. Then the Leray–Schauder degree $\deg(I - C, G)$ is well defined and we set

$$\deg(B, M; G) = \deg(I - C, G).$$

If $G = \varnothing$, then let $\deg(B, M; G) = 0$.

29.2e.* Show that $\deg(B, M; G)$ is independent of P, Q, and J, i.e., the coincidence degree $\deg(B, M; G)$ depends only on the orientation chosen in both $N(B)$ and $Y/R(B)$. If we change the orientation of either $N(B)$ and $Y/R(B)$, then the coincidence degree changes its sign.
Hint: Use the representation

$$C = KPM + J^{-1}P^{\perp}M + Q$$

given in Proposition 29.3, and use the topological invariance (T) of the Leray–Schauder degree in Section 13.7. Cf. Gaines and Mawhin (1977), p. 19.

29.3. *Properties of the coincidence degree.*

29.3a. *Existence principle.* Show that if $\deg(B, M; G) \neq 0$, then the equation $Bu = Mu$, $u \in G \cap D(B)$, has a solution.
Solution: By hypothesis, $\deg(I - C, G) \neq 0$. Hence the equation $Cu = u$, $u \in G$, has a solution. This equation is equivalent to $Bu = Mu$, $u \in G \cap D(B)$.

29.3b. *Additivity property*. Let

$$\bar{G} = \bigcup_{i=1}^{n} \bar{G}_i,$$

where $\{G_i\}$ is a family of pairwise disjoint bounded open sets. Let $\{B, M\}$ be admissible on $\bar{G}$ and let

$$Bu \neq Mu \quad \text{on } \partial G_i \qquad \text{for all } i.$$

Show that

$$\deg(B, M; G) = \sum_{i=1}^{n} \deg(B, M; G_i).$$

Solution: This follows from the corresponding property of the Leray–Schauder degree.

29.3c. *Homotopy invariance*. Suppose that the map

$$H\colon \bar{G} \times [0,1] \to X$$

has the property that $\{B, H(\cdot, \lambda)\}$ is admissible on $\bar{G}$ for all $\lambda \in [0,1]$ and the operator H is B-compact, i.e., $KPH\colon \bar{G} \times [0,1] \to X$ is compact. Show that

$$\deg(B, H(\cdot, \lambda); G) = \text{const} \qquad \text{for all} \quad \lambda \in [0,1].$$

Solution: This is a consequence of the homotopy invariance of the Leray–Schauder degree.

29.3d. *Continuation principle*. Suppose that:
(i) $\{B, M\}$ is admissible on $\bar{G}$.
(ii) $Bu \neq \lambda Mu$ for all $(u, \lambda) \in \partial G \times]0,1]$.
(iii) $\deg(J^{-1}P^{\perp}M, G \cap N(B)) \neq 0$.
Show that the equation $Bu = Mu$, $u \in G \cap D(B)$, has a solution.
Hint: Cf. Gaines and Mawhin (1977). pp. 29 and 40.

29.3e. *Reduction principle*. Let $\{B, M\}$ admissible with $R(M) \subseteq R(B)^{\perp}$. Show that

$$|\deg(B, M; G)| = |\deg(J^{-1}M, G \cap N(B))|.$$

Solution: We have

$$C = KPM + J^{-1}P^{\perp}M + Q = J^{-1}M + Q.$$

Hence $I - C = -J^{-1}M$ on $N(B)$. By the reduction principle (R) of the Leray–Schauder degree in Section 13.6, we obtain that

$$\begin{aligned}\deg(B, M; G) &= \deg(I - C, G) = \deg(I - C, G \cap N(B)) \\ &= \deg(-J^{-1}M, G \cap N(B)) = (-1)^{\dim N(B)} \deg(J^{-1}M, G \cap N(B)).\end{aligned}$$

29.4. *Application to periodic solutions of differential equations*. Let $0 < T < \infty$ be fixed. We consider the problem

$$\begin{aligned} u'(t) &= f(t, u(t)) \quad \text{on } [0, T], \\ u(0) &= u(T), \end{aligned} \tag{161}$$

i.e., we seek periodic solutions $u\colon [0, T] \to \mathbb{R}^N$.

29.4a. *Operator equation.* Show that this problem can be reduced to the equation

$$Bu = Mu, \qquad u \in G. \tag{162}$$

Solution: We set

$$Bu = u', \qquad (Mu)(t) = f(t, u(t))$$

and

$$X = Y = C([0, T], \mathbb{R}^N),$$
$$D(B) = \{u \in C^1([0, T], \mathbb{R}^N) \colon u(0) = u(T)\}.$$

Then

$$N(B) = \{u \in X \colon u = \text{const}\},$$
$$R(B) = \left\{y \in Y \colon \int_0^T y(t)\,dt = 0\right\}.$$

The projection operators $Q\colon X \to N(B)$ and $P\colon Y \to R(B)$ can be defined by

$$Qu = u(0), \qquad (Py)(t) = y(t) - T^{-1} \int_0^T y(t)\,dt.$$

Then $N(B)^\perp = (I - Q)(X)$ and $R(B)^\perp = (I - P)(Y)$, i.e.,

$$N(B)^\perp = \{u \in X \colon u(0) = 0\},$$
$$R(B)^\perp = \{y \in Y \colon y = \text{const}\}.$$

The operator $B\colon N(B)^\perp \cap D(B) \to R(B)$ is bijective, and the inverse operator $K\colon R(B) \to N(B)^\perp \cap D(B)$ is given by

$$(Ky)(t) = \int_0^t y(s)\,ds.$$

The operator K is compact, i.e., M is B-compact. The bijective linear operator $J\colon N(B) \to R(B)^\perp$ can be defined by

$$Ju = u \quad \text{on } N(B).$$

29.4b. *Compute the coincidence degree of* (162).
Solution: By Problem 29.2, equation (162) is equivalent to

$$u = Cu, \qquad u \in G, \tag{162*}$$

with

$$C = KPM + J^{-1}(I - P)M + Q.$$

Hence

$$(Cu)(t) = \int_0^t (f(s, u(s)) - a)\,ds + a + u(0)$$

with $a = T^{-1} \int_0^T f(s, u(s))\,ds$ and

$$\deg(B, M; G) = \deg(I - C, G).$$

29.4c. *An important property of the Brouwer degree for potential operators.* Suppose that:

(H) The function $V: \mathbb{R}^N \to \mathbb{R}$ is C^1 and weakly coercive, i.e., $V(x) \to +\infty$ as $|x| \to \infty$. Furthermore, there is an $r > 0$ such that $V'(x) \neq 0$ for all x: $|x| \geq r$.

Let $U(0,r) = \{x \in \mathbb{R}^N: |x| < r\}$. Show that

$$\deg(V', U(0,r)) = 1.$$

Solution: This is a special case of Problem 14.4d.

29.4d. *Gradient systems.* We consider (161) with $f(t,u) = -V'(u)$, i.e., we consider the gradient system

$$\begin{aligned} u' &= -V'(u) \quad \text{on } [0,T], \\ u(0) &= u(T). \end{aligned} \tag{163}$$

Suppose that condition (H) holds and set $G = \{u \in X: \|u\| < r\}$. Show that

$$|\deg(B, M; G)| = \deg(V', U(0,r)) = 1. \tag{164}$$

This relation is the key to the general existence theorem in Problem 29.4e below.

Moreover show that the function u is a solution of (163) iff $u = \text{const} = c$ and $V'(c) = 0$. Since $\deg(V', U(0,r)) = 1$ by Problem 29.4c, there exists at least one such solution of (163) with $c \in U(0,r)$.

Solution: We set $P^\perp = I - P$ and use the homotopy

$$\begin{aligned} u' &= -\lambda V'(u) - (1-\lambda)T^{-1}\int_0^T V'(u(t))\,dt, \\ u(0) &= u(T), \end{aligned} \tag{165}$$

where $0 \leq \lambda \leq 1$. This equation corresponds to

$$Bu = \lambda Mu + (1-\lambda)P^\perp Mu, \qquad u \in \bar{G}, \tag{166}$$

where we use the notation of Problem 29.4a.

(I) We show that this is a homotopy. Let u be a solution of (166), i.e., $\|u\| \leq r$. Multiplication of (165) with u' and subsequent integration yield

$$\int_0^T \langle u'(t)|u'(t)\rangle\,dt = 0.$$

In this connection, note that $V(u(t))' = \langle V'(u(t))|u'(t)\rangle$ and $u(0) = u(T)$. Thus, $u = \text{const}$, and (165) yields $u'(0) = 0$ and hence $V'(u(0)) = 0$. By (H), $|u(0)| < r$. That means $u \in G$.

(II) The homotopy invariance of the coincidence degree yields that, for $\lambda = 1$ and $\lambda = 0$,

$$\deg(B, M; G) = \deg(B, P^\perp M, G),$$

and it follows from the reduction property (Problem 29.3e) that

$$|\deg(B, P^\perp M, G)| = |\deg(J^{-1}P^\perp M, G \cap N(B))|.$$

On $N(B)$, the operator $J^{-1}P^\perp M$ is identical to $V': N(B) \to N(B)$. Because of $N(B) = \mathbb{R}^N$, we obtain

$$\deg(J^{-1}P^\perp M, G \cap N(B)) = \deg(V', U(0,r)).$$

By Problem 29.4c, $\deg(V', U(0,r)) = 1$.

29.4e. *A general existence theorem.* Suppose that:

(i) The function $f\colon [0, T] \times \mathbb{R}^N \to \mathbb{R}^N$ is continuous.

(ii) There exists a C^1-function $V\colon \mathbb{R}^N \to \mathbb{R}$ with $V(x) \to +\infty$ as $|x| \to \infty$, and there exists a function $g \in C[0, T]$ such that

$$\langle V'(x)|f(t, x)\rangle \leq g(t) \qquad \text{for all} \quad (x, t) \in \mathbb{R}^N \times [0, T].$$

(iii) There exist a number $r > 0$ and a C^1-function $W\colon \mathbb{R}^N - U(0, r) \to \mathbb{R}$ such that

$$\langle V'(x)|W'(x)\rangle > 0$$

for all x with $|x| \geq r$, and

$$\int_0^T \langle W'(u(t))|f(t, u(t))\rangle\, dt \leq 0$$

for all C^1-functions $u\colon [0, T] \to \mathbb{R}^N$ with $u(0) = u(T)$ and $|u(t)| \geq r$ on $[0, T]$.

Show that the original problem (161) has a solution.
Hint: Use the homotopy

$$\begin{aligned} u' &= (1 - \lambda) f(t, u) - \lambda V'(u) \quad \text{on } [0, T], \\ u(0) &= u(T). \end{aligned} \tag{167}$$

Then the homotopy invariance of the coincidence degree and (164) yield

$$|\deg(B, M; G)| = 1,$$

and the assertion follows from the existence principle of the coincidence degree.

In order to show that (167) is a homotopy, one needs *a priori* estimates by means of the function W. Cf. Mawhin (1979, L), p. 64.

29.4f. *A special case.* Let $f\colon \mathbb{R} \to \mathbb{R}$ be a monotone increasing continuous function. Show that the problem

$$\begin{aligned} u' &= f(u) \quad \text{on } [0, T], \\ u(0) &= u(T), \end{aligned}$$

has a C^1-solution iff there exists a function $v \in C[0, T]$ with

$$\int_0^T f(v(x))\, dx = 0.$$

Solution: Use Problem 29.4e with

$$V(x) = \int_0^{x^2} (1 + y^{1/2})^{-1}\, dy,$$

$$W(x) = \int_{r^2}^{x^2} y^{-1/2}\, dy.$$

Cf. Mawhin (1979), p. 66.

29.5. *Condition* (S) *and linear Fredholm operators of index zero.*
Let $B\colon X \to X^*$ be a continuous linear operator satisfying condition (S) on the real reflexive B-space X.

Show that B is a Fredholm operator of index zero.

Solution:

(I) We show that $R(B)$ is closed. Since B is linear it is sufficient to prove that $B(M)$ is closed for all closed bounded subsets M of X (cf. Dunford and Schwartz (1958, M), Vol. I, Theorem VI, 6.5).

Let (u_n) be a sequence in the closed bounded set M and let

$$Bu_n \to v \qquad \text{as} \quad n \to \infty.$$

Since X is reflexive, there is a subsequence, again denoted by (u_n), such that

$$u_n \rightharpoonup u \qquad \text{as} \quad n \to \infty.$$

Thus, $\langle Bu_n, u_n \rangle \to \langle v, u \rangle$ as $n \to \infty$. By Figure 27.1, (S) implies $(S)_0$. Consequently,

$$u_n \to u \qquad \text{as} \quad n \to \infty.$$

Hence $u \in M$ and $Bu = v$, i.e., $v \in B(M)$.

(II) We show that $\dim N(B) < \infty$. The set $N(B) = \{u \in X: Bu = 0\}$ is a closed linear subspace of X. Set

$$M = \{u \in N(B): \|u\| \leq 1\}.$$

Letting $v = 0$, we obtain from (I) that each sequence in M has a convergent subsequence, i.e., M is compact. By Theorem 21.C, $\dim N(B) < \infty$.

(III) Let $n = \dim N(B)$ and $n^* = \dim N(B)$. Suppose first that $n \geq 1$. We show that $n^* \leq n$.

Otherwise, $n^* > n$. Let $\{u_1, \ldots, u_n\}$ be a basis in $N(B)$, and let $u_1^*, \ldots, u_n^*$ be linearly independent elements in $N(B^*)$. Note that B^* maps X into X^* since $X^{**} = X$. Hence $u_i^* \in X$. By $A_1(51)$, there exist elements $v_i, v_i^* \in X^*$ such that

$$\langle v_i^*, u_j \rangle_X = \langle v_i, u_j^* \rangle_X = \delta_{ij}, \qquad i, j = 1, \ldots, n.$$

As in Section 29.1, we construct the operator

$$Su = Bu + \sum_{k=1}^{n} \langle v_k^*, u \rangle v_k. \tag{168}$$

(III-1) We show that $Su = 0$ implies $u = 0$. Indeed, $Su = 0$ implies $\langle Su, u_i^* \rangle = 0$. Since $B^* u_i^* = 0$,

$$\langle Bu, u_i^* \rangle = \langle u, B^* u_i^* \rangle = 0.$$

Applying u_i^* to (168) we get $\langle v_i^*, u \rangle = 0$ for all i. Hence $Su = Bu = 0$, i.e., $u \in \operatorname{span}\{u_1, \ldots, u_n\}$. This yields $u = 0$, since $\langle v_i^*, u_j \rangle = \delta_{ij}$.

(III-2) We show that $R(S)$ is closed. The linear continuous operator $S: X \to X^*$ is a strongly continuous perturbation of the (S)-operator B. By Figure 27.1, the operator S satisfies condition (S). Now, the same argument as in (I) above shows that $R(S)$ is closed.

(III-3) By (III-1), the operator $S: X \to R(B)$ is bijective. By the open mapping theorem $A_1(36)$, the operator $S^{-1}: R(S) \to X$ is linear and continuous.

(III-4) Let $S_t u = Su - tb$. For all u with $\|u\| = R$ and sufficiently large R and all $t \in [0, 1]$, we obtain

$$\|S_t u\| \geq \|S^{-1}\|^{-1}\|u\| - \|b\| > 0.$$

By Theorem 29.B, $R(S) = X^*$.

(III-5) Because of $n^* > n$, there exists a $u^* \in N(B^*)$ with $u^* \neq 0$ and

$$\langle v_i, u^* \rangle = 0, \qquad i = 1, \ldots, n.$$

By the Hahn–Banach theorem, there exists a $b_0 \in X^*$ with $\langle b_0, u^* \rangle_X = \|u^*\|$. Since $R(S) = X^*$, there is a u_0 with $Su_0 = b_0$. Hence

$$\|u^*\| = \langle Su_0, u^* \rangle = \langle Bu_0, u^* \rangle = \langle u_0, B^* u^* \rangle = 0.$$

This contradicts $u^* \neq 0$.

(IV) Let $n = 0$, i.e., $N(B) = \{0\}$. Letting $S = B$ and using the same argument as in (III), we again obtain that $n^* \leq n$.

(V) We show that $n \leq n^*$. Since X is reflexive, $(B^*)^* = B$. The definition of (S) in Section 27.1 shows that B^* satisfies (S) if B has this property. Now, replace B by B^* and B^* by $(B^*)^* = B$ in (III) and (IV).

(VI) It follows from (III) through (V) that $n = n^*$, i.e., $\dim N(B) = \dim N(B^*)$.

29.6.* *The singular values of functionals.* Let $F: X \to \mathbb{R}$ be a C^k-functional on the real, separable, reflexive B-space X with $k \geq 2$ and $k \geq \dim N(F''(u))$ for all $u \in X$. Suppose that $F': X \to X^*$ is a Fredholm operator.

Show that the set of singular values of F has zero Lebesgue measure in $\mathbb{R}$.

Hint: Cf. Berger (1977, M), p. 127. Recall that b is a singular value of F iff there is a $u \in X$ such that $F(u) = b$ and $F'(u) = 0$.

29.7. *A bifurcation theorem in $\mathbb{R}^N$ of variational type.*

29.7a. Let $N \geq 1$. We consider the equation

$$F_u(u, \lambda) = 0, \qquad u \in \mathbb{R}^N, \quad \lambda \in \mathbb{R}, \tag{169}$$

where

$$F(u, \lambda) = \tfrac{1}{2}\lambda|u|^2 + r(u),$$

and the function $r: U(0) \subseteq \mathbb{R}^N \to \mathbb{R}$ is C^2 on a neighborhood of $u = 0$ with $r(u) = o(|u|^2)$, $u \to 0$. Show that $(0, 0)$ is a bifurcation point of equation (169).

Solution: By the Taylor theorem (Theorem 4.A), it follows that $r'(0) = 0$, $r''(0) = 0$, and hence $r'(u) = o(|u|)$, $u \to 0$. The two problems

$$r(u) = \min!, \qquad \langle u|u \rangle = \rho,$$

$$r(u) = \max!, \qquad \langle u|u \rangle = \rho,$$

have two different solutions u_1 and u_2 for each sufficiently small $\rho > 0$. Note that $r(\cdot)$ cannot be constant on small spheres, since $r(u) = o(|u|^2)$, $u \to 0$.

By the well-known classical Lagrange multiplier rule (see Section 43.10), there exists a real number λ_i such that $r'(u_i) = -\lambda_i u_i$, and hence

$$\lambda_i = -\frac{\langle r'(u_i)|u_i \rangle}{|u_i|^2} \to 0 \qquad \text{as} \quad \rho \to 0.$$

From $F_u(u, \lambda) = \lambda u + r'(u)$ it follows that

$$F_u(u_i, \lambda_i) = 0, \qquad i = 1, 2.$$

Obviously, $u_i \to 0$ as $\rho \to 0$.

29.7b.* *Proof of* (iii) *in Proposition* 29.25(H). If $r(\cdot)$ does not depend on the parameter λ, then the proof follows from Problem 29.7a by using a new scalar product on $X = \mathbb{R}^N$ induced by $a(\cdot, \cdot)$.

If $r(\cdot)$ depends on λ, then the sophisticated proof is based on a detailed study of dynamical systems by using arguments of the Ljusternik–Schnirelman theory (cf. Part III) and the theory of isolating blocks of Conley.

Hint: Use the dynamical system

$$u'(t) = -F_u(u(t), \lambda)$$

and observe that $u'(t_0) = 0$ implies $F_u(u(t_0), \lambda) = 0$. Cf Rabinowitz (1977) and Chow and Hale (1982, M), p. 159.

29.8. *A variant of the implicit function theorem* (*blowing-up technique*). We consider the equation

$$\lambda B(v, v) = C(v, \lambda), \qquad v \in X, \quad \lambda \in \mathbb{K}, \tag{170}$$

in a neighborhood of the point $(0, 0)$. The prototype for (170) is the real equation $\lambda v^2 = g(\lambda)v^3(1 + O(v))$ with the solution $\lambda = v$ in the special case $\lambda v^2 = v^3$. Suppose that:

(i) $B: X \times X \to \mathbb{R}$ is bilinear and bounded on the B-space X over $\mathbb{K} = \mathbb{R}, \mathbb{C}$, and $|B(v, v)| \geq c\|v\|^2$ for all $v \in X$ and fixed $c > 0$.

(ii) $C: U(0, 0) \subseteq X \times \mathbb{R} \to \mathbb{R}$ is C^k in a neighborhood of $(0, 0)$, where $1 \leq k \leq \infty$. Moreover, as $(v, \lambda) \to (0, 0)$, we have

$$C(v, \lambda) = o(\|v\|^2) \qquad \text{and} \qquad C_\lambda(v, \lambda) = o(\|v\|^2).$$

Show that, in a neighborhood of the point $(0, 0)$, equation (170) has a unique solution curve $\lambda = \lambda(v)$ through the point $(0, 0)$. This curve is continuous on a neighborhood of $v = 0$ and C^k away from $v = 0$.

Proof. Set

$$F(v, \lambda) = \begin{cases} \lambda - \dfrac{C(v, \lambda)}{B(v, v)} & \text{if} \quad v \neq 0, \\ \lambda & \text{if} \quad v = 0, \end{cases}$$

and write equation (170) in the form

$$F(v, \lambda) = 0.$$

Then F and F_λ are continuous in a neighborhood of $(0, 0)$ and $F_\lambda(0, 0) = 1$. The assertion now follows from the implicit function theorem (Theorem 4.B).

29.9.* *A multiplicity theorem for semilinear equations.* We want to study the boundary value problem

$$\begin{aligned} -\Delta u &= F(u) + f \quad \text{on } G, \\ u &= 0 \quad \text{on } \partial G. \end{aligned} \tag{171}$$

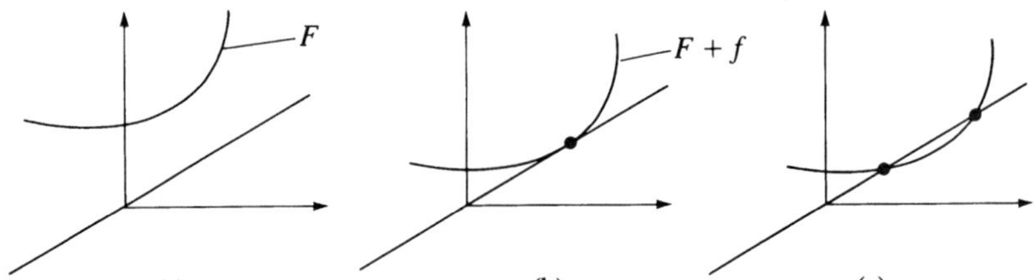

(a) (b) (c)

Figure 29.25

To this end, we consider the linear eigenvalue problem

$$\begin{aligned} -\Delta u &= \mu u \quad \text{on } G, \\ u &= 0 \quad \text{on } \partial G \end{aligned} \tag{172}$$

with the eigenvalues $0 < \mu_1 < \mu_2 \leq \cdots$. We set

$$X = \{u \in C^{2,\alpha}(\bar{G}): u = 0 \text{ on } \partial G\}, \qquad Y = C^\alpha(\bar{G}),$$

and we assume:

(H1) G is a bounded region in $\mathbb{R}^M$ with $\partial G \in C^\infty$ and $M \geq 1, 0 < \alpha < 1$.

(H2) The C^3-function $F: \mathbb{R} \to \mathbb{R}$ is strictly monotone increasing and convex (Fig. 29.25(a)). More precisely, let $F''(0) > 0$ and

$$F'(u) > 0 \quad \text{and} \quad F''(u) \geq 0 \qquad \text{for all} \quad u \in \mathbb{R},$$

$$0 < \lim_{u \to -\infty} F'(u) < \mu_1 < \lim_{u \to +\infty} F'(u) < \mu_2.$$

Show that there exists a closed connected C^1-manifold C of codimension 1 in Y such that $Y - C$ consists of exactly two connected components T, N, and equation (171) has

(i) one solution $u \in X$ for $f \in C$,

(ii) two solutions $u \in X$ for $f \in T$, and

(iii) no solution for $f \in N$.

Hint: Use Theorem 29.N. Cf. Ambrosetti and Prodi (1972), Berger and Podolak (1975), and Nirenberg (1974, L) p. 117. If we consider the real equation

(E) $$-\Delta u = F(u) + f, \qquad u \in \mathbb{R},$$

with $-\Delta > 0$ and $f \in \mathbb{R}$, then Figure 29.25 gives an intuitive interpretation of the result above. Note that the intersection points in Figure 29.25 correspond to the solutions of (E).

More general results can be found in the survey articles by Ambrosetti (1983) and Ruf (1986).

29.10.* *Resonance at the first eigenvalue of the linearized problem.* We consider the semilinear boundary value problem

$$\begin{aligned} -\Delta u(x) - a(x, u(x))u(x) &= f(x, u(x)) \quad \text{on } G, \\ u &= 0 \quad \text{on } \partial G, \end{aligned} \tag{173}$$

along with the linearized eigenvalue problem

$$\begin{aligned} -\Delta u - \mu u &= 0 \quad \text{on } G, \\ u &= 0 \quad \text{on } \partial G. \end{aligned} \tag{174}$$

Let G be a bounded region in $\mathbb{R}^N$ with $\partial G \in C^\infty$, $N \geq 1$, and let the functions a, $f: \bar{G} \times \mathbb{R} \to \mathbb{R}$ be C^∞. Moreover, let μ_1 be the smallest eigenvalue of (174) with the corresponding eigenfunction u_1. Prove the following.

(a) *Nonresonce case.* If

$$a(x, u) < \mu_1 \qquad \text{for all} \quad (x, u) \in \bar{G} \times \mathbb{R},$$

and if f is bounded on $\bar{G} \times \mathbb{R}$, then the original problem (173) has a solution.

(b) *Resonance case.* If

$$a(x, u) \equiv \mu_1 \qquad \text{for all} \quad (x, u) \in \bar{G} \times \mathbb{R},$$

and if $f_u(x, u) \leq 0$ for all $(x, u) \in \bar{G} \times \mathbb{R}$, then (173) has a solution iff there exists a function $v \in C^\infty(\bar{G})$ such that $v = 0$ on ∂G and

$$\int_G f(x, v(x)) u_1(x)\, dx = 0. \tag{175}$$

In the case where the function f does not depend on u, condition (175) represents the well-known classical solvability condition for (173).

Hint: Cf. Kazdan and Warner (1975). The proof is based on the method of subsolutions and supersolutions from Chapter 7.

Further important material about semilinear elliptic differential equations can be found in the problems section of Chapter 49 (variational methods). Cf. also Problem 29.14.

29.11. *Stable homotopy (generalized mapping degree) and the solvability of nonlinear operator equations.* In the following we will use tools from Section 16.8. Let N, $M \geq 1$.

29.11a. *Essential maps.* We consider the equation

$$F(x) = 0, \qquad x \in B^N, \tag{176}$$

where B^N denotes the closed unit ball in $\mathbb{R}^N$. As usual let $S^{N-1} = \partial B^N$. The continuous map

$$f: \partial B^N \to \mathbb{R}^M \tag{177}$$

is called *essential* with respect to B^N iff $f(x) \neq 0$ on ∂B^N and equation (176) has a solution for each continuous map

$$F: B^N \to \mathbb{R}^M \qquad \text{with} \quad F = f \quad \text{on } \partial B^N.$$

From Chapter 16 we obtain the following results:

(i) For $N = M = 1$, the map f in (177) is essential with respect to $B^N = [-1, 1]$ iff

$$f(1) f(-1) \neq 0.$$

(ii) For $N = M$, the continuous map f in (177) is essential with respect to B^N

iff $f(x) \neq 0$ on ∂B^N and

$$\deg(f, \operatorname{int} B^N) \neq 0.$$

(iii) For $M > N$, each continuous map (177) is nonessential with respect to B^N.

The following methods can be applied if $M \leq N$.

29.11b. *Nullhomotopic maps.* The continuous map

$$g: S^{N-1} \to S^{M-1}$$

is called *nullhomotopic* (or homotopically trivial) iff it can be deformed into a constant map, i.e., there is a continuous map

$$H: S^{N-1} \times [0, 1] \to S^{M-1}$$

such that $H(x, 0) = g(x)$ and $H(x, 1) = \text{const}$ for all $x \in S^{N-1}$.

According to Example 16.24, the continuous map $f: \partial B^N \to \mathbb{R}^M - \{0\}$ is essential with respect to B^N iff the normalized map

$$\frac{f}{|f|}: \partial B^N \to \partial B^M$$

is not nullhomotopic.

29.11c. *Nontrivial stable homotopy.* Let $N \geq M \geq 1$. We consider the continuous map

$$g: S^{N-1} \to S^{M-1} \tag{178}$$

along with the Kth *suspension*

$$S^K g: S^{N-1+K} \to S^{M-1+K}. \tag{179}$$

The suspension operator S has been introduced in Section 16.8. An important topological result tells us the following.

For $K \geq N - M$, the map $S^K g$ in (179) *is not nullhomotopic iff $S^{K+1} g$ is not nullhomotopic.*

Let $B_r^N = \{x \in \mathbb{R}^N: |x| \leq r\}$, and let $g(x) = f(x)/|f(x)|$. The continuous map

$$f: B_r^N \to \mathbb{R}^M \tag{180}$$

is said to have *stable homotopy* iff $f(x) \neq 0$ on ∂B_r^N and all the suspensions $S^K g$ in (179) with $K \geq 1$ are not nullhomotopic.

(i) For $N = M$, the continuous map f in (180) has stable homotopy iff it is essential with respect to B_r^N, i.e., $f(x) \neq 0$ on ∂B_r^N and

$$\deg(f, \operatorname{int} B_r^N) \neq 0.$$

In the case $N = M = 1$, this means $f(1)f(-1) \neq 0$.

(ii) For $N > M$, the continuous map f in (180) has stable homotopy iff $f(x) \neq 0$ on ∂B_r^N and all the suspensions $S^K g$ in (179) with $1 \leq K \leq N - M$ are not nullhomotopic.

29.11d. *The key result.* Let $1 \leq d^* \leq d < N$. We set

$$F_i(\xi_1, \ldots, \xi_N) = \begin{cases} f_i(\xi_1, \ldots, \xi_d) & \text{if } 1 \leq i \leq d^*, \\ \xi_i & \text{if } i \geq d + 1, \end{cases}$$

where the map

$$f: B^d \to \mathbb{R}^{d^*}$$

is continuous with $f(y) \neq 0$ on ∂B^d. Show that the map

$$F: \partial B^N \to \mathbb{R}^{N-d+d^*} \tag{181}$$

is essential with respect to B^N if the map f has stable homotopy.

Solution: It is not difficult to show that the normalized map

$$\frac{F}{|F|}: S^{N-1} \to S^{N-d+d^*-1}$$

is homotopic to the $(N-d)$-fold suspension of the map

$$\frac{f}{|f|}: S^{d-1} \to S^{d^*-1},$$

and hence $F/|F|$ is *not* nullhomotopic, since f has stable homotopy. By Problem 29.11b, the map F is essential.

29.11e.* *The abstract main theorem on semilinear operator equations with bounded nonlinearities.* The equation

$$Lu = Fu, \qquad u \in X, \tag{182}$$

has a solution in the case where the following conditions are satisfied:

(i) *Linear operator.* The linear continuous map $L: X \to Y$ is a Fredholm operator of index $\operatorname{ind} L \geq 0$, where X and Y are real B-spaces. We choose fixed topological direct sums

$$X = N(L) \oplus N(L)^\perp \qquad \text{and} \qquad Y = R(L) \oplus R(L)^\perp$$

along with the corresponding linear projection operator $P: Y \to R(L)^\perp$.

(ii) *Nonlinear perturbation.* The map $F: X \to Y$ is *compact*. There are positive constants r and ρ such that

$$\|(I-P)Fu\| \leq \rho \qquad \text{for all} \quad u \in X, \tag{183}$$

and

$$F(v+w) \notin R(L) \tag{184}$$

for all $v \in N(L)$, $w \in N(L)^\perp$ with $\|v\| \geq r$ and $\|w\| \leq \|L_0^{-1}\|\rho$, where L_0 denotes the restriction of L to $N(L)^\perp$.

(iii) *Stable homotopy.* Let $B_r = \{u \in N(L): \|u\| \leq r\}$. The finite-dimensional map

$$PF: B_r \subseteq N(L) \to R(L)^\perp$$

has stable homotopy.

Remark. Condition (183) is satisfied in the case where F is *bounded*, i.e.,

$$\sup_{u \in X} \|Fu\| < \infty.$$

If $\operatorname{ind} L = 0$ and $\dim N(L) > 0$, then condition (iii) holds true iff

$$\deg(PF, \operatorname{int} B_r) \neq 0.$$

If ind $L > 0$, then it is not possible to formulate condition (iii) in terms of the mapping degree, since the map PF lowers dimension.

Hint: Cf. Cronin (1973), Nirenberg (1974, L), p. 134, Berger and Podolak (1974), (1975), and Berger (1977, M), p. 279.

29.11f.* *Application to general semilinear elliptic boundary value problems with nonegative index (the generalized Landesman–Lazer theorem).* We consider the boundary value problem

$$\begin{aligned} Lu(x) &= f(x, u(x)) \quad \text{on } G, \\ Bu &= 0 \quad \text{on } \partial G, \end{aligned} \tag{185}$$

where

$$Lu = \sum_{|\alpha| \leq 2m} a_\alpha D^\alpha u$$

and $Bu = (B_1 u, \ldots, B_K u)$ with $B_k u = \sum_{|\beta| < 2m} b_{k,\beta} D^\beta u$.

Show that problem (185) has a classical solution u in the case where the following hold.

(H1) G is a bounded region in $\mathbb{R}^N$ with $\partial G \in C^\infty$, where $N, m \geq 1$. All the functions $a_\alpha, b_{k,\beta}: \bar{G} \to \mathbb{R}$ are C^∞.

(H2) The linear differential operator L is elliptic, i.e.,

$$\sum_{|\alpha| = 2m} a_\alpha(x) D^\alpha \neq 0 \qquad \text{for all} \quad x \in \bar{G}, \quad D \in \mathbb{R}^N - \{0\},$$

where $D = (D_1, \ldots, D_N)$ and $D^\alpha = D_1^{\alpha_1} D_2^{\alpha_2} \cdots D_N^{\alpha_N}$.

(H3) The boundary operator B satisfies the algebraic complementing condition. According to Problem 6.8, large classes of boundary conditions satisfy this condition. For example, we can choose the boundary condition $Bu = 0$ in the form $D^\gamma u = 0$ on ∂G for all γ: $|\gamma| \leq m - 1$.

(H4) Let $g = 0$. The only solution of

$$\begin{aligned} Lu &= g \quad \text{on } G, \\ Bu &= 0 \quad \text{on } \partial G, \end{aligned} \tag{186}$$

which vanishes on a set of positive measure, is $u = 0$.

(H5) Let $\{u_1, \ldots, u_d\}$ be a basis of the solution set of (186) with $g = 0$, and let $\{u_1, \ldots, u_{d^*}\}$ be a cobasis, i.e., for given $g \in C^\infty(\bar{G})$, problem (186) has a solution u iff

$$\int_G g(x) u_i^*(x)\, dx = 0 \qquad \text{for} \quad i = 1, \ldots, d^*.$$

Suppose that $1 \leq d^* \leq d$.

(H6) The function $f: \bar{G} \times \mathbb{R} \to \mathbb{R}$ is C^∞, and

$$|f(x, u)| \leq \text{const} \qquad \text{for all} \quad x \in \bar{G}, \quad u \in \mathbb{R}.$$

The limit

$$h(x, v) = \lim_{r \to +\infty} f(x, rv)$$

exists uniformly for all $x \in \bar{G}$ and $v \in \mathbb{R}$ with $|v| = 1$.

(H7) *The key condition.* We set $F = (F_1, \ldots, F_{d^*})$ with

$$F_i(c) = \int_G u_i^*(x) h\left(x, \sum_{j=1}^{d} c_j u_j(x)\right) dx$$

and $c = (c_1, \ldots, c_d)$, where c_j is real for all j. Suppose that the map

$$F\colon B^d \to \mathbb{R}^{d^*}$$

has *stable homotopy.*

In particular, if $d = d^*$, then (H7) holds in the case where

$$\deg(F, \operatorname{int} B^d) \neq 0.$$

If $d = d^* = 1$, then (H7) is equivalent to

$$F_1(1)F_1(-1) < 0.$$

Hint: Cf. Nirenberg (1970), (1974, L), p. 140.

A variant of this theorem can be found in Schechter (1973) by replacing condition (H7) with inequalities.

29.12. *A general model for beams and its stability.* We consider a beam of length l as pictured in Figure 29.26. In an idealized form, let the deformed beam be described by the curve

$$\xi = \alpha + a(\alpha), \qquad \zeta = c(\alpha), \qquad 0 \le \alpha \le l, \tag{187}$$

where ξ, η, ζ are Cartesian coordinates, and $u = (a, c)$ is the displacement vector in the (ξ, ζ)-plane. More precisely, the undeformed beam is given by the set

$$\{(\xi, \eta, \zeta) \in \mathbb{R}^3 \colon 0 \le \xi \le l, (\eta, \zeta) \in Q(\xi)\},$$

where $Q(\xi)$ denotes the cross setion of the beam. To simplify our considerations, let us assume that the cross section Q is constant, i.e., it is independent

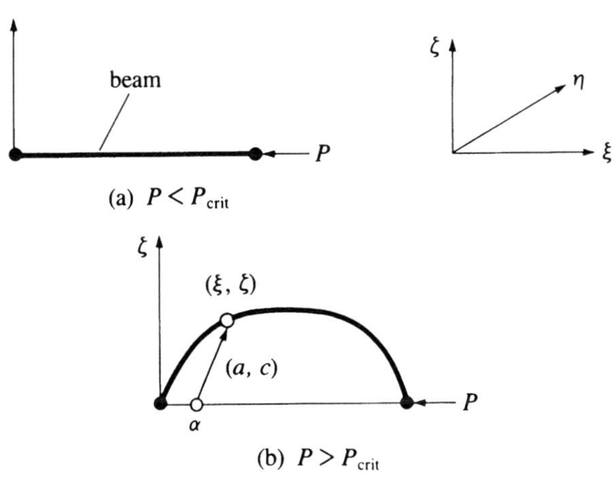

Figure 29.26

of ξ. Variable cross sections will be studied in Section 64.7. The number $P > 0$ is the strength of an outer force acting in the direction of the negative ξ-axis. We want to motivate the following *variational principle* for the functions a, c:

$$E_{\text{pot}} \stackrel{\text{def}}{=} \int_0^l \frac{A}{2}k^2 + \frac{B}{2}(\tau' - 1)^2 \, d\xi + Pa(l) = \min!.$$

$$a(0) = c(0) = 0, \qquad c(l) = 0. \tag{188}$$

In this connection,

$$k = \frac{((1 + a')c'' - a''c')^2}{(1 + a'^2 + c'^2)^3}$$

is the *curvature* of (187) and

$$\tau' = \sqrt{1 + a'^2 + c'^2}$$

is the derivative of the arclength of (187). The boundary condition corresponds to Figure 29.26.

Motivation. According to Section 61.6, the potential energy E_{pot} of an elastic body under the displacement u is given by

$$E_{\text{pot}}(u) = U(u) - W_{\text{out}}(u),$$

where U is the elastic potential energy and $W_{\text{out}}(u)$ is the work of the outer forces with respect to the displacement u. In our special case, we obtain that

$$W_{\text{out}}(u) = -Pa(l), \qquad u = (a, c),$$

since work = force times displacement, and the force acts in direction of the negative ξ-axis.

We now assume that the elastic potential energy U depends on

(i) the curvature k of the beam, and on
(ii) the length contraction of the beam.

Ad(i). Suppose that U has the form

$$U = \int_0^l f(k) \, dk.$$

For small curvature k, Taylor expansion yields

$$f(k) = f(0) + f'(0)k + \tfrac{1}{2}f''(0)k^2 + \cdots.$$

It is natural to assume that $k = 0$ implies $f(k) = 0$ and hence $f(0) = 0$. Moreover, we assume that f does not depend on the sign of the curvature. This yields the second-order approximation

$$f(k) = \frac{A}{2}k^2,$$

where A depends on the material and the cross section. In Section 64.7 we shall motivate that

$$A = EJ, \qquad J = \int_Q \zeta^2 \, d\eta \, d\zeta,$$

where E is the so-called elasticity module related to Hooke's law (189) below.

Ad(ii). We consider two points $(\xi, 0, 0)$ and $(\xi + \Delta\xi, 0, 0)$ of the undeformed beam. The relative change of length of this element of the beam is equal to

$$\gamma = (\Delta\tau - \Delta\xi)/\Delta\xi,$$

where τ denotes the arclength of the deformed beam. By Hooke's law, in Section 60.1, the corresponding tension is given by

$$\sigma = E\gamma. \tag{189}$$

We now consider a volume element $\Delta V = m(Q)\Delta\xi$ of the underformed beam, where $m(Q)$ denotes the measure of the constant cross section Q. According to Section 61.7, the elastic potential energy

$$\tfrac{1}{2}\sigma\gamma\Delta V$$

corresponds to the strain γ. This leads to the term $\left(\frac{B}{2}\right)(\tau' - 1)^2\,d\xi$ in (188) with

$$B = m(Q)E.$$

Study the stability of the trivial solution $a(\xi) \equiv 0$, $c(\xi) \equiv 0$, and determine the buckling force P_{crit} (Fig. 29.26).

Hint: Use similar arguments as in Section 29.13 to obtain that

$$P_{\text{crit}} = \min\left(\frac{A\pi^2}{l^2}, B\right).$$

If the length of the beam l is sufficiently large, i.e., $A\pi^2/l^2 < B$, then we obtain the classical Euler buckling force $P_{\text{crit}} = A\pi^2/l^2$. Note that our model (188) is more general than (97). Cf. Klötzler (1971, M), p. 147.

29.13.* *An existence theorem for semilinear operator equations in* H-*spaces*. Show that the equation

$$Lu = F'(u), \qquad u \in X,$$

has a solution if the following conditions are satisfied:

(i) The operator $L: D(L) \subseteq X \to X$ is self-adjoint on the real H-space X, and the range of L is closed.

(ii) The linearized equation $Lu = 0$ has a nontrivial solution $u \neq 0$.

(iii) The functional $F: X \to \mathbb{R}$ is convex, continuous, and G-differentiable.

(iv) For all $u \in X$, we have the growth condition

$$F(u) \leq \frac{b}{2}\|u\|^2 + \text{const},$$

where the constant b satisfies the inequality $0 < b < \lambda_1$. Here, λ_1 denotes the smallest positive eigenvalue of L.

(v) $$F(w) \to \infty \qquad \text{if} \quad \|w\| \to \infty,$$

where w lives in the null space of L.

Hint: The proof is based on a dual variational principle. Cf. Mawhin (1986a).

29.14. *Positive solutions of semilinear elliptic differential equations.* We consider the following boundary value problem:

$$\begin{aligned} -\Delta u &= f(u) \quad \text{in } G, \qquad u = 0 \quad \text{on } \partial G, \\ u &> 0 \qquad \text{in } G, \qquad u \in C^2(\bar{G}). \end{aligned} \tag{190}$$

Let

$$F(u) = \int_0^u f(v)\,dv.$$

Then the first line of problem (190) corresponds to the variational problem

$$\begin{aligned} &\int_G (\tfrac{1}{2}|Du|^2 - F(u))\,dx = \text{stationary!}, \\ &u = 0 \quad \text{on } \partial G, \end{aligned} \tag{190*}$$

where $Du = (D_1 u, \ldots, D_N u)$. We assume:

(H) Let G be a bounded region in $\mathbb{R}^N$, $N \geq 2$, with smooth boundary, i.e., $\partial G \in C^\infty$.

A typical special case of (190) is given by the following problem:

$$\begin{aligned} -\Delta u &= \mu u^q \quad \text{in } G, \qquad u = 0 \quad \text{on } \partial G, \\ u &> 0 \qquad \text{in } G, \qquad u \in C^2(\bar{G}), \end{aligned} \tag{191}$$

where $\mu > 0$ is fixed. If $q > 1$ (resp. $0 < q < 1$), then problem (191) is called *superlinear* (resp. sublinear). We introduce the critical exponent

$$q_{\text{crit}} = \begin{cases} (N+2)/(N-2) & \text{for } \quad N \geq 3, \\ +\infty & \text{for } \quad N = 1, 2. \end{cases}$$

29.14a. *Typical special case.* From our general results below we obtain the following special statements. Let $N \geq 3$ and assume (H). Moreover, let $\mu > 0$ be given. Then:

(i) If $1 < q < q_{\text{crit}}$, then problem (191) has a solution.

(ii) If $q \geq q_{\text{crit}}$ and G is *star-shaped* (e.g., G is convex), then problem (191) has *no* solution.

(iii) If $q = q_{\text{crit}}$ and G has *nontrivial topology* (e.g., G is *not* contractible to a point if $N = 3$), then problem (191) has a solution.

Now, assume (H) with $N = 2$. Then $q_{\text{crit}} = \infty$ and problem (191) has a solution for *each* q with $1 < q < q_{\text{crit}}$.

Finally, let G be an open ball around the origin in $\mathbb{R}^N$, $N \geq 1$, and let $q \geq 1$. Then, each solution u of (191) is *radially symmetric*.

Lack of compactness. The results above show clearly that the critical exponent q_{crit} plays a fundamental role. Note the following crucial fact:

(C) *Under the assumption* (H) *with* $N \geq 3$, *the embedding*

$$\mathring{W}_2^1(G) \subseteq L_{q+1}(G)$$

is compact for $1 \leq q < q_{\text{crit}}$ *and not compact for* $q = q_{\text{crit}}$.

Roughly speaking, this lack of compactness is responsible for (i) and (ii) above. In particular, if we consider problem (191), then $f(u) = |u|^q$ and

hence

$$F(u) = \frac{1}{q+1}(\operatorname{sgn} u)|u|^{q+1}.$$

In this case, the functional $u \mapsto \int_G F(u)\,dx$ is *not* weakly sequentially continuous on the Sobolev space $\mathring{W}_2^1(G)$ if $q = q_{\text{crit}}$. This fact is related to (C) (cf. Step 2 of the existence proof from Problem 29.14g). In Part III we will study two standard methods for solving variational problems, namely, the Ljusternik–Schnirelman theory and the Morse theory. If $q = q_{\text{crit}}$, then it is not possible to apply straightforwardly these theories to problem (190*), since the crucial Palais–Smale compactness condition is violated, by (C).

This lack of compactness only appears in $\mathbb{R}^N$ for $N \geq 3$. In fact, let G be a bounded region in $\mathbb{R}^N$. If $N = 2$ (resp. $N = 1$), then the embedding (E) above is compact for all q with $1 \leq q < \infty$ (resp. $1 \leq q \leq \infty$).

Critical exponents in differential geometry. It is quite remarkable that important nonlinear differential equations in differential geometry are related to situation (C) above for $q = q_{\text{crit}}$. For example, this concerns:

(a) the Yamabe problem (cf. Schoen (1984) and the survey article by Lee and Parker (1987)); and

(b) the Yang–Mills equations in gauge field theory (cf. Taubes (1982), (1982a), (1986, S)).

Problems (a) and (b) will be discussed in Chapters 85 and 96, respectively. Concerning critical exponents for semilinear partial differential equations, we also recommend the survey article by Brézis (1986).

29.14b*. *Existence theorem for superlinear problems in the noncritical case.* Let $N \geq 3$. Along with (H) above, we make the following assumptions:

(H1) *Superlinearity.* The function $f\colon \mathbb{R}_+ \to \mathbb{R}_+$ is locally Lipschitz continuous and superlinear, i.e., more precisely, we have $f(0) = 0$,

$$\lim_{u \to +0} \frac{f(u)}{u} = 0 \qquad \text{and} \qquad \lim_{u \to +\infty} \frac{f(u)}{u} = +\infty.$$

(H2) *Subcritical growth.* The function $u \mapsto f(u)$ does not grow faster than $u \mapsto u^{q_{\text{crit}}}$ as $u \to +\infty$. More precisely,

$$\lim_{u \to +\infty} \frac{f(u)}{u^{q_{\text{crit}}}} = 0,$$

and $u \mapsto f(u)/u^{q_{\text{crit}}}$ is nonincreasing on $]0, \infty[$.

(H3) There exists a number q with $-1 \leq q < q_{\text{crit}}$ such that

$$uf(u) \leq (q+1)F(u) \qquad \text{for all} \quad u \geq u_0 \quad \text{and fixed } u_0.$$

Then:

(a) *Existence.* The original problem (190) has a solution.

(b) *A priori estimates.* There exists a constant $C > 0$ such that $\|u\|_\infty \leq C$ for each solution u of (190).

Corollary (The simpler two-dimensional case). Assume (H) with $N = 2$ and (H1). Then statements (a) and (b) remain true.

Example. Let $\mu > 0$. The assumptions made above are satisfied for the function

$$f(u) = \mu u^q,$$

where $1 < q < q_{\text{crit}}$ for $N \geq 2$. Note that $F(u) = \mu u^{q+1}/(q+1)$ for $u \geq 0$ in (H3).

Hint: Cf. De Figueiredo, Lions, and Nussbaum (1982).

29.14c. *The nonexistence theorem of Pohožaev* (1965). Let G be a bounded region in $\mathbb{R}^N$, $N \geq 1$, with $\partial G \in C^\infty$ (i.e., G is a bounded open interval for $N = 1$). Let $f: \mathbb{R}_+ \to \mathbb{R}$ be continuous. Show that the following hold.

(i) If u is a solution of the original problem (190), then

$$2N \int_G F(u)\,dx - (N-2)\int_G uf(u)\,dx = \int_{\partial G} (x-y)n(x)\left|\frac{\partial u}{\partial n}\right|^2 dO, \tag{192}$$

for each fixed $y \in \mathbb{R}^N$. Here, $n(x)$ denotes the outer unit normal at the point $x \in \partial G$. Recall that $F(u) = \int_0^u f(v)\,dv$.

(ii) Let $N \geq 3$. If the region G is *star-shaped* (e.g., G is convex) and if

$$(q_{\text{crit}} + 1)F(u) \leq uf(u) \qquad \text{for all} \quad u \geq 0, \tag{193}$$

where $f(u) \geq 0$ for all $u \geq 0$, then the original problem (190) has *no* solution u.

Example. Let $f(u) = \mu u^q$ for fixed $q \geq q_{\text{crit}}$, $\mu > 0$, and all $u \geq 0$. Then $F(u) = \mu u^{q+1}/(q+1)$ for all $u \geq 0$, and hence condition (193) is satisfied.

Solution: Ad(i). Multiplying the original equation $-\Delta u = f(u)$ by $(\xi_i - \eta_i)D_i u$, we get

$$\text{(R)} \qquad -\int_G (\xi_i - \eta_i)D_i u \Delta u\,dx = \int_G (\xi_i - \eta_i)D_i u f(u)\,dx.$$

Here, we sum over $i, j = 1, \ldots, N$, and we set $x = (\xi_1, \ldots, \xi_N)$, $y = (\eta_1, \ldots, \eta_N)$, $n = (n_1, \ldots, n_N)$, and $D_i = \partial/\partial\xi_i$. We want to show that (R) implies the desired relation (192) by simply using integration by parts. In fact, since $u = 0$ on ∂G, integration by parts yields

$$\begin{aligned}\int_G (\xi_i - \eta_i)D_i u f(u)\,dx &= \int_G (\xi_i - \eta_i)D_i F(u)\,dx \\ &= \int_{\partial G} (\xi_i - \eta_i)n_i F(u(x))\,dO - \int_G (D_i(\xi_i - \eta_i))F(u)\,dx \\ &= -\int_G NF(u)\,dx,\end{aligned}$$

since $F(0) = 0$. Moreover, using $\Delta u = D_j(D_j u)$, integration by parts yields

$$-\int_G (\xi_i - \eta_i)D_i u D_j(D_j u)\,dx = A + B,$$

where

$$A = -\int_{\partial G} (\xi_i - \eta_i)D_i u(n_j D_j u)\,dO,$$

$$B = \int_G D_j[(\xi_i - \eta_i)D_i u]D_j u\,dx.$$

Since $u = 0$ on ∂G, the tangential derivatives of u vanish on ∂G, and hence $Du = |Du|n$, i.e.,

$$D_j u = |Du| n_j \quad \text{on } \partial G \qquad \text{for all } j. \tag{194}$$

This implies

$$\begin{aligned} A &= -\int_{\partial G} (\xi_i - \eta_i) n_i |Du|^2 \, dO \\ &= -\int_{\partial G} (x - y) n |Du|^2 \, dO. \end{aligned}$$

By the product rule,

$$B = \int_G \delta_{ij} D_i u D_j u \, dx + C, \qquad \text{where} \quad C = \int_G (\xi_i - \eta_i) D_i D_j u D_j u \, dx.$$

Integration by parts yields

$$C = \int_G (\xi_i - \eta_i) n_i |Du|^2 \, dO - \int_G N |Du|^2 \, dx - C,$$

and hence

$$C = \frac{1}{2} \int_G (x - y) n |Du|^2 \, dO - \frac{N}{2} \int_G |Du|^2 \, dx.$$

Finally, integration by parts yields

$$\begin{aligned} \int_G |Du|^2 \, dx &= \int_G D_j u D_j u \, dx \\ &= \int_{\partial G} n_j u D_j u \, dx - \int_G u \Delta u \, dx = \int_G u f(u) \, dx. \end{aligned}$$

Putting all these relations together, we get the claimed inequality (192). Note that $|Du|^2 = |\partial u / \partial n|^2$ on ∂G, by (194).

Ad(ii). If the region G is *star-shaped*, then there exists a point $y \in G$ such that

$$(x - y) n(x) > 0 \qquad \text{for all} \quad x \in \partial G. \tag{195}$$

Suppose that u is a solution of the original problem (190). Since $f(u) \geq 0$ for all $u \geq 0$, we get

$$\Delta u \leq 0 \quad \text{in } G,$$

and $u = 0$ on ∂G as well as $u > 0$ in G. Thus, the C^2-function $u \colon \bar{G} \to \mathbb{R}$ attains its minimum at each point of the boundary ∂G. By the strong maximum principle for elliptic equations applied to $-u$, this implies

$$\frac{\partial u}{\partial n} < 0 \quad \text{on } \partial G$$

(cf. Problem 7.2 from Part II). Noting that $q_{\text{crit}} + 1 = 2N/(N - 2)$, it follows from the decisive inequality (192) along with (193) and (195) that $\partial u/\partial n = 0$ on ∂G. That is a contradiction.

29.14d*. *The impact of topology on semilinear elliptic equations with critical exponent.* We consider the problem

$$\begin{aligned} -\Delta u &= \mu u^{q_{\mathrm{crit}}} \quad \text{in } G, \qquad u = 0 \quad \text{on } \partial G, \\ u &> 0 \quad \text{in } G, \qquad u \in C^2(\bar{G}), \end{aligned} \tag{196}$$

where $\mu > 0$ is fixed, and G is a bounded region in $\mathbb{R}^N$, $N \geq 3$, with $\partial G \in C^\infty$. We say that G has *nontrivial topology* iff there exists an integer $k \geq 1$ such that

$$H_{2k-1}(G, \mathbb{Q}) \neq 0 \qquad \text{or} \qquad H_{2k-1}(G, \mathrm{mod}\, 2) \neq 0.$$

Here, $H_m(G, \mathbb{Q})$ and $H_m(G, \mathrm{mod}\, 2)$ denotes the mth homology group of G with coefficients in $\mathbb{Q}$ (the field of rational numbers) and $\mathbb{Z}_2 = \mathbb{Z}/2\mathbb{Z}$, respectively (cf. Chapter 97). For $N = 3$, G has nontrivial topology iff G is not contractible to a point. For $N \geq 4$, a large variety of regions G has nontrivial topology. For example, this is the case if G is homeomorphic to a solid torus or G has a hole.

Show that, for each $\mu > 0$, problem (196) has a solution if G has nontrivial topology.

Hint: Cf. Bahri and Coron (1988).

29.14e*. *Existence theorem in the sublinear case.* We consider the problem

$$\begin{aligned} -\Delta u &= \mu f(u) \quad \text{in } G, \qquad u = 0 \quad \text{on } \partial G, \\ u &> 0 \quad \text{in } G, \qquad u \in C^2(\bar{G}), \end{aligned} \tag{197}$$

where $\mu > 0$. Let S denote the set of all solution pairs (μ, u) of (197). Suppose that $f: \mathbb{R} \to \mathbb{R}$ is C^1 and sublinear, i.e.,

$$\lim_{u \to 0} \frac{f(u)}{u} = +\infty.$$

Show that the set $S \cup (0, 0)$ contains an unbounded component Σ in the space $\mathbb{R}_+ \times C^2(\bar{G})$ with $(0, 0) \in \Sigma$.

Hint: Cf. Turner (1974).

29.14f*. *Radially symmetric solutions.* Let G be an open ball around the origin in $\mathbb{R}^N$, $N \geq 1$, and let $f: \mathbb{R} \to \mathbb{R}$ be C^1.

Show that each solution u of (197) is radially symmetric.

Hint: Cf. Gidas, Ni, and Nirenberg (1979). The proof is based on the maximum principle and on an important device of *moving parallel planes* to a critical position and then showing that the solution is symmetric about the limiting plane (cf. also Kawohl (1985, L)).

29.14g*. *The importance of best Sobolev constants.* Finally, we study the following problem:

$$\begin{aligned} -\Delta u - a(x)u &= u^q \quad \text{in } G, \qquad u = 0 \quad \text{on } \partial G, \\ u &> 0 \quad \text{in } G, \qquad u \in C^2(\bar{G}). \end{aligned} \tag{198}$$

In the special case where $N \geq 3$ and $q = q_{\mathrm{crit}}$, this problem is closely related to the famous Yamabe problem in differential geometry (cf. Chapter 85).

(H1) Let G be a bounded region in $\mathbb{R}^N$, $N \geq 2$, with $\partial G \in C^\infty$. Let $a \in C^\infty(\bar{G})$.

(H2) The smallest eigenvalue μ_1 of the linearized eigenvalue problem

$$-\Delta u - a(x)u = \mu u \quad \text{in } G, \qquad u = 0 \quad \text{on } \partial G, \tag{199}$$

is positive. Recall that

$$\mu_1 = \inf_{u \in \mathring{W}_2^1(G)-\{0\}} \frac{\int_G (|Du|^2 - au^2)\,dx}{\|u\|_2^2}. \tag{199*}$$

In the case where $N \geq 3$, we also introduce the so-called *best Sobolev constant*

$$S = \inf_{u \in \mathring{W}_2^1(G)-\{0\}} \frac{\int_G |Du|^2\,dx}{\|u\|_{q_{\text{crit}}+1}^2}.$$

Note that the embedding $\mathring{W}_2^1(G) \subseteq L_{q_{\text{crit}}+1}(G)$ is continuous. Hence the number $S > 0$ is the smallest constant such that

$$\|u\|_{q_{\text{crit}}+1}^2 \leq S\|u\|_{1,2,0}^2 \qquad \text{for all} \quad u \in \mathring{W}_2^1(G).$$

Moreover, we define

$$J = \inf_{u \in \mathring{W}_2^1(G)-\{0\}} \frac{\int_G (|Du|^2 - au^2)\,dx}{\|u\|_{q_{\text{crit}}+1}^2}.$$

Finally, for $N = 3$, let K denote the Green function with respect to the linearized problem for (198), i.e., for each fixed $y \in G$, we have

$$K(x, y) = \frac{1}{4\pi|x - y|} + k(x, y),$$

where

$$-\Delta_x K - aK = \delta_y \quad \text{in } G,$$

$$K(x, y) = 0 \qquad \text{for all} \quad x \in \partial G.$$

Here, δ_y denotes Dirac's delta distribution at the point y (cf. $A_2(64)$). Observe that the function $k(\cdot, \cdot)$ is continuous on $G \times G$.

Remark (Necessary condition). Let $q \geq 1$. Then, under the assumption (H1) above, condition (H2) is necessary for the solvability of the original problem (198).

In fact, let u be a solution of (198), and let (u_1, μ_1) be an eigensolution of (199). Then $u > 0$ and $u_1 > 0$ on G. Multiplying (198) with u_1 and integrating by parts, we get

$$\int_G u^p u_1\,dx = \int_G (-\Delta u_1 - au_1)u\,dx = \mu_1 \int_G u_1 u\,dx,$$

and hence $\mu_1 > 0$.

Existence theorem. Show that, under the assumptions (H1) and (H2), the following hold:

(i) If $N \geq 2$ and $1 < q < q_{\text{crit}}$, then the original problem (198) has a solution.

(ii)* If $N = 3$ and $q = q_{\text{crit}}$, then (198) has a solution provided $k(x, x) > 0$ for some $x \in G$.

(iii)* If $N \geq 4$ and $q = q_{\text{crit}}$, then (198) has a solution provided $a(x) > 0$ for some $x \in G$.

Corollary. Assume (H1) and (H2). Let $N \geq 3$. Then:

(a) $0 < J \leq S$.

(b) The best Sobolev constant S is not achieved in any bounded region G. Moreover, S is independent of G; it depends only on N.

(c) If J is achieved, then the original problem (198) has a solution.

(d) $J < S \Rightarrow J$ is achieved.

(e) If $N = 3$, then:

$$k(x, x) > 0 \quad \text{for some} \quad x \in G \Rightarrow J < S.$$

(f) If $N \geq 4$, then:

$$a(x) > 0 \quad \text{for some} \quad x \in G \quad \Leftrightarrow \quad J < S \quad \Leftrightarrow \quad J \text{ is achieved.}$$

Hint: Cf. the survey article by Brézis (1986). The main idea is to solve the following constrained minimum problem:

$$\begin{aligned} \min f(u) &= J_q, \qquad u \in \mathring{W}_2^1(G), \\ g(u) &= 1, \end{aligned} \tag{200}$$

where

$$f(u) = \int_G (|Du|^2 - au^2)\,dx, \qquad g(u) = \int_G |u|^{q+1}\,dx.$$

Obviously,

$$J_q = \inf_{u \in \mathring{W}_2^1(G) - \{0\}} \frac{\int_G (|Du|^2 - au^2)\,dx}{\|u\|_{q+1}^2},$$

and $J = J_{q_{\text{crit}}}$.

Let us prove statement (i). Let $X = \mathring{W}_2^1(G)$.

Step 1. Let $N \geq 2$ and $1 < q < q_{\text{crit}}$. Suppose that J_q is achieved, i.e., the minimum problem (200) has a solution. We want to show that this implies the solvability of the original problem (198).

We first show that $u \mapsto f(u)^{1/2}$ is an equivalent norm on X. Since $\mu_1 > 0$,

$$f(u) \geq \mu_1 \|u\|_2^2 \qquad \text{for all} \quad u \in X.$$

For sufficiently small $\varepsilon > 0$ and all $u \in X$,

$$\varepsilon f(u) \geq \varepsilon \int_G |Du|^2\,dx - \varepsilon \|u\|_2^2 \max_{x \in \bar{G}} |a(x)|.$$

Thus, there is a constant $c > 0$ such that

$$f(u) \geq c \int_G (|Du|^2 + u^2)\,dx \qquad \text{for all} \quad u \in X.$$

We now show that $J_q > 0$. Otherwise, there would exist a sequence (u_n) in X such that

$$f(u_n) \to 0 \quad \text{as } n \to \infty \qquad \text{and} \qquad g(u_n) = 1 \quad \text{for all } n.$$

Hence $u_n \to 0$ in X as $n \to \infty$. Since the embedding $X \subseteq L_{q+1}(G)$ is *continuous*, we get $g(u_n) \to 0$ as $n \to \infty$. This is a contradiction.

Suppose now that u is a solution of (200). We may assume that $u \geq 0$ on $\bar{G}$. Otherwise, we pass from u to $|u|$, by using Problem 21.3g. According to the *Lagrange multiplier rule* (Proposition 43.6 from Part III), there exists a number $\mu > 0$ such that

$$f'(u)h = \mu g'(u)h \qquad \text{for all} \quad h \in X, \tag{201}$$

where

$$f'(u)h = 2\int_G (DuDh - auh)\,dx, \qquad g'(u)h = (q+1)\int_G u^q h\,dx.$$

In this connection, note that $g'(u)u > 0$. Letting $h = u$, it follows from (201) that $\mu > 0$. Furthermore, by (201), u is a generalized solution of the following boundary value problem:

$$-\Delta u - au = \tfrac{1}{2}\mu(q+1)u^q \quad \text{in } G, \qquad u = 0 \quad \text{on } \partial G. \tag{202}$$

By regularity theory, u is also a classical solution of (202), where $u \in C^\infty(\bar{G})$, since $a \in C^\infty(\bar{G})$. Furthermore, since $u \geq 0$ on $\bar{G}$ and $u = 0$ on ∂G, it follows from the strong maximum principle (Problem 7.2) that $u > 0$ in G. Otherwise, we would have $u = 0$ on $\bar{G}$ which contradicts $g(u) = 1$.

Consequently, it follows from (202) that the function tu is a solution of the original problem (198) if we choose an appropriate number $t > 0$.

Step 2. Again let $N \geq 2$ and $1 < q < q_{\text{crit}}$. By using a standard compactness argument, we want to show that the minimum problem (200) has a solution.

Let (u_n) be a minimal sequence in X corresponding to (200), i.e., we have

$$f(u_n) \to J_q \quad \text{as } n \to \infty \qquad \text{and} \qquad g(u_n) = 1 \quad \text{for all } n.$$

Since $u \mapsto f(u)^{1/2}$ is an equivalent norm on X, the sequence (u_n) is bounded in X. Thus, there exists a subsequence, again denoted by (u_n), such that

$$u_n \rightharpoonup u \quad \text{in } X \quad \text{as } u \to \infty \tag{203}$$

and

$$f(u) \leq \varliminf_{n \to \infty} f(u_n).$$

Since the embedding $X \subseteq L_{q+1}(G)$ is *compact* it follows from (203) that

$$g(u_n) \to g(u) \qquad \text{as} \quad n \to \infty.$$

Hence $f(u) = J_q$ and $g(u) = 1$, i.e., u is a solution of (200).

From Steps 1 and 2 we immediately obtain assertion (i).

In the critical case $N \geq 3$ and $q = q_{\text{crit}}$, Step 1 above remains valid, since the embedding $X \subseteq L_{q+1}(G)$ is continuous for $q = q_{\text{crit}}$. Consequently, in order to prove assertions (ii) and (iii) above, it is sufficient to show that J is achieved. In this connection, we refer to Brézis (1986, S).

References to the Literature

Classical works: Fredholm (1900) (integral equations), Smale (1965) (nonlinear Fredholm operators), Pohožaev (1967) (nonresonance case), Landesman and Lazer (1969), Ambrosetti and Prodi (1972) (global multiplicity theorem), Ambrosetti and Rabinowitz (1973) (variational methods).

Introduction: Nirenberg (1974, L), (1981, S), Berger (1977, M), Fučik and Kufner (1980, M), Fučik (1981, M).

Nonlinear Fredholm alternatives: Pohožaev (1967), Landesman and Lazer (1969), Kačurovskii (1970), Nirenberg (1970), (1974, L), Hess (1972), (1974a), Cronin (1973), Schechter (1973), (1978), Fučik, Nečas, and Souček (1973, L), Nečas (1973), (1983, L), Berger and Podolak (1974), (1975), Kazdan and Warner (1975), Petryshyn (1975, S), (1980), (1981), (1986, S), Browder (1978), (1986, P), Pascali and Sburlan (1978, M).

Nonlinear Fredholm alternatives for general semilinear elliptic equations: Nirenberg (1974, L) (recommended as an introduction), Nirenberg (1970), Schechter (1973), Berger and Schechter (1977).

Stable homotopy and nonlinear Fredholm alternatives: Nirenberg (1974, L) (recommended as an introduction), Nirenberg (1970), Cronin (1973), Berger and Podolak (1974), (1975), Berger (1977, M).

Stable homotopy and the topological classification of nonlinear Fredholm maps: Švarc (1964), Berger (1977, M), Booss and Bleecker (1985, M) (K-theory and the Atiyah–Jänich theorem).

Nonlinear Fredholm maps: Berger (1977, M), Borisovič (1977, S).

Bifurcation theory for potential operators: Böhme (1972), Rabinowitz (1974, S), (1977), Chow and Hale (1982, M) (see also the References to the Literature for Chapter 45).

Stability in nonlinear elasticity: Beckert (1976), (1984, S).

Semicoercive problems: Hess (1974a).

Variational methods for semilinear elliptic equations: Rabinowitz (1986, L).

Positivity methods for semilinear elliptic equations: Pohožaev (1965) (nonexistence theorem), Turner (1974), Amann (1976, S) and Lions (1982, S) (important survey articles), De Figueiredo, Lions, and Nussbaum (1982).

Critical Sobolev exponents and positive solutions of semilinear elliptic equations: Brézis (1986, S) (important survey article), Schoen (1984), Bahri and Coron (1988) (the impact of topology).

Radially symmetric solutions of semilinear elliptic equations: Gidas, Ni, and Nirenberg (1979).

Symmetrization and rearrangements: Kawohl (1985, L).

Existence theorems for quasi-linear elliptic and parabolic equations via subsolutions and supersolutions: Deuel and Hess (1975), Deuel (1978).

Combination of different methods: Hofer (1982).

Multiplicity results for semilinear elliptic equations: Ambrosetti (1983, S) and Ruf (1986, S) (recommended as an introduction), Ambrosetti and Prodi (1972), Ambrosetti and Rabinowitz (1973), Nirenberg (1974, L), Rabinowitz (1974, S), (1986, L), Kazdan and Warner (1975), Amann (1976, S), Hess (1981), Hofer (1982), Lions (1982, S), De Figueiredo and Solimini (1984), Berger (1985), Benkert (1985), (1989), Rabier (1986).

Range of sum operators and applications to semilinear equations: Brézis and Haraux (1976), Brézis and Nirenberg (1978).

Alternative problems, coincidence degree, and ordinary differential equations: Cesari (1964), (1976, S), Mawhin (1972), (1979, L), Gaines and Mawhin (1977, L, B, H).

(Cf. also the References to the Literature for Chapters 7 and 49 concerning semilinear differential equations.)

Nonlinear differential equations in differential geometry: Yau (1986, S) (fundamental survey article), Yau (1978) (complex Monge–Ampère equations and the Calabi problem), Schoen and Yau (1979) (positive energy theorem in general relativity), Sacks and Uhlenbeck (1981) (minimal immersions of 2-spheres), Hamilton (1982a); Taubes (1982), (1982a), (1986), Uhlenbeck (1982), as well as Freed and Uhlenbeck (1984, L) (Yang–Mills equations in gauge field theory); Schoen (1984), (1986, S) and Lee (1987, S) (solution of the Yamabe problem—conformal deformation of a Riemannian metric to a constant scalar curvature), Kazdan (1985, L) (prescribing the curvature of a Riemannian manifold), Hildebrandt (1984a, S), (1986, S) (minimal surfaces), (1985, S) (harmonic mappings between Riemannian manifolds), Wente (1986) (counterexample to the Hopf conjecture), Jost (1984, L), (1985, L), (1988, L), (1988a, S) (harmonic mappings and minimal surfaces).

Physics and geometry: Witten (1986, S) (string theory), Piran and Weinberg (1988, L) (superstrings), Atiyah and Hitchin (1988, L) (magnetic monopoles).

(Cf. also the References to the Literature for Chapter 85 (Riemannian geometry) and Chapter 96 (gauge field theory).)

GENERALIZATION TO NONLINEAR NONSTATIONARY PROBLEMS

The strides that have been made recently, in the theory of nonlinear partial differential equations, are as great as in the linear theory. Unlike the linear case, no wholesale liquidation of broad classes of problems has taken place; rather, it is steady progress on old fronts and on some new ones, the complete solution of some special problems, and the discovery of some brand new phenomena. The old tools—variational methods, fixed-point theorems, mapping degree, and other topological tools have been augmented by some new ones. Preeminent for discovering new phenomena is numerical experimentation; but it is likely that in the future numerical calculations will be parts of proofs.

Peter Lax (1983)

In Chapters 30 through 33 we want to apply the following methods to nonlinear evolution equations:

(i) the Galerkin method;
(ii) the method of nonlinear semigroups; and
(iii) the method of maximal monotone operators.

Chapter 32 is devoted to a detailed study of the properties of maximal monotone operators, which play a key role in the theory of monotone operators.

The study of abstract nonlinear evolution equations will be continued in Part V in connection with applications to important problems in mathematical physics.

CHAPTER 30

First-Order Evolution Equations and the Galerkin Method

Use several function spaces for the same problem.

The modern strategy for nonlinear partial differential equations

In this chapter, we generalize the Hilbert space methods in Chapter 23, for the investigation of linear parabolic differential equations, to nonlinear problems. In this connection, as an essential auxiliary tool, we use the Galerkin method.

30.1. Equivalent Formulations of First-Order Evolution Equations

The goal of this section is to present various equivalent formulations of first-order evolution equations. This is useful with a view to both abstract existence proofs and applications to nonlinear parabolic differential equations. Let "$V \subseteq H \subseteq V^*$" be an evolution triple.

30.1a. The Original Problem

We consider the following basic *initial value problem*:

$$u'(t) + A(t)u(t) = b(t) \qquad \text{for almost all} \quad t \in]0, T[, \tag{1a}$$

$$u(0) = u_0 \in H, \tag{1b}$$

$$u \in W_p^1(0, T; V, H), \qquad 1 < p < \infty. \tag{1c}$$

For each $t \in]0, T[$, let $b(t) \in V^*$ and let $A(t)$ be an operator of the form

$$A(t): V \to V^*.$$

For given $u_0 \in H$, we seek a function $u: [0, T] \to V$ such that (1) is satisfied. Recall, from Section 23.6, that condition (1c) means that

$$u \in L_p(0, T; V), \qquad u' \in L_q(0, T; V^*), \qquad p^{-1} + q^{-1} = 1,$$

where the derivative u' is to be understood in the generalized sense.

The initial condition (1b) is meaningful because it follows from Proposition 23.23 that the embedding $W_p^1(0, T; V, H) \subseteq C([0, T], H)$ is continuous. More precisely, after a modification of the function $u \in W_p^1(0, T; V, H)$ on a subset of $[0, T]$ of measure zero, if necessary, we obtain a uniquely determined continuous function $u: [0, T] \to H$. Condition (1b) is to be understood in this sense.

30.1b. The Functional Equation

Equation (1a) is equivalent to

$$\langle u'(t), v\rangle_V + \langle A(t)u(t), v\rangle_V = \langle b(t), v\rangle_V, \tag{2}$$

for all $v \in V$ and almost all $t \in]0, T[$. To be precise, we assume that there exists a subset Z of $]0, T[$ of measure zero such that (2) is valid for all $v \in V$ and for all $t \in]0, T[- Z$. Note that Z is independent of v.

For all $u, v \in V$, $t \in]0, T[$, we set

$$a(t; u, v) = \langle A(t)u, v\rangle_V, \qquad b(t; v) = \langle b(t), v\rangle_V. \tag{3}$$

By Proposition 23.20, the original problem (1) is *equivalent* to the following problem. We seek a function u such that, for all $v \in V$ and almost all $t \in]0, T[$,

$$\frac{d}{dt}(u(t)|v)_H + a(t; u, v) = b(t, v), \tag{4a}$$

$$u(0) = u_0 \in H, \tag{4b}$$

$$u \in W_p^1(0, T; V, H), \qquad 1 < p < \infty. \tag{4c}$$

In (4a), d/dt is to be understood as the generalized derivative, i.e., to be precise, (4a) means

$$-\int_0^T (u(t)|v)_H \varphi'(t)\,dt + \int_0^T a(t; u(t), v)\varphi(t)\,dt$$
$$= \int_0^T b(t; v)\varphi(t)\,dt \qquad \text{for all} \quad \varphi \in C_0^\infty(0, T), \qquad v \in V.$$

One can easily carry over concrete nonlinear parabolic equations by means of integration by parts into problems of the form (4). We discuss this in Section

30.4. In this connection, there will result immediately $a(t; u, v)$ and $b(t, v)$ from the differential equation under consideration. By means of (3), one has then to construct $A(t)$ and $b(t)$. The explicit time-dependence of $a(\cdot)$ and hence of $A(\cdot)$ is due to the time-dependent coefficients, and $b(\cdot)$ results from the right member of the differential equation.

30.1c. The Galerkin Equations

Let $\{w_1, w_2, \ldots\}$ be a basis in V. We set

$$u_n(t) = \sum_{k=1}^{n} c_{kn}(t) w_k.$$

Motivated by equation (2), we define the *Galerkin equations* in the following way. For almost all $t \in]0, T[$, we seek a function u_n such that

$$\langle u_n'(t), w_j \rangle_V + \langle A(t)u_n(t), w_j \rangle_V = \langle b(t), w_j \rangle_V, \qquad j = 1, \ldots, n, \tag{5a}$$

$$u_n(0) = u_{n0} \in H_n, \tag{5b}$$

$$u_n \in L_p(0, T; H_n), \qquad u_n' \in L_q(0, T; H_n), \tag{5c}$$

where $H_n = \operatorname{span}\{w_1, \ldots, w_n\}$, and the derivative u_n' is to be understood in the generalized sense.

Using the relations (3) and (4), the Galerkin equations (5) can be written in the following *explicit* form:

$$\sum_{k=1}^{n} c_{kn}'(t)(w_k | w_j)_H + a\left(t; \sum_{k=1}^{n} c_{kn}(t) w_k, w_j \right) = b(t, w_j), \tag{6a}$$

$$c_{jn}(0) = \alpha_{j0}, \qquad j = 1, \ldots, n, \tag{6b}$$

for almost all $t \in]0, T[$, where the initial values α_{j0} are given.

Since the $w_1, \ldots, w_n$ are linearly independent, we get

$$\det((w_k | w_j)_H) \neq 0, \qquad k, j = 1, \ldots, n.$$

Therefore, the system (6a) can be solved for the derivatives c_{kn}'. Hence, the Galerkin equations (6) represent a first-order system of *ordinary differential equations* for the real functions $t \mapsto c_{kn}(t)$ on $]0, T[$, where $k = 1, \ldots, n$.

In Section 30.4 we shall encounter the Galerkin method in the form (6), in connection with nonlinear parabolic equations. However, the coordinate-free form (5) is more convenient for the proof of convergence for the Galerkin method.

30.1d. The Final Operator Equation

We set

$$X = L_p(0, T; V),$$

and hence $X^* = L_q(0, T; V^*)$, by Section 23.3. Moreover, we set

$$(Au)(t) = A(t)u(t) \qquad \text{for all} \quad t \in]0, T[.$$

In Theorem 30.A below, we shall show that this defines an operator of the form

$$A: X \to X^*,$$

and the original problem (1) is *equivalent* to the following operator equation:

$$u' + Au = b, \tag{7a}$$

$$u(0) = u_0, \tag{7b}$$

$$u \in X, \qquad u' \in X^*, \tag{7c}$$

where $u_0 \in H$ and $b \in X^*$ are given. Recall from Section 23.6 that (7c) is equivalent to $u \in W_p^1(0, T; V, H)$.

Problem (7) is undoubtedly the most elegant form of a first-order evolution equation. However, for *concrete* parabolic differential equations, one arrives at (7) only via (1)–(4). For that reason, we have devoted so much attention to the various formulations of the original problem (1).

30.2. The Main Theorem on Monotone First-Order Evolution Equations

We summarize our assumptions:

(H1) *Evolution triple.* Let "$V \subseteq H \subseteq V^*$" be an evolution triple with $\dim V = \infty$, and let $\{w_1, w_2, \ldots\}$ be a basis in V. We set

$$H_n = \operatorname{span}\{w_1, \ldots, w_n\}$$

and introduce on the n-dimensional space H_n the scalar product of the H-space H. Note that $H_n \subseteq V \subseteq H$.

Let the sequence (u_{n0}) be an approximation of the given initial value $u_0 \in H$, i.e., suppose that

$$u_{n0} \to u_0 \quad \text{in } H \qquad \text{as} \quad n \to \infty.$$

Let $0 < T < \infty$, $1 < p < \infty$, and $p^{-1} + q^{-1} = 1$.

(H2) *Monotonicity.* For each $t \in]0, T[$, the operator

$$A(t): V \to V^*$$

is monotone and hemicontinuous.

(H3) *Coerciveness.* For each $t \in]0, T[$, the operator $A(t)$ is coercive, i.e., more precisely, there exist constants $c_1 > 0$ and $c_2 \geq 0$ such that

$$\langle A(t)v, v\rangle_V \geq c_1 \|v\|_V^p - c_2 \qquad \text{for all} \quad v \in V, \quad t \in]0, T[.$$

(H4) *Growth condition.* There exist a nonnegative function $c_3 \in L_q(0, T)$ and a constant $c_4 > 0$ such that

$$\|A(t)v\|_{V^*} \leq c_3(t) + c_4\|v\|_V^{p-1} \qquad \text{for all} \quad v \in V, \quad t \in]0, T[.$$

(H5) *Measurability.* The function $t \mapsto A(t)$ is weakly measurable, i.e., the function

$$t \mapsto \langle A(t)u, v\rangle_V$$

is measurable on $]0, T[$, for all $u, v \in V$.

(H6) Let $u_0 \in H$ and $b \in L_q(0, T; V^*)$ be given.

In Section 30.4 we shall show that the assumptions in the present form can easily be verified for nonlinear parabolic equations.

Theorem 30.A. *Assume* (H1)–(H6). *Then*:

(a) Existence and uniqueness. *The original initial value problem* (1) *has a unique solution u.*

(b) Convergence of the Galerkin method. *For each* $n = 1, 2, \ldots,$ *the Galerkin equation* (5) *has a unique solution* u_n. *The sequence* (u_n) *converges to the solution u of* (1) *in the following sense*:

$$u_n \rightharpoonup u \quad \text{in } L_p(0, T; V) \qquad \text{as} \quad n \to \infty, \tag{8}$$

$$\max_{0 \leq t \leq T} \|u_n(t) - u(t)\|_H \to 0 \qquad \text{as} \quad n \to \infty. \tag{9}$$

(c) Equivalent operator equation. *Let* $X = L_p(0, T; V)$. *If we set*

$$(Au)(t) = A(t)u(t) \qquad \text{for all} \quad t \in]0, T[,$$

then the operator $A: X \to X^*$ *is monotone, hemicontinuous, coercive, and bounded. The original initial value problem* (1) *is equivalent to the operator equation* (7).

30.3. Proof of the Main Theorem

If a man's wit be wandering, let him study mathematics, for in demonstration, if his wit be called away ever so little he must begin again.

Lord Bacon (1561–1626)

We first summarize several auxiliary tools. In order to simplify the writing of formulas, we agree to use the following abbreviations:

$$\langle u, v\rangle = \langle u, v\rangle_V, \qquad \|u\| = \|u\|_V,$$

$$(u|v) = (u|v)_H, \qquad |u| = \|u\|_H.$$

By (23.25), the integration by parts formula

$$(u(t)|v(t)) - (u(0)|v(0)) = \int_0^t \langle u'(s), v(s)\rangle + \langle v'(s), u(s)\rangle\, ds \tag{10}$$

holds for all $u, v \in W_p^1(0, T; V, H)$ and for all $t \in [0, T]$. This formula forms the decisive *basis* for our proof together with the *monotonicity* trick in Lemma 30.6 below.

Furthermore, for all $u \in X^*$, $v \in X$, and $0 \leq t \leq T$, we have that

$$\begin{aligned} \|u\|_{X^*}^q &= \int_0^T \|u(t)\|_{V^*}^q\, dt, \qquad \|v\|_X^p = \int_0^T \|v(t)\|^p\, dt, \\ \langle u, v\rangle_X &= \int_0^T \langle u(t), v(t)\rangle\, dt \end{aligned} \tag{11}$$

and

$$\begin{aligned} \left|\int_0^t \langle u(s), v(s)\rangle\, ds\right| &\leq \int_0^t |\langle u(s), v(s)\rangle|\, ds \\ &\leq \|u\|_{X^*}\|v\|_X, \end{aligned} \tag{12}$$

where (12) follows from the Hölder inequality. In the sense of the identification of Convention 23.14, we obtain that, for all $u \in H$, $v \in V$,

$$\begin{aligned} &\langle u, v\rangle = (u|v), \\ &\|v\|_{V^*} \leq \text{const}|v| \leq \text{const}\|v\|. \end{aligned} \tag{13}$$

Lemma 30.1. *Suppose that either*

$$u_n \to u \quad \textit{in } X^* \qquad \textit{and} \qquad v_n \rightharpoonup v \quad \textit{in } X \qquad \textit{as} \quad n \to \infty,$$

or

$$u_n \rightharpoonup u \quad \textit{in } X^* \qquad \textit{and} \qquad v_n \to v \quad \textit{in } X \qquad \textit{as} \quad n \to \infty.$$

Then, for all $t \in [0, T]$, as $n \to \infty$:

$$\int_0^t \langle u_n(s), v_n(s)\rangle\, ds \to \int_0^t \langle u(s), v(s)\rangle\, ds. \tag{14}$$

This follows from Proposition 23.9. Furthermore, for each $n \in \mathbb{N}$,

$$H_n \subseteq V \subseteq H \subseteq V^*,$$

and

$$\|v\|, \|v\|_{V^*} \leq c|v| \qquad \text{for all} \quad v \in H_n \quad \text{and fixed} \quad c > 0, \tag{15}$$

since $\dim H_n < \infty$, and all norms are equivalent on finite-dimensional B-spaces. Therefore, it follows from

$$u_n \in L_p(0, T; H_n), \qquad u_n' \in L_q(0, T; H_n),$$

that

$$u_n \in L_p(0, T; V), \qquad u_n' \in L_q(0, T; V^*), \tag{16}$$

and hence $u_n \in W_p^1(0, T; V, H)$.

30.3a. Proof of Uniqueness

We first prove the uniqueness of the solution u of the original problem (1). Let $u, v \in W_p^1(0, T; V, H)$ be two solutions of (1). Using integration by parts and the *monotonicity* of the operator $A(s)$ for all $s \in]0, T[$, we obtain that, for all $t \in [0, T]$,

$$|u(t)-v(t)|^2 - |u(0)-v(0)|^2 = 2\int_0^t \langle u'(s)-v'(s), u(s)-v(s)\rangle\, ds$$

$$= -2\int_0^t \langle A(s)u(s)-A(s)v(s), u(s)-v(s)\rangle\, ds \leq 0.$$

Thus, from $u(0) = v(0)$ we get $u(t) = v(t)$ for all $t \in [0, T]$.

We now prove the uniqueness of the solution u_n of the Galerkin equation (5). To this end, let u_n and v_n be two solutions of (5). Then, $u_n, v_n \in W_p^1(0, T; V, H)$, and

$$\langle u_n'(t) - v_n'(t), u_n(t) - v_n(t)\rangle = -\langle A(t)u_n(t) - A(t)v_n(t), u_n(t) - v_n(t)\rangle.$$

As above, this implies $u_n(t) = v_n(t)$ for all $t \in [0, T]$. □

30.3b. The Equivalent Operator Equation

We want to prove Theorem 30.A(c). Recall that

$$X = L_p(0, T; V), \qquad X^* = L_q(0, T; V^*).$$

For each $u \in X$, we set

$$(Au)(t) = A(t)u(t) \qquad \text{for all} \quad t \in]0, T[.$$

Lemma 30.2. *For all $u \in X$, $v \in V$, the real function*

$$t \mapsto \langle A(t)u(t), v\rangle$$

is measurable on $]0, T[$.

The proof will be given in Problem 30.1. Since "$V \subseteq H \subseteq V^*$" is an evolution triple, the B-space V is reflexive. Therefore, Lemma 30.2 and the Pettis theorem $A_2(10)$ imply that, for each $u \in X$, the function

$$t \mapsto A(t)u(t)$$

is *measurable* from $]0, T[$ to V^*.

(I) *Boundedness*. We show that the operator $A: X \to X^*$ is bounded. Let $u \in X$. Noting $p/q = p - 1$ and the inequality $A_2(30b)$, it follows from the growth condition (H4) that

$$\|A(t)u(t)\|_{V^*}^q \le \text{const}(|c_3(t)|^q + \|u(t)\|^p) \qquad \text{for all} \quad t \in]0, T[. \quad (17)$$

By $A_2(9)$, the real function $t \mapsto \|A(t)u(t)\|_{V^*}^q$ is measurable on $]0, T[$. Since $c_3 \in L_q(0, T)$ and $u \in L_p(0, T; V)$, the function on the right-hand side of (17) is integrable over $]0, T[$. By integration, we obtain from (17) that

$$\|Au\|_{X^*} \le \text{const}(\|c_3\|_q + \|u\|_X^{p/q}) \qquad \text{for all} \quad u \in X,$$

recalling that $\|Au\|_{X^*}^q = \int_0^T \|A(t)u(t)\|_{V^*}^q \, dt$.

(II) *Monotonicity*. We show that $A: X \to X^*$ is monotone. Let $u, v \in X$. Since $Au \in X^*$, the Hölder inequality (Proposition 23.6) tells us that the real function

$$t \mapsto \langle A(t)u(t), v(t) \rangle$$

is integrable over $]0, T[$. From the monotonicity of $A(t): V \to V^*$ for each $t \in]0, T[$, it now follows immediately that

$$\langle Au - Av, u - v \rangle_X = \int_0^T \langle A(t)u(t) - A(t)v(t), u(t) - v(t) \rangle \, dt \ge 0,$$

for all $u, v \in X$.

(III) *Coerciveness*. The operator $A: X \to X^*$ is coercive, since

$$\begin{aligned} \langle Au, u \rangle_X &= \int_0^T \langle A(t)u(t), u(t) \rangle \, dt \\ &\ge c_1 \|u\|_X^p - c_2 T \qquad \text{for all} \quad u \in X, \end{aligned} \quad (18)$$

by assumption (H3). This implies that

$$\langle Au, u \rangle_X / \|u\|_X \to \infty \qquad \text{as} \quad \|u\|_X \to \infty.$$

(IV) *Hemicontinuity*. Finally, we show that $A: X \to X^*$ is hemicontinuous. Let $u, v, w \in X$ and $0 \le \lambda, \mu \le 1$. For all $t \in]0, T[$, it follows from (17) and the inequality $A_2(30b)$ that

$$\begin{aligned} &|\langle A(t)(u(t) + \lambda v(t)), w(t) \rangle| \\ &\qquad \le \text{const}(|c_3(t)| + \|u(t) + \lambda v(t)\|^{p/q}) \|w(t)\| \le m(t), \end{aligned}$$

where

$$m(t) = \text{const}(|c_3(t)| + \|u(t)\|^{p/q} + \|v(t)\|^{p/q}) \|w(t)\|.$$

Because of $c_3 \in L_q(0, T)$, and hence $c_3 \in L_1(0, T)$, and because of $u, v, w \in L_p(0, T; V)$ it follows that

$$\|u(\cdot)\|^{p/q}, \|v(\cdot)\|^{p/q} \in L_q(0, T) \qquad \text{and} \qquad \|w(\cdot)\| \in L_p(0, T).$$

By the Hölder inequality, the majorant function $m(\cdot)$ belongs to $L_1(0, T)$.

Therefore, it follows from the principle of *majorized convergence* $A_2(19)$ that

$$\lim_{\lambda \to \mu} \langle A(u + \lambda v), w \rangle_X = \lim_{\lambda \to \mu} \int_0^T \langle A(t)(u(t) + \lambda v(t)), w(t) \rangle_X \, dt$$
$$= \langle A(u + \mu v), w \rangle_X,$$

which shows the hemicontinuity of A.

(V) *Equivalence.* The equivalence between the original initial value problem (1) and the operator equation (7) follows immediately from our considerations in Section 30.1.

The proof of Theorem 30.A(c) is complete. □

30.3c. Proof of Existence

Step 1: *A priori* estimates for the solutions u_n of the Galerkin equations (5).

Lemma 30.3. *There exists a constant* $C > 0$, *independent of* n, *such that, for all* $n \in \mathbb{N}$ *and all solutions* u_n *of* (5), *the following are valid*:

$$\|u_n\|_X \leq C, \tag{19}$$

$$\|Au_n\|_{X^*} \leq C, \tag{20}$$

$$\max_{0 \leq t \leq T} |u_n(t)| \leq C. \tag{21}$$

PROOF. Ad(19). Since $A(t)$ is monotone, we obtain that

$$\langle A(t)u_n(t) - A(t)(0), u_n(t) \rangle \geq 0 \qquad \text{for all} \quad t \in \left]0, T\right[.$$

Integration by parts yields

$$\begin{aligned} \tfrac{1}{2}(|u_n(t)|^2 - |u_n(0)|^2) &= \int_0^t \langle u_n'(s), u_n(s) \rangle \, ds \\ &= \int_0^t \langle b(s) - A(s)u_n(s), u_n(s) \rangle \, ds \qquad (22) \\ &\leq \int_0^t \langle b(s) - A(s)(0), u_n(s) \rangle \, ds \qquad \text{for all} \quad t \in \left]0, T\right[, \end{aligned}$$

noting (10) and $u_n \in W_p^1(0, T; V, H)$. Because $u_n(0) \to u_0$ in H as $n \to \infty$, there is a number a such that

$$\tfrac{1}{2}|u_n(0)|^2 \leq a \qquad \text{for all} \quad n \in \mathbb{N}.$$

Therefore, by (22),

$$-a \leq \tfrac{1}{2}(|u_n(T)|^2 - |u_n(0)|^2) = \langle b - Au_n, u_n \rangle_X.$$

By the coerciveness condition (18), this implies

$$-a \le \|b\|_{X^*}\|u_n\|_X - c_1\|u_n\|_X^p + Tc_2,$$

and hence we obtain the assertion (19), since $p > 1$ and the real function

$$g(\xi) = \|b\|_{X^*}\xi - c_1|\xi|^p + Tc_2$$

goes to $-\infty$ as $\xi \to +\infty$.

Ad(20). This follows from (19) and the boundedness of the operator $A: X \to X^*$.

Ad(21). Noting (22) and the Hölder inequality (12), we get

$$\tfrac{1}{2}(|u_n(t)|^2 - |u_n(0)|^2) \le \|b - A(0)\|_{X^*}\|u_n\|_X.$$

This implies (21), since $|u_n(0)|^2 \le 2a$ for all n. □

Lemma 30.4. *For each $n \in \mathbb{N}$, the Galerkin equation* (5) *has a unique solution u_n.*

This follows from the existence theorem of Carathéodory for ordinary differential equations in $\mathbb{R}^n$ (cf. $A_2(61)$). The details will be given in Problem 30.3.

Step 2: Weak convergence of a *subsequence* of the Galerkin sequence (u_n) in the space X to a solution u of the original problem (1).

The spaces X, X^*, and H are reflexive. Therefore, by the *a priori* estimates (19)–(21), there exists a subsequence, again denoted by (u_n), such that

$$u_n \rightharpoonup u \quad \text{in } X \qquad \text{and} \qquad Au_n \rightharpoonup w \quad \text{in } X^* \qquad \text{as} \quad n \to \infty, \tag{23}$$

$$u_n(T) \rightharpoonup z \quad \text{in } H \qquad \text{as} \quad n \to \infty. \tag{24}$$

Recall that, by assumption (H1),

$$u_n(0) \to u_0 \quad \text{in } H \qquad \text{as } n \to \infty.$$

Lemma 30.5. *The limit elements u, w, and z satisfy*:

$$u' + w = b, \qquad u \in W_p^1(0, T; V, H), \tag{25}$$

$$u(0) = u_0, \qquad u(T) = z. \tag{26}$$

Proof. The point of departure is the formula

$$(z|\psi(T)v) - (u_0|\psi(0)v) = \int_0^T \langle b(t) - w(t), \psi(t)v\rangle + \langle \psi'(t)v, u(t)\rangle\, dt, \tag{27}$$

for all $\psi \in C^\infty[0, T]$, $v \in V$, which we shall prove below, by using integration by parts and a limiting process.

Ad(25). As a special case of (27), we obtain that

$$\int_0^T \langle b(t) - w(t), v\rangle\psi(t)\, dt = -\int_0^T (u(t)|v)\psi'(t)\, dt, \tag{28}$$

for all $\psi \in C_0^\infty(0, T)$, $v \in V$. Note that $\langle v, u(t) \rangle = (v|u(t))$ by (13). According to Proposition 23.20, equation (28) tells us that the generalized derivative u' exists and $u' = b - w$. Since $b, w \in X^*$, we get $u' \in X^*$. Moreover, from $u \in X$ and $u' \in X^*$ it follows that $u \in W_p^1(0, T; V, H)$.

Ad(26). The integration by parts formula (10) yields

$$(u(T)|\psi(T)v) - (u(0)|\psi(0)v) = \int_0^T \langle u'(t), \psi(t)v \rangle + \langle \psi'(t)v, u(t) \rangle \, dt,$$

for all $\psi \in C^\infty[0, T]$, $v \in V$. Since $u' = b - w$, it follows from (27) that

$$(u(T)|\psi(T)v) - (u(0)|\psi(0)v) = (z|\psi(T)v) - (u_0|\psi(0)v) \quad \text{for all} \quad \psi \in C^\infty[0, T], \quad v \in V.$$

In particular, if we choose a function ψ with $\psi(T) = 1$ and $\psi(0) = 0$, then

$$(u(T) - z|v) = 0 \qquad \text{for all} \quad v \in V.$$

Since V is dense in H, this implies $u(T) = z$.

Analogously, if we choose ψ with $\psi(T) = 0$ and $\psi(0) = 1$, then we get $u(0) = u_0$.

Ad(27). First let $\psi \in C^\infty[0, T]$ and $v \in H_m$. Then $v \in W_p^1(0, T; V, H)$. For $n \geq m$, the integration by parts formula (10) yields

$$(u_n(T)|\psi(T)v) - (u_n(0)|\psi(0)v) = \int_0^T \langle u_n'(t), \psi(t)v \rangle + \langle \psi'(t)v, u_n(t) \rangle \, dt.$$

By the Galerkin equation (5),

$$\begin{aligned} (u_n(T)|\psi(T)v) - (u_n(0)|\psi(0)v) &= \int_0^T \langle b(t) - A(t)u_n(t), \psi(t)v \rangle + \langle \psi'(t)v, u_n(t) \rangle \, dt \\ &= \langle b - Au_n, \psi v \rangle_X + \langle \psi' v, u_n \rangle_X. \end{aligned}$$

Letting $n \to \infty$ and noting (23), (24), we obtain that

$$(z|\psi(T)v) - (u_0|\psi(0)v) = \langle b - w, \psi v \rangle_X + \langle \psi' v, u \rangle_X \qquad \text{for all} \quad v \in \bigcup_m H_m.$$

Since $\bigcup_m H_m$ is dense in V by (H1), we get (27) for all $v \in V$. In this connection, note the following. For each $v \in V$, there exists a sequence (v_k) in $\bigcup_m H_m$ such that

$$v_k \to v \quad \text{in } V \qquad \text{as} \quad n \to \infty.$$

This implies $\psi v_k \to \psi v$ in $X = L_p(0, T; V)$ and $\psi' v_k \to \psi' v$ in $X^* = L_q(0, T; V^*)$ as $k \to \infty$, since the embedding $V \subseteq V^*$ is continuous. □

Lemma 30.6. *The limit elements u and w from* (23) *satisfy the equation $Au = w$.*

PROOF. We will use a *typical* argument of the theory of *monotone operators*. By (23).

$$u_n \rightharpoonup u \quad \text{in } X \qquad \text{as} \quad n \to \infty,$$

$$Au_n \rightharpoonup w \quad \text{in } X^* \qquad \text{as} \quad n \to \infty.$$

Below we show that

$$\overline{\lim_{n\to\infty}} \, \langle Au_n, u_n \rangle_X \le \langle w, u \rangle_X. \tag{29}$$

The operator $A: X \to X^*$ is monotone and hemicontinuous. Therefore, it follows from the fundamental *monotonicity trick* (25.4) that

$$Au = w.$$

We now prove (29). The integration by parts formula (10) yields

$$\begin{aligned}\tfrac{1}{2}(|u_n(T)|^2 - |u_n(0)|^2) &= \int_0^T \langle u_n'(t), u_n(t) \rangle \, dt \\ &= \int_0^T \langle b(t) - A(t)u_n(t), u_n(t) \rangle \, dt,\end{aligned}$$

noting the Galerkin equation (5). Hence

$$\langle Au_n, u_n \rangle_X = \langle b, u_n \rangle_X + \tfrac{1}{2}(|u_n(0)|^2 - |u_n(T)|^2. \tag{30}$$

By Lemma 30.5,

$$u_n(0) \to u(0) \qquad \text{and} \qquad u_n(T) \rightharpoonup u(T) \quad \text{in } H \qquad \text{as} \quad n \to \infty,$$

and hence

$$|u(T)| \le \underline{\lim_{n\to\infty}} \, |u_n(T)|.$$

Since $u_n \rightharpoonup u$ in X as $n \to \infty$, $\langle b, u_n \rangle_X \to \langle b, u \rangle_X$. From (30) we get

$$\overline{\lim_{n\to\infty}} \, \langle Au_n, u_n \rangle_X \le \langle b, u \rangle_X + \tfrac{1}{2}(|u(0)|^2 - |u(T)|^2). \tag{31}$$

Finally, integration by parts yields

$$\begin{aligned}\tfrac{1}{2}(|u(0)|^2 - |u(T)|^2) &= -\int_0^T \langle u'(t), u(t) \rangle \, dt \\ &= -\langle u', u \rangle_X = \langle w - b, u \rangle_X,\end{aligned} \tag{32}$$

by Lemma 30.5. From (31) and (32) we get (29). □

From Lemmas 30.5 and 30.6 we obtain that the limit element u in (23) satisfies the following operator equation:

$$u' + Au = b, \qquad u \in W_p^1(0, T; V, H),$$

$$u(0) = u_0.$$

This equation is equivalent to the original equation (1).

Therefore, the proof of Theorem 30.A(a) is complete. □

Step 3: Weak convergence of the *total* Galerkin sequence (u_n) in the space $X = L_p(0, T; V)$ to the solution u of (1).

According to Step 2, it always follows from the convergence

$$u_{n'} \rightharpoonup v \quad \text{in } X \qquad \text{as} \quad n \to \infty,$$

of an arbitrary subsequence of (u_n) that v is a solution of the original problem (1). Since this solution is uniquely determined, we get $v = u$. By Proposition 21.23(i), the total sequence (u_n) converges, i.e.,

$$u_n \rightharpoonup u \quad \text{in } X \qquad \text{as} \quad n \to \infty.$$

Step 4: Convergence of (u_n) in the space $C([0, T]; H)$.

Lemma 30.7. *As* $n \to \infty$, $\max_{0 \le t \le T} |u_n(t) - u(t)| \to 0$.

The proof of this lemma will be given in Problem 30.4. This finishes the proof of Theorem 30.A. □

30.4. Application to Quasi-Linear Parabolic Differential Equations of Order $2m$

Let $Q_T = G \times]0, T[$. We consider the following initial–boundary value problem:

$$\begin{gathered} u_t(x,t) + L(t)u(x,t) = f(x,t) \quad \text{on } Q_T, \\ D^\beta u(x,t) = 0 \quad \text{on } \partial G \times [0,T] \qquad \text{for all} \quad \beta\colon |\beta| \le m-1, \\ u(x,0) = u_0(x) \quad \text{on } G, \end{gathered} \tag{33}$$

where

$$L(t)u(x,t) = \sum_{|\alpha| \le m} (-1)^{|\alpha|} D^\alpha A_\alpha(x, Du(x,t); t)$$

and $Du = (D^\beta u)_{|\beta| \le m}$. Note that the derivatives D^α and D^β refer to the spatial variable x. Thus, for each fixed time t, the differential operator $L(t)$ is of order $2m$ with respect to x.

Our goal is to prove the existence of a generalized solution for (33). In Section 26.5, we proved the existence of generalized solutions for quasi-linear elliptic equations. Roughly speaking, we will suppose in this section that, for each fixed time t, the operator $L(t)$ is a quasi-linear elliptic operator of the form considered in Section 26.5. To be precise, we assume:

(A1) Let G be a bounded region in $\mathbb{R}^N$, $N \ge 1$. We set

$$V = \mathring{W}_p^m(G), \qquad H = L_2(G), \qquad 2 \le p < \infty.$$

Moreover, let $0 < T < \infty$, $m = 1, 2, \ldots, p^{-1} + q^{-1} = 1$.

(A2) For each fixed $t \in]0, T[$, the differential operator $L(t)$ satisfies the assumptions (H1) through (H4) of Proposition 26.12, where we assume that all the constants and the real functions g and h that appear in (H1) through (H4) are *independent* of t.

Recall that (H1) through (H4) concern the Carathéodory condition, the growth condition, the monotonicity condition, and the coerciveness condition for the functions A_α.

(A3) The real function

$$t \mapsto a(t; v, w),$$

introduced in Definition 30.9 below, is measurable on $]0, T[$, for all v, $w \in V$.

This weak assumption means that the functions A_α depend on time t in a reasonable manner.

Because of the growth condition (H2) assumed in (A2), the assumption (A3) is satisfied if all the functions A_α are measurable, that is, for each α, the real function

$$(x, D, t) \mapsto A_\alpha(x, D, t)$$

is measurable on $G \times \mathbb{R}^d \times]0, T[$.

(A4) Let $f \in L_q(Q_T)$ and $u_0 \in L_2(G)$ be given.

(A5) Let $\{w_1, w_2, \ldots\}$ be a basis in $V = \mathring{W}_p^m(G)$. We set

$$H_n = \operatorname{span}\{w_1, \ldots, w_n\}.$$

Moreover, let (u_{n0}) be a sequence in $H = L_2(G)$ such that $u_{n0} \to u_0$ in H as $n \to \infty$, i.e.,

$$\int_G (u_{n0}(x) - u_0(x))^2 \, dx \to 0 \qquad \text{as} \quad n \to \infty.$$

EXAMPLE 30.8. In the case where $m = 1$, the assumptions (A1)–(A3) are satisfied for the following generalized heat equation:

$$\begin{aligned} u_t - \sum_{i=1}^N D_i(|D_i u|^{p-2} D_i u) &= f \quad \text{on } Q_T, \\ u &= 0 \quad \text{on } \partial G \times [0, T], \\ u(x, 0) &= u_0(x) \quad \text{on } G, \end{aligned} \tag{34}$$

where $x = (\xi_1, \ldots, \xi_N)$ and $D_i = \partial/\partial\xi_i$. By Definition 30.9 below,

$$a(t; w, v) = \int_G \sum_{i=1}^N |D_i w(x)|^{p-2} D_i w(x) D_i v(x) \, dx,$$

for all v, $w \in V$. Here, $a(\cdot)$ is independent of t.

PROOF. This follows from the proof of Proposition 26.10. □

Definition 30.9. The *generalized problem* associated with (33) reads as follows. Let $f \in L_q(Q_T)$ and $u_0 \in L_2(G)$ be given. We seek a function $u \in W_p^1(0, T; V, H)$

such that, for all $v \in V$ and almost all $t \in]0, T[$,

$$\frac{d}{dt}(u(t)|v)_H + a(t; u(t), v) = b(t; v),$$
$$u(0) = u_0, \tag{35}$$

where

$$a(t; w, v) = \int_G \sum_{|\alpha| \le m} A_\alpha(x, Dw(x); t) D^\alpha v(x)\, dx,$$

$$b(t; v) = \int_G f(x, t) v(x)\, dx \qquad \text{for all} \quad v, w \in V, \quad t \in]0, T[.$$

Recall that

$$(u(t)|v)_H = \int_G u(x, t) v(x)\, dx.$$

In (35), d/dt denotes the generalized derivative on $]0, T[$. Therefore, equation (35) means explicitly

$$-\int_0^T (u(t)|v)_H \varphi'(t)\, dt + \int_0^T a(t; u(t), v)\varphi(t)\, dt$$
$$= \int_0^T b(t; v)\varphi(t)\, dt \qquad \text{for all} \quad \varphi \in C_0^\infty(0, T).$$

Formally, one *obtains* the generalized problem (35) by multiplying the original problem (33) by the function $v \in C_0^\infty(G)$ and then integrating by parts with respect to the spatial variable x.

The *Galerkin method* for the original problem (33) follows immediately from the generalized problem (35). We seek a function

$$u_n(t) = \sum_{k=1}^n c_{kn}(t) w_k$$

such that, for almost all $t \in]0, T[$,

$$\frac{d}{dt}(u_n(t)|w_j)_H + a(t; u_n(t), w_j) = b(t; w_j), \qquad j = 1, \ldots, n,$$
$$u_n(0) = u_{n0}, \tag{36}$$
$$u_n \in L_p(0, T; H_n), \qquad u_n' \in L_q(0, T; H_n).$$

If we set

$$u_{n0} = \sum_{k=1}^n \alpha_{kn} w_k.$$

then (36) represents a first-order system of *ordinary differential equations* for the unknown real functions $t \mapsto c_{kn}(t)$ with the initial condition $c_{kn}(0) = \alpha_{kn}$ for $k = 1, \ldots, n$.

Proposition 30.10 (Solution of the Initial–Boundary Value Problem (33)). *Assume* (A1)–(A5). *Then the generalized problem* (35) *corresponding to the original problem* (33) *is equivalent to equation* (1), *and all the assumptions of Theorem* 30.A *are fulfilled. Therefore, all the assertions of Theorem* 30.A *are also valid. In particular, the following hold*:

(a) *Problem* (35) *has a unique solution* u.
(b) *For each* $n \in \mathbb{N}$, *the Galerkin equation* (36) *has a unique solution* u_n, *and* (u_n) *converges to* u *in the sense of*

$$\max_{0 \le t \le T} \int_G (u_n(x,t) - u(x,t))^2 \, dx \to 0 \qquad \text{as} \quad n \to \infty.$$

Corollary 30.11 (Properties of the Solution u). *By Proposition* 23.23, *the solution* $u \in W_p^1(0, T; V, H)$ *belongs to the space* $C([0, T]; H)$, *that is,*

$$\lim_{t \to s} \int_G (u(x,t) - u(x,s))^2 \, dx = 0 \qquad \text{for each} \quad s \in [0, T].$$

Therefore, the function $t \mapsto u(x,t)$ *is continuous on* $[0, T]$ *in the mean.*

By the construction of the space $W_p^1(0, T; V, H)$, *the function* $x \mapsto u(x,t)$ *belongs to the Sobolev space* $V = \mathring{W}_p^m(G)$, *for almost all* $t \in]0, T[$, *i.e., this function has generalized derivatives up to order* m *with respect to the spatial variable* x.

Proof of Proposition 30.10. As in the proof of Proposition 26.12, for each $t \in]0, T[$, there exists an operator $A(t): V \to V^*$ such that

$$\langle A(t)w, v \rangle_V = a(t; w, v) \qquad \text{for all} \quad v, w \in V.$$

By assumptions (A2), (A3), and the proof of Proposition 26.12, the assumptions regarding $A(t)$ in Theorem 30.A are fulfilled.

As in Example 23.4, because $f \in L_q(Q_T)$ there results the existence of a function $b \in L_q(0, T; V^*)$ such that

$$\langle b(t), v \rangle_V = b(t; v) \qquad \text{for all} \quad v \in V, \quad t \in]0, T[.$$

It follows from Section 30.1b that problem (35) is identical to (4), and hence (35) is equivalent to (1). □

Corollary 30.12. *Let* $X = L_p(0, T; V)$. *By Theorem* 30.A, *the generalized problem* (35) *is equivalent to the operator equation*

$$u' + Au = b, \qquad u \in W_p^1(0, T; V, H),$$
$$u(0) = u_0,$$

where the operator $A: X \to X^*$ *is monotone, hemicontinuous, coercive, and bounded.*

In particular, all the results stated above hold true for problem (34) in Example 30.8. In the special case of Example 30.8, the operator $A: X \to X^*$ is also *uniformly monotone*. In fact, by Proposition 26.10, we have

$$a(u, u - v) - a(v, u - v) \geq c\|u - v\|_V^p \qquad \text{for all} \quad u, v \in V,$$

and thus

$$\begin{aligned} \langle Au - Av, u - v \rangle_X &= \int_0^T \langle A(t)u(t) - A(t)v(t), u(t) - v(t) \rangle \, dt \\ &= \int_0^T a(u(t), u(t) - v(t)) - a(v(t), u(t) - v(t)) \, dt \\ &\geq \int_0^T c\|u(t) - v(t)\|_V^p \, dt \\ &= c\|u - v\|_X^p \qquad \text{for all} \quad u, v \in X. \end{aligned}$$

30.5. The Main Theorem on Semibounded Nonlinear Evolution Equations

We want to study the nonlinear evolution equation

$$\begin{gathered} u'(t) + A(u(t), t) = 0 \qquad \text{for all} \quad t \in [0, T], \\ u(0) = u_0 \in H, \end{gathered} \tag{37}$$

where $0 < T < \infty$. Parallel to (37) we consider the problem

$$\frac{d}{dt}(u(t)|v)_H + \langle A(u(t), t), v \rangle = 0 \quad \text{on } [0, T] \tag{37*}$$

for all $v \in V$, and $u(0) = u_0$.

The point is that the operator

$$A: H \times [0, T_0] \to V^+$$

satisfies the *semiboundedness* condition

$$\langle A(v, t), v \rangle \geq -c(\|v\|_H^2) \tag{38}$$

for all $(v, t) \in V \times [0, T_0]$. With respect to applications to nonlinear partial differential equations, it is important that we use three different B-spaces V, H, V^+ with

$$V \subseteq H \subseteq V^+.$$

In order to obtain a great flexibility in applications, we do not assume that V^+ is the dual space V^* to V. We only assume that $\{V, V^+\}$ forms a *dual pair* in the sense of Section 27.6. The following Theorem 30.B is closely related to

Theorem 27.B (the main theorem on locally coercive operators), where we also made use of dual pairs. Recall that dual pairs also play an important role in formulating general Fredholm alternatives. This can be found in the Appendix to Part I (see $A_1(53)$). The simple idea of proof of Theorem 30.B below is to use a Galerkin method for (37) in the "very nice" space V. The point is that we obtain *a priori* estimates for the Galerkin equations only in the "nice" space H. Thus, a subsequence (u_n) of the Galerkin solutions converges weakly in H and the properties of A imply that the sequence $(A(u_n(t), t))$ converges weakly in the "bad" space V^+.

Definition 30.13. We understand an *admissible triplet* "$V \subseteq H \subseteq V^+$" to be the following:

(i) H is a real separable H-space with scalar product $(\cdot|\cdot)_H$.
(ii) $\{V, V^+\}$ is a dual pair of real separable B-spaces with the corresponding bilinear form $\langle \cdot, \cdot \rangle$.
(iii) The embeddings $V \subseteq H \subseteq V^+$ are continuous and dense.
(iv) For all $h \in H$, $v \in V$,

$$\langle h, v \rangle = (h|v)_H. \tag{39}$$

EXAMPLE 30.14. Each evolution triple is an admissible triplet. This follows from Problem 24.1.

Let $\{V, V^+\}$ be a dual pair. By definition, the *weak$^+$ convergence*

$$w_n \overset{+}{\rightharpoonup} w \quad \text{in } V^+ \qquad \text{as} \quad n \to \infty$$

means

$$\langle w_n, v \rangle \to \langle w, v \rangle \qquad \text{as} \quad n \to \infty \qquad \text{for all} \quad v \in V.$$

If V^+ is equal to the dual space V^* of V, then weak$^+$ convergence is identical to weak* convergence. The weak$^+$ limit on V^+ is unique, since $\langle w, v \rangle = 0$ for all $v \in V$ implies $w = 0$.

Definition 30.15. The function $u\colon [0, T] \to V^+$ has the *weak$^+$-derivative* $u'(s) = w$ at the point s iff

$$\frac{u(s+h) - u(s)}{h} \overset{+}{\rightharpoonup} w \qquad \text{as} \quad h \to 0.$$

For $s = 0$ and $s = T$, this is to be understood as a one-sided derivative.

This is equivalent to

$$\frac{d}{dt}\langle u(t), v \rangle|_{t=s} = \langle w, v \rangle \qquad \text{for all} \quad v \in V.$$

Let Y be a B-space. Then $C_w([0, T], Y)$ denotes the set of all functions $u\colon [0, T] \to Y$, which are *weakly continuous*, i.e., $t \mapsto \langle y^*, u(t) \rangle$ is continuous on $[0, T]$ for all $y^* \in Y^*$.

We make the following assumptions:

(H1) "$V \subseteq H \subseteq V^+$" is an admissible triplet.

(H2) For fixed $T_0 > 0$, the operator $A: H \times [0, T_0] \to V^+$ is weakly sequentially continuous, i.e.,

$$u_n \rightharpoonup u \quad \text{in } H \qquad \text{and} \qquad t_n \to t \quad \text{in } [0, T_0] \qquad \text{as} \quad n \to \infty$$

implies

$$A(u_n, t_n) \rightharpoonup A(u, t) \quad \text{in } V^+ \qquad \text{as} \quad n \to \infty.$$

(H3) The operator A is *semibounded*, i.e., there exists a monotone increasing C^1-function $c: \mathbb{R}_+ \to \mathbb{R}_+$ such that the *key* condition (38) above holds true.

(H4) Let $u_0 \in H$ be given.

Theorem 30.B (Kato and Lai (1984)). *Assume* (H1) *through* (H4). *Then there is a* $T > 0$ *such that the original initial value problem* (37) *has a solution* u *with*

$$u \in C_w([0, T], H), \qquad u' \in C_w([0, T], V^+), \tag{40}$$

where u' *denotes a weak*$^+$*-derivative. Moreover, we have* $u(t) \to u_0$ *in* H *as* $t \to +0$.

At the same time, u *is also a solution of* (37*).

Corollary 30.16 (Majorization). *For fixed* $T_1 \in]0, T_0]$, *let* $g: [0, T_1] \to \mathbb{R}_+$ *be a solution of the ordinary differential equation*

$$\begin{aligned} g'(t) &= 2c(g(t)) \quad \textit{on } [0, T_1], \\ g(0) &= \|u_0\|_H^2. \end{aligned} \tag{41}$$

Then Theorem 30.B *holds with* $T = T_1$ *and we obtain*

$$\|u(t)\|_H^2 \le g(t) \quad \textit{on } [0, T] \tag{42}$$

for the solution u *of* (37).

We now consider the *reversible* case, i.e., we assume that the operator A in (37) does *not* depend on time t and the semiboundedness condition (38) is replaced with the following stronger condition:

$$|\langle Av, v\rangle| \le c(\|v\|_H^2) \qquad \text{for all} \quad v \in V. \tag{43}$$

More precisely, we replace (H1)–(H4) with the following assumptions.

(H1*) "$V \subseteq H \subseteq V^+$" is an admissible triplet.

(H2*) The operator $A: H \to V^+$ is weakly sequentially continuous and continuous.

(H3*) There exists a monotone increasing C^1-function $c: \mathbb{R}_+ \to \mathbb{R}_+$ such that the *key* condition (43) above holds true.

(H4*) Uniqueness. For each $u_0 \in H$, there is a $T_0 > 0$ depending on u_0 such that problem (37) has at most one solution u with (40) and $T = T_0$.

Furthermore, we assume that the same uniqueness assumption is valid for the modified problem (37) which is obtained from (37) by replacing the operator A with $-A$.

Corollary 30.17 (Uniqueness and Regularity of the Solution). *Assume* (H1*) *through* (H4*). *Let $u_0 \in H$ be given. Then there is a $T > 0$ such that the original initial value problem* (37) *has a unique solution*

$$u \in C^1([0, T], V^+) \cap C([0, T], H).$$

Note that this is a much stronger result than Theorem 30.B. In particular, we obtain that the problem

$$u' + Au = 0 \quad \text{on } [0, T],$$

$$u(0) = u_0 \in H,$$

has a unique classical solution, i.e., the derivative $u'(t)$ exists in the usual classical sense on $[0, T]$ with respect to the convergence on the space V^+ and both u': $[0, T] \to V^+$ and u: $[0, T] \to H$ are continuous.

In the next section we will apply Corollary 30.17 to the generalized Korteweg–de Vries equation.

The following proof of Theorem 30.B is much simpler than the proof of Theorem 30.A, since we now have more regularity of the operator A at hand. In addition, note that the semiboundedness condition (38) is more flexible than the global monotonicity condition for A in Theorem 30.A.

Proof of Theorem 30.B. We set $(u|v) = (u|v)_H$ and $\|u\| = \|u\|_H$.

Step 1: Projection operator P_n.

We choose a sequence $V_1 \subseteq V_2 \subseteq \cdots \subseteq V$ of finite-dimensional subspaces of V so that $\bigcup_n V_n$ is dense in V. Then $\bigcup_n V_n$ is also dense in H. Let $\{e_1, \ldots, e_n\}$ be an orthonormal basis in V_n with respect to $(\cdot|\cdot)$. We set

$$P_n u = \sum_{i=1}^{n} \langle u, e_i \rangle e_i \qquad \text{for all} \quad u \in V^+.$$

Then the operator P_n: $V^+ \to V$ is linear and continuous, and we have

$$P_n u = \sum_{i=1}^{n} (u|e_i) e_i \qquad \text{for all} \quad u \in H,$$

i.e., P_n: $H \to V_n$ is an orthogonal projection operator onto V_n. Obviously, we have

$$\begin{aligned} \langle v, P_n u \rangle &= \langle u, P_n v \rangle \qquad \text{for all} \quad u, v \in V^+, \\ P_n u &= u \qquad \text{for all} \quad u \in V_n. \end{aligned} \tag{44}$$

Step 2: Solution of the Galerkin equations.

We consider the Galerkin equation

$$\begin{aligned} u_n'(t) + P_n A(u_n(t), t) &= 0 \quad \text{on } [0, T], \\ u_n(0) &= P_n u_0. \end{aligned} \tag{45}$$

We seek a solution $u_n: [0, T] \to V_n$ of this ordinary differential equation for suitable $T > 0$ independent of n. Since $\dim V_n < \infty$, the operator $P_nA: V_n \times [0, T_0] \to V_n$ is continuous. By the Peano theorem (Theorem 3.B), there is a $T^* > 0$ and a C^1-solution u_n of (45) on $[0, T^*]$.

Step 3: *A priori* estimates for the Galerkin solutions.

Our goal is to obtain the *a priori* estimates (47) below by using the *majorant* equation (46b) below. From (44) and (45) we obtain

$$\frac{1}{2}\frac{d}{dt}\|u_n(t)\|^2 = (u_n(t)|u_n'(t)) = -\langle u_n(t), P_nA(u_n(t), t)\rangle$$
$$= -\langle A(u_n(t), t), u_n(t)\rangle \le c(\|u_n(t)\|^2),$$

and $\|u_n(0)\| = \|P_nu_0\| \le \|u_0\|$. Thus, if u_n is a solution of (45), then

$$\frac{d}{dt}\|u_n(t)\|^2 \le 2c(\|u_n(t)\|^2), \qquad \|u_n(0)\|^2 \le \|u_0\|^2, \tag{46a}$$

and

$$\frac{d}{dt}g(t) = 2c(g(t)), \qquad g(0) = \|u_0\|^2. \tag{46b}$$

This implies the basic *a priori* estimates

$$\|u_n(t)\|^2 \le g(t) \quad \text{on } [0, T] \qquad \text{for all } n. \tag{47}$$

More precisely, we have the following situation. If we extend the given function $c: \mathbb{R}_+ \to \mathbb{R}_+$ to a monotone increasing C^1-function $c: \mathbb{R} \to \mathbb{R}$, then it follows from the Picard–Lindelöf theorem (Proposition 1.8) that there is a $T_1 > 0$ such that the differential equation (46b) has a unique solution g on $[0, T_1]$. Let $t \mapsto \|u_n(t)\|^2$ be a solution of (46a) on $[0, T]$, where $0 < T \le T_1$. By using a standard argument on differential inequalities (cf. Proposition 33.11), we obtain (47) from (46a) and (46b).

By the *a priori estimate* (47) and by the continuation principle (cf. Problem 30.2), we can extend the solution u_n in Step 2 to the interval $[0, T_1]$. The point is that T_1 *does* not depend on n. In what follows we set $T = T_1$.

Step 4: Convergence of the Galerkin method.

Let D be a dense countable set in $[0, T]$. By (47) there exists a constant C such that

$$\|u_n(t)\| \le C \qquad \text{for all} \quad t \in [0, T] \qquad \text{and all } n. \tag{48}$$

The H-space H is reflexive. Using a diagonal procedure, we can find a subsequence, again denoted by (u_n), such that

$$u_n(t) \rightharpoonup u(t) \quad \text{in } H \qquad \text{as} \quad n \to \infty \qquad \text{for all} \quad t \in D. \tag{49}$$

Let $v \in V_k$ for fixed k and let $n \ge k$. By (44) and (45),

$$\frac{d}{dt}(u_n(t)|v) = -\langle v, P_nA(u_n(t), t)\rangle$$
$$= -\langle A(u_n(t), t), v\rangle.$$

Integration yields the *key* relation

$$(u_n(t)|v) - (u_0|v) = -\int_0^t \langle A(u_n(s), s), v\rangle\, ds \tag{50}$$

for all $v \in V_k$ and all $n \geq k$. Note that $(u_n(0)|v) = (P_n u_0|v) = (u_0|v)$.

(I) We want to show that

$$u_n(t) \rightharpoonup u(t) \quad \text{in } H \qquad \text{as} \quad n \to \infty \qquad \text{for } all \quad t \in [0, T]. \tag{51}$$

The operator $A: H \times [0, T] \to V^+$ is weakly sequentially continuous. Since H is reflexive, A is bounded. Hence there is a constant $K(v)$ such that, for each $v \in V$,

$$|\langle A(u_n(t), t), v\rangle| \leq K(v) \tag{52}$$

holds for all $t \in [0, T]$ and all n. We set $f_n(t) = (u_n(t)|v)$ for fixed $v \in V_k$. By (50) and (52), the sequence $\{f_n\}$ is equicontinuous on $[0, T]$. According to Problem 19.14c, the limit

$$\lim_{n\to\infty} (u_n(t)|v) = f(t) \tag{53}$$

exists for all $t \in [0, T]$ and all $v \in V_k$. This is true for all k, i.e., for all $v \in \bigcup_k V_k$. Moreover, since $\bigcup_k V_k$ is dense in H and $\|u_n(t)\| \leq C$, the limit (53) exists for all $v \in V$. This is (51).

(II) We show that

$$(u(t)|v) - (u_0|v) = -\int_0^t \langle A(u(s), s), v\rangle\, ds \tag{54}$$

for all $t \in [0, T]$, $v \in V$. In fact, for all $(u, v) \in V^+ \times V$, we have

$$|\langle u, v\rangle| \leq \text{const}\|u\|_{V^+}\|v\|_V.$$

It follows that $u \mapsto \langle u, v\rangle$ is a linear continuous functional on V^+ for all $v \in V$. Relation (51) implies

$$A(u_n(t), t) \rightharpoonup A(u(t), t) \quad \text{in } V^+ \qquad \text{as} \quad n \to \infty$$

and hence

$$\langle A(u_n(t), t), v\rangle \to \langle A(u(t), t), v\rangle \qquad \text{as} \quad n \to \infty$$

for all $t \in [0, T]$, $v \in V$. Letting $n \to \infty$, equation (50) implies (54) for all $v \in \bigcup_k V_k$. Since $\bigcup_k V_k$ is dense in V and the bilinear form $\langle \cdot, \cdot \rangle$ is continuous on $V^+ \times V$, we obtain that (54) holds for all $v \in V$.

(III) From (47) and (51) it follows that

$$\|u(t)\| \leq \underline{\lim}_{n\to\infty} \|u_n(t)\| \leq g(t)^{1/2} \quad \text{on } [0, T] \tag{55}$$

and hence $\|u(t)\| \leq \text{const}$ on $[0, T]$.

(IV) Weak continuity of $u(\cdot)$. Equation (54) shows that $t \mapsto (u(t)|v)$ is continuous on $[0, T]$ for all $v \in V$. Since V is dense in H and $\|u(t)\| \leq \text{const}$ on $[0, T]$, we obtain that $t \mapsto (u(t)|v)$ is continuous on $[0, T]$ for *all* $v \in H$, i.e., $u \in C_w([0, T], H)$.

(V) Weak$^+$-differentiability of u. From (54) we obtain

$$\frac{d}{dt}(u(t)|v) = -\langle A(u(t), t), v\rangle \quad \text{on } [0, T]$$

for all $v \in V$, i.e., there exists the weak$^+$-derivative

$$u'(t) = -A(u(t), t) \quad \text{on } [0, T].$$

(VI) Weak continuity of u'. If $t \to s$ on $[0, T]$, then $u(t) \rightharpoonup u(s)$ in H and hence $A(u(t), t) \rightharpoonup A(u(s), s)$ in V^+; therefore, $u' \in C_w([0, T], V^+)$.

(VII) Equation (54) shows that $(u(0)|v) = (u_0|v)$ for all $v \in V$. Since V is dense in H, we get $u(0) = u_0$.

(VIII) From

$$u(t) \rightharpoonup u_0 \quad \text{in } H \qquad \text{as} \quad t \to +0$$

and $\|u(t)\|^2 \leq g(t)$ with $g(0) = \|u_0\|^2$ we obtain

$$\|u_0\| \leq \varliminf_{t \to +0} \|u(t)\| \leq \varlimsup_{t \to +0} \|u(t)\| \leq \|u_0\|.$$

This implies

$$\|u(t)\| \to \|u_0\| \qquad \text{as} \quad t \to +0.$$

Since H is an H-space, this yields $u(t) \to u_0$ in H as $t \to +0$. □

Corollary 30.16 follows from (55).

PROOF OF COROLLARY 30.17.

(I) By Theorem 30.B and the uniqueness assumption (H4*), problem (37) has a unique solution $u \in C_w([0, T], H)$ with the weak$^+$-derivative $u' \in C_w([0, T], V^+)$. Furthermore, we have

$$u(t) \to u(0) \quad \text{in } H \qquad \text{as} \quad t \to +0.$$

(II) Letting $v(t) = u(t + t_0)$ for fixed $t_0 \in [0, T[$, v is a solution of (37) with $v(0) = u(t_0)$. This solution is unique. Applying (I) to v, it follows that $v(t) \to v(0)$ in H as $t \to +0$, and hence

$$u(t) \to u(t_0) \quad \text{in } H \qquad \text{as} \quad t \to t_0 + 0 \qquad \text{for each} \quad t_0 \in [0, T[.$$

(III) Letting $w(t) = u(-t + T)$, w is a solution of

$$w' - Aw = 0 \quad \text{on } [0, T], \qquad w(0) = u(T). \tag{56}$$

By (H4*), this solution is unique. Assumption (H3*) implies that the key condition (38) holds for both the operators A and $-A$. Therefore, we can apply the argument (I), (II) to equation (56). Hence

$$u(t) \to u(t_0) \quad \text{in } H \qquad \text{as} \quad t \to t_0 - 0 \qquad \text{for each} \quad t_0 \in]0, T].$$

(IV) To finish the proof note the following. By (I)–(III), the function $u: [0, T] \to H$ is continuous, i.e., $u \in C([0, T], H)$. Since $A: H \to V^+$ is continuous, the function $t \mapsto Au(t)$ belongs to $C([0, T], V^+)$. Therefore, noting (39), it follows from (54) that

$$u(t) - u_0 = -\int_0^t Au(s)\,ds \qquad \text{for all} \quad t \in [0, T].$$

This implies $u'(t) = -Au(t)$ for all $t \in [0, T]$, and hence $u' \in C([0, T], V^+)$. □

30.6. Application to the Generalized Korteweg–de Vries Equation

We want to apply the results of Section 30.5 to the following initial value problem for the generalized Korteweg–de Vries equation:

$$\begin{gathered} u_t + u_{xxx} + a(u)u_x = 0, \qquad x \in \mathbb{R}, \quad t > 0, \\ u(x, 0) = u_0(x). \end{gathered} \tag{57*}$$

We are looking for the function $u = u(x, t)$. Introducing the operator

$$Au = u_{xxx} + a(u)u_x, \qquad u \in H^3,$$

problem (57*) can be written in the form

$$\begin{gathered} u'(t) + Au(t) = 0, \qquad t > 0, \\ u(0) = u_0. \end{gathered} \tag{57}$$

For simplifying notation, we set

$$H^m = W_2^m(\mathbb{R}), \qquad m = 0, 1, \ldots.$$

In particular, $H^0 = L_2(\mathbb{R})$. In the following, we will consider the case $u_0 \in H^m$, $m \geq 3$.

Proposition 30.18. *Suppose that the function $a: \mathbb{R} \to \mathbb{R}$ is C^3. Let $u_0 \in H^3$ be given. Then there is a $T > 0$ such that the initial value problem (57) has a unique solution*

$$u \in C^1([0, T], H^0) \cap C([0, T], H^3).$$

The proof will be given below. This is a very natural result, since we shall show below that Au is well defined for $u \in H^3$, i.e., the function u has generalized derivatives up to order three with respect to the spatial coordinate x. Then $Au \in H^0$. More precisely, our proof will show that the operator

$$A: H^3 \to H^0$$

is continuous (and also weakly sequentially continuous). Therefore, by (57), we expect that if $u_0 \in H^3$, then $u(t) \in H^3$ and hence $u'(t) \in H^0$ for all $t \in [0, T]$, since $Au(t) \in H^0$. Indeed, Proposition 30.18 justifies this formal consideration.

The following stronger result shows that the solution $u(t)$, $t > 0$, has the same regularity (in terms of the Sobolev spaces H^m, $m \geq 3$) as the initial value $u(0) = u_0$.

Proposition 30.19 (Regularity). *Suppose that* $a: \mathbb{R} \to \mathbb{R}$ *is* C^∞. *Let* $u_0 \in H^m$ *be given for fixed* $m \geq 3$. *Then there exists a* $T > 0$ *only depending on the norm* $\|u_0\|_{H^2}$ *such that the initial value problem* (57) *has a unique solution*

$$u \in C^1([0, T], H^{m-3}) \cap C([0, T], H^m).$$

In addition, the solution u at time $t > 0$ *depends continuously on the initial value* u_0 *at time* $t = 0$ *with respect to the Sobolev space* H^m. *More precisely, the map*

$$u_0 \mapsto u$$

is continuous from each ball B in H^m *into the space* $C([0, T], H^m)$, *where T depends on B.*

The continuous dependence of the solution u on the initial value u_0 can also be expressed as follows. Let (u_{0n}) be a sequence of initial values such that

$$u_{0n} \to u_0 \quad \text{in } H^m \qquad \text{as} \quad n \to \infty$$

for fixed $m \geq 3$. Then there is an n_0 and a $T > 0$ such that, for all $n \geq n_0$, the corresponding solutions of problem (57) satisfy the condition

$$u_n(t) \to u(t) \quad \text{in } H^m \qquad \text{as} \quad n \to \infty,$$

uniformly with respect to all $t \in [0, T]$.

Corollary 30.20 (C^∞-Solutions). *Suppose that* $a: \mathbb{R} \to \mathbb{R}$ *is* C^∞. *Let* u_0 *be given such that*

$$u_0 \in C^\infty(\mathbb{R}) \cap H^m \qquad \textit{for all} \quad m = 3, 4, \ldots, \tag{58}$$

e.g., $u_0 \in C_0^\infty(\mathbb{R})$. *Then the solution u from Proposition* 30.19 *satisfies*

$$u(t) \in C^\infty(\mathbb{R}) \qquad \textit{for all} \quad t \in [0, T].$$

PROOF OF COROLLARY 30.20. From (58) it follows that $u_0 \in H^m$ for all $m \geq 3$. By Proposition 30.19, $u(t) \in H^m$ for all $t \in [0, T]$ and all $m \geq 3$. By the Sobolev embedding theorem, $H^m \subseteq C^{m-1}(\mathbb{R})$ for all $m \geq 1$. Hence $u(t) \in C^\infty(\mathbb{R})$. □

Proposition 30.21 (Global Solutions for Small Initial Values). *Suppose that* $a: \mathbb{R} \to \mathbb{R}$ *is* C^∞. *Then there is a number* $\gamma > 0$ *depending only on* $a(\cdot)$ *such that, for given* $u_0 \in H^m$ *with fixed* $m \geq 3$ *and*

$$\|u_0\|_{H^2} \leq \gamma,$$

the initial value problem (57) *has a unique solution*

$$u \in C^1([0, \infty[, H^{m-3}) \cap C([0, \infty[, H^m).$$

The following global existence theorem holds for arbitrarily large initial values u_0 in the case where the function $a(\cdot)$ satisfies the following condition

$$\overline{\lim_{|\lambda| \to \infty}} |\lambda|^{-6} \int_0^\lambda (\lambda - \mu) a(\mu) \, d\mu \leq 0. \tag{59}$$

Theorem 30.C (Global Solutions of the Generalized Korteweg–de Vries Equation). *Suppose that* $a: \mathbb{R} \to \mathbb{R}$ *is* C^∞ *and suppose that condition* (59) *holds. Let* $u_0 \in H^m$ *be given for fixed* $m \geq 3$. *Then the original initial value problem* (57) *has a unique solution*

$$u \in C^1([0, \infty[, H^{m-3}) \cap C([0, \infty[, H^m).$$

Remark 30.22 (Special Korteweg–de Vries Equation and the Method of the Inverse Spectral Transform). In the special case

$$a(u) = -6u, \tag{60}$$

the original equation (57*) is called the (special) Korteweg–de Vries equation. The physical meaning of this important equation and its history will be discussed in Chapter 71 in connection with nonlinear wave processes (cf. Problem 71.7). Note that all the preceding results are valid for the special Korteweg–de Vries equation, since condition (59) is fulfilled for (60).

For the special Korteweg–de Vries equation (57*) with (60), it is possible to construct solutions of the initial value problem in an explicit manner by using the method of inverse spectral transform. This method was discovered by Gardner, Green, Kruskal, and Miura (1967). In fact, this was an important discovery in modern mathematical physics. Before this discovery, mathematicians and physicists believed that it would be extremely difficult to construct explicit solutions for nonlinear evolution equations. Nowadays, we know a number of important nonlinear evolution equations in physics which allow the construction of explicit solutions via the inverse spectral transform. This will be discussed in Problem 30.7. The method of inverse spectral transform represents a sophisticated nonlinear variant of the classical Fourier transform.

However, note that, in contrast to our general results above, the method of inverse spectral transform is only applicable to equation (57*) in the case where the function $a(\cdot)$ has a *special* form.

PROOF OF PROPOSITION 30.18. We will use Theorem 30.B and Corollary 30.17.

Step 1: Formal motivation of the choice of the spaces $V \subseteq H \subseteq V^+$.

Let u be a sufficiently smooth solution of the original problem (57*), that is, $u = u(x,t)$ and

$$\begin{gathered} u_t + u_{xxx} + a(u)u_x = 0, \qquad x \in \mathbb{R}, \quad t > 0, \\ u(x,0) = u_0(x). \end{gathered} \tag{61}$$

Our goal is to write this in the form (37*), that is,

$$\begin{gathered} \frac{d}{dt}(u(t)|v)_H + \langle Au(t), v\rangle = 0 \quad \text{on } [0,T] \qquad \text{for all} \quad v \in V, \\ u(0) = u_0. \end{gathered} \tag{62}$$

To this end, we set

$$V = H^6, \qquad H = H^3, \qquad V^+ = H^0.$$

Recall that $H^m = W_2^m(\mathbb{R})$ and $H^0 = L_2(\mathbb{R})$. Let $v \in C_0^\infty(\mathbb{R})$. Multiplying (61) by both v and $-\partial^6 v/\partial x^6$ and using subsequent integration by parts, we obtain (62) from (61) in the case where we set

$$\begin{aligned} Au &= u_{xxx} + a(u)u_x, \\ \langle w, v\rangle &= \int_{\mathbb{R}} \left(wv - w\frac{\partial^6 v}{\partial x^6}\right) dx \qquad \text{for all} \quad v \in V, \quad w \in V^+, \\ (u|v)_H &= \int_{\mathbb{R}} (uv + u_{xxx}v_{xxx})\, dx \qquad \text{for all} \quad u, v \in H. \end{aligned} \tag{63}$$

In the following we set $\partial^k u/\partial x^k = u^{(k)}$ and $\int_{\mathbb{R}} = \int$.

Step 2: Properties of the spaces H^m.
Recall that

$$\|u\|_{H^m} = \left(\int \sum_{j=0}^m |u^{(j)}|^2\, dx\right)^{1/2}, \qquad m = 0, 1, \ldots,$$

and note that $C_0^\infty(\mathbb{R})$ is dense in H^m, $m \geq 0$. The easiest way to investigate the spaces H^m is to use the Fourier transform as in Section 21.20. For example, if $\hat{u}$ denotes the Fourier transform of u, then

$$\int |u^{(k)}|^2\, dx = \int |\xi^k \hat{u}(\xi)|^2\, d\xi \tag{64}$$

for all $u \in C_0^\infty(\mathbb{R})$ and $k = 0, 1, \ldots$. In particular, it follows from $|\xi|^2 \leq \text{const}(1 + |\xi|^4)$ for all $\xi \in \mathbb{R}$ that

$$\int u'^2\, dx \leq \text{const} \int (u^2 + u''^2)\, dx \qquad \text{for all} \quad u \in C_0^\infty(\mathbb{R}). \tag{65}$$

(I) Equivalent scalar product on the space H. Analogously to (65), it follows from (64) and

$$\|u\|_H^2 = (u|u)_H = \int (u^2 + u'''^2)\, dx$$

that

$$\|u\|_{H^3} \leq \text{const}\|u\|_H \qquad \text{for all} \quad u \in C_0^\infty(\mathbb{R}). \tag{66}$$

Since $C_0^\infty(\mathbb{R})$ is dense in $H = H^3$, relation (66) also holds true for all $u \in H$. Therefore, the space H is an H-space with respect to the scalar product $(u|v)_H$ introduced in (63).

(II) Sobolev embedding theorem. Let $C_b^k(\mathbb{R})$ denote the B-space of all C^k-functions $u: \mathbb{R} \to \mathbb{R}$ with

$$\|u\|_{C^k} \stackrel{\text{def}}{=} \sum_{j=0}^{k} \sup_{x \in \mathbb{R}} |u^{(j)}(x)| < \infty$$

for fixed $k = 0, 1, \ldots$. By Proposition 21.70, it follows via Fourier transform that the embedding

$$H^m \subseteq C_b^{m-1}(\mathbb{R}), \qquad m = 1, 2, \ldots,$$

is continuous.

In particular, since the embedding $H \subseteq C_b^2(\mathbb{R})$ is continuous, we get

$$\|u^{(k)}\|_C \leq \text{const}\|u\|_H \qquad \text{for all} \quad u \in H, \quad k = 0, 1, 2. \tag{67}$$

Step 3: We show that $V \subseteq H \subseteq V^+$ is an admissible triplet.
Using integration by parts, it follows from (63) that

$$\langle w, v\rangle = (w|v)_H \qquad \text{for all} \quad w, v \in C_0^\infty(\mathbb{R}).$$

Since $C_0^\infty(\mathbb{R})$ is dense in both H and V^+, we obtain the *compatibility* condition

$$\langle w, v\rangle = (w|v)_H \qquad \text{for all} \quad w \in H, \quad v \in V. \tag{68}$$

In addition, we have to show that $\{V, V^+\}$ represents a dual pair with respect to $\langle \cdot, \cdot \rangle$. In fact, for all $w \in V^+$, $v \in V$, we obtain from (63), by the Hölder inequality, that

$$\begin{aligned} |\langle w, v\rangle|^2 &\leq \int w^2\, dx \int (v^2 + |v^{(6)}|^2)\, dx \\ &\leq \|w\|_{V^+}^2 \|v\|_V^2. \end{aligned}$$

Furthermore, let $v \in V$ be given such that

$$\langle w, v\rangle = 0 \qquad \text{for all} \quad w \in V^+.$$

By (68), $(v|v)_H = 0$, and hence $v = 0$.

Finally, let $w \in V^+$ be given such that

$$\langle w, v\rangle = 0 \qquad \text{for all} \quad v \in V.$$

We have to show that $w = 0$. To this end, we consider the equation

$$v_0 - v_0^{(6)} = w, \qquad v_0 \in V. \tag{69}$$

Recall that $V^+ = L_2(\mathbb{R})$ and $V = W_2^6(\mathbb{R})$. It follows from the regularity theory for elliptic equations that problem (69) has a solution v_0 (cf. Problem 83.1).

This implies

$$0 = \langle w, v_0 \rangle = \int w^2\,dx,$$

and hence $w = 0$.

Step 4: The operator $A: H \to V^+$ is continuous and weakly sequentially continuous.

In fact, it is obvious that the operator

$$u \mapsto u'''$$

is linear and continuous from $H = H^3$ into $V^+ = H^0$, since $\|u'''\|_{H^0} \le \|u\|_{H^3}$ for all $u \in H^3$. By Proposition 21.81, each linear continuous operator between B-spaces is also weakly continuous. In addition, it follows from Proposition 21.84 via *Moser-type calculus* that the operator

$$u \mapsto a(u)u'$$

is continuous and weakly sequentially continuous from H into V^+.

Step 5: We prove the following *key* condition.

There exists a monotone increasing C^1-function $c: \mathbb{R}_+ \to \mathbb{R}_+$ such that

$$|\langle Av, v \rangle| \le c(\|v\|_H^2) \qquad \text{for all} \quad v \in V. \tag{70}$$

To prove this, we set

$$\langle Av, v \rangle = B(v, v) + C(v, v) + D(v, v),$$

where

$$B(v, v) = \int (v'''v - v'''v^{(6)})\,dx,$$

$$C(v, v) = \int a(v)v'v\,dx, \qquad D(v, v) = -\int a(v)v'v^{(6)}\,dx.$$

Recall that $H = H^3$. Thus, in order to obtain (70), we have to reduce B, C, D to terms which only contain derivatives up to order three. This will be done by using integration by parts.

(I) The first trick. Integration by parts yields

$$\int v'''v\,dx = -\int vv'''\,dx \qquad \text{for all} \quad v \in C_0^\infty(\mathbb{R})$$

and hence $B(v, v) = -B(v, v)$ for all $v \in C_0^\infty(\mathbb{R})$. This implies

$$B(v, v) = 0 \qquad \text{for all} \quad v \in V.$$

(II) The second trick. Note that

$$v'''v'''' = \frac{1}{2}\frac{d}{dx}v'''^2. \tag{71}$$

Integration by parts yields

$$D(v, v) = \int (a(v)v')'''v''' \, dx \qquad \text{for all} \quad v \in C_0^\infty(\mathbb{R}).$$

By (71),

$$(a(v)v')'''v''' = [a(v)'''v' + 3a(v)''v'' + 3a(v)'v''']v''' + 2^{-1}a(v)(v'''^2)'. \quad (72)$$

(III) The third trick. Let $b\colon \mathbb{R} \to \mathbb{R}$ be continuous. Then there exists a monotone increasing C^1-function $d\colon \mathbb{R} \to \mathbb{R}$ such that

$$|b(v)| \le d(|v|) \qquad \text{for all} \quad v \in \mathbb{R}.$$

Therefore, for all $v \in C_b(\mathbb{R})$,

$$\sup_{x \in \mathbb{R}} |b(v(x))| \le d(\|v\|_C).$$

Moreover, for all $v \in V$, we obtain

$$\sup_{x \in \mathbb{R}} |b(v(x))| \le d(K\|v\|_H), \tag{73}$$

where K denotes a constant. In this connection, observe that the embeddings $V \subseteq H \subseteq C_b(\mathbb{R})$ are continuous.

(IV) Proof of (70). Inequality (70) now follows easily from the tricks (I)–(III) above and from the Sobolev embedding theorem (67). For example, for all $v \in C_0^\infty(\mathbb{R})$, we get

$$\begin{aligned}\left|\int a(v)(v'''^2)' \, dx\right| &= \left|-\int a(v)'v'''^2 \, dx\right| \\ &= \left|\int a'(v)v'v'''^2 \, dx\right| \le d(K\|v\|_H) \int v'v'''^2 \, dx \\ &\le d(K\|v\|_H)\|v'\|_C\|v\|_H^2 \le \text{const}\, d(K\|v\|_H)\|v\|_H^3.\end{aligned}$$

In this connection, we use integration by parts as well as (67) and (73).

Similarly, using (72) we obtain

$$|D(v, v) + C(v, v)| \le c(\|v\|_H^2) \tag{74}$$

for all $v \in C_0^\infty(\mathbb{R})$, where $c\colon \mathbb{R}_+ \to \mathbb{R}_+$ is an appropriate monotone increasing C^1-function. Since $C_0^\infty(\mathbb{R})$ is dense in V, a simple limiting argument shows that (74) remains true for all $v \in V$. This finishes the proof of (70).

Step 6: Existence.

It follows from Theorem 30.B that, for given $u_0 \in H$, there is a $T > 0$ such that the original problem (57) has a solution

$$u \in C([0, T], H), \qquad u' \in C([0, T], V^+), \tag{75}$$

where the derivative $u'(t)$ is to be understood in the sense of weak$^+$ convergence, i.e., for all $v \in V$,

$$\left\langle \frac{u(t+h)-u(t)}{h}, v \right\rangle \to \langle u'(t), v\rangle \qquad \text{as} \quad h \to 0.$$

From (69) it follows that, for each element w in $V^+ = L_2(\mathbb{R})$, there is a v_0 in V such that $\int uw\,dx = \int u(v_0 - v_0^{(6)})\,dx$, i.e.,

$$(u|w)_{V^+} = \langle u, v_0\rangle \qquad \text{for all} \quad u \in V^+.$$

Therefore, weak$^+$ convergence and weak convergence coincide on the space V^+. Hence the derivative $u'(t)$ also exists in the sense of weak convergence on V^+.

Step 7: Uniqueness.
We show that the solution u of problem (57) is unique in the sense of Step 6.

(I) Let $u_0 \in H$ be given, and let u and v be solutions of (57) such that u and v satisfy condition (75). By Step 6, the derivatives $u'(t)$ and $v'(t)$ exist on $[0, T]$ in the sense of weak convergence on V^+. According to Problem 30.9, this implies the continuity of $u, v\colon [0, T] \to V^+$ and

$$\frac{d}{dt}\|u(t)-v(t)\|_{V^+}^2 = 2(u'(t)-v'(t)|u(t)-v(t)) \quad \text{on } [0, T], \tag{76}$$

where $(\cdot|\cdot)$ denotes the scalar product on $V^+ = L_2(\mathbb{R})$, i.e.,

$$(u|v) = \int uv\,dx \qquad \text{for all} \quad u, v \in V^+.$$

From $u' + Au = 0$ and $v' + Av = 0$ we get

$$\frac{d}{dt}\|u(t)-v(t)\|_{V^+}^2 = -2(Au(t)-Av(t)|u(t)-v(t)) \tag{77}$$

for all $t \in [0, T]$. We shall show below that there is a constant $C > 0$ such that

$$|(Au(t)-Av(t)|u(t)-v(t))| \le C\|u(t)-v(t)\|_{V^+}^2 \tag{78}$$

for all $t \in [0, T]$. From (77) and (78) we obtain

$$\|u(t)-v(t)\|_{V^+}^2 \le \|u(0)-v(0)\|_{V^+}^2 + \int_0^t 2C\|u(s)-v(s)\|_{V^+}^2\,ds$$

for all $t \in [0, T]$. Note that the integration of equation (77) is allowed, since the right-hand side of (76), (77) is continuous on $[0, T]$, by (75). From the *Gronwall lemma* in Section 3.5, we obtain

$$\|u(t)-v(t)\|_{V^+} \le \|u(0)-v(0)\|_{V^+}e^{CT} \qquad \text{for all} \quad t \in [0, T].$$

Since $u(0) = v(0)$, this implies the desired relation

$$u(t) = v(t) \qquad \text{for all} \quad t \in [0, T].$$

(II) Proof of (78). By (75), the functions u, v: $[0, T] \to H$ are weakly continuous. Thus, we get

$$\sup_{0 \le t \le T} \|u(t)\|_H + \|v(t)\|_H < \infty$$

(cf. Problem 30.9c). Let B be a closed ball in H. In order to prove (78) it is sufficient to show that

$$|(Au - Av|u - v)| \le C\|u - v\|_{V^+}^2 \qquad \text{for all} \quad u, v \in B, \tag{79}$$

where C depends on B. In what follows, note that the continuity of the embedding $H \subseteq C_b^1(\mathbb{R})$ implies that

$$\sup_{x \in \mathbb{R}} |w(x)| + |w'(x)| \le \text{const}\|w\|_H \qquad \text{for all} \quad w \in H. \tag{80}$$

Let u, $v \in C_0^\infty(\mathbb{R}) \cap B$. Since $Au = u''' + a(u)u'$, we have the decomposition

$$(Au - Av|u - v) = (u''' - v'''|u - v)$$
$$+ (a(u)(u' - v')|u - v) + ((a(u) - a(v))v'|u - v).$$

(II-1) By the first trick from Step 5, integration by parts yields

$$(u''' - v'''|u - v) = 0.$$

(II-2) From

$$2a(u)(u' - v')(u - v) = ((u - v)^2 a(u))' - (u - v)^2 a'(u)u'$$

and (80) we get

$$\begin{aligned} &|2(a(u)(u' - v')|u - v)| \\ &= \left|2 \int a(u)(u' - v')(u - v)\, dx\right| \\ &= \left|\int a'(u)u'(u - v)^2\, dx\right| \\ &\le \text{const} \int (u - v)^2\, dx = \text{const}\|u - v\|_{V^+}^2. \end{aligned}$$

(II-3) By the mean value theorem, there is a $\vartheta \in [0, 1]$ such that

$$a(u(x)) - a(v(x)) = a'(u(x) + \vartheta(v(x) - u(x)))[u(x) - v(x)],$$

and hence it follows from (80) that

$$\begin{aligned} |((a(u) - a(v))v'|u - v)| &= \left|\int (a(u) - a(v))v'(u - v)\, dx\right| \\ &\le \text{const} \int |u - v|^2\, dx = \text{const}\|u - v\|_{V^+}^2. \end{aligned}$$

Therefore, relation (79) holds for all $u, v \in C_0^\infty(\mathbb{R}) \cap B$. Since $C_0^\infty(\mathbb{R})$ is dense in the spaces H and V^+, and since the operator $A: H \to V^+$ is continuous, relation (79) also holds for all $u, v \in B$.

Step 8: By Step 7, the same uniqueness result holds if we replace A by $-A$. Consequently, Proposition 30.18 follows from Corollary 30.17. □

Remark 30.23. In order to give an elegant proof of Propositions 30.19–30.21 and Theorem 30.C, we need *nonlinear interpolation theory* which will be considered in Chapter 83 (cf. also Problem 30.11). We therefore postpone the corresponding proofs to Chapter 83. At this place, we only mention the following ideas of proof:

(i) The weak sequential continuity of the operator $A: H^m \to H^{m-3}$, $m \geq 3$, follows from Section 21.23 (Moser-type calculus).
(ii) The continuous dependence of the solution u on the initial values u_0 follows from nonlinear interpolation theory which ensures the *continuity* of appropriate nonlinear operators on interpolation spaces.
(iii) We use the method of *norm compression* in order to obtain the existence of the solution on an interval $[0, T]$ being independent of the space H^m, $m \geq 3$.
(iv) As usual, the existence of global solutions follows from suitable *a priori* estimates.

Problems

30.1. *Proof of Lemma* 30.2.

Solution: We want to use the substitution theorem $A_2(12)$. Let $w \in V$ be fixed. We set

$$f(t, v) = \langle A(t)v, w\rangle \qquad \text{for all} \quad t \in]0, T[, \quad v \in V.$$

By assumption (H5) in Section 30.2, for each $v \in V$, the function

$$t \mapsto f(t, v)$$

is measurable on $]0, T[$. Moreover, for each $t \in]0, T[$, the function

$$v \mapsto f(t, v)$$

is continuous on V, since the operator $A(t): V \to V^*$ is monotone and hemicontinuous and hence demicontinuous.

Now let $u \in L_p(0, T; V)$ be given, $1 < p < \infty$. Then the function $t \mapsto u(t)$ is measurable on $]0, T[$. Therefore, it follows from $A_2(12)$ that the function

$$t \mapsto f(t, u(t))$$

is measurable on $]0, T[$.

30.2. *Continuation of the solutions of ordinary differential equations in* B-*spaces*. We consider the initial value problem

$$\begin{aligned} u'(t) &= F(u(t), t) \quad \text{on } J, \\ u(t_0) &= u_0, \end{aligned} \tag{81}$$

where J is a real finite or infinite interval with $t_0 \in J$, i.e., $J \subseteq \mathbb{R}$, and the map F is of the form

$$F\colon X \times J \to X,$$

where X is a B-space. Let $u_0 \in X$ be given. Suppose that the following *a priori* estimate holds:

There is a positive number c such that if $u\colon J_0 \to X$ *is a solution of* (81) *on an arbitrary subinterval* J_0 *of* J, *then*

$$\|u(t)\| \le c \quad \text{on } J_0. \tag{82}$$

Show that problem (81) has a solution on J in the case where one of the following four conditions is met:

(i) F is C^1 and bounded on bounded sets.

(ii) F is continuous, bounded on bounded sets, and locally Lipschitz continuous with respect to u, i.e., for each $(u, t) \in X \times J$, there is a neighborhood $U(u, t)$ in $X \times J$ and a number L such that

$$\|F(v, s) - F(w, s)\| \le L\|v - w\| \qquad \text{for all} \quad (v, s), (w, s) \in U.$$

(iii) F is compact (e.g., F is continuous on $X \times \mathbb{R}$ and $\dim X < \infty$).

(iv) Let $X = \mathbb{R}^N$ and $F = (F_1, \ldots, F_N)$, and let J be closed (e.g., $J = \mathbb{R}$). There is a real integrable function $f \in L_1(J)$ such that

$$\|F(u, t)\| \le f(t) \qquad \text{for all} \quad (u, t) \in B \times J, \tag{83}$$

where $B = \{u \in X\colon \|u\| \le 2c\}$. Moreover, F satisfies the Carathéodory condition on $B \times J$, i.e., for all i,

$$\begin{aligned} t &\mapsto F_i(u, t) \quad \text{is measurable on } J \qquad \text{for all} \quad u \in B; \\ u &\mapsto F_i(u, t) \quad \text{is continuous on } B \qquad \text{for almost all} \quad t \in J. \end{aligned}$$

In the cases (i) or (ii), the solution of (81) is unique on J.

In the case (iv), a solution is understood to be a continuous function $u\colon J \to X$ such that the differential equation (81) holds for *almost* all $t \in J$.

Remark. In order to obtain *a priori* estimates of the form (82) one can use integral inequalities (the Gronwall lemma in Section 3.5 or the generalized Gronwall lemma in Section 7.3), as well as differential inequalities (see Propositions 33.11, 33.13, and 33.15).

Solution: We use similar arguments as in Section 3.3.

Step 1: Let J be a finite interval, e.g., let $J = [t_0, t_1]$.

Ad(i), (ii). By Theorem 3.A, there exists a $T > 0$ such that equation (81) has a unique solution on $[t_0, t_0 + T]$.

Let $J_* = [t_0, t_*[$ be the maximal half-open subinterval of J such that a solution $u = u(t)$ of (81) exists on J_*. The definition of J_* makes sense, since the solution of (81) is locally unique, by Theorem 3.A, and hence also globally unique.

Integrating equation (81), we obtain that

$$u(t) - u(s) = \int_s^t F(u(\tau), \tau)\, d\tau \tag{84}$$

for all $t_0 \leq s < t < t_*$. Since F is bounded on bounded sets,

$$M \stackrel{\text{def}}{=} \sup_{\|u\| \leq 2c,\, t \in J} \|F(u, t)\| < \infty.$$

By the *a priori* estimate (82), we obtain that

$$\|u(t) - u(s)\| \leq \int_s^t M\, d\tau = (t - s)M,$$

and hence the limit $u(t) \to u_*$ exists as $t \to t_* - 0$. This implies $t_* = t_1$. Otherwise, we could solve locally the new initial value problem

$$u'(t) = F(u(t), t), \qquad u(t_*) = u_*,$$

by using Theorem 3.A. This way, we could continue the solution $u = u(t)$ of (81) to a larger interval than J_*, which contradicts the maximality of J_*.

Ad(iii). By Theorem 3.B, there is a number $T > 0$ such that the equation (81) has a solution $u = u(t)$ on $[t_0, t_0 + T]$. Note the important fact that T depends *only* on the numbers c, M and $t_1 - t_0$. Hence, the solution $u = u(t)$ can be continued to the interval $J = [t_0, t_1]$ after a *finite* number of steps.

Ad(iv). We change F outside of the set $B \times J$ in such a way that condition (iv) holds for $B = X$. To this end, we set

$$\tilde{F}(u, t) = F(u/2c\|u\|, t) \qquad \text{if} \quad \|u\| > 2c.$$

Now apply the theorem of Carathéodory, $A_2(61)$, to the changed equation corresponding to (81). Because of the *a priori* estimate (82) and the construction of the set B, the solutions of the changed equation corresponding to (81) are also solutions of the original equation (81).

Step 2: Let J be an infinite interval, e.g., let $J = \mathbb{R}$.

Let $T > 0$ be fixed. By Step 1, we can construct a solution $u = u(t)$ of (81) on $[-T, T]$, and we can continue this solution successively to the intervals $[-nT, nT]$, where $n = 2, 3, \ldots$.

30.3. *Proof of Lemma* 30.4. Use Problem 30.2 to show that the Galerkin equation (5) has a unique solution for each $n \in \mathbb{N}$.

Solution: By (6), the Galerkin equation (5) represents a system of ordinary differential equations.

The uniqueness of the solution follows from Section 30.3a. The existence of a solution of (5) on $[0, T]$ follows from Problem 30.2(iv) above. The assumptions (H1)–(H5) of Theorem 30.A imply condition (iv) of Problem 30.2. In this connection, note the following. From (21) we obtain the *a priori* estimate

$$|u_n(t)| \leq \text{const} \quad \text{on } [0, T] \qquad \text{for all } n.$$

By (H2), the operator $A(t)\colon V \to V^*$ is demicontinuous, and hence the function

$$(u, t) \mapsto \langle A(t)u, w_j \rangle$$

satisfies the *Carathéodory condition* on $H_n \times [0, T]$, according to (H5). Finally, from (H4) we obtain the estimate

$$|\langle A(t)u, w_j\rangle| \le \text{const}(c_3(t) + \|u\|^{p/q})\|w_j\| \tag{85}$$

for all $(u, t) \in H_n \times [0, T]$ and $j = 1, \ldots, n$.

By the Carathéodory theorem, $A_2(61)$, the solution u_n: $[0, T] \to H_n$ of (5) is continuous, and the derivative $u_n'(t)$ exists for almost all $t \in [0, T]$. Hence

$$u_n \in L_p(0, T; H_n).$$

By (85) and $c_3 \in L_q(0, T)$, the function

$$t \mapsto \langle A(t)u_n(t), w_j\rangle$$

belongs to $L_q(0, T)$. Moreover, it follows from $|\langle b(t), w_j\rangle|^q \le \|b(t)\|^q\|w_j\|^q$ and $b \in L_q(0, T; V^*)$ that the function

$$t \mapsto \langle b(t), w_j\rangle$$

also belongs to $L_q(0, T)$. Finally, by the Galerkin equation (5a), we get

$$u_n' \in L_q(0, T; H_n).$$

30.4. *Proof of Lemma* 30.7.

Solution: We will use a similar argument as in the proof of Theorem 23.A.

(I) Let $u \in W_p^1(0, T; V, H)$ be the solution of the original problem (1), and let u_n be the solution of the Galerkin equation (5). By Step 5 of the proof of Theorem 23.A, there exists a sequence (p_n) of polynomials p_n: $[0, T] \to H_n$ such that

$$p_n \to u \quad \text{in } W_p^1(0, T; V, H) \qquad \text{as} \quad n \to \infty,$$

and hence

$$p_n \to u \quad \text{in } C([0, T], H) \qquad \text{as} \quad n \to \infty. \tag{86}$$

By Lemma 30.5, $u_n(0) \to u(0)$ in H as $n \to \infty$, and hence

$$p_n(0) - u_n(0) \to 0 \quad \text{in } H \qquad \text{as} \quad n \to \infty. \tag{87}$$

Below we shall show that

$$\max_{0 \le t \le T} |u_n(t) - p_n(t)|_H^2 - |u_n(0) - p_n(0)|_H^2 \to 0 \qquad \text{as} \quad n \to \infty. \tag{88}$$

This yields the desired result, namely,

$$u_n \to u \quad \text{in } C([0, T], H) \qquad \text{as} \quad n \to \infty,$$

by (86) and (87).

(II) We prove (88). From the original problem (1),

$$u'(t) + A(t)u(t) - b(t) = 0,$$

and from the Galerkin equation (5),

$$\langle u_n'(t) + A(t)u_n(t) - b(t), v\rangle = 0 \qquad \text{for all} \quad v \in H_n,$$

we obtain

$$\begin{aligned}\langle u_n'(t), u_n(t) - p_n(t)\rangle &= \langle b(t) - A(t)u_n(t), u_n(t) - p_n(t)\rangle \\ &= \langle u'(t) + A(t)u(t) - A(t)u_n(t), u_n(t) - p_n(t)\rangle.\end{aligned}$$

By the monotonicity of $A(t)$,

$$\langle A(t)u(t) - A(t)u_n(t), u_n(t) - u(t)\rangle \le 0.$$

Finally, for all $t \in [0, T]$, the integration by parts formula (10) yields the following *key* relation:

$$\begin{aligned}
&\tfrac{1}{2}(|u_n(t) - p_n(t)|_H^2 - |u_n(0) - p_n(0)|_H^2) \\
&\quad = \int_0^t \langle u_n'(s) - p_n'(s), u_n(s) - p_n(s)\rangle\, ds \\
&\quad = \int_0^t \langle u' + Au - Au_n - p_n', (u_n - u) + (u - p_n)\rangle\, ds \\
&\quad \le \int_0^t \langle u' - p_n', u_n - p_n\rangle + \langle Au - Au_n, u - p_n\rangle\, ds \\
&\quad \le \|u' - p_n'\|_{X^*}\|u_n - p_n\|_X + \|Au - Au_n\|_{X^*}\|u - p_n\|_X \overset{\text{def}}{=} d_n.
\end{aligned}$$

Since $u_n \to u$ and $p_n \to u$ in X as $n \to \infty$, the sequences (u_n) and (p_n) are bounded in X. Because $A: X \to X^*$ is bounded, the sequence (Au_n) is bounded in X^*. Therefore, we obtain that

$$d_n \le \text{const}\|u - p_n\|_{W_p^1} \to 0 \qquad \text{as} \quad n \to \infty.$$

This implies (88).

30.5. *The spectral transform.* The following considerations will be used basically in Problem 30.7. The connection with scattering theory in quantum mechanics will be explained in Problem 30.8 below.

We consider the stationary Schrödinger equation

$$-\psi''(x) + u(x)\psi(x) = k^2\psi(x). \tag{89}$$

We assume that the real C^∞-potential u vanishes sufficiently fast at infinity, i.e.,

$$\int_{-\infty}^{\infty} |u(x)|(1 + |x|)\, dx < \infty.$$

We are looking for complex-valued eigenfunctions ψ. The spectrum of (89) has the following structure:

(i) *Continuous spectrum.* For each real $k \neq 0$, the number k^2 is a double eigenvalue of (89) with the two linearly independent eigenfunctions ψ_1 and ψ_2, which are uniquely characterized by the following asymptotic behavior:

$$\begin{aligned}
\psi_1(x) &= e^{-ikx} + o(1), \qquad x \to -\infty \\
\psi_2(x) &= e^{ikx} + o(1), \qquad x \to -\infty.
\end{aligned} \tag{90}$$

Additionally, we obtain

$$\begin{aligned}
\psi_1(x) &= a(k)e^{-ikx} + b(k)e^{ikx} + o(1), \qquad x \to +\infty, \\
\psi_2(x) &= \overline{b(k)}e^{-ikx} + \overline{a(k)}e^{ikx} + o(1), \qquad x \to +\infty.
\end{aligned} \tag{91}$$

The matrix

$$T = \begin{pmatrix} a(k) & b(k) \\ \overline{b(k)} & \overline{a(k)} \end{pmatrix} \tag{92}$$

is called the transition matrix.

(ii) *Discrete spectrum.* Equation (89) has either no negative eigenvalues or a finite number of negative eigenvalues

$$-\infty < k_1^2 < k_2^2 < \cdots < k_N^2 < 0.$$

All these eigenvalues are simple. Letting $k_j = iq_j$ with $q_j > 0$, the corresponding eigenfunctions $\psi^{(j)}, j = 1, \ldots, N$, are uniquely characterized by the following asymptotic behavior:

$$\psi^{(j)}(x) = e^{q_j x} + o(e^{q_j x}), \qquad x \to -\infty. \tag{93}$$

Additionally, we have

$$\psi^{(j)}(x) = c_j e^{-q_j x} + o(e^{-q_j x}), \qquad x \to +\infty,$$

where c_j is real.

The mapping

$$u \mapsto (a(k), b(k), q_j, c_j), \tag{94}$$

with $k \in \mathbb{R}$ and $j = 1, \ldots, N$, is called the *spectral transform.*

In terms of quantum mechanics, (i) and (ii) above correspond to scattered particles and to bound states of particles, respectively. In the special case $u(x) \equiv 0$, there are no negative eigenvalues, and (90) holds with $o(1) = 0$.

30.6. *The inverse spectral transform.* Let all the scattering data $a(k)$, $b(k)$, q_j and c_j be given. We want to construct the corresponding potential u. To this end, we set

$$F(x) = \sum_{j=1}^{N} \frac{c_j e^{-q_j x}}{ia'(iq_j)} + \frac{1}{2\pi} \int_{-\infty}^{\infty} \frac{b(k)}{a(k)} e^{ikx}\, dk,$$

and we consider the so-called *Gelfand–Levitan–Marčenko equation*

$$K(x, y) + F(x + y) + \int_x^{\infty} K(x, z) F(z + y)\, dz = 0.$$

If we know a solution K of this linear integral equation, then we obtain the unknown potential u by the relation

$$u(x) = -2 \frac{d}{dx} K(x, x).$$

The details can be found in Novikov (1980, M) and Levitan (1984, M).

30.7. *Construction of solutions of the Korteweg–de Vries equation via the inverse spectral transform.* We consider the *nonlinear* Korteweg–de Vries equation

$$\begin{gathered} u_t + u_{xxx} - 6uu_x = 0, \qquad -\infty < x < \infty, \quad t > 0, \\ u(x, 0) = u_0(x) \end{gathered} \tag{95}$$

together with the *linear* Schrödinger equation

$$-\psi''(x) + u(x, t)\psi(x) = 0 \tag{95*}$$

for fixed, but otherwise arbitrary t. Then we have the following important result:

(R) *Let $u = u(x,t)$ be a solution of* (95), *which vanishes sufficiently fast as $|x| \to \infty$. Then the spectral transform of u satisfies the following very simple linear equation*:

$$\begin{aligned} \dot{a}(k,t) &= 0, & \dot{b}(k,t) &= 8ik^3 b(k,t), \\ \dot{q}_j &= 0, & \dot{c}_j &= 8q_j^3 c_j. \end{aligned} \tag{96}$$

The dot denotes the t-derivative.

According to (R), we can use the following elegant procedure in order to solve the initial value problem (95).

(i) For given initial values $u = u_0(x)$, we compute the spectral transform $a(k,0)$, $b(k,0)$, $q_j(0)$, $c_j(0)$.

(ii) We solve equation (96). This yields

$$\begin{aligned} a(k,t) &= a(k,0), & q_j(t) &= q_j(0), \qquad t > 0, \\ b(k,t) &= b(k,0)e^{8ik^3 t}, & c_j(t) &= c_j(0)e^{8q_j^3(0)t}. \end{aligned}$$

(iii) We get the solution $u = u(x,t)$ of the original problem (95) by using the inverse spectral transform

$$(a(k,t), b(k,t), q_j(t), c_j(t)) \mapsto u(x,t),$$

which we considered in Problem 30.6.

30.7a. *Motivation of* (R). In order to display the simple idea of proof as clearly as possible, we restrict ourselves to formal considerations.

(I) Discrete spectrum. We introduce the following two differential operators:

$$L(t)\psi = -\frac{d^2}{dx^2}\psi(x) + u(x,t)\psi(x),$$

$$A(t)\psi = 4\frac{d^3}{dx^3}\psi(x) - 3u(x,t)\frac{d}{dx}\psi(x) - \frac{\partial}{\partial x}(u(x,t)\psi(x)),$$

where the fixed function u satisfies the Korteweg–de Vries equation (95). From (95), we get the key relation

$$L_t = LA - AL. \tag{97}$$

Hence

$$L(t) = e^{-At}L(0)e^{At}. \tag{97*}$$

The unitary group $\{e^{At}\}$ is to be understood in the sense of a skew-adjoint extension of the skew-symmetric operator A in $L_2^{\mathbb{C}}(\mathbb{R})$ with $D(A) = C_0^\infty(\mathbb{R})$. By (97*), the operator $L(t)$ has the same eigenvalues as $L(0)$, i.e., we obtain the crucial relation

$$q_j(t) = q_j(0) \qquad \text{for all } t,$$

and hence $\dot{q}_j = 0$. The pair $\{L, A\}$ with (97) is called a *Lax pair*. Let $\psi = \psi(x,t)$ and let

$$L(t)\psi = -q^2\psi \tag{98}$$

for fixed t, where $q = q_j$. By Problem 30.5, for fixed t, the eigenfunction ψ

can be characterized by the following asymptotic behavior:

$$\psi(x,t) = e^{qx} + o(e^{qx}), \qquad x \to -\infty. \tag{99a}$$

In addition,

$$\psi(x,t) = c(t)e^{-qx} + o(e^{-qx}), \qquad x \to +\infty. \tag{99b}$$

Differentiation of (98) with respect to time t yields

$$L_t\psi + L\psi_t = -q^2\psi_t.$$

Note that $q = q_j$ is independent of t. By (97) and (98), $(L + q^2)(\psi_t + A\psi) = 0$. Hence we obtain

$$L(t)\varphi = -q^2\varphi,$$

where $\varphi = \psi_t + A\psi$. By (99a),

$$\varphi = 4q^3e^{qx} + o(e^{qx}), \qquad x \to -\infty.$$

Hence $\varphi = 4q^3\psi$, i.e., we obtain the *key* equation

$$\psi_t + A\psi = 4q^3\psi. \tag{100}$$

In this connection, note that u is rapidly vanishing as $|x| \to \infty$, i.e., we may put $A = 4d^3/dx^3$ as $|x| \to \infty$. Finally, it follows from (99b) and (100) that

$$\dot{c}(t) = 8q^3c(t).$$

(II) Continuous spectrum. Let $k \geq 0$ be fixed and let ψ denote the eigenfunction of the equation

$$L(t)\psi = k^2\psi \tag{101}$$

for fixed t, where $\psi = \psi(x,t)$ is characterized by the following asymptotic behavior:

$$\psi(x,t) = e^{-ikx} + o(1), \qquad x \to -\infty. \tag{102a}$$

In addition,

$$\psi(x,t) = a(k,t)e^{-ikx} + b(k,t)e^{ikx} + o(1), \qquad x \to +\infty. \tag{102b}$$

As above, differentiation of (101) with respect to time t yields

$$L(t)\varphi = k^2\varphi,$$

where $\varphi = \psi_t + A\psi$. By (102a),

$$\varphi = 4ik^3e^{-ikx} + o(1), \qquad x \to -\infty.$$

Hence $\varphi = 4ik^3\psi$, i.e., we obtain the key equation

$$\psi_t + A\psi = 4ik^3\psi. \tag{103}$$

Using (102b) we get

$$\begin{aligned}\dot{a}e^{-ikx} + \dot{b}e^{ikx} &= \left(-4\frac{d^3}{dx^3} + 4ik^3\right)(ae^{-ikx} + be^{ikx}) + o(1)\\ &= 8ik^3be^{ikx} + o(1), \qquad x \to +\infty.\end{aligned}$$

Hence $\dot{a} = 0$ and $\dot{b} = 8ik^3b$. This finishes the motivation of (R).

For more details compare Novikov (1980, M) and Levitan (1984, M).

30.7b. *Discussion of the method of the inverse spectral transform* ($\mathscr{S}$).

(i) *Explicit solutions.* The method ($\mathscr{S}$) reduces the integration of a *nonlinear* partial differential equation to the integration of *linear* ordinary differential equations. In a certain sense, one obtains the solutions in an explicit manner in contrast to general existence proofs via the Galerkin method. There exists a number of important nonlinear evolution equation in physics, which can be treated similarly. A list of 37 equations, which can be solved by the inverse spectral transform method, can be found in Calogero and Degasperis (1982, M). Compare also Mihailov, Šabat and Jamilov (1987). We also recommend Bullough and Caudrey (1980, S), Novikov (1980, M), Ablowitz and Segur (1981, M), Faddeev and Takhtadjan (1987, M).

(ii) *Nonlinear Fourier transform.* The method ($\mathscr{S}$) represents a nonlinear variant of the classical Fourier transform. To explain this, consider the linearized Korteweg–de Vries equation

$$u_t + u_{xxx} = 0.$$

Using the Fourier transform

$$u(x,t) = \int_{-\infty}^{\infty} b(k,t)e^{ikx}\,dk,$$

we get

$$\int_{-\infty}^{\infty} (\dot{b} - ik^3 b)e^{ikx}\,dk = 0.$$

This implies $\dot{b} = ik^3 b$ and hence

$$b(k,t) = e^{ik^3 t} b(k,0).$$

(iii) *Infinite-dimensional integrable Hamiltonian system.* The Hamiltonian system

$$\dot{p}_j = -\frac{\partial H(p,q)}{\partial q_j}, \qquad \dot{q}_j = \frac{\partial H(p,q)}{\partial p_j}, \qquad j = 1, \ldots, N, \tag{104}$$

is called integrable if there exists a sufficiently regular transformation $P = P(p,q)$, $Q = Q(p,q)$ such that (104) is transformed into the simple equation

$$\dot{P}_j = -\frac{\partial \tilde{H}(P)}{\partial Q_j}, \qquad \dot{Q}_j = \frac{\partial \tilde{H}(P)}{\partial P_j}, \qquad j = 1, \ldots, N. \tag{105}$$

Equation (105) has the solution

$$P_j(t) = \text{const}, \qquad Q_j(t) = \frac{\partial \tilde{H}(P)}{\partial P_j} t + \text{const}. \tag{106}$$

Transforming back to the variables q and p, we obtain the solution of (104).

In terms of classical mechanics, the Hamiltonian H frequently represents the energy of the system, where q_j and p_j correspond to a position coordinate and a generalized momentum, respectively.

Note the crucial fact that P_j, $j = 1, \ldots, N$, are conserved quantities by (106), i.e., P_j is constant along each solution of (104). Moreover, H is also a conserved quantity (conservation of energy). In fact, it follows from (104) that

$$\frac{d}{dt}H(p(t), q(t)) = H_p\dot{p} + H_q\dot{q} = \dot{q}\dot{p} - \dot{p}\dot{q} = 0 \qquad \text{for all } t.$$

The Korteweg–de Vries equation (95) and equation (96) correspond to (104) and (105), respectively. Moreover, the canonical transformation $P = P(p, q)$, $Q = Q(p, q)$ corresponds to the spectral transform. More precisely, set

$$H(u) = \int_{-\infty}^{\infty} (2^{-1}u_x^2 + u^3)\,dx. \tag{107}$$

For all $h \in C_0^\infty(\mathbb{R})$, we obtain the first variation

$$\begin{aligned} H'(u)h &= \frac{d}{dt}H(u + th)|_{t=0} \\ &= \int_{-\infty}^{\infty} (u_x h_x + 3u^2 h)\,dx = \int_{-\infty}^{\infty} (3u^2 - u_{xx})h\,dx. \end{aligned}$$

We define the variational derivative $\delta H/\delta u(x)$ by the relation

$$H'(u)h = \int_{-\infty}^{\infty} \frac{\delta H}{\delta u(x)} h\,dx \qquad \text{for all} \quad h \in C_0^\infty(\mathbb{R}).$$

Hence $\delta H/\delta u(x) = 3u^2 - u_{xx}$. Thus, the Korteweg–de Vries equation $u_t + u_{xxx} - 6uu_x = 0$ may be written as the following infinite-dimensional Hamiltonian system

$$u_t = \frac{\partial}{\partial x}\frac{\delta H}{\delta u(x)}, \tag{107*}$$

which corresponds to (104). The important Hamiltonian approach to nonlinear evolution equations can be found in Faddeev and Takhtadjan (1987, M). Physicists like to use the Hamiltonian approach in order to get maximal insight and to find the quantized form of physical theories (cf. Part V).

(iv) *Conservation of energy and an infinite number of conservation laws.* As we mentioned above, the function H from (104) represents the energy of systems in classical mechanics. In Problem 71.7 we shall show that the Korteweg–de Vries equation (95) describes the motion of an infinite number of nonlinearly coupled oscillators (nonlinear oscillations of a system with an infinite number of degrees of freedom). We therefore expect that $H(u)$ from (107) represents the energy of the oscillating system which is a conserved quantity (conservation of energy). More precisely, we shall show the following. If $u = u(x, t)$ is a solution of the Korteweg–de Vries equation (95), where $u \in C^4(\mathbb{R}^2)$ and $u(\cdot, t) \in C_0^\infty(\mathbb{R})$ for all $t \in \mathbb{R}$, then

$$H(u(t)) = \text{const} \qquad \text{for all} \quad t \in \mathbb{R}.$$

The same remains true if, for each fixed t, the function $u = u(x, t)$ and its x-derivatives vanish sufficiently rapidly as $|x| \to \infty$.

In fact, we obtain from (95) that $u_t = 6uu_x - u_{xxx}$ and $u_{xt} = 6u_x^2 + 6uu_{xx} -$

u_{xxxx}. Hence

$$\frac{d}{dt}H(u(t)) = \int_{-\infty}^{\infty} (u_x u_{xt} + 3u^2 u_t)\,dx$$

$$= \int_{-\infty}^{\infty} \frac{\partial}{\partial x}(6uu_x^2 - 3u^2 u_{xx} - \tfrac{1}{2}u_{xxx}^2 + \tfrac{9}{2}u^2)\,dx = 0.$$

More generally, one can show that there exists an infinite number of conserved quantities for the Korteweg–de Vries equation. Roughly speaking, this reflects the fact that (95) can be regarded as an integrable Hamiltonian system with an infinite number of degrees of freedom. Recall that the classical system (104), with N degrees of freedom, has the N conserved quantities P_j, $j = 1, \ldots, N$.

Generally, conservation laws play an important role in the theory of nonlinear evolution equations, since they represent *a priori* estimates for the solutions, which are necessary for proving global existence (cf. Chapter 83).

(v) *Solitons*. The simple spectral transform

$$b(k,0) = 0, \qquad a(k,0) = \frac{k - iq}{k + iq}, \; q, c$$

corresponds to the solution

$$u(x,t) = -\frac{2q^2}{\cosh^2 q(x - 4q^2 t - \alpha)}$$

of the Korteweg–de Vries equation (95), where $\alpha = (1/2q)\ln c$ (cf. Novikov (1980, M)). This solution is called a solitary wave or a soliton. Note that the shape of this propagating wave remains constant for all times t (Fig. 30.1).

(vi) *Collision between solitons*. Using the spectral transform

$$b(k,0) = 0, \qquad a(k,0) = \prod_{j=1}^{N} \frac{k - iq_j}{k + iq_j}, \; q_j, c_j,$$

one can construct solutions of the Korteweg–de Vries equation which correspond to N solitons (cf. Novikov (1980, M)). It is remarkable that such solitons behave elastically in collisions, i.e., collisions between solitons do *not* change the speed and the shape of the solitons (Fig. 30.2). This behavior of solitons is very interesting from the physical point of view. Roughly speaking, in many fields of physics, solitons describe the strictly localized transport of energy, where the transported energy behaves like a particle. This represents a basic phenomenon in nature. This is why physicists are very interested in the theory of solitons.

The method of inverse spectral transform goes back to Gardner, Green, Kruskal, and Miura (1967). This method represents an important discovery in mathematical physics. More about the interesting history of this method can be found in Problem 71.7.

A general representation formula for explicit solutions of the Korteweg–de Vries equation via *theta functions* and its discussion can be found in Its and Matveev (1975), Matveev (1986, S), Bobenko and Bordag (1987), and in the monograph by Bobenko, Its, and Matveev (1991). In this connection, it is quite

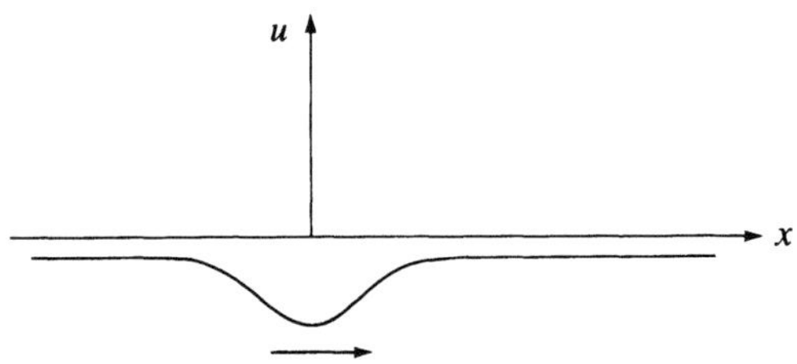

Figure 30.1

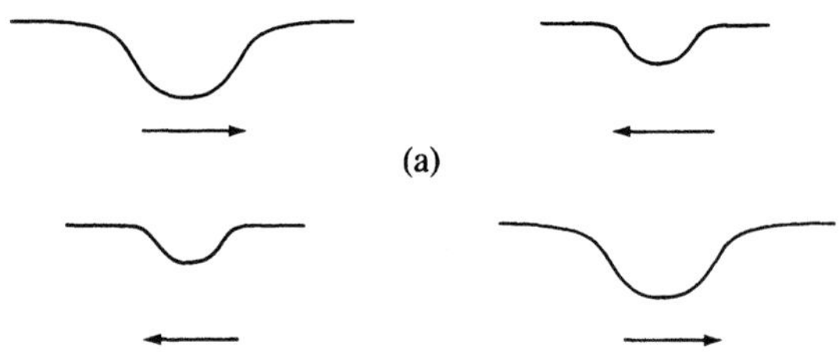

(a)

(b)

Figure 30.2

remarkable that the explicit solutions of nonlinear evolution equations are parametrized, in some sense, by Riemannian surfaces. The theory of theta functions can be found in Mumford (1983, L). Cf. also Toda (1981, M).

30.8. *Scattering of particles in quantum mechanics and the physical background of the spectral transform.*

30.8a. *Classical mechanics.* The equation

$$m\ddot{x} = -V'(x) \tag{108}$$

describes the motion $x = x(t)$ of a particle of mass m on $\mathbb{R}$. Here V is called a potential. The momentum vector is given by $p = m\dot{x}e$, where e denotes the unit vector in the direction of the x-axis. Equation (108) implies conservation of energy, i.e., for each solution of (108), there exists a constant E such that, for all t,

$$E = \frac{p^2}{2m} + V(x(t)). \tag{109}$$

Here E is called the total energy. Consequently, the motion of the particle is only possible in such regions where x satisfies the condition

$$E - V(x) \geq 0.$$

Figure 30.3(a) and (b) corresponds to a bounded and an unbounded motion, respectively. In terms of our solar system, the motion of planets and comets corresponds to bounded and unbounded motions, respectively.

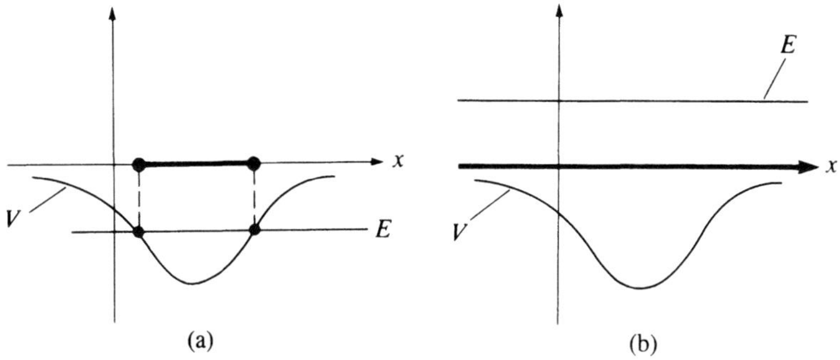

Figure 30.3

30.8b. *Quantum mechanics.* In terms of quantum mechanics, the motion of a particle of mass m on $\mathbb{R}$ is described by the Schrödinger equation

$$i\hbar\varphi_t = -\frac{\hbar^2}{2m}\varphi_{xx} + V\varphi, \tag{110}$$

where $\hbar = h/2\pi$, and h denotes the Planck quantum of action. Letting

$$\rho = \varphi\bar{\varphi}, \qquad v = \frac{i\hbar}{2m\rho}(\varphi\bar{\varphi}_x - \bar{\varphi}\varphi_x)e,$$

we obtain from (110) the continuity equation

$$\rho_t + \operatorname{div} \rho v = 0. \tag{111}$$

Each physical quantity in classical mechanics corresponds to an operator in quantum mechanics. For example, energy E and momentum vector p correspond to the following operators:

$$E \Rightarrow i\hbar\frac{\partial}{\partial t}, \qquad p \Rightarrow -i\hbar e\frac{\partial}{\partial x}.$$

This way, we obtain the Schrödinger equation (110) from the classical energy equation (109) (quantization of classical mechanics).

(i) *Bound states.* Let φ be a solution of (110) with $\varphi \in L_2^{\mathbb{C}}(\mathbb{R})$ normalized by $\int_{-\infty}^{\infty} \rho\, dx = 1$. Then, by definition, the number

$$\int_a^b \rho\, dx$$

is equal to the probability for finding the particle in the interval $[a, b]$. Such so-called bound states correspond to bounded motions in classical mechanics.

(ii) *Unbound states.* Let φ be a solution of (110) with $\int_{-\infty}^{\infty} \rho\, dx = \infty$. Such so-called unbound states correspond to unbounded motions in classical mechanics. In this case, the function φ describes a stream of particles, where ρ represents the particle density (i.e., the number of particles per volume) and v represents the velocity vector. The continuity equation (111) above describes the conservation of the number of particles (cf. Section 69.1).

Let ψ be a nonzero solution of the stationary Schrödinger equation

$$E\psi = -\frac{\hbar^2}{2m}\psi_{xx} + V\psi, \tag{112}$$

where E is a fixed real number. Then the function

$$\varphi(x,t) = e^{-iEt/\hbar}\psi(x)$$

is a solution of the Schrödinger equation (110). If $\int_{-\infty}^{\infty} |\psi|^2\,dx = 1$, then φ corresponds to a bound state of energy E.

In contrast to this situation, the function

$$\varphi_\pm(x,t) = e^{-iEt/\hbar}e^{\pm ikx}, \qquad E = \frac{\hbar^2 k^2}{2m},$$

satisfies the Schrödinger equation (110) with vanishing potential $V = 0$. Since $\int_{-\infty}^{\infty} |\varphi_\pm|^2\,dx = \infty$, the function $\varphi_\pm$ describes a stream of free particles with particle density $\rho = |\varphi_\pm|^2 = 1$ and velocity vector

$$v = \pm\frac{\hbar k}{m}e.$$

The momentum vector p and the kinetic energy E_{kin} of one particle are given by the relations

$$p = mv = \pm\hbar k e, \qquad E_{\text{kin}} = \frac{mv^2}{2} = E.$$

30.8c. *The spectral transform.* By Problem 30.8b, the spectral transform

$$u \mapsto (a(k), b(k), q_j, c_j)$$

of Problem 30.5 allows the following physical interpretation:

(i) The function u in (89) corresponds to the potential $V = \hbar^2 u/2m$ of the Schrödinger equation (112).

(ii) The value q_j corresponds to an eigenvalue $E = -q_j^2\hbar^2/2m$ of (112), i.e., q_j corresponds to a bound state of energy E.

(iii) The function ψ_1 with the asymptotic behavior

$$\psi_1(x) = e^{-ikx} + o(1), \qquad x \to -\infty,$$

$$\psi_1(x) = a(k)e^{-ikx} + b(k)e^{ikx} + o(1), \qquad x \to +\infty,$$

corresponds to the scattering of a particle stream on the x-axis caused by the potential V. More precisely, the original particle stream $a(k)e^{-ikx}$ at $x = +\infty$ with velocity vector $v = -\hbar k e/m$ splits into:

(a) the reflected particle stream $b(k)e^{ikx}$ at $x = +\infty$ with velocity vector $v = \hbar k e/m$ and

(b) the transmitted particle stream e^{-ikx} at $x = -\infty$ with velocity vector $v = -\hbar k e/m$ (Fig. 30.4).

Scattering theory in quantum physics will be studied in greater detail in Part V (cf. Chapters 89 and 92 on quantum mechanics and quantum field theory, respectively).

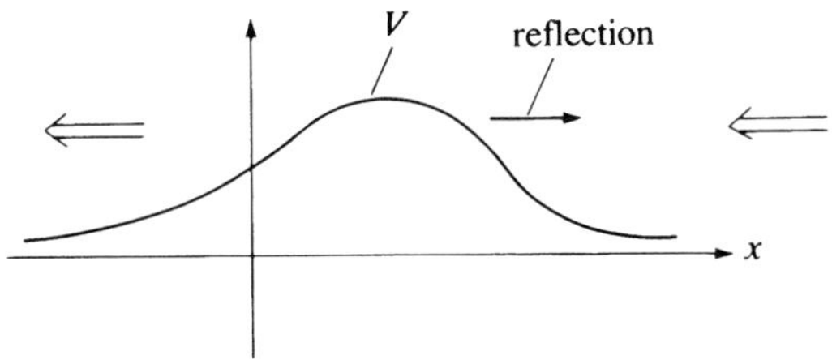

Figure 30.4

30.9. *Weak derivatives.* Let $u: [0, T] \to X$ be a given function, where X is a real H-space and $0 < T < \infty$.

30.9a. For fixed $t \in [0, T]$, suppose that the derivative $u'(t)$ exists in the sense of *weak* convergence on X, i.e.,

$$\frac{u(t+h) - u(t)}{h} \rightharpoonup u'(t) \quad \text{in } X \qquad \text{as} \quad h \to 0. \tag{113}$$

Note that if $t = 0$ or $t = T$, then this limit is to be understood as a one-sided limit. Show that u is continuous at t and that

$$\frac{d}{dt}(u(t)|u(t)) = 2(u'(t)|u(t)). \tag{114}$$

Solution: By the uniform boundedness theorem $A_1(35)$, relation (113) implies

$$\sup_{|h| \le h_0} \left\| \frac{u(t+h) - u(t)}{h} \right\| < \infty.$$

Thus, u is continuous at t. Consequently, the sequence

$$\frac{(u(t+h)|u(t+h)) - (u(t)|u(t))}{h} = \left(\frac{u(t+h) - u(t)}{h} \middle| u(t) + u(t+h) \right)$$

goes to $2(u'(t)|u(t))$ as $h \to 0$.

30.9b. Suppose that, for each $t \in [0, T]$, the derivative $u'(t)$ exists in the sense of weak convergence on X and $u' \in C_w([0, T], X)$, i.e., $u': [0, T] \to X$ is weakly continuous. Show that $u: [0, T] \to X$ is Lipschitz continuous.

Solution: The set $\{u'(t): t \in [0, T]\}$ is bounded in X, by Problem 30.9c.

For fixed $v \in X$, let $\varphi(t) = (u(t)|v)$. By the mean value theorem, $\varphi(t) - \varphi(s) = \varphi'(\xi)(t - s)$, and hence

$$\begin{aligned} |(u(t)|v) - (u(s)|v)| &= |(u'(\xi)|v)(t - s)| \\ &\le \left(\sup_{0 \le \xi \le T} \|u'(\xi)\| \right) \|v\| \, |t - s|. \end{aligned}$$

This implies

$$\|u(t) - u(s)\| \le \text{const}\,|t - s| \qquad \text{for all} \quad t, s \in [0, T].$$

° 30.9c. Let $u: [0, T] \to X$ be weakly continuous. Show that

$$\sup_{0 \le t \le T} \|u(t)\| < \infty.$$

Solution: If the assertion is not true, then there exists a sequence (t_n) in $[0, T]$ such that $\|u(t_n)\| \to \infty$ as $n \to \infty$. Since $[0, T]$ is compact, we may assume that $t_n \to t$ as $n \to \infty$, and hence $u(t_n) \rightharpoonup u(t)$ as $n \to \infty$. Thus, $(u(t_n))$ is bounded in X. This is a contradiction.

30.10. *Approximation method of Rothe.* With this method, the time derivatives of nonlinear parabolic differential equations are discretized. Study Kačur (1985, M) and Schumann (1987). Cf. also Chapter 83.

30.11.* *Nonlinear interpolation theory and the continuity of nonlinear operators on interpolation spaces.* In the Appendix, $A_2(112)$, we summarize basic results of linear interpolation theory. We want to generalize this to nonlinear operators. To this end, we consider schematically the following situation:

$$\begin{array}{ccccc} X & \subseteq & Y & \subseteq & Z \\ \downarrow & & \downarrow & & \downarrow \quad S \\ \tilde{X} & \subseteq & \tilde{Y} & \subseteq & \tilde{Z}. \end{array} \tag{115}$$

Let $X, \tilde{X}$ and $Z, \tilde{Z}$ be real B-spaces, where the embeddings $X \subseteq Z$ and $\tilde{X} \subseteq \tilde{Z}$ are continuous. For fixed $0 < \theta < 1$ and $1 \le p < \infty$, we set

$$Y = [Z, X]_{\theta, p}, \qquad \tilde{Y} = [\tilde{Z}, \tilde{X}]_{\theta, p},$$

i.e., Y and $\tilde{Y}$ are interpolation spaces in the sense of the K-method (cf. $A_2(112)$). In applications, the space X is nicer than Z, i.e., the functions in X are smoother than those in Z, etc. We are given the *nonlinear* operator

$$S: M \subseteq Z \to \tilde{Z}, \tag{116}$$

where M is an open subset of Y (e.g., $M = Y$). Show that $S(M) \subseteq \tilde{Y}$ and that

$$S: M \subseteq Y \to \tilde{Y} \tag{117}$$

is *continuous* with respect to the nice spaces $Y, \tilde{Y}$ in the case where the following two conditions are satisfied:

(i) The operator S is *Lipschitz continuous* with respect to the "bad" spaces $Z, \tilde{Z}$, i.e., there is a constant $L > 0$ such that

$$\|Su - Sv\|_{\tilde{Z}} \le L\|u - v\|_Z \qquad \text{for all} \quad u, v \in M.$$

(ii) The operator S is *bounded* with respect to the "very nice" spaces $X, \tilde{X}$, i.e., there is a constant $C > 0$ such that $S(M \cap X) \subseteq \tilde{X}$ and

$$\|Su\|_{\tilde{X}} \le C(1 + \|u\|_X) \qquad \text{for all} \quad u \in M \cap X.$$

Hint: This is a special case of a more general result due to Günther (1987), which we will prove in Chapter 83.

Special case of linear operators. Let the operators $S: Z \to \tilde{Z}$ and $S: X \to \tilde{X}$ be linear and continuous. Then conditions (i) and (ii) are satisfied. Hence the operator $S: Y \to \tilde{Y}$ is also linear and continuous.

Important applications to nonlinear evolution equations:

$$\begin{aligned} u'(t) + Bu(t) &= 0, \qquad 0 < t < T, \\ u(0) &= u_0. \end{aligned} \tag{118}$$

Such applications will be considered in Chapter 83. The main idea is the following. Let $u = u(t)$ be the unique solution of (118). We set

$$Su_0 = u(t)$$

for fixed time t. From our continuity result above we obtain that the solution of (118) depends continuously on the initial data. More precisely, we shall show that this continuous dependence is uniform with respect to all $t \in [0, T]$ in the case where the constant C in (ii) above is independent of t (cf. Chapter 83).

For example, in Chapter 83, we shall apply the above result to the generalized Korteweg–de Vries equation by letting

$$X = \tilde{X} = W_2^m(\mathbb{R}), \qquad Z = \tilde{Z} = L_2(\mathbb{R}),$$

and

$$Y = \tilde{Y} = W_2^k(\mathbb{R}), \qquad 1 \le k < m.$$

Here, condition (i) (i.e., the Lipschitz continuity of the evolution operator S) follows from the Gronwall lemma similarly as in the proof of Proposition 30.18, and condition (ii) (i.e., the boundedness of S) can be verified easily.

References to the Literature

General expositions: Lions (1969, M) (numerous methods), Caroll (1969, M), Gajewski, Gröger, and Zacharias (1974, M), Barbu (1976, M), Tanabe (1979, M), Henry (1981, L), Haraux (1981, L), Višik and Fursikov (1981, M), Pazy (1983, M), Smoller (1983, M), Majda (1984, L), v. Wahl (1985, M), Browder (1986, P), Pavel (1987, M), Bęnilan, Crandall, and Pazy (1989, M), Christodoulou and Klainerman (1990, M).

Survey articles: Dubinskii (1968), (1976), v. Wahl (1978), (1982), Kato (1986), Crandall (1986).

Semibounded evolution equations and the Euler flow for incompressible inviscid fluids: Kato and Lai (1984), Kato (1986, S).

Quasi-linear parabolic differential equations: Oleinik and Kružkov (1961, S), Višik (1962), Ladyženskaja (1967, M, B, H), Friedman (1964, M), (1969, M), Lions (1969, M), Dubinskii (1968, S), (1976, S), Amann (1986), (1986a–c), (1988), (1988a–c).

Semilinear parabolic equations and reaction–diffusion processes: Henry (1981, L) (recommended as an introduction), Fife (1979, L) (applications to mathematical biology), Smoller (1983, M), Rothe (1984, L), Amann (1984), (1985), (1986c), (1988), (1988a), Browder (1986, P).

Semilinear evolution equations and existence and nonexistence of global solutions in reactor dynamics: Pao (1978), (1979), (1980).

Pseudoparabolic equations: Showalter (1977, M) (linear equations), (1978, M) (non-Newtonian fluids), Gajewski, Gröger, and Zacharias (1974, M), Caroll and Showalter (1976, M), Böhm and Showalter (1985) (nonlinear diffusion), Kačur (1985, L) (Rothe's method for pseudoparabolic equations).

Scales of B-spaces, hard implicit function theorem, and evolution equations of the Cauchy–Kovalevskaja type: Nirenberg (1972), Ovsjannikov (1976) (applications to the shallow water theory), Nishida (1977), Hamilton (1982, S), Deimling (1985, M), Walter (1985).

The Korteweg–de Vries equation, solitons, and the inverse spectral transform: Novikov (1980, M) and Ablowitz and Segur (1981, M, H) (recommended as an introduction), Gardner, Green, Kruskal, and Miura (1967) (classical paper), Lax (1968), Bullough and Caudrey (1980, P) (applications to physics), Calogero and Degasperis (1982, M), Lee (1981, M) and Rajaraman (1982, M) (solitons, instantons, and elementary particles), Davydov (1984, M) (solitons in molecular systems), Knörrer (1986, S) (solitons, integrable Hamiltonian systems, and algebraic geometry), Faddeev and Takhtadjan (1987, M) (infinite-dimensional Hamiltonian systems, solitons, the Riemann–Hilbert problem, and inverse scattering theory), Bobenko and Bordag (1987) (multi-cnoidal waves), Konopelchenko (1987, M) (nonlinear integrable systems in physics), Newell (1987, S).

Nonlinear lattices in physics: Toda (1981, M).

General solutions of the Korteweg–de Vries equation and of other nonlinear evolution equations via multi-dimensional theta functions and Riemannian surfaces: Its and Matveev (1975), Toda (1981, M), Matveev (1986, S), Bobenko and Bordag (1987), Bobenko, Its, and Matveev (1991, M).

Theory of theta functions: Mumford (1983, L).

Complete classification of integrable systems: Mihailov, Šabat, and Jamilov (1987, S).

Magnetic monopoles: Atiyah and Hitchin (1988, M).

Collection of classical papers on solitons: Rebbi (1984).

Scattering processes in quantum mechanics: Schechter (1981, M) (introduction), Reed and Simon (1971, M), Vol. 3, Berezin and Shubin (1983, M) (cf. also the References to the Literature for Chapter 89).

Inverse spectral theory: Pöschel and Trubowitz (1987, M).

The generalized Korteweg–de Vries equation: Bona and Smith (1975), Bona and Scott (1976), Kato (1983, S), (1986), Kružkov and Faminskii (1985), Günther (1987) (applications of nonlinear interpolation theory; cf. Chapter 83).

Numerical methods

Handbook of numerical analysis: Ciarlet and Lions (1988, M), Vols. 1ff.

Finite elements and the Galerkin method: Thomée (1984, L) (recommended as an introduction), Douglas and Dupont (1970, S), Glowinski, Lions, and Trémolières (1976, M), Ciarlet (1977, M), Temam (1977, M), Fletcher (1984, M), Gekeler (1984, M), Reinhardt (1985, M), Girault and Raviart (1986, M).

Fractional step method: Raviart (1967), Janenko (1969, L), Temam (1968), (1977, M), Lions (1969, M).

Regularization methods: Lions (1969, M).

Approximation method of Rothe: Kačur (1985, M), Schumann (1987).

Difference methods: Meis and Marcowitz (1978, M), Marchuk and Shaidurov (1983, M), Reinhardt (1985, M).

Multigrid methods: Hackbusch and Trottenberg (1982, P), Hackbusch (1985, M).

Supercomputer methods: Murman (1985, P), Lichnewsky and Saguez (1987, S), Martin (1988, S).

(Cf. also the References to the Literature for Chapters 31–33.)

CHAPTER 31

Maximal Accretive Operators, Nonlinear Nonexpansive Semigroups, and First-Order Evolution Equations

The analytical theory of semigroups is a recent addition to the ever-growing list of mathematical disciplines .... I hail a semigroup when I see one and I seem to see them everywhere! Friends have observed, however, that there are mathematical objects which are not semigroups.

Einar Hille (1948)

The importance of the class of nonexpansive mappings lies neither in its trivial generalization of a Lipschitz condition, nor in a comparable durability or fruitfulness, but in two key observations: first, nonexpansive mappings are intimately tied to the monotonicity methods developed since the early 1960's, and constitute one of the first classes of mappings for which fixed-point results were obtained by using the fine geometric structure of the underlying Banach space instead of compactness properties. Second, they appear in applications as the shift operator for initial value problems of differential inclusions of the form

$$u'(t) + A(t)u(t) \ni 0,$$

where the operators $A(t)$ are in general multivalued, and in some sense positive (accretive) and only minimally continuous.

Ronald Bruck (1983)

The condition of maximal accretivity is enjoyed by many important operators in applications.

Michael Crandall (1986)

In Chapter 30 we considered first-order evolution equations of the form

$$u'(t) + A(t)u(t) = b(t), \qquad u(0) = u_0 \in H, \tag{1}$$

with the operators $A(t)\colon V \to V^*$ and $b(t) \in V^*$ for all $t \in]0, T[$. In this connection, "$V \subseteq H \subseteq V^*$" is an evolution triple. In this chapter, we investigate problems of the form

$$u'(t) + Au(t) = 0, \qquad u(0) = u_0 \in H, \tag{2}$$

with the operator $A: D(A) \subseteq H \to H$. Whereas we use the three spaces V, H, V^* in (1) in an essential way, we work in (2) only in the H-space H. However, in this connection, A is not necessarily defined on the whole space H. The approach in Chapter 30 was a generalization of the Galerkin method for linear evolution equations in Chapter 23. In this chapter, we want to generalize the theory of linear nonexpansive semigroups studied in Chapter 19 via the Yosida approximation. In Theorem 19.E we showed that a linear densely defined operator $-A$ on an H-space is the generator of a linear nonexpansive semigroup iff A is maximal accretive. As we will show, the theory of nonlinear nonexpansive semigroups on H-spaces and B-spaces is based on nonlinear *maximal accretive* operators. We suppose that A in (2) is maximal accretive. Let $u = u(t)$ be the unique solution of (2). If we set

$$u(t) = S(t)u_0, \tag{3}$$

then $\{S(t)\}$ is a nonlinear nonexpansive semigroup. In Chapter 55 (resp. Chapter 57), we shall consider the more general problem

$$\begin{gathered} u'(t) + Au(t) \ni b(t), \\ u(0) = u_0, \end{gathered} \tag{4}$$

where $A: X \to 2^X$ is a multivalued maximal accretive operator on the H-space X (resp. B-space X). If $b(t) \equiv 0$, then the solutions of (4) generate a nonlinear nonexpansive semigroup $\{S(t)\}$ according to (3). Thus, in contrast to linear nonexpansive semigroups, nonlinear nonexpansive semigroups can be generated by multivalued operators. If $A: D(A) \subseteq X \to X$ is single-valued, then we obtain the corresponding multivalued operator $A: X \to 2^X$ by setting $Au = \{Au\}$ for $u \in D(A)$ and $Au = \emptyset$ for $u \notin D(A)$. Then (4) goes over into the equation

$$\begin{gathered} u'(t) + Au(t) = b(t), \\ u(0) = u_0. \end{gathered} \tag{4*}$$

In an H-space one has the following fundamental equivalence:

$$A \text{ is maximal accretive} \quad \Leftrightarrow \quad A \text{ is maximal monotone.}$$

This is the source for the close connection between nonexpansive semigroups, first-order evolution equations, and the theory of monotone operators.

In this chapter the main idea is the following. Besides the original equation

$$u'(t) + Au(t) = 0, \qquad u(0) = u_0,$$

we consider the well-behaved approximate equation

$$u_\mu'(t) + A_\mu u_\mu(t) = 0, \qquad u_\mu(0) = u_0,$$

where $A_\mu = A(I + \mu A)^{-1}$ is the *Yosida approximation* of the operator A. Since A_μ is Lipschitz continuous, the approximate equation can easily be solved by using the Picard–Lindelöf theorem (Theorem 3.A). The point is to consider the passage to the limit $\mu \to +0$. In this connection, the maximal accretivity of A plays a fundamental role. More precisely, we use the fact that the maximal accretivity of A implies the maximal monotonicity of A. Hence we can apply the trick of maximal monotonicity.

The same method can be used to prove the existence of solutions for equation (4) in case X is an H-space (cf. Chapter 55). In Chapter 57, we will use a completely new technique in order to obtain generalized solutions for the general class (4) of multivalued evolution equations in B-spaces, where A is maximal accretive.

31.1. The Main Theorem

We consider the equation

$$\begin{aligned} u'(t) + Au(t) = 0, &\qquad 0 < t < \infty, \\ u(0) = u_0. & \end{aligned} \tag{5}$$

By definition, the derivative $u'(t)$ exists in the sense of weak convergence iff

$$\frac{u(t+h) - u(t)}{h} \rightharpoonup u'(t) \quad \text{in } H \qquad \text{as} \quad h \to 0.$$

If we replace "$\rightharpoonup$" by "$\to$", then we obtain the derivative in the usual sense.

Theorem 31.A (Komura (1967)). *Let $A: D(A) \subseteq H \to H$ be an operator on the real* H-*space H with the following two properties*:

(H1) *A is monotone, i.e., $(Au - Av|u - v) \geq 0$ for all $u, v \in D(A)$.*
(H2) *$R(I + A) = H$.*

Then, for each given $u_0 \in D(A)$, there exists exactly one continuous function $u: [0, \infty[\to H$ such that equation (5) holds for all $t \in]0, \infty[$, where the derivative $u'(t)$ is to be understood in the sense of weak convergence.

Moreover, this unique solution has the following additional properties:

(a) *$u(t) \in D(A)$ for all $t \geq 0$.*
(b) *$u(\cdot)$ is Lipschitz continuous on $[0, \infty[$.*
(c) *For almost all $t \in]0, \infty[$, the derivative $u'(t)$ exists in the usual sense and satisfies equation (5). Furthermore, we have $\|u'(t)\| \leq \|Au_0\|$.*
(d) *The function $t \mapsto u'(t)$ is the generalized derivative of the function $t \mapsto u(t)$ on $]0, \infty[$. Moreover, $u' \in C_w([0, \infty[, H)$, i.e., $u'(\cdot): [0, \infty[\to H$ is weakly continuous.*
(e) *For all $t \geq 0$, there exists the derivative $u'_+(t)$ from the right and*

$$u'_+(t) + Au(t) = 0, \qquad u(0) = u_0.$$

In the next section we will prove the following equivalences:

(H1), (H2) $\Leftrightarrow$ A is maximal accretive $\Leftrightarrow$ A is maximal monotone.

Corollary 31.1 (Connection with Nonexpansive Semigroups). *Let $u = u(t)$ be the solution of equation* (5). *We set*

$$S(t)u_0 = u(t) \qquad \textit{for all} \quad t \geq 0 \quad \textit{and} \quad u_0 \in D(A). \tag{6}$$

Then, $\{S(t)\}$ is a nonexpansive semigroup on $D(A)$ which can be uniquely extended to a nonexpansive semigroup on $\overline{D(A)}$.

The generator of $\{S(t)\}$ on $\overline{D(A)}$ is $-A$.

Definition 31.2. If we set $u(t) = S(t)u_0$ for $u_0 \in \overline{D(A)}$, then $u(\cdot)$ is called the generalized solution of (5).

It is possible to give a complete characterization of the class of all nonlinear nonexpansive semigroups on H-spaces. To this end, one needs equation (5), where A is a *multivalued* maximal monotone operator. This will be considered in Section 32.24.

31.2. Maximal Accretive Operators

In order to make the proof of Theorem 31.A, in the next section, as transparent as possible, we first prove some results of general interest.

Definition 31.3. Let $A: D(A) \subseteq H \to H$ be an operator on the real H-space H.

(i) A is called *monotone* iff $(Au - Av|u - v) \geq 0$ for all $u, v \in D(A)$.
(ii) A is called *maximal monotone* iff A is monotone and

$$(b - Av|u - v) \geq 0 \qquad \text{for all} \quad v \in D(A)$$

implies $Au = b$, i.e., A has no proper monotone extension.
(iii) A is called *accretive* iff $(I + \mu A): D(A) \to H$ is injective and $(I + \mu A)^{-1}$ is nonexpansive for all $\mu > 0$.
(iv) A is called *maximal accretive* (m-accretive) iff A is accretive and $(I + \mu A)^{-1}$ exists on H for all $\mu > 0$.

For historical reasons, the use of the term "accretive" is not uniform in the literature. We follow the modern theory of nonexpansive semigroups. Maximal accretive operators are sometimes called "hypermaximal accretive" or "hyperaccretive."

Proposition 31.4. *The operator $A: D(A) \subseteq H \to H$ on the real* H-*space H is monotone iff it is accretive.*

PROOF. For all $u, v \in D(A)$ and $\mu > 0$,

$$\|(u + \mu Au) - (v + \mu Av)\|^2 = \|u - v\|^2 + 2\mu(Au - Av|u - v) + \mu^2 \|Au - Av\|^2.$$

Hence we have

$$\|(u + \mu Au) - (v + \mu Av)\|^2 \geq \|u - v\|^2 \qquad \text{for } all \quad \mu > 0$$

iff $(Au - Av|u - v) \geq 0$. □

Proposition 31.5. *Let $A: D(A) \subseteq H \to H$ be an operator on the real* H-*space H. Then the following three properties of A are mutually equivalent*:

(i) *A is monotone and $R(I + A) = H$.*
(ii) *A is maximal accretive.*
(iii) *A is maximal monotone.*

We prove (i) ⇒ (ii) ⇒ (iii). Note that only these implications are needed in the proof of Theorem 31.A below. The implication (iii) ⇒ (i) is a famous result of Minty (1962) which stood at the very beginning of the modern theory of monotone operators. In Section 32.4 we shall prove the main theorem on multivalued maximal monotone operators in B-spaces via a Galerkin method. Then the implication (iii) ⇒ (i) is a simple consequence of this main theorem (cf. Proposition 32.8).

Proof. (i) ⇒ (ii). This follows from the Banach fixed-point theorem. We give the simple proof in Problem 31.1.

(ii) ⇒ (iii). Let A be maximal accretive. Then A is monotone by Proposition 31.4. Suppose that

$$(b - Av|u - v) \geq 0 \qquad \text{for all} \quad v \in D(A). \tag{7}$$

We set $v_t = (I + A)^{-1}(u + b + tz)$ for $t > 0$ and $z \in H$. Addition of (7) with $v = v_t$ and $(u - v_t|u - v_t) \geq 0$ yields

$$(b + u - Av_t - v_t|u - v_t) \geq 0.$$

Hence $(z|u - v_t) \leq 0$. Letting $t \to 0$ we obtain

$$(z|u - (I + A)^{-1}(b + u)) \leq 0 \qquad \text{for all} \quad z \in H,$$

i.e., $u = (I + A)^{-1}(b + u)$. Hence $Au = b$. □

Proposition 31.6 (Convergence Trick of Maximal Monotonicity). *Let A: $D(A) \subseteq H \to H$ be maximal monotone on the real* H-*space H. Then, it follows from either*

$$Au_n \rightharpoonup b \qquad as \quad n \to \infty,$$

$$u_n \to u \qquad as \quad n \to \infty,$$

or

$$Au_n \to b \qquad as \quad n \to \infty,$$

$$u_n \rightharpoonup u \qquad as \quad n \to \infty,$$

that $Au = b$.

PROOF. From $(Au_n - Av|u_n - v) \geq 0$ for all n it follows that $(b - Av|u - v) \geq 0$ for all $v \in D(A)$. Hence $Au = b$. □

31.3. Proof of the Main Theorem

31.3a. Proof of Uniqueness

Let $u: [0, \infty[\rightarrow H$ be a solution of (5), where u is continuous and $u'(t)$ exists for all $t \in]0, \infty[$ in the sense of weak convergence. By Problem 30.9,

$$\frac{d}{dt}\|u(t)\|^2 = 2(u'(t)|u(t)) \qquad \text{for all} \quad t \in]0, \infty[.$$

Let v be another solution of (5). Then, for all $t \in]0, \infty[$,

$$\begin{aligned}\frac{d}{dt}\|u(t) - v(t)\|^2 &= 2(u'(t) - v'(t)|u(t) - v(t)) \\ &= -(Au(t) - Av(t)|u(t) - v(t)) \leq 0,\end{aligned}$$

since A is monotone. Hence

$$\|u(t) - v(t)\| \leq \|u(0) - v(0)\| \qquad \text{for all} \quad t \geq 0. \tag{8}$$

Therefore, since $u(0) = v(0)$, we get $u(t) = v(t)$ for all $t \geq 0$.

31.3b. Proof of Existence

The following proof of Theorem 31.A generalizes the proof of Theorem 19.E.

Step 1: The nonlinear resolvent $R_\mu = (I + \mu A)^{-1}$ for all $\mu > 0$.

It follows from (H1), (H2) in Theorem 31.A and from Proposition 31.5 that A is maximal accretive. Hence the operator R_μ exists for all $\mu > 0$ and

$$R_\mu: H \rightarrow D(A)$$

is bijective and nonexpansive, i.e.,

$$\|R_\mu u - R_\mu v\| \leq \|u - v\| \qquad \text{for all} \quad u, v \in H, \quad \mu > 0. \tag{9}$$

Step 2: The nonlinear Yosida approximation $A_\mu = \mu^{-1}(I - R_\mu)$ for all $\mu > 0$.

Lemma 31.7. *For all $\mu > 0$ and $u, v \in H$ we have:*

$$A_\mu u = AR_\mu u, \tag{10}$$

$$\|A_\mu u - A_\mu v\| \leq 2\mu^{-1}\|u - v\|, \tag{11}$$

$$(A_\mu u - A_\mu v|u - v) \geq 0. \tag{12}$$

That means, A_μ is Lipschitz continuous and monotone.

For all $\mu > 0$ *and* $u \in D(A)$, *we have* $R_\mu Au = A_\mu u$ *and*

$$\|A_\mu u\| \leq \|Au\|. \tag{13}$$

PROOF. Equation (10) follows from $A_\mu R_\mu^{-1} = \mu^{-1}(R_\mu^{-1} - I) = A$. Relation (9) implies that

$$\|A_\mu u - A_\mu v\| = \mu^{-1}\|(u - v) - (R_\mu u - R_\mu v)\| \leq 2\mu^{-1}\|u - v\|,$$

$$\|A_\mu u\| = \mu^{-1}\|u - R_\mu u\| = \mu^{-1}\|R_\mu(I + \mu A)u - R_\mu u\| \leq \|Au\|.$$

The operator R_μ is nonexpansive. Thus, $I - R_\mu$ is monotone (cf. Problem 31.3), and hence A_μ is monotone. □

Step 3: The operator $\bar{A}: D(\bar{A}) \subseteq X \to X$ with $X = L_2(0, T; H)$ for fixed $T > 0$.

We want to extend the operator $A: D(A) \subseteq H \to H$ to the H-space X in a natural way. To this end, we set

$$(\bar{A}u)(t) = Au(t) \qquad \text{for almost all} \quad t \in [0, T]. \tag{14}$$

To be precise, we define the domain $D(\bar{A})$ to be the set of all $u \in X$ with the property that

$$u(t) \in D(A)$$

holds for almost all $t \in [0, T]$ and $t \mapsto Au(t)$ belongs to X.

Lemma 31.8. *The operator* $\bar{A}: D(\bar{A}) \subseteq X \to X$ *is maximal accretive and maximal monotone.*

This follows from the maximal accretiveness of A. We give the simple proof in Problem 31.2.

Step 4: The solution of the *auxiliary problem*:

$$\begin{gathered} u_\mu'(t) + A_\mu u_\mu(t) = 0, \qquad 0 < t < \infty, \\ u_\mu(0) = u_0 \in D(A). \end{gathered} \tag{15}$$

The operator A_μ is Lipschitz continuous on H. By the global Picard–Lindelöf theorem (Corollary 3.8), problem (15) has exactly one C^1-solution $u: \mathbb{R} \to H$.

Let u_μ and v_μ be two C^1-solutions of (15). As in the uniqueness proof above, it follows from the monotonicity of A_μ that

$$\|u_\mu(t) - v_\mu(t)\| \leq \|u_\mu(0) - v_\mu(0)\| \qquad \text{for all} \quad t \geq 0. \tag{16}$$

Step 5: Estimates for u_μ.

Lemma 31.9. *For all* $t, s \geq 0$ *and all* $\mu, \lambda > 0$, *the following hold*:

$$\|u_\mu'(t)\| = \|A_\mu u_\mu(t)\| \leq \|Au_0\|, \tag{17}$$

$$\|u_\mu(t) - u_\mu(s)\| \leq \|Au_0\|\,|t - s|, \tag{18}$$

$$\|u_\mu(t) - R_\mu u_\mu(t)\| \le \mu \|Au_0\|, \tag{19}$$

$$\|u_\mu(t) - u_\lambda(t)\| \le 2\sqrt{\mu + \lambda}(t\|Au_0\|). \tag{20}$$

PROOF. Ad(17). Along with $t \mapsto u_\mu(t)$, the function $t \mapsto u_\mu(t+h)$ with $h > 0$ is also a solution of (15) with appropriate initial values. It follows from (16) that

$$\|u_\mu(t) - u_\mu(t+h)\| \le \|u_\mu(0) - u_\mu(h)\|.$$

Dividing by h and then letting $h \to +0$, we get, by (13), that

$$\|u'_\mu(t)\| \le \|u'_\mu(0)\| = \|A_\mu u_0\| \le \|Au_0\|.$$

Ad(18). This follows from the mean value theorem (Proposition 3.5) and (17).

Ad(19). Note that $A_\mu = \mu^{-1}(I - R_\mu)$ and use (17).

Ad(20). We set

$$\Delta = -(A_\mu u_\mu(t) - A_\lambda u_\lambda(t) | R_\lambda u_\lambda(t) - u_\lambda(t) + u_\mu(t) - R_\mu u_\mu(t)).$$

By (17) and (19),

$$|\Delta| \le 2(\mu + \lambda)\|Au_0\|^2.$$

Taking into consideration the differential equation (15), $A_\mu = AR_\mu$, and the monotonicity of A, we have that

$$\begin{aligned}
\frac{1}{2}\frac{d}{dt}\|u_\mu(t) - u_\lambda(t)\|^2 &= (u'_\mu(t) - u'_\lambda(t) | u_\mu(t) - u_\lambda(t)) \\
&= -(A_\mu u_\mu(t) - A_\lambda u_\lambda(t) | u_\mu(t) - u_\lambda(t)) \\
&= -(AR_\mu u_\mu(t) - AR_\lambda u_\lambda(t) | R_\mu u_\mu(t) - R_\lambda u_\lambda(t)) + \Delta \\
&\le \Delta.
\end{aligned}$$

Now, (20) results from this, after integration over $[0, t]$. In this connection, note that $u_\mu(0) - u_\lambda(0) = 0$. $\square$

Step 6: Convergence in H as $\mu \to +0$.

By (20), $u_\mu(t)$ converges in H as $\mu \to +0$ to a certain $u(t)$ and indeed uniformly with respect to all compact t-intervals. Inequality (18) yields

$$\|u(t) - u(s)\| \le \|Au_0\| \, |t - s| \quad \text{for all} \quad t, s \ge 0. \tag{21}$$

Step 7: Convergence in $X = L_2(0, T; H)$ as $\mu \to +0$.

The uniform convergence in the preceding step yields

$$u_\mu \to u \quad \text{in } X \qquad \text{as} \quad \mu \to +0.$$

By (21), the function $t \mapsto u(t)$ is Lipschitz continuous on $\mathbb{R}_+$. According to Corollary 23.22, the derivative $u'(t)$ exists for almost all $t \in \mathbb{R}_+$. Furthermore, it follows, from (21), that

$$\|u'(t)\| \le \|Au_0\| \qquad \text{for almost all} \quad t \in \mathbb{R}_+,$$

i.e., $u' \in X$. Moreover, u' is the generalized derivative of u on each interval $]0, T[$. By (17), there exists a constant c such that

$$\|u'_\mu\|_X \leq c \qquad \text{for all} \quad \mu > 0.$$

As an H-space, X is reflexive. Therefore, by passing to a suitable subsequence, we obtain that

$$u_\mu \to u \quad \text{in } X \qquad \text{and} \qquad u'_\mu \rightharpoonup w \quad \text{in } X \qquad \text{as} \quad \mu \to +0.$$

By Proposition 23.19, it follows that

$$u' = w.$$

Because of

$$u'_\mu(t) = -A_\mu u_\mu(t) = -AR_\mu u_\mu(t)$$

and (19), the following hold:

$$R_\mu u_\mu \to u \quad \text{in } X \qquad \text{as} \quad \mu \to +0,$$
$$-\bar{A}R_\mu u \rightharpoonup w \quad \text{in } X \qquad \text{as} \quad \mu \to +0.$$

The operator $\bar{A}$ is maximal monotone. Hence the *convergence trick of maximal monotonicity* (Proposition 31.6) yields the fundamental result that $u \in D(\bar{A})$ and

$$w = -\bar{A}u,$$

i.e., $u' = -\bar{A}u$. This implies $u'(t) = -Au(t)$ for almost all $t \in \mathbb{R}_+$.

Let $t \geq 0$. From $\|AR_\mu u_\mu(t)\| \leq \|Au_0\|$, by (17), and from (19) in connection with Step 6, it follows, for an appropriate subsequence, that

$$R_\mu u_\mu(t) \to u(t) \quad \text{in } H \qquad \text{as} \quad \mu \to +0,$$
$$AR_\mu u_\mu(t) \rightharpoonup z \quad \text{in } H \qquad \text{as} \quad \mu \to +0.$$

Since A is maximal monotone, we obtain $u(t) \in D(A)$ and $Au(t) = z$. Consequently,

$$\|Au(t)\| \leq \varliminf_{\mu \to +0} \|AR_\mu u_\mu(t)\| \leq \|Au_0\|.$$

If $t \mapsto u(t)$ is a solution of (5), then $t \mapsto u(t+s)$ is also a solution of (5) with initial value $u(s)$. Hence

$$\|Au(t)\| \leq \|Au(s)\| \qquad \text{for all} \quad t \geq s \geq 0,$$

i.e., the function $t \mapsto \|Au(t)\|$ is monotone decreasing on $\mathbb{R}_+$.

Step 8: The function $t \mapsto Au(t)$ is continuous from the right on $\mathbb{R}_+$.

Let $t_n \searrow t$ as $n \to \infty$. Then $\|A(u(t_n))\| \leq \|Au(t)\|$ for all n. Thus, there exists a subsequence, again denoted by $(Au(t_n))$, such that

$$Au(t_n) \rightharpoonup z \qquad \text{as} \quad n \to \infty,$$
$$u(t_n) \to u(t) \qquad \text{as} \quad n \to \infty.$$

Since A is maximal monotone, $Au(t) = z$. The sequence $(\|Au(t_n)\|)$ is monotone decreasing and hence convergent. Furthermore, $Au(t_n) \rightharpoonup Au(t)$ as $n \to \infty$ implies

$$\|Au(t)\| \leq \varliminf_{n\to\infty} \|Au(t_n)\| \leq \|Au(t)\|,$$

and hence $\|Au(t_n)\| \to \|Au(t)\|$. Thus, $Au(t_n) \to Au(t)$ as $n \to \infty$, by Proposition 21.23(d). The convergence principle (Proposition 10.13) implies $Au(s) \to Au(t)$ as $s \to t + 0$.

Step 9: Derivative from the right.

Let $h > 0$. Integration of $u'(t) + Au(t) = 0$ yields

$$\frac{u(t+h) - u(t)}{h} = -\frac{1}{h}\int_t^{t+h} Au(s)\,ds.$$

By Step 8,

$$\frac{d^+}{dt}u(t) = -Au(t) \qquad \text{for all} \quad t \geq 0.$$

Step 10: Weak continuity.

The same argument as in Step 8 shows that $t_n \to t$ as $n \to \infty$ implies $Au(t_n) \rightharpoonup Au(t)$ in H as $n \to \infty$, i.e., for each $w \in H$, the function

$$t \mapsto (Au(t)|w)$$

is continuous on $[0, \infty[$.

Step 11: Weak derivative.

In what follows let $t, s \geq 0$, $\mu > 0$, and $w \in H$. From (15) we get

$$(u_\mu(t)|w) = (u_0|w) - \int_0^t (A_\mu u_\mu(s)|w)\,ds.$$

By (17),

$$|(A_\mu u_\mu(s)|w)| \leq \|Au_0\|\,\|w\|.$$

Noting that $A_\mu = AR_\mu$, we obtain from Step 7 that

$$A_\mu u_\mu(t) \rightharpoonup Au(t) \quad \text{in } H \qquad \text{as} \quad \mu \to +0.$$

By Step 6, $u_\mu(t) \to u(t)$ in H as $\mu \to +0$. Therefore, as $\mu \to +0$, majorized convergence yields

$$(u(t)|w) = (u_0|w) - \int_0^t (Au(s)|w)\,ds.$$

Since the integrand is continuous by Step 10, we obtain

$$\lim_{h\to 0} \frac{(u(t+h)|w) - (u(t)|w)}{h} = -(Au(t)|w).$$

That means

$$u'(t) = -Au(t),$$

where the derivative $u'(t)$ is to be understood in the sense of weak convergence on H.

The proof of Theorem 31.A is complete. The proof of Corollary 31.1 will be given in Problem 31.4. □

31.4. Application to Monotone Coercive Operators on B-Spaces

Proposition 31.10. *Let $A: D(A) \subseteq H \to H$ be a linear operator on the real* H-*space H. Then A is maximal monotone iff $-A$ is maximal dissipative.*

PROOF. This follows immediately from Definition 19.44 and Proposition 31.5. □

We now want to explain the connection between maximal monotone operators and the Friedrichs extension.

EXAMPLE 31.11. Let $A: D(A) \subseteq H \to H$ be a linear symmetric and strongly monotone operator on the real H-space H, i.e.,

$$(Au|u) \geq c\|u\|^2 \qquad \text{for all} \quad u \in D(A) \quad \text{and fixed} \quad c > 0. \tag{22}$$

Let $A_F: D(A_F) \subseteq H \to H$ be the Friedrichs extension of A with the corresponding energetic space H_E. By Section 19.10, there exists an energetic extension A_E of A such that

$$A \subseteq A_F \subseteq A_E$$

with $D(A) \subseteq D(A_F) \subseteq H_E \subseteq H \subseteq H_E^*$, and the operator

$$A_E: H_E \to H_E^*$$

is a linear, strongly monotone homeomorphism. Moreover, the Friedrichs extension A_F is the restriction of A_E to $D(A_F) = A_E^{-1}(H)$. Then (22) also holds for A_F, and the operator A_F is self-adjoint. By construction of A_F, $R(I + A_F) = H$. Hence A_F is maximal monotone.

Consequently, the *Friedrichs extension* of A is a *maximal monotone* extension of A.

Proposition 31.12. *Let $A: D(A) \subseteq H \to H$ be a linear symmetric monotone operator on the real* H-*space H, i.e.,*

$$(Au|u) \geq 0 \qquad \textit{for all} \quad u \in D(A).$$

Then A is maximal monotone iff it is self-adjoint.

PROOF. Let A be self-adjoint. By Problem 19.7, all $\lambda < 0$ belong to the resolvent set of A. In particular, $R(I + A) = H$, i.e., A is maximal monotone.

Let A be maximal monotone. We set $B = I + A$. Then B is strongly monotone. Let B_F be the Friedrichs extension of B. Then $A \subseteq B_F - I$. The operator $B_F - I$ is self-adjoint. Since A is maximal monotone, there is no proper monotone extension of A. Hence $A = B_F - I$, i.e., A is self-adjoint. $\square$

We now want to generalize the situation of the Friedrichs extension in Example 31.11 to nonlinear operators. At the same time, we want to explain the connection with evolution equations. To this end, we consider the two evolution equations

$$\begin{aligned} u'(t) + A_E u(t) = 0, \qquad 0 < t < \infty, \\ u(0) = u_0, \end{aligned} \tag{23}$$

where the derivative $u'(t)$ is to be understood in the sense of the convergence in H, for almost all $t \in]0, \infty[$, and

$$\begin{aligned} u'(t) + Au(t) = 0, \qquad 0 < t < \infty, \\ u(0) = u_0. \end{aligned} \tag{24}$$

Furthermore, we consider the equation

$$\begin{aligned} \frac{d}{dt}(u(t)|w) + a(u(t), w) = 0 \qquad \text{for all} \quad w \in H, \quad 0 < t < \infty, \\ u(0) = u_0. \end{aligned} \tag{25}$$

Here, we set $a(u, v) = \langle A_E u, v \rangle_V$.

We make the following assumptions:

(H1) "$V \subseteq H \subseteq V^*$" is an evolution triple. In particular, this means that V is a B-space, H is an H-space, and

$$\langle h, v \rangle_V = (h|v)_H \qquad \text{for all} \quad h \in H, \quad v \in V.$$

(H2) The operator $A_E: V \to V^*$ is monotone, hemicontinuous, and coercive.

By Theorem 26.A, the operator A_E is surjective.

(H3) We set $D(A) = \{u \in V: Au \in H\}$ and $Au = A_E u$ for all $u \in D(A)$.

Proposition 31.13. *Suppose that* (H1) *through* (H3) *hold. Then*:

(a) *The operator* $A: D(A) \subseteq H \to H$ *is maximal monotone.*
(b) *Each solution of equation* (23) *is also a solution of* (24). *For every* $u_0 \in D(A)$, *equation* (24) *has a unique solution in the sense of Theorem* 31.A. *This solution is also a solution of* (25).
(c) *For each* $u_0 \in \overline{D(A)}$, *equation* (24) *has the generalized solution* $u(t) = S(t)u_0$, *where* $\{S(t)\}$ *is the nonexpansive semigroup generated by* $-A$.

Proof. The operator A_E is monotone, i.e.,

$$\langle A_E u - A_E v, u - v \rangle_V \geq 0 \qquad \text{for all} \quad u, v \in V.$$

Since $A_E u = Au \in H$ for all $u \in D(A)$, we obtain

$$(Au - Av|u - v) \geq 0 \qquad \text{for all} \quad u, v \in H.$$

Let I denote the identity operator on H. Then

$$\langle (A_E + I)u, u \rangle_V = \langle A_E u, u \rangle_V + (u|u) \qquad \text{for all} \quad u \in H.$$

Thus, the operator $A_E + I: V \to V^*$ is monotone, hemicontinuous, coercive, and hence surjective by Theorem 26.A. This implies $R(A + I) = H$, i.e., A is maximal monotone.

Let $u = u(t)$ be a solution of (23). Then $u(t) \in V$, i.e., $u(t) \in H$ and $u'(t) \in H$. This implies $A_E u(t) \in H$. Thus, $u(t) \in D(A)$ and $A_E u(t) = Au(t)$. Consequently, (23) is a solution of (24) and hence of (25).

The remaining assertions are obvious. □

31.5. Application to Quasi-Linear Parabolic Differential Equations

We want to solve the initial–boundary value problem

$$\begin{aligned} u_t + Lu &= 0 && \text{on } G \times]0, \infty[, \\ D^\beta u(x,t) &= 0 && \text{on } \partial G \times]0, \infty[\quad \text{for all} \quad \beta: |\beta| \leq m - 1, \\ u(x,0) &= u_0(x) && \text{on } G, \end{aligned} \tag{26}$$

with

$$Lu(x,t) = \sum_{|\alpha| \leq m} (-1)^{|\alpha|} D^\alpha A_\alpha(x, Du(x,t))$$

in an appropriate generalized sense. To this end, we will apply Proposition 31.13. We make the following assumptions:

(H1) G is a bounded region in $\mathbb{R}^N$, $N \geq 1$. Let $m \geq 1$.

We set $H = L_2(G)$ and $V = \mathring{W}_p^m(G)$ with $2 \leq p < \infty$.

(H2) The differential operator L satisfies the assumptions (H1)–(H4) of Proposition 26.12 (Carathéodory condition, growth condition, monotonicity and coerciveness condition).

We set

$$a(u, v) = \int_G \sum_{|\alpha| \leq m} A_\alpha(x, Du(x)) D^\alpha v(x)\, dx.$$

As usual, integration by parts yields the generalized problem for (26):

$$\begin{gathered} \frac{d}{dt}(u(t)|w)_H + a(u(t), w) = 0 \qquad \text{for all} \quad w \in V, \quad 0 < t < \infty, \\ u(0) = u_0. \end{gathered} \tag{27}$$

By the proof of Proposition 26.12, there exists a monotone, continuous, and coercive operator $A_E: V \to V^*$ with

$$\langle A_E u, v\rangle_V = a(u,v) \qquad \text{for all} \quad u, v \in V.$$

Furthermore, let $D(A) = A_E^{-1}(H)$ and $Au = A_E u$ on $D(A)$. Motivated by Proposition 31.13 we consider, instead of (27), the operator equation

$$\begin{aligned} u'(t) + Au(t) &= 0, \qquad 0 < t < \infty, \\ u(0) &= u_0. \end{aligned} \tag{28}$$

Proposition 31.14. *Suppose that* (H1) *and* (H2) *hold. Then*:

(a) *For each* $u_0 \in D(A)$, *equation* (28) *has a unique solution in the sense of Theorem* 31.A.
(b) *For each* $u_0 \in \overline{D(A)}$, *equation* (28) *has the generalized solution* $u(t) = S(t)u_0$, *where* $\{S(t)\}$ *is the nonexpansive semigroup generated by* $-A$.

PROOF. This follows from Proposition 31.13. □

Corollary 31.15. *If all the* A_α *are* C^∞*-functions, then* $C_0^\infty(G) \subseteq D(A)$ *and* $\overline{D(A)} = L_2(G)$.

The proof will be given in Problem 31.5.

31.6. A Look at Quasi-Linear Evolution Equations

We consider the quasi-linear evolution equation

$$\begin{aligned} u'(t) + A(u(t))u(t) &= F(u(t)) \qquad \text{for all} \quad t \in [0, T], \\ u(0) &= u_0, \end{aligned} \tag{29}$$

where $T > 0$ is sufficiently small and the operator $A(u)$: $D(A(u))$: $X \to X$ is linear and densely defined on the real B-space X. More precisely, we assume:

(H1) Let X and Y be real reflexive B-spaces such that the embedding $Y \subseteq X$ is continuous and Y is dense in X. Furthermore, there exists a linear isometric homeomorphism $S: Y \to X$.
(H2) Maximal accretivity. There is an open subset U of Y and a number $\beta \geq 0$ such that, for each $u \in U$, the linear, densely defined operator

$$A(u) + \beta I: D(A(u)) \subseteq X \to X$$

is maximal accretive, and $Y \subseteq D(A(u))$.
(H3) Contractivity. For all $u, v \in U$ and $y \in Y$,

$$\|A(u)y - A(v)y\|_X \leq \text{const}\, \|u - v\|_X \|y\|_Y,$$

$$\|A(u)y\|_X \leq \text{const}\, \|y\|_Y.$$

(H4) For each $u \in U$, there is a linear continuous operator $L(u)\colon X \to X$ such that $\sup_{u \in U} \|L(u)\| < \infty$ and

$$SA(u)S^{-1} = A(u) + L(u).$$

(H5) The operator $F\colon U \to Y$ is bounded and, for all $u, v \in U$,

$$\|F(u) - F(v)\|_X \leq \text{const}\, \|u - v\|_X.$$

(H6) For all $u, v \in U$, $w \in X$,

$$\|L(u)w - L(v)w\|_X \leq \text{const}\, \|u - v\|_Y \|w\|_X,$$

$$\|F(u) - F(v)\|_Y \leq \text{const}\, \|u - v\|_Y.$$

Theorem 31.B (Kato (1976)). *If* (H1)–(H5) *hold then, for each* $u_0 \in U$, *there is a* $T > 0$ *such that the original problem* (29) *has a solution*

$$u \in C([0, T], Y) \cap C^1([0, T], X).$$

Corollary 31.16 (Continuous Dependence of the Solution on the Initial Values). *If, in addition, condition* (H6) *holds, then the map* $u_0 \mapsto u$ *is continuous from* U *into* $C([0, T], Y)$, *where* T *is locally constant on* U.

An "elementary" proof of this important result can be found in Crandall and Sougandis (1986). The idea is to prove the convergence of the difference method

$$\begin{gathered} \frac{v_{i+1} - v_i}{\Delta t} + A(v_i)v_{i+1} = F(v_i), \qquad i = 0, 1, \ldots, n, \\ v_0 = u_0, \end{gathered} \tag{30}$$

where v_i denotes an approximation for $u(i\Delta t)$. For simplicity, let $\beta = 0$. Since, by hypothesis, the operator $A(v_i)$ is maximal accretive, the inverse operator $(I + \Delta t\, A(v_i))^{-1}$ exists on X and we obtain

$$v_{i+1} = (I + \Delta t\, A(v_i))^{-1}(v_i + \Delta t\, F(v_i)).$$

The convergence proof for (30) is related to our proof of Theorem 57.A for multivalued, maximal accretive evolution equations.

The original proof of Kato for Theorem 31.B was based on the iteration method

$$u'_{n+1}(t) + A(u_n(t))u_{n+1}(t) = F(u_n(t)) \quad \text{on } [0, T]$$

for $n = 0, 1, \ldots$. In order to prove the convergence of (u_n) via the Banach fixed-point theorem, one needs sharp results on linear semigroups. Generalizations and important applications of Theorem 31.B to nonlinear partial differential equations in mathematical physics can be found in Kato (1975), (1975a), (1976, L), (1985, L), (1986, S) and in Kato, Marsden, and Hughes (1977) (second-order evolution equations). In this connection, quasi-linear symmetric hyperbolic systems play an important role. This class of problems will be studied in Chapter 83 by using another approach.

31.7. A Look at Quasi-Linear Parabolic Systems Regarded as Dynamical Systems

Many time-dependent problems in physics, chemistry, and biology can be reduced to the following abstract quasi-linear first-order evolution equation:

$$\begin{gathered} u'(t) + A(t,u)u = F(t,u), \qquad 0 < t \leq T, \\ u(0) = u_0, \end{gathered} \tag{31}$$

where the operator $w \mapsto A(u,t)w$ is linear. An important goal of modern analysis is to develop a *qualitative theory* (or dynamical theory) for such processes, parallel to the classical theory of dynamical systems with a finite number of degrees of freedom (ordinary differential equations). The main ingredients of such a theory should be the following:

(a) Existence and uniqueness of classical solutions.
(b) Existence of global smooth semiflows on appropriate state spaces in the autonomous case (i.e., A and F are independent of time t).
(c) Blowing-up of solutions.
(d) Existence of global solutions for all times $t \geq 0$.
(e) Stability of the time-evolution.
(f) Structural stability of the semiflow.
(g) Bifurcation.

For example, we are interested in the following questions with respect to (e)–(g):

(i) Stability of equilibrium points.
(ii) Existence of limit cycles (stable periodic solutions).
(iii) Existence of attractors and repellers and their structure (e.g., their Hausdorff dimension), and their stability under external influences.
(iv) Existence of stable, unstable, and center manifolds.
(v) Classification of structural stable semiflows.
(vi) Bifurcation of invariant sets under external influences (e.g., an equilibrium point looses its stability and passes to a stable periodic solution (Hopf bifurcation)).
(vii) Classification of the roads to chaotic behavior (turbulence).
(viii) Classification of shock waves.

Simple typical examples for these phenomena have been considered in Sections 3.7 and 3.8 in terms of ordinary differential equations.

Until today, a general theory is not available. The creation of such a theory will be a huge *program* for the mathematics of the future. However, for general processes of the reaction–diffusion type, we are able today to complete the first important step towards a general theory. To this end, we consider fairly general quasi-linear parabolic systems of the following form:

$$\begin{aligned} u_t + \mathscr{A}(x,t,u)u &= f(x,t,u) && \text{on } G \times]0,\infty[, \\ \mathscr{B}(x,t,u)u &= \alpha g(x,t,u) && \text{on } \partial G \times]0,\infty[, \\ u(x,0) &= u_0(x) && \text{on } G. \end{aligned} \tag{32}$$

We want to show that, for given u_0, there exists a unique maximal classical solution, which forms a local *semiflow* in the autonomous case. A typical application to reaction–diffusion processes will be considered in Example 31.17 below. In the following we *sum* over two equal indices j, k from 1 to N. We set

$$u = (u_1, \ldots, u_m)$$

and $x = (\xi_1, \ldots, \xi_N)$ with $D_i = \partial/\partial\xi_i$ as well as

$$\mathscr{A}(x,t,u)u = -D_j(a_{jk}(x,t,u)D_k u) + a_j(x,t,u)D_j u$$

and either

$$\mathscr{B}(x,t,u)u = a_{jk}(x,t,u)n_j D_k u, \qquad \alpha = 1,$$

(Neumann boundary condition) or

$$\mathscr{B}(x,t,u)u = u, \qquad \alpha = 0,$$

(Dirichlet boundary condition). As usual, $n = (n_1, \ldots, n_N)$ denotes the outer unit normal to ∂G.

We assume:

(H1) Let G be a bounded region in $\mathbb{R}^N$ with $\partial G \in C^\infty$ and $N \geq 1$.

(H2) All the coefficients are C^∞ for $x \in \bar{G}$ (resp. $x \in \partial G$) and $t \geq 0$. Explicitly, this means that, for all j and k,

$$a_{jk}, a_j \in C^\infty(\bar{G} \times \mathbb{R}_+ \times \mathbb{R}^m, \mathbb{R}^{m^2}),$$

$$f \in C^\infty(\bar{G} \times \mathbb{R}_+ \times \mathbb{R}^m, \mathbb{R}^m), \qquad g \in C^\infty(\partial G \times \mathbb{R}_+ \times \mathbb{R}^m, \mathbb{R}^m),$$

where $g(x,t,0) \equiv 0$. Note that a_{jk} and a_j are real $(m \times m)$-matrices for each fixed argument (x,t,u), whereas f and g are m-vectors.

(H3) Ellipticity of $\mathscr{A}$. For each $(x,t,u) \in \bar{G} \times \mathbb{R}_+ \times \mathbb{R}^m$ and $D \in \mathbb{R}^N$ with $|D| = 1$, all the eigenvalues of the $(m \times m)$-matrix

$$a_{jk}(x,t,u)D_j D_k$$

have positive real part.

Recall that we sum over j, $k = 1, \ldots, N$.

If a_{jk}, a_j, f, and g are independent of time t, then we have the *autonomous* case at hand.

As the *state space* we use

$$X = \{u \in W_p^1(G)^m : (1 - \alpha)u = 0 \text{ on } \partial G\},$$

where $N < p < \infty$. Recall that $u \in W_p^1(G)^m$ means that $u_i \in W_p^1(G)$ for all i.

Theorem 31.C (Amann (1988)). *Assume* (H1)–(H3). *Then*:

(a) Existence and uniqueness. *For each* $u_0 \in X$, *the original problem* (32) *has a unique maximal classical solution* $u = u(t; u_0)$. *Let* $[0, t_e[$ *be the maximal half-open interval of existence, where* t_e *depends on* u_0.

(b) Global existence. *If we have the a priori estimate*

$$\sup_{t \in J} \|u(t; u_0)\|_X < \infty$$

for each bounded interval $J = [0, T[$ *with* $T \leq t_e$, *then* $t_e = \infty$.

(c) Continuous dependence on the data. *The solution* $u(t)$, $t > 0$, *depends in the topology of* X *smoothly upon* u_0, a_{jk}, a_j, f, *and* g. *For each* $h \in X$, *the function*

$$t \mapsto \frac{\partial u(t; u_0)}{\partial u_0} h$$

is equal to the solution of the initial–boundary value problem which is obtained by differentiating the original problem (32) *formally with respect to* u_0 *in the direction* h.

(d) Local semiflow. *In the autonomous case, the map*

$$(t, u_0) \mapsto u(t; u_0)$$

represents a local semiflow on X.

(e) Bounded orbits. *Suppose that there is a number* $t_0 \in]0, t_e[$ *such that*

$$\sup_{t_0 \leq t < t_e} \|u(t; u_0)\|_Y < \infty,$$

where $Y = W_p^l(G)^m$ *for some fixed* $l > 1$. *Then* $t_e = \infty$, *and the orbit*

$$\{u(t; u_0) \colon 0 \leq t < \infty\}$$

is relatively compact and hence bounded in the state space X.

Set $u(t; u_0) = S(t)u_0$. Recall that the map in (d) is called a *local semiflow* iff $S(0) = I$ and

$$S(t + s)u_0 = S(t)S(s)u_0$$

for all $u_0 \in X$ and for all $t, s \geq 0$ for which these expressions exist. In addition, we obtain that the map $u_0 \mapsto u(t; u_0)$ in (d) is C^∞ on X for $0 \leq t < t_e(u_0)$.

In very rough terms, the proof of this deep result proceeds along the lines of the proof for Theorem 19.I, i.e., the abstract evolution equation (31) is reduced to an integral equation via semigroup theory. In this connection, it is very important, that the operator $-A(u, t)$ generates an analytic semigroup for each fixed argument (u, t). This follows from (H3) and reflects the parabolicity of the problem. The detailed proof of Theorem 31.C is complex. It is based on sophisticated results for linear elliptic and parabolic differential equations, the theory of analytic semigroups and time-dependent first-order evolution equations, interpolation theory, and the regularity theory for quasi-linear parabolic equations. See Amann (1988), (1988a)).

EXAMPLE 31.17 (Chemotaxis Problems for Two Components). Let $u = (u_1, u_2)$ and $x \in \mathbb{R}^3$. The results of Theorem 31.C can be applied to the following semilinear parabolic system:

$$\begin{aligned} u_t - a\Delta u &= f(x,u) \quad \text{on } G \times]0,\infty[, \\ -a\frac{\partial u}{\partial n} &= g(x,u) \quad \text{on } \partial G \times]0,\infty[, \\ u(x,0) &= u_0(x) \quad \text{on } G, \end{aligned} \tag{33}$$

where

$$a = \begin{pmatrix} \alpha_1 & \beta\alpha_2 \\ 0 & \alpha_2 \end{pmatrix}.$$

Such equations describe reaction–diffusion processes in physics, chemistry, biology, and population dynamics. The positive numbers α_1 and α_2 are the so-called diffusion coefficients, and the real number β describes the drift of the quantity u_1 caused by the gradient of u_2.

In order to obtain an interpretation in terms of chemistry, let M_i be the mass of the ith substance, and let

$$u_i = M_i/(M_1 + M_2)$$

be the *concentration* of the ith substance. The so-called *current density vector* of diffusion for the ith substance is given by

$$j_i = -\alpha_i \operatorname{grad} u_i.$$

An equation of the form

$$\frac{\partial u_i}{\partial t} + \operatorname{div} j_i = h_i$$

means

$$\frac{d}{dt}\int_H u_i(x,t)\,dx = -\int_{\partial H} j_i n\,dO + \int_H h_i\,dx, \tag{34}$$

for each reasonable subregion H of G. This follows after integration by parts. Relation (34) describes the change of the mass of the ith substance in the region H by diffusion via j_i and by a source h_i, which corresponds to chemical reactions. A more detailed discussion can be found in Section 69.1.

The original equation (33) can now be written as

$$\begin{aligned} \frac{\partial u_1}{\partial t} + \operatorname{div} j_1 &= f_1 - \beta \operatorname{div} j_2, \\ \frac{\partial u_2}{\partial t} + \operatorname{div} j_2 &= f_2 \quad \text{on } G \times]0,\infty[, \end{aligned}$$

with the boundary condition

$$\begin{aligned} j_1 n &= g_1 - \beta j_2 n, \\ j_2 n &= g_2 \quad \text{on } \partial G \times]0,\infty[. \end{aligned}$$

and the initial condition

$$u_i(x,0) = u_{0i}(x) \quad \text{on } G, \qquad i = 1, 2.$$

The boundary condition describes diffusion through the boundary ∂G.

PROBLEMS

31.1. *Maximal accretive operators.* Let $A: D(A) \subseteq H \to H$ be a monotone operator with $R(I + \lambda A) = H$ for fixed $\lambda > 0$, where H is a real H-space. Show that A is maximal accretive.

Solution: By Proposition 31.4, the operator A is accretive. Hence the operator $R_\lambda = (I + \lambda A)^{-1}$ is nonexpansive and it is sufficient to show that $R(I + \mu A) = H$ for all $\mu > 0$. The equation

$$u + \mu Au = w, \qquad u \in H,$$

is equivalent to

$$u = L_w u, \qquad u \in H, \tag{35}$$

with $L_w u = R_\lambda((1 - \mu^{-1}\lambda)u + \mu^{-1}\lambda w)$. For $\mu > \lambda/2$, $|1 - \mu^{-1}\lambda| < 1$ and

$$\|L_w u - L_w v\| \leq |1 - \mu^{-1}\lambda| \, \|u - v\| \qquad \text{for all} \quad u, v \in H.$$

By the Banach fixed-point theorem (Theorem 1.A), equation (35) has a unique solution, i.e., $R(I + \mu A) = H$ for all $\mu > \lambda/2$. An n-fold repetition of this argument yields

$$R(I + \mu A) = H \qquad \text{for all} \quad \mu > \lambda/2^n \quad \text{and all} \quad n.$$

31.2. *Proof of Lemma* 31.8.

Solution: Let $X = L_2(0, T; H)$. The operator $\bar{A}$ is monotone. In fact, for all $u, v \in D(\bar{A})$, the monotonicity of A implies

$$(\bar{A}u - \bar{A}v | u - v)_X = \int_0^T (Au(t) - Av(t) | u(t) - v(t)) \, dt \geq 0.$$

We show that $R(I + \bar{A}) = X$. Let $w \in X$. We set

$$u(t) = (I + A)^{-1} w(t), \qquad u_0 = (I + A)^{-1}(0).$$

Since A is maximal accretive, the operator $(I + A)^{-1}$ is nonexpansive, i.e., $\|u(t) - u_0\|^2 \leq \|w(t)\|^2$. Integration over $[0, T]$ yields $u - u_0 \in X$, i.e., $u \in X$.

By Proposition 31.5, the operator $\bar{A}$ is maximal accretive and maximal monotone.

31.3. *Nonexpansive operators and monotone operators.* Let $C: D(C) \subseteq H \to H$ be a nonexpansive operator, where H is a real H-space. Show that $I - C$ is monotone.

Solution: For all $u, v \in D(C)$,

$$\begin{aligned}((u - Cu) - (v - Cv) | u - v) &= (u - v | u - v) - (Cu - Cv | u - v) \\ &\geq \|u - v\|^2 - \|Cu - Cv\| \, \|u - v\| \geq 0.\end{aligned}$$

31.4. *Proof of Corollary* 31.1.

Solution:

(I) Let $u_0, v_0 \in D(A)$ and $t, s \geq 0$. We show that

$$S(0)u_0 = u_0, \tag{36}$$

$$S(t+s)u_0 = S(t)S(s)u_0, \tag{37}$$

$$\|S(t)u_0 - S(t)v_0\| \leq \|u_0 - v_0\|, \tag{38}$$

$$t \mapsto S(t)u_0 \text{ is continuous on } [0, \infty[. \tag{39}$$

Let $u(t) = S(t)u_0$. Then, by definition, $u(\cdot)$ is the unique solution of the equation

$$u'(t) + Au(t) = 0, \qquad 0 < t < \infty, \quad u(0) = u_0.$$

For fixed $s \geq 0$ we set $v(t) = u(t+s)$. Then, v is a solution of the equation

$$v'(t) + Av(t) = 0, \qquad 0 < t < \infty, \quad v(0) = u(s).$$

Hence $v(t) = S(t)u(s)$. This is (37). Inequality (38) follows from (8), and (39) follows from the continuity of $u = u(t)$.

(II) Since $S(t)\colon D(A) \to D(A)$ is nonexpansive for all $t \geq 0$, there exists a unique extension $S(t)\colon \overline{D(A)} \to \overline{D(A)}$ such that (36)–(39) remain true. In fact, let $u \in \overline{D(A)}$ and let (u_n) be a sequence in $D(A)$ with $u_n \to u$ as $n \to \infty$. Then (u_n) is a Cauchy sequence. By (38), the sequence $(S(t)u_n)$ is also Cauchy. We set

$$S(t)u = \lim_{n\to\infty} S(t)u_n.$$

Inequality (38) shows the independence of this limiting value of the chosen sequence (u_n) and the uniformity of the limit process for all $t \geq 0$. Therefrom follow (36) through (39) also for the extended operators.

(III) By Step 9 in Section 31.3,

$$\lim_{h\to+0} \frac{S(h)u_0 - u_0}{h} = -Au_0 \qquad \text{for all} \quad u_0 \in D(A).$$

Let $-B$ the generator of $\{S(t)\}$ on $\overline{D(A)}$. Then $A \subseteq B$. By Problem 31.3,

$$(u - S(t)u - (v - S(t)v)|u - v) \geq 0 \qquad \text{for all} \quad u, v \in \overline{D(A)}.$$

Differentiation at $t = 0$ yields $(Bu - Bv|u - v) \geq 0$. Hence B is monotone. Since A is maximal monotone, $B = A$.

31.5. *Proof of Corollary* 31.15.

Solution: If all A_α are C^∞-functions, then for $u \in C_0^\infty(G)$, after integration by parts, the inequality

$$|a(u,v)| = \left|\int_G vLu\,dx\right| \leq \text{const}\,\|v\|_H$$

holds for all $v \in V$. Therefore, there exists a $w \in H$ with

$$a(u,v) = (w|v)_H \qquad \text{for all} \quad v \in V.$$

Hence

$$a(u,v) = \langle w, v\rangle_V \qquad \text{for all} \quad v \in V.$$

On the other hand, $a(u, v) = \langle A_E u, v \rangle_V$ for all $v \in V$. Hence $A_E u \in H$, i.e., $u \in D(A)$.

31.6. *The generic property of existence of solutions for differential equations on* B-*spaces.* Let X and Y be B-spaces over $\mathbb{K} = \mathbb{R}, \mathbb{C}$. Let

$$F: G \subseteq X \to Y$$

be continuous on the nonempty open subset G of X. Then there exists a sequence (F_n) of locally Lipschitz continuous operators

$$F_n: G \subseteq X \to Y$$

such that $\sup_{u \in G} \|Fu - F_n u\| \to 0$ as $n \to \infty$.

Hint: Use a partition of unity and set $F_n u = \sum_j \psi_j(u) F u_j$, where ψ_j is an appropriate real function. Cf. Lasota and Yorke (1973).

Discussion. This result has the following important consequence. Along with the original initial value problem

$$\begin{aligned} u'(t) &= Fu(t), \qquad t \in U(t_0), \\ u(t_0) &= u_0, \end{aligned} \tag{40}$$

we consider the approximate problem

$$\begin{aligned} u'(t) &= F_n u(t), \qquad t \in U(t_0), \\ u(t_0) &= u_0. \end{aligned} \tag{41}$$

Since F_n is locally Lipschitz continuous, the approximate problem (41) has locally a unique solution by Theorem 3.A. By Problem 31.6, the approximation of F by F_n is arbitrarily close. Roughly speaking, this means the following.

Generically, the initial value problem (40) *has always a solution.*

31.7.* *Nonexpansive semigroups on* B-*spaces and initial value problems.* We consider the initial value problem

$$\begin{aligned} u'(t) + Au(t) &= 0, \qquad t \geq 0, \\ u(0) &= u_0. \end{aligned} \tag{42}$$

We assume that the operator $A: D(A) \subseteq X \to X$ is accretive on the B-space X over $\mathbb{K} = \mathbb{R}, \mathbb{C}$, and that there is a $\lambda_0 > 0$ such that

$$D(A) \subseteq R(I + \lambda A) \qquad \text{for all} \quad 0 < \lambda \leq \lambda_0.$$

For example, this condition is satisfied if A is maximal accretive. Then:

(a) The operator A generates a nonexpansive semigroup $\{S(t)\}$ on $D(A)$ by means of

$$S(t)u = \lim_{n \to \infty} \left(1 + \frac{t}{n} A\right)^{-n} u \qquad \text{for all} \quad t \geq 0, \tag{43}$$

and all $u \in D(A)$. This semigroup can be uniquely extended to a nonexpansive semigroup on $\overline{D(A)}$.

(b) If X is reflexive, then, for each $u_0 \in D(A)$, the initial value problem (42) has the unique solution $u(t) = S(t)u_0$, where $u'(t)$ exists for almost all $t \in \mathbb{R}_+$.

This fundamental result underlines the importance of maximal accretive operators for the theory of nonexpansive semigroups on B-spaces. A more general result for *multivalued* maximal accretive operators will be considered in Chapter 57. If $X = \mathbb{R}$, then formula (43) means that $S(t) = e^{-tA}$.

Hint: Ad(a). Cf. Crandall and Liggett (1971).

Ad(b). Cf. Brézis and Pazy (1970).

More recent results can be found in Crandall (1986, S) and Benilan, Crandall, and Pazy (1989, M). Cf. also Pavel (1987, M).

References to the Literature

Classical papers: Komura (1967), Crandall and Pazy (1969), Crandall and Evans (1975).

Introduction to the theory of linear semigroups: Pazy (1983, M).

Linear semigroups, interpolation theory, and approximation theory: Butzer and Berens (1967, M).

Linear semigroups of positive operators: Nagel (1986, L).

Survey of the theory of nonlinear nonexpansive semigroups: Crandall (1986).

Nonlinear semigroups: Brézis (1973, L), Barbu (1976, M), Deimling (1985, M), Clément (1987, M), Pavel (1987, M), Benilan, Crandall, and Pazy (1989, M).

Quasi-linear evolution equations: Kato (1975), (1975a), (1976), (1985, L), (1986, S), Crandall and Sougandis (1986), Amann (1988), (1988a) (dynamic theory).

Quasi-linear evolution equations of hyperbolic type and mathematical physics: Kato(1975), Hughes, Kato, and Marsden (1977), Marsden and Hughes (1983, M), Majda (1984, L), Christodoulou and Klainerman (1990, M).

Quasi-linear evolution equations of parabolic type and reaction–diffusion processes: Henry (1981, L), Amann (1984), (1985), (1986), (1986a–c), (1988), (1988a–c).

Evolution equations, dimension of attractors, and Ljapunov exponents: Babin and Višik (1983), (1983a, S), (1986, S), Temam (1986, S), (1988, M), Ladyženskaja (1987, S), Hale (1988, M).

CHAPTER 32

Maximal Monotone Mappings

Each symmetric operator can be extended to a maximal symmetric operator. To a given maximal symmetric operator, there belongs none or exactly one spectral family (i.e., the operator is not self-adjoint or it is self-adjoint).

John von Neumann (1932)
(*Mathematical Foundations of Quantum Mechanics*)

Let X be a Hilbert space, with real or complex scalars, and with inner product $(u|v)$. For $D \subseteq X$, we call (a not necessarily linear) operator $A: D \to X$ a monotone operator provided that, for all $u, v \in D$,

(C) $$\operatorname{Re}(Au - Av|u - v) \geq 0.$$

We prove several properties of such operators, the most important of which are that the "integral equation of the second kind"

$$Au + u = b,$$

under a few additional hypotheses, always has a solution u for given b, and that the solution depends continuously on b.

It is the author's feeling that some problems of mathematical physics could be better reformulated in terms of the "graphs" of the operators—i.e., in terms of a subset of the product space $X \times X$....

The term "monotone" is preferred here because the theorems of this paper resemble theorems on monotone nondecreasing real functions of a real variable.

George Minty (1962)

The reader should not be surprised at the appearance of multivalued operators in the theory of maximal monotone operators. They play an important role for the following two reasons:

(i) A coherent theory of nonlinear nonexpansive semigroups necessarily demands multivalued operators.
(ii) Certain classes of boundary value problems, particularly variational inequalities, are related to nondifferentiable convex functionals and these equations can be described very conveniently by means of multivalued operators.

Haïm Brézis (1973)

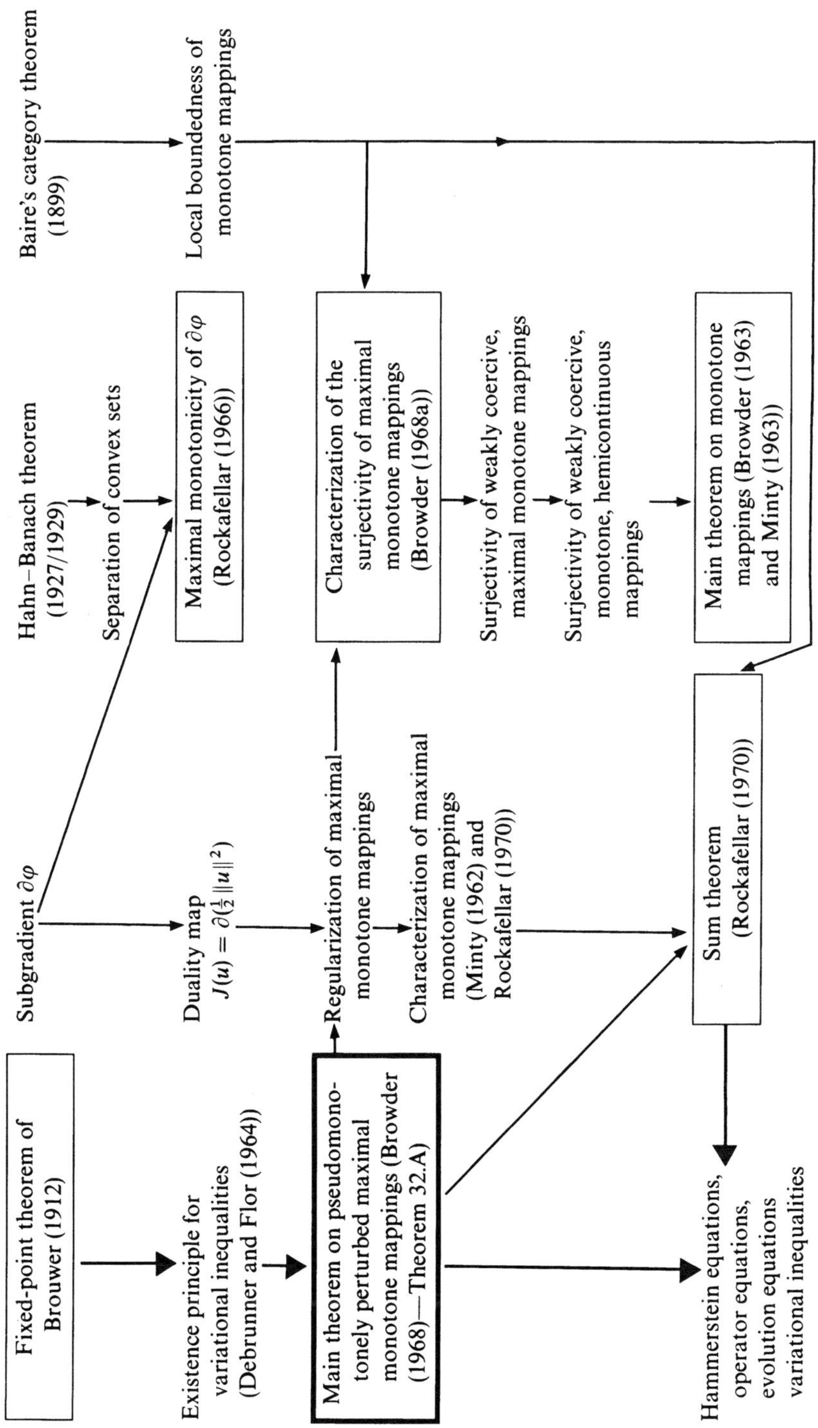
Fixed-point theorem of Brouwer (1912)
Existence principle for variational inequalities (Debrunner and Flor (1964))
Main theorem on pseudomonotonely perturbed maximal monotone mappings (Browder (1968)—Theorem 32.A)
Hammerstein equations, operator equations, evolution equations variational inequalities
Subgradient ∂φ
Duality map J(u) = ∂(½ ‖u‖²)
Regularization of maximal monotone mappings
Characterization of maximal monotone mappings (Minty (1962) and Rockafellar (1970))
Sum theorem (Rockafellar (1970))
Hahn–Banach theorem (1927/1929)
Separation of convex sets
Maximal monotonicity of ∂φ (Rockafellar (1966))
Characterization of the surjectivity of maximal monotone mappings (Browder (1968a))
Surjectivity of weakly coercive, maximal monotone mappings
Surjectivity of weakly coercive, monotone, hemicontinuous mappings
Main theorem on monotone mappings (Browder (1963) and Minty (1963))
Baire's category theorem (1899)
Local boundedness of monotone mappings

Figure 32.1

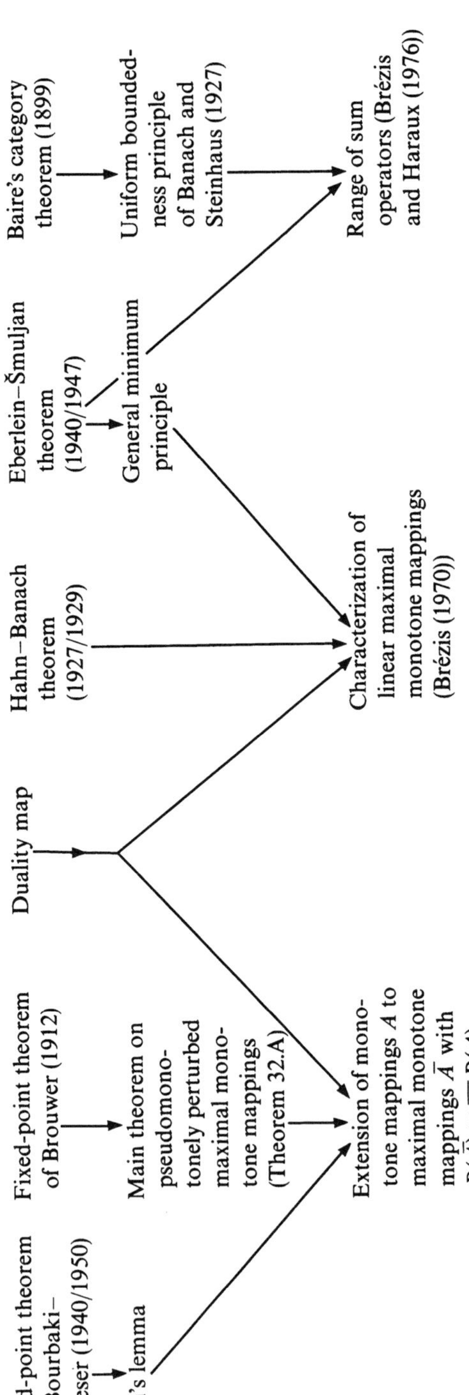
Fixed-point theorem of Bourbaki–Kneser (1940/1950)
Zorn's lemma
Fixed-point theorem of Brouwer (1912)
Main theorem on pseudomonotonely perturbed maximal monotone mappings (Theorem 32.A)
Duality map
Extension of monotone mappings A to maximal monotone mappings $\bar{A}$ with $D(\bar{A}) \subseteq \overline{\text{co}}\, D(A)$
Hahn–Banach theorem (1927/1929)
Characterization of linear maximal monotone mappings (Brézis (1970))
Eberlein–Šmuljan theorem (1940/1947)
General minimum principle
Baire's category theorem (1899)
Uniform boundedness principle of Banach and Steinhaus (1927)
Range of sum operators (Brézis and Haraux (1976))

Figure 32.2

The logical structure of this chapter is represented in Figures 32.1 and 32.2. The *key* to our approach is the main theorem on pseudomonotone perturbations of maximal monotone mappings due to Browder (1968) (Theorem 32.A in Section 32.4). This theorem will be proved via the *Galerkin method*. Throughout this chapter, as an important auxiliary tool, we shall use the *method of regularization* based on the duality map J. Here, the idea is to replace the original equation

$$Au = b, \qquad u \in X,$$

with

$$Au + \varepsilon Ju = b, \qquad u \in X,$$

and to consider the limit $\varepsilon \to +0$. Figures 32.1 and 32.2 show clearly that the theory of maximal monotone operators is based on a number of fundamental principles of functional analysis. Typical applications of this theory concern operator equations, evolution equations of first and second order, Hammerstein equations, and variational inequalities.

32.1. Basic Ideas

The notion of a *maximal monotone* operator is the *most important concept* in the theory of monotone operators. Each monotone operator possesses a maximal monotone extension. The first basic result on maximal monotone operators was proved by Minty (1962). This paper marks the beginning of the modern theory of monotone operators. Minty proved that a monotone operator

$$A\colon D(A) \subseteq X \to X \tag{1}$$

on the real H-space X is maximal monotone iff

$$R(A + I) = X.$$

In particular, each continuous monotone operator $A\colon X \to X$ on the real H-space X is maximal monotone, and $A + I\colon X \to X$ is a homeomorphism, i.e., the equation

$$Au + u = b, \qquad u \in X, \tag{2}$$

has a unique solution u for each given $b \in X$, and u depends continuously on b.

32.1a. Typical Existence Theorems

The two main existence theorems of this chapter are due to Browder (1968), (1968a), namely:

(i) the main theorem on pseudomonotone perturbations of maximal monotone operators (Theorem 32.A); and

(ii) the surjectivity theorem for maximal monotone operators (Theorem 32.G).

In the special case of single-valued operators on H-spaces, we obtain from (i) and (ii) the following two results:

(i*) Let $A: D(A) \subseteq X \to X$ be maximal monotone on the real H-space X, and let $B: X \to X$ be pseudomonotone, demicontinuous, and bounded. Moreover, let B be coercive with respect to A, i.e., there is a $u_0 \in D(A)$ such that

$$\lim_{\|u\| \to \infty} \frac{(Bu|u - u_0)}{\|u\|} = +\infty.$$

Then, for each given $b \in X$, the equation

$$Au + Bu = b, \qquad u \in X, \tag{3}$$

has a solution.

If we set $B = I$, then we obtain equation (2). Consequently, theorem (i) generalizes the theorem of Minty mentioned above.

(ii*) Let $A: D(A) \subseteq X \to X$ be maximal monotone. Then A is surjective iff $A^{-1}: X \to 2^X$ is locally bounded, i.e., for each $u \in X$, there is a neighborhood U such that $A^{-1}(U)$ is bounded.

Note that the surjectivity of A means that, for each $b \in X$, the equation

$$Au = b, \qquad u \in X, \tag{4}$$

has a solution.

In particular, if $A: D(A) \subseteq X \to X$ is maximal monotone and

$$\|Au\| \to \infty \qquad \text{as} \quad \|u\| \to \infty,$$

then A is surjective.

This generalizes the following simple result. The monotone continuous function $A: \mathbb{R} \to \mathbb{R}$ is surjective iff $|Au| \to \infty$ as $|u| \to \infty$ (Fig. 32.3).

For a linear operator $A: D(A) \subseteq X \to X$ on a real H-space X, the following two statements are equivalent:

(a) A is maximal monotone.
(b) A and A^* are monotone, $D(A)$ is dense in X, and A is graph closed.

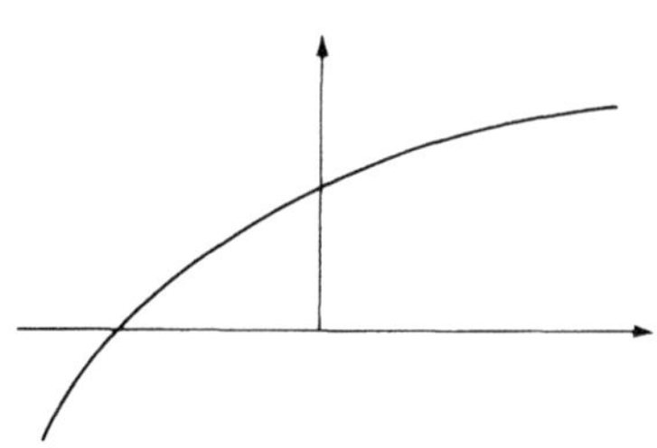

Figure 32.3

From Proposition 31.12 we obtain the following special case:

The linear symmetric monotone operator $A: D(A) \subseteq X \to X$ *on the real* H-*space* X *is maximal monotone iff it is self-adjoint.*

Thus, the theory of maximal monotone operators *generalizes* the classical theory of self-adjoint operators due to John von Neumann. For example, this theory can be found in v. Neumann (1932, M) and in Riesz and Nagy (1956, M). In Chapter 19 we have shown that linear self-adjoint operators allow important applications to variational problems and to differential equations of elliptic, parabolic, and hyperbolic type.

As we shall show in this chapter, the same is true for nonlinear maximal monotone operators.

32.1b. Typical Maximal Monotone Mappings

Let X be a real reflexive B-space. The following examples show that many important operators are maximal monotone:

(a) *Prototype.* If $A: X \to X^*$ is monotone and hemicontinuous, then A and $A^{-1}: X^* \to 2^X$ are maximal monotone.

(b) *Gradients.* If the C^1-functional $f: X \to \mathbb{R}$ is convex, then the derivative $f': X \to X^*$ is maximal monotone.

(c) *Subgradients.* If $f: X \to]-\infty, \infty]$ is convex and lower semicontinuous with $f \not\equiv +\infty$, then the subgradient $\partial f: X \to 2^{X^*}$ is maximal monotone (Theorem of Rockafellar (1966)).

In particular, if $f: X \to \mathbb{R}$ is G-differentiable, then $\partial f(u) = \{f'(u)\}$ for all $u \in X$, i.e., the subgradient generalizes the G-derivative.

(d) The *duality map* $J: X \to 2^{X^*}$ defined through

$$Ju = \partial(2^{-1}\|u\|^2) \qquad \text{for all} \quad u \in X,$$

is maximal monotone.

In particular, if X^* is strictly convex (e.g., X is an H-space), then $J = f'$ where $f(u) = 2^{-1}\|u\|^2$.

(e) *Time derivatives.* We consider functions of the form $u: [0, T] \to V$, where $0 < T < \infty$. Let

$$X = L_p(0, T; V), \qquad 1 < p < \infty,$$

where "$V \subseteq H \subseteq V^*$" is an evolution triple. We define the operator

$$Lu = u',$$

where either

$$D(L) = \{u \in W_p^1(0, T; V, H): u(0) = 0\}$$

or

$$D(L) = \{u \in W_p^1(0, T; V, H): u(0) = u(T) = 0\}.$$

Then the operator

$$L: D(L) \subseteq X \to X^*$$

is maximal monotone. Recall that $X^* = L_q(0, T; V^*)$, where $p^{-1} + q^{-1} = 1$.

(f) *Regularization and generators of semigroups.* Let X and X^* be strictly convex, and let $A: X \to 2^{X^*}$ be monotone. Then, A is maximal monotone iff

$$R(A + \lambda J) = X \qquad \text{for all} \quad \lambda > 0 \tag{5}$$

(Theorem of Rockafellar (1970)). This result is the basis of an important regularization technique to be discussed below.

In the special case of an H-space X, we may identify $X^* = X$. Then $J = I$ (the identity on X). Thus, (5) generalizes the Theorem of Minty mentioned in (1) above.

By Theorem 31.A, each maximal monotone operator $A: D(A) \subseteq X \to X$ on a real H-space X has the property that $-A$ generates a nonexpansive semigroup $\{S(t)\}$ on $\overline{D(A)}$. Letting

$$u(t) = S(t)u_0,$$

we obtain generalized solutions of the initial value problem

$$u' + Au = 0, \qquad u(0) = u_0. \tag{6}$$

In order to obtain a complete characterization of nonexpansive (nonlinear) semigroups on H-spaces, one needs multivalued maximal monotone operators. The precise formulation can be found in Section 32.24.

(g) *Sum rule.* Let $A, B: X \to 2^{X^*}$ be maximal monotone and suppose that

$$D(A) \cap \operatorname{int} D(B) \neq \emptyset.$$

Then the mapping $A + B: X \to 2^{X^*}$ is maximal monotone (Theorem of Rockafellar (1970)).

32.1c. Typical Applications

In this chapter we study the following applications:

Hammerstein operator equations and Hammerstein integral equations;
evolution equations of first and second order (initial value problems as well as periodic problems);
variational inequalities of elliptic and parabolic type.

Applications of variational inequalities to elasticity, plasticity, hydrodynamics, gas dynamics, optimal control, etc., will be considered in Parts III through V.

Let us explain some basic ideas. The details will be studied below in this chapter. The main theorem on perturbed maximal monotone mappings of Browder in Section 32.4 concerns the solvability of the *fundamental equation*

$$b \in Au + Bu, \qquad u \in C, \tag{7}$$

where $A: C \to 2^{X^*}$ is maximal monotone and $B: C \to 2^{X^*}$ is pseudomonotone. Moreover, C is a nonempty closed convex subset of X. The proof of this fundamental result is based on the existence principle for inequalities in Section 2.11 (the extension theorem of Debrunner and Flor (1964) for monotone sets). This principle follows from the Brouwer fixed-point theorem. In the following, we want to show that completely different problems can be reduced to equation (7).

Hammerstein Equations

To solve the equation

$$u + KFu = b, \qquad u \in X, \tag{8}$$

we use the multivalued inverse mapping K^{-1}. This way we obtain the equivalent problem

$$0 \in K^{-1}(u - b) + Fu, \qquad u \in X, \tag{8*}$$

which is of the form (7).

First-Order Evolution Equations

The initial value problem

$$u' + Bu = b, \qquad u(0) = 0, \tag{9}$$

can be written in the form

$$Au + Bu = b, \qquad u \in X, \tag{9*}$$

where $Au = u'$ with $D(A) = \{u \in W_p^1(0, T; V, H): u(0) = 0\}$ and $X = L_p(0, T; V)$.

Similarly, the following evolution equation with periodicity condition

$$u' + Bu = b, \qquad u(0) = u(T) = 0, \tag{10}$$

leads to the problem

$$Au + Bu = b, \qquad u \in X, \tag{10*}$$

where $Au = u'$ and $D(A) = \{u \in W_p^1(0, T; V, H): u(0) = u(T) = 0\}$.

Reduction of Second-Order Evolution Equations to First-Order Evolution Equations

We consider the initial value problem

$$u'' + Nu' + Lu = b, \qquad u(0) = u'(0) = 0. \tag{11}$$

Letting $v = u'$, we obtain the *first equivalent problem*

$$v' + Nv + L\left(\int_0^t v(s)\,ds\right) = b, \qquad v(0) = 0. \tag{11*}$$

Again letting $v = u'$, we obtain from (11) the *second equivalent problem*

$$\begin{aligned} u' - v &= 0, \\ v' + Nv + Lu &= b, \\ u(0) = 0, \qquad v(0) &= 0. \end{aligned} \tag{11**}$$

Letting $w = (u, v)$, problem (11**) can be written in the form

$$w' + Bw = (0, b), \qquad w(0) = 0.$$

Note that (11*) and (11**) represent first-order evolution equations. Equation (11*) (resp. (11**)) will be used in Chapters 32 and 33 (resp. Chapter 56).

Free Minimum Problems

The minimum problem

$$f(u) = \min!, \qquad u \in X, \tag{12}$$

for the functional $f: X \to \mathbb{R}$ is equivalent to the Euler equation

$$0 \in \partial f(u), \qquad u \in X. \tag{12*}$$

This problem is of the form (7) with $A = \partial f$.

Constrained Minimum Problems

Let C be a nonempty closed convex subset of the real reflexive B-space X, and let $f: X \to \mathbb{R}$ be convex and lower semicontinuous. Define

$$\chi(u) = \begin{cases} 0 & \text{if } u \in C, \\ +\infty & \text{if } u \notin C. \end{cases}$$

Then the minimum problem

$$f(u) = \min!, \qquad u \in C, \tag{13}$$

is equivalent to the Euler equation

$$0 \in \partial f(u) + \partial\chi(u), \qquad u \in X. \tag{13*}$$

By the Theorem of Rockafellar (c) above, the mappings $\partial f, \partial\chi: X \to 2^{X^*}$ are maximal monotone. By the sum rule (g) above, the mapping $A = \partial f + \partial\chi$ is maximal monotone if $\operatorname{int} C \neq \emptyset$. Hence problem (13*) is of the form (7).

If $f: X \to \mathbb{R}$ is G-differentiable, then (13*) is equivalent to the variational inequality

$$\langle f'(u), v - u \rangle \geq 0 \qquad \text{for all} \quad v \in C. \tag{13**}$$

Here we seek $u \in C$. Problems of the form (13) will be studied in detail in Chapter 47 in the context of convex analysis.

Elliptic Variational Inequalities

Instead of (13**), we consider the more general problem

$$\langle Bu - b, v - u \rangle \geq 0, \qquad \text{for all} \quad v \in C, \tag{14}$$

where we seek $u \in C$. This is equivalent to the equation

$$b \in \partial\chi(u) + Bu, \qquad u \in C, \tag{14*}$$

which is again of the form (7) with $A = \partial\chi$. In the special case $C = X$, the variational inequality (14) for the operator $B\colon X \to X^*$ is equivalent to the operator equation

$$Bu = b, \qquad u \in X.$$

Evolution Variational Inequalities

We set $Lu = u'$, where $D(L)$ is given as in (e) above and $X = L_2(0, T; V)$. If we replace the operator B in (14) with $Lu + Bu$, then we obtain a so-called variational inequality:

$$\langle Lu + Bu - b, v - u \rangle \geq 0 \qquad \text{for all} \quad v \in C, \tag{15}$$

where we seek $u \in C$. This problem is equivalent to the equation

$$b \in \partial\chi(u) + Lu + Bu, \qquad u \in C. \tag{15*}$$

The operator $L\colon D(L) \subseteq X \to X^*$ is maximal monotone. By the sum rule (g) above, the mapping $A = \partial\chi + L$ is maximal monotone if $\operatorname{int} C \neq \varnothing$.

More General Variational Inequalities

Let $\varphi\colon X \to]-\infty, \infty]$ be a convex lower semicontinuous function with $\varphi \not\equiv +\infty$. We consider the variational inequality

$$\langle b - Bu, v - u \rangle + \varphi(u) \leq \varphi(v) \qquad \text{for all} \quad v \in X, \tag{16}$$

where we seek $u \in C$. This problem is equivalent to the equation

$$b \in \partial\varphi(u) + Bu, \qquad u \in C. \tag{16*}$$

In the special case $\varphi = \chi$, problem (16) is equivalent to (14).

Method of Regularization

Along with the original problem

$$b \in Au + Bu, \qquad u \in X, \tag{17}$$

we study the regularized problems

$$b \in Au_\varepsilon + \varepsilon Ju_\varepsilon + Bu_\varepsilon, \qquad u_\varepsilon \in X \tag{17*}$$

for small $\varepsilon > 0$. Let $A: X \to 2^{X^*}$ be maximal monotone and let $B: X \to X^*$ be pseudomonotone. Suppose that X and X^* are strictly convex. Then the inverse operator $(A + \varepsilon J)^{-1}: X^* \to X$ is single-valued, and (17*) is equivalent to the equation

$$u_\varepsilon = (A + \varepsilon J)^{-1}(b - Bu_\varepsilon), \qquad u_\varepsilon \in X. \tag{17**}$$

Suppose that, for each small $\varepsilon > 0$, equation (17**) has a unique solution u_ε (e.g., $B = 0$). In Section 32.18 we shall formulate conditions which guarantee that, as $\varepsilon \to +0$, the sequence (u_ε) converges to a solution u of the original problem (17). Summarizing, we obtain:

The method of regularization allows us to solve nonuniquely solvable equations by means of a family of uniquely solvable equations.

32.2. Definition of Maximal Monotone Mappings

We first consider some basic notions for multivalued mappings.

Definition 32.1. Let

$$A: M \to 2^Y$$

be a *multivalued mapping*, i.e., A assigns to each point $u \in M$ a subset Au of Y.

(i) The set

$$D(A) = \{u \in M: Au \neq \emptyset\}$$

is called the *effective domain* of A.

(ii) The set

$$R(A) = \bigcup_{u \in M} Au$$

is called the *range* of A.

(iii) The set

$$G(A) = \{(u, v) \in M \times Y: u \in D(A), v \in Au\}$$

is called the *graph* of A. Instead of $(u, v) \in G(A)$ we briefly write

$$(u, v) \in A.$$

The inverse mapping

$$A^{-1}: Y \to 2^M$$

is defined naturally enough by

$$A^{-1}(v) = \{u \in M: v \in Au\}.$$

Obviously, we have $D(A^{-1}) = R(A)$ and

$$(u, v) \in A \quad \text{iff} \quad (v, u) \in A^{-1}.$$

Let X and Y be linear spaces over $\mathbb{K}$ and let $M \subseteq X$. For given maps

$$A, B: M \to 2^Y$$

and for fixed $\alpha, \beta \in \mathbb{K}$, we define the linear combination

$$\alpha A + \beta B: M \to 2^Y$$

through

$$(\alpha A + \beta B)(u) = \begin{cases} \alpha Au + \beta Bu & \text{if } u \in D(A) \cap D(B), \\ \varnothing & \text{otherwise.} \end{cases}$$

Obviously, $D(\alpha A + \beta B) = D(A) \cap D(B)$.

In terms of set theory, the multivalued map $A: M \to 2^Y$ is a subset of $M \times Y$. In this sense, the graph $G(A)$ is identical to the subset A of $M \times Y$.

Each single-valued map

$$A: D(A) \subseteq M \to Y$$

can be identified with a multivalued map

$$\bar{A}: M \to 2^Y$$

by setting

$$\bar{A}u = \begin{cases} \{Au\} & \text{if } u \in D(A), \\ \varnothing & \text{otherwise.} \end{cases}$$

Then $D(\bar{A}) = D(A)$ and $R(\bar{A}) = R(A)$. In the following, single-valued maps will always be regarded as multivalued maps. Instead of $\bar{A}$ we will briefly write A.

The map $B: M \to 2^Y$ is called an *extension* of $A: M \to 2^Y$ iff $G(A) \subseteq G(B)$.

The following definition is basic.

Definition 32.2. We consider the multivalued map

$$A: M \to 2^{X^*}, \tag{18}$$

where M is a subset of the real B-space X.

(a) A subset S of $M \times X^*$ is called *monotone* iff

$$\langle u^* - v^*, u - v \rangle_X \geq 0 \quad \text{for all} \quad (u, u^*), (v, v^*) \in S.$$

(b) A subset S of $M \times X^*$ is called *maximal monotone* iff it is monotone and there is *no* proper monotone extension in $M \times X^*$.

(c) The map A in (18) is called *monotone* iff the graph $G(A)$ is a monotone set in $M \times X^*$, i.e.,

$$\langle u^* - v^*, u - v \rangle_X \geq 0 \qquad \text{for all} \quad (u, u^*), (v, v^*) \in A.$$

(d) The map A in (18) is called *maximal monotone* iff the graph $G(A)$ is a maximal monotone set in $M \times X^*$, i.e., A is monotone and it follows from $(u, u^*) \in M \times X^*$ and

$$\langle u^* - v^*, u - v \rangle_X \geq 0 \qquad \text{for all} \quad (v, v^*) \in A$$

that $(u, u^*) \in A$.

An operator

$$A: D(A) \subseteq X \to X^* \tag{19}$$

is to be understood as a multivalued map $A: X \to 2^{X^*}$. Thus, A in (19) is called monotone iff

$$\langle Au - Av, u - v \rangle_X \geq 0 \qquad \text{for all} \quad u, v \in D(A). \tag{20}$$

Furthermore, A in (19) is called maximal monotone iff A is monotone and it follows from $(u, u^*) \in X \times X^*$ and

$$\langle u^* - Av, u - v \rangle_X \geq 0 \qquad \text{for all} \quad v \in D(A) \tag{21}$$

that $u \in D(A)$ and $u^* = Au$.

If X is an H-*space*, then we identify X^* with X in the sense of the Identification Principle 21.18. Thus, we have to replace $\langle \cdot, \cdot \rangle_X$ by the scalar product $(\cdot | \cdot)$ on X. For example, a set S in $M \times X$ is called monotone iff

$$(u^* - v^* | u - v) \geq 0 \qquad \text{for all} \quad (u, u^*), (v, v^*) \in S.$$

The map

$$A: M \to 2^X$$

with $M \subseteq X$ is called maximal monotone iff

$$(u^* - v^* | u - v) \geq 0 \qquad \text{for all} \quad (u, u^*), (v, v^*) \in A,$$

and it follows from $(u, u^*) \in M \times X$ and

$$(u^* - v^* | u - v) \geq 0 \qquad \text{for all} \quad (v, v^*) \in A,$$

that $(u, u^*) \in A$.

EXAMPLE 32.3. Consider the H-space $X = \mathbb{R}$. A set S in $X \times X = \mathbb{R}^2$ is monotone iff

$$(u^* - v^*)(u - v) \geq 0 \qquad \text{for all} \quad (u, u^*), (v, v^*) \in S,$$

i.e.,

$$(u, u^*), (v, v^*) \in S \qquad \text{and} \qquad u < v \qquad \text{implies} \qquad u^* \leq v^*.$$

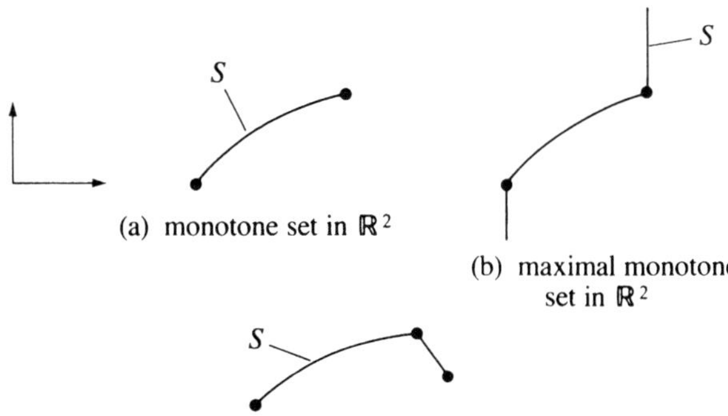

(a) monotone set in $\mathbb{R}^2$

(b) maximal monotone set in $\mathbb{R}^2$

(c) nonmonotone set in $\mathbb{R}^2$

Figure 32.4

The set S in Figure 32.4(a) is monotone in $\mathbb{R}^2$, whereas S in Figure 32.4(c) is *not* monotone in $\mathbb{R}^2$. Furthermore, S in Figure 32.4(b) is a maximal monotone set in $\mathbb{R}^2$, since there is no proper monotone extension of S in $\mathbb{R}^2$.

EXAMPLE 32.4. We consider the function

$$f: \mathbb{R} \to \mathbb{R} \tag{22}$$

on the H-space $X = \mathbb{R}$. Then:

(a) If f is monotone increasing and continuous, then f is maximal monotone (Fig. 32.5(a)). Indeed, it follows from $(u, u^*) \in \mathbb{R}^2$ and

$$(u^* - f(v))(u - v) \geq 0 \qquad \text{for all} \quad v \in \mathbb{R},$$

that $u^* = f(u)$. Argue by contradiction.

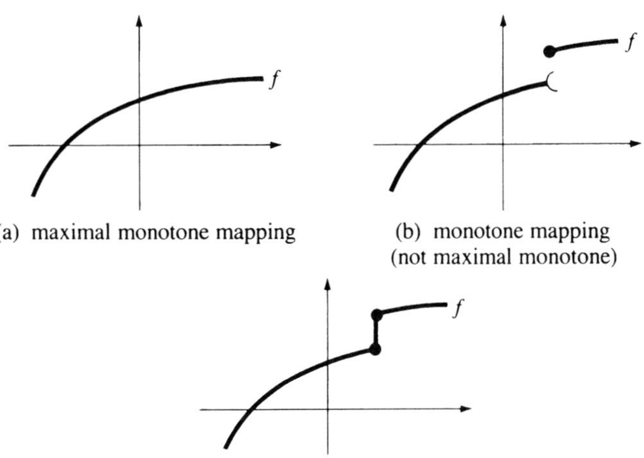

(a) maximal monotone mapping

(b) monotone mapping (not maximal monotone)

(c) maximal monotone mapping

Figure 32.5

(b) However, the monotone increasing *discontinuous* function f in Figure 32.5(b) is a monotone map, but *not* maximal monotone, since the graph of f possesses a proper monotone extension in $\mathbb{R}^2$ pictured in Figure 32.5(c).
(c) The map $f: \mathbb{R} \to 2^{\mathbb{R}}$ pictured in Figure 32.5(c) is maximal monotone.

Proposition 32.5. *The map $A: X \to 2^{X^*}$ on the real reflexive* B-*space X is maximal monotone iff the inverse map $A^{-1}: X^* \to 2^X$ is maximal monotone.*
Here, we identify X^{**} with X.

Proof. This follows from

$$G(A^{-1}) = \{(u^*, u) \in X^* \times X: (u, u^*) \in G(A)\}$$

and from $\langle u^*, u\rangle_X = \langle u, u^*\rangle_{X^*}$ for all $(u, u^*) \in X \times X^*$. □

Proposition 32.6. *If $A: X \to 2^{X^*}$ is maximal monotone on the real* B-*space X, then the set Au is convex and weakly* closed in X^* for each $u \in X$.*

Proof. Let $u_0^*, u_1^* \in Au$. Set $u_t^* = (1-t)u_0^* + t u_1^*, 0 \le t \le 1$. For all $(v, v^*) \in A$,

$$\begin{aligned}\langle u_t^* - v^*, u - v\rangle &= (1-t)\langle u_0^* - v^*, u - v\rangle + t\langle u_1^* - v^*, u - v\rangle \\ &\ge 0.\end{aligned}$$

Hence $u_t^* \in Au$.

Let (u_α^*) be an M–S sequence in Au such that $u_\alpha^* \overset{*}{\rightharpoonup} u^*$ in X^*. It follows from

$$\langle u_\alpha^* - v^*, u - v\rangle_X \ge 0 \qquad \text{for all} \quad (v, v^*) \in A,$$

that $\langle u^* - v^*, u - v\rangle_X \ge 0$ for all $(v, v^*) \in A$. Hence $u^* \in Au$. □

32.3. Typical Examples for Maximal Monotone Mappings

32.3a. Single-Valued Monotone Operators

Proposition 32.7 (Prototype). *Let $A: X \to X^*$ be a monotone hemicontinuous operator on the real reflexive* B-*space X. Then A and $A^{-1}: X^* \to 2^X$ are maximal monotone.*
Here, we identify X^{**} with X.

Proof. The maximal monotonicity of A follows from the monotonicity trick (25.4), and the maximal monotonicity of A^{-1} follows from Proposition 32.5. □

Proposition 32.8 (Minty (1962)). *A monotone operator* $A: D(A) \subseteq X \to X$ *on the real* H*-space* X *is maximal monotone iff* $R(A + I) = X$.

PROOF. This is a special case of Theorem 32.A below. More precisely, if $R(A + I) = X$, then A is maximal monotone by Proposition 31.5. Conversely, if A is maximal monotone, then $R(A + I) = X$ by Corollary 32.25. □

Corollary 32.9. *If the operator* $A: X \to X$ *is monotone and continuous on the real* H*-space* X, *then* $A + I: X \to X$ *is a homeomorphism.*

PROOF. By Proposition 32.7, A is maximal monotone. Hence $R(A + I) = X$. By Proposition 31.5, the operator $(A + I)^{-1}: X \to X$ is nonexpansive. □

32.3b. Time-Derivatives

We define

$$L_1 u = u', \qquad D(L_1) = \{u \in W_p^1(0, T; V, H): u(0) = 0\}, \tag{23}$$

$$L_2 u = u', \qquad D(L_2) = \{u \in W_p^1(0, T; V, H): u(0) = u(T)\}. \tag{24}$$

Proposition 32.10. *Let "*$V \subseteq H \subseteq V^*$*" be an evolution triple and let*

$$X = L_p(0, T; V),$$

where $1 < p < \infty$ *and* $0 < T < \infty$. *Then the two linear operators*

$$L_i: D(L_i) \subseteq X \to X^*, \qquad i = 1, 2,$$

are maximal monotone.

Recall that $X^* = L_q(0, T; V^*)$, where $q^{-1} + p^{-1} = 1$.

PROOF. We make essential use of the integration by parts formula

$$\int_0^T \langle u'(t), u(t)\rangle_V \, dt = 2^{-1}(\|u(T)\|_H^2 - \|u(0)\|_H^2), \tag{25}$$

for all $u \in W_p^1(0, T; V, H)$.

(I) Obviously, L_i is linear.

(II) L_i is monotone. Indeed, it follows from (25) that

$$\langle L_i u, u\rangle_X = 2^{-1}(\|u(T)\|_H^2 - \|u(0)\|_H^2) \tag{26}$$

for all $u \in D(L_i)$. Hence $\langle L_i u, u\rangle_X \geq 0$ for all $u \in D(L_i)$.

(III) L_i is maximal monotone. To prove this, suppose that $(v, w) \in X \times X^*$ and

$$0 \leq \langle w - L_i u, v - u\rangle_X \qquad \text{for all} \quad u \in D(L_i). \tag{27}$$

We have to show that $v \in D(L_i)$ and $w = L_i v$, i.e., $w = v'$.

To this end, choose

$$u = \varphi z, \qquad \text{where} \quad \varphi \in C_0^\infty(0, T) \quad \text{and} \quad z \in V.$$

Then $u' = \varphi' z$ and $u \in D(L_i)$. By (26), $\langle L_i u, u \rangle = 0$. From (27) it follows that

$$0 \leq \langle w, v \rangle_X - \int_0^T \langle \varphi'(t)v(t) + \varphi(t)w(t), z \rangle_V \, dt, \tag{28}$$

for all $z \in V$. In this connection, note that $\langle a, b \rangle_V = \langle b, a \rangle_V$ for all $a, b \in V$ by (23.17). From (28) we get

$$\int_0^T \varphi'(t)v(t) + \varphi(t)w(t) \, dt = 0 \qquad \text{for all} \quad \varphi \in C_0^\infty(0, T).$$

Hence

$$v' = w$$

and $v \in W_p^1(0, T; V, H)$, since $w \in X^*$.

It remains to show that $v \in D(L_i)$. Using integration by parts, we obtain from (27) that

$$\begin{aligned} 0 &\leq \langle v' - u', v - u \rangle_X \\ &= 2^{-1}(\|v(T) - u(T)\|_H^2 - \|v(0) - u(0)\|_H^2). \end{aligned} \tag{29}$$

Ad L_1. Choose a sequence (a_n) in V with $Ta_n \to v(T)$ in H as $n \to \infty$. Set $u(t) = ta_n$. Then $u \in D(L_1)$. By (29), $v(0) = u(0) = 0$. Hence $v \in D(L_1)$.

Ad L_2. From (29) we obtain

$$0 \leq \|v(T)\|_H^2 - \|v(0)\|_H^2 + 2(u(0) | v(0) - v(T))_H,$$

for all $u \in D(L_2)$. Note that $u(0) = u(T)$. In particular, we can choose $u(t) \equiv a$ for arbitrary $a \in V$. Hence $v(0) = v(T)$, i.e., $v \in D(L_2)$. Note that V is dense in H. □

32.3c. Subgradients

Subgradients generalize the classical concept of a derivative. In this connection, the following formula

$$f(v) \geq f(u) + \langle u^*, v - u \rangle_X \qquad \text{for all} \quad v \in X \tag{30}$$

is crucial.

Definition 32.11. Let $f: X \to [-\infty, \infty]$ be a functional on the real B-space X.

The functional u^* in X^* is called a *subgradient* of f at the point u iff $f(u) \neq \pm\infty$ and (30) holds.

The set of all subgradients of f at u is called the *subdifferential* $\partial f(u)$ at u. If

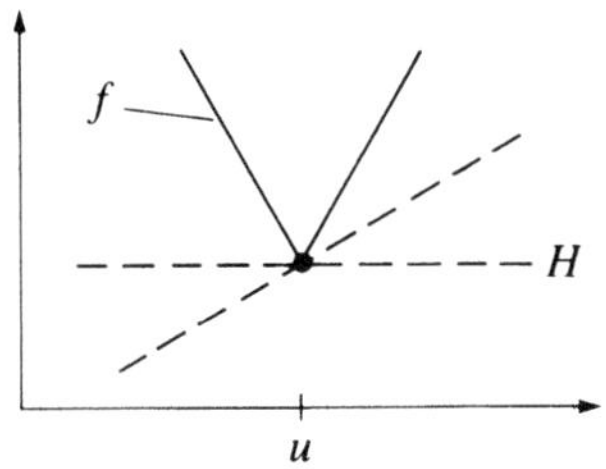

Figure 32.6

no subgradients exist, then we set $\partial f(u) = \emptyset$. In particular, let $\partial f(u) = \emptyset$ if $f(u) = \pm\infty$.

If $\partial f(u) \neq \emptyset$, then $f(v) > -\infty$ for all $v \in X$, by (30). Subgradients will be considered in detail in Part III.

Example 32.12 (Real Functions). For a function $f: \mathbb{R} \to \mathbb{R}$, the subdifferential at u equals the set of all slopes $u^* \in \mathbb{R}$ of straight lines through the point $(u, f(u))$ which lie below the graph of f (the generalized tangents in Figure 32.6). If the derivative $f'(u)$ exists, then $\partial f(u)$ is single-valued, i.e.,

$$\partial f(u) = \{f'(u)\}. \tag{31}$$

We now want to generalize relation (31) to B-spaces.

Proposition 32.13. *Let $f: X \to \mathbb{R}$ be a functional on the real* B-*space X. Then:*

(a) *If f is convex and if f possesses the* G-*derivative $f'(u)$ at the point u, then* (31) *holds.*
(b) *Conversely, if $\partial f: X \to X^*$ is single-valued and hemicontinuous, then f is* G-*differentiable on X* and (31) *holds for all $u \in X$.*

Proof. Ad(a). Set $\varphi(t) = f(u + th)$. Since $\varphi: \mathbb{R} \to \mathbb{R}$ is convex, $\varphi(1) - \varphi(0) \geq \varphi'(0)$. Hence

$$f(u + h) - f(u) \geq \langle f'(u), h\rangle \qquad \text{for all} \quad h \in X,$$

i.e., $f'(u) \in \partial f(u)$.

Conversely, if $u^* \in \partial f(u)$, then

$$f(u + th) - f(u) \geq \langle u^*, th\rangle \qquad \text{for all} \quad h \in X, \quad t > 0,$$

and hence $\langle f'(u), h\rangle \geq \langle u^*, h\rangle$ for all $h \in X$. This implies $u^* = f'(u)$.

Ad(b). Let $t > 0$. From

$$f(u + th) - f(u) \geq \langle \partial f(u), th\rangle,$$

$$f(u) - f(u + th) \geq -\langle \partial f(u + th), th\rangle,$$

we obtain

$$\begin{aligned}\langle \partial f(u), h\rangle &\leq \underline{\lim}_{t\to+0} \frac{f(u+th)-f(u)}{t}\\ &\leq \overline{\lim}_{t\to+0} \frac{f(u+th)-f(u)}{t} \leq \langle \partial f(u), h\rangle,\end{aligned}$$

for all $h \in X$, i.e., $f'(u) = \partial f(u)$. □

Proposition 32.14 (Minimum Principle). *Let $f\colon X \to]-\infty, \infty]$ be a functional on the real* B*-space X with $f \not\equiv +\infty$. Then, u is a solution of the minimum problem*

$$f(u) = \min!, \qquad u \in X, \tag{32}$$

iff

$$0 \in \partial f(u). \tag{33}$$

PROOF. This follows immediately from (30). □

Intuitively, the Euler equation (33) means that there exists a *horizontal* hyperplane H through the minimal point $(u, f(u))$ such that the graph of f lies above H (Fig. 32.6).

EXAMPLE 32.15 (Support Functionals of Convex Sets). Let C be a nonempty closed convex set in the real B-space X with the *indicator function* χ, i.e.,

$$\chi(u) = \begin{cases} 0 & \text{if} \quad u \in C \\ +\infty & \text{if} \quad u \in X - C. \end{cases}$$

By a *support functional* to C at the point u we understand a functional u^* in X^* such that

$$\langle u^*, u - v\rangle \geq 0 \qquad \text{for all} \quad v \in C. \tag{34}$$

Obviously, if u is an interior point of C, then (34) is equivalent to $u^* = 0$. Generally, the equation

$$\langle u^*, u - v\rangle = 0, \qquad v \in X,$$

describes a closed hyperplane H in X through the point u. Thus, relation (34) means that the set C lies on one side of the hyperplane H (Fig. 32.7). According to (30), we obtain:

$$\partial\chi(u) = \begin{cases} \text{set of all support functionals to } C \text{ at } u & \text{if} \quad u \in C; \\ \varnothing \quad \text{if} \quad u \in X - C. \end{cases}$$

In particular,

$$\partial\chi(u) = \{0\} \qquad \text{if} \quad u \in \operatorname{int} C.$$

Moreover, we have $0 \in \partial\chi(u)$ if $u \in C$.

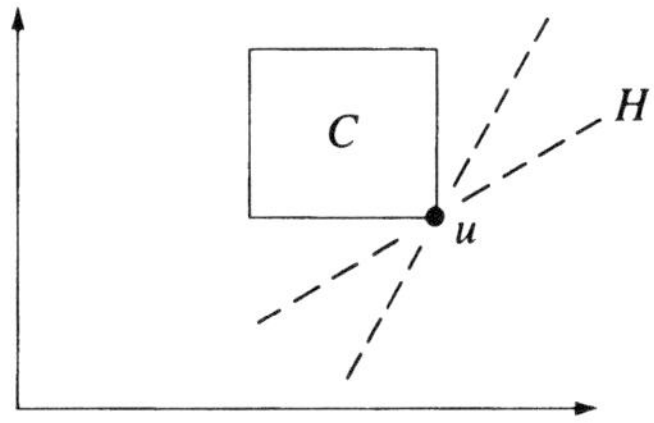

Figure 32.7

The mappings

$$\partial\chi: C \to 2^{X^*} \tag{35}$$

and

$$\partial\chi: X \to 2^{X^*} \tag{36}$$

are maximal monotone.

PROOF. Ad(35). We show first that $\partial\chi: C \to 2^{X^*}$ is monotone. Let $u^* \in \partial\chi(u)$ and $v^* \in \partial\chi(v)$, where $u, v \in C$. Then, for all $w \in C$,

$$\langle u^*, u - w\rangle \geq 0 \qquad \text{and} \qquad \langle v^*, v - w\rangle \geq 0.$$

Hence $\langle u^* - v^*, u - v\rangle \geq 0$.

Next we show that $\partial\chi: C \to 2^{X^*}$ is maximal monotone. Suppose that $(u, u^*) \in C \times X^*$ and

$$\langle u^* - v^*, u - v\rangle \geq 0 \qquad \text{for all} \quad (v, v^*) \in \partial\chi, \tag{37}$$

i.e., $v \in C$ and $v^* \in \partial\chi(v)$. Note that $0 \in \partial\chi(v)$ for each $v \in C$. Letting $v^* = 0$ in (37), we obtain

$$\langle u^*, u - v\rangle \geq 0 \qquad \text{for all} \quad v \in C, \quad \text{i.e.,} \quad u^* \in \partial\chi(u).$$

Ad(36). This is a special case of Proposition 32.17 below. □

Definition 32.16. The mapping $A: X \to 2^{X^*}$ on the real B-space X is called *cyclic monotone* iff the inequality

$$\langle u_1^*, u_1 - u_2\rangle + \langle u_2^*, u_2 - u_3\rangle + \cdots + \langle u_n^*, u_n - u_{n+1}\rangle \geq 0$$

holds for all $(u_i, u_i^*) \in A$, $i = 1, \ldots, n$, and all $n \in \mathbb{N}$, where we set $u_{n+1} = u_1$.

Furthermore, A is called *maximal cyclic monotone* iff A is cyclic monotone and there is *no* cyclic monotone mapping $B: X \to 2^{X^*}$ such that $G(A) \subset G(B)$.

We make the following assumption:

(H) Let $f: X \to\,]-\infty, \infty]$ be convex and lower semicontinuous on the real B-space X and let $f \not\equiv +\infty$.

Recall that f is called *convex* iff

$$f((1 - t)u + tv) \leq (1 - t)f(u) + tf(v)$$

for all $u, v \in X, t \in]0, 1[$. Moreover, f is called *lower semicontinuous* iff the set $\{u \in X: f(u) \leq r\}$ is closed for all $r \in \mathbb{R}$.

For example, condition (H) is satisfied for each continuous convex functional $f: X \to \mathbb{R}$.

Proposition 32.17 (Rockafellar (1966)). *If* (H) *holds, then the subgradient* $\partial f: X \to 2^{X^*}$ *is maximal monotone.*

Corollary 32.18 (Rockafellar (1970b)). *For a mapping* $A: X \to 2^{X^*}$ *on the real* B*-space* X, *the following two assertions are equivalent*:

(i) $A = \partial f$ *and* f *satisfies* (H).
(ii) A *is maximal cyclic monotone.*

PROOF. This follows from Theorem 47.F, which we prove in Section 47.11 by using the separation of convex sets. □

Corollary 32.18 reflects the fact that *not* all maximal monotone operators are subgradients.

32.3d. Duality Map

The duality map represents an important auxiliary tool in the theory of maximal monotone operators.

Definition 32.19. Set $\varphi(u) = 2^{-1}\|u\|^2$ for all $u \in X$, where X is a real B-space. The duality map

$$J: X \to 2^{X^*}$$

of X is defined to be $J = \partial\varphi$.

EXAMPLE 32.20 (Prototype). Let X be a real H-space. Then the duality map $J: X \to X^*$ is single-valued and

$$\langle Ju, v\rangle = (u|v) \qquad \text{for all} \quad u, v \in X.$$

PROOF. Set $\psi(t) = \varphi(u + th) = 2^{-1}(u + th|u + th)$, $t \in \mathbb{R}$. Then $\psi'(0) = (u|h)$ and hence $\langle\varphi'(u), h\rangle = (u|h)$ for all $h \in X$. By Proposition 32.13(a), $J = \varphi'$. □

Example 32.20 shows that, in the case of H-spaces, Definition 32.19 is equivalent to our earlier definition given in Section 21.6.

Proposition 32.21. *The duality map* $J: X \to 2^{X^*}$ *of a real* B*-space* X *is maximal monotone and cyclic monotone.*

PROOF. This follows from Proposition 32.17 and Corollary 32.18. □

Proposition 32.22 (Nice Properties of the Duality Map). *Let X be a real reflexive* B-*space, and let the dual space X^* be strictly convex. Set $\varphi(u) = 2^{-1}\|u\|^2$ for all $u \in X$. Then:*

(a) *The duality map*
$$J: X \to X^*$$
is single-valued, surjective, odd, demicontinuous, maximal monotone, bounded, and coercive.

For all $u \in X$, we have
$$Ju = \varphi'(u), \tag{38}$$
where φ' denotes the G-*derivative. Thus, J is a potential operator with potential φ.*

The norm $u \mapsto \|u\|$ is G-*differentiable on $X - \{0\}$. If we set $\psi(u) = \|u\|$, then*
$$\psi'(u) = \frac{Ju}{\|u\|} \qquad \text{for all} \quad u \in X - \{0\}.$$

Moreover, for each $u \in X$, there exists exactly one functional $u^* \in X^*$ such that
$$\langle u^*, u\rangle_X = \|u^*\|\,\|u\| \qquad \text{and} \qquad \|u^*\| = \|u\|. \tag{39*}$$
The functional u^* is equal to Ju. Thus, for all $u \in X$,
$$\langle Ju, u\rangle_X = \|u\|^2 \qquad \text{and} \qquad \|Ju\| = \|u\|. \tag{39}$$

Moreover, J is positively homogeneous, i.e.,
$$J(\lambda u) = \lambda Ju \qquad \text{for all} \quad \lambda > 0, \quad u \in X.$$

(b) *Let X be strictly convex. Then the operator $J: X \to X^*$ is strictly monotone and bijective. The inverse operator*
$$J^{-1}: X^* \to X$$
is equal to the duality map of the dual space X^* provided we identify X^{**} with X.

Furthermore, it follows from
$$\langle Ju_n - Ju, u_n - u\rangle_X \to 0 \qquad \text{as} \quad n \to \infty, \tag{40}$$
that $u_n \rightharpoonup u$ in X as $n \to \infty$. If, in addition, X is locally uniformly convex, then (40) *implies $u_n \to u$ as $n \to \infty$, i.e., J satisfies condition* (S).

(c) *Let X^* be locally uniformly convex. Then $J: X \to X^*$ is continuous. Furthermore, the norm $u \mapsto \|u\|$ is* F-*differentiable on $X - \{0\}$.*

(d) *Let X and X^* be locally uniformly convex. Then the duality map $J: X \to X^*$ is a homeomorphism.*

(e) *Let X^* be uniformly convex. Then $J: X \to X^*$ is uniformly continuous on bounded subsets of X.*

The F*-derivative of the norm* $u \mapsto \|u\|$ *is uniformly continuous on bounded subsets of* X *that lie outside some neighborhood of the point* $u = 0$.

Recall that, for each B-space Y, we have the following implications:

$$\begin{array}{ccc} Y \text{ is uniformly convex} & \rightarrow & Y \text{ is locally uniformly convex} \\ \downarrow & & \downarrow \\ Y \text{ is reflexive} & & Y \text{ is strictly convex.} \\ \updownarrow & & \\ Y^* \text{ is reflexive} & & \end{array} \tag{41}$$

The corresponding definitions may be found in Section 21.7. In this connection, the following deep result of the geometry of B-spaces is important.

Proposition 32.23 (Kadec (1959) and Troyanski (1971)). *In every reflexive* B-space X, *an equivalent norm can be introduced so that* X *and* X^* *are locally uniformly convex and thus also strictly convex with respect to the new norms on* X *and* X^*.

PROOF. Cf. Ciorănescu (1974, M), p. 98. □

Combining Propositions 32.22 and 32.23, we obtain the following result which we will use frequently:

Corollary 32.24. *Let* X *be a real reflexive* B-*space. Then we can introduce an equivalent norm on* X *so that, with respect to the new norms on* X *and* X^*, *the following holds true. The duality map*

$$J: X \rightarrow X^*$$

is an odd homeomorphism. Moreover, J *is strictly monotone, maximal monotone, bounded, coercive, and* J *satisfies condition* (S). *For all* $u, v \in X$, *the relations* (38) *and* (39) *above hold. The inverse operator* $J^{-1}: X^* \rightarrow X$ *is the duality map of the dual space* X^*.

Consequently, in considerations that are independent relative to passing to an equivalent norm on X, one can always assume that the duality map J possesses the propitious properties given in Corollary 32.24.

PROOF OF PROPOSITION 32.22(a).

(I) We define the map $A: X \rightarrow 2^{X^*}$ through

$$Au = \{u^* \in X^*: \langle u^*, u\rangle_X = \|u\|^2, \|u^*\| = \|u\|\}.$$

We first show that $Au \neq \emptyset$. Let $X_0 = \operatorname{span}\{u\}$. We set

$$f(tu) = t\|u\|^2 \qquad \text{for all} \quad t \in \mathbb{R}.$$

Hence, $f: X_0 \rightarrow \mathbb{R}$ is a linear functional with norm $\|f\| = \|u\|$. By the

Hahn–Banach theorem, we may extend f to a linear continuous functional $u^*: X \to \mathbb{R}$ with $\|u^*\| = \|u\|$. Obviously, $\langle u^*, u\rangle = f(u) = \|u\|^2$.

Next we show that A is single-valued. To this end, let $u_i^* \in Au$, $i = 1, 2$, i.e.,

$$\langle u_i^*, u\rangle = \|u\|^2 = \|u_i^*\|^2, \qquad i = 1, 2.$$

Hence

$$\begin{aligned} 2\|u_1^*\|\,\|u\| &\leq \|u_1^*\|^2 + \|u_2^*\|^2 = \langle u_1^* + u_2^*, u\rangle \\ &\leq \|u_1^* + u_2^*\|\,\|u\| \qquad \text{for all} \quad u \in X. \end{aligned}$$

This implies $\|u_1^*\| \leq 2^{-1}\|u_1^* + u_2^*\|$. Since $\|u_1^*\| = \|u_2^*\|$ and X^* is strictly convex, $u_1^* = u_2^*$.

(II) We show that $A: X \to X^*$ is demicontinuous. Let

$$u_n \to u \quad \text{in } X \qquad \text{as} \quad n \to \infty.$$

Hence $\|Au_n\| = \|u_n\| \to \|u\|$ as $n \to \infty$. Since (Au_n) is bounded and X^* is reflexive, there exists a subsequence, again denoted by (u_n), such that

$$Au_n \rightharpoonup u^* \quad \text{in } X^* \qquad \text{as} \quad n \to \infty.$$

For all $v \in X$,

$$\langle u^*, v\rangle = \lim_{n\to\infty} \langle Au_n, v\rangle \leq \lim_{n\to\infty} \|u_n\|\,\|v\| = \|u\|\,\|v\|.$$

Moreover,

$$\langle u^*, u\rangle = \lim_{n\to\infty} \langle Au_n, u_n\rangle = \lim_{n\to\infty} \|u_n\|^2 = \|u\|^2.$$

Hence $u^* = Au$. The convergence principle (Proposition 21.23(i)) shows that the entire sequence (Au_n) converges weakly to $u^* = Au$.

(III) We prove that

$$\lim_{t\to 0} \frac{2^{-1}\|u + th\|^2 - 2^{-1}\|u\|^2}{t} = \langle Au, h\rangle \qquad \text{for all} \quad h \in X. \tag{42}$$

Indeed, for all $u, v \in X$,

$$\begin{aligned} \langle Av, v - u\rangle &\geq \|v\|^2 - \|v\|\,\|u\| \\ &\geq \|v\|^2 - 2^{-1}(\|u\|^2 + \|v\|^2) \\ &= 2^{-1}\|v\|^2 - 2^{-1}\|u\|^2 \geq \|u\|\,\|v\| - \|u\|^2 \\ &\geq \langle Au, v - u\rangle. \end{aligned}$$

Letting $v = u + th$, $h \in X$, $t \in \mathbb{R}$, we obtain

$$t\langle Au, h\rangle \leq 2^{-1}\|u + th\|^2 - 2^{-1}\|u\|^2 \leq t\langle A(u + th), h\rangle.$$

This implies (42), since $\langle A(u + th), h\rangle \to \langle Au, h\rangle$ as $t \to 0$ by (II).

(IV) It follows from (42) that $Au = \varphi'(u)$. By Proposition 32.13(a), $\partial\varphi = \varphi'$. Hence $J = A$.

(V) Since $\|u\| = (2\varphi(u))^{1/2}$, it follows from the chain rule in Section 4.3 that $u \mapsto \|u\|$ is G-differentiable on $X - \{0\}$.

(VI) Since φ is even, the G-derivative $J = \varphi'$ is odd. From $\langle Ju, u\rangle = \|u\|^2$ and $\|Ju\| = \|u\|$, it follows that J is coercive and bounded.

(VII) J is monotone. Indeed, for all $u, v \in X$,

$$\begin{aligned}\langle Ju - Jv, u - v\rangle &= \langle Ju, u\rangle + \langle Jv, v\rangle - \langle Ju, v\rangle - \langle Jv, u\rangle \\ &\geq \|u\|^2 + \|v\|^2 - 2\|u\|\,\|v\|.\end{aligned}$$

Hence

$$\langle Ju - Jv, u - v\rangle \geq (\|u\|^2 - \|v\|^2)^2 \qquad \text{for all} \quad u, v \in X. \tag{43}$$

(VIII) Since $J\colon X \to X^*$ is monotone, coercive, and demicontinuous, it follows from Theorem 26.A that J is surjective.

(IX) By Proposition 32.7, J is maximal monotone.

Proof of Proposition 32.22(b). Suppose that X is strictly convex.

(X) We show that $J\colon X \to X^*$ is strictly monotone. From

$$\langle Ju - Jv, u - v\rangle = 0$$

and (43) we obtain

$$\begin{aligned}0 &= \left\langle Ju - J\left(\frac{u+v}{2}\right), \frac{u-v}{2}\right\rangle + \left\langle J\left(\frac{u+v}{2}\right) - Jv, \frac{u-v}{2}\right\rangle \\ &\geq \left(\|u\| - \left\|\frac{u+v}{2}\right\|\right)^2 + \left(\left\|\frac{u+v}{2}\right\| - \|v\|\right)^2.\end{aligned}$$

Hence $\|u\| = \|2^{-1}(u+v)\| = \|v\|$. Since X is strictly convex, $u = v$. Consequently, J is strictly monotone.

(XI) Since the operator $J\colon X \to X^*$ is strictly monotone and surjective, it is also bijective.

(XII) We show that $J^{-1}\colon X^* \to X$ is the duality map of X^*. By hypothesis, X is reflexive, i.e., $X^{**} = X$. Let

$$\bar{J}\colon X^* \to X$$

denote the duality map of X^*. By Proposition 32.22(a), the operator $\bar{J}$ is characterized through

$$\langle \bar{J}u^*, u^*\rangle_{X^*} = \|u^*\|^2 \qquad \text{and} \qquad \|\bar{J}u^*\| = \|u^*\|,$$

for all $u^* \in X^*$. Since $\langle u, u^*\rangle_{X^*} = \langle u^*, u\rangle_X$ for all $u \in X$, $u^* \in X^*$, it follows from (39*) that $\bar{J}u^* = J^{-1}(u^*)$. Hence $J^{-1} = \bar{J}$.

(XIII) Let

$$\langle Ju_n - Ju, u_n - u\rangle \to 0 \qquad \text{as} \quad n \to \infty.$$

We want to show that $u_n \rightharpoonup u$ in X as $n \to \infty$. A simple calculation shows

that

$$\langle Ju_n - Ju, u_n - u\rangle = (\|u_n\| - \|u\|)^2 + (\|u_n\|\,\|u\| - \langle Ju_n, u\rangle)$$
$$+ (\|u_n\|\,\|u\| - \langle Ju, u_n\rangle).$$

Since each of the three terms on the right-hand side is nonnegative, we obtain

$$\|u_n\| \to \|u\| \quad \text{and} \quad \langle Ju, u_n\rangle \to \|u\|^2 \quad \text{as} \quad n \to \infty.$$

Since X is reflexive, there exists a subsequence, again denoted by (u_n), such that

$$u_n \rightharpoonup v \quad \text{as} \quad n \to \infty. \tag{44}$$

We have to show that $u = v$. Then we are done by using the convergence principle (Proposition 21.23(i)). Indeed, by (44),

$$\langle Ju, u_n\rangle \to \langle Ju, v\rangle \quad \text{as} \quad n \to \infty.$$

This implies $\langle Ju, v\rangle = \|u\|^2$, i.e., $\|Ju\|\,\|v\| \geq \|u\|^2$. Hence $\|v\| \geq \|u\|$. On the other hand,

$$\|v\| \leq \varliminf_{n\to\infty} \|u_n\| = \|u\|.$$

Hence

$$\langle Ju, v\rangle_X = \|u\|^2 \quad \text{and} \quad \|v\| = \|Ju\|.$$

By (XII), $v = \bar{J}(Ju)$. Since $\bar{J} = J^{-1}$, $v = u$.

If, in addition, X is locally uniformly convex, then it follows from

$$u_n \rightharpoonup u \quad \text{and} \quad \|u_n\| \to \|u\| \quad \text{as} \quad n \to \infty$$

that $u_n \to u$ as $n \to \infty$ (cf. Proposition 21.23(d)).

Proof of Proposition 32.22(c). Suppose that X^* is locally uniformly convex.

(XIV) We want to show that $J: X \to X^*$ is continuous. To this end, let

$$u_n \to u \quad \text{in } X \quad \text{as} \quad n \to \infty.$$

This implies $\|Ju_n\| \to \|Ju\|$ as $n \to \infty$. By Proposition 32.22(a), J is demicontinuous. Hence

$$Ju_n \rightharpoonup Ju \quad \text{as} \quad n \to \infty.$$

Since X is locally uniformly convex, $Ju_n \to Ju$ as $n \to \infty$.

(XV) Since $J: X \to X^*$ is continuous and $J = \varphi'$, we obtain from Proposition 4.8(c) that the G-derivative φ' is actually an F-derivative. By the chain rule in Section 4.3, it follows from $\|u\| = (2\varphi(u))^{1/2}$ that the norm $u \mapsto \|u\|$ is F-differentiable on $X - \{0\}$.

Proof of Proposition 32.22(d). This follows from (b) and (c).

PROOF OF PROPOSITION 32.22(e). Let X^* be uniformly convex. We want to show that $J: X \to X^*$ is uniformly continuous on bounded sets.

Let $S = \{u \in X: \|u\| = 1\}$. We first show that J is uniformly continuous on S. Otherwise, there exist an $\varepsilon > 0$ and two sequences (u_n) and (v_n) with $\|u_n\| = \|v_n\| = 1$ for all n such that

$$\|u_n - v_n\| \to 0 \quad \text{as} \quad n \to \infty \quad \text{and} \quad \|Ju_n - Jv_n\| \geq \varepsilon \quad \text{for all } n.$$

For all $u, v \in X$,

$$\begin{aligned}\|Ju + Jv\| \, \|u\| &\geq \langle Ju + Jv, u \rangle \\ &= \langle Ju, u \rangle + \langle Jv, v \rangle + \langle Jv, u - v \rangle \\ &\geq \|u\|^2 + \|v\|^2 - \|v\| \, \|u - v\|.\end{aligned}$$

Letting $u = u_n$ and $v = v_n$, we get

$$\|2^{-1}(Ju_n + Jv_n)\| \geq 1 - 2^{-1}\|u_n - v_n\|.$$

This contradicts the uniform convexity of X^*. In this connection, note that $\|Ju_n\| = \|Jv_n\| = 1$ for all n.

Now observe that $J(\lambda w) = \lambda Jw$ for all $w \in X$, $\lambda > 0$. From

$$\begin{aligned}\|Ju - Jv\| &= \big\| \|u\| J(\|u\|^{-1}u) - \|v\| J(\|v\|^{-1}v) \big\| \\ &\leq \|u\| \, \|J(\|u\|^{-1}u) - J(\|v\|^{-1}v)\| \\ &\quad + \big| \|u\| - \|v\| \big| \, \|J(\|v\|^{-1}v)\|,\end{aligned}$$

and from the uniform continuity of J on S, it follows that J is uniformly continuous on bounded sets that lie outside some neighborhood of the point $u = 0$. Hence, J is uniformly continuous on each bounded set, since J is continuous at $u = 0$ and $J(0) = 0$.

If we set $\psi(u) = \|u\|$, and $\varphi(u) = 2^{-1}\|u\|^2$, then $\psi(u) = (2\varphi(u))^{1/2}$ and $\varphi'(u) = Ju$ for each $u \in X$. By the chain rule,

$$\psi'(u) = \frac{Ju}{\|u\|} \quad \text{for all} \quad u \neq 0.$$

Hence the F-derivative ψ' is uniformly continuous on each subset of X which lies outside some neighborhood of the point $u = 0$.

The proof of Proposition 32.22 is complete. □

32.4. The Main Theorem on Pseudomonotone Perturbations of Maximal Monotone Mappings

Our objective is to solve the basic equation

$$b \in Au + Bu, \qquad u \in C, \tag{45}$$

where $A: C \subseteq X \to 2^{X^*}$ is maximal monotone and $B: C \to X^*$ is pseudomonotone. As we shall show in the next sections, equations of type (45) allow many applications. Explicitly, equation (45) means the following. For given $b \in X^*$, we seek a $u \in C$ such that

$$b = v + w, \qquad \text{where} \quad v \in Au \quad \text{and} \quad w \in Bu.$$

If A and B are single-valued, then (45) is equivalent to the operator equation

$$b = Au + Bu, \qquad u \in C.$$

We assume:

(H1) C is a nonempty closed convex set in the real reflexive B-space X.
(H2) The mapping $A: C \to 2^{X^*}$ is maximal monotone.
(H3) The mapping $B: C \to X^*$ is pseudomonotone, bounded, and demicontinuous.

Recall that $B: C \to X^*$ is called pseudomonotone iff the following holds true. For each sequence (u_n) in C, it follows from $u_n \rightharpoonup u$ as $n \to \infty$ and

$$\overline{\lim_{n\to\infty}} \langle Bu_n, u_n - u\rangle \leq 0$$

that

$$\langle Bu, u - v\rangle \leq \underline{\lim}_{n\to\infty} \langle Bu_n, u_n - v\rangle \qquad \text{for all} \quad v \in C.$$

(H4) If the set C is unbounded, then the operator B is A-coercive with respect to the fixed element $b \in X^*$, i.e., there exists a point $u_0 \in C \cap D(A)$ and a number $r > 0$ such that

$$\langle Bu, u - u_0\rangle > \langle b, u - u_0\rangle \qquad \text{for all} \quad u \in C \quad \text{with} \quad \|u\| > r. \tag{46}$$

Theorem 32.A (Browder (1968)). *Let $b \in X^*$ be given and assume* (H1) *through* (H4). *Then the original problem* (45) *has a solution.*

This theorem represents a fundamental result in the theory of monotone operators. In Problem 32.4 we will show that this theorem remains true if $B: C \to 2^{X^*}$ is multivalued. In this connection, we will use a regularization argument. Before proving Theorem 32.A let us prove some important corollaries.

Corollary 32.25. *Suppose that* (H1) *through* (H3) *hold, and suppose that one of the following two conditions is satisfied*:

(i) *C is bounded.*
(ii) *C is unbounded, and B is A-coercive, i.e., there is a $u_0 \in C \cap D(A)$ such that*

$$\frac{\langle Bu, u - u_0\rangle}{\|u\|} \to +\infty \qquad \text{as} \quad \|u\| \to \infty \quad \text{in } C. \tag{47}$$

Then, $R(A + B) = X^*$. That is, for each $b \in X^*$, the original equation $b \in Au + Bu$, $u \in C$, has a solution.

Proof. For each $b \in X^*$, the assumption (H4) above is satisfied. □

Corollary 32.26. *Suppose that:*

(i) *The mapping $A: X \to 2^{X^*}$ is maximal monotone on the real reflexive* B-*space* X.
(ii) *The operator $B: X \to X^*$ is monotone, hemicontinuous, and bounded.*
(iii) *B is A-coercive, i.e., there exists a $u_0 \in D(A)$ such that*

$$\lim_{\|u\| \to \infty} \frac{\langle Bu, u - u_0 \rangle}{\|u\|} = +\infty.$$

Then, $R(A + B) = X^*$.

Proof. By Proposition 27.6, the operator B is pseudomonotone. The assertion now follows from Corollary 32.25. □

We now study the unperturbed equation

$$b \in Au, \qquad u \in C. \tag{48}$$

Corollary 32.27. *Suppose that:*

(i) *C is a nonempty closed convex subset of the real reflexive* B-*space* X.
(ii) *The mapping $A: C \to 2^{X^*}$ is maximal monotone.*
(iii) *If the set C is unbounded, then the operator A is coercive with respect to the fixed element $b \in X^*$, i.e., there exist a $u_0 \in D(A)$ and an $r > 0$ such that*

$$\langle u^*, u - u_0 \rangle > \langle b, u - u_0 \rangle \qquad \text{for all} \quad (u, u^*) \in A \quad \text{with} \quad \|u\| > r.$$

Then, equation (48) *has a solution.*

Proof. We use a *regularization method.* In place of (48), we study the regularized equations

$$b \in Au_n + \varepsilon_n J(u_n - u_0), \qquad u_n \in C, \quad n = 1, 2, \ldots, \tag{49}$$

where $J: X \to 2^{X^*}$ denotes the duality map $J: X \to X^*$, and (ε_n) is a sequence of positive real numbers with $\varepsilon_n \to 0$ as $n \to \infty$. According to Proposition 32.23, we introduce an equivalent norm on X such that both X and X^* are locally uniformly convex. By Corollary 32.24, the duality map $J: X \to X^*$ is single-valued, continuous, bounded, coercive, and strictly monotone. Notice that A remains maximal monotone and coercive with respect to b, when passing to equivalent norms.

(I) *Solution of the regularized equation* (49). From $\langle J(u - u_0), u - u_0 \rangle = \|u - u_0\|^2$ it follows that

$$\lim_{\|u\| \to \infty} \frac{\langle J(u - u_0), u - u_0 \rangle}{\|u\|} = +\infty.$$

According to Corollary 32.26, for each n, equation (49) has a solution u_n.

(II) *A priori estimates*. By (49), for each n, there exists a $u_n^* \in Au_n$ such that

$$b = u_n^* + \varepsilon_n J(u_n - u_0). \tag{50}$$

Hence

$$\begin{aligned}\langle b, u_n - u_0\rangle &= \langle u_n^*, u_n - u_0\rangle + \varepsilon_n\langle J(u_n - u_0), u_n - u_0\rangle \\ &= \langle u_n^*, u_n - u_0\rangle + \varepsilon_n\|u_n - u_0\|^2 \end{aligned}\tag{51}$$

and $(u_n, u_n^*) \in A$ for all n. By hypothesis (iii), the sequence (u_n) is bounded.

(III) *Weak convergence of the regularization method*. Since (u_n) is bounded, there exists a subsequence, again denoted by (u_n), such that

$$u_n \rightharpoonup u \quad \text{as} \quad n \to \infty.$$

Hence $u \in C$.

Since A is monotone, it follows from $(u_n, u_n^*) \in A$ that

$$\langle u_n^* - v^*, u_n - v\rangle \geq 0 \qquad \text{for all} \quad (v, v^*) \in A.$$

By (50),

$$\langle b - \varepsilon_n J(u_n - u_0) - v^*, u_n - v\rangle \geq 0 \qquad \text{for all} \quad (v, v^*) \in A \tag{52}$$

and all n. Notice that

$$\|\varepsilon_n J(u_n - u_0)\| = \varepsilon_n\|u_n - u_0\| \to 0 \qquad \text{as} \quad n \to \infty.$$

Letting $n \to \infty$, it follows from (52) that

$$\langle b - v^*, u - v\rangle \geq 0 \qquad \text{for all} \quad (v, v^*) \in A.$$

Since A is maximal monotone, $b \in Au$. □

PROOF OF THEOREM 32.A. The idea of proof is the following:

(a) We use a Galerkin method. In contrast to the proof of the main theorem on monotone operators (Theorem 26.A), we now work with *Galerkin inequalities*.
(b) The Galerkin inequalities are solved by means of the existence principle for inequalities in $\mathbb{R}^n$ (Proposition 2.17). In this connection, we use a truncation method.
(c) Coerciveness implies *a priori* estimates for the solutions of the Galerkin inequalities.
(d) The convergence of the Galerkin method is based on the pseudomonotonicity of the operator B.

Since $A: C \to 2^{X^*}$ is maximal monotone, the mapping A is not empty, i.e., there exists a $(u_0, u_0^*) \in A$. Without loss of generality, we may assume that $u_0 = 0$ and $u_0^* = 0$, i.e.,

$$(0, 0) \in A.$$

Otherwise, we pass from the equation $b \in Au + Bu$, $u \in C$, to the modified equation

$$b - u_0^* \in \bar{A}v + \bar{B}v, \qquad v \in C - u_0,$$

where we set $\bar{A}v = A(v + u_0)$ and $\bar{B}v = B(v + u_0) - u_0^*$ for all $v \in C - u_0$.

Step 1: Equivalent variational inequality.

We seek a $u \in C$ such that

$$\langle b - Bu - v^*, u - v\rangle_X \geq 0 \qquad \text{for all} \quad (v, v^*) \in A. \tag{53}$$

This problem is equivalent to the original problem (45). Indeed, $u \in C$ is a solution of (53) iff $b - Bu \in Au$, since A is *maximal monotone.*

Step 2: *A priori* estimates.

Let u be a solution of (53). Since $(0, 0) \in A$, we obtain

$$\langle b - Bu, u\rangle \geq 0.$$

By the *coerciveness* condition (46) with $u_0 = 0$, we get $\|u\| \leq r$ for fixed $r > 0$.

Step 3: The Galerkin inequalities.

Let $\mathscr{L}$ denote the set of all finite-dimensional subspaces Y of the original B-space X, i.e.,

$$Y \in \mathscr{L} \qquad \text{implies} \qquad Y \subseteq X \quad \text{and} \quad \dim Y < \infty.$$

We fix $Y \in \mathscr{L}$. In place of the variational inequality (53), we consider the *approximate problem*

$$\langle b - Bu_Y - v^*, u_Y - v\rangle_Y \geq 0, \tag{54}$$

for all $(v, v^*) \in A$ with $v \in C \cap Y$, where we seek $u_Y \in C \cap Y$.

In this connection, note that $X^* \subseteq Y^*$. This is to be understood in the sense of

$$\langle u^*, u\rangle_Y = \langle u^*, u\rangle_X \qquad \text{for all} \quad (u^*, u) \in X^* \times Y.$$

Step 4: Solution of (54) by a truncation method.

(IV-1) *The truncated problem.* Let $R > 0$. For fixed $Y \in \mathscr{L}$, we replace (54) by the truncated problem

$$\langle b - Bu_R - v^*, u_R - v\rangle_Y \geq 0 \qquad \text{for all} \quad (v, v^*) \in G_R \tag{55}$$

and unknown $u_R \in K_R$. Here, we set

$$K_R = \{v \in C \cap Y\colon \|v\| \leq R\},$$
$$G_R = \{(v, v^*) \in A\colon v \in K_R\}.$$

(IV-2) *Solution of* (55) *via the existence principle for finite-dimensional variational inequalities.* By Proposition 2.17, problem (55) has a solution $u_R \in K_R$.

(IV-3) *Solution of* (54) *via the finite intersection property.* We fix $Y \in \mathscr{L}$. Let $\mathscr{S}_R$ denote the set of all the solutions $u_R \in K_R$ of (55), i.e.,

$$\mathscr{S}_R \subseteq K_R \qquad \text{for all} \quad R > 0.$$

As in Step 2 above, we obtain the *a priori* estimate

$$\|u_R\| \le r \qquad \text{for all} \quad u_R \in \mathscr{S}_R \quad \text{and all} \quad R > 0,$$

where r is independent of R and Y. Since B is *demicontinuous*, the set $\mathscr{S}_R$ is closed. Moreover, $\mathscr{S}_R$ lies in the compact set

$$\{u \in C \cap Y: \|u\| \le r\}.$$

If $r \le R \le R'$, then $G_R \subseteq G_{R'}$. Hence

$$\mathscr{S}_R \supseteq \mathscr{S}_{R'} \qquad \text{for all } R, R' \text{ with} \quad R' \ge R \ge r.$$

Therefore, by the finite intersection principle $A_1(12g)$, there exists an element u_Y such that

$$u_Y \in \bigcap_{R \ge r} \mathscr{S}_R.$$

Obviously, this element u_Y is a solution of our approximate problem (54). In addition, we obtain that

$$\|u_Y\| \le r \qquad \text{and} \qquad u_Y \in C \cap Y. \tag{56}$$

Step 5: Convergence of the Galerkin method via the finite intersection principle.

For each $Y \in \mathscr{L}$, the approximate problem (54) has a solution u_Y. We want to show that (u_Y) "converges" in some sense to a solution u of the original problem (53). To this end, we will use the maximal monotonicity of A and the pseudomonotonicity of B. Since the defintion of pseudomonotonicity is based on sequences and not on more general M–S sequences, we need an additional approximation argument, which will be used in (V-3) below.

(V-1) *The finite intersection principle.* Let $Y, Z \in \mathscr{L}$. We set

$$M_Z = \{(u_Y, Bu_Y) \in X \times X^*: u_Y \text{ is a solution of (54) with } Y \supseteq Z\}.$$

We want to show that there exists an ordered pair (u, u^*) such that

$$(u, u^*) \in \bigcap_{Z \in \mathscr{L}} \overline{M_Z}, \tag{57}$$

where $\overline{M_Z}$ denotes the closure of M_Z with respect to the weak topology on the product space $X \times X^*$. Furthermore, we shall show below that u is the desired solution of the original problem (53).

We now prove (57). According to (56), there exists a closed ball K in $X \times X^*$ such that

$$\bigcup_{Z \in \mathscr{L}} M_Z \subseteq K.$$

In this connection, note that the operator B is *bounded*.

Since X is *reflexive*, so are X^* and $X \times X^*$. Consequently, K is weakly compact (see $A_1(38)$). Since $\overline{M_Z}$ is a weakly closed subset of the weakly compact set K, the set $\overline{M_Z}$ is weakly compact and

$$\bigcup_{Z \in \mathscr{L}} \overline{M_Z} \subseteq K.$$

Let $Y, Z \in \mathscr{L}$ and set $S = \operatorname{span}\{Y, Z\}$. Then

$$M_Y \cap M_Z \supseteq M_S.$$

This implies

$$\overline{M_Y} \cap \overline{M_Z} \neq \emptyset \qquad \text{for all} \quad Y, Z \in \mathscr{L}.$$

The same argument shows that the intersection of finitely many sets $\overline{M_Y}$ is not empty, i.e.,

$$\overline{M_{Y_1}} \cap \cdots \cap \overline{M_{Y_n}} \neq \emptyset \qquad \text{for all} \quad Y_1, \ldots, Y_n \in \mathscr{L}.$$

By the finite intersection principle $A_1(12g)$, there exists a (u, u^*) such that (57) holds.

(V-2) *Construction of a special ordered pair via the maximal monotonicity of A*. There exists a $(v_0, v_0^*) \in A$ such that

$$\langle b - u^* - v_0^*, u - v_0 \rangle_X \leq 0. \tag{58}$$

Otherwise, we have

$$\langle b - u^* - v^*, u - v \rangle_X > 0 \qquad \text{for all} \quad (v, v^*) \in A.$$

This implies $b - u^* \in Au$, because A is maximal monotone. Letting $v = u$ and $v^* = b - u^*$, we get $(v, v^*) \in A$ and $\langle b - u^* - v^*, u - v \rangle_X = 0$. This is a contradiction.

(V-3) *A special approximation argument*. We fix $Y \in \mathscr{L}$. The set $\overline{M_Y}$ is weakly closed in the B-space $X \times X^*$, and we have $(u, u^*) \in \overline{M_Y}$, according to (57). It follows from Problem 32.1 that there exists a sequence of elements (u_n, u_n^*) in M_Y such that

$$(u_n, u_n^*) \rightharpoonup (u, u^*) \quad \text{in } X \times X^* \qquad \text{as} \quad n \to \infty.$$

By construction of M_Y, $u_n^* = Bu_n$. Consequently, there exists a sequence (u_n) in C such that

$$u_n \rightharpoonup u \quad \text{in } X \qquad \text{and} \qquad Bu_n \rightharpoonup u^* \quad \text{in } X^* \qquad \text{as} \quad n \to \infty \tag{59}$$

and

$$\langle b - Bu_n - v^*, u_n - v \rangle_Y \geq 0 \qquad \text{for all} \quad (v, v^*) \in A \quad \text{with} \quad v \in Y, \tag{60}$$

according to (54). Since C is closed and convex, we obtain

$$u \in C.$$

(V-4) *Pseudomonotonicity of B*. We want to show that

$$\langle Bu, u - v \rangle \leq \langle v^* - b, v - u \rangle \quad \text{for all} \quad (v, v^*) \in A \text{ with } v \in Y. \tag{61}$$

In the following proof of (61), we consider only such $Y \in \mathscr{L}$ with $v_0 \in Y$, where the fixed element v_0 satisfies (58). Notice that

$$\bigcup_{Y \in \mathscr{L}, v_0 \in Y} Y = X. \tag{62}$$

We now fix $Y \in \mathscr{L}$. From (60) it follows that

$$\langle Bu_n, u_n - w \rangle \le \langle v^* - b, v - u_n \rangle + \langle Bu_n, v - w \rangle, \tag{63}$$

for all $w \in C$, $(v, v^*) \in A$ and $n \in \mathbb{N}$. Letting $w = v$, we obtain

$$\langle Bu_n, u_n - v \rangle \le \langle v^* - b, v - u_n \rangle \qquad \text{for all} \quad (v, v^*) \in A \quad \text{with} \quad v \in Y, \tag{64}$$

and for all $n \in \mathbb{N}$. Choosing $w = u$, $v = v_0$, and $v^* = v_0^*$ in (63), we get

$$\overline{\lim_{n \to \infty}} \langle Bu_n, u_n - u \rangle \le \langle v_0^* - b + u^*, v_0 - u \rangle,$$

since $Bu_n \rightharpoonup u^*$ as $n \to \infty$.

By (58), $\langle v_0^* - b + u^*, v_0 - u \rangle \le 0$ and hence

$$\overline{\lim_{n \to \infty}} \langle Bu_n, u_n - u \rangle \le 0.$$

Moreover, we have $u_n \rightharpoonup u$ as $n \to \infty$. Since $B: C \to X^*$ is *pseudomonotone*, we obtain

$$\langle Bu, u - v \rangle \le \underline{\lim_{n \to \infty}} \langle Bu_n, u_n - v \rangle \qquad \text{for all} \quad v \in C. \tag{65}$$

Now, relation (64) implies (61).

(V-5) *Solution of the original problem* (53). From (61) and (62), it follows that

$$\langle Bu, u - v \rangle \le \langle v^* - b, v - u \rangle \qquad \text{for all} \quad (v, v^*) \in A,$$

i.e., u is a solution of (53).

This completes the proof of Theorem 32.A. □

32.5. Application to Abstract Hammerstein Equations

We consider the equation

$$u + KFu = 0, \qquad u \in X^*. \tag{66}$$

Theorem 32.B. *Suppose that:*

(i) *The operator $K: X \to X^*$ is linear and monotone on the real reflexive* B-*space X.*
(ii) *The operator $F: X^* \to X$ is pseudomonotone, bounded, and coercive.*

Then, equation (66) has a solution, and this solution is unique if one of the following two additional conditions is satisfied:

(a) *F is strictly monotone.*
(b) *K is strictly monotone, and F is monotone.*

If X is an H-space, then we can replace assumption (ii) with the following condition: The operator $F: X^* \to X$ is monotone, hemicontinuous, and 3*-

monotone. This follows from Theorem 32.O in Section 32.22. The advantage of this modification is that F need not be coercive. In Section 32.21 we show that large classes of monotone operators are 3*-monotone (e.g., monotone Nemyckii operators on $L_2(G)$ are 3*-monotone in the case where G is a bounded region).

PROOF.

(I) Existence. Problem (66) is equivalent to the equation

$$0 \in K^{-1}u + Fu, \qquad u \in X^*. \tag{67}$$

By Proposition 26.4, K is continuous, and according to Propositions 32.5 and 32.7, the mappings K and K^{-1} are maximal monotone. Moreover, by Proposition 27.7, F is demicontinuous. Now, it follows from Corollary 32.25 that equation (67) has a solution.

(II) Uniqueness. Let

$$u_i + KFu_i = 0, \qquad u_i \in X^*, \qquad i = 1, 2. \tag{68}$$

Ad(a). Since K is monotone, it follows that

$$\langle u_1 - u_2, Fu_1 - Fu_2\rangle = -\langle KFu_1 - KF_2u, Fu_1 - Fu_2\rangle \leq 0.$$

Hence $u_1 = u_2$, according to the strict monotonicity of F.

Ad(b). Since F is monotone, it follows from (68) that

$$0 \leq \langle u_1 - u_2, Fu_1 - Fu_2\rangle = -\langle KFu_1 - KFu_2, Fu_1 - Fu_2\rangle.$$

Because of the strict monotonicity of K, we obtain $Fu_1 = Fu_2$. By (68), $u_1 = u_2$. □

32.6. Application to Hammerstein Integral Equations

Theorem 32.B allows applications to Hammerstein integral equations

$$u(x) + \int_G k(x, y)f(y, u(y))\,dy = 0 \quad \text{on } G, \qquad u \in L_q(G),$$

where G is a bounded region in $\mathbb{R}^N$ and $1 < q < \infty$. Such applications can be found in Proposition 28.4 in connection with other existence results for Hammerstein integral equations.

32.7. Application to Elliptic Variational Inequalities

By an elliptic variational inequality, we understand the following problem. For given $b \in X^*$, we seek a $u \in C$ such that

$$\langle b - Au, u - v\rangle \geq 0 \qquad \text{for all} \quad v \in C. \tag{69}$$

By means of the indicator function χ of C, problem (69) can be written in the equivalent form

$$b \in \partial\chi + Au, \qquad u \in C, \tag{70}$$

where we use the subgradient $\partial\chi$ from Example 32.15.

Theorem 32.C. *Suppose that:*

(i) *C is a nonempty closed convex subset of the real reflexive* B-*space X.*
(ii) *The operator $A: C \to X^*$ is pseudomonotone, demicontinuous, and bounded.*
(iii) *In the case where C is unbounded, there exists a point $u_0 \in C$ such that*

$$\frac{\langle Au, u - u_0\rangle}{\|u\|} \to +\infty \quad \text{as} \quad \|u\| \to \infty \quad \text{in } C.$$

Then the following hold true:

(a) Existence. *For each $b \in X^*$, the variational inequality* (69) *has a solution $u \in C$.*
(b) Solution set. *If $A: C \to X^*$ is monotone, then the set of all the solutions $u \in C$ of* (69) *is closed and convex.*
(c) Uniqueness. *If $A: C \to X^*$ is strictly monotone, then the solution $u \in C$ of* (69) *is unique.*

Corollary 32.28. *Assumption* (ii) *holds true in the case where one of the following two conditions is satisfied:*

(α) *The operator $A: X \to X^*$ is monotone, hemicontinuous, and bounded.*
(β) *The operator $A: X \to X^*$ is pseudomontone and bounded.*

Variational inequalities and their numerous applications will be studied in greater detail in Parts III through V.

Proof. Ad(a). By Example 32.15, the mapping $\partial\chi: C \to 2^{X^*}$ is maximal monotone, and $\partial\chi(u_0) \neq \varnothing$. Corollary 32.25 yields the assertion.

Ad(b). Let $A: C \to X^*$ be monotone. Then, the original problem (69) is equivalent to

$$\langle b - Av, u - v\rangle \geq 0 \quad \text{for all} \quad v \in C, \tag{71}$$

where we seek $u \in C$. Indeed, if $u \in C$ is a solution of (69), then

$$\begin{aligned}\langle Av, v - u\rangle &= \langle Au, v - u\rangle + \langle Av - Au, v - u\rangle \\ &\geq \langle Au, v - u\rangle \geq \langle b, v - u\rangle \quad \text{for all} \quad v \in C,\end{aligned}$$

i.e., u is a solution of (71). Conversely, let $u \in C$ be a solution of (71). We set

$$v = (1 - t)u + tw, \qquad 0 < t < 1, \quad w \in C.$$

Since C is convex, $v \in C$. By (71),

$$\langle b - A(u + t(w - u)), u - w\rangle \geq 0 \quad \text{for all} \quad w \in C.$$

As $t \to +0$, we obtain (69), since A is demicontinuous.

Let $\mathscr{S}$ denote the set of all the solutions $u \in C$ of (71). From (71) it follows that

$$u, \bar{u} \in \mathscr{S} \quad \text{implies} \quad (1-t)u + t\bar{u} \in \mathscr{S} \quad \text{for} \quad 0 \le t \le 1,$$

and

$$u_n \in \mathscr{S} \quad \text{for all } n \quad \text{and} \quad \lim_{n\to\infty} u_n = u \quad \text{implies} \quad u \in \mathscr{S}.$$

Thus, $\mathscr{S}$ is convex and closed.

Ad(c). Let u and $\bar{u}$ be solutions of (69), i.e., for all $v \in C$,

$$\langle b - Au, u - v\rangle \ge 0 \quad \text{and} \quad \langle b - A\bar{u}, \bar{u} - v\rangle \ge 0.$$

Letting $v = \bar{u}$ and $v = u$, we obtain

$$\langle Au - A\bar{u}, u - \bar{u}\rangle \le 0.$$

Since A is strictly monotone, $u = \bar{u}$. □

Corollary 32.28 follows from Section 27.2.

32.8. Application to First-Order Evolution Equations

We study the initial value problem

$$\begin{aligned} u'(t) + Au(t) &= b(t), \qquad 0 < t < T, \\ u(0) &= 0, \end{aligned} \tag{72*}$$

and the periodic problem

$$\begin{aligned} u'(t) + Au(t) &= b(t), \qquad 0 < t < T, \\ u(0) &= u(T). \end{aligned} \tag{73*}$$

In order to obtain the corresponding operator equations, we set

$$L_1 u = u', \qquad D(L_1) = \{u \in W_p^1(0, T; V, H): u(0) = 0\},$$

$$L_2 u = u', \qquad D(L_2) = \{u \in W_p^1(0, T; V, H): u(0) = u(T)\}.$$

We now replace (72*) and (73*) with the operator equation

$$L_1 u + Au = b, \qquad u \in D(L_1), \tag{72}$$

and

$$L_2 u + Au = b, \qquad u \in D(L_2), \tag{73}$$

respectively.

Theorem 32.D. *Let "$V \subseteq H \subseteq V^*$" be an evolution triple, and let $X = L_p(0, T; V)$, where $1 < p < \infty$ and $0 < T < \infty$. Suppose that the* *operator*

$$A: X \to X^*$$

is pseudomonotone, coercive, and bounded.

Then, for each $b \in X^*$, problems (72) *and* (73) *have solutions. If, in addition, A is strictly monotone, then the corresponding solutions are unique.*

PROOF.

(I) Existence. By Proposition 32.10, the operators $L_j: D(L_j) \subseteq X \to X^*$, $j = 1, 2$, are maximal monotone. The assertion then follows from Corollary 32.25 with $C = X$ and $u_0 = 0$.

(II) Uniqueness. Since L_j is monotone, it follows from

$$L_j u_k + A u_k = b, \qquad j, k = 1, 2,$$

that

$$\begin{aligned} 0 &= \langle L_j(u_1 - u_2), u_1 - u_2 \rangle + \langle Au_1 - Au_2, u_1 - u_2 \rangle \\ &\geq \langle Au_1 - Au_2, u_1 - u_2 \rangle. \end{aligned}$$

Since A is strictly monotone, $u_1 = u_2$. □

32.9. Application to Time-Periodic Solutions for Quasi-Linear Parabolic Differential Equations

Let $Q_T = G \times]0, T[$. We consider the time-periodic problem

$$\begin{gathered} u_t(x,t) + Lu(x,t) = f(x,t) \quad \text{on } Q_T, \\ D^\beta u(x,t) = 0 \quad \text{on } \partial G \times [0,T] \qquad \text{for all } \beta\colon |\beta| \leq m-1, \qquad (74^*) \\ u(x,0) = u(x,T) \quad \text{on } G, \end{gathered}$$

where

$$Lu(x,t) = \sum_{|\alpha| \leq m} (-1)^{|\alpha|} D^\alpha A_\alpha(x, Du(x,t), t)$$

and $Du = (D^\beta u)_{|\beta| \leq m}$, i.e., for each fixed t, the differential operator L is of order $2m$ with respect to the spatial variable x, where $m = 1, 2, \ldots$. Note that the derivatives D^α and D^β refer to x. The condition "$u(x,0) = u(x,T)$ on G" means that we are looking for solutions of time-period T.

We want to replace this problem with the *generalized problem*

$$u' + Au = b, \qquad u(0) = u(T), \qquad u \in W_p^1(0, T; V, H). \tag{74}$$

To this end, we make the following assumptions:

(A1) G is a bounded region in $\mathbb{R}^N$, $N \geq 1$. Let $0 < T < \infty$, $2 \leq p < \infty$, and $q^{-1} + p^{-1} = 1$. We set

$$V = \mathring{W}_p^m(G), \qquad H = L_2(G), \qquad X = L_p(0, T; V).$$

(A2) Let $f \in L_q(Q_T)$ be given.

(A3) For each fixed $t \in]0, T[$, the differential operator L satisfies the assumptions (H1) through (H4) of Proposition 26.12 (Carathéodory condition, growth condition, monotonicity, and coerciveness). In this connection, the constants and the functions g and h that appear in (H1) through (H4) should be *independent* of time t. Moreoever, the real function

$$t \mapsto \int_G \sum_{|\alpha| \leq m} A_\alpha(x, Du(x), t) D^\alpha v(x)\, dx$$

is measurable on $]0, T[$ for all $u, v \in V$.

By Section 30.4, there exists an operator $A(t)\colon V \to V^*$ such that

$$\langle A(t)u, v\rangle_V = \int_G \sum_{|\alpha| \leq m} A_\alpha(x, Du(x), t) D^\alpha v(x)\, dx,$$

for all $u, v \in V, t \in]0, T[$. Letting $w = Au$, where $w(t) = A(t)u(t)$, we obtain an operator

$$A\colon X \to X^*,$$

which is monotone, hemicontinuous, bounded, and coercive, according to Section 30.4 and Theorem 30.A(c). Moreover, there exists a functional $b \in X^*$ such that

$$\langle b(t), v\rangle_V = \int_G f(x, t) v(x)\, dx \qquad \text{for all} \quad v \in V.$$

Proposition 32.29. *Under the assumptions* (A1) *through* (A3), *the generalized problem* (74) *corresponding to* (74*) *has a solution. If A is strictly monotone, then the solution is unique.*

PROOF. This follows from Theorem 32.D. □

Let $m = 1$. As a special case of (74*), we consider the following problem:

$$\begin{aligned} u_t(x,t) - \sum_{i=1}^{N} D_i(|D_i u(x,t)|^{p-2} D_i u(x,t)) &= f(x,t) \quad \text{on } Q_T, \\ u(x,t) &= 0 \qquad \text{on } \partial G \times [0, T], \\ u(x,0) &= u(x,T) \quad \text{on } G, \end{aligned} \tag{75}$$

where $x = (\xi_1, \ldots, \xi_N)$ and $D_i = \partial/\partial\xi_i$.

Let G be a bounded region in $\mathbb{R}^N$, $N \geq 1$, and let $0 < T < \infty$, $2 \leq p < \infty$, and $q^{-1} + p^{-1} = 1$. Set

$$V = \mathring{W}_p^1(G), \qquad H = L_2(G), \qquad X = L_p(0, T; V).$$

By Proposition 26.10, the assumptions of Proposition 32.29 are fulfilled. In addition, the operator $A\colon X \to X^*$ is uniformly monotone, according to

Corollary 30.12. Explicitly, we have

$$\langle A(t)u, v\rangle_V = \int_G \sum_{i=1}^N |D_i u(x)|^{p-2} D_i u(x) D_i v(x)\, dx,$$

for all $u, v \in V$.

According to Proposition 32.29, for each given $f \in L_q(Q_T)$, the generalized problem (74) corresponding to (75) has a unique solution.

32.10. Application to Second-Order Evolution Equations

We now study the following initial value problem:

$$\begin{gathered} u'' + Nu' + Lu = b, \\ u(0) = u'(0) = 0, \\ u \in C([0, T], V), \qquad u' \in W_p^1(0, T; V, H). \end{gathered} \tag{76}$$

We set

$$X = L_p(0, T; V).$$

Then $X^* = L_q(0, T; V^*)$, where $p^{-1} + q^{-1} = 1$. If $2 \le p < \infty$, then $q \le p$ and hence $X \subseteq X^*$. The condition $u' \in W_p^1(0, T; V, H)$ means that

$$u' \in X \qquad \text{and} \qquad u'' \in X^*.$$

Let $u' \in W_p^1(0, T; V, H)$. After changing the function $t \mapsto u'(t)$ on a subset of $]0, T[$ of measure zero, we obtain a uniquely determined function

$$u' \in C([0, T], H),$$

i.e., $u'\colon [0, T] \to H$ is continuous. The initial condition $u'(0) = 0$ above is to be understood in the sense of $u' \in C([0, T], H)$. We assume:

(H1) Let $2 \le p < \infty$ and $0 < T < \infty$, and let "$V \subseteq H \subseteq V^*$" be an evolution triple.

(H2) The operator $N\colon X \to X^*$ is monotone, hemicontinuous, coercive, and bounded.

(H3) The operator $L\colon V \to V^*$ is linear, monotone, and symmetric, i.e.,

$$\langle Lv, w\rangle_V = \langle Lw, v\rangle_V \qquad \text{for all} \quad v, w \in V.$$

For $u \in X$, we set $(Lu)(t) = Lu(t)$.

Theorem 32.E. *Under the assumptions* (H1) *through* (H3), *for each given* $b \in X^*$, *the initial value problem* (76) *has a solution. If N is strictly monotone, then this solution is unique.*

Applications of this theorem to nonlinear hyperbolic differential equations will be considered in the next chapter.

The idea of proof is to introduce the operator S by

$$(Sv)(t) = \int_0^t v(s)\, ds.$$

Letting $u = Sv$, the original problem (76) is reduced to the first-order evolution equation

$$\begin{gathered} v' + Nv + LSv = b, \qquad v \in W_p^1(0, T; V, H), \\ v(0) = 0. \end{gathered} \tag{77}$$

PROOF.

(I) Equivalence of (76) and (77). Let u be a solution of (76). We set

$$v = u' \qquad \text{and} \qquad w = Sv.$$

From $v \in C([0, T], H)$ it follows that $w \in C^1([0, T], H)$. This implies $w' = v$. Since $u(0) = w(0) = 0$, we obtain $w = u$. Hence, v is a solution of (77).

Conversely, let v be a solution of (77). Hence $v \in C([0, T]; H)$. We set

$$u = Sv.$$

Thus, $u' = v$. Since $v \in L_p(0, T; V)$, we obtain $u \in C([0, T], V)$. Consequently, u is a solution of (76).

(II) We want to show that the operator $LS: X \to X^*$ is linear and continuous.

(II-1) The operator $S: X \to X$ is continuous. Indeed, for all $v \in X$, the Hölder inequality implies

$$\begin{aligned} \|Sv\|_X^p &= \int_0^T \left\| \int_0^t v(s)\, ds \right\|^p dt \\ &\leq T\left(\int_0^T \|v(s)\|\, ds \right)^p \\ &\leq \text{const} \int_0^T \|v(s)\|^p\, ds = \text{const}\, \|v\|_X^p. \end{aligned}$$

(II-2) By assumption, the operator $L: V \to V^*$ is linear and monotone. By Proposition 26.4, $L: V \to V^*$ is continuous. Thus, for all $v \in X$, we obtain

$$\begin{aligned} \|Lv\|_{X^*}^q &= \int_0^T \|Lv(s)\|_{V^*}^q\, ds \\ &\leq \text{const} \int_0^T \|v(s)\|_V^q\, ds \\ &\leq \text{const} \left(\int_0^T \|v(s)\|_V^p\, ds \right)^{q/p} = \text{const}\, \|v\|_X^q, \end{aligned}$$

since $1 < q \leq p$. Hence $L: X \to X^*$ is linear and continuous.

(III) The operator $LS: X \to X^*$ is *monotone*. To prove this, let P denote the set of all polynomials $p: [0, T] \to V$ with $p(0) = 0$, where the coefficients of p live in V. Since P is dense in X, it is sufficient to show that

$$\langle LSp, p\rangle_X \geq 0 \qquad \text{for all} \quad p \in P. \tag{78}$$

Let $q \in P$. Using the product rule and the *symmetry* of $L: V \to V^*$, we obtain

$$\begin{aligned}\frac{d}{dt}\langle Lq(t), q(t)\rangle_V &= \langle Lq'(t), q(t)\rangle_V + \langle Lq(t), q'(t)\rangle_V \\ &= 2\langle Lq(t), q'(t)\rangle_V,\end{aligned}$$

and hence

$$2\int_0^t \langle Lq(s), q'(s)\rangle_V \, ds = \langle Lq(t), q(t)\rangle_V \geq 0, \tag{79}$$

for all $t \in [0, T]$. For $q = Sp$, $p \in P$, this means

$$2\int_0^t \langle L(Sp)(s), p(s)\rangle_V \, ds \geq 0. \tag{79*}$$

Letting $t = T$ we get (78).

(IV) Existence. Since $LS: X \to X^*$ is linear, continuous, and monotone, the operator

$$N + LS: X \to X^*$$

is monotone, hemicontinuous, coercive, and bounded. By Theorem 32.D, the initial value problem (77) has a solution.

(V) Uniqueness. If $N: X \to X^*$ is strictly monotone, then so is $N + LS$. According to Theorem 32.D, the solution of (77) is unique. □

Since P is dense in X, it follows from (79*) that

$$\int_0^t \langle L(Sv)(s), v(s)\rangle_V \, ds \geq 0 \tag{80}$$

for all $v \in X$ and all $t \in [0, T]$. This inequality will be used in the proof of Theorem 33.A.

32.11. Regularization of Maximal Monotone Operators

The following theorem characterizes maximal monotone operators in terms of the duality map.

Theorem 32.F (Rockafellar (1970)). *Let X be a real reflexive* B-*space, where X*

and X^* are strictly convex. Then the monotone mapping

$$A: X \to 2^{X^*}$$

is maximal monotone iff $R(A + J) = X^*$.

Here, $J: X \to X^*$ denotes the duality map of the space X.

Corollary 32.30 (Regularization). *Let C be a nonempty closed convex subset of the real reflexive* B-*space X, where X and X^* are strictly convex. Suppose that the mapping $A: C \to 2^{X^*}$ is maximal monotone.*

Then, for each $\lambda > 0$, the inverse operator

$$(A + \lambda J)^{-1}: X^* \to X$$

is single-valued, demicontinuous, and maximal monotone.

Proof of Corollary 32.30. Let $\lambda > 0$.

(I) Surjectivity. By Proposition 32.22, the duality map $J: X \to X^*$ is single-valued, strictly monotone, demicontinuous, and bounded. From $\langle Ju, u - u_0 \rangle \geq \|u\|^2 - \|u\| \, \|u_0\|$ it follows that

$$\lim_{\|u\| \to \infty} \frac{\langle Ju, u - u_0 \rangle}{\|u\|} = +\infty \qquad \text{for each} \quad u_0 \in X.$$

By Corollary 32.25, we get $R(A + \lambda J) = X^*$.

(II) Single-valuedness of $(A + \lambda J)^{-1}$. Let $u, v \in (A + \lambda J)^{-1}(b)$. We set

$$c = b - \lambda Ju \qquad \text{and} \qquad d = b - \lambda Jv.$$

Hence $c \in Au$ and $d \in Av$. By the monotonicity of A,

$$\begin{aligned} 0 = \langle b - b, u - v \rangle &= \langle c + \lambda Ju - (d + \lambda Jv), u - v \rangle \\ &= \langle c - d, u - v \rangle + \lambda \langle Ju - Jv, u - v \rangle \\ &\geq \lambda \langle Ju - Jv, u - v \rangle. \end{aligned}$$

Since J is strictly monotone, $u = v$.

(III) The operator $A + \lambda J: X \to 2^{X^*}$ is monotone, since A and J are monotone. This implies the monotonicity of the inverse mapping $(A + \lambda J)^{-1}$: $X^* \to X$. By Proposition 26.4, $(A + \lambda J)^{-1}$ is locally bounded.

(IV) The operator $(A + \lambda J)^{-1}: X^* \to X$ is demicontinuous. To prove this, let

$$b_n \to b \quad \text{in } X^* \qquad \text{as} \quad n \to \infty.$$

We set $u_n = (A + \lambda J)^{-1} b_n$ and $u = (A + \lambda J)^{-1} b$. This implies

$$Au_n + \lambda Ju_n \to Au + \lambda Ju \qquad \text{as} \quad n \to \infty.$$

Since $(A + \lambda J)^{-1}$ is locally bounded, the sequence (u_n) is bounded.

Hence

$$\langle Au_n + \lambda Ju_n - (Au + \lambda Ju), u_n - u\rangle$$

$$= \langle Au_n - Au, u_n - u\rangle + \lambda\langle Ju_n - Ju, u_n - u\rangle \to 0 \quad \text{as} \quad n \to \infty.$$

Since the two right-hand terms are nonnegative, it follows that

$$\langle Ju_n - Ju, u_n - u\rangle \to 0 \qquad \text{as} \quad n \to \infty.$$

By (40), this implies $u_n \rightharpoonup u$ as $n \to \infty$.

(V) According to Proposition 32.7, the monotone demicontinuous operator $(A + \lambda J)^{-1}: X^* \to X$ is maximal monotone. □

PROOF OF THEOREM 32.F.

(I) If $A: X \to 2^{X^*}$ is maximal monotone, then it follows from Corollary 32.30 that $R(A + J) = X^*$.

(II) Conversely, let $A: X \to 2^{X^*}$ be monotone and suppose that $R(A + J) = X^*$. The proof of Corollary 32.30 shows that $(A + J)^{-1}: X^* \to X$ is monotone and demicontinuous, and hence maximal monotone. By Proposition 32.5, the inverse mapping

$$A + J: X \to 2^{X^*}$$

is also maximal monotone. To prove the maximal monotonicity of A, let $(u, u^*) \in X \times X^*$ and suppose that

$$\langle u^* - v^*, u - v\rangle \geq 0 \qquad \text{for all} \quad (v, v^*) \in A.$$

This implies

$$\langle u^* - (w^* - Jv), u - v\rangle \geq 0 \qquad \text{for all} \quad (v, w^*) \in A + J.$$

From $\langle Ju - Jv, u - v\rangle \geq 0$ we get

$$\langle u^* + Ju - w^*, u - v\rangle \geq 0 \qquad \text{for all} \quad (v, w^*) \in A + J.$$

Since $A + J$ is maximal monotone, we obtain $u^* + Ju \in (A + J)u$, and hence $u^* \in Au$. □

32.12. Regularization of Pseudomonotone Operators

Proposition 32.31. *Let X be a real reflexive* B-*space, where X and X^* are locally uniformly convex. Further let*

$$A: C \subseteq X \to X^*$$

be a pseudomonotone operator on the nonempty closed convex set C.

Then, for each $\lambda > 0$, the operator $A + \lambda J: C \to X^*$ satisfies condition $(S)_+$.

PROOF. Let $\lambda > 0$, and set $A_\lambda = A + \lambda J$. Suppose that $u_n \rightharpoonup u$ in C as $n \to \infty$, and

$$\overline{\lim_{n\to\infty}}\, \langle A_\lambda u_n - A_\lambda u, u_n - u\rangle \le 0.$$

We have to show that $u_n \to u$ as $n \to \infty$. Indeed, from the decomposition

$$\langle A_\lambda u_n - A_\lambda u, u_n - u\rangle = \langle Au_n - Au, u_n - u\rangle + \lambda\langle Ju_n - Ju, u_n - u\rangle \tag{81}$$

and $\langle Ju_n - Ju, u_n - u\rangle \ge 0$, it follows that

$$\overline{\lim_{n\to\infty}}\, \langle Au_n - Au, u_n - u\rangle \le 0. \tag{82}$$

Since A is pseudomonotone, this implies

$$\langle Au, u - v\rangle \le \underline{\lim_{n\to\infty}}\, \langle Au_n, u_n - v\rangle \qquad \text{for all} \quad v \in C.$$

Letting $v = u$, we obtain

$$\underline{\lim_{n\to\infty}}\, \langle Au_n, u_n - u\rangle \ge 0.$$

Hence

$$\underline{\lim_{n\to\infty}}\, \langle Au_n - Au, u_n - u\rangle \ge 0. \tag{83}$$

By (82) and (83), $\lim_{n\to\infty} \langle Au_n - Au, u_n - u\rangle = 0$. From (81) we obtain

$$\lim_{n\to\infty} \langle Ju_n - Ju, u_n - u\rangle = 0.$$

According to Proposition 32.22, this implies $u_n \to u$ as $n \to \infty$. □

32.13. Local Boundedness of Monotone Mappings

Definition 32.32. Let Y and Z be B-spaces. The mapping $A: Y \to 2^Z$ is called *locally bounded* at the point y iff there exists a neighborhood U of y such that the set

$$A(U) = \bigcup_{u \in U} Au$$

is bounded. Moreover, A is locally bounded on the set D iff A is locally bounded at each point of D.

The local boundedness of monotone operators is a fundamental property of these operators. The following result will be used critically in the next section.

Proposition 32.33. *A monotone mapping $A: X \to 2^{X^*}$ on the real* B-*space X is locally bounded at each interior point of $D(A)$.*

PROOF. Our proof makes use of the Baire category.

(I) Let (u_n^*) be a sequence in X^* with

$$\inf_{n,u} \langle u_n^*, u\rangle > -\infty \qquad \text{for all } u \text{ on a fixed open set.}$$

Then

$$\sup_n \|u_n^*\| < \infty. \tag{84}$$

Indeed, it follows first that there exist an $r > 0$ and a ball $B = \{v \in X: \|v\| < r\}$ such that

$$\inf_{n,v} \langle u_n^*, u_0 + v\rangle > -\infty \qquad \text{for all} \quad v \in B \quad \text{and fixed } u_0.$$

Since $0 \in B$, this implies $\inf_{n,v} \langle u_n^*, \pm v\rangle > -\infty$ for all $v \in B$, and hence

$$\sup_{n,v} |\langle u_n^*, v\rangle| < \infty \qquad \text{for all} \quad v \in B.$$

This yields (84).

(II) Let (u_n) and (u_n^*) be sequences in X and X^*, respectively, such that

$$u_n \to 0 \qquad \text{and} \qquad \|u_n^*\| \to \infty \qquad \text{as} \quad n \to \infty.$$

Then, for each $r > 0$, there exists a $v \in X$ with $\|v\| \le r$ such that

$$\varliminf_{n\to\infty} \langle u_n^*, u_n - v\rangle = -\infty.$$

Otherwise, there exists an $r > 0$ such that

$$\varliminf_{n\to\infty} \langle u_n^*, u_n - v\rangle > -\infty,$$

for each v in the ball $B = \{v \in X: \|v\| \le r\}$. This implies

$$B = \bigcup_{k=1}^{\infty} B_k.$$

where $B_k = \{v \in B: \langle u_n^*, u_n - v\rangle \ge -k \text{ for all } n\}$. The set B_k is closed for each k. Since the closed ball B in the B-space X has second Baire category (cf. A_1(65a)), there exists at least one B_k which has nonempty interior. This implies

$$\inf_{n,u} \langle u_n^*, u\rangle > -\infty \qquad \text{for all } u \text{ in a small ball.}$$

By (I), the sequence (u_n^*) is bounded in X^*. This is a contradiction.

(III) We now prove the statement of Proposition 32.33 by contradiction. If A is not locally bounded at a fixed point $u \in \operatorname{int} D(A)$, then there exists a sequence of points $(u_n, u_n^*) \in A$ such that

$$u_n \to u \qquad \text{and} \qquad \|u_n^*\| \to \infty \qquad \text{as} \quad n \to \infty.$$

We may assume that $u = 0$, since the monotonicity of A is invariant under translation. Choose an $r > 0$ such that the ball $B = \{v \in X: \|v\| \le r\}$

lies in $D(A)$. According to (II), there exists a $v \in B$ such that

$$\varliminf_{n\to\infty} \langle u_n^*, u_n - v\rangle = -\infty. \tag{85}$$

For fixed $(v, v^*) \in A$, the monotonicity of A implies

$$\langle u_n^*, u_n - v\rangle \ge \langle v^*, u_n - v\rangle \qquad \text{for all } n.$$

This contradicts (85). □

32.14. Characterization of the Surjectivity of Maximal Monotone Mappings

By definition of the range of A, $R(A)$, the equation

$$b \in Au, \qquad u \in X, \tag{86}$$

has a solution for each given $b \in X^*$ iff $R(A) = X^*$.

Theorem 32.G (Browder (1968a)). *Let $A: X \to 2^{X^*}$ be a maximal monotone mapping on the real reflexive* B-space *X. Then, $R(A) = X^*$ iff $A^{-1}: X^* \to 2^X$ is locally bounded.*

The proof will be given below.

In order to formulate important consequences of this theorem, we need the following notions.

Definition 32.34. Let $A: X \to 2^{X^*}$ be a mapping on the real B-space X.

(a) A is called *coercive* iff either $D(A)$ is bounded or $D(A)$ is unbounded and

$$\frac{\inf_{u^* \in Au} \langle u^*, u\rangle}{\|u\|} \to +\infty \qquad \text{as} \quad \|u\| \to \infty, \quad u \in D(A).$$

(b) A is called *weakly coercive* iff either $D(A)$ is bounded or $D(A)$ is unbounded and

$$\inf_{u^* \in Au} \|u^*\| \to +\infty \qquad \text{as} \quad \|u\| \to \infty, \quad u \in D(A).$$

Since $\langle u^*, u\rangle \le \|u^*\|\,\|u\|$, we have:

A is coercive implies A is weakly coercive.

In particular, let A be single-valued, i.e., we consider the operator $A: D(A) \subseteq X \to X^*$. Then A is coercive iff either $D(A)$ is bounded or $D(A)$ is unbounded and

$$\frac{\langle Au, u\rangle}{\|u\|} \to +\infty \qquad \text{as} \quad \|u\| \to \infty, \quad u \in D(A).$$

Moreover, A is weakly coercive iff either $D(A)$ is bounded or $D(A)$ is unbounded and

$$\|Au\| \to \infty \qquad \text{as} \quad \|u\| \to \infty, \quad u \in D(A).$$

Corollary 32.35. *Let $A: X \to 2^{X^*}$ (resp. $A: D(A) \subseteq X \to X^*$) be a maximal monotone and weakly coercive mapping on the real reflexive* B*-space X. Then $R(A) = X^*$.*

Proof. The weak coerciveness of A implies the local boundedness of A^{-1}. Theorem 32.G yields the assertion. □

Theorem 32.H (Main Theorem on Weakly Coercive Monotone Operators). *Let $A: X \to X^*$ be a monotone, hemicontinuous, and weakly coercive operator on the real reflexive* B*-space X.*

Then $R(A) = X^*$.

Proof. By Proposition 32.7, the operator A is maximal monotone. Corollary 32.35 yields the assertion. □

Proof of Theorem 32.G. The key to the proof is the method of regularization in (II-3) below.

(I) Let $R(A) = X^*$. Since A is maximal monotone, so is A^{-1}. According to Proposition 32.33, the mapping $A^{-1}: X^* \to 2^X$ is locally bounded.

(II) Conversely, let $A^{-1}: X^* \to 2^X$ be locally bounded. We want to show that $R(A) = X^*$. To do this, it is sufficient to prove that $R(A)$ is nonempty, closed, and open.

(II-1) $R(A)$ is nonempty. Indeed, it follows from the maximal monotonicity of A that $A \neq \varnothing$ and hence $R(A) \neq \varnothing$.

(II-2) $R(A)$ is closed. To prove this, let $u_n^* \in Au_n$ for all n and

$$u_n^* \to u^* \qquad \text{as} \quad n \to \infty.$$

By the monotonicity of A, we obtain

$$\langle u_n^* - v^*, u_n - v\rangle \geq 0 \qquad \text{for all} \quad (v, v^*) \in A \quad \text{and all } n. \tag{87}$$

Since A^{-1} is *locally bounded*, the sequence (u_n) is bounded. Thus, there exists a subsequence, again denoted by (u_n), such that

$$u_n \rightharpoonup u \qquad \text{as} \quad n \to \infty.$$

From (87), we get

$$\langle u^* - v^*, u - v\rangle \geq 0 \qquad \text{for all} \quad (v, v^*) \in A.$$

Since A is maximal monotone, this implies $u^* \in Au$, i.e., $u^* \in R(A)$.

(II-3) $R(A)$ is open. To show this, let

$$u^* \in R(A),$$

i.e., $u^* \in Au$ for some u. Since the maximal monotonicity of A remains invariant under a translation, we may assume that $u = 0$. We now choose an $r > 0$ such that A^{-1} is bounded on the ball

$$B = \{v^* \in X^*: \|v^* - u^*\| < r\}.$$

Our goal is to show that

$$\|v^* - u^*\| < r/2 \qquad \text{implies} \qquad v^* \in R(A). \tag{88}$$

This yields the openness of $R(A)$.

In order to prove (88), we use the *method of regularization*. We first equip the B-space X with an equivalent norm such that both X and X^* are strictly convex. Let v^* be given such that

$$\|v^* - u^*\| < r/2.$$

By Corollary 32.30, for each $\lambda > 0$, the equation

$$v_\lambda^* + \lambda J u_\lambda = v^*, \qquad v_\lambda^* \in Au_\lambda, \tag{89}$$

has a solution u_λ. This existence result is the *key* to our proof. By the monotonicity of A, we obtain

$$\langle v^* - \lambda J u_\lambda - u^*, u_\lambda - u\rangle \geq 0 \qquad \text{for all} \quad \lambda > 0,$$

where $u = 0$. This implies $\|v^* - u^*\| \|u_\lambda\| - \lambda \|u_\lambda\|^2 \geq 0$ and hence

$$\lambda \|u_\lambda\| \leq \|v^* - u^*\| < r/2 \qquad \text{for all} \quad \lambda > 0.$$

By (89),

$$\|v^* - v_\lambda^*\| = \lambda \|J u_\lambda\| = \lambda \|u_\lambda\| < r/2, \tag{90}$$

and thus

$$\|v_\lambda^* - u^*\| < r \qquad \text{for all} \quad \lambda > 0. \tag{91}$$

Since A^{-1} is bounded on the ball B, it follows from (91) and $u_\lambda \in A^{-1} v_\lambda^*$ that (u_λ) is bounded. By (90),

$$v_\lambda^* \to v^* \qquad \text{as} \quad \lambda \to 0.$$

Since $R(A)$ is closed, we obtain $v^* \in R(A)$. □

32.15. The Sum Theorem

Theorem 32.I (Rockafellar (1970)). *Let the mappings $A, B: X \to 2^{X^*}$ be maximal monotone on the real reflexive* B-*space X and let*

$$D(A) \cap \operatorname{int} D(B) \neq \varnothing.$$

Then, the sum $A + B: X \to 2^{X^*}$ is also maximal monotone.

Applications of this important result will be considered in the next three sections.

Proof. The main idea of our proof is to use Section 32.11 on the regularization of maximal monotone operators (Theorem 32.F). Furthermore, we use the main theorem on maximal monotone operators in Section 32.4 and a truncation argument based on the maximal monotonicity of subgradients.

Passing to an equivalent norm on X, we may assume that X and X^* are strictly convex. Moreover, using a translation, we may assume that

$$0 \in D(A) \cap \operatorname{int} D(B). \tag{92}$$

Finally, replacing $u \mapsto Au$ with $u \mapsto Au + c$ for fixed c, we may assume that

$$0 \in A(0). \tag{93}$$

Step 1: Suppose that $D(B)$ is bounded.

The mapping $A + B\colon X \to 2^{X^*}$ is monotone. By Theorem 32.F, the mapping $A + B$ is maximal monotone iff $R(A + B + J) = X^*$. Therefore, it is sufficient to prove that, for each $b^* \in X^*$, the equation

$$b^* \in Au + Bu + Ju, \qquad u \in X,$$

has a solution. Replacing $u \mapsto Bu$ with $u \mapsto Bu - b^*$, it is sufficient to prove that the equation

$$0 \in Au + Bu + Ju, \qquad u \in X, \tag{94}$$

has a solution. To obtain a solution u of (94), it is sufficient to find a (u, b) such that

$$\begin{aligned} -b &\in (A + 2^{-1}J)u, \qquad (u, b) \in X \times X^*, \\ b &\in (B + 2^{-1}J)u. \end{aligned} \tag{95}$$

To this end, we set

$$\begin{aligned} Eb &= -(A + 2^{-1}J)^{-1}(-b), \\ Fb &= (B + 2^{-1}J)^{-1}(b). \end{aligned}$$

By Corollary 32.30, the operators

$$E, F\colon X^* \to X$$

are monotone and demicontinuous. Moreover, we have

$$R(F) = D(B + 2^{-1}J) = D(B),$$

and hence

$$R(F) \text{ is bounded and } 0 \in \operatorname{int} R(F). \tag{96}$$

Now, to solve problem (95), it is sufficient to solve the equation

$$Eb + Fb = 0, \qquad b \in X^*. \tag{97}$$

In the following we want to solve (97) by means of the main theorem on maximal monotone operators in Section 32.4. Our considerations above show that the solvability of (97) implies the desired maximal monotonicity of $A + B$.

(I) The operator $E + F: X^* \to X$ is monotone and demicontinuous. Thus, $E + F$ is maximal monotone.

(II) Since $J(0) = 0$ and hence $0 \in (A + 2^{-1}J)(0)$, we get $E(0) = 0$. From the monotonicity of E, it follows that

$$\langle Ev, v\rangle \geq 0 \qquad \text{for all} \quad v \in X^*. \tag{98a}$$

(III) We shall prove below that there is an $r > 0$ such that

$$\langle Fv, v\rangle > 0 \qquad \text{for all} \quad v \in X^* \quad \text{with} \quad \|v\| > r. \tag{98b}$$

Adding (98a) and (98b), we get

$$\langle (E + F)v, v\rangle > 0 \qquad \text{for all} \quad v \in X^* \quad \text{with} \quad \|v\| > r,$$

i.e., $E + F$ is coercive with respect to 0. By the *main theorem* on maximal monotone mappings (Corollary 32.27), equation (97) above has a solution.

(IV) We finish our argument by proving (98b) by means of (96). From the monotonicity of F, it follows that $\langle Fv - Fw, v - w\rangle \geq 0$ and hence

$$\langle Fv, v\rangle \geq \langle Fw, v\rangle + \langle Fv - Fw, w\rangle \qquad \text{for all} \quad v, w \in X^*. \tag{99}$$

Since $R(F)$ is bounded, there is an $R > 0$ such that

$$|\langle Fv - Fw, w\rangle| \leq R\|w\| \qquad \text{for all} \quad v, w \in X^*. \tag{100}$$

By Proposition 32.33, the monotone mapping $F^{-1}: X \to 2^{X^*}$ is *locally bounded* at 0. In this connection, note that $0 \in \operatorname{int} R(F)$. Thus, there exist positive numbers δ_0 and ε such that

$$\|Fw\| \leq \delta_0 \qquad \text{implies} \qquad \|w\| \leq \varepsilon.$$

Since $0 \in \operatorname{int} R(F)$, there exists a δ such that $0 < \delta \leq \delta_0$ and the equation

$$u = Fw, \qquad w \in X,$$

has a solution for each given $u \in X^*$ with $\|u\| \leq \delta$. From (99) and (100), we obtain

$$\begin{aligned}\langle Fv, v\rangle &\geq \sup_{\|u\|=\delta} \langle u, v\rangle + \langle Fv - Fw, w\rangle \\ &\geq \delta\|v\| - R\varepsilon \qquad \text{for all} \quad v \in X^*.\end{aligned}$$

This implies (98b).

Step 2: Suppose that $D(B)$ is unbounded.

We want to reduce this case to Step 1 by using a *truncation argument*. We may assume that

$$0 \in D(A) \cap \operatorname{int} D(B) \qquad \text{and} \qquad 0 \in A(0), \quad 0 \in B(0).$$

We set

$$\chi_r(u) = \begin{cases} 0 & \text{if} \quad \|u\| \le r, \\ +\infty & \text{if} \quad \|u\| > r, \end{cases}$$

i.e., χ_r is the indicator function of the closed ball $\{u \in X: \|u\| \le r\}$. For the corresponding subgradient, we obtain

$$\partial\chi_r(u) = \begin{cases} \{0\} & \text{if} \quad \|u\| < r, \\ \{\lambda Ju: \lambda \ge 0\} & \text{if} \quad \|u\| = r, \\ \emptyset & \text{if} \quad \|u\| > r. \end{cases}$$

In this connection, note that $u^* \in \partial\chi_r(u)$ iff

$$\chi_r(v) \ge \chi_r(u) + \langle u^*, v - u \rangle \qquad \text{for all} \quad v \in X.$$

For example, let $\|u\| = r$. Then $u^* \in \partial\chi_r(u)$ iff

$$\langle u^*, v - u \rangle \le 0 \qquad \text{for all} \quad v: \|v\| \le r.$$

This implies $\langle u^*, u \rangle = \sup_{\|v\|=r} \langle u^*, v \rangle$ and hence

$$\langle u^*, u \rangle = \|u^*\| r = \|u^*\| \, \|u\|.$$

By (39*), we obtain that $u^* \in \partial\chi_r(u)$ iff $u^* = \lambda Ju$ for some $\lambda \ge 0$.

By Proposition 32.17, the subgradient $\partial\chi_r: X \to 2^{X^*}$ is maximal monotone.

(I) We show that the mapping

$$A + (B + \partial\chi_r): X \to 2^{X^*}$$

is maximal monotone for each $r > 0$. To this end, we use Step 1.

(I-1) The mappings $B, \partial\chi_r: X \to 2^{X^*}$ are maximal monotone and

$$0 \in D(B) \cap \operatorname{int} D(\partial\chi_r).$$

Since $D(\partial\chi_r)$ is bounded, it follows from Step 1 that $B + \partial\chi_r: X \to 2^{X^*}$ is maximal monotone.

(I-2) The mappings $A, B + \partial\chi_r: X \to 2^{X^*}$ are maximal monotone. We have

$$D(B + \partial\chi_r) = \{u \in D(B): \|u\| \le r\}.$$

From $0 \in D(A) \cap \operatorname{int} D(B)$ we get

$$0 \in D(A) \cap \operatorname{int} D(B + \partial\chi_r).$$

Since $D(B + \partial\chi_r)$ is bounded, it follows from Step 1 that $A + (B + \partial\chi_r)$: $X \to 2^{X^*}$ is maximal monotone.

(II) We show that $A + B: X \to 2^{X^*}$ is maximal monotone. To this end, we set $C = A + B$. By Theorem 32.F, it is sufficient to prove that

$$R(C + J) = X^*. \tag{101}$$

Since $C + \partial\chi_r$ is maximal monotone, we have $R(C + \partial\chi_r + J) = X^*$, according to Theorem 32.F. Thus, for each $b \in X^*$, the equation

$$b \in Cu + \partial\chi_r(u) + Ju, \qquad u \in X,$$

has a solution. We choose an $r > 0$ such that $r > \|b\|$. The explicit form of $\partial\chi_r$ yields

$$b \in Cu + (1 + \lambda)Ju, \tag{102}$$

where $\|u\| \le r$ and

$$\lambda = \begin{cases} 0 & \text{if} \quad \|u\| < r, \\ \ge 0 & \text{if} \quad \|u\| = r. \end{cases}$$

To prove (101), it is sufficient to show that $\|u\| < r$. Indeed, it follows from (102) that there exists a $u^* \in Cu$ such that

$$b = u^* + (1 + \lambda)Ju,$$

This implies

$$\langle b, u\rangle = \langle u^*, u\rangle + (1 + \lambda)\langle Ju, u\rangle.$$

Since C is monotone and $0 \in C(0)$, we get $\langle u^*, u\rangle \ge 0$. Noting that $\langle Ju, u\rangle = \|u\|^2$, we obtain

$$\langle b, u\rangle \ge (1 + \lambda)\|u\|^2,$$

and hence $(1 + \lambda)\|u\| \le \|b\| < r$. Therefore, $\|u\| < r$.

The proof of Theorem 32.I is complete. □

32.16. Application to Elliptic Variational Inequalities

We want to solve the variational inequality

$$\langle Au, v - u\rangle \ge \varphi(u) - \varphi(v) \qquad \text{for all} \quad v \in X, \tag{103}$$

where we seek $u \in D(A) \cap D(\partial\varphi)$. By definition of the subgradient $\partial\varphi$ in Section 32.3c, this inequality is equivalent to $-Au \in \partial\varphi(u)$. Consequently, problem (103) is equivalent to the equation

$$0 \in Au + \partial\varphi(u), \qquad u \in X. \tag{104}$$

In particular, if φ is the indicator function of the nonempty closed convex set C in the B-space X, i.e.,

$$\varphi(u) = \begin{cases} 0 & \text{if} \quad u \in C, \\ +\infty & \text{if} \quad u \notin C, \end{cases}$$

then problem (103) is equivalent to

$$\langle Au, v - u\rangle \ge 0 \qquad \text{for all} \quad v \in C,$$

where we seek $u \in C$.

In this special case, assumption (H2) below is satisfied. We assume:

(H1) The mapping $A: X \to 2^{X^*}$ is monotone on the real reflexive B-space X.
(H2) The functional $\varphi: X \to]-\infty, \infty]$ is convex, lower semicontinuous, and $\varphi \not\equiv +\infty$.
(H3) One of the following three conditions is satisfied:
  (i) $A: X \to X^*$ is single-valued and hemicontinuous.
  (ii) A is maximal monotone and $\operatorname{int} D(A) \cap D(\partial\varphi) \neq \varnothing$.
  (iii) A is maximal monotone and $D(A) \cap \operatorname{int} D(\partial\varphi) \neq \varnothing$.
(H4) The sum $A + \partial\varphi: X \to 2^{X^*}$ is coercive with respect to 0, i.e., there are an $r > 0$ and a $u_0 \in D(A) \cap D(\partial\varphi)$ such that

$$\langle u^*, u - u_0 \rangle > 0 \qquad \text{for all} \quad (u, u^*) \in A + \partial\varphi \quad \text{with} \quad \|u\| > r.$$

Proposition 32.36. *Assume* (H1) *through* (H4). *Then, the original variational inequality* (103) *has a solution.*

Proof. By Proposition 32.17, the subgradient $\partial\varphi: X \to 2^{X^*}$ is maximal monotone.

Ad(i). This is a special case of (ii).

Ad(ii), (iii). According to Theorem 32.I, the sum $A + \partial\varphi: X \to 2^{X^*}$ is maximal monotone. Hence it follows from (H4) and Corollary 32.27 that the equation (104) has a solution. This equation is equivalent to (103). □

32.17. Application to Evolution Variational Inequalities

We want to solve the variational inequality

$$\langle Lu + Au, v - u \rangle \geq \varphi(u) - \varphi(v) \qquad \text{for all} \quad v \in X, \tag{105}$$

where we seek $u \in D(L) \cap D(A) \cap D(\partial\varphi)$. By definition of the subgradient $\partial\varphi$ in Section 32.3c, problem (105) is equivalent to the equation

$$0 \in Lu + Au + \partial\varphi(u), \qquad u \in X. \tag{106}$$

We assume:

(H1) The linear operator $L: D(L) \subseteq X \to X^*$ is maximal monotone on the real reflexive B-space X.
(H2) The mapping $A: X \to 2^{X^*}$ is monotone.
(H3) The functional $\varphi: X \to]-\infty, \infty]$ is convex, lower semicontinuous, and $\varphi \not\equiv +\infty$.
(H4) One of the following three conditions is satisfied:
  (i) $A: X \to X^*$ is single-valued and hemicontinuous.
  (ii) A is maximal monotone and $\operatorname{int} D(A) \cap D(\partial\varphi) \neq \varnothing$.
  (iii) A is maximal monotone and $D(A) \cap \operatorname{int} D(\partial\varphi) \neq \varnothing$.

(H5) The sum $L + A + \partial\varphi: X \to 2^{X^*}$ is coercive with respect to 0, i.e., there are an $r > 0$ and a $u_0 \in D(L) \cap D(A) \cap D(\partial\varphi)$ such that

$$\langle u^*, u - u_0 \rangle > 0 \quad \text{for all} \quad (u, u^*) \in L + A + \partial\varphi \quad \text{with} \quad \|u\| > r.$$

(H6) $D(L) \cap \operatorname{int}(A + \partial\varphi) \neq \emptyset$.

Theorem 32.J. *Assume* (H1) *through* (H6). *Then the original variational inequality* (105) *has a solution.*

PROOF. As in the proof of Proposition 32.36, we obtain that $A + \partial\varphi: X \to 2^{X^*}$ is maximal monotone. It follows from (H6) and Theorem 32.I that $L + (A + \partial\varphi): X \to 2^{X^*}$ is maximal monotone. By (H5) and Corollary 32.27, problem (106) has a solution. This problem is equivalent to (105). □

EXAMPLE 32.37. (Evolution Variational Inequalities). We consider the special case, where $X = L_p(0, T; V)$, $1 < p < \infty$, and either

$$Lu = u', \qquad D(L) = \{u \in W_p^1(0, T; V, H): u(0) = 0\},$$

or

$$Lu = u', \qquad D(L) = \{u \in W_p^1(0, T; V, H): u(0) = u(T)\}.$$

Here, "$V \subseteq H \subseteq V^*$" is an evolution triple and $0 < T < \infty$.

Then, by Proposition 32.10, assumption (H1) above is fulfilled. In this case, the original problem (105) is called an evolution variational inequality.

32.18. The Regularization Method for Nonuniquely Solvable Operator Equations

We want to solve the equation

$$b \in Au, \qquad u \in X, \tag{107}$$

by means of the family of regularized equations

$$b \in Au_n + \varepsilon_n Bu_n, \qquad u_n \in X, \qquad n = 1, 2, \ldots, \tag{108}$$

where

$$\varepsilon_n > 0 \quad \text{for all} \quad n \quad \text{and} \quad \varepsilon_n \to 0 \quad \text{as} \quad n \to \infty. \tag{109}$$

For given fixed $b \in X^*$, let $\mathcal{S}$ denote the set of all the solutions u of (107). Our goal is to show that

$$u_n \to u \qquad \text{as} \quad n \to \infty,$$

where u solves the original equation (107) and, at the same time, u is the unique solution of the variational inequality

$$\langle Bu, v - u \rangle \geq 0 \quad \text{for all} \quad v \in \mathcal{S} \quad \text{and fixed} \quad u \in \mathcal{S}. \tag{110}$$

Notice that we do not assume that the original equation (107) possesses a unique solution. However, the regularized equation (108) is uniquely solvable for each n. We make the following assumptions:

(H1) The mapping $A: X \to 2^{X^*}$ is maximal monotone on the real reflexive B-space X (e.g., $A: X \to X^*$ is monotone and hemicontinuous).

(H2) A is coercive with respect to the given fixed $b \in X^*$, i.e., there is an $r > 0$ such that $0 \in D(A)$ and

$$\langle u^* - b, u \rangle > 0 \qquad \text{for all} \quad (u, u^*) \in A \quad \text{with} \quad \|u\| > r.$$

(H3) The operator $B: X \to X^*$ is strictly monotone, hemicontinuous, and bounded. Moreover, $B(0) = 0$.

(H4) The given sequence (ε_n) of real numbers satisfies (109).

Theorem 32.K (Convergence of the Method of Regularization). *Assume* (H1) *through* (H4). *Let* $b \in X^*$ *be given. Then*:

(a) *The solution set* $\mathscr{S}$ *of the original problem* (107) *is not empty, convex, and closed in* X.

(b) *For each* $n \in \mathbb{N}$, *the regularized equation* (108) *has a unique solution* u_n.

(c) *As* $n \to \infty$, *the sequence* (u_n) *converges weakly in* X *to a solution* u *of the original equation* (107), *where* u *is the unique solution of the variational inequality* (110).

(d) *If the operator* B *satisfies condition* (S), *then* (u_n) *converges strongly in* X *to* u.

Proof. For simplicity of notation, the elements u^* of Au are sometimes denoted briefly by Au.

Ad(a). By Corollary 32.27, the solution set $\mathscr{S}$ of (107) is not empty. Since A is maximal monotone, we have $u \in \mathscr{S}$ iff

$$\langle b - v^*, u - v \rangle \geq 0 \qquad \text{for all} \quad (v, v^*) \in A.$$

Hence $\mathscr{S}$ is convex and closed in X.

Ad(b). The operator $B: X \to X^*$ is maximal monotone. According to Theorem 32.I, the sum $A + \varepsilon_n B: X \to 2^{X^*}$ is maximal monotone. Since $B(0) = 0$, the monotonicity of B implies

$$\langle Bu, u \rangle \geq 0.$$

From (H2) we obtain

$$\langle u^* + \varepsilon_n Bu - b, u \rangle > 0 \qquad \text{for all} \quad (u, u^*) \in A \quad \text{with} \quad \|u\| > r.$$

Hence $0 \in D(A + \varepsilon_n B)$ and

$$\langle v^* - b, u \rangle > 0 \qquad \text{for all} \quad (u, v^*) \in A + \varepsilon_n B \quad \text{with} \quad \|u\| > r.$$

By Corollary 32.27, the regularized equation (108) has a solution u_n for each n.

To prove the uniqueness of u_n, we suppose that u_n and v_n solve (108) for fixed n, i.e.,

$$Au_n + \varepsilon_n Bu_n = b \qquad \text{and} \qquad Av_n + \varepsilon_n Bv_n = b.$$

Hence

$$\begin{aligned} 0 = \langle b - b, u_n - v_n \rangle &= \langle Au_n - Av_n, u_n - v_n \rangle + \varepsilon_n \langle Bu_n - Bv_n, u_n - v_n \rangle \\ &\geq \varepsilon_n \langle Bu_n - Bv_n, u_n - v_n \rangle, \end{aligned}$$

by the monotonicity of A. Since B is strictly monotone, $u_n = v_n$.

Ad(c).

(I) We prove first that the variational inequality (110) has at most one solution. Indeed, let u and w be solutions of (110). Setting $v = w$ and $v = u$, we obtain

$$\langle Bu - Bw, u - w \rangle \leq 0.$$

The strict monotonicity of B implies $u = w$.

(II) *A priori* estimates. We want to show that

$$\|u_n\| \leq r \qquad \text{for all } n$$

and all solutions u_n of (108). Otherwise, there is a solution u_n of (108) with $\|u_n\| > r$. From (108) and the coerciveness condition (H2), we obtain

$$\langle Au_n - b, u_n \rangle = \langle -\varepsilon_n Bu_n, u_n \rangle > 0.$$

On the other hand, it follows from $B(0) = 0$ and from the monotonicity of B that $\langle Bu_n, u_n \rangle \geq 0$. This is a contradiction.

(III) Since (u_n) is bounded, there exists a sequence, again denoted by (u_n), such that

$$u_n \rightharpoonup u \qquad \text{as} \quad n \to \infty.$$

We want to show that $u \in \mathscr{S}$. Using the monotonicity of A, we obtain, for all $(v, v^*) \in A$,

$$\begin{aligned} 0 &\leq \lim_{n\to\infty} \langle Au_n - v^*, u_n - v \rangle \\ &= \lim_{n\to\infty} \langle b - \varepsilon_n Bu_n - v^*, u_n - v \rangle \\ &= \langle b - v^*, u - v \rangle. \end{aligned}$$

In this connection, note that (Bu_n) is bounded, since B is bounded. Because of the maximal monotonicity of A, we get $b \in Au$, i.e., $u \in \mathscr{S}$.

(IV) We want to prove that u is a solution of the variational inequality (110). To this end, we set

$$u_t = u + t(v - u) \qquad \text{for} \quad 0 < t \leq 1 \quad \text{and} \quad v \in \mathscr{S},$$

where t and v are fixed. Since $\mathscr{S}$ is convex, we obtain $u_t \in \mathscr{S}$ and hence $b \in Au_t$. From the regularized equation (108) it follows that

$$\begin{aligned} 0 &= \langle Au_n - Au_t, u_n - u_t \rangle + \varepsilon_n \langle Bu_n, u_n - u_t \rangle \\ &\geq \varepsilon_n \langle Bu_n, u_n - u_t \rangle \qquad \text{for all } n. \end{aligned}$$

Hence

$$0 \geq \langle Bu_n, u_n - u_t \rangle = \langle Bu_n - Bu_t, u_n - u_t \rangle + \langle Bu_t, u_n - u_t \rangle$$
$$\geq \langle Bu_t, u_n - u_t \rangle.$$

Letting $n \to \infty$, we get

$$0 \geq \langle Bu_t, u - u_t \rangle = t\langle Bu_t, u - v \rangle.$$

Dividing by t and letting $t \to +0$, we obtain

$$\langle Bu, v - u \rangle \geq 0 \qquad \text{for all} \quad v \in \mathcal{S},$$

i.e., u is a solution of (110).

(V) According to our considerations above, each subsequence of (u_n) has, in turn, a subsequence which converges weakly to the unique solution u of (110). By the convergence principle (Proposition 21.23(i)), the entire sequence (u_n) converges weakly to u.

Ad(d). From (108) and $b \in Au$, we obtain

$$0 = \langle Au_n - b + \varepsilon_n Bu_n, u_n - u \rangle \geq \varepsilon_n \langle Bu_n, u_n - u \rangle$$

and hence

$$0 \geq \langle Bu_n, u_n - u \rangle = \langle Bu_n - Bu, u_n - u \rangle + \langle Bu, u_n - u \rangle.$$

Noting that $u_n \rightharpoonup u$ as $n \to \infty$, this implies

$$0 \leq \langle Bu_n - Bu, u_n - u \rangle \to 0 \quad \text{as} \quad n \to \infty.$$

Since B satisfies condition (S), we obtain that $u_n \to u$ as $n \to \infty$. □

32.19. Characterization of Linear Maximal Monotone Operators

Theorem 32.L (Brézis (1970)). *For a linear operator $L: D(L) \subseteq X \to X^*$ on the real reflexive* B*-space X, the following two statements are equivalent:*

(i) *L is maximal monotone.*
(ii) *$D(L)$ is dense in X, L and L^* are monotone, and L is graph closed.*

Proof. To prove that (ii) implies (i), we use a minimum problem and the duality maps on X and X^*.

(i) ⇒ (ii).

(I) $D(L)$ is dense in X. Otherwise, it follows from the *Hahn–Banach theorem* that there exists a functional $b \in X^*$ such that $b \neq 0$ and

$$\langle b, u \rangle = 0 \qquad \text{for all} \quad u \in \overline{D(L)}.$$

Hence

$$\langle Lu - b, u \rangle = \langle Lu, u \rangle \geq 0 \qquad \text{for all} \quad u \in D(L).$$

Since L is maximal monotone, this implies $b = L(0) = 0$, which is a contradiction.

(II) L is graph closed. To prove this, let

$$u_n \to u \quad \text{and} \quad Lu_n \to b \quad \text{as} \quad n \to \infty.$$

From

$$\langle Lu_n - Lv, u_n - v \rangle \geq 0 \quad \text{for all} \quad v \in D(L) \quad \text{and all } n,$$

we obtain

$$\langle b - Lv, u - v \rangle \geq 0 \quad \text{for all} \quad v \in D(L)$$

and hence $b = Lu$.

(III) L^* is monotone. Since L^* is linear, it is sufficient to prove that

$$\langle L^*u, u \rangle \geq 0 \quad \text{for all} \quad u \in D(L^*).$$

First, let $u \in D(L) \cap D(L^*)$. Then

$$\langle L^*u, u \rangle = \langle u, Lu \rangle \geq 0.$$

Secondly, let $u \in D(L^*)$ and $u \notin D(L)$. Since L is maximal monotone, there exists a $v \in D(L)$ such that

$$\langle Lv - b, v - u \rangle < 0,$$

where $b = -L^*u$. Hence

$$\langle Lv, v \rangle < \langle Lv, u \rangle - \langle L^*u, v \rangle + \langle L^*u, u \rangle = \langle L^*u, u \rangle.$$

Since $\langle Lv, v \rangle \geq 0$, we obtain $\langle L^*u, u \rangle > 0$.

(ii) $\Rightarrow$ (i). Passing to an equivalent norm on X, we may assume that X and X^* are strictly convex. We have to show that

$$\langle Lu - b, u - v \rangle \geq 0 \quad \text{for all} \quad u \in D(L) \tag{111}$$

implies $b = Lv$. To prove this, we consider the minimum problem

$$g(u, u^*) = \min !, \qquad (u, u^*) \in G(L), \tag{112}$$

where $(v, b) \in X \times X^*$ is fixed and

$$g(u, u^*) = \langle u^* - b, u - v \rangle + \|u^* - b\|_{X^*}^2 + \|u - v\|_X^2.$$

By (ii), L is graph closed, i.e., $G(L)$ is a closed linear subspace of $X \times X^*$. The functional $g\colon G(L) \to \mathbb{R}$ is convex and continuous. From the well-known inequality $\pm ab \leq \frac{1}{2}a^2 + \frac{1}{2}b^2$, we obtain

$$g(u, u^*) \geq \tfrac{1}{2}\|u^* - b\|^2 + \tfrac{1}{2}\|u - v\|^2,$$

and hence $g(u, u^*) \to +\infty$ as $\|u\| + \|u^*\| \to \infty$. According to Theorem 25.E, the minimum problem (112) has a solution $(u, u^*) \in G(L)$, which satisfies the Euler equation

$$g'(u, u^*)(h, h^*) = 0 \quad \text{for all} \quad (h, h^*) \in G(L).$$

Explicitly, that means

$$\langle h^*, u - v\rangle + \langle u^* - b, h\rangle + 2\bar{J}(u^* - b)h^* + 2J(u - v)h = 0, \tag{113}$$

for all $(h, h^*) \in G(L)$, where $J: X \to X^*$ and $\bar{J}: X^* \to X$ denotes the duality map on X and X^*, respectively. The condition (u, u^*), $(h, h^*) \in G(L)$ means that $u^* = Lu$ and $h^* = Lh$. To simplify notation, we set

$$x = u - v \qquad \text{and} \qquad y = Lu - b.$$

By (113),

$$\langle Lh, x + 2\bar{J}y\rangle = -\langle 2Jx + y, h\rangle \qquad \text{for all} \quad h \in D(L).$$

Hence $L^*(x + 2\bar{J}y) = -(2Jx + y)$. This implies

$$-\langle 2Jx + y, x + 2\bar{J}y\rangle \geq 0,$$

since L^* is monotone. By (111), $\langle y, x\rangle \geq 0$. Hence

$$\langle Jx, x\rangle + \langle y, \bar{J}y\rangle + 2\langle Jx, \bar{J}y\rangle \leq 0.$$

Noting that $\langle Jx, x\rangle = \|x\|^2$ and $\langle Jx, \bar{J}y\rangle \geq -\|Jx\|\,\|\bar{J}y\| = -\|x\|\,\|y\|$, we obtain

$$\|x\|^2 + \|y\|^2 - 2\|x\|\,\|y\| \leq 0,$$

and hence $x = y = 0$, i.e., $u = v$ and $Lv = b$. □

32.20. Extension of Monotone Mappings

Proposition 32.38. *Let C be a nonempty closed convex subset of the real reflexive* B-*space X, and let the mapping*

$$A: C \to 2^{X^*}$$

be maximal monotone. If we set

$$\bar{A}u = \begin{cases} Au & \text{if} \quad u \in C, \\ \varnothing & \text{if} \quad u \in X - C, \end{cases}$$

then $\bar{A}: X \to 2^{X^*}$ is also maximal monotone.

Proof. Passing to an equivalent norm on X, we may assume that both X and X^* are strictly monotone. By Proposition 32.22, the duality map $J: X \to X^*$ is single-valued, monotone, bounded, and demicontinous, and hence J is pseudomonotone. For each $u_0 \in X$,

$$\lim_{\|u\| \to \infty} \frac{\langle Ju, u - u_0\rangle}{\|u\|} \geq \lim_{\|u\| \to \infty} \|u\| - \|u_0\| = +\infty.$$

By Corollary 32.25, $R(A + J) = X^*$.

Hence $R(\bar{A} + J) = X^*$. Thus, by Theorem 32.F, the mapping $\bar{A}: X \to 2^{X^*}$ is maximal monotone. $\square$

The following theorem shows that each monotone mapping can be extended to a maximal monotone mapping.

Theorem 32.M (Extension Theorem). *Let*

$$A: X \to 2^{X^*}$$

be a monotone mapping on the real reflexive B-*space X with $D(A) \neq \emptyset$, i.e., A is not the empty map. Then there exists a maximal monotone mapping*

$$\bar{A}: X \to 2^{X^*}$$

such that $G(A) \subseteq G(\bar{A})$ and $D(\bar{A}) \subseteq \overline{\text{co}}\, D(A)$.

PROOF. We use Zorn's lemma. To this end, set $C = \overline{\text{co}}\, D(A)$. Denote by $\mathscr{S}$ the set of all monotone mappings $B: C \to 2^{X^*}$ such that B is an extension of A. For $B_1, B_2 \in \mathscr{S}$, we write

$$B_1 \leq B_2 \qquad \text{iff} \qquad G(B_1) \subseteq G(B_2).$$

This way, $\mathscr{S}$ becomes an ordered set. Let $\mathscr{C}$ be a chain, i.e., $\mathscr{C}$ is nonempty and $B_1, B_2 \in \mathscr{C}$ implies $B_1 \leq B_2$ or $B_2 \leq B_1$. If we set

$$Su = \bigcup_{B \in \mathscr{C}} Bu \qquad \text{for all} \quad u \in C,$$

then the mapping $S: C \to 2^{X^*}$ is monotone and

$$B \leq S \qquad \text{for all} \quad B \in \mathscr{C},$$

i.e., S is a upper bound for the chain $\mathscr{C}$.

By Zorn's lemma, there is a maximal element A_0 in $\mathscr{S}$, i.e., $A_0 \leq B$ and $B \in \mathscr{S}$ implies $B = A_0$ (cf. Problem 11.1b).

Obviously, the mapping $A_0: C \to 2^{X^*}$ is maximal monotone. We set $\bar{A}u = A_0 u$ if $u \in C$, and $\bar{A}u = \emptyset$ if $u \in X - C$. By Proposition 32.38, the mapping $\bar{A}: X \to 2^{X^*}$ is maximal monotone. From $A \leq A_0$, we obtain

$$G(A) \subseteq G(A_0) = G(\bar{A}) \qquad \text{and} \qquad D(\bar{A}) = D(A_0) \subseteq C. \qquad \square$$

EXAMPLE 32.39. Let $X = \mathbb{R}$. We consider the monotone mapping

$$A: \mathbb{R} \to 2^{\mathbb{R}}$$

with $D(A) =]a, b[$ as pictured in Figure 32.8(a). Then the mapping

$$\bar{A}: \mathbb{R} \to 2^{\mathbb{R}},$$

pictured in Figure 32.8(b), is the unique maximal monotone extension of A with the property $D(\bar{A}) = \overline{\text{co}}\, D(A) = [a, b]$.

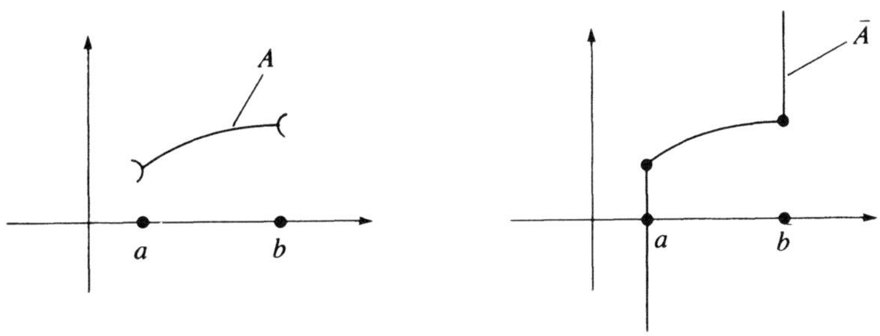

(a) monotone mapping (b) maximal monotone extension

Figure 32.8

32.21. 3-Monotone Mappings and Their Generalizations

Let us recall that, by definition, a mapping $A: X \to 2^{X^*}$ on the real B-space X is *cyclic monotone* iff the condition

$$\langle u_1^*, u_1 - u_2 \rangle + \langle u_2^*, u_2 - u_3 \rangle + \cdots + \langle u_n^*, u_n - u_{n+1} \rangle \geq 0 \qquad (114)$$

holds for all $(u_i, u_i^*) \in A$, $i = 1, \ldots, n$, and all $n = 2, 3, \ldots$, where we set $u_{n+1} = u_1$. In particular, the single-valued mapping $A: D(A) \subseteq X \to X^*$ is cyclic monotone iff

$$\langle Au_1, u_1 - u_2 \rangle + \langle Au_2, u_2 - u_3 \rangle + \cdots + \langle Au_n, u_n - u_{n+1} \rangle \geq 0 \quad (114^*)$$

holds for all $u_i \in D(A)$, $i = 1, \ldots, n$, and all $n = 2, 3, \ldots$, where we set $u_{n+1} = u_1$.

Definition 32.40. Let $A: X \to 2^{X^*}$ be a mapping on the real B-space X.

(a) A is called *n-monotone* iff condition (114) holds for fixed $n \geq 2$. In particular, A is *3-monotone* iff

$$\langle v^* - u^*, w - v \rangle \leq \langle u^* - w^*, u - w \rangle$$

holds for all (u, u^*), (v, v^*), $(w, w^*) \in A$.

(b) A is called *3-σ-monotone* iff A is monotone and there is a number $\sigma > 0$ such that

$$\langle v^* - u^*, w - v \rangle \leq \sigma \langle u^* - w^*, u - w \rangle$$

holds for all (u, u^*), (v, v^*), $(w, w^*) \in A$.

(c) A is called *3*-monotone* iff A is monotone and

$$\sup_{(v, v^*) \in A} \langle v^* - u^*, w - v \rangle < \infty$$

holds for all $w \in D(A)$, $u^* \in R(A)$.

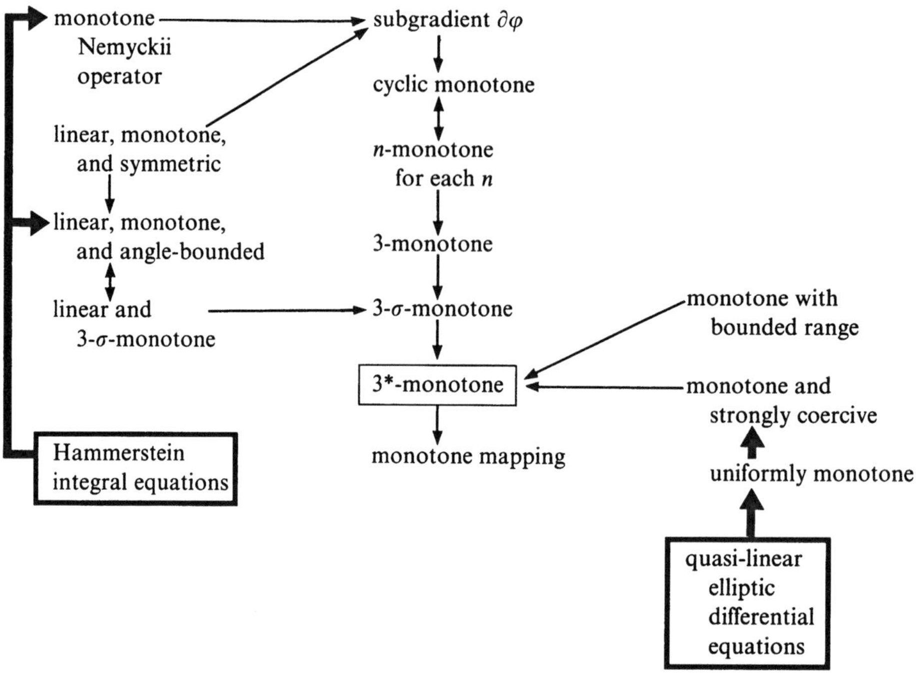

Figure 32.9

Figure 32.9 shows important interrelations which will be proved below. The notion of *3*-monotonicity* will play a decisive role in the next section.

Proposition 32.41 (3*-monotonicity). *Let $A: X \to 2^{X^*}$ be a mapping on the real reflexive* B*-space X.*

(a) *If A is n-monotone, then A is monotone.*
(b) *If A is 3-monotone, then A is 3-σ-monotone with $\sigma = 1$.*
(c) *If A is 3-σ-monotone, then A is 3*-monotone.*
(d) *If A is monotone and strongly coercive (e.g., $A: X \to X^*$ is uniformly monotone), then A is 3*-monotone.*
(e) *If A is monotone and the range $R(A)$ is bounded, then A is 3*-monotone. If, in addition, A is maximal monotone, then $D(A) = X$.*
(f) *If A is 3*-monotone, then so is $A^{-1}: X^* \to 2^X$.*

By definition, the mapping $A: X \to 2^{X^*}$ on the real B-space X is *strongly coercive* iff either $D(A)$ is bounded or $D(A)$ is unbounded and

$$\frac{\langle v^*, v - w\rangle}{\|v\|} \to +\infty \qquad \text{as} \quad \|v\| \to \infty, \quad (v, v^*) \in A, \tag{115}$$

for each $w \in D(A)$. More precisely, this means that, for each given $R > 0$ and

$w \in D(A)$, there exists an $r(w) > 0$ such that

$$\frac{\langle v^*, v - w\rangle}{\|v\|} \geq R \qquad \text{for all} \quad (v, v^*) \in A \quad \text{with} \quad \|v\| \geq r(w).$$

For example, suppose that the operator $A: X \to X^*$ is uniformly monotone, i.e.,

$$\langle Av - Aw, v - w\rangle \geq a(\|v - w\|)\,\|v - w\| \qquad \text{for all} \quad v, w \in X,$$

where the continuous function $a: \mathbb{R}_+ \to \mathbb{R}_+$ is strictly monotone with $a(0) = 0$ and $a(t) \to +\infty$ as $t \to +\infty$. Then A is strongly coercive. This follows from

$$\langle Av, v - w\rangle \geq a(\|v - w\|)\,\|v - w\| - \|Aw\|\,\|v - w\|.$$

PROOF. Ad(a), (b), (c), (f). This is an immediate consequence of the corresponding definitions.

Ad(d). Let $w \in D(A)$ and $u^* \in R(A)$. We choose a fixed $w^* \in Aw$. It follows from (115) that there is an $r > 0$ such that

$$\Delta \stackrel{\text{def}}{=} \langle v^* - u^*, w - v\rangle \leq 0,$$

for all $(v, v^*) \in A$ with $\|v\| \geq r$. If $\|v\| < r$, then the monotonicity of A implies

$$\langle v^* - u^*, w - v\rangle \leq \langle w^* - u^*, w - v\rangle \qquad \text{for all} \qquad (v, v^*) \in A,$$

and hence

$$\Delta \leq (\|w^*\| + \|u^*\|)(\|w\| + r),$$

for all $(v, v^*) \in A$ with $\|v\| < r$. Thus $\sup_{(v,v^*)\in A} \Delta < \infty$.

Ad(e). Let $w \in D(A)$ and $u^* \in R(A)$. We choose w^* and u such that (w, w^*), $(u, u^*) \in A$. From the monotonicity of A it follows that

$$\begin{aligned}\langle v^* - u^*, w - v\rangle &\leq \langle v^* - u^*, w - v\rangle + \langle v^* - u^*, v - u\rangle \\ &= \langle v^* - u^*, w - u\rangle \qquad \text{for all} \quad (v, v^*) \in A.\end{aligned}$$

Since $R(A)$ is bounded, the mapping A is 3*-monotone.

If A is maximal monotone, then so is A^{-1}. Since $R(A)$ is bounded, the mapping A is bounded. By Theorem 32.G, the mapping A^{-1} is surjective, i.e., $D(A) = X$. □

Proposition 32.42. *Let $\varphi: X \to]-\infty, \infty]$ be a functional on the real* B-*space X. Then the subdifferential*

$$\partial\varphi: X \to 2^{X^*}$$

is cyclic monotone, i.e., $\partial\varphi$ is n-monotone for each $n \geq 2$.

In particular, $\partial\varphi$ is 3-monotone.*

PROOF. If $(u_i, u_i^*) \in \partial\varphi$ for $i = 1, \ldots, n$, then

$$\langle u_i^*, u_{i+1} - u_i\rangle \leq \varphi(u_{i+1}) - \varphi(u_i), \qquad i = 1, \ldots, n,$$

where $u_{n+1} = u_1$. Addition shows that $\partial\varphi$ is n-monotone. □

Proposition 32.43. *Let $K: X \to X^*$ be a linear monotone operator on the real* B*-space X. Then the following two conditions are equivalent:*

(i) *K is angle-bounded, i.e., there is a number $a \geq 0$ such that*

$$|\langle Ku, v\rangle - \langle Kv, u\rangle|^2 \leq 4a^2\langle Ku, u\rangle\langle Kv, v\rangle \qquad \textit{for all} \quad u, v \in X. \tag{116}$$

(ii) *K is 3-σ-monotone, i.e., there is a number $\sigma > 0$ such that*

$$\langle Kv - Ku, w - v\rangle \leq \sigma\langle Ku - Kw, u - w\rangle \quad \textit{for all} \quad u, v, w \in X. \tag{117}$$

PROOF. We first show that (117) is equivalent to the inequality

$$\langle Ku, v\rangle^2 \leq 4\sigma\langle Ku, u\rangle\langle Kv, v\rangle \qquad \text{for all} \quad u, v \in X. \tag{118}$$

In fact, letting $x = u - w$ and $y = v - w$ and noting the linearity of K, we obtain that (117) is equivalent to

$$\langle Kx, y\rangle - \langle Ky, y\rangle \leq \sigma\langle Kx, x\rangle \qquad \text{for all} \quad x, y \in X. \tag{118*}$$

Replacing x by tx, we get

$$\sigma\langle Kx, x\rangle t^2 - \langle Kx, y\rangle t + \langle Ky, y\rangle \geq 0 \qquad \text{for all} \quad t \in \mathbb{R}.$$

Consequently, (118*) is equivalent to (118).

We now set

$$[u, v]_\pm = \tfrac{1}{2}(\langle Ku, v\rangle \pm \langle Kv, u\rangle) \qquad \text{for all} \quad u, v \in X.$$

Since K is monotone, we have $[u, u]_+ \geq 0$ for all $u \in X$. By the generalized Schwarz inequality (Problem 21.16),

$$[u, v]_+^2 \leq [u, u]_+[v, v]_+ \qquad \text{for all} \quad u, v \in X.$$

Obviously,

$$\langle Ku, v\rangle = [u, v]_+ + [u, v]_- \qquad \text{for all} \quad u, v \in X. \tag{119}$$

(I) If K is angle-bounded, then

$$[u, v]_-^2 \leq a^2[u, u]_+[v, v]_+ \qquad \text{for all} \quad u, v \in X.$$

By (119) and $(\alpha + \beta)^2 \leq 2\alpha^2 + 2\beta^2$ for real α, β, we obtain

$$\langle Ku, v\rangle^2 \leq (2 + 2a^2)[u, u]_+[v, v]_+ \qquad \text{for all} \quad u, v \in X.$$

This is (118), i.e., K is 3-σ-monotone.

(II) Conversely, if K is 3-σ-monotone, then (118) holds. From (119) we obtain

$$[u, v]_- = \langle Ku, v\rangle - [u, v]_+$$

and hence

$$[u, v]_-^2 \leq (8\sigma + 2)[u, u]_+^2[v, v]_+^2 \qquad \text{for all} \quad u, v \in X.$$

This is (116), i.e., K is angle-bounded. $\square$

Finally, we consider the *Nemyckii operator* $F: X \to X^*$, where

$$(Fu)(x) = f(x, u(x)) \qquad \text{for all} \quad x \in G$$

and $X = L_p(G)$. Our assumptions are the following:

(H1) Let G be a nonempty bounded measurable set in $\mathbb{R}^N$, $N \geq 1$, and let $1 < p, q < \infty$ with $p^{-1} + q^{-1} = 1$.

(H2) *Monotonicity*. The function $f: G \times \mathbb{R} \to \mathbb{R}$ satisfies the Carathéodory condition (e.g., f is continuous), and $u \mapsto f(x, u)$ is monotone increasing on $\mathbb{R}$ for almost all $x \in G$.

(H3) *Growth condition*. There is a function $a \in L_q(G)$ and a number $b > 0$ such that

$$|f(x, u)| \leq |a(x)| + b|u|^{p-1} \qquad \text{for all} \quad x \in G, \quad u \in \mathbb{R}.$$

Proposition 32.44. *Assume* (H1) *through* (H3). *Then the Nemyckii operator* $F: X \to X^*$ *is cyclic monotone, i.e., F is n-monotone for each $n \geq 2$.*

In particular, F is 3-monotone.*

In Proposition 41.10 we will prove the stronger result that $F: X \to X^*$ is a continuous potential operator.

PROOF. We set

$$g(x, v) = \int_0^v f(x, u)\, du$$

and

$$h(u) = \int_G g(x, u(x))\, dx.$$

By (H2), the function $v \mapsto g(x, v)$ is convex and C^1 on $\mathbb{R}$ for almost all $x \in G$. This implies

$$g(x, v) - g(x, w) \geq (v - w)g_v(x, w) = (v - w)f(x, w),$$

for all $v, w \in \mathbb{R}$ and almost all $x \in G$. By Proposition 26.7, the operator $F: X \to X^*$ is continuous. Recall that $X = L_p(G)$ and $X^* = L_q(G)$. Since

$$|g(x, v)| \leq |a(x)||v| + bp^{-1}|v|^p \qquad \text{for all} \quad x \in G, \quad v \in \mathbb{R},$$

the integral $\int_G g(x, v(x))\, dx$ exists for all $v \in X$, by the Hölder inequality. Therefore, integration yields

$$h(v) - h(w) \geq \langle Fw, v - w \rangle \qquad \text{for all} \quad v, w \in X.$$

Letting

$$w = u_i, \qquad v = u_{i+1}, \qquad i = 1, \ldots, n, \quad u_{n+1} = u_1,$$

and adding the corresponding inequalities, we obtain that F is n-monotone for each n. □

32.22. The Range of Sum Operators

Our goal is to prove that

$$R(A + B) \simeq R(A) + R(B), \tag{120}$$

i.e., the range $R(A + B)$ of the sum operator $A + B$ is "almost equal" to $R(A) + R(B)$. More precisely, we will use the symbol "$\simeq$" in the following sense.

Definition 32.45. Let S and T be subsets of a topological space X (e.g., X is a B-space). We write

$$S \simeq T$$

iff

$$\operatorname{int} S = \operatorname{int} T \qquad \text{and} \qquad \bar{S} = \bar{T}.$$

In this case, we say that the set S is *almost equal* to the set T.

Recall that $\operatorname{int} S$ and $\bar{S}$ denotes the interior and the closure of the set S, respectively.

For example, if T is a B-space, then $S \simeq T$ means $S = T$. We now want to study the equation

$$b \in Au + Bu, \qquad u \in X. \tag{121}$$

Theorem 32.N (Brézis and Haraux (1976)). *Assume*:

(i) *The operators A, $B: X \to 2^X$ are maximal monotone and 3*-monotone on the real* H-*space X.*
(ii) *The sum operator $A + B: X \to 2^X$ is maximal monotone (e.g., $D(A) \cap \operatorname{int} D(B) \neq \emptyset$).*

Then

$$R(A + B) \simeq R(A) + R(B).$$

In particular, if $R(A) = X$ or $R(B) = X$, then $R(A + B) = X$, i.e., equation (121) *has a solution for each given $b \in X$.*

Corollary 32.46. *The assertion remains true if we replace the* 3*-*monotonicity of the operator A by the condition $D(A) \subseteq D(B)$.*

The proof will be based on the following crucial result.

Lemma 32.47. *Let $C: X \to 2^X$ be a maximal monotone operator on the real* H-*space X. Let S be a subset of X such that, for each $u^* \in S$, there exists a $w \in X$ with*

$$\sup_{(v, v^*) \in C} (v^* - u^* | w - v) < \infty. \tag{122}$$

Then, $\operatorname{co} S \subseteq \overline{R(C)}$ *and* $\operatorname{int}(\operatorname{co} S) \subseteq R(C)$.

PROOF OF LEMMA 32.47. We will use critically the uniform boundedness principle of Banach and Steinhaus.

(I) We show that $S \subseteq \overline{R(C)}$. Let $u^* \in S$. Since C is maximal monotone, the equation

$$Cv_\varepsilon + \varepsilon v_\varepsilon = u^*, \qquad v_\varepsilon \in D(C) \tag{123}$$

has a solution for each $\varepsilon > 0$. In this connection, note that C is maximal accretive by Proposition 31.5. By assumption (122), there are a $w \in X$ and a real number c such that

$$(v^* - u^* | w - v) \leq c \qquad \text{for all} \quad (v, v^*) \in C.$$

We set $v = v_\varepsilon$ and $v^* = u^* - \varepsilon v_\varepsilon$. Hence

$$(\varepsilon v_\varepsilon | v_\varepsilon - w) \leq c \qquad \text{for all} \quad \varepsilon > 0.$$

This implies

$$\varepsilon \|v_\varepsilon\|^2 \leq \varepsilon \|v_\varepsilon\| \, \|w\| + c \leq \tfrac{1}{2}\varepsilon^2 \|v_\varepsilon\|^2 + \tfrac{1}{2}\|w\|^2 + c.$$

Thus the sequence $(\sqrt{\varepsilon}\|v_\varepsilon\|)$ is bounded as $\varepsilon \to 0$. By equation (123),

$$Cv_\varepsilon \to u^* \qquad \text{as} \quad \varepsilon \to 0, \quad \text{i.e.,}\ u^* \in \overline{R(C)}.$$

(II) We show that $\operatorname{int} S \subseteq R(C)$. Let $u^* \in \operatorname{int} S$. We choose $r > 0$ such that $u^* + w^* \in S$ for all w^* with $\|w^*\| < r$. By assumption (122), for each $w^* \in X$ with $\|w^*\| < r$, there are a $w \in X$ and a real number c such that

$$(v^* - u^* - w^* | w - v) \leq c \qquad \text{for all} \quad (v, v^*) \in C.$$

We set $v = v_\varepsilon$ and $v^* = u^* - \varepsilon v_\varepsilon$. Hence

$$(\varepsilon v_\varepsilon + w^* | v_\varepsilon - w) \leq c \qquad \text{for all} \quad \varepsilon > 0.$$

By (I), the sequence $(\varepsilon \|v_\varepsilon\|^2)$ is bounded as $\varepsilon \to 0$. Therefore, the sequence $\{(w^* | v_\varepsilon)\}$ is bounded as $\varepsilon \to 0$ for each $w^* \in X$ with $\|w^*\| < r$. By the *uniform boundedness theorem* $A_1(35)$, the sequence $(\|v_\varepsilon\|)$ is bounded as $\varepsilon \to 0$. Consequently, there exists a subsequence with

$$v_\varepsilon \rightharpoonup v \qquad \text{as} \quad \varepsilon \to 0.$$

From equation (123) it follows that $u^* - \varepsilon v_\varepsilon \in Cv_\varepsilon$ for all $\varepsilon > 0$ and

$$u^* - \varepsilon v_\varepsilon \to u^* \qquad \text{as} \quad \varepsilon \to 0.$$

Since C is maximal monotone, we obtain that $u^* \in Cv$, by Proposition 31.6. Thus $u^* \in R(C)$.

(III) We show that the assumption (122) is also satisfied for the convex hull $\operatorname{co} S$. In fact, if $u^* \in \operatorname{co} S$, then

$$u^* = \sum_i t_i u_i^*, \qquad \sum_i t_i = 1,$$

where $u_i^* \in S$ and $t_i \geq 0$ for all i. By assumption (122), there are elements

w_i and numbers c_i such that

$$(v^* - u_i^*|w_i - v) \le c_i \qquad \text{for all} \quad (v, v^*) \in C \quad \text{and all } i.$$

Letting $w = \sum_i t_i w_i$ and noting that $\sum_i t_i = 1$, we obtain

$$(v^* - u^*|w - v) \le \sum_i t_i c_i + \sum_i t_i(u_i^*|w_i) - (u^*|w),$$

for all $(v, v^*) \in C$. This is (122) for $\operatorname{co} S$.

(IV) According to (III), we can replace the set S by $\operatorname{co} S$ in (I) and (II). This yields the assertion. □

PROOF OF THEOREM 32.N. Obviously,

$$R(A + B) \subseteq R(A) + R(B).$$

Hence, it is sufficient to show that

$$R(A) + R(B) \subseteq \overline{R(A + B)} \qquad \text{and} \qquad \operatorname{int}(R(A) + R(B)) \subseteq R(A + B).$$

To prove this, we want to apply Lemma 32.47 to the operator

$$C = A + B$$

and to the set $S = R(A) + R(B)$. Thus it remains to check the assumption (122). Let $u^* \in S$, i.e., $u^* = u_A^* + u_B^*$ with $u_A^* \in R(A)$ and $u_B^* \in R(B)$. Choose a fixed $w \in D(A) \cap D(B)$. Note that $D(A) \cap D(B) \neq \emptyset$, since $A + B$ is maximal monotone and hence $A + B \neq \emptyset$. Since A and B are 3*-monotone, we have

$$\sup_{(v,v^*) \in A} (v^* - u_A^*|w - v) < \infty, \tag{124a}$$

$$\sup_{(v,v^*) \in B} (v^* - u_B^*|w - v) < \infty. \tag{124b}$$

This implies

$$\sup_{(v,v^*) \in C} (v^* - u^*|w - v) < \infty, \tag{125}$$

i.e., condition (122) is satisfied. □

PROOF OF COROLLARY 32.46. In contrast to the preceding proof, we now choose w such that $w \in D(A)$ and $u_A^* \in Aw$. Since $D(A) \subseteq D(B)$, we again have $w \in D(A) \cap D(B)$. From the monotonicity of A it follows that

$$(v^* - u_A^*|w - v) \le 0 \qquad \text{for all} \quad (v, v^*) \in A.$$

This is (124a). From (124a) and (124b) we again obtain (125). □

32.23. Application to Hammerstein Equations

We consider the abstract Hammerstein equation

$$b \in KFu + u, \qquad u \in X. \tag{126}$$

Theorem 32.O. *Suppose that the mappings $K, F: X \to 2^X$ are maximal monotone on the real* H-*space X with $D(K) = D(F) = X$, and at least one of these two mappings is* 3*-*monotone.*

Then, for each $b \in X$, equation (126) *has a solution.*

For example, the assumptions of Theorem 32.O are satisfied if the following conditions hold:

(i) The operator $K: X \to X$ is linear and monotone.
(ii) The operator $F: X \to X$ is continuous and monotone.
(iii) The operator K is angle-bounded or F is a potential operator or F is strongly coercive or the range of F is bounded.

In contrast to Theorem 32.B, the theorem above also applies to operators F which are not coercive.

PROOF. We apply Corollary 32.46.

(I) Suppose that K is 3*-monotone. We consider the equation

$$b \in Av + Bv, \qquad v \in X, \tag{126*}$$

with $A = F^{-1}$ and $B = K$. We have $R(A) = D(F) = X$ and $D(B) = X$. By Corollary 32.46, $R(A + B) = X$, i.e., (126*) has a solution v for each $b \in X$. Choosing $u \in Av$, we obtain a solution u of the original equation (126).

(II) Suppose that F is 3*-monotone. Then equation (126) is equivalent to

$$0 \in Au + Bu, \qquad u \in X,$$

where $Au = -K^{-1}(b - u)$ and $Bu = Fu$. Since K is maximal monotone, so is A. Moreover, we have $R(A) = D(K) = X$ and $D(B) = X$. By Corollary 32.46, $R(A + B) = X$. □

An application of this theorem to Hammerstein integral equations in $L_2(G)$ has already been formulated in Proposition 28.4. In this connection, observe Proposition 32.44 on the Nemyckii operator.

32.24. The Characterization of Nonexpansive Semigroups in H-spaces

We consider the following multivalued initial value problem

$$\begin{aligned} u' + Au &\ni 0, \\ u(0) &= u_0. \end{aligned} \tag{127}$$

We assume:

(H) The operator $A: X \to 2^X$ is maximal monotone on the real H-space X.

Definition 32.48. Let $u_0 \in D(A)$. By a *generalized solution* of (127), we understand a Lipschitz continuous function

$$u\colon [0, \infty[\to D(A)$$

such that $u(0) = u_0$ and

$$u'(t) + Au(t) \ni 0 \qquad \text{for almost all} \quad t \in]0, \infty[,$$

where the derivative $u'(t)$ is to be understood in the classical sense (cf. Definition 3.4).

If $u = u(t)$ is a generalized solution of (127), then we set

$$u(t) = S(t)u_0 \qquad \text{for all} \quad t \geq 0.$$

Under the assumption (H), the set Au is nonempty, closed, and convex for each $u \in D(A)$, by Proposition 32.6. Thus, the minimum problem

$$\min_{v \in Au} \|v\| = \|v_0\|$$

has a unique solution v_0, by Theorem 25.E. We set

$$A_0 u = v_0.$$

This way we obtain the single-valued operator $A_0\colon D(A) \subseteq X \to X$ (Fig. 32.10).

The following theorem shows that, on real H-spaces, there exists a one-to-one correspondence between maximal monotone operators and nonexpansive semigroups.

Theorem 32.P (Crandall and Pazy (1969)). *Let X be a real* H-*space.*

(a) *If the mapping $A\colon X \to 2^X$ is maximal monotone, then the operator $-A_0$ is the generator of a uniquely determined nonexpansive semigroup $\{S(t)\}$ on $\overline{D(A)}$.*

More precisely, for each $u_0 \in D(A)$, the initial value problem (127) *has a unique generalized solution $u = u(t)$, and there exists a unique nonexpansive*

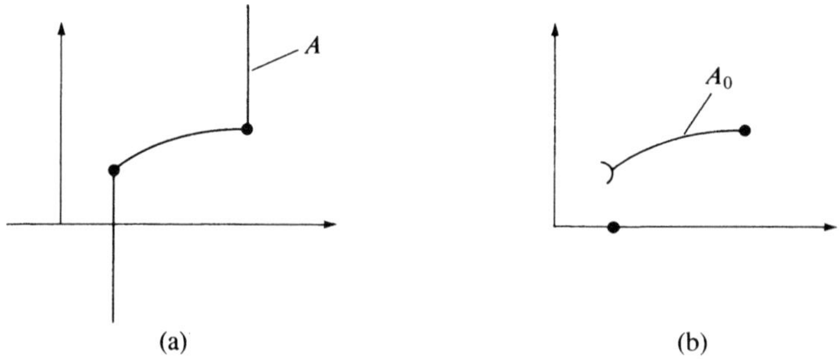

Figure 32.10

semigroup $\{S(t)\}$ *on* $\overline{D(A)}$ *such that*

$$u(t) = S(t)u_0 \qquad \text{for all} \quad t \geq 0.$$

(b) *Conversely, if* $\{S(t)\}$ *is a nonexpansive semigroup on the nonempty closed convex subset* C *of* X, *then there exists a uniquely determined maximal monotone mapping* $A\colon X \to 2^X$ *such that* $C = \overline{D(A)}$, *and* $-A_0$ *is the generator of* $\{S(t)\}$.

More precisely, the semigroup $\{S(t)\}$ *can be constructed as in* (a).

This theorem generalizes the main theorem on linear nonexpansive semigroups (Theorem 19.E). Moreover, this theorem generalizes Theorem 31.A to multivalued mappings. The proof of Theorem 32.P will be discussed in Problem 32.8.

PROBLEMS

32.1. *Characterization of the weak closure by means of sequences.* Let M be a bounded subset of the reflexive B-space X, and let u be a point in the weak closure of M. By A_1(17c), there exists an M–S-sequence (u_α) in M such that

$$u_\alpha \rightharpoonup u \quad \text{in } X.$$

Prove that there exists a *sequence* $(u_n)_{n\in\mathbb{N}}$ in M such that

$$u_n \rightharpoonup u \quad \text{in } X \qquad \text{as} \quad n \to \infty. \tag{128}$$

This result shows that:

$$\text{weak closure of } M = \text{sequentially weak closure of } M.$$

Hint: Use the following general results.

(i) The weak topology on a weakly compact subset of a separable B-space is a metric topology.

(ii) The weak topology on a bounded closed convex subset of a reflexive separable B-space is a metric topology.

The proof of (i) can be found in Dunford and Schwartz (1958, M), Vol. 1, p. 434. According to the Eberlein–Šmuljan theorem, statement (ii) is a special case of (i).

Solution:

(I) Since X is reflexive, so is X^*. Moreover, each finite product $X^* \times \cdots \times X^*$ is also a reflexive B-space. Finally, each closed linear subspace X_0 of X is reflexive.

(II) Let u be fixed. We prove first that there exists an at most countable subset M_0 of M such that u lies in the weak closure of M_0 with respect to the weak topology on X. Let B^n be the product of n copies of the closed unit ball B in X^*. Choose fixed natural numbers m and n. For $(f_1,\ldots,f_n) \in B^n$, we set

$$U(u) = \left\{v \in X\colon |\langle f_j, v - u\rangle| < \frac{1}{m}, j = 1,\ldots,n\right\}.$$

Since u lies in the weak closure of M, there exists a $v \in M$ such that

$v \in U(u)$, i.e.,

$$|\langle f_j, v - u\rangle| < \frac{1}{m}, \qquad j = 1, \ldots, n. \tag{129}$$

For fixed $v \in M$, the set

$$V = \{(f_1, \ldots, f_n) \in B^n\colon \text{inequality (129) holds}\}$$

represents a weakly open subset of B^n. By the Eberlein–Šmuljan theorem, B is weakly compact in the reflexive B-space X^*. Moreover, B^n is weakly compact in the reflexive B-space $(X^*)^n$. All the weakly open sets V cover B^n. Since B^n is weakly compact, B^n can be covered by finitely many sets of type V. This means that there exists a *finite* subset S_{nm} of M such that, for each $(f_1, \ldots, f_n) \in B^n$, there exists a $v \in S_{nm}$ satisfying (129).

We now set

$$M_0 = \bigcup_{n,m \in \mathbb{N}} S_{nm}.$$

Then, for each n, $m \in \mathbb{N}$ and each $(f_1, \ldots, f_n) \in B^n$, there exists a $v \in M_0$ such that (129) holds, i.e., the point u lies in the weak closure of M_0 with respect to the weak topology on X.

(III) Let X_0 denote the smallest closed linear subspace of X which contains M_0 and u. Then X_0 is a separable and reflexive B-space by (I). According to (II), the point u lies in the weak closure of M_0 with respect to the weak topology on X. By the Hahn–Banach theorem, each functional $f \in X_0^*$ can be extended to a functional $f \in X^*$. Therefore, u also lies in the weak closure of M_0 with respect to the weak topology on X_0.

Now, choose a closed ball K in X_0 such that $M_0 \subseteq K$. According to (ii) above, the weak X_0-topology on K corresponds to a metric. Thus, there exists a sequence (u_n) in M_0 such that

$$u_n \rightharpoonup u \quad \text{in } X_0 \qquad \text{as} \quad n \to \infty.$$

This implies the assertion (128), since $X^* \subseteq X_0^*$.

32.2. *Continuity properties of multivalued maps.* We consider the map

$$A\colon C \to 2^Y, \tag{130}$$

where C and Y are topological spaces. Then A is called *upper semicontinuous* at the point u iff, for each open neighborhood V of the set Au, there exists a neighbhorhood U of the point u such that

$$A(U) \subseteq V.$$

Properties of such maps have been studied in Section 9.2.

Let X and Y be B-spaces over $\mathbb{K} = \mathbb{R}, \mathbb{C}$, and let $C \subseteq X$. Then the map A in (130) is called *demicontinuous* at u iff A is upper semicontinuous at u from the strong topology on C into the weak topology on Y.

The map A in (130) is called *strongly continuous* at u iff A is upper semicontinuous at u from the weak topology on C into the strong topology on Y.

Let the set C be convex. The map A in (130) is called *hemicontinuous* at u iff, for each $v \in C$, the map

$$t \mapsto A(tu + (1 - t)v)$$

is upper semicontinuous from the interval $[0, 1]$ into the space Y equipped with the weak topology.

The map A in (130) is called upper semicontinuous (resp. demicontinuous, strongly continuous, hemicontinuous) iff A has the corresponding property at each point u in $D(A)$.

The map A is called *bounded* iff the image $A(M)$ is bounded for each bounded subset M of $D(A)$.

32.2a. *Hemicontinuity and demicontinuity*. Obviously, each demicontinuous map is also hemicontinuous.

Conversely, let

$$A: X \to 2^{X^*} \tag{131}$$

be hemicontinuous and monotone on the real reflexive B-space X. Let U be an open subset of $D(A)$ and suppose that, for each $u \in U$, the set Au is convex and weakly closed in X^*. Show that A is demicontinuous on U.

Hint: Use a similar argument as in the proof of Proposition 26.4. Cf. Kluge (1979, M), p. 32.

32.2b. *General monotonicity trick*. Consider A and U as in Problem 32.2a. Show that it follows from $(u, u^*) \in U \times X^*$ and

$$\langle u^* - v^*, u - v\rangle_X \geq 0 \qquad \text{for all} \quad (v, v^*) \in A$$

that $(u, u^*) \in A$.

This result generalizes the fundamental monotonicity trick (25.4).

Hint: Assume $(u, u^*) \notin A$ and use the separation theorem for convex sets in Section 39.1. Cf. Kluge (1979, M), p. 32.

32.3. *Pseudomonotone multivalued maps*. Let C be a closed convex subset of the real reflexive B-space X. The map

$$A: C \to 2^{X^*} \tag{132}$$

is called *pseudomonotone* iff the following holds. Let $(u_n, u_n^*) \in A$ for all $n \in \mathbb{N}$, and suppose that $u_n \rightharpoonup u$ as $n \to \infty$ such that

$$\overline{\lim_{n\to\infty}}\, \langle u_n^*, u_n - u\rangle \leq 0.$$

Then, for each $v \in C$, there is a u_v^* such that $(v, u_v^*) \in A$ and

$$\langle u_v^*, u - v\rangle \leq \underline{\lim_{n\to\infty}}\, \langle u_n^*, u_n - v\rangle.$$

32.3a. Show that each strongly continuous map $A: C \to X^*$ is pseudomonotone.

Solution: If $u_n \rightharpoonup u$ as $n \to \infty$, then $Au_n \to Au$.

32.3b. Show that each continuous operator $A: C \to X^*$ is pseudomonotone in the case $\dim X < \infty$.

Solution: This is a special case of Problem 32.3a, since weak convergence and strong convergence coincide on X.

32.3c. Let $A: C \to 2^{X^*}$ be monotone and hemicontinuous on the nonempty closed convex subset C of the real reflexive B-space X. In addition, suppose that, for each $u \in C$, the set Au is nonempty, convex, and closed in X^*. Show that A is pseudomonotone.

Hint: Use a similar argument as in the proof of Proposition 27.6. Cf. Kluge (1979, M), p. 193.

32.3d. *Additivity*. Let the maps $A, B: X \to 2^{X^*}$ be pseudomonotone on the real reflexive B-space X. Show that the sum

$$A + B: X \to 2^{X^*}$$

is pseudomonotone.

Hint: Use a similar argument as in the proof of Proposition 27.6(e).

32.4.* *Generalization of the main theorem on pseudomonotone perturbations of maximal monotone mappings* (*Theorem* 32.A). We consider the multivalued operator equation

$$b \in Au + Bu, \qquad u \in C, \tag{133}$$

where, in contrast to Theorem 32.A, both the maps A and B are multivalued. We assume:

(i) C is a nonempty closed convex subset of the real reflexive B-space X.
(ii) The map $A: C \to 2^{X^*}$ is maximal monotone, and $B: C \to 2^{X^*}$ is pseudomonotone and bounded.
(iii) B is A-coercive with respect to the fixed $b \in X^*$, i.e., there is a $u_0 \in C \cap D(A)$ and a number $r > 0$ such that

$$\langle u^*, u - u_0 \rangle > \langle b, u - u_0 \rangle,$$

for all $(u, u^*) \in B$ with $\|u\| > r$.
(iv) For each $u \in C$, the set Bu is a nonempty closed convex subset of X^*.
(v) B is demicontinuous on simplices, i.e., for each finite subset F of C, the map B: co $F \to 2^{X^*}$ is demicontinuous.

Show that problem (133) has a solution.

Hint: Passing to an equivalent norm on X, if necessary, we may assume that both X and X^* are strictly convex.

(I) Regularization. Instead of (133), we start with the regularized equation

$$b \in (Au + \varepsilon Ju + Bu), \tag{134}$$

where $\varepsilon > 0$, and J denotes the duality map of X. Since A is maximal monotone, we have $R(A + \varepsilon J) = X^*$.
(II) Galerkin method. Now apply the proof idea of Theorem 32.A to (134).
(III) Study the limiting process $\varepsilon \to 0$.

A detailed proof can be found in Browder (1968/76, M), p. 92 and in Kluge (1979, M), p. 195.

32.5. *A special class of maximal monotone mappings*. Let

$$A: X \to 2^{X^*}$$

be monotone and hemicontinuous on the real reflexive B-space X, and suppose that, for each $u \in X$, the set Au is nonempty, closed, and convex in X^*. Show that A is maximal monotone.

Hint: Use the monotonicity trick from Problem 32.2b.

32.6. *Properties of maximal monotone mappings*. Let the map

$$A: X \to 2^{X^*}$$

be maximal monotone on the real reflexive B-space X. We summarize a number of important properties of A.

(a) Let $(u_n, u_n^*) \in A$ for all $n \in \mathbb{N}$. If

$$u_n \to u \quad \text{in } X \qquad \text{and} \qquad u_n^* \rightharpoonup u^* \quad \text{in } X^* \qquad \text{as} \quad n \to \infty,$$

then $(u, u^*) \in A$.

(b) Let $(u_n, u_n^*) \in A$ for all $n \in \mathbb{N}$. If

$$u_n \rightharpoonup u \quad \text{in } X \qquad \text{and} \qquad u_n^* \to u^* \quad \text{in } X^* \qquad \text{as} \quad n \to \infty,$$

then $(u, u^*) \in A$.

(c) For each $u \in X$, the set Au is convex and weakly closed.

(d) The map A is demicontinuous and and locally bounded on $\operatorname{int} D(A)$.

(e)* The sets $\operatorname{int} D(A)$, $\overline{D(A)}$, and $\overline{R(A)}$ are convex.

(f)* If $\operatorname{int} D(A) \neq \emptyset$, then $\overline{\operatorname{int} D(A)} = \overline{D(A)}$.

(g) If $D(A) = X$, then A is pseudomonotone.

(h) The inverse map $A^{-1}: X^* \to 2^X$ is maximal monotone.

(i) The map A is surjective iff A^{-1} is locally bounded.

(j)* *Theorem of Zarantonello* (1973) *and Kenderov* (1976). If $\operatorname{int} D(A) \neq \emptyset$, then there exists a *dense* subset S of $D(A)$ such that the map A is *single-valued* on S.

The set S is the intersection of a countable family of open sets. If $\dim X < \infty$, then the set $D(A) - S$ has Lebesgue measure zero.

This theorem tells us that multivalued maximal monotone mappings are single-valued up to a "small" singular set $D(A) - S$.

Hint: Ad(a), (b). The map A is monotone. From

$$\langle u_n^* - v^*, u_n - v \rangle \geq 0 \qquad \text{for all} \quad (v, v^*) \in A, \quad n \in \mathbb{N},$$

it follows that $\langle u^* - v^*, u - v \rangle \geq 0$ for all $(v, v^*) \in A$, and hence $(u, u^*) \in A$.

Ad(c), (h), (i). Cf. Propositions 32.5 and 32.6, and Theorem 32.G.

Ad(d). The local boundedness of A follows from Proposition 32.33. The demicontinuity of A follows similarly as in the proof of Proposition 26.4. Cf. Kluge (1979, M), p. 33.

Ad(e), (f). Cf. Rockafellar (1968), (1969).

Ad(g). Cf. Pascali and Sburlan (1978, M), p. 106. Use (d).

Ad(j). Cf. Nirenberg (1974, L), p. 183, and Deimling (1985, M), p. 292.

32.7. *Characterization of maximal monotone mappings by skew-symmetric saddle functions.*

32.7a. *Definitions.* A function $f: X \to [-\infty, \infty]$ on the real B-space X is *convex* iff the epigraph,

$$\operatorname{epi} f = \{(u, t) \in X \times \mathbb{R}: f(u) \leq t\},$$

is a convex set (see Section 47.1). Let $M(u)$ be the set of all continuous affine minorants of f at the point u, i.e., $g \in M(u)$ iff

$$f(v) \geq g(v) \qquad \text{for all} \quad v \in X,$$

where $g = u^* + \text{const}$ with $u^* \in X^*$. The closure of f is defined by

$$(\operatorname{cl} f)(u) = \sup_{g \in M(u)} g(u).$$

In particular, if $M(u) = \emptyset$, then $(\operatorname{cl} f)(u) = -\infty$.

The function $f\colon X \to [-\infty, \infty]$ is called *concave* iff $-f$ is convex.

By a *skew-symmetric saddle function*

$$L\colon X \times X \to [-\infty, \infty], \tag{135}$$

we understand a function which has the following properties:

(i) $u \mapsto L(u, v)$ is concave on X for each $v \in X$.

(ii) $v \mapsto L(u, v)$ is convex on X for each $u \in X$.

(iii) $\mathrm{cl}_2\, L(u, v) = \mathrm{cl}_1(-L(v, u))$ for all $u, v \in X$, where cl_i refers to the ith variable.

32.7b.* *A general theorem.* Let X be a real reflexive B-space. Let L in (135) be a skew-symmetric saddle function. We define the map

$$A_L\colon X \to 2^{X^*}$$

through $(u, u^*) \in A_L$ iff

$$u^*(h) \le L(u, u + h) \qquad \text{for all} \quad h \in X.$$

Show the following:

(i) The map A_L is monotone.

(ii) If $\mathrm{cl}_2\, L = L$, then A is maximal monotone.

(iii) Conversely, each maximal monotone mapping $A\colon X \to 2^{X^*}$ can be obtained this way, i.e., there exists a skew-symmetric saddle function $L\colon X \times X \to [-\infty, \infty]$ such that $\mathrm{cl}_2\, L = L$ and $A_L = A$.

Moreover, we have

$$0 \in Au$$

iff (u, u) is a saddle point of L.

This theorem allows a pseudovariational approach to maximal monotone mappings. Moreover, this theorem explains why maximal monotone mappings behave, in many respects, like subgradients of convex functions.

32.7c. *Simple example.* Let $f\colon X \to \mathbb{R}$ be convex. We set

$$L(u, v) = f(v) - f(u).$$

Then $L\colon X \times X \to \mathbb{R}$ is a skew-symmetric saddle function, and

$$A_L = \partial f.$$

In particular, if f is C^1, then the operator

$$A_L = f'$$

is maximal monotone on X, since A_L is monotone and continuous.

Hint: Cf. Krauss (1985b), (1986). These papers also contain interesting applications of this approach.

32.8.* *Proof of Theorem* 32.P.

Hint: The proof of Theorem 32.P(a) proceeds similarly to the proof of Theorem 31.A. In this connection, use the properties of the *Yosida approximation* for multivalued maximal monotone operators (cf. Chapter 55).

Let $\{S(t)\}$ be a nonexpansive semigroup on the nonempty closed convex subset C of the real H-space X. In order to prove Theorem 32.P(b), let

$$Bu = \lim_{t \to +0} \frac{u - S(h)u}{h},$$

where $u \in D(B)$ iff the right-hand limit exists. The operator B is monotone. Let A be a maximal monotone extension of B. Hence $D(A) \subseteq \overline{\text{co}}\, D(B) \subseteq C$. According to Theorem 32.P(a), the operator A generates a nonexpansive semigroup $\{S_A(t)\}$ by solving $u' + Au \ni 0$ and letting $u(t) = S_A(t)u(0)$. Show that $S(t) = S_A(t)$ for all $t \geq 0$.

A detailed proof can be found in Brézis (1973, L), p. 114.

32.9.* *The range of sum operators.* We assume:

(H1) The linear operator $A: D(A) \subseteq X \to X$ is graph closed on the real H-space X, where $D(A)$ is dense in X and $R(A)$ is closed.

(H2) Let $N(A) = N(A^*)$, where A^* denotes the adjoint operator. This is equivalent to $R(A) = N(A)^\perp$. Consequently, the operator $A: D(A) \cap R(A) \to R(A)$ is bijective.

(H3) The inverse operator $A^{-1}: R(A) \to D(A) \cap R(A)$ is compact.

(H4) Let α be the largest positive number such that

$$(Au|u) \geq -\frac{1}{\alpha}\|Au\|^2 \qquad \text{for all} \quad u \in D(A).$$

In particular, if $(Au|u) \geq 0$ for all $u \in D(A)$, then we set $\alpha = +\infty$.

In this connection, observe the following. We have the orthogonal decomposition

$$X = N(A) \oplus R(A).$$

Let P be the orthogonal projection operator from X onto $R(A)$. From (H3) it follows that there is a number $c > 0$ such that, for all $u \in D(A)$, we obtain

$$\|Au\| = \|APu\| \geq c\|Pu\|,$$

and hence

$$(Au|u) = (Au|Pu) \geq -\|Au\|\,\|Pu\| \geq -c^{-1}\|Au\|^2.$$

Consequently, condition (H4) makes sense.

(H5) The nonlinear operator $B: X \to X$ is monotone and hemicontinuous.

32.9a. Assume (H1) through (H5) and prove the following four statements:

(i) If $R(B)$ is bounded, then

$$R(A + B) \simeq R(A) + \text{co}\, R(B) \tag{136}$$

(see Definition 32.45).

(ii) Suppose that there is a number β with $0 < \beta < \alpha$ such that

$$(Bu - Bv|u) \geq \frac{1}{\beta}\|Bu\|^2 - \gamma(v) \qquad \text{for all} \quad u, v \in X,$$

where the real number γ depends only on v. Then relation (136) holds. If, in addition, $N(A) \subseteq R(B)$ (e.g., B is weakly coercive), then

$$R(A + B) = X.$$

(iii) Suppose that $A + \lambda I: X \to X$ is bijective for some fixed $\lambda > 0$, and suppose that

$$\lim_{\|u\| \to \infty} \frac{\|Bu - \lambda u\|}{\|u\|} = 0. \tag{137}$$

Then $R(A + B) = X$.

(iv) Suppose that $R(B) = X$ (e.g., B is weakly coercive), and suppose that there is a number β with $0 < \beta < \alpha$ such that

$$(Bu - Bv|u - v) \geq \frac{1}{\beta}\|Bu - Bv\|^2 \qquad \text{for all} \quad u, v \in X.$$

Then $R(A + B) = X$. For each $b \in X$, the equation

$$Au + Bu = b, \qquad u \in X, \tag{138}$$

has a unique solution mod $N(A)$, i.e., if u and $\bar{u}$ are solutions of (138), then $u - \bar{u} \in N(A)$. If $B: X \to X$ is bijective (e.g., B is weakly coercive and strictly monotone), then the solution of (138) is unique.

32.9b. Assume (H1) through (H5) and assume that $B = \varphi'$, where the function $\varphi: X \to \mathbb{R}$ is convex, continuous, and G-differentiable. Prove the following two statements:

(i) If

$$\overline{\lim_{\|u\| \to \infty}} \frac{\|Bu\|}{\|u\|} < \frac{\alpha}{2},$$

then $R(A + B) \simeq R(A) + \text{co}\, R(B)$.

(ii) If $R(B) = X$ (e.g., B is weakly coercive), and if there is a number β with $0 < \beta < \alpha$ such that

$$(Bu - Bv|u - v) \leq \beta\|u - v\|^2 \qquad \text{for all} \quad u, v \in X,$$

then $R(A + B) \simeq R(A) + \text{co}\, R(B)$.

In (ii), the compactness condition (H3) drops out.

Hint: Cf. Brézis and Nirenberg (1978). In this paper, one also finds applications to nonlinear elliptic, parabolic, and hyperbolic equations.

References to the Literature

Classical works: Minty (1962), (1963), Browder (1963), (1968), (1968a), Rockafellar (1966), (1968), (1969), (1970), Brézis and Haraux (1976).

Monograph on maximal monotone operators in H-spaces: Brézis (1973).

Monographs on maximal monotone operators in B-spaces: Browder (1968/76), Pascali and Sburlan (1978), Kluge (1979).

Range of sum operators: Brézis and Haraux (1976), Brézis and Nirenberg (1978).

Single-valuedness of maximal monotone mappings: Zarantonello (1973), Kenderov (1976), Fitzpatrick (1977), Deimling (1985, M).

Maximal monotone mappings and skew-symmetric saddle functions: Krauss (1985b), (1986).

CHAPTER 33

Second-Order Evolution Equations and the Galerkin Method

I attended the Technological Institute in Graz (Austria). One of the most worthy representatives of structural engineering there gave a lecture on the most reasonable installations of loo furnishings. One day, during a lecture, he drew a circle on the seat of a chair with chalk. Then he sat down, stood up again, and turned his colossal backside towards us. Thus he showed us the most convenient choice of the toilet seat. This experience also contributed to me devoting myself to the pure science of mathematics.

Wilhelm Blaschke (1957)

Two men travel in a balloon above the earth. "Where are we?", asks one of them. Thereupon the other thinks for a long time and then answers: "We are in a basket under a flying balloon."

This must have been a mathematician, because he has thought for such a long time, the answer is correct and useless.

Folclore

At the beginning of this century, Dedekind (1831–1916) opened a calender and read: "Richard Dedekind. Died in Braunschweig on September 4, 1899." Dedekind then wrote to the publisher of the calender: "Dear Sir. Please allow me to draw your attention to the circumstance that the date of my death is in error at least as far as the year is concerned."

Folclore

One day a man came to Bertrand Russell (1872–1970) and asked him: "You say that a wrong hypothesis implies everything. Please, prove to me from a wrong hypothesis that I am the Emperor of Japan." "Well," Russell said,

"0 = 1 implies that 1 = 2."

"The Emperor of Japan and you are two, two equals one, and we are done."

Folclore

In order to solve nonlinear second-order evolution equations of the form

$$u'' + B(t, u, u') = 0, \tag{E}$$

one can reduce them to first-order evolution equations. In this connection, one has the following two possibilities:

(i) If we set

$$u(t) = \int_0^t v(s)\,ds,$$

then we obtain from (E) the equation

$$v' + B\left(t, \int_0^t v(s)\,ds, v\right) = 0.$$

(ii) If we set

$$v = u',$$

then we obtain from (E) the system

$$v' + B(t, u, v) = 0,$$
$$u' - v = 0.$$

Both methods will be used in this chapter. By using (i), Theorem 33.A in Section 33.3 justifies the Galerkin method for a class of second-order evolution equations which contain monotone operators. The proof of Theorem 33.A will be based on Theorem 32.E which follows from the theory of maximal monotone operators. In Section 33.5 we consider applications of Theorem 33.A to a class of quasi-linear hyperbolic differential equations, which are nonlinear with respect to the first time derivative.

In Sections 33.6–33.11, we study the following topics:

(a) Symmetric hyperbolic systems and systems of conservation laws.
(b) Formation of shocks by intersection of characteristics.
(c) General blowing-up effects.
(d) Blow-up of solutions for semilinear wave equations.
(e) Global solutions for semilinear wave equations and the semilinear Klein–Gordon equation.
(f) Generalized (viscosity) solutions of the Hamilton–Jacobi equation.

These topics will be investigated in greater detail in Part V in connection with basic problems in mathematical physics. For example, in Chapter 83 we will explain the fundamental role which is played by quasi-linear symmetric hyperbolic systems in mathematical physics.

Roughly speaking, all processes in nature which correspond to the *propagation of waves* lead to *hyperbolic* partial differential equations. This explains the importance of such equations. The mathematical theory for hyperbolic equations encounters a number of typical *difficulties* which are mainly based on the possible formation of *shocks* (the appearance of discontinuites after a finite time despite smooth initial data) and on *blowing-up* effects. This will be discussed in Section 33.7.

33.1. The Original Problem

Let "$V \subseteq H \subseteq V^*$" be an evolution triple. We want to solve the following initial value problem:

$$u''(t) + A(t)u'(t) + Lu(t) = b(t) \qquad \text{for almost all} \quad t \in]0, T[,$$

$$u(0) = u'(0) = 0, \tag{1}$$

$$u \in C([0, T], V), \qquad u' \in W_p^1(0, T; V, H), \qquad 2 \leq p < \infty.$$

For each $t \in]0, T[$, we are given $b(t) \in H$ and the operators

$$L,\ A(t)\colon V \to V^*.$$

Moreover, let $0 < T < \infty$ and $q^{-1} + p^{-1} = 1$. The precise assumptions for problem (1) will be formulated in Section 33.3. We set

$$X = L_p(0, T; V).$$

Then $X^* = L_q(0, T: V^*)$. Since $1 < q \leq 2 \leq p$,

$$X \subseteq L_2(0, T; H) \subseteq X^*.$$

Recall that the condition $u' \in W_p^1(0, T; V, H)$ means that

$$u' \in X \qquad \text{and} \qquad u'' \in X^*.$$

Let $u' \in W_p^1(0, T; V, H)$. After changing the function $t \mapsto u'(t)$ on a subset of $[0, T]$ of measure zero, if necessary, we obtain a uniquely determined function

$$u' \in C([0, T], H),$$

i.e., $u'\colon [0, T] \to H$ is continuous. The initial condition $u'(0) = 0$ is to be understood in this sense.

33.2. Equivalent Formulations of the Original Problem

33.2a. The Functional Equation

Problem (1) is equivalent to the following equation. For all $v \in V$ and almost all $t \in]0, T[$, let

$$\frac{d^2}{dt^2}(u(t)|v)_H + a(t; u'(t), v) + c(u(t), v) = b(t; v),$$

$$u(0) = u'(0) = 0, \tag{2}$$

$$u \in C([0, T], V), \qquad u' \in W_p^1(0, T; V, H),$$

where d^2/dt^2 denotes the second generalized derivative on $]0, T[$, and we set

$$\begin{aligned} a(t; v, w) &= \langle A(t)v, w\rangle_V, \\ c(v; w) &= \langle Lv, w\rangle_V, \\ b(t; v) &= \langle b(t), v\rangle_V \qquad \text{for all} \quad v, w \in V. \end{aligned}$$

Indeed, it follows from (1) that

$$\langle u''(t), v\rangle_V + \langle A(t)u'(t), v\rangle_V + \langle Lu(t), v\rangle_V = \langle b(t), v\rangle_V \qquad \text{for all} \quad v \in V. \tag{2*}$$

According to Proposition 23.20,

$$\langle u''(t), v\rangle_V = \frac{d^2}{dt^2}\langle u(t), v\rangle_V \qquad \text{for all} \quad v \in V.$$

Furthermore, since $u(t), v \in V$, we obtain that

$$\langle u(t), v\rangle_V = (u(t)|v)_H.$$

In Section 33.5, we shall show that initial value problems for hyperbolic differential equations can be easily formulated in the form (2).

33.2b. The Galerkin Method

Let $\{w_1, w_2, \ldots\}$ be a basis in V. We set

$$u_n(t) = \sum_{k=1}^{n} c_{kn}(t) w_k.$$

Motivated by (2*), we define the *Galerkin equations* in the following way. For $j = 1, \ldots, n$ and almost all $t \in\,]0, T[$, we consider:

$$\begin{gathered} \langle u_n''(t), w_j\rangle_V + \langle A(t)u_n'(t), w_j\rangle_V + \langle Lu_n(t), w_j\rangle_V = \langle b(t), w_j\rangle_V, \\ u_n(0) = u_n'(0) = 0, \\ u_n \in C([0, T], V_n), \qquad u_n' \in W_p^1(0, T; V_n, H_n). \end{gathered} \tag{3}$$

Here, we set

$$V_n = H_n = \operatorname{span}\{w_1, \ldots, w_n\},$$

and we equip V_n and H_n with the norm of V and the scalar product of H, respectively.

Using (2), the Galerkin equations (3) may be written in the following equivalent form:

$$\begin{gathered} \sum_{k=1}^{n} c_{kn}''(t)(w_k|w_j)_H + a\left(t; \sum_{k=1}^{n} c_{kn}'(t)w_k, w_j\right) + c\left(\sum_{k=1}^{n} c_{kn}(t)w_k, w_j\right) = b(t; w_j), \\ c_{jn}(0) = c_{jn}'(0) = 0, \qquad j = 1, \ldots, n, \end{gathered} \tag{4}$$

for almost all $t \in\,]0, T[$.

In Section 33.5, we shall show that the Galerkin method for quasi-linear hyperbolic differential equations leads directly to (4). Equation (4) represents a second-order system of ordinary differential equations. Since $w_1, \ldots, w_n$ are linearly independent, we have

$$\det((w_k|w_j)_H) \neq 0, \qquad j, k = 1, \ldots, n,$$

and hence equation (4) can be solved for the second time-derivatives c''_{kn}. In the standard textbooks of numerical analysis, one finds well-elaborated methods for solving equation (4) on computers.

33.2c. The Equivalent Operator Equation

The original problem (1) is equivalent to the following equation:

$$\begin{gathered} u'' + Au' + Lu = b, \\ u(0) = u'(0) = 0, \\ u \in C([0, T], V), \qquad u' \in W_p^1(0, T; V, H), \end{gathered} \tag{5}$$

where we set $X = L_p(0, T; V)$, and the operators

$$A, L\colon X \to X^*$$

are defined through

$$(Au)(t) = A(t)u(t) \qquad \text{and} \qquad (Lu)(t) = Lu(t) \qquad \text{for all} \quad t \in]0, T[.$$

Here, let $b \in L_2(0, T; H)$ be given. Hence $b \in X^*$.

33.3. The Existence Theorem

We make the following assumptions:

(H1) Let "$V \subseteq H \subseteq V^*$" be an evolution triple with $\dim V = \infty$. Let $\{w_1, w_2, \ldots\}$ be a basis in V. Moreover, let $2 \leq p < \infty$, $q^{-1} + p^{-1} = 1$, and $0 < T < \infty$.

(H2) For each $t \in]0, T[$, the operator

$$A(t)\colon V \to V^*$$

is monotone and hemicontinuous with $A(t)(0) = 0$.

(H3) For each $t \in]0, T[$, the operator $A(t)$ is coercive, i.e., there are constants $c_1 > 0$ and $c_2 \geq 0$ such that

$$\langle A(t)v, v\rangle_V \geq c_1 \|v\|_V^p - c_2 \qquad \text{for all} \quad v \in V, \quad t \in]0, T[.$$

(H4) For each $t \in]0, T[$, the operator $A(t)$ is bounded, i.e., there exists a nonnegative function $c_3 \in L_q(0, T)$ and a constant $c_4 > 0$ such that

$$\|A(t)v\|_{V^*} \leq c_3(t) + c_4 \|v\|_V^{p/q} \qquad \text{for all} \quad v \in V, \quad t \in]0, T[.$$

(H5) The map $t \mapsto A(t)$ is weakly measurable, i.e., for each $u, v \in V$, the function

$$t \mapsto \langle A(t)u, v\rangle_V$$

is measurable on $]0, T[$.

(H6) The operator $L: V \to V^*$ is linear, *symmetric*, and strongly monotone.

(H7) Let $b \in L_2(0, T; H)$ be given.

Theorem 33.A. *Assume* (H1) *through* (H7). *Then*:

(a) Existence and uniqueness. *The original problem* (1) *has a unique solution* u.

(b) Convergence of the Galerkin method. *For each* $n \in \mathbb{N}$, *the Galerkin equation* (3) *has a unique solution* u_n. *As* $n \to \infty$, *the sequence* (u_n) *converges to the solution* u *of the original problem. More precisely, we have that*

$$\max_{0 \le t \le T} \|u_n(t) - u(t)\|_V \to 0 \qquad as \quad n \to \infty,$$

and

$$\max_{0 \le t \le T} \|u_n'(t) - u'(t)\|_H \to 0 \qquad as \quad n \to \infty.$$

(c) Equivalent operator equation. *The original problem* (1) *is equivalent to the operator equation* (5), *where the operator* $A: X \to X^*$ *is monotone, hemicontinuous, coercive, and bounded. Moreover, the operator* $L: X \to X^*$ *is linear, symmetric, and strongly monotone.*

Applications to hyperbolic differential equations will be considered in Section 33.5.

33.4. Proof of the Existence Theorem

We will use Theorem 32.E. To simplify notation, we write

$$\|v\| = \|v\|_V, \qquad |v| = \|v\|_H, \qquad (u|v) = (u|v)_H,$$
$$\langle u, v\rangle = \langle u, v\rangle_V.$$

Let us recall that

$$(u|v) = \langle u, v\rangle \qquad \text{for all} \quad u \in H, \quad v \in V,$$

according to (23.17).

Step 1: Existence proof for the original problem (1).

We set $(Au)(t) = (Au)(t)$ and $(Lu)(t) = Lu(t)$. The proof of Theorem 30.A shows that the operator

$$A: X \to X^*$$

is monotone, hemicontinuous, coercive, and bounded. Obviously, the original problem (1) is equivalent to problem (5). According to Theorem 32.E, problem (5) possesses a solution.

Step 2: Uniqueness proof for (1).

We set

$$(Sv)(t) = \int_0^t v(s)\,ds.$$

The proof of Theorem 32.E shows that problem (5) is equivalent to the first-order evolution equation

$$v' + Av + LSv = b, \qquad v \in W_p^1(0, T; V, H), \qquad v(0) = 0. \tag{6}$$

Moreover, by (32.80),

$$\int_0^t \langle L(Sv)(s), v(s)\rangle\,ds \geq 0,$$

for all $t \in [0, T]$ and all $v \in W_p^1(0, T; V, H)$. Using integration by parts, the same argument as in the uniqueness proof for Theorem 30.A yields the unique solvability of (6). Thus, the original problem (1) has a unique solution.

Step 3: Unique solvability of the Galerkin equation (3).

For $v \in V$ and fixed $t \in \,]0, T[$, we consider the following functionals on V:

$$A(t)v,\ Lv,\ b(t) \in V^*. \tag{7}$$

The restrictions of these functionals to the linear subspace V_n of V are denoted by

$$A_n(t)v,\ L_n v,\ b_n(t) \in V_n^*, \tag{7*}$$

respectively. Using this notation, the Galerkin equation (3) can be written in the following form:

$$\begin{gathered} u_n''(t) + A_n(t)u_n'(t) + L_n u_n(t) = b_n(t) \qquad \text{for almost all} \quad t \in \,]0, T[, \\ u_n(0) = u_n'(0) = 0, \\ u_n \in C([0, T], V_n), \qquad u_n' \in W_p^1(0, T; V_n, H_n). \end{gathered} \tag{8}$$

Replacing V and H with V_n and H_n, respectively, the same proof as for (1) shows that problem (8) has a unique solution.

Step 4: *A priori* estimates for the Galerkin solutions.

For each $n \in \mathbb{N}$, we have $V_n = H_n$, according to the definition of these spaces. Since $\dim V_n < \infty$, we may identify V_n^* with V_n, i.e.,

$$V_n = H_n = V_n^*.$$

Moreover, since all norms are equivalent on finite-dimensional B-spaces, we obtain that

$$W_p^1(0, T; V_n, H_n) \subseteq W_p^1(0, T; V, H).$$

Lemma 33.1. *For all $n \in \mathbb{N}$ and all $t \in [0, T]$, we have the following estimates*:

$$|u_n'(t)| \leq \text{const}, \tag{9}$$

$$\|u_n'\|_X \leq \text{const}, \tag{10}$$

$$\|u_n(t)\| \leq \text{const}, \tag{11}$$

$$\|Au_n'\|_{X^*} \leq \text{const}, \tag{12}$$

$$\|Lu_n\|_{X^*} \leq \text{const}. \tag{13}$$

PROOF. The proofs of (9), (10), and (11) will be given in Problem 33.1. Since the operators $A, L: X \to X^*$ are bounded, the estimates (12) and (13) follow from (10) and (11), respectively. □

Step 5: Convergence of the Galerkin method.
Let u denote the solution of the original problem (1). Then $u' \in W_p^1(0, T; V, H)$.

(I) Preparations. We will use the following approximation argument.

Lemma 33.2. *For each $n \in \mathbb{N}$, there is a polynomial $q_n: [0, T] \to V_n$ with coefficients in V_n such that $q_n(0) = 0$ and*:

$$\|q_n' - u'\|_X + \|q_n'' - u''\|_{X^*} \to 0 \quad \text{as} \quad n \to \infty, \tag{14}$$

$$q_n \to u \quad \text{in } C([0, T], V) \quad \text{as} \quad n \to \infty, \tag{15}$$

$$q_n' \to u' \quad \text{in } C([0, T], H) \quad \text{as} \quad n \to \infty. \tag{16}$$

In particular, it follows from (16) that

$$q_n'(0) \to u'(0) \quad \text{in } H \quad \text{as} \quad n \to \infty. \tag{17}$$

The proof of Lemma 33.2 will be given in Problem 32.2.

Lemma 33.3. *For all $n \in \mathbb{N}$ and all $t \in [0, T]$, we have*

$$\int_0^t \langle Au_n' - Au', u_n' - u' \rangle \, ds \geq 0, \tag{18}$$

$$\begin{aligned} 2\int_0^t \langle Lu_n - Lu, u_n' - u' \rangle \, ds &= \langle Lu_n(t) - Lu(t), u_n(t) - u(t) \rangle \\ &\geq c\|u_n(t) - u(t)\|^2, \end{aligned} \tag{19}$$

where $c > 0$ is fixed.

PROOF. Ad(18). This follows from the monotonicity of $A(s)$ for all s.
Ad(19). This follows from (32.79) and from the strong monotonicity of L. □

(II) They key relation. Using the original problem (1), $u'' + Au' + Lu = b$, and the Galerkin equation (3), integration by parts yields the following *key relation*:

$$
\begin{aligned}
\Delta_n \stackrel{\text{def}}{=} \tfrac{1}{2}|u_n'(t) - q_n'(t)|^2 - \tfrac{1}{2}|u_n'(0) - q_n'(0)|^2 \\
&= \int_0^t \langle u_n'' - q_n'', u_n' - q_n' \rangle \, ds \\
&= \int_0^t \langle -Au_n' - Lu_n + (u'' + Au' + Lu) - q_n'', u_n' - u' + u' - q_n' \rangle \, ds \\
&= -\int_0^t \langle Au_n' - Au', u_n' - u' \rangle + \langle Lu_n - Lu, u_n' - u' \rangle \, ds \\
&\quad + \int_0^t \langle u'' - q_n'', u_n' - q_n' \rangle - \langle Au_n' - Au' + Lu_n - Lu, u' - q_n' \rangle \, ds \\
&\leq \int_0^t \langle u'' - q_n'', u_n' - q_n' \rangle - \langle Au_n' - Au' + Lu_n - Lu, u' - q_n' \rangle \, ds \\
&\leq \|u'' - q_n''\|_{X^*} \|u_n' - q_n'\|_X \\
&\quad + \|Au_n' - Au' + Lu_n - Lu\|_{X^*} \|u' - q_n'\|_X \stackrel{\text{def}}{=} \delta_n,
\end{aligned} \tag{20}
$$

where

$$\delta_n \to 0 \quad \text{as} \quad n \to \infty, \tag{20*}$$

by Lemmas 33.1 and 33.2. In particular, note that

$$\|u'' - q_n''\|_{X^*} \to 0 \quad \text{and} \quad \|u' - q_n'\|_X \to 0 \quad \text{as} \quad n \to \infty,$$

and note that the following sequences

$$(\|u_n'\|_X), (\|q_n'\|_X), (\|Au_n'\|_{X^*}), (\|Lu_n\|_{X^*})$$

are bounded.

(III) We want to show that

$$u_n' \to u' \quad \text{in } C([0, T], H) \quad \text{as} \quad n \to \infty. \tag{21}$$

Indeed, from $u_n'(0) = 0$, $u'(0) = 0$, and

$$q_n'(0) \to u'(0) \quad \text{in } H \quad \text{as} \quad n \to \infty$$

it follows that

$$u_n' - q_n' \to 0 \quad \text{in } C([0, T], H) \quad \text{as} \quad n \to \infty,$$

according to the key relation (20), (20*). By (16), we obtain (21).

(IV) We want to show that

$$u_n \to u \quad \text{in } C([0, T], V) \quad \text{as} \quad n \to \infty. \tag{22}$$

Indeed, consider the key relation (20), (20*) and notice that

$$\Delta_n \to 0 \qquad \text{as} \quad n \to \infty$$

and

$$\left| \int_0^t \langle u'' - q_n'', u_n' - q_n' \rangle - \langle Au_n' - Au' + Lu_n - Lu, u' - q_n' \rangle \, ds \right|$$
$$\leq \delta_n \to 0 \qquad \text{as} \quad n \to \infty.$$

Thus, it follows from Lemma 33.3 and (20) that

$$\max_{0 \leq t \leq T} \int_0^t \langle Lu_n - Lu, u_n' - u' \rangle \, ds \to 0 \qquad \text{as} \quad n \to \infty.$$

Again by Lemma 33.3,

$$\frac{c}{2} \|u_n(t) - u(t)\|^2 \leq \int_0^t \langle Lu_n - Lu, u_n' - u' \rangle \, ds.$$

This yields (22). □

33.5. Application to Quasi-Linear Hyperbolic Differential Equations

Let $Q_T = G \times]0, T[$. We consider the following boundary–initial value problem corresponding to a wave equation with *nonlinear friction*:

$$\begin{gathered} u_{tt} - \Delta u - \sum_{i=1}^N D_i(\alpha(|Du_t|^2) D_i u_t) = f \quad \text{on } Q_T, \\ u(x,t) = 0 \quad \text{on } \partial G \times [0, T], \\ u(x,0) = u_t(x,0) = 0 \quad \text{on } G. \end{gathered} \tag{23}$$

We make the following assumptions:

(H1) Let G be a bounded region in $\mathbb{R}^N$, $N \geq 1$. Let $x = (\xi_1, \ldots, \xi_N)$, $D_i = \partial/\partial \xi_i$. We set

$$|Dv(x)|^2 = \sum_{i=1}^N |D_i v(x)|^2.$$

(H2) Let $0 < T < \infty$ and $f \in L_2(Q_T)$.

(H3) The function α: $[0, \infty[\to [0, \infty[$ is continuous. There are positive constants M and d such that

$$0 \leq \alpha(\sigma^2) \leq M \qquad \text{for all} \quad \sigma \geq 0,$$

$$\alpha(\tau^2)\tau - \alpha(\sigma^2)\sigma \geq d(\tau - \sigma) \qquad \text{for all real} \quad \tau \geq \sigma \geq 0.$$

(H4) Let $\{w_1, w_2, \ldots\}$ be a basis in $\mathring{W}_2^1(G)$.

Definition 33.4. Let $V = \mathring{W}_2^1(G)$ and $H = L_2(G)$. The *generalized problem* corresponding to (23) reads as follows. We are given $f \in L_2(Q_T)$. We seek a function $t \mapsto u(t)$ such that, for all $v \in V$ and almost all $t \in]0, T[$,

$$\frac{d^2}{dt^2}(u(t)|v)_H + a(u'(t), v) + c(u(t), v) = b(t; v),$$

$$u(0) = u'(0) = 0, \tag{24}$$

$$u \in C([0, T], V), \qquad u' \in W_2^1(0, T; V, H),$$

where we set

$$a(u, v) = \int_G \sum_{i=1}^N \alpha(|Du|^2) D_i u D_i v \, dx,$$

$$c(u, v) = \int_G \sum_{i=1}^N D_i u D_i v \, dx, \qquad b(t, v) = \int_G f(x, t) v(x) \, dx,$$

$$(u|v)_H = \int_G uv \, dx,$$

for all $u, v \in V$ and all $t \in]0, T[$. Here, d^2/dt^2 is to be understood as a second generalized derivative on $]0, T[$.

Problem (24) follows formally from the original problem by multiplying (23) with $v \in C_0^\infty(G)$ and by using subsequent integration by parts.

To formulate the Galerkin method, we set

$$u_n(t) = \sum_{k=1}^n c_{kn}(t) w_k.$$

Then, for each $n \in \mathbb{N}$, the *Galerkin equation* reads as follows:

$$\sum_{k=1}^n c''_{kn}(t)(w_k|w_j)_H + a\left(\sum_{k=1}^n c'_{kn}(t) w_k, w_j\right) + c\left(\sum_{k=1}^n c_{kn}(t) w_k, w_j\right)$$

$$= b(t, w_j) \quad \text{on }]0, T[, \tag{25}$$

$$c_{jn}(0) = c'_{jn}(0) = 0, \qquad j = 1, \ldots, n.$$

Proposition 33.5. *Assume* (H1) *through* (H4). *Then*:

(a) *For each* $f \in L_2(Q_T)$, *the generalized problem* (24) *corresponding to* (23) *has a unique solution* u.

(b) *For each* $n \in \mathbb{N}$, *the Galerkin equation* (25) *has a unique solution* u_n *with*

$$u_n, u'_n \in L_2(0, T; V).$$

As $n \to \infty$, *the sequence* (u_n) *converges to* u. *More precisely, we have*

$$\max_{0 \le t \le T} \int_G |u_n(x, t) - u(x, t)|^2 + \sum_{i=1}^N |D_i u_n(x, t) - D_i u(x, t)|^2 \, dx \to 0,$$

and

$$\max_{0 \le t \le T} \int_G |D_t u_n(x,t) - D_t u(x,t)|^2 \, dx \to 0 \qquad as \quad n \to \infty,$$

where $D_t = \partial/\partial t$.

PROOF. According to (25.87*), there exists an operator $A: V \to V^*$ such that

$$\langle Au, v \rangle_V = a(u, v) \qquad \text{for all} \quad u, v \in V,$$

where A is continuous, strongly monotone, and bounded. By Section 22.2a, there exists an operator $L: V \to V^*$ such that

$$\langle Lu, v \rangle_V = c(u, v) \qquad \text{for all} \quad u, v \in V,$$

where L is linear, symmetric, and strongly monotone.

Finally, it follows from $f \in L_2(Q_T)$ that there exists a $b \in L_2(0, T; H)$ such that

$$\langle b(t), v \rangle = b(t; v) \qquad \text{for all} \quad v \in V,$$

and almost all $t \in \,]0, T[$, according to Example 23.4.

The assertion now follows from Theorem 33.A. □

33.6. Strong Monotonicity, Systems of Conservation Laws, and Quasi-Linear Symmetric Hyperbolic Systems

> The part of my work that was most strongly influenced by von Neumann's work concerned spectral theory and partial differential equations. In it, I have restricted myself to linear equations and also to systems of first order. The latter restriction is convenient and not severe, since equations of higher order can—in general—be reduced to systems of the first order.
>
> I have also always restricted myself to dealing with symmetric systems. The reason for this restriction was that almost all nondegenerate partial differential equations of classical mathematical physics appear either naturally as symmetric equations, or can be reduced to symmetric equations. Also, fortunately, it is much easier to solve symmetric equations than nonsymmetric ones.
>
> Kurt Otto Friedrichs (1980)

We consider the following initial value problem for a system of conservation laws:

$$\begin{aligned} u_t + \sum_{j=1}^{N} D_j F_j(u) &= S(x, t, u) \quad \text{on } \mathbb{R}^N \times [0, T], \\ u(x, 0) &= u_0(x) \quad \text{on } \mathbb{R}^N, \end{aligned} \tag{26}$$

where $u = (u_1, \ldots, u_M)$ and $x = (\xi_1, \ldots, \xi_N)$ with $D_j = \partial/\partial \xi_j$. Note that $F_j(u)$ is a real $(M \times M)$-matrix, and $S(x, t, u)$ is a real M-vector. We assume:

(H1) Let F_j and S be smooth, i.e.,

$$F_j \in C^\infty(\mathbb{R}^M, \mathbb{R}^M \times \mathbb{R}^M), \qquad S \in C^\infty(\mathbb{R}^{N+M+1}, \mathbb{R}^M),$$

where $j = 1, \ldots, N$ and $N, M \geq 1$.

(H2) Let $u_0 \in W_2^m$ be given with $m > 1 + N/2$.
Here, we set $W_2^m = W_2^m(\mathbb{R}^N, \mathbb{R}^M)$, i.e., all the components of u_0 belong to the Sobolev space $W_2^m(\mathbb{R}^N)$.

(H3) Let $u_0(x) \in U_0$ for all $x \in \mathbb{R}^N$, where U_0 is a bounded open subset of a given open set U in $\mathbb{R}^N$ with $\bar{U}_0 \subset U$.

(H4) The linearized system

$$u_t + \sum_{j=1}^N A_j(v) D_j u = 0$$

with $A_j(v) = F_j'(v)$ is *symmetric hyperbolic* on U, i.e., there exists a C^∞-matrix function

$$u \mapsto B(u)$$

from U into $\mathbb{R}^M \times \mathbb{R}^M$ such that the matrices

$$B(u) \qquad \text{and} \qquad B(u)A_j(u)$$

are symmetric for all j, and for each compact subset C of U, there is a constant $c > 0$ such that

$$\langle B(u)x|x\rangle \geq c|x|^2 \qquad \text{for all} \quad u \in C, \quad x \in \mathbb{R}^N,$$

i.e., $B(u)$ is *strongly monotone* (uniformly with respect to compact u-sets).

Theorem 33.B (Majda (1984)). *Assume* (H1) *through* (H4). *Then there is a time interval* $[0, T]$ *with* $T > 0$, *so that the original problem* (26) *has a unique classical solution*

$$u \in C^1(\mathbb{R}^N \times [0, T], \mathbb{R}^M)$$

with the property that $u(x, t)$ *lies in a compact subset of* U *for all* $x \in \mathbb{R}^N$, $t \in [0, T]$.

Moreover, we have

$$u \in C([0, T], W_2^m) \cap C^1([0, T], W_2^{m-1}),$$

and T *depends only on* $\|u_0\|_{W_2^m}$ *and the set* U_0.

The proof of this important theorem can be based on the iteration method of Theorem 21.H (see Majda (1984, L), p. 30) or on the Galerkin method of Theorem 30.B along the lines of the proof for the generalized Korteweg–de Vries equation in Chapter 30. In Chapter 83 we will prove a general existence theorem for *nonlinear symmetric hyperbolic systems* which contains Theorem 33.B above as a special case. In Chapter 83 we will also discuss in detail the physical background of equation (26) and the physical meaning of the corresponding *a priori* estimates (energy estimates). Those estimates concern the quantity

$$E(t) = \int_{\mathbb{R}^N} \langle B(u)u|u\rangle \, dx$$

which frequently corresponds to the physical energy of the system. Many important problems in mathematical physics, which describe the propagation of waves, can be reduced either to problem (26) or to more general quasi-linear symmetric hyperbolic systems (e.g., the basic equations in fluid dynamics, gas dynamics, electromagnetism, relativistic quantum mechanics, general relativity).

If u is a solution of (26), then integration by parts yields

$$\frac{d}{dt}\int_G u(x,t)\,dx = -\int_{\partial G} \sum_{j=1}^{N} F_j(u)n_j\,dO + \int_G S\,dx \tag{27}$$

for each bounded region G in $\mathbb{R}^N$ with $\partial G \in C^{0,1}$, where $n = (n_1, \ldots, n_N)$ denotes the unit normal to ∂G. As in Example 31.17, relation (27) motivates the designation "conservation law" for (26).

The theory of linear symmetric hyperbolic systems was created by Friedrichs (1954), who discovered the crucial role played by such systems in mathematical physics.

As a special application of Theorem 33.B to second-order *semilinear wave equations*, we consider the following initial value problem:

$$\begin{aligned} v_{tt} - \Delta v &= F(x, t, D_1 v, \ldots, D_N v, v_t, v) \quad \text{on } \mathbb{R}^N \times [0, T], \\ v(x, 0) &= a(x) \quad \text{on } \mathbb{R}^N, \\ v_t(x, 0) &= b(x) \quad \text{on } \mathbb{R}^N. \end{aligned} \tag{28}$$

We set $x = (\xi_1, \ldots, \xi_N)$, $D_j = \partial/\partial\xi_j$, and we are looking for a real function $v = v(x, t)$.

Proposition 33.6 (Semilinear Wave Equation). *Let $F\colon \mathbb{R}^{2N+3} \to \mathbb{R}$ be C^∞. Then, for given real functions*

$$a \in W_2^{m+1}(\mathbb{R}^N) \qquad \text{and} \qquad b \in W_2^m(\mathbb{R}^N)$$

with $m > 1 + N/2$, there exists a $T > 0$ such that the original problem (28) *has a unique classical solution $v \in C^2(\mathbb{R}^N \times [0, T])$.*

PROOF. Let $N = 2$. The general case proceeds analogously. We set

$$u = (u_1, u_2, u_3, u_4) = (D_1 v, D_2 v, v_t, v)$$

and $u_0 = (D_1 a, D_2 a, b, a)$. Using

$$(D_j v)_t = D_j v_t,$$

we obtain from (28) the following *first-order* system:

$$\begin{aligned} (u_1)_t - D_1 u_3 &= 0, \\ (u_2)_t - D_2 u_3 &= 0, \\ (u_3)_t - D_1 u_1 - D_2 u_2 &= F(x, t, u), \\ (u_4)_t &= u_3. \end{aligned} \tag{29*}$$

This can be written as

$$\begin{aligned} u_t + A_1 D_1 u + A_2 D_2 u = S \quad & \text{on } \mathbb{R}^N \times [0, T], \\ u(x, 0) = u_0(x) \quad & \text{on } \mathbb{R}^N, \end{aligned} \tag{29}$$

where

$$A_1 = \begin{pmatrix} 0 & 0 & -1 & 0 \\ 0 & 0 & 0 & 0 \\ -1 & 0 & 0 & 0 \\ 0 & 0 & 0 & 0 \end{pmatrix}, \qquad A_2 = \begin{pmatrix} 0 & 0 & 0 & 0 \\ 0 & 0 & -1 & 0 \\ 0 & -1 & 0 & 0 \\ 0 & 0 & 0 & 0 \end{pmatrix},$$

$$S = \begin{pmatrix} 0 \\ 0 \\ F(x, t, u) \\ u_3 \end{pmatrix}.$$

Since the matrices A_1 and A_2 are symmetric, the system (29) is symmetric hyperbolic with $B = I$, the unit matrix.

Obviously, the original problem (28) is equivalent to (29). In this connection, observe the following. Suppose that u is a classical solution of (29). Letting $v = u_4$, it follows from (29*) that

$$(D_j v)_t = D_j u_3 = (u_j)_t \qquad \text{for} \quad t \geq 0,$$

and $D_j v = u_j$ for $t = 0$. Hence $D_j v = u_j$ for $t \geq 0$.

The assertion now follows from Theorem 33.B. In this connection, note that the functions a and b are continuous and bounded on $\mathbb{R}^N$, by the Sobolev embedding theorems (see $A_2(74l)$). $\square$

Remark 33.7 (Energy). Since $B = I$, we obtain for the energy the following well-known classical expression:

$$E(t) = \int_{\mathbb{R}^N} \langle Bu | u \rangle \, dx = \int_{\mathbb{R}^N} \sum_{j=1}^{N} (D_j v)^2 + v_t^2 + v^2 \, dx.$$

33.7. Three Important General Phenomena

We want to discuss the following three phenomena:

(i) the appearance of shock waves (discontinuities of the solution);
(ii) the blow-up of solutions;
(iii) the crucial difference between parabolic equations (reaction–diffusion processes) and hyperbolic equations (wave propagation) with respect to the smoothness properties of the solutions.

Generally, it is not possible to prove the existence of smooth solutions u of equation (26) above for all times $t > 0$. This is caused by (i) or (ii). Roughly speaking, shocks correspond to the collision of waves which propagate with different speed. Such shocks play a crucial role in gas dynamics, where they correspond to discontinuities of physical quantities (e.g., density, pressure, velocity). The simplest model for this effect will be considered in Example 33.9. For example, supersonic aircraft cause shock waves which one hears as sharp cracks. Today, there are no general existence theorems available for the equations of gas dynamics. The mathematical difficulties are mainly related to the appearance of complex shock waves.

By a blowing-up effect we understand that the solution goes to infinity after a finite time. Roughly speaking, such effects occur if too much energy is added to the system. The simplest model for this will be considered in Example 33.10.

With respect to (iii), roughly speaking, one has the following general principle:

(P) *In contrast to hyperbolic equations, the solutions of parabolic equations at time $t > 0$ are, as a rule, smoother than the initial data at time $t = 0$.*

In terms of physics, *hyperbolic* equations describe the *propagation of waves*, which does *not* lead to an increase of smoothness for increasing time. *Parabolic* equations describe *diffusion* processes. Such processes try to remove irregular situations, which *increases* the smoothness.

Mathematically, let us look at symmetric hyperbolic systems (Theorem 33.B) and parabolic systems (Theorem 31.C) in $\mathbb{R}^N$. In Theorem 33.B, we need initial data of the form

$$u_0 \in W_2^m(\mathbb{R}^N, \mathbb{R}^M), \qquad m > 1 + N/2, \tag{30}$$

in order to obtain classical C^1-solutions. Note that, by the Sobolev embedding theorems, relation (30) implies that $u_0 \in C^1$. In Theorem 31.C, we only need initial data of the form

$$u_0 \in W_p^1(\mathbb{R}^N, \mathbb{R}^M), \qquad p > N,$$

in order to obtain a classical solution for $t > 0$.

EXAMPLE. 33.8. In order to explain principle (P) above in the simplest way, we consider the simplest special cases of Theorem 33.B and Theorem 31.C. First we study the hyperbolic equation

$$\begin{aligned} u_t + cu_x &= 0 \qquad x \in \mathbb{R}, \quad t > 0, \\ u(x, 0) &= u_0(x), \end{aligned} \tag{31}$$

where c is a positive number. This equation has the solution

$$u(x, t) = u_0(x - ct),$$

which corresponds to the propagation of a wave with velocity c. Obviously, the solution u has the same smoothness properties as the initial values u_0. If u_0 is continuous, but not differentiable, then so is u, etc.

In contrast to (31), we now consider the parabolic equation

$$\begin{aligned} u_t - u_{xx} &= 0, \qquad x \in \mathbb{R}, \quad t > 0, \\ u(x, 0) &= u_0(x) \end{aligned} \tag{32}$$

with the solution

$$u(x, t) = \frac{1}{2\sqrt{\pi t}} \int_{-\infty}^{\infty} e^{-|x-y|^2/4t} u_0(y)\, dy.$$

Observe that u is C^∞ for $t > 0$ if u_0 is continuous with compact support or, more generally, $u_0 \in L_1(-\infty, \infty)$ (see A_2(25b)).

33.8. The Formation of Shocks

STANDARD EXAMPLE 33.9. Let $u = u(x, t)$ be a real function. We consider the following initial value problem:

$$\begin{aligned} u_t + f(u)_x &= 0, \qquad x \in \mathbb{R}, \quad t > 0, \\ u(x, 0) &= u_0(x), \end{aligned} \tag{33}$$

which represents the simplest nonlinear example for Theorem 33.B. Let $f: \mathbb{R} \to \mathbb{R}$ be C^2. It is obvious that (33) is symmetric hyperbolic. However, in this special case, we can construct classical solutions in an explicit manner by using the so-called method of characteristics, which will be explained below. We write (33) in the form

$$u_t + c(u)u_x = 0, \qquad \text{where} \quad c(u) = f'(u). \tag{33*}$$

(i) *Necessary condition.* Let $u = u(x, t)$ be a C^1-solution of (33). The solutions $x = x(t)$ of the equation

$$x'(t) = c(u(x(t), t)) \tag{34}$$

are called *characteristics* of (33). Differentiation shows that

$$\frac{d}{dt} u(x(t), t) = u_x c(u) + u_t = 0.$$

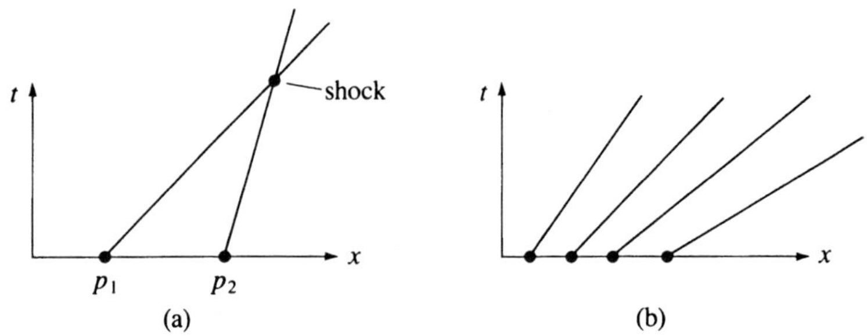

Figure 33.1

Thus, we obtain that:

$$u \text{ is constant along each characteristic.} \tag{35}$$

From (34) and (35) it follows that the characteristics are straight lines, and hence they have the form

$$x = c(u_0(p))t + p, \tag{36}$$

where p is a parameter with $x(0) = p$.

(ii) *Shocks.* Suppose that there are two points p_1 and p_2 such that $p_1 < p_2$ and

$$c(u_0(p_1)) > c(u_0(p_2)). \tag{37}$$

In this case, the characteristics intersect at a point (x, T) (Fig. 33.1(a)). Since u is constant along the characteristics, we obtain that

$$u(x, T) = u_0(p_1) \qquad \text{and} \qquad u(x, T) = u_0(p_2).$$

But it follows from (37) that $u_0(p_1) \neq u_0(p_2)$, and hence the classical solution u can only exist for

$$t < T.$$

At least at time $t = T$, a discontinuity appears. The following observation is important:

The appearance of discontinuities of the solution does not depend on the smoothness of the initial data.

The condition (37) is satisfied in most cases. Thus, as a rule, problem (33) does *not* have global classical solutions for all $t > 0$. This is a quite remarkable fact.

(iii) *Physical interpretation.* Regard $u(x, t)$ as the *mass density* at the point x at time t. Then equation (33) describes the conservation of mass (see Section 69.1). The characteristics $x = x(t)$ are the trajectories which transport the mass particles. Discontinuities of the mass density u appear if two trajectories intersect which transport *different* densities.

(iv) *Construction of classical solutions.* Let u_0 be given as a C^1-function. By (35) and (36), a solution of the original problem (33) must have the form

$$u(x, t) = u_0(p),$$

where p is given through (36). Differentiation shows immediately that u is a solution in the case where equation (36) can be solved for p, i.e., we obtain a classical C^1-solution u in each region where the characteristics do not intersect (Fig. 33.1(b)).

(v) *Generalized solutions.* We postpone the consideration of generalized solutions of (33) to Chapter 86, where we will explain the connection with gas dynamics. At this point, let us only make some general remarks.

(a) By a generalized solution of (33), we understand a solution in the sense of distributions.
(b) Generally, problem (33) has *several* generalized solutions.
(c) However, if we add the so-called *entropy condition*, then it is possible to prove the existence of a *unique* global generalized solution for all times $t > 0$. This has been done by Oleinik (1957).
(d) Roughly speaking, the *entropy condition* guarantees that the solution does *not* violate the *second law of thermodynamics.* Thus, among all the possible generalized solutions of (33), there is precisely one which is meaningful from the physical point of view. This is a very remarkable fact.

In this connection, we recommend Smoller (1983, M) and the survey article Lax (1984).

33.9. Blowing-Up Effects

STANDARD EXAMPLE 33.10. Let $a \in \mathbb{R}$, and let $u = u(t)$ be a solution of the ordinary differential equation

$$u' = f(u). \tag{38}$$

If $f(u) = au$, then the solution $u(t) = e^{at}u(0)$ exists for *all* times $t \in \mathbb{R}$. However, if $a \neq 0$ and

$$f(u) = a(1 + u^2)^\alpha, \qquad \alpha > 1/2, \tag{39}$$

then the solutions *blow up*, i.e., they go to infinity after a finite time. For example, if $\alpha = 1$ and $a > 0$, $u(0) = 0$, then

$$u(t) = \tan at,$$

i.e., the solution exists only in the interval $]-\pi/2a, \pi/2a[$ (Fig. 33.2).

PROOF. If $u = u(t)$ is a solution of (38) with (39), then

$$\int_{u(0)}^{u} \frac{dv}{a(1 + v^2)^\alpha} = t.$$

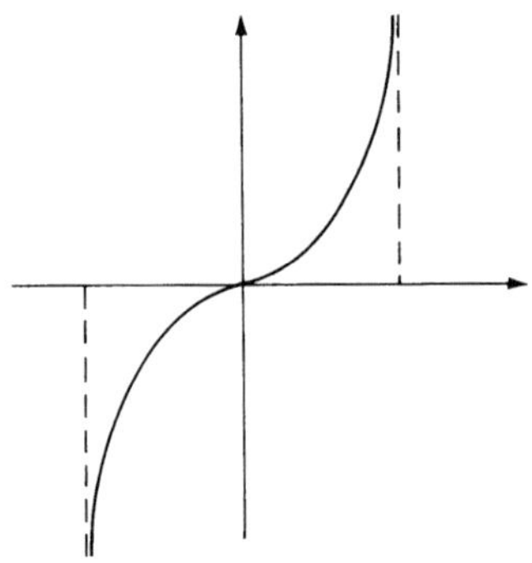

Figure 33.2

Since the integral converges as $u \to \pm\infty$, the function $u = u(t)$ can only exist on a finite time interval.

Since f is C^1, the solution $u = u(t)$ can be continued as long as it remains bounded, by Section 3.3. Thus, the solution must blow up after a finite time. $\square$

The same argument justifies the following general principle. Let $f: \mathbb{R} \to \mathbb{R}$ be C^1 with $f(u) \neq 0$ for all $u \in \mathbb{R}$.

The solutions of (38) *blow up if f grows stronger than linearly as $u \to +\infty$.*

This principle is a special case of a general result which we will prove below in Theorem 33.C. As a preparation, we need the following well-known result on differential inequalities which allows to estimate the *life-span* of solutions. We want to show that

$$\begin{aligned} u' &\leq f(u) \quad \text{on } [0, T], \\ v' &= f(v) \quad \text{on } [0, T], \\ u(0) &\leq v(0), \end{aligned} \tag{40}$$

implies that

$$u(t) \leq v(t) \quad \text{on } [0, T]. \tag{41}$$

It is important that f is monotone increasing.

Proposition 33.11 (Differential Inequalities). *Let $0 < T < \infty$, and let $f: \mathbb{R} \to \mathbb{R}$ be a monotone increasing C^1-function, i.e., $u \leq v$ implies $f(u) \leq f(v)$. Suppose that the C^1-functions $u, v: [0, T] \to \mathbb{R}$ satisfy* (40).

Then inequality (41) *holds.*

Corollary 33.12. *The same assertion remains true if we replace "$\leq$" everywhere with "$\geq$".*

PROOF. From (40) it follows that

$$v(t) = v(0) + \int_0^t f(v(s))\,ds \quad \text{on } [0, T],$$

$$u(t) \le u(0) + \int_0^t f(u(s))\,ds.$$

We now consider the iterative method

$$v_{n+1}(t) = v(0) + \int_0^t f(v_n(s))\,ds, \qquad n = 0, 1, \ldots, \tag{41*}$$

with $v_0 = u$. Obviously, $u \le v_1$ on $[0, T]$. By induction, we obtain that

$$u(t) \le v_n(t) \quad \text{on } [0, T] \qquad \text{for all } n.$$

By the proof of Theorem 3.A, there is a $T_1 > 0$ such that

$$v_n(t) \to v(t) \quad \text{on } [0, T_1] \qquad \text{as} \quad n \to \infty.$$

This implies

$$u(t) \le v(t) \quad \text{on } [0, T_1].$$

A simple continuation argument shows that the latter inequality is valid on $[0, T]$ (cf. Problem 30.2). □

PROOF OF COROLLARY 33.12. In contrast to the proof above, we now use

$$u(t) \ge u(0) + \int_0^t f(u(s))\,ds \quad \text{on } [0, T]$$

and the iterative method

$$v_{n+1}(t) = v(0) + \int_0^t f(v_n(s))\,ds, \qquad n = 0, 1, \ldots,$$

with $v_0 = u$. Note that $v(0) \le u(0)$, by assumption. Hence, by induction,

$$v_n(t) \le u(t) \quad \text{on } [0, T] \qquad \text{for all } n.$$

As in the proof above, this implies $v(t) \le u(t)$ on $[0, T]$. □

We now consider the case where f is only monotone increasing on $\mathbb{R}_+$, e.g., $f(u) = 1 + u^2$.

Proposition 33.13. *Let $0 < T < \infty$, and let $f\colon \mathbb{R}_+ \to \mathbb{R}_+$ be a monotone increasing C^1-function, i.e., $0 \le u \le v$ implies $0 \le f(u) \le f(v)$. Then:*

(a) *If the C^1-functions $u, v\colon [0, T] \to \mathbb{R}$ satisfy*

$$u' \le f(u) \quad \text{on } [0, T],$$

$$v' = f(v) \quad \text{on } [0, T],$$

$$u(0) \le v(0),$$

and if $u(t) \geq 0$ *on* $[0, T]$, *then*

$$u(t) \leq v(t) \quad \text{on } [0, T].$$

(b) *If the* C^1*-functions* $u, v\colon [0, T] \to \mathbb{R}$ *satisfy*

$$u' \geq f(u) \quad \text{on } [0, T],$$

$$v' = f(v) \quad \text{on } [0, T],$$

$$0 \leq v(0) \leq u(0),$$

and if $u(t) \geq 0$ *on* $[0, T]$, *then*

$$0 \leq v(t) \leq u(t) \quad \text{on } [0, T].$$

PROOF. We use the same argument as in the proofs of Proposition 33.11 and Corollary 33.12 above. Note that, in addition, $v_n(t) \geq 0$ on $[0, T]$ for all n. This follows from $u(t) \geq 0$ on $[0, T]$ and from $v(0) \geq 0$ in (b). □

More general results on differential inequalities can be found in Walter (1964, M).

EXAMPLE 33.14 (Quadratic Differential Equations). Let $u = u(t)$ be a solution of the differential equation

$$u'(t) = \alpha(t)u(t)^2 \quad \text{on } [0, T[,$$

where $0 < T < \infty$ and $u(0) > 0$. Suppose that there are numbers β and γ such that

$$0 < \beta \leq \alpha(t) \leq \gamma \quad \text{on } [0, T[,$$

and the function $\alpha\colon [0, \infty[\to \mathbb{R}$ is continuous.

Then, for the *life-span* of the solution u, we obtain the estimate

$$(\gamma u(0))^{-1} \leq T \leq (\beta u(0))^{-1}.$$

PROOF. For $\mu > 0$, the differential equation

$$v' = \mu v^2, \qquad v(0) = u(0),$$

has the solution $v_\mu(t) = u(0)/(1 - \mu u(0)t)$. By Proposition 33.13,

$$v_\beta(t) \leq u(t) \leq v_\gamma(t) \quad \text{on } [0, T[. \qquad \square$$

We now want to apply Proposition 33.13 to the differential equation

$$\begin{aligned} u'(t) &= F(u(t), t), \qquad t \in \mathbb{R}, \\ u(0) &= u_0 \end{aligned} \tag{42}$$

in a real H-space X. We will use estimates of the following form:

(i) *Sublinear growth of F as* $\|u\| \to \infty$:

$$\|F(u,t)\| \le c\|u\| + d \qquad \text{for all} \quad t \in \mathbb{R}, \quad u \in X. \tag{43}$$

(ii) *Superlinear growth of F as* $\|u\| \to \infty$:

$$\begin{aligned} (F(u,t)|u) &\ge 0 && \text{for all} \quad t \in \mathbb{R}, \quad u \in X, \\ (F(u,t)|u) &\ge b\|u\|^\beta && \text{for all} \quad t \in \mathbb{R}, \quad \|u\| \ge \|u_0\| > 0. \end{aligned} \tag{44}$$

Here, $\beta > 2$ and b, c, d are positive numbers.

Theorem 33.C (Global Existence and Blowing Up). *Let $u_0 \in X$ be given, and let*

$$F: U \times \mathbb{R} \to X$$

be C^1, where U is an open subset of the real B-*space X with $u_0 \in U$. Moreover, let F be bounded on bounded sets. Then*:

(a) Existence and uniqueness. *There exists a maximal open interval $J =]S, T[$ with $0 \in J$ such that the initial value problem* (42) *has a unique solution $u = u(t)$ with $u(t) \in U$ for all $t \in J$.*
(b) Continuation. *The solution $u = u(t)$ of* (42) *can be uniquely continued as long as it remains bounded and it does not reach the boundary of U (Fig.* 33.3).
(c) Global existence. *Let $U = X$, where X is a real* H-*space. If* (43) *holds, then $J = \mathbb{R}$.*
(d) Blowing up. *Let $U = X$, where X is a real* H-*space. If* (44) *holds, then $T < \infty$ and*

$$\lim_{t \to T-0} \|u(t)\| = \infty.$$

Proof. Ad(a), (b). Cf. Problem 30.2.

Ad(c). Since $(u(t)|u(t))' = 2(u'(t)|u(t))$, we obtain from (42) the key relation

$$\frac{d}{dt}\|u(t)\|^2 = 2(F(u(t),t)|u(t)) \quad \text{on } J. \tag{45}$$

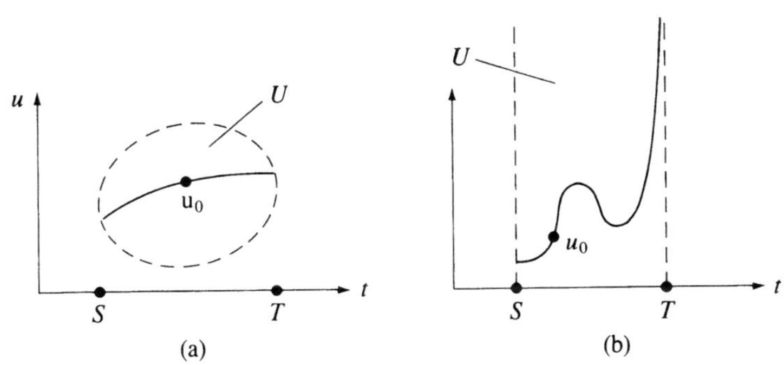

Figure 33.3

If (43) holds, then there is a number $a > 0$ such that

$$|(F(u,t)|u)| \leq c\|u\|^2 + d\|u\| \leq 2^{-1}a(1 + \|u\|^2)$$

for all $t \in \mathbb{R}$, $u \in X$. Hence

$$\frac{d}{dt}\|u(t)\|^2 \leq a(1 + \|u(t)\|^2) \quad \text{on } J.$$

Since the majorant equation

$$v' = a(1 + v), \qquad v(0) = \|u(0)\|^2,$$

has the solution $v(t) = v(0)e^{at} - 1$, we obtain from Proposition 33.13 the *a priori* estimate

$$\|u(t)\|^2 \leq v(t) \qquad \text{for} \quad t \geq 0.$$

An analogous argument yields the same estimate for $u(-t)$. By (b), $J = \mathbb{R}$.

Ad(d). Since $(F(u,t)|u) \geq 0$, we obtain from (45) that $\|u(t)\| \geq \|u(0)\|$ for $t \geq 0$. By (44) and (45), there is a number $a > 0$ such that

$$\frac{d}{dt}\|u(t)\|^2 \geq a(1 + \|u(t)\|^4)^{\beta/4} \qquad \text{for} \quad t \geq 0,$$

where $\beta/4 > \frac{1}{2}$. The assertion now follows from Example 33.10 and Proposition 33.13. □

Finally, we want to use the method of differential inequalities in order to prove the existence of a unique *global* solution for the following initial value problem:

$$\text{(E)} \qquad \begin{aligned} u'(t) &= h(u(t),t) \quad \text{on } [0,T], \\ u(0) &= u_0. \end{aligned}$$

To this end, we consider the following two differential equations:

$$\text{(E*)} \qquad \begin{aligned} v'(t) &= f(v(t),t) \quad \text{on } [0,T], \\ w'(t) &= g(w(t),t) \quad \text{on } [0,T], \end{aligned}$$

along with the decisive *majorant condition*

$$\text{(M)} \qquad \begin{aligned} g(u,t) \leq h(u,t) \leq f(u,t) \qquad &\text{for all} \quad u \in \mathbb{R}_+, \quad t \in [0,T], \\ w(0) \leq u_0 \leq v(0).& \end{aligned}$$

Our objective is the following inequality:

$$\text{(I)} \qquad w(t) \leq u(t) \leq v(t) \quad \text{on } [0,T].$$

Recall that $\mathbb{R}_+ = \{u \in \mathbb{R}: u \geq 0\}$.

By definition, a function $h: \mathbb{R}_+ \times [0,T] \to \mathbb{R}$ satisfies the condition (L) iff h is continuous on $\mathbb{R}_+ \times [0,T]$ and h is locally Lipschitz continuous with respect to u, i.e., for each point $(u,t) \in \mathbb{R}_+ \times [0,T]$, there exists a neighborhood

U in $\mathbb{R}^2$ and a constant $L \geq 0$ such that

$$|h(v, s) - h(w, s)| \leq L|v - w|$$

for all points (v, s) and (w, s) in the set $U \cap (\mathbb{R}_+ \times [0, T])$.

For example, the condition (L) is satisfied if

$$h\colon \mathbb{R}_+ \times [0, T] \to \mathbb{R} \quad \text{is } C^1.$$

Proposition 33.15 (Majorant Method). *The original problem* (E) *has a unique C^1-solution u, and this solution satisfies the inequality* (I) *provided the following conditions are satisfied:*

(i) *Let $0 < T < \infty$. The two C^1-functions v, w: $[0, T] \to \mathbb{R}$ satisfy the differential equations* (E*).
(ii) *The functions f, g, h: $\mathbb{R}_+ \times [0, T] \to \mathbb{R}_+$ satisfy the condition* (L) *and the majorant condition* (M).
(iii) *The functions f and g are monotone increasing with respect to the first variable, i.e.,*

$$f(u, t) \leq f(v, t) \qquad \text{and} \qquad g(u, t) \leq g(v, t)$$

for all $u, v \in \mathbb{R}_+$ and $t \in [0, T]$ with $u \leq v$.
(iv) *We are given $u_0 \in \mathbb{R}_+$ such that the majorant condition* (M) *is satisfied, and $w(0) \in \mathbb{R}_+$.*

Corollary 33.16 (General Case). *The assertions of Proposition* 33.15 *remain valid if we replace $\mathbb{R}_+$ with $\mathbb{R}$ everywhere in the assumptions* (i) *through* (iv) *above, and in conditions* (L) *and* (M) *above.*

Proof of Proposition 33.15. We first prove an *a priori* estimate for u. Suppose that u is a C^1-solution of the equation (E) on the interval $[0, S]$, where $0 < S \leq T$. We consider the iterative method

$$u_{n+1}(t) = u(0) + \int_0^t h(u_n(s), s)\, ds, \qquad n = 0, 1, \ldots,$$

with $u_0(t) = u(0)$ on $[0, S]$. By induction,

$$u_n(t) \geq 0 \quad \text{on } [0, S],$$

since $u(0) \geq 0$ and $h(u, s) \geq 0$ for all $u \geq 0$, $s \in [0, T]$. By Theorem 3.A, there is a number $S_0 > 0$ such that $u_n(t) \to u(t)$ on $[0, S_0]$ as $n \to \infty$, and hence

$$u(t) \geq 0 \quad \text{on } [0, S_0].$$

The continuation argument from Problem 30.2 yields $u(t) \geq 0$ on $[0, S]$.

Now it is possible to use the same argument as in the proof of Proposition 33.13. Hence we obtain the following *a priori* estimate for the solution u:

$$0 \leq w(t) \leq u(t) \leq v(t) \quad \text{on } [0, S].$$

By Problem 30.2, this *a priori* estimate guarantees the *existence* of a unique solution u for the original problem (E) on the interval $[0, S]$.

Suppose that $S < T$, otherwise we are done. Applying the argument above to the initial value problem for the differential equation (E) at the new point $t = S$, we obtain the existence of a unique solution u on some larger interval $[0, S + \Delta]$ such that $0 \le w(t) \le u(t) \le v(t)$ on $[0, S + \Delta]$.

Finally, using the continuation argument from Problem 30.2, we obtain the existence of a unique solution u for (E) on the interval $[0, T]$ such that $0 \le v(t) \le u(t) \le w(t)$ on $[0, T]$. □

The same way we prove Corollary 33.16.

33.10. Blow-Up of Solutions for Semilinear Wave Equations

The reason for the improved behavior of solutions of semilinear wave equations in higher dimensions was beautifully demonstrated by Fritz John (1976) with the following quotation from Shakespeare, Henry VI:

"Glory is like a circle in the water,
Which never ceaseth to enlarge itself,
Till by broad spreading it disperses to naught."

The case of dimension $N = 3$ (three space variables), which nature gives preference, is not only the most important, but also the most challenging.

Sergiu Klainerman (1983)

In this section we summarize a number of deep results on classical solutions for semilinear wave equations and semilinear Klein–Gordon equations. We first consider the semilinear *wave equation*

$$\begin{gathered}\Box u = F(u, u', u'') \quad \text{on } \mathbb{R}^N \times [0, T[, \\ u(x, 0) = \varepsilon u_0(x) \qquad \text{and} \qquad u_t(x, 0) = \varepsilon v_0(x) \quad \text{on } \mathbb{R}^N\end{gathered} \tag{46}$$

for the real function $u = u(x, t)$, where ε denotes a small positive parameter. Here, $x = (\xi_1, \ldots, \xi_N)$, $t = \xi_{N+1}$, $D_j = \partial/\partial\xi_j$, and

$$\Box u = \frac{1}{c^2} u_{tt} - \sum_{j=1}^{N} D_j^2 u.$$

It is well known that the positive number c describes the speed of the propagation of waves in the case of the linear wave equation (46) with $F \equiv 0$. Moreover, let u' and u'' denote the first and second partial derivatives of u, i.e.,

$$u' = (D_1 u, \ldots, D_{N+1} u) \qquad \text{and} \qquad u'' = (D_i D_j u)_{i,j=1,\ldots,N+1}.$$

We assume:

(H1) The function $F\colon \mathbb{R}^M \to \mathbb{R}$ is C^∞, and F is independent of u_{tt}. We are given $u_0, v_0 \in C_0^\infty(\mathbb{R}^N)$.

(H2) F vanishes together with all its first derivatives at $(u, u', u'') = (0, 0, 0)$.
(H3) F is independent of u.

Definition 33.17. By the *life-span* $T_{\text{crit}}(\varepsilon)$, we understand the supremum over all $T \geq 0$ such that equation (46) has a C^∞-solution u on $\mathbb{R}^N \times [0, T[$.

In particular, $T_{\text{crit}}(\varepsilon) = \infty$ means that there exists a global C^∞-solution for all times $t \geq 0$.

To begin with, we first formulate the fundamental classical local existence theorem.

Proposition 33.18 (Schauder (1935)). *Assume* (H1) *and* (H2) *with* $N \geq 1$ *for equation* (46). *Then there are positive constants* ε_0, *and* A *such that*

$$T_{\text{crit}}(\varepsilon) \geq \frac{A}{\varepsilon} \qquad \textit{for all} \quad \varepsilon \in \,]0, \varepsilon_0[.$$

The following example shows that this result is sharp in the case where $N = 1$.

EXAMPLE 33.19 (Lax (1964), Klainerman and Majda (1980)). Let $N = 1$ and let

$$F = f(u_x)u_{xx}.$$

Here, $f: \mathbb{R} \to \mathbb{R}$ is C^∞ with

$$f(0) = f'(0) = \cdots = f^{(k-1)}(0) = 0, \qquad f^{(k)}(0) \neq 0,$$

where $k \geq 1$. In this case, equation (46) corresponds to the nonlinear *string equation*

$$\begin{gathered} u_{tt} = c^2(1 + f(u_x))u_{xx}, \qquad x \in \mathbb{R}, \quad t \geq 0, \\ u(x, 0) = \varepsilon u_0(x), \qquad u_t(x, 0) = \varepsilon v_0(x), \end{gathered} \tag{46*}$$

where $u = u(x, t)$ describes the shape of the string at time t (Fig. 33.4). Let u_0, $v_0 \in C_0^\infty(\mathbb{R})$ be given. Then the solution of (46*) blows up after the finite time

$$T_{\text{crit}}(\varepsilon) = O(\varepsilon^{-k}) \qquad \text{as} \quad \varepsilon \to 0.$$

The blow-up occurs in the second derivative of u, i.e., u_{xx} becomes infinite

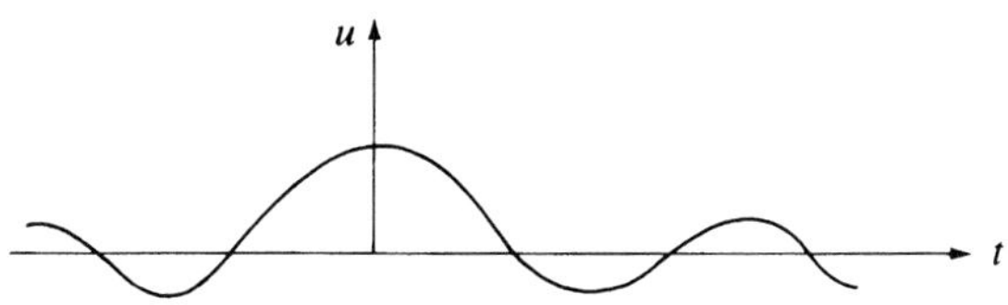

Figure 33.4

while u_x and u_t remain bounded. Such blow-ups are typical of shock formations.

An explicit example will be considered in Problem 33.9.

In the case where $N \geq 3$, Proposition 33.18 can be improved substantially.

Theorem 33.D (Klainerman (1985)). *Assume* (H1), (H2), *and* (H3). *Then there are positive constants* ε_0 *and* A *such that, for all* $\varepsilon \in \,]0, \varepsilon_0[$, *we obtain*

$$T_{\text{crit}}(\varepsilon) \geq \begin{cases} \infty & \text{if} \quad N > 3 \text{ (global solution)}, \\ e^{A/\varepsilon} & \text{if} \quad N = 3 \text{ (almost global solution)}. \end{cases} \tag{47}$$

Corollary 33.20 (Klainerman (1983)). *Assume* (H1), (H2), *and* (H3), *and let*

$$F(u', u'') = O((|u'| + |u''|)^p) \qquad \text{as} \quad |u'| + |u''| \to 0,$$

where $p > 1 + 2/(N - 1)$ *and* $N \geq 2$. *Then there is an* $\varepsilon_0 > 0$ *such that, for each* $\varepsilon \in \,]0, \varepsilon_0[$, *the original problem* (46) *has a unique global classical solution* $u \in C^\infty(\mathbb{R}^N \times [0, \infty[)$.

The proofs of these important results are based on sharp *a priori* estimates for linear and nonlinear wave equations. In this connection, the invariance of the linear wave equation under the Lorentz group (cf. Section 75.7) plays a fundamental role. The estimate $T_{\text{crit}}(\varepsilon) \geq e^{A/\varepsilon}$ in the case where $N = 3$ tells us that the solution exists for "a long time."

We now study the semilinear *Klein–Gordon equation*

$$\begin{gathered} \Box u + \frac{m^2 c^2}{\hbar^2} u = F(u, u', u''), \qquad x \in \mathbb{R}^3, \quad t > 0, \\ u(x, 0) = \varepsilon u_0(x), \qquad u_t(x, 0) = \varepsilon v_0(x). \end{gathered} \tag{48}$$

Here, the positive number m can be regarded as the mass of a meson, where c denotes the velocity of light, and $\hbar = h/2\pi$, where h denotes the Planck quantum of action (cf. Chapter 91 in Part V). It is quite remarkable that, in contrast to the case $m = 0$ (wave equation), we obtain global solutions if $m > 0$.

Theorem 33.E (Klainerman (1985a)). *Assume* (H1), (H2), *and let* $m > 0$. *Then there is an* $\varepsilon_0 > 0$ *such that, for each* $\varepsilon \in \,]0, \varepsilon_0[$, *problem* (48) *has a unique global classical solution* $u \in C^\infty(\mathbb{R}^3 \times [0, \infty[)$, *i.e.*,

$$T_{\text{crit}}(\varepsilon) = \infty \qquad \text{for all} \quad \varepsilon \in \,]0, \varepsilon_0[,$$

with respect to (48). *The solution decays as* $t^{-5/4}$ *for large times* t, *i.e.*,

$$\sup_{x \in \mathbb{R}^3} |u(x, t)| \leq \frac{\text{const}}{t^{5/4}} \qquad \text{for all} \quad t > 0.$$

The techniques of proof for the above results can also be applied to the initial value problem for the Einstein equations in general relativity, and to more general equations of mathematical physics. This can be found in the monograph by Christodoulou and Klainerman (1990).

33.11. A Look at Generalized Viscosity Solutions of Hamilton–Jacobi Equations

> The notion of viscosity solution of Hamilton–Jacobi equations admits nowhere differentiable functions and permits a good existence and uniqueness theory. It is akin to the standard distribution theory, but "integration by parts" is replaced by "differentiation by parts" and is done "inside" the nonlinearity. It is extremely convenient (as in the distribution theory) for passages to limits. Classical solutions are viscosity solutions. Complete consistency of the classical and viscosity notions requires that a viscosity solution which happens to be C^1 will also be a classical solution. This is indeed the case.
>
> Michael Crandall and Pierre-Louis Lions (1983)

33.11a. Motivation

We consider the Hamilton–Jacobi equation

$$S_t + H(q, t, S_q) = 0 \tag{49}$$

along with the canonical equations

$$\dot{p} = -H_q(q, t, p), \qquad \dot{q} = H_p(q, t, p), \tag{50}$$

where $q, p \in \mathbb{R}^N$, and the dot denotes the derivative with respect to time t. In classical mechanics, one uses the solutions $S = S(q, t)$ of (49) in order to obtain solutions $q = q(t)$, $p = p(t)$, of the equations of motion (50). Conversely, one can solve the partial differential equation (49) by constructing appropriate families of solutions of the ordinary differential equation (50). This classical procedure and its physical interpretation will be considered in detail in Sections 58.24 and 58.25. The Hamilton–Jacobi equation (49) is closely related to the variational problem:

$$\begin{gathered} \int_{t_0}^{t_1} L(q(t), \dot{q}(t), t)\, dt = \min!, \\ q(t_0) = q_0, \qquad q(t_1) = q_1. \end{gathered} \tag{51}$$

For fixed q_0, t_0, we set

$$S(q_1, t_1) = \textit{minimal value of } (51).$$

Then the function S satisfies equation (49) if the situation is sufficiently regular (see Section 37.4i). Here, we set

$$H(q,t,p) = \langle p|q\rangle - L(q,\dot{q}(p,t),t),$$

where $\dot{q} = \dot{q}(p,t)$ follows from the equation

$$p = L_{\dot{q}}(q,\dot{q},t),$$

which describes the so-called Legendre transformation. In classical mechanics, the functions L and H are called Lagrangian and Hamiltonian, respectively. Moreover, we have

$$S = \text{action}.$$

The minimum problem (51) corresponds to the *principle of least action*. In many cases, we have:

$$L = \text{kinetic energy} - \text{potential energy},$$

$$H = \text{total energy} = \text{kinetic energy} + \text{potential energy}.$$

The Hamilton–Jacobi equation plays a fundamental role in the calculus of variations and in modern control theory. This will be studied in detail in Part III. Concerning control problems in economics, we have:

$$S(q,t) = \text{minimal costs of the optimal process},$$

where q and t are external parameters.

The classical method of solving the Hamilton–Jacobi equation (49) via (50) frequently runs into trouble, since singularities may appear. In order to motivate this in terms of geometrical optics, we set

$$L = n(t,q)c^{-1}\sqrt{1+\dot{q}^2},$$

where c is the velocity of light and $n(t,q)$ is the index of refraction at the point $(t,q) \in \mathbb{R}^2$. The solutions $q = q(t)$ of (51) are light rays in the (t,q)-plane. Note that t and q are space variables. In this special case, the minimum problem (51) represents *Fermat's principle of least time* and we have:

$$S = \text{minimal value of (51)} = \text{running time of light}.$$

We expect that S loses its smoothness if light rays bifurcate as in Figure 33.5.

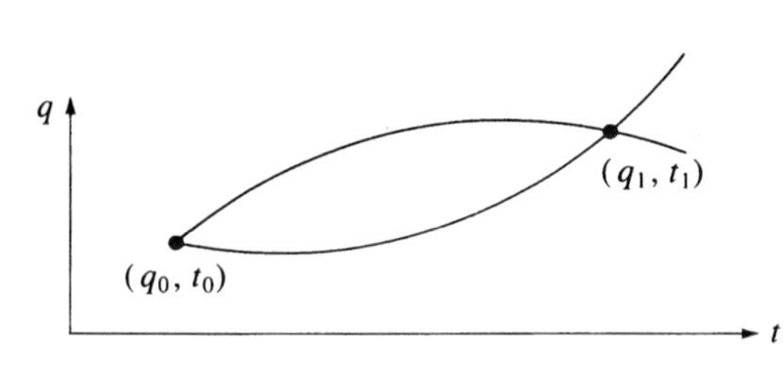

Figure 33.5

Our goal is to introduce a notion of generalized solutions of Hamilton–Jacobi equations (so-called viscosity solutions) which allows general existence and uniqueness theorems. Observe that viscosity solutions are always continuous, but not necessarily differentiable. This is reasonable from the point of view of geometrical optics, since we expect that the running time S depends continuously on the distance if the index of refraction is sufficiently smooth. The designation "viscosity solution" is motivated by the following important fact. Instead of the Hamilton–Jacobi equation (49), we consider the *regularized* parabolic equation

$$S_t - \eta \Delta S + H(q, t, S_q) = 0, \tag{52}$$

where η is a small positive number called *artificial viscosity*. As we will show in Section 70.3, the Navier–Stokes equations for viscous fluids are of the form (52), where η denotes the viscosity. In fluid dynamics, the limiting process $\eta \to 0$ corresponds to a passage from viscous fluids to inviscid fluids. This motivates the following basic idea of our approach:

(a) For each small $\eta \to 0$, we determine a solution $S^{(\eta)}$ of (52).
(b) We consider the limiting process $S^{(\eta)} \to S$ as $\eta \to 0$.
(c) S is a viscosity solution of (52) for $\eta = 0$.

This procedure makes sense if we are able to show the following:

(i) The notion of viscosity solution can be introduced independently of the approximation process above.
(ii) There exist uniqueness theorems for viscosity solutions.
(iii) Classical solutions are viscosity solutions.
(iv) Conversely, C^1-viscosity solutions are classical solutions.

This method is called *parabolic regularization.*

If the Hamiltonian H is independent of time t, we can also consider the *stationary* Hamilton–Jacobi equation

$$H(q, S_q) = 0. \tag{53}$$

In this case, we use the method of *elliptic regularization*

$$-\eta \Delta S + H(q, S_q) = 0, \tag{54}$$

which leads to viscosity solutions of (53).

In what follows we use the following notation. Let M be a subset of $\mathbb{R}^K$. We set:

$$C_0^\infty(M)_+ = \{u \in C_0^\infty(M): u \geq 0 \text{ on } M\},$$

$$C(M) = \{u: M \to \mathbb{R}, \text{ continuous}\},$$

$$C_b(M) = \{u: M \to \mathbb{R}, \text{ continuous and bounded}\},$$

$$C_{ub}(M) = \{u: M \to \mathbb{R}, \text{ uniformly continuous and bounded.}\}.$$

In the spaces $C(M)$, $C_b(M)$, and $C_{ub}(M)$, we introduce the norm

$$\|u\| = \sup_{x \in M} |u(x)|.$$

33.11b. The Basic Definition

Instead of equation (49), we consider the more general equation

$$F(x, S, S') = 0 \quad \text{on } G, \tag{55}$$

where G is an open subset of $\mathbb{R}^K$. We are looking for a function $S = S(x)$ from G into $\mathbb{R}$, where $S' = (D_1 S, \ldots, D_K S)$ denotes the tupel of the first partial derivatives. The Hamilton–Jacobi equation (49) corresponds to (55) with $x = (q, t)$, $K = N + 1$, and $S' = (S_q, S_t)$. The stationary Hamilton–Jacobi equation (53) corresponds to (55) with $x = q$, $K = N$, and $S' = S_q$.

Definition 33.21. Let $\varphi \in C_0^\infty(\mathbb{R}^N)_+$ and $c \in \mathbb{R}$. The *viscosity derivative* of the function $S: G \to \mathbb{R}$ at the point x with respect to φ and c is defined through

$$S'_{\varphi,c}(x) = \frac{c - S(x)}{\varphi(x)} \varphi'(x),$$

where we assume that $\varphi(x) \neq 0$.

Remark 33.22 (Motivation). Let $S \in C^1(G)$, where G is open. Suppose that x is a minimal point of

$$\min_{y \in G} \varphi(y)(S(y) - c) = \varphi(x)(S(x) - c) < 0, \tag{56}$$

or a maximal point of

$$\max_{y \in G} \varphi(y)(S(y) - c) = \varphi(x)(S(x) - c) > 0. \tag{57}$$

Then the viscosity derivative is equal to the classical derivative, i.e.,

$$S'_{\varphi,c}(x) = S'(x).$$

This follows from $(\varphi(S - c))' = 0$ at x, and hence

$$(S(x) - c)\varphi'(x) + \varphi(x)S'(x) = 0.$$

Definition 33.23. A *viscosity solution* of equation (55) is a *continuous* function $S: G \to \mathbb{R}$ such that for *every* $\varphi \in C_0^\infty(G)_+$ and $c \in \mathbb{R}$ the following hold:

(i) If the solution set of (56) is not empty, then there exists a minimal point x of (56) such that

$$F(x, S(x), S'_{\varphi,c}(x)) \geq 0.$$

(ii) If the solution set of (57) is not empty, then there exists a maximal point x of (57) such that

$$F(x, S(x), S'_{\varphi,c}(x)) \leq 0.$$

33.11c. Properties of Viscosity Solutions

According to Remark 33.22, each classical C^1-solution of (55) on an open set G is also a viscosity solution. Conversely, one can prove the following.

Proposition 33.24 (Consistency). *Let $F: G \times \mathbb{R}^{K+1} \to \mathbb{R}$ be continuous, where G is an open subset of $\mathbb{R}^K$, $K \geq 1$. If $S: G \to \mathbb{R}$ is a viscosity solution of the original equation* (55), *and if the classical derivative $S'(x)$ exists, then the function S satisfies* (55) *at the point x.*

The methods of parabolic and elliptic regularization described above are based on the following result.

Proposition 33.25 (Stability). *Let $F_k: G \times \mathbb{R}^{K+1} \to \mathbb{R}$ be continuous for all $k \in \mathbb{N}$, where G is an open subset of $\mathbb{R}^K$, $K \geq 1$. For each k, let $S_k \in C(G)$ be a viscosity solution of the equation*

$$F_k(x, S_k, S'_k) = 0 \quad \text{on } G.$$

As $k \to \infty$, let

$$F_k \to F \quad \text{in } C(G \times \mathbb{R}^{K+1}), \tag{58}$$

$$S_k \to S \quad \text{in } C(G). \tag{59}$$

Then S is a viscosity solution of the equation

$$F(x, S, S') = 0 \quad \text{on } G.$$

The convergence in (58) and (59) means uniform convergence on each compact subset of $G \times \mathbb{R}^{K+1}$ and G, respectively.

33.11d. The Initial Value Problem

We consider the following initial value problem:

$$S_t + H(S_q) = 0 \qquad \text{on } \mathbb{R}^N \times]0, \infty[, \tag{60a}$$

$$S(q, 0) = S_0(q) \quad \text{on } \mathbb{R}^N. \tag{60b}$$

Theorem 33.F (Crandall and Lions (1983)). *Let $H \in C(\mathbb{R}^N)$, $N \geq 1$. For each given $S_0 \in C_{ub}(\mathbb{R}^N)$, there is a unique function S which has the following* *properties*:

(i) $S \in C(\mathbb{R}^N \times [0, \infty[) \cap C_{ub}(\mathbb{R}^N \times [0, T])$ *for all* $T > 0$.
(ii) *S is a viscosity solution of equation* (60a), *and the initial condition* (60b) *is satisfied.*
(iii) $\lim\limits_{t \to +0} \sup\limits_{q \in \mathbb{R}^N} |S(q, t) - S_0(q)| = 0$.

Morever, if we set $\mathscr{S}(t)S_0(q) = S(q, t)$, *then* $\{\mathscr{S}(t)\}$ *is a nonexpansive semigroup on* $C_{ub}(\mathbb{R}^N)$.

The proof uses the method of parabolic regularization.

33.11e. The Stationary Case

We now study the following equation

$$S + H(S_q) = f(q) \quad \text{on } \mathbb{R}^N. \tag{61}$$

Proposition 33.26. *Let* $H \in C(\mathbb{R}^N)$ *and let* $f \in C_{ub}(\mathbb{R}^N)$. *Then equation* (61) *has a unique viscosity solution* $S \in C_b(\mathbb{R}^N)$.

The proof uses the method of elliptic regularization.

Example 33.27. The equation

$$S + |S_q| = 1 \quad \text{on } \mathbb{R} \tag{62}$$

has the following infinite family of Lipschitz continuous, bounded solutions:

$$S(q) = \begin{cases} 1 - Ce^{q} & \text{if } q \le 0, \\ 1 - Ce^{-q} & \text{if } q \ge 0, \end{cases}$$

where $C \ge 0$ is an arbitrary constant. The uniquely determined *viscosity solution* of (62) corresponds to $C = 0$. If $C > 0$, then we obtain solutions which satisfy equation (62) except at the point $q = 0$.

The regularized equation

$$-\eta S_{qq} + S + |S_q| = 1 \quad \text{on } \mathbb{R}$$

has the classical solution

$$S(q) = 1 + e^{-\alpha q/2\eta}, \qquad \alpha = 1 + \sqrt{1 + 4\eta},$$

which goes to the viscosity solution $S \equiv 1$ of (62) as $\eta \to +0$.

The proofs to the results mentioned above and more general theorems can be found in Crandall and Lions (1983) and Lions (1983, L). We also recommend the survey article by Lions (1983a).

Problems

33.1. *Proof of Lemma* 33.1.

Solution: Noting $A(s)(0) = 0$ as well as the monotonicity and coerciveness of A, we obtain that

$$\int_0^t \langle A(s)u_n'(s), u_n'(s)\rangle\, ds \geq 0 \qquad \text{for all} \quad t \in [0, T], \tag{63}$$

and

$$\int_0^T \langle A(s)u_n'(s), u_n'(s)\rangle\, ds \geq c_1 \|u_n'\|_X^p - c_2 T. \tag{64}$$

By (32.79) and the strong monotonicity of L, there is a $c > 0$ such that

$$\begin{aligned} 2\int_0^t \langle Lu_n(s), u_n'(s)\rangle &= \langle Lu_n(t), u_n(t)\rangle \\ &\geq c\|u_n(t)\|^2 \qquad \text{for all} \quad t \in [0, T]. \end{aligned} \tag{65}$$

Moreover, since $b \in L_2(0, T; H)$, we obtain that

$$\begin{aligned} \int_0^t (b(s)|u_n'(s))\, ds &\leq \int_0^t |b(s)||u_n'(s)|\, ds \\ &\leq \frac{1}{2}\int_0^t |b(s)|^2 + |u_n'(s)|^2\, ds \\ &\leq \text{const}\left(1 + \int_0^t |u_n'(s)|^2\, ds\right) \qquad \text{for all} \quad t \in [0, T]. \end{aligned}$$

Since $u_n' \in W_p^1(0, T; V, H)$ and $u_n'(0) = 0$, integration by parts yields the *key estimate*:

$$\begin{aligned} \tfrac{1}{2}|u_n'(t)|^2 &= \int_0^t \langle u_n''(s), u_n'(s)\rangle\, ds \\ &= \int_0^t - \langle Au_n', u_n'\rangle - \langle Lu_n, u_n'\rangle + (b|u_n')\, ds \\ &\leq \text{const}\left(1 + \int_0^t |u_n'(s)|^2\, ds\right) \qquad \text{for all} \quad t \in [0, T]. \end{aligned} \tag{66}$$

In this connection, we use the Galerkin equation (3) and (63)–(65).

(I) Using (66) and the *Gronwall lemma* from Section 3.5, we obtain that

$$|u_n'(t)| \leq \text{const} \qquad \text{for all} \quad t \in [0, T] \quad \text{and all } n.$$

(II) From (66) with $t = T$ and (64), (65), we get

$$\begin{aligned} 0 &\leq \int_0^T - \langle Au_n', u_n'\rangle - \langle Lu_n, u_n'\rangle + (b|u_n')\, ds \\ &\leq -c_1\|u_n'\|_X^p + c_2 T + \text{const}\left(1 + \int_0^T |u_n'(s)|^2\, ds\right) \\ &\leq -c_1\|u_n'\|_X^p + \text{const}, \end{aligned}$$

where $c_1 > 0$. Hence

$$\|u_n'\|_X \le \text{const} \qquad \text{for all } n.$$

(III) Finally, we obtain from (63)–(66) that

$$0 \le \int_0^t -\langle Lu_n(s), u_n'(s)\rangle + (b(s)|u_n'(s))\,ds$$
$$\le -c\|u_n(t)\|^2 + \text{const},$$

and hence

$$\|u_n(t)\| \le \text{const} \qquad \text{for all} \quad t \in [0, T] \quad \text{and all } n.$$

33.2. *Proof of Lemma* 33.2.

Solution: Let $u' \in W_p^1(0, T; V, H)$ and $u(0) = 0$. According to Step 5 of the proof of Theorem 23.A, for each $n \in \mathbb{N}$, there exists a polynomial p_n: $[0, T] \to V_n$ with coefficients in V_n such that

$$p_n \to u' \quad \text{in } W_p^1(0, T; V, H) \qquad \text{as} \quad n \to \infty$$

and $p_n(0) \to u'(0)$ in H as $n \to \infty$. That means

$$\|p_n - u'\|_X + \|p_n' - u''\|_{X^*} \to 0 \qquad \text{as} \quad n \to \infty.$$

We set

$$q_n(t) = \int_0^t p_n(s)\,ds \qquad \text{for all} \quad t \in [0, T].$$

Hence $q_n(0) = 0$ and $q_n' = p_n$. The continuity of the embedding $W_p^1(0, T; V, H) \subseteq C([0, T], H)$ implies that

$$p_n \to u' \quad \text{in } C([0, T], H) \qquad \text{as} \quad n \to \infty.$$

Finally, it follows from the Hölder inequality that, for all $t \in [0, T]$,

$$\|q_n(t) - u(t)\|_V \le \left\| \int_0^t (q_n' - u')\,ds \right\|$$
$$\le \text{const}\,\|q_n' - u'\|_X \to 0 \qquad \text{as} \quad n \to \infty,$$

and hence $q_n \to u$ in $C([0, T], V)$ as $n \to \infty$.

33.3.* *A semilinear wave equation.* We consider the following initial value problem:

$$u_{tt} - \Delta u + |u|^{p-2}u = f(x, t) \quad \text{on } G \times]0, T[, \tag{67a}$$

$$u(x, 0) = u_0(x) \qquad \text{and} \qquad u_t(x, 0) = v_0(x) \quad \text{on } G \tag{67b}$$

along with the boundary condition

$$u(x, t) = 0 \quad \text{on } \partial G \times [0, T].$$

Let G be a bounded region in $\mathbb{R}^N$, and let $0 < T < \infty$, $2 < p < \infty$, $N \ge 1$. We are given

$$u_0 \in Y, \qquad v_0 \in L_2(G), \qquad f \in L_2(G \times]0, T[),$$

where $Y = \mathring{W}_2^1(G) \cap L_p(G)$. We equip Y with the norm $\max\{\|u\|_{1,2}, \|u\|_p\}$.

Show that problem (67) has a solution

$$u \in L_\infty(0, T; Y), \qquad u_t \in L_\infty(0, T; L_2(G)). \tag{68}$$

Moreover, show that it follows from (67a) and (68) that

$$u \in C([0, T], L_2(G)), \qquad u_t \in C([0, T]; Y^*).$$

In this connection use Problem 23.13. Thus, the initial condition (67b) makes sense. Finally, show that the solution is unique if $p - 2 > 0$ is sufficiently small.

The solution is to be understood in the generalized sense as in Section 33.5, i.e., all the derivatives are to be understood in the sense of distributions.

Hint: Use a Galerkin method. See Lions (1969, M), Section 1.3. Observe that $Y^* = \mathring{W}_2^1(G)^* + L_p(G)^*$ (see Problem 23.14).

33.4. *Conservation laws and shocks.* In connection with Section 33.8, study Smoller (1983, M) and Lax (1984, S).

33.5. *Semilinear wave equations.* In connection with Section 33.10, study the fundamental papers by John (1985), (1987), (1988), Klainerman (1985a), (1985), and the monograph by Christodoulou and Klainerman (1990).

33.6. *Semilinear hyperbolic systems in two variables and the Colombeau algebra of generalized distributions.* Study Oberguggenberger (1987). In this paper, semilinear first-order hyperbolic systems in two variables (i.e., one space variable) are considered, whose nonlinearity satisfies a global Lipschitz condition. It is shown that such systems admit unique global generalized solutions in the so-called Colombeau algebra $\mathscr{G}(\mathbb{R}^2)$ of distributions. In particular, this provides unique generalized solutions for arbitrary distributions as initial data. The generalized solutions are limits of smooth approximate solutions. Note that the Colombeau algebra contains the set of distributions $\mathscr{D}'(\mathbb{R}^N)$, and $C^\infty(\mathbb{R}^N)$ is a subalgebra of $\mathscr{G}(\mathbb{R}^N)$ (see Colombeau (1984, M), (1985, M)).

Further interesting results on semilinear hyperbolic equations in two variables can be found in Rauch and Reed (1980), (1981), (1982), (1988).

33.7. *Explicit solutions of evolution equations.* Cf. Problem 30.7.

33.8. *Generalized solutions of the Hamilton–Jacobi equations.* In connection with Section 33.11, study Lions (1983, L) and Crandall and Lions (1983).

33.9.* *Blowing-up effects for nonlinear strings.* We consider the nonlinear string equation

$$\begin{gathered} u_{tt} = g(u_x)u_{xx}, \qquad 0 \le x \le \pi, \quad t > 0, \\ u(x,0) = u_0(x), \qquad u_t(x,0) = 0, \qquad 0 \le x \le \pi, \\ u(0,t) = u(\pi,t) = 0, \qquad t \ge 0, \end{gathered} \tag{69}$$

with

$$g = c^2(1 + au_x)^2, \qquad u_0(x) = \varepsilon \sin x,$$

where c, a, $\varepsilon > 0$ and $a\varepsilon$ is small. Here, $u = u(x,t)$ describes the shape of the string at time t.

Reduce equation (69) to a first-order system and use Example 33.14 in order to show that the classical solution of (69) breaks down after the time

$$T_{\text{crit}} \cong 4/c\, a\varepsilon.$$

This formula is asymptotically true for $a\varepsilon \to 0$.

Hint: Cf. Lax (1964). This paper contains general results for quasi-linear hyperbolic first-order systems in two variables.

References to the Literature

Classical existence proofs for quasi-linear hyperbolic equations of second order: Schauder (1935), Leray, (1952, M) (systems of equations).

Introduction to nonlinear hyperbolic equations: Courant and Hilbert (1953, M), Vol. 2, Smoller (1983, M), Majda (1984, L).

Monographs containing important results on nonlinear hyperbolic equations: Leray (1952), Lichnerowicz (1967), Lions (1969), Strauss (1969), Hawking and Ellis (1973), Gajewski, Gröger, and Zacharias (1974), Barbu (1976), Jeffrey (1976), Reed (1976), Haraux (1981), Choquet–Bruhat (1982a), Marsden and Hughes (1983), Pazy (1983) (theory of semigroups), John (1985) (collected papers), Christodoulou and Klainerman (1990).

Nonlinear wave equations in mathematical physics: Courant and Hilbert (1953, M), Vol. 2, Jeffrey and Taniuti (1964, M) and Lichnerowicz (1967, M) (magnetohydrodynamics), Whitham (1974, M) (nonlinear waves), Kato (1975, S) (many important examples), (1986, S), Reed (1976, L) (quantum theory), Marsden and Hughes (1983, M) and John (1985) (elasticity), Majda (1984, L) (fluid dynamics).

Einstein equations of general relativity and the Einstein–Yang–Mills equations: Hawking and Ellis (1973, M), Choquet–Bruhat and Yorke (1980, S), Choquet–Bruhat and Christodoulou (1981), (1982), Arms, Marsden, and Moncrieff (1982), Eardley and Moncrieff (1982), Klainerman (1987, S), Christodoulou and Klainerman (1990, M) (cf. also the References to the Literature for Chapter 76).

Gas dynamics: Smoller (1983, M) (cf. also the References to the Literature for Chapter 86).

Vlasov–Maxwell equations in plasma physics: Wollman (1984).

Boltzmann equation in molecular gas kinetics: Albeverio (1986, M) (cf. also the References to the Literature for Chapter 86).

Mathematical viscoelasticity: Renardy, Hrusa, and Nohel (1987, M) (fundamental monograph), Nohel, Rogers, and Tzavaras (1988).

Solitons: Cf. the References to the Literature for Chapter 30.

Linear symmetric hyperbolic systems: Friedrichs (1954) (classical paper), (1980) (collected papers), Courant and Hilbert (1953, M), Vol. 2, John (1982, M).

Quasi-linear symmetric hyperbolic systems: Kato (1975, S), (1975a), (1986, S), Hughes, Kato, and Marsden (1977), Majda (1984, L) (cf. also the References to the Literature for Chapter 83).

Conservation laws: Lax (1973, S), (1984, S), Bardos (1982, L), Smoller (1983, M), Majda (1984, L).

The interaction of nonlinear hyperbolic waves: Glimm (1988, S).

Blow-up of the solutions for nonlinear wave equations: Lax (1964), John (1976), (1985) (collected papers), (1987), Klainerman (1983, S) (recommended as an introduction), Shatah (1982), Klainerman and Ponce (1983), John and Klainerman (1984), Klainerman (1985), (1985a), Christodoulou (1986), Christodoulou and Klainerman (1990, M).

Almost global existence of elastic waves: John (1988).

Self-similar blow-up: Bebernes and Eberly (1988).

Rescaling algorithm for the numerical calculation of blow-up: Berger and Kohn (1988).

Generalized solutions for semilinear hyperbolic systems in two variables and the propagation of singularities: Rauch and Reed (1980), (1981), (1982), (1988), Oberguggenberger (1986), (1987), (1988) (Colombeau algebra of generalized distributions; the initial value data are distributions).

Colombeau algebra of generalized distributions: Colombeau (1984, M), (1985, M).

Propagation of singularities and paradifferential operators: Bony (1981), (1984, S), Runst (1985), Lebeau (1986).

Generalized solutions (viscosity solutions) of Hamilton–Jacobi equations: Lions (1983, L), (1983a, S), Crandall and Lions (1983), (1985).

Periodic solutions for the nonlinear string equation: Rabinowitz (1978), (1984a), Brézis and Nirenberg (1978a), Brézis, Coron, and Nirenberg (1980).

Blow-up for the nonlinear string equation: Lax (1964), Klainerman and Majda (1980).

Conferences on nonlinear hyperbolic partial differential equations: Carasso (1987, P), Ballman (1988, P).

(Cf. also the References to the Literature for Chapter 30).

GENERAL THEORY OF DISCRETIZATION METHODS

> The interplay between generality and individuality, deduction and construction, logic and imagination—this is the profound essence of live mathematics. Any one or another of these aspects can be at the center of a given achievement. In a far-reaching development all of them will be involved. Generally speaking, such a development will start from the "concrete ground," then discard ballast by abstraction and rise to the lofty layers of thin air where navigations and observations are easy: after this flight comes the crucial test of landing and reaching specific goals in the newly surveyed low plains of individual "reality." In brief, the flight into abstract generality must start from and return to the concrete and the specific.
>
> Richard Courant (1888–1972)

We want to explain some basic ideas of general discretization methods. Suppose that we are given the operator equation

$$\text{(E)} \qquad Au = b, \qquad u \in X,$$

which may correspond to a differential or integral equation. In order to solve (E) by means of an approximation method, we replace the operator $A: X \to Y$ by an appropriate approximation $A_n: X_n \to Y_n$, where X_n and Y_n are finite-dimensional spaces, and we replace the right-hand side $b \in Y$ by $b_n \in Y_n$. This way we obtain the approximate problems

$$(\text{E}_n) \qquad A_n u_n = b_n, \qquad u_n \in X_n, \qquad n = 1, 2, \ldots,$$

which correspond to linear or nonlinear systems of equations with finitely many unknowns. Such problems can be solved on computers. For example, equations of type (E_n) arise in connection with the Galerkin method and the difference method. Note that we do *not* assume that $X_n \subseteq X$ and $Y_n \subseteq Y$. For example, this is important for describing difference methods by (E_n). In this case, X and Y are spaces of functions defined on the region G, whereas X_n and Y_n are spaces of functions defined on the set of grid points.

For numerical mathematics it is of great importance to answer the following two questions.

PROBLEM 1. Which conditions must be satisfied by X, Y, X_n, Y_n, A, A_n in order to obtain the following? If the original equation (E) has a unique solution, then the approximate problems (E_n) have unique solutions and the approximation method converges.

This problem can be formulated briefly like this:

When is it possible to convert a nonconstructive existence proof into a constructive existence proof?

PROBLEM 2. Which conditions must be satisfied by X, Y, X_n, Y_n, A, A_n in order to obtain the following? If the approximate problems (E_n) are uniquely solvable, then the approximation method converges to a solution of the original problem (E).

We shall give a final answer to Problems 1 and 2. Roughly speaking, we shall prove the following *main result* of numerical functional analysis.

If the approximation method is consistent and stable, then the following three conditions are equivalent.

(C1) Solvability. *The original problem* (E) *has a solution, i.e., we know an abstract existence proof for* (E).
(C2) Unique approximation-solvability. *The approximate problems* (E_n) *have unique solutions and these solutions converge strongly to the unique solution of the original problem* (E), *i.e., we obtain a constructive existence proof.*
(C3) A-properness. *The operator* $A: X \to Y$ *is* A*-proper.*

This underlines the fundamental importance of A-proper operators for the general theory of approximation methods. This class of operators was introduced and studied in detail by Petryshyn (1967), (1968ff). An extensive survey of the general theory of A-proper operators and its applications can be found in Petryshyn (1975).

For example, the following operators A are A-proper:

(i) $A = B + C$, where B is uniformly monotone and continuous, and C is compact.
(ii) $A = \pm I + K + C$, where K is k-contractive and C is compact.
(iii) $A = B + C$, where B is strongly stable (e.g., strongly monotone) and continuous, and C is compact.

Therefore, the theory of A-proper operators unifies the following fundamental results under a *constructive* aspect:

The fixed-point theorem of Banach ($A = I - K$).
The fixed-point theorem of Schauder ($A = I - C$).
The existence theorems on monotone and pseudomonotone operator equations.

The next two results can be used to prove the A-properness of perturbed A-proper operators:

If the operator A is A-proper and if C is compact, then $A + C$ is also A-proper.
If A is A-proper on a normed space over $\mathbb{K}$, then λA is also A-proper for each nonzero $\lambda \in \mathbb{K}$.

Our plan is the following. In Chapter 34 we consider so-called inner approximation schemes which correspond to the Galerkin method.
In Chapter 35 we consider external approximation schemes which correspond to the difference method.
In Chapter 36 we introduce a mapping degree for A-proper operators.

CHAPTER 34

Inner Approximation Schemes, A-Proper Operators, and the Galerkin Method

For what type of a linear or nonlinear mapping A is it possible to construct a solution u of the equation

$$Au = b$$

as a strong limit of solutions u_n of the simpler finite-dimensional equations

$$A_n u_n = Q_n b?$$

In a series of papers the author studied this problem, and the notion which evolved from these investigations is that of an A-proper mapping. It turned out that the A-properness of A is not only intimately connected with the approximation-solvability of the equation $Au = b$ but, in view of the fact that the class of A-proper mappings is quite extensive, the theory of A-proper mappings extends and unifies earlier results concerning Galerkin type methods for linear and nonlinear operator equations with the more recent results in the theory of strongly monotone and accretive operators, operators of type (S), P_γ-compact, ball-condensing and other mappings.

W. V. Petryshyn (1975)

34.1. Inner Approximation Schemes

Along with the operator equation

$$Au = b, \qquad u \in X, \tag{1}$$

we consider the approximate equations

$$A_n u_n = Q_n b, \qquad u_n \in X_n, \qquad n = 1, 2, \ldots, \tag{2}$$

which correspond to the following approximation scheme:

$$\begin{array}{ccc} X & \xrightarrow{A} & Y \\ {\scriptstyle R_n}\downarrow\uparrow{\scriptstyle E_n} & & \downarrow{\scriptstyle Q_n} \\ X_n & \xrightarrow{A_n} & Y_n \end{array}, \qquad A_n = Q_n A E_n. \tag{3}$$

The operator $R_n: X \to X_n$ is called a *restriction* operator, and $E_n: X_n \to X$ is called an *extension* operator. An important role will be played by the so-called *compatibility condition*:

$$\lim_{n\to\infty} \|E_n R_n u - u\|_X = 0 \qquad \text{for all} \quad u \in X. \tag{4}$$

EXAMPLE 34.1 (Galerkin Method in H-Spaces). Let (e_i) be a complete orthonormal system in the separable infinite-dimensional H-space X. We set

$$X_n = \operatorname{span}\{e_1, \ldots, e_n\}, \qquad n = 1, 2, \ldots.$$

Let $P_n: X \to X_n$ be the orthogonal projection operator from X onto X_n, i.e.,

$$P_n u = \sum_{i=1}^{n} (e_i|u)e_i \qquad \text{for all} \quad u \in X.$$

Moreover, let

$$E_n: X_n \to X$$

be the embedding operator corresponding to $X_n \subseteq X$. In order to obtain (3), we set

$$Y = X, \qquad Y_n = X_n, \qquad R_n = Q_n = P_n.$$

It is easy to check that this approximation scheme is an admissible inner approximation scheme in the sense of Definition 34.2 below. In this connection, note that $\|P_n\| = 1$ for all n and that $P_n u \to u$ as $n \to \infty$ for all $u \in X$.

Let the operator $A: X \to X$ be given and let

$$A_n = P_n A E_n.$$

Then the approximate problems (2) correspond to a projection method (Galerkin method). For $n = 1, 2, \ldots$, equation (2) is equivalent to the following Galerkin equations:

$$(Au_n|e_j) = (b|e_j), \qquad u_n \in X_n, \qquad j = 1, \ldots, n.$$

Definition 34.2. The tupel $\{X, X_n, Y, Y_n, R_n, E_n, Q_n\}_{n\in\mathbb{N}}$ represented in (3) is called an *admissible inner approximation* scheme iff the following hold.

(i) X and Y are infinite-dimensional normed spaces over $\mathbb{K} = \mathbb{R}, \mathbb{C}$.
(ii) X_n and Y_n are normed spaces over $\mathbb{K}$ with $\dim X_n = \dim Y_n < \infty$ for all n.
(iii) For all n, the operators $E_n: X_n \to X$ and $Q_n: Y \to Y_n$ are linear and continuous with

$$\sup_n \|E_n\| < \infty \qquad \text{and} \qquad \sup_n \|Q_n\| < \infty.$$

(iv) The compatibility condition (4) is satisfied.

In contrast to Example 34.1, we do not postulate that $X_n \subseteq X$ and $Y_n \subseteq Y$, i.e., we allow the original problem (1) and the approximate problems (2) to live in completely different spaces. However, we postulate $A_n = Q_n A E_n$, i.e., the diagram (3) is commutative. In the next chapter, this condition will drop out by using external approximation schemes.

In connection with diagram (3), there are two different possibilities of defining the convergence of the approximation method, namely, we can use either

$$\lim_{n\to\infty} \|E_n u_n - u\|_X = 0,$$

or

$$\lim_{n\to\infty} \|u_n - R_n u\|_{X_n} = 0.$$

In this chapter (resp. in the next chapter), we will use the first (resp. second) possibility.

34.2. The Main Theorem on Stable Discretization Methods with Inner Approximation Schemes

Our goal is to investigate the connection between the following conditions (C1) through (C3).

(C1) *Solvability.* For each $b \in Y$, the original equation $Au = b$, $u \in X$, has a solution.

(C2) *Unique approximation-solvability.* The original equation is uniquely approximation-solvable, i.e., for each $b \in Y$, the following hold:
  (i) The equation $Au = b$, $u \in X$, has a unique solution.
  (ii) For each $n \geq n_0$, the approximate equation $A_n u_n = Q_n b$, $u_n \in X_n$, has a unique solution.
  (iii) The sequence (u_n) converges to the solution u of the equation $Au = b$ in the sense of $\lim_{n\to\infty} \|E_n u_n - u\|_X = 0$.

(C3) *A-properness.* The operator $A: X \to Y$ is A-proper with respect to the approximation scheme (3). That is, by definition, the following holds. Let (n') be any subsequence of the sequence of natural numbers. If $(u_{n'})$ is a sequence with $u_{n'} \in X_{n'}$ for all n' and if
  (i) $\lim_{n\to\infty} \|A_{n'} u_{n'} - Q_{n'} b\|_{Y_{n'}} = 0$ for fixed $b \in Y$, and
  (ii) $\sup_{n'} \|u_{n'}\|_{X_{n'}} < \infty$,
  then there exists a subsequence $(u_{n''})$ such that

$$\lim_{n\to\infty} \|E_{n''} u_{n''} - u\|_X = 0$$

  and $Au = b$.

Examples for A-proper operators will be given in Sections 34.4 and 34.5. The designation "A-proper" stands for "approximation-proper." Note that

A-proper operators represent a modification of proper continuous operators. To show this let $A: X \to Y$ be proper. This means that the preimages $A^{-1}(K)$ of compact sets K are again compact. Thus, it follows from

$$Au_{n'} \to b \qquad \text{as} \quad n \to \infty$$

that there is a subsequence $(u_{n''})$ such that

$$u_{n''} \to u \qquad \text{as} \quad n \to \infty.$$

If, in addition, the operator A is continuous, then $Au = b$.

The following theorem is the main result of this chapter.

Theorem 34.A (Petryshyn (1970)). *The three conditions* (C1) *through* (C3) *are mutually equivalent in the case where the following hold.*

(H1) Approximation scheme. *The diagram* (3) *represents an admissible inner approximation scheme. All the operators $A_n: X_n \to Y_n$ are continuous.*

(H2) Consistency. *For all $u \in X$,*

$$\lim_{n \to \infty} \|Q_n Au - A_n R_n u\|_{Y_n} = 0. \tag{5}$$

(H3) Stability. *There is an n_0 such that*

$$\|A_n u - A_n v\|_{Y_n} \geq a(\|u - v\|_{X_n}), \tag{6}$$

for all $u, v \in X_n$ and all $n \geq n_0$. Here, the given function $a: \mathbb{R}_+ \to \mathbb{R}_+$ is continuous and strictly monotone increasing with $a(0) = 0$ and $a(t) \to +\infty$ as $t \to +\infty$ (Fig. 34.1).

Roughly speaking, this theorem tells us the following:

If the approximation method is consistent and stable, then the original equation $Au = b$, $u \in X$, is uniquely approximation-solvable iff the operator A is A-proper.

Before proving Theorem 34.A, we consider some variants of this theorem.

Corollary 34.3 (Sufficient Condition for Consistency). *If the operator $A: X \to Y$ is continuous, then* (H1) *implies* (H2).

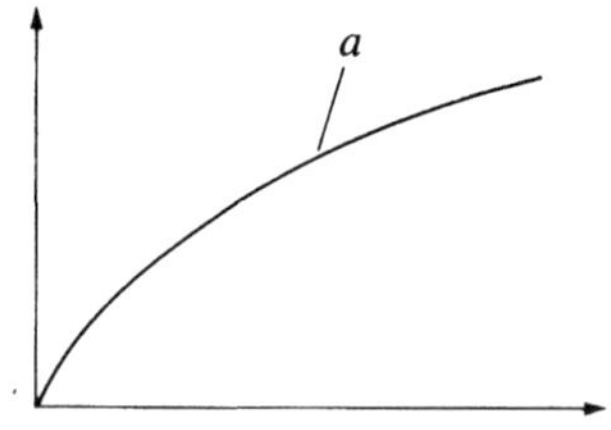

Figure 34.1

Corollary 34.4 (Lack of Consistency). *Suppose that the assumptions* (H1) *and* (H3) *hold and suppose that* $A: X \to Y$ *is A-proper. Then*:

(a) *For each* $b \in Y$, *the equation* $Au = b$, $u \in X$, *has a solution.*
(b) *For each* $b \in Y$ *and each* $n \geq n_0$, *the approximate equation* $A_n u_n = Q_n b$, $u_n \in X_n$, *has a unique solution.*
(c) *The sequence* (u_n) *has a subsequence* $(u_{n'})$ *which converges to a solution* u *of the original equation* $Au = b$, $u \in X$, *in the sense of*

$$\lim_{n \to \infty} \|E_{n'} u_{n'} - u\|_X = 0.$$

(d) *If, for fixed* $b \in Y$, *the equation* $Au = b$, $u \in X$, *has a unique solution, then the entire sequence* (u_n) *converges to* u, *i.e.*,

$$\lim_{n \to \infty} \|E_n u_n - u\|_X = 0.$$

Corollary 34.5 (Compensation for the Lack of Stability by *a priori* Estimates). *Suppose that*:

(i) *The operator* $A: X \to Y$ *is A-proper with respect to the admissible inner approximation scheme* (3).
(ii) *For fixed* $b \in Y$ *and all* $n \geq n_0$, *the approximate equation* $A_n u_n = Q_n b$, $u_n \in X_n$, *has a unique solution, and we have the a priori estimate* $\sup_n \|u_n\|_{X_n} < \infty$. *Then*:
  (a) *There exists a subsequence* $(u_{n'})$ *such that*

$$\lim_{n \to \infty} \|E_{n'} u_{n'} - u\|_X = 0 \qquad \text{and} \qquad Au = b.$$

  (b) *If the equation* $Au = b$, $u \in X$, *has a unique solution, then*

$$\lim_{n \to \infty} \|E_n u_n - u\|_X = 0.$$

Corollary 34.6 (Compact Perturbations). *If the operator* $A: X \to Y$ *is A-proper with respect to the admissible inner approximation scheme* (3), *and if* $C: X \to Y$ *is compact, then* $A + C: X \to Y$ *is also A-proper.*

If $A: X \to Y$ *is A-proper, then so is* λA *for all nonzero* $\lambda \in \mathbb{K}$, *where* X *and* Y *are normed spaces over* $\mathbb{K}$.

Remark 34.7. Let the above assumptions (H1) through (H3) be satisfied. Then there are two different possibilities of using Theorem 34.A.

(i) *Abstract existence theorems imply the unique approximation-solvability.* Suppose that we know that the equation $Au = b$, $u \in X$, has a solution, i.e., condition (C1) holds. Then Theorem 34.A says that this equation is uniquely approximation-solvable, and the operator $A: X \to Y$ is A-proper.

By Corollary 34.6, the compact perturbation $A + C: X \to Y$ is also A-proper.

(ii) *A-properness implies the unique approximation-solvability.* Conversely, suppose that we can prove the A-properness of $A: X \to Y$ by means of a direct argument. Then Theorem 34.A tells us that the equation $Au = b$, $u \in X$, is uniquely approximation-solvable.

34.3. Proof of the Main Theorem

We prove Theorem 34.A. The main idea of proof is to use the theorem on the *invariance of the domain* from Section 16.4, which follows from the antipodal theorem.

Step 1: (C3) implies (C2).

(I) We show that, for fixed $b \in Y$ and each $n \geq n_0$, the approximate equation $A_n u_n = Q_n b$, $u_n \in X_n$, has a unique solution. To this end, we will use the invariance of domain theorem.

(I-1) The set $A_n(X_n)$ is open. Indeed, the operator $A_n: X_n \to Y_n$ is injective, by the stability condition (H3). The invariance of domain theorem (Theorem 16.C) tells us that $A_n(X_n)$ is open.

(I-2) The set $A_n(X_n)$ is closed. To prove this let $A_n u_k \to z$ as $k \to \infty$. Thus $(A_n u_k)$ is a Cauchy sequence. By (H3), (u_k) is also a Cauchy sequence in X_n. Hence $u_k \to u$ as $k \to \infty$. Since A_n is continuous, we get $A_n u = z$, i.e., $z \in A_n(X_n)$.

(I-3) Since the nonempty set $A_n(X_n)$ is open and closed in Y_n, we obtain $A_n(X_n) = Y_n$ (cf. A_1(13f)).

(II) We show that, for fixed $b \in Y$, the equation $Au = b$, $u \in X$, has at most one solution. To this end, let $Au = Av$. By consistency (H2) and stability (H3),

$$a(\|R_n u - R_n v\|) \leq \|A_n R_n u - A_n R_n v\|$$

$$\leq \|A_n R_n u - Q_n Au\| + \|Q_n Av - A_n R_n v\| \to 0 \quad \text{as} \quad n \to \infty.$$

Hence $\|R_n u - R_n v\| \to 0$ as $n \to \infty$. Since $\sup_n \|E_n\| < \infty$,

$$\|E_n R_n u - E_n R_n v\|_X \to 0 \quad \text{as} \quad n \to \infty.$$

Now, the compatibility condition (4) yields $u = v$.

(III) We prove that $\|E_n u_n - u\|_X \to 0$ as $n \to \infty$ and $Au = b$. For fixed $b \in Y$ and $n \geq n_0$, we solve the equation $A_n u_n = Q_n b$, $u_n \in X_n$, according to (I). From

$$A_n(0) = Q_n A E_n(0) = Q_n A(0)$$

and stability (H3) it follows that

$$a(\|u_n\|) \leq \|A_n u_n - A_n(0)\| \leq \|Q_n\|(\|b\| + \|A(0)\|).$$

Since $\sup_n \|Q_n\| < \infty$, we obtain $\sup_n a(\|u_n\|) < \infty$ and hence the *a priori* estimate:

$$\sup_n \|u_n\| < \infty.$$

Obviously, $\|A_n u_n - Q_n b\| \to 0$ as $n \to \infty$. The operator A is A-proper. Thus, there is a subsequence $(u_{n'})$ such that

$$\|E_{n'} u_{n'} - u\|_X \to 0 \qquad \text{as} \quad n \to \infty. \tag{7}$$

and $Au = b$.

The same argument shows that each subsequence of $(E_n u_n)$ has, in turn, a subsequence $(E_{n'} u_{n'})$ satisfying (7). By the unique solvability of $Au = b$, the limit u in (7) is the same for all subsequences. Therefore, it follows from the convergence principle (Proposition 10.13(1)) that relation (7) holds for the entire sequence $(E_n u_n)$.

Step 2: (C2) implies (C1). This is obvious.

Step 3: (C1) implies (C3).

For given $b \in Y$, we solve the equation $Au = b$, $u \in X$. To prove that $A: X \to Y$ is A-proper, assume that $\sup_{n'} \|u_{n'}\| < \infty$ and

$$\lim_{n \to \infty} \|A_{n'} u_{n'} - Q_{n'} b\| = 0.$$

For brevity we will write u_n instead of $u_{n'}$. By consistency (H2) and stability (H3),

$$\begin{aligned} a(\|u_n - R_n u\|) &\le \|A_n u_n - A_n R_n u\| \\ &\le \|A_n u_n - Q_n b\| + \|Q_n A u - A_n R_n u\| \to 0 \qquad \text{as} \quad n \to \infty. \end{aligned}$$

Hence $\|u_n - R_n u\| \to 0$ as $n \to \infty$. Since $\sup_n \|E_n\| < \infty$,

$$\|E_n u_n - E_n R_n u\| \to 0 \qquad \text{as} \quad n \to \infty.$$

The compatibility condition (4) yields

$$\|E_n R_n u - u\| \to 0 \qquad \text{as} \quad n \to \infty.$$

Hence $\|E_n u_n - u\| \to 0$ as $n \to \infty$, i.e., A is A-proper.

This finishes the proof of Theorem 34.A. The proofs for Corollaries 34.3 through 34.6 will be given in the problems section to this chapter.

34.4. Inner Approximation Schemes in H-Spaces and the Main Theorem on Strongly Stable Operators

We consider the operator equation

$$Au = b, \qquad u \in X, \tag{8}$$

together with the approximate equations

$$P_n A u_n = P_n b, \qquad u_n \in X_n, \qquad n = 1, 2, \ldots, \tag{9}$$

corresponding to the following approximation scheme:

$$\begin{array}{ccc} X & \xrightarrow{A} & X \\ {\scriptstyle P_n}\downarrow\uparrow{\scriptstyle E_n} & & \downarrow{\scriptstyle P_n}, \\ X_n & \xrightarrow{A_n} & X_n \end{array} \qquad A_n = P_n A E_n. \tag{10}$$

We make the following assumptions.

(A1) X is a separable H-space over $\mathbb{K} = \mathbb{R}, \mathbb{C}$ with $\dim X = \infty$, and (e_i) is a complete orthonormal system in X. Let

$$X_n = \operatorname{span}\{e_1, \ldots, e_n\},$$

and let $P_n\colon X \to X_n$ be the orthogonal projection operator from X onto X_n, i.e.,

$$P_n u = \sum_{i=1}^{n} (e_i|u)e_i.$$

Let $E_n\colon X_n \to X$ be the embedding operator corresponding to $X_n \subseteq X$.

(A2) The operator $A\colon X \to X$ is continuous and *strongly stable*, i.e., there is an $a > 0$ such that

$$|(Au - Av|u - v)| \geq a\|u - v\|^2 \qquad \text{for all} \quad u, v \in X.$$

(A3) Let $A = B + L$, where $B\colon X \to X$ is continuous and strongly monotone, and $L\colon X \to X$ is Lipschitz continuous (e.g., $L = 0$). More precisely, we assume that there are numbers c, d with $0 \leq d < c$ such that, for all $u, v \in X$,

$$\operatorname{Re}(Bu - Bv|u - v) \geq c\|u - v\|^2,$$

and

$$\|Lu - Lv\| \leq d\|u - v\|.$$

(A4) Let $A = \pm I + K$, where $K\colon X \to X$ is k-contractive.

Obviously, (A3) and (A4) are special cases of (A2). For example, consider the operator A in (A3). Then, for all $u, v \in X$,

$$\begin{aligned} |(Au - Av|u - v)| &\geq \operatorname{Re}(Au - Av|u - v) \\ &= \operatorname{Re}(Bu - Bv|u - v) + \operatorname{Re}(Lu - Lv|u - v) \\ &\geq (c - d)\|u - v\|^2, \end{aligned}$$

i.e., A satisfies (A2). In case (A4) we obtain

$$\begin{aligned} |(Au - Av|u - v)| &= |(u - v|u - v) \pm (Ku - Kv|u - v)| \\ &\geq (1 - k)\|u - v\|^2 \qquad \text{for all} \quad u, v \in X, \end{aligned}$$

since $\|Ku - Kv\| \leq k\|u - v\|$ for all $u, v \in X$ with $0 \leq k < 1$, i.e., A satisfies (A2).

By Example 34.1, the diagram (10) represents an admissible inner approximation scheme. For $n = 1, 2, \ldots$, the approximate equation (9) is equivalent to the following Galerkin equations:

$$(Au_n|e_j) = (b|e_j), \qquad u_n \in X_n, \qquad j = 1, \ldots, n.$$

Theorem 34.B. *Assume* (A1), (A2). *Then, for each* $b \in X$, *the equation* $Au = b$, $u \in X$, *is uniquely approximation-solvable.*

The operator A *is A-proper with respect to the approximation scheme* (10). *If* $C: X \to X$ *is compact, then* $A + C: X \to X$ *is also A-proper with respect to* (10).

Corollary 34.8. *Assume* (A1). *Let* $A: X \to X$ *be continuous and strongly monotone, and let* $C: X \to X$ *be compact. Then for each nonzero* $\lambda \in \mathbb{K}$, *the operator*

$$\lambda(A + C): X \to X$$

is A-proper with respect to (10). *In particular, the operator*

$$\pm I + K + C: X \to X$$

is A-proper with respect to (10) *if* $K: X \to X$ *is k-contractive and* $C: X \to X$ *is compact* (*e.g.*, $K = 0$ *or* $C = 0$).

Proof. We use Theorem 34.A. We shall prove the A-properness of A by a direct argument.

(I) Consistency. Since $A: X \to X$ is continuous, the consistency follows from Corollary 34.3.

(II) Stability. We want to show that

$$\|P_nAu - P_nAv\| \geq a\|u - v\| \qquad \text{for all} \quad u, v \in X_n.$$

Indeed, for all $u, v \in X_n$, we obtain

$$\begin{aligned} a\|u - v\|^2 &\leq |(P_nu - P_nv|Au - Av)| \\ &= |(u - v|P_nAu - P_nAv)| \leq \|u - v\|\,\|P_nAu - P_nAv\|. \end{aligned}$$

(III) The operator $A: X \to X$ is A-proper. To show this, let $\sup_n\|u_n\| < \infty$ and

$$\|P_nAu_n - P_nb\| \to 0 \qquad \text{as} \quad n \to \infty,$$

where $u_n \in X_n$ for all n. Since X is reflexive, there exists a subsequence, again denoted by (u_n), such that

$$u_n \rightharpoonup u \quad \text{in } X \qquad \text{as} \quad n \to \infty.$$

We have to show that

$$u_n \to u \quad \text{in } X \qquad \text{as} \quad n \to \infty$$

and $Au = b$.

(III-1) We will show in Problem 34.5 that, as $n \to \infty$,

$$u_n \rightharpoonup u \qquad \text{implies} \qquad u_n - P_n u \rightharpoonup 0 \tag{11}$$

and

$$v_n \to v \qquad \text{implies} \qquad P_n v_n \to v. \tag{12}$$

Thus, we obtain that, as $n \to \infty$,

$$P_n b \to b, \qquad P_n u \to u, \qquad AP_n u \to Au,$$
$$P_n A P_n u \to Au,$$

and hence

$$P_n(Au_n - AP_n u) = (P_n Au_n - P_n b) + (P_n b - P_n AP_n u) \to b - Au. \tag{13}$$

(III-2) From (11) and (13) it follows that, as $n \to \infty$,

$$a\|u_n - P_n u\|^2 \le |(Au_n - AP_n u | u_n - P_n u)|$$
$$= |(P_n(Au_n - AP_n u)) | u_n - P_n u)| \to 0.$$

Therefore, $u_n - P_n u \to 0$ as $n \to \infty$ and hence $u_n - u \to 0$ as $n \to \infty$. Thus, $Au = b$ by the continuity of A.

Theorem 34.B follows now from Theorem 34.A.
Corollary 34.8 follows from Corollary 34.6. □

34.5. Inner Approximation Schemes in B-Spaces

We consider the operator equation

$$Au = b, \qquad u \in X, \tag{14}$$

together with the approximate equations

$$E_n^* A E_n u_n = E_n^* b, \qquad u_n \in X_n, \qquad n = 1, 2, \ldots, \tag{15}$$

corresponding to the following approximation scheme:

$$\begin{array}{ccc} X & \xrightarrow{\;A\;} & X^* \\ R_n \downarrow \uparrow E_n & & \downarrow E_n^* \\ X_n & \xrightarrow{\;A_n\;} & X_n^* \end{array}, \qquad A_n = E_n^* A E_n. \tag{16}$$

We make the following assumptions.

(A1) X is a real separable reflexive B-space with $\dim X = \infty$. Let (X_n) be a

Galerkin scheme in X with

$$X_n = \operatorname{span}\{w_{1n}, \dots, w_{n'n}\}, \qquad n = 1, 2, \dots.$$

Let $E_n: X_n \to X$ be the embedding operator corresponding to $X_n \subseteq X$.

By Problem 34.6, for each $u \in X$, there exists at least an element $R_n u \in X_n$ such that

$$\|u - R_n u\|_X = \operatorname{dist}(u, X_n).$$

This way we construct the operator $R_n: X \to X_n$, by choosing a fixed solution $R_n u$.

For $n = 1, 2, \dots$, the approximate equation (15) is equivalent to the Galerkin equations:

$$\langle Au_n, w_{jn} \rangle = \langle b, w_{jn} \rangle, \qquad u_n \in X_n, \qquad j = 1, \dots, n'.$$

We will show below that the diagram (16) represents an admissible inner approximation scheme.

(A2) The operator $A: X \to X^*$ is uniformly monotone and continuous.

Proposition 34.9. *Assume* (A1) *and* (A2). *Then, for each* $b \in X^*$, *the original equation* (14) *is uniquely approximation-solvable.*

Corollary 34.10. *Assume* (A1). *Let* $A: X \to X^*$ *be uniformly monotone and continuous, and let* $C: X \to X^*$ *be compact. Then, for each real* $\lambda \neq 0$, *the operator*

$$\lambda(A + C): X \to X^*$$

is A-proper.

PROOF.

(I) Admissible inner approximation scheme. From $\|E_n\| = 1$ it follows that $\|E_n^*\| = 1$ for all n. Since (X_n) is a Galerkin scheme, $\operatorname{dist}(u, X_n) \to 0$ as $n \to \infty$ for all $u \in X$. This implies $\|R_n u - u\| \to 0$ as $n \to \infty$. That is the compatibility condition (4).

(II) Consistency. Since A is continuous, the consistency follows from Corollary 34.3.

(III) Stability. By (A2), for all $u, v \in X$,

$$\langle Au - Av, u - v \rangle \geq a(\|u - v\|)\|u - v\|.$$

For all $u, v \in X_n$, this implies

$$\begin{aligned}\|A_n u - A_n v\| \, \|u - v\| &\geq \langle A_n u - A_n v, u - v \rangle \\ &= \langle Au - Av, E_n u - E_n v \rangle = \langle Au - Av, u - v \rangle \\ &\geq a(\|u - v\|)\|u - v\|.\end{aligned}$$

Hence

$$\|A_n u - A_n v\| \geq a(\|u - v\|) \qquad \text{for all} \quad u, v \in X_n.$$

(IV) Solvability. It follows from Theorem 26.A that, for each $b \in X$, the equation $Au = b$, $u \in X$, has a solution.
(V) By Theorem 34.A, the operator $A: X \to X^*$ is A-proper and the equation $Au = b$, $u \in X$, is uniquely approximation-solvable.

Corollary 34.10 follows from Corollary 34.6. □

34.6. Application to the Numerical Range of Nonlinear Operators

We want to apply the main theorem on strongly stable operators (Theorem 34.B) to the operator equation

$$Au - \lambda u = b, \qquad u \in X, \tag{17}$$

for fixed $\lambda \in \mathbb{K}$. Our objective is to find a set $\mathcal{M}$ such that, for each $\lambda \in \mathcal{M}$ and each $b \in X$, equation (17) has a unique solution. In this connection, the numerical range of A plays a fundamental role.

Definition 34.11. Let $A: X \to X$ be an operator on the H-space X over $\mathbb{K} = \mathbb{R}$, $\mathbb{C}$. The set

$$\mathcal{N}(A) = \left\{\frac{(u - v|Au - Av)}{\|u - v\|^2} : u, v \in X, u \neq v\right\}$$

is called the *numerical range* of A. Obviously, $\mathcal{N}(A)$ is a subset of $\mathbb{K}$.

Example 34.12. Let $A: X \to X$ be linear and continuous. Then:

(a) $\mathcal{N}(A)$ contains the eigenvalues of A.
(b) The closure of $\mathcal{N}(A)$ contains the spectrum of A in the case where $\mathbb{K} = \mathbb{C}$.
(c) If $\lambda \in \mathcal{N}(A)$, then $|\lambda| \leq \|A\|$.
(d) The set $\mathcal{N}(A)$ is convex (Theorem of Hausdorff).

Proof. Ad(a). If $Au = \lambda u$ and $u \neq 0$, then $(u|Au)/\|u\|^2 = \lambda$ and hence $\lambda \in \mathcal{N}(A)$.

Ad(b). Cf. Corollary 34.14(a) below.

Ad(c). Cf. Example 34.13 below.

Ad(d). A proof of this classical result can be found in Stone (1932a, M). □

Example 34.13. Let $A: X \to X$ be Lipschitz continuous, i.e.,

$$\|Au - Av\| \leq L\|u - v\| \qquad \text{for all} \quad u, v \in X.$$

Then $\lambda \in \mathcal{N}(A)$ implies $|\lambda| \leq L$.

PROOF. This follows from

$$|(u - v|Au - Av)| \le \|u - v\| \, \|Au - Av\| \le L\|u - v\|^2. \qquad \square$$

Theorem 34.C (Zarantonello (1964)). *Suppose that:*

(i) *The operator* $A: X \to X$ *is continuous on the separable* H-*space* X *over* $\mathbb{K} = \mathbb{R}, \mathbb{C}$. *We set* $A_\lambda = A - \lambda I$.
(ii) *The number* $\lambda \in \mathbb{K}$ *has a positive distance from the numerical range of* A. *We set* $d = \operatorname{dist}(\lambda, \mathcal{N}(A))$, *i.e.,* $d > 0$.

Then, for each $b \in X$, *the original equation* (17) *has a unique solution.*

Corollary 34.14. *Assume* (i) *and* (ii). *Then:*

(a) *The inverse operator* $A_\lambda^{-1}: X \to X$ *is Lipschitz continuous, i.e.,*

$$\|A_\lambda^{-1}b - A_\lambda^{-1}\bar{b}\| \le d^{-1}\|b - \bar{b}\| \qquad \text{for all} \quad b, \bar{b} \in X. \tag{18}$$

(b) *If* $A(0) = 0$, *then* $\mathcal{N}(A)$ *contains the eigenvalues of* A.
(c) *If, in addition,* $\dim X = \infty$, *then, for each* $b \in X$, *the original equation* (17) *is uniquely approximation-solvable.*

PROOF. We use Theorem 34.B. For all $u, v \in X$ with $u \ne v$, we obtain the *key* inequality

$$|(u - v|A_\lambda u - A_\lambda v)| = \left|\lambda - \frac{(u - v|Au - Av)}{\|u - v\|^2}\right| \|u - v\|^2 \ge d\|u - v\|^2, \tag{19}$$

i.e., $A_\lambda: X \to X$ is *strongly stable*. This implies

$$\|A_\lambda u - A_\lambda v\| \ge d\|u - v\| \qquad \text{for all} \quad u, v \in X. \tag{20}$$

(I) Existence and uniqueness.
Let $\dim X = \infty$. Then $A_\lambda u = b$ is uniquely approximation-solvable by Theorem 34.B.
Let $\dim X < \infty$. As in Step 1 of Section 34.3, it follows from (20) and the invariance of domain theorem that $A_\lambda: X \to X$ is bijective.
(II) Lipschitz continuity of A_λ^{-1}. Relation (18) follows from (20).
(III) If $Au = \lambda u$, $u \ne 0$ and $A(0) = 0$, then $(u|Au)/\|u\|^2 = \lambda$, i.e., $\lambda \in \mathcal{N}(A)$.

$\square$

PROBLEMS

34.1. *Proof of Corollary* 34.3.
Solution: By the compatibility condition (4),

$$\|E_n R_n u - u\| \to 0 \qquad \text{as} \quad n \to \infty.$$

By assumption, the operator A is continuous. Hence

$$\|AE_n R_n u - Au\| \to 0 \qquad \text{as} \quad n \to \infty.$$

Now, from $\sup_n \|Q_n\| < \infty$, we obtain the consistency condition

$$\|Q_n Au - Q_n AE_n R_n u\| \le \|Q_n\| \, \|Au - AE_n R_n u\| \to 0 \qquad \text{as} \quad n \to \infty.$$

34.2. *Proof of Corollary* 34.4.
Solution: Use (I) and (III) in Step 1 of Section 34.3.

34.3. *Proof of Corollary* 34.5.
Solution: Use (III) in Step 1 of Section 34.3.

34.4. *Proof of Corollary* 34.6.
Solution: For brevity, we write n instead of n'. Let $\sup_n \|u_n\| < \infty$ and let

$$\lim_{n \to \infty} \|Q_n(A + C)E_n u_n - Q_n b\| = 0.$$

Since $\sup_n \|E_n\| < \infty$, the sequence $(E_n u_n)$ is bounded. By assumption, the operator C is compact. Thus, there is a subsequence, again denoted by (u_n), such that

$$CE_n u_n \to z \quad \text{in } Y \qquad \text{as} \quad n \to \infty. \tag{21}$$

From $\sup_n \|Q_n\| < \infty$ it follows that $\|Q_n CE_n u_n - Q_n z\| \to 0$ as $n \to \infty$. This implies

$$\lim_{n \to \infty} \|Q_n AE_n u_n - Q_n(b - z)\| = 0.$$

By assumption, the operator A is A-proper. Thus, there is a subsequence, again denoted by (u_n), such that

$$E_n u_n \to u \quad \text{in } X \qquad \text{as} \quad n \to \infty \tag{22}$$

and $Au = b - z$. By (21) and (22), $Cu = z$. Therefore, $(A + C)u = b$, i.e., $A + C$ is A-proper.

34.5. *Convergence of projection operators.* Let $X_1 \subseteq X_2 \subseteq \cdots$ be a sequence of linear closed subspaces X_n of the H-space X such that $X = \overline{\bigcup_n X_n}$. Let $P_n \colon X \to X_n$ be the orthogonal projection operator from X onto X_n.

34.5a. Show that, as $n \to \infty$, $u_n \rightharpoonup u$ implies $u_n - P_n u \rightharpoonup 0$.
Solution: Let $u_n \rightharpoonup u$. For all $w \in X$,

$$(u_n - P_n u | w) = (u_n | w) - (P_n u | w) \to 0,$$

since $P_n u \to u$. This implies $u_n - P_n u \rightharpoonup 0$.

34.5b. Show that, as $n \to \infty$, $u_n \to u$ implies $P_n u_n \to u$.
Solution: Let $u_n \to u$. Then

$$\begin{aligned}\|P_n u_n - u\| &\le \|P_n u_n - P_n u\| + \|P_n u - u\| \\ &\le \|P_n\| \, \|u_n - u\| + \|P_n u - u\| \to 0.\end{aligned}$$

Note that $\|P_n\| = 1$ if $X \ne \{0\}$.

34.6. *An extremal problem.* Let X_n be a finite-dimensional linear subspace of the B-space X. Show that, for fixed $u \in X$, the minimum problem

$$\|u - u_n\| = \min!, \qquad u_n \in X_n, \tag{23}$$

has a solution.

Solution: Since $0 \in X_n$, it is sufficient to consider problem (22) for all $u_n \in X_n$ with $\|u_n\| \leq \|u\|$. This modified problem has a solution by the classical Weierstrass theorem, since $\dim X_n < \infty$.

References to the Literature

Classical survey article on the applications of functional analysis to numerical mathematics: Kantorovič (1948).

Classical papers on A-proper maps: Petryshyn (1967), (1968), (1970), Browder and Petryshyn (1969).

Extensive survey article on A-proper maps and their applications: Petryshyn (1975, B).

Survey on the historical development: Petryshyn (1975), Schumann and Zeidler (1979).

A-proper maps, fixed-point theory, and eigenvalue problems: Petryshyn (1975, S), Fitzpatrick and Petryshyn (1975), (1978), (1979), (1982), Petryshyn (1987).

A-proper maps and the numerical stability of the Galerkin method: Petryshyn (1975).

A-proper maps and nonlinear elliptic differential equations: Petryshyn (1975), (1976), (1980), (1980a), (1981).

A-proper maps and periodic solutions of ordinary differential equations: Petryshyn (1986, S), (1986a), Petryshyn and Yu (1985) (periodic motions of satellites).

Approximation schemes and numerical functional analysis: Céa (1964), Browder (1967), (1968/76, M), Michlin (1969, M) (numerical stability of the Ritz method), Stummel (1970/72), Temam (1970, M), (1977, M), Anselone (1971, M), Aubin (1972, M), Grigorieff (1973), Glowinski, Lions, and Trémolières (1976, M), Ciarlet (1977, M), Vainikko (1976, L), (1978, S), (1979, S), Glowinski (1984, M), Deimling (1985, M), Reinhardt (1985, M).

Nonstationary problems: Raviart (1967), Temam (1968), (1977, M), Girault and Raviart (1986, M).

Collectively compact operators and integral equations: Anselone (1971, M), Fenyö and Stolle (1982, M), Vol. 4.

Projection methods: Krasnoselskii (1973, M).

Eigenvalue problems: Krasnoselskii (1973, M), Osborn (1975), Vainikko (1976, L), (1977), (1979, S).

Variational inequalities: Glowinski, Lions, and Trémolières (1976, M).

Applications to mathematical physics: Glowinski, Lions, and Trémolieres (1976, M), Ciarlet (1977, M), Temam (1977, M), Glowinski (1984, M), Girault and Raviart (1986, M).

The functional analysis of discretization methods: Vainikko (1976, L, H), (1978, S), (1979, S).

Numerical range: Bonsall and Duncan (1971, M) (recommended as an introduction), Stone (1932a, M), Zarantonello (1964), Rhodius (1977).

CHAPTER 35

External Approximation Schemes, A-Proper Operators, and the Difference Method

The most practical solution is a good theory.

Albert Einstein

The devil rides high on detail.

In order to elucidate the basic ideas, we consider the boundary value problem for a quasi-linear elliptic differential equation of order $2m$:

$$\begin{gathered}\sum_{|\alpha|\leq m}(-1)^{|\alpha|}D^{\alpha}A_{\alpha}(x, Du(x)) = f(x) \quad \text{on } G,\\ D^{\beta}u(x) = 0 \quad \text{on } \partial G \qquad \text{for all} \quad \beta\colon |\beta| \leq m-1.\end{gathered} \tag{1}$$

The corresponding *difference method* reads as follows:

$$\begin{gathered}\sum_{|\alpha|\leq m}(-1)^{|\alpha|}\nabla_{-}^{\alpha}A_{\alpha}(P, \nabla_{+}u(P)) = \bar{f}_h(P) \qquad \text{for all} \quad P \in \mathcal{G}_{h,m},\\ u_h(P) = 0 \qquad \text{for all} \quad P \notin \mathcal{G}_{h,m}.\end{gathered} \tag{2}$$

In contrast to (1), the partial derivatives D^{α} are replaced by difference quotients $\nabla_{\pm}^{\alpha}$.

Here, G denotes a bounded region in $\mathbb{R}^N$. In connection with the difference method, we consider a lattice having grid width h. The precise definition of $\mathcal{G}_{m,h}$ will be given in Section 35.4. Roughly speaking, $\mathcal{G}_{m,h}$ consists of all *interior* lattice points of G that are at a distance greater than or equal to mh from the boundary lattice points corresponding to the boundary ∂G of G. Thus it immediately follows from

$$u_h(P) = 0 \qquad \text{for all} \quad P \notin \mathcal{G}_{m,h}$$

that

$$\nabla_{+}^{\beta}u_h(P) = 0 \qquad \text{for all boundary lattice points and for all} \quad \beta\colon |\beta| \leq m-1.$$

This condition corresponds to the boundary condition in (1).

One obtains $\bar{f}_h$ by using an integral mean value.

The appearance of the distinguishable difference quotients ∇_+ and ∇_- is explained in a natural way if one goes over to the generalized statement of the problem.

(A) *The generalized differential equation problem.* Let $X = \mathring{W}_p^m(G)$, i.e., X is a Sobolev space. We seek a function $u \in X$ such that

$$\int_G \sum_{|\alpha| \le m} A_\alpha(x, Du(x)) D^\alpha v(x)\, dx = \int_G f(x) v(x)\, dx \qquad \text{for all} \quad v \in X. \quad (1^*)$$

(B) *The generalized difference equation problem.* Let $X_h = \mathring{\mathscr{W}}_p^m(\mathscr{G}_h)$, i.e., X_h is a so-called discrete Sobolev space. We seek a lattice function $u_h \in X_h$ such that

$$\int \sum_{|\alpha| \le m} A_\alpha(P, \nabla u_h(P)) \nabla^\alpha v_h(P)\, dx_h = \int \bar{f}_h(P) v_h(P)\, dx_h \qquad \text{for all} \quad v_h \in X_h. \quad (2^*)$$

We abbreviate ∇_+ to ∇. Here, $\int \ldots dx_h$ means discrete integration, i.e., one sums over all lattice points P and $dx_h = h^N$, where N is the dimension of G.

One notes here that (2*) follows from (1*) if one systematically makes the following very natural substitutions:

(i) The region G goes over into the set of lattice points.
(ii) The functions u on G are replaced by the lattice functions u_h that are defined in the lattice points.
(iii) Differential quotients D^α are replaced by difference quotients ∇^α.
(iv) The integrals $\int_G \ldots dx$ go over into the discrete integrals $\int \ldots dx_h$.
(v) f is replaced by appropriate mean values $\bar{f}_h$.
(vi) Sobolev spaces go over into discrete Sobolev spaces.

As we shall see in Section 35.5, problem (2) follows from (2*) by discrete integration by parts. For that reason, we need both ∇_- and ∇_+ in (2).

In Section 26.5 we have proved that, under suitable assumptions, there exists exactly one solution of (1*). The problem arises of showing the *convergence* of the corresponding difference method.

We solve the problem of convergence in this chapter by proving, parallel to the preceding chapter, an abstract main theorem for external approximation schemes (Theorem 35.A) and then applying this theorem to the difference method (2*). In this connection, it is important that the differential equation problem (1*) leads to an operator equation

$$Au = b, \qquad u \in X,$$

where the operator $A: X \to X^*$ is uniformly monotone.

Another possibility for an approximate solution of the differential equation (1*) consists in using a Galerkin method with finite elements as basis functions. For a suitable choice of the finite elements according to $A_2(60)$, the Galerkin equations have the form (2*) in special cases. However, in the general case, the simple difference method (2*) does not arise here.

The main ingredients of our approach in this chapter are the synchronization condition (7) below and the discrete convergence $u_n \xrightarrow{d} u$ as well as the discrete* convergence $u_n^* \xrightarrow{d^*} u^*$ of functionals.

35.1. External Approximation Schemes

Together with the initial problem

$$Au = b, \qquad u \in X, \tag{3}$$

we consider the discretized problem

$$A_n u_n = b_n, \qquad u_n \in X_n, \qquad n = 1, 2, \ldots, \tag{4}$$

with the corresponding approximation scheme:

$$\begin{array}{ccccc} F & \xleftarrow{\ \omega\ } & X & \xrightarrow{\ A\ } & X^* \\ & \nwarrow^{E_n} & \downarrow R_n & & \\ & & X_n & \xrightarrow{\ A_n\ } & X_n^*. \end{array} \tag{5}$$

The operator R_n: $X \to X_n$ is called a *restriction* operator and E_n: $X_n \to F$ is called an *extension* operator. We call ω a *synchronization* operator. The elements of F that lie in $\omega(X)$ are said to be synchronized. The so-called *compatibility condition* plays an essential role:

$$E_n R_n u \to \omega(u) \quad \text{in } F \qquad \text{as} \quad n \to \infty \qquad \text{and for all} \quad u \in X, \tag{6}$$

and also the *synchronization condition*:

> The weak limits in F of the sequence $\{E_n u_n\}$ and their subsequences are synchronized, i.e., if
>
> $$E_{n'} u_{n'} \rightharpoonup g \quad \text{in } F \qquad \text{as} \quad n \to \infty, \tag{7}$$
>
> then $g \in \omega(X)$.

EXAMPLE 35.1. We show how to realize the *external approximation scheme* (5) with respect to our concrete problem (1*), (2*). The statements will be made precise in Section 35.6.

Let (h_n) be a sequence of positive numbers with $h_n \to 0$ as $n \to \infty$. Here, h_n denotes the grid width of the lattice. We choose:

(a) X is the Sobolev space $\mathring{W}_p^m(G)$.
(b) X_n is the discrete Sobolev space $\mathring{\mathscr{W}}_p^m(\mathscr{G}_{h_n})$.
(c) We assign to each $u \in X$ the tuple

$$\omega(u) = (D^\alpha u)_{|\alpha| \le m}$$

that consists of u and all partial derivatives of u up to order m. Here,

$$D^\alpha u \in L_p(G) \qquad \text{for all} \quad \alpha\colon |\alpha| \le m.$$

Therefore, $\omega(u)$ lies in the corresponding product space, which we denote by F, i.e., we have $\omega(u) \in F$, where

$$F = \prod_{|\alpha| \leq m} L_p(G).$$

Consequently, the space F consists of all the tuples

$$(u_\alpha)_{|\alpha| \leq m} \qquad \text{with} \quad u_\alpha \in L_p(G) \qquad \text{for all } \alpha.$$

The *synchronized* elements $\omega(u)$ of F have the special property that

$$u_\alpha = D^\alpha u \qquad \text{for all} \quad \alpha\colon |\alpha| \leq m \qquad \text{and fixed} \quad u \in X.$$

(d) By means of an averaging process, the restriction operator R_n assigns to each $u \in X$ a lattice function $u_n \in X_n$ on the set $\mathcal{G}_{h_n}$ of grid points.
(e) The extension operator E_n describes an extension process for lattice functions. In order to explain this, let $u_n \in X_n$ be a lattice function. We first construct the tuple of difference quotients

$$(\nabla^\alpha u_n)_{|\alpha| \leq m}$$

and extend all $\nabla^\alpha u_n$ to $L_p(G)$-functions u_α on G. Then we define

$$E_n u_n = (u_\alpha)_{|\alpha| \leq m}.$$

We use the so-called *external space* F in order to be able to choose this extension process *simply*. It is essential that we do *not* require that $E_n u_n = (u_\alpha)_{|\alpha| \leq m}$ is synchronized.

Let u_n be a solution of the difference equation (2*). The role of the synchronization condition (7) consists in that the *weak limit* of $(E_n u_n)$ is synchronized and yields a solution of the corresponding differential equation (1*). In this connection, we will use the following important fact:

The extended difference quotients converge weakly in $L_p(G)$ to the corresponding generalized derivatives.

This principle is the *key* to our approach.

The approximation scheme described in Example 35.1 turns out, in Section 35.6, to be an admissible external approximation scheme in the sense of the following definition.

Definition 35.2. The approximation scheme (5) is called an *admissible external* approximation scheme iff the following hold:

(i) X, F, X_n are real B-spaces with $\dim X_n < \infty$ for all $n \in \mathbb{N}$, where F is reflexive.
(ii) The operator $\omega\colon X \to F$ is linear, continuous, and injective.
(iii) All the operators $R_n\colon X \to X_n$ and $E_n\colon X_n \to F$ are linear and continuous with $\sup_n \|R_n\| < \infty$ and $\sup_n \|E_n\| < \infty$.
(iv) The compatibility condition (6) and the synchronization condition (7) above hold.

An inner approximation scheme of the form (34.16) arises as a special case when $F = X$ and ω is the identity.

In Theorem 35.A in the following section, we will require that the operators A_n and A of the external scheme (5) are related to each other by a weak consistency condition. However, in contrast to the inner approximation scheme in Chapter 34, we do *not* require that the equation $A_n = E_n^* A E_n$ holds.

To describe the convergence of the difference method we use the following discrete convergence:

$$u_n \xrightarrow{d} u \qquad \text{iff} \qquad \|u_n - R_n u\|_{X_n} \to 0 \qquad \text{as} \quad n \to \infty. \tag{8a}$$

To be precise, we have the following definition.

Definition 35.3. Let (u_n) be a sequence of elements with $u_n \in X_n$ for all $n \in \mathbb{N}$. The sequence (u_n) converges *discretely* to $u \in X$ iff

$$\lim_{n \to \infty} \|u_n - R_n u\|_{X_n} = 0.$$

We write $u_n \xrightarrow{d} u$.

The following convergence concept for functionals is decisive for the proof of convergence for the difference method:

$$u_n^* \xrightarrow{d*} u^* \qquad \text{iff} \qquad \langle u_n^*, u_n \rangle_{X_n} \to \langle u^*, u \rangle_X \qquad \text{as} \quad n \to \infty. \tag{8b}$$

To be more precise, the following definition of discrete* convergence holds.

Definition 35.4. Let (u_n^*) be a sequence of functionals with $u_n^* \in X_n^*$ for all $n \in \mathbb{N}$. The sequence (u_n^*) converges *discretely** to $u^* \in X^*$ iff

$$\lim_{n \to \infty} \langle u_n^*, u_n \rangle_{X_n} = \langle u^*, u \rangle_X$$

holds for all sequences (u_n), $u_n \in X_n$, with $\sup_n \|u_n\|_{X_n} < \infty$ and

$$E_n u_n \rightharpoonup \omega(u) \quad \text{in } F \qquad \text{as} \quad n \to \infty.$$

We write $u_n^* \xrightarrow{d*} u^*$.

35.2. Main Theorem on Stable Discretization Methods with External Approximation Schemes

We investigate the connection between the following conditions.

(C1) *Solvability*. For each $b \in X$, the original equation

$$Au = b, \qquad u \in X,$$

has a solution.

(C2) *Unique approximation-solvability.* The following hold for all $b \in X^*$:
 (i) The equation $Au = b$ has a unique solution $u \in X$.
 (ii) For each $b_n \in X_n^*$ and all $n \geq n_0$, the approximate equation

$$A_n u_n = b_n$$

has a unique solution $u_n \in X_n$.
 (iii) As $n \to \infty$,

$$b_n \overset{d^*}{\to} b \qquad \text{implies} \qquad u_n \overset{d}{\to} u$$

and $E_n u_n \to \omega(u)$ in F.

(C3) *A-properness.* The operator $A: X \to X^*$ is A-proper with respect to the external approximation scheme (5), i.e., by definition the following holds:

$$A_{n'} u_{n'} \overset{d^*}{\to} b$$

and $\sup_{n'} \|u_{n'}\|_{X_{n'}} < \infty$ imply the existence of a subsequence $(u_{n''})$ with

$$u_{n''} \overset{d}{\to} u \qquad \text{and} \qquad Au = b.$$

In this connection (n') denotes an arbitrary subsequence of the sequence of natural numbers and we suppose that $u_{n'} \in X_{n'}$ for all n'.

The following theorem is the main result of this chapter.

Theorem 35.A (Schumann (1979)). *The three conditions* (C1), (C2), *and* (C3) *are mutually equivalent in the case where the following hold*:

(H1) Approximation scheme. *We are given the admissible external approximation scheme* (5). *All the operators A_n in* (5) *are continuous.*

(H2) Weak consistency. *For all $u \in X$,*

$$A_n R_n u \overset{d^*}{\to} Au.$$

(H3) Stability. *For all $u, v \in X_n$ and all $n \geq n_0$,*

$$\|A_n u - A_n v\|_{X_n^*} \geq a(\|u - v\|_{X_n}),$$

where the given function $a: [0, \infty[\to \mathbb{R}$ is strictly monotone increasing and continuous with $a(0) = 0$ and

$$a(t) \to +\infty \qquad \text{as} \quad t \to +\infty.$$

(H4) Approximation of the right-hand term b. *For each $b \in X^*$, there exists a sequence (b_n) with*

$$b_n \overset{d^*}{\to} b,$$

where $b_n \in X_n^*$ for all $n \geq n_0$.

The significance of this theorem for the applications in Section 35.5 consists in that, for the difference method, conditions (H1) through (H4) can easily be verified for quasi-linear elliptic differential equations. In this connection, (H2) results from the convergence theorems for the Lebesque integral. It is important that *no a priori* estimates whatsoever are needed for the difference quotients.

35.3. Proof of the Main Theorem

The following preparatory lemma is proved in Problem 35.1. The synchronization condition is used here in (ii) and (iii).

Lemma 35.5. *The following hold for an admissible external approximation scheme as $n \to \infty$:*

(i) $u_n \xrightarrow{d} u$ *implies* $E_n u_n \to \omega(u)$ *in* F.
(ii) $u_n^* \xrightarrow{d^*} u^*$ *implies* $\sup_n \|u_n^*\|_{X_n^*} < \infty$.
(iii) $u_n^* \xrightarrow{d^*} 0$ *implies* $\|u_n^*\|_{X_n^*} \to 0$.
(iv) $E_n R_n u \to \omega(u)$ *for all* $u \in U$ *implies* $E_n R_n u \to \omega(u)$ *for all* $u \in X$ *provided* U *is dense in* X.

We now give a proof of Theorem 35.A that is parallel to that of Theorem 34.A.

Step 1: (C3) implies (C2).

(I) For fixed $b_n \in X_n^*$, equation $A_n u_n = b_n$ has exactly one solution $u_n \in X_n$ for all $n \geq n_0$. This follows analogously to Section 34.3.

(II) We show that, for fixed $b \in X^*$, equation $Au = b$ has at most one solution $u \in X$.

Let $Au = Av$. It follows from (H3) that

$$a(\|R_n u - R_n v\|) \leq \|A_n R_n u - A_n R_n v\| \qquad \text{for all } n. \tag{9}$$

By (H2),

$$A_n R_n u - A_n R_n v \xrightarrow{d^*} 0.$$

Lemma 35.5(iii) yields $\|A_n R_n u - A_n R_n v\| \to 0$. Therefore, by (9),

$$\|R_n u - R_n v\| \to 0 \qquad \text{as} \quad n \to \infty.$$

From this it follows that

$$\|E_n R_n u - E_n R_n v\| \leq (\sup_n \|E_n\|) \|R_n u - R_n v\| \to 0 \qquad \text{as} \quad n \to \infty.$$

The compatibility condition (6) yields $\omega(u - v) = 0$; therefore, $u = v$.

(III) We show that, for each $b \in X^*$, the equation $Au = b$ has exactly one solution $u \in X$. To this end, we choose a sequence $b_n \xrightarrow{d^*} b$ according to (H4). Furthermore, we choose u_n with $A_n u_n = b_n$ according to (I).

By (H2), it follows from $A_n R_n(0) \xrightarrow{d^*} A(0)$ and Lemma 35.5(ii) that $\sup_n \|A_n R_n(0)\| < \infty$. Because $R_n(0) = 0$, the following holds for all n:

$$\begin{aligned} \|b_n\| = \|A_n u_n\| &\geq \|A_n u_n - A_n(0)\| - \|A_n R_n(0)\| \\ &\geq a(\|u_n\|) - \|A_n R_n(0)\|. \end{aligned}$$

Therefore, $\sup_n a(\|u_n\|) < \infty$ and hence $\sup_n \|u_n\| < \infty$. The A-properness of the operator A ensures the existence of a subsequence $(u_{n'})$ with

$$u_{n'} \xrightarrow{d} u \qquad \text{and} \qquad Au = b.$$

(IV) We show that $b_n \xrightarrow{d^*} b$ and $A_n u_n = b_n$ imply $u_n \xrightarrow{d} u$ and $E_n u_n \to \omega(u)$ in F.

By (III), we obtain: Each subsequence $(u_{n'})$ of (u_n) has another subsequence $(u_{n''})$ with $u_{n''} \xrightarrow{d} u$ and $Au = b$. The limit element u is the same for all subsequences since $Au = b$ has exactly one solution u. This implies the convergence of the entire sequence, i.e., we get $u_n \xrightarrow{d} u$. Indeed, this follows by using an analogous argument as in the proof of the convergence principle (Proposition 10.13(1)). By Lemma 35.5(i), $u_n \xrightarrow{d} u$ implies $E_n u_n \to \omega(u)$.

Step 2: (C2) implies (C1): This is trivial.

Step 3: (C1) implies (C3).

For brevity, we write n instead of n'. Let

$$A_n u_n \xrightarrow{d^*} b$$

with $\sup_n \|u_n\| < \infty$. We choose a point u with $Au = b$ according to (C1). We want to show that

$$u_n \xrightarrow{d} u.$$

Then the operator A is A-proper with respect to the approximation scheme (5), i.e., condition (C3) holds. Indeed, by (H2),

$$A_n R_n u \xrightarrow{d^*} Au.$$

This implies $A_n u_n - A_n R_n u \xrightarrow{d^*} 0$. Lemma 35.5(iii) yields

$$\|A_n u_n - A_n R_n u\| \to 0 \qquad \text{as} \quad n \to \infty.$$

By (H3), $a(\|u_n - R_n u\|) \leq \|A_n u_n - A_n R_n u\| \to 0$ as $n \to \infty$. This implies $\|u_n - R_n u\| \to 0$, i.e., $u_n \xrightarrow{d} u$.

The proof of Theorem 35.A is complete. □

35.4. Discrete Sobolev Spaces

We now make ready the applications of Theorem 35.A to partial differential equations.

To this end, we choose a cube lattice in $\mathbb{R}^N$ with the grid mesh h. The coordinates of the lattice points P are integer multiples of h. To each lattice point P we assign a cube $c_h(P)$ with edges parallel to the axes having edge length h and P as midpoint. In this connection, we choose $c_h(P)$ to be, say, half-open and in fact so that the union of all cubes yields a disjoint decomposition of $\mathbb{R}^N$. Let G be a bounded region in $\mathbb{R}^N$ with sufficiently smooth boundary, i.e., $\partial G \in C^{0,1}$.

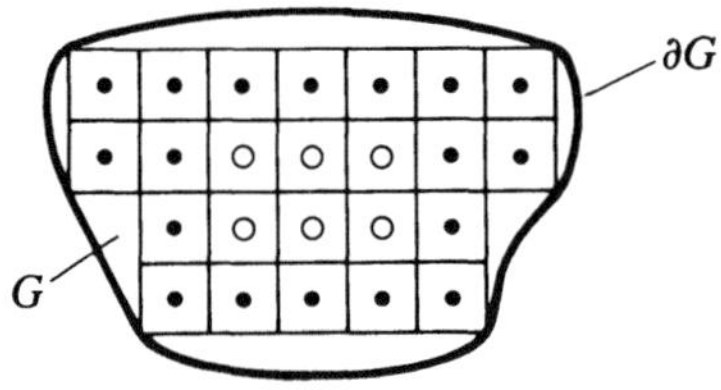

Figure 35.1

Definition 35.6. We understand $\mathscr{G}_h$ to be the set of all lattice points P with $\overline{c_h(P)} \subseteq \bar{G}$. These lattice points are called *interior lattice points* of G. The cubes belonging to $\mathscr{G}_h$ approximate G from the interior.

By the set of *boundary lattice points* $\partial\mathscr{G}_h$ we understand all lattice points of $\mathscr{G}_h$ that belong to the boundary cubes. These are in a natural way the cubes whose closure does not lie entirely in the interior of the cube set belonging to $\mathscr{G}_h$.

For $m = 1, 2, \ldots$, by $\mathscr{G}_{h,m}$ we understand the set of all lattice points of $\mathscr{G}_h$ that have distance from the boundary lattice points greater than or equal to mh.

In Figure 35.1, the points and the small circles mean the lattice points belonging to $\mathscr{G}_h$ while the points mark the boundary lattice points of $\partial\mathscr{G}_h$. Moreover, the small circles mark the lattice points of $\mathscr{G}_{h,1}$.

By a *lattice function* u_h we understand a function that assigns a real number to each lattice point of $\mathbb{R}^N$.

Definition 35.7. We define the difference quotient $\nabla_i^\pm u_h(P)$ to be

$$\nabla_i^\pm u_h(P) = \frac{u_h(P \pm he_i) - u_h(P)}{\pm h}. \tag{10}$$

In this connection, e_i is the unit vector in the direction of the ith coordinate. If P has the coordinates $(hg_1, \ldots, hg_N)$ where the $g_1, \ldots, g_N$ are integers, then there results the point $P \pm he_i$ if one replaces g_i by $g_i \pm 1$.

The difference operator $\nabla_i^\pm$ corresponds to the differential operator $D_i = \partial/\partial\xi_i$. Analogous to the differential operator

$$D^\alpha = D_1^{\alpha_1} \cdots D_N^{\alpha_N},$$

we define the difference operator

$$\nabla_\pm^\alpha = (\nabla_1^\pm)^{\alpha_1} \cdots (\nabla_N^\pm)^{\alpha_N}$$

for $\alpha = (\alpha_1, \ldots, \alpha_N)$. For brevity, we agree to write

$$\nabla_i = \nabla_i^+, \qquad \nabla^\alpha = \nabla_+^\alpha.$$

We now define discrete Lebesgue spaces and Sobolev spaces parallel to $L_p(G)$ and $\mathring{W}_p^m(G)$, respectively. In this connection, we make use of the discrete integral

$$\int u_h\,dx_h \overset{\text{def}}{=} \sum_P u_h(P)h^N.$$

Here, the summation is over all lattice points P of $\mathbb{R}^N$. The sum always exists since the lattice functions to be considered are different from zero only in finitely many lattice points.

Definition 35.8. Let $1 \leq p < \infty$, $m = 1, 2, \ldots$. By the *discrete Lebesgue space* $\mathscr{L}_p(\mathscr{G}_h)$ we understand the set of all lattice functions that vanish identically outside $\mathscr{G}_h$. We choose the norm to be

$$|u_h|_p = \left(\int |u_h|^p\,dx_h\right)^{1/p}.$$

By the *discrete Sobolev space* $\mathring{\mathscr{W}}_p^m(\mathscr{G}_h)$ we understand the set of all lattice functions u_h that vanish identically outside $\mathscr{G}_{m,h}$. We choose the norm to be

$$|u_h|_{m,p} = \left(\int \sum_{|\alpha| \leq m} |\nabla^\alpha u_h|^p\,dx_h\right)^{1/p}.$$

Moreover, we define

$$|u_h|_{m,p,0} = \left(\int \sum_{|\alpha| = m} |\nabla^\alpha u_h|^p\,dx_h\right)^{1/p}.$$

Proposition 35.9 *With the given norms, $\mathscr{L}_p(\mathscr{G}_h)$ and $\mathring{\mathscr{W}}_p^m(\mathscr{G}_h)$ are real* B-*spaces. The norms $|\cdot|_{m,p}$ and $|\cdot|_{m,p,0}$ are equivalent on $\mathring{\mathscr{W}}_p^m(\mathscr{G}_h)$.*

We deal with the proof in Problem 35.2. Note that the norms are defined in a way completely parallel to the norms on $L_p(G)$ and $\mathring{W}_p^m(G)$. Parallel to integration by parts there is also a formula for *discrete integration by parts*:

$$\int u\nabla_\pm^\alpha v\,dx_h = (-1)^{|\alpha|}\int (\nabla_\mp^\alpha u)v\,dx_h. \tag{11}$$

Proposition 35.10. *Formula* (11) *holds for all $u, v \in \mathring{\mathscr{W}}_p^m(\mathscr{G}_h)$, where $|\alpha| \leq m$.*

We recommend that the reader do the proof as a simple exercise. In conclusion, we mention an extension method that will play an important role in the next section.

Definition 35.11. Let u_h be a lattice function. By a lattice function *extended* to the region G, which we shall also designate by u_h, we mean:

$$u_h(x) = \begin{cases} u_h(P) & \text{for} \quad x \in c_h(P), \quad P \in \mathscr{G}_h, \\ 0 & \text{otherwise.} \end{cases}$$

This way u_h is extended in a natural way as a constant to each cube that belongs to an interior lattice point, i.e., u_h is piecewise constant. Obviously,

$$\int_G u_h \, dx = \int u_h \, dx_h.$$

In the following, the integral appearing on the left-hand side of this equation is always to be understood in this sense.

35.5. Application to Difference Methods

In order not to obscure the simple basic idea by purely technical details, we consider only a simple example. We treat the general situation of quasi-linear elliptic differential equations of order $2m$ in Problem 35.4.

We investigate the boundary value problem

$$\begin{gathered} -\sum_{i=1}^{N} D_i(|D_i u|^{p-2} D_i u) + su = f \quad \text{on } G, \\ u = 0 \quad \text{on } \partial G, \end{gathered} \tag{12}$$

with the corresponding difference equations

$$\begin{gathered} -\sum_{i=1}^{N} \nabla_i^-(|\nabla_i u_h(P)|^{p-2} \nabla_i u_h(P)) + su_h(P) = \bar{f}_h(P) \qquad \text{for all} \quad P \in \mathcal{G}_h, \\ u_h(P) = 0 \qquad \text{for all} \quad P \in \partial\mathcal{G}_h. \end{gathered} \tag{13}$$

We make the following assumptions.

(A) G is a bounded region in $\mathbb{R}^N$, $N \geq 1$, with sufficiently smooth boundary, i.e., $\partial G \in C^{0,1}$, and s is a nonnegative real number. We choose a sufficiently small positive number h_0 so that the set $\mathcal{G}_h$ of interior lattice points is not empty for all h, $0 < h \leq h_0$. We understand $\bar{f}_h(P)$ to be the integral mean value of f over the cube $c_h(P)$ belonging to P, i.e.,

$$\bar{f}_h(P) = h^{-N} \int_{c_h(P)} f(x) \, dx. \tag{14}$$

Definition 35.12. Let $X = \mathring{W}_p^1(G)$, $X_h = \mathring{\mathscr{W}}_p^1(\mathcal{G}_h)$, $2 \leq p < \infty$, $p^{-1} + q^{-1} = 1$.

The *generalized problem* for (12) reads as follows: For a given function $f \in L_q(G)$, we seek a function $u \in X$ such that

$$a(u, v) = b(v) \qquad \text{for all} \quad v \in X, \tag{12*}$$

with

$$a(u, v) = \int_G \left(\sum_{i=1}^{N} |D_i u|^{p-2} D_i u D_i v + suv \right) dx,$$

$$b(v) = \int_G fv \, dx.$$

The generalized problem for (13) reads as follows: We seek a lattice function $u_h \in X_h$ such that

$$a_h(u_h, v_h) = b_h(v_h) \qquad \text{for all} \quad v_h \in X_h, \tag{13*}$$

with

$$a_h(u_h, v_h) = \int \left(\sum_{i=1}^{N} |\nabla_i u_h|^{p-2} \nabla_i u_h \nabla_i v_h + s u_h v_h \right) dx_h,$$

$$b_h(v_h) = \int \bar{f}_h v_h \, dx_h.$$

Proposition 35.13. *Under the assumption* (A) *above, the following three assertions are valid*:

(a) Differential equation. *The generalized problem for* (12) *has exactly one solution* u.
(b) Difference equation. *The generalized problem for* (13) *has exactly one solution* u_h *for all* h, $0 < h \leq h_0$.
(c) Convergence. *As* $h \to +0$, *the sequence* (u_h) *converges to* u *in the following sense*:

$$\int_G \left(\sum_{i=1}^{N} |D_i u - \nabla_i u_h|^p + |u - u_h|^p \right) dx \to 0, \tag{15}$$

$$\sum_{P \in \mathscr{G}_{h,1}} \left(\sum_{i=1}^{N} |\nabla_i \bar{u} - \nabla_i u_h|^p + |\bar{u} - u_h|^p \right) h^N \to 0. \tag{16}$$

In this connection, (15) and (16) describe the convergence of the difference method on the region G and on the lattice, respectively. In (15), u_h and $\nabla_i u_h$ denote the lattice functions extended to G in the sense of Definition 35.11. In (16), for brevity, we write u_h and $\bar{u}$ instead of $u_h(P)$ and $\bar{u}(P)$, where $\bar{u}(P)$ is the mean value of u at the point P in the sense of (14). Moreover, $\nabla_i \bar{u}$ and $\nabla_i u_h$ stand for the difference quotients $\nabla_i \bar{u}(P)$ and $\nabla_i u_h(P)$ of the lattice functions $P \mapsto \bar{u}(P)$ and $P \mapsto u_h(P)$, respectively.

Corollary 35.14 (Equivalence). *The difference equation* (13) *and the corresponding generalized problem* (13*) *are mutually equivalent.*

Proof of Corollary 35.14. Multiplication of (13) by v_h, discrete integration, and discrete integration by parts yield (13*). Furthermore, (13) follows from (13*) by a reversal of this procedure. □

Therefore, instead of (13*), we need to solve only the nonlinear system of equations (13). This happens with the aid of the following iteration method for $k = 0, 1, \ldots$:

$$\begin{aligned} u_h^{(k+1)}(P) &= u_h^{(k)}(P) - t g_P(u_h^{(k)}) \qquad \text{for} \quad P \in \mathscr{G}_{h,1}, \\ u_h^{(k+1)}(P) &= 0 \qquad \text{for} \quad P \in \partial \mathscr{G}_h, \end{aligned} \tag{17}$$

with the starting elements $u_h^{(0)}(P) \equiv 0$, where

$$g_P(u_h) = -\sum_{i=1}^{N} \nabla_i^-(|\nabla_i u_h(P)|^{p-2}\nabla_i u_h(P)) + su_h(P) - \bar{f}_h(P).$$

Corollary 35.15. *Let* $0 < h \le h_0$ *and* $s > 0$. *Then, for sufficiently small* $t > 0$, *the iteration method* (17) *converges as* $k \to \infty$ *to the solution* u_h *of the difference equation* (13).

By Theorem 26.B, one can specify explicitly the domain of t values. The proof will be given in Problem 35.7.

35.6. Proof of Convergence

We want to prove Proposition 35.13. To this end, we apply Theorem 35.A and simply verify the hypotheses (H1)–(H4) and (C1). Then the statement of Proposition 35.13 is precisely (C2) in Theorem 35.A.

Step 1: Condition (H1).

We construct the following approximation scheme:

$$\begin{array}{ccccc} F & \xleftarrow{\ \omega\ } & X & \xrightarrow{\ A\ } & X^* \\ & {\scriptstyle E_n}\nwarrow & \downarrow {\scriptstyle R_n} & & \\ & & X_n & \xrightarrow{\ A_n\ } & X_n. \end{array} \tag{18}$$

To this end, let (h_n) be a sequence of positive numbers with $h_n \to 0$ as $n \to \infty$ and $0 < h_n \le h_0$ for all $n \in \mathbb{N}$. We set

$$X = \mathring{W}_p^1(G), \qquad X_n = \mathring{\mathscr{W}}_p^1(\mathscr{G}_{h_n}),$$

$$F = \prod_{i=1}^{N} L_p(G), \qquad 2 \le p < \infty.$$

Moreover, we set $u_n = u_{h_n}$. We equip X_n with the norm $\|\cdot\| = |\cdot|_{1,p,0}$.

The *synchronization* operator ω: $X \to F$ is defined by

$$\omega(u) = (u, D_1 u, \ldots, D_N u).$$

The *restriction* operator R_n: $X \to X_n$ results from forming the mean value, that is, we set $k = h_n$ and we define

$$(R_n u)(P) = \begin{cases} k^{-N} \displaystyle\int_{c_k(P)} u(x)\,dx & \text{for} \quad P \in \mathscr{G}_{k,1}, \\ 0 & \text{for} \quad P \notin \mathscr{G}_{k,1}. \end{cases}$$

Furthermore, we define the *extension* operator E_n: $X_n \to F$ by

$$E_n u_n = (u_n, \nabla_1 u_n, \ldots, \nabla_N u_n),$$

where the lattice functions extended to the region G occur on the right (cf. Definition 35.11). This way we obtain $E_n u_n \in F$.

Lemma 35.16. *The diagram* (18) *represents an admissible external approximation scheme.*

The proof which uses standard arguments on discrete Sobolev spaces can be found in Schumann and Zeidler (1979). In this connection, by Lemma 35.5(iv), one needs to verify the compatibility condition

$$E_n R_n u \to \omega(u) \quad \text{in } F \qquad \text{as} \quad n \to \infty$$

only for all $u \in C_0^\infty(G)$ since $C_0^\infty(G)$ is dense in X.

The synchronization condition is obtained as follows. Let

$$E_n u_n \rightharpoonup g \quad \text{in } F \qquad \text{as} \quad n \to \infty$$

with $g = (u, U_1, \ldots, U_N)$. From this it follows that

$$u_n \rightharpoonup u \quad \text{in } L_p(G) \qquad \text{as} \quad n \to \infty,$$

$$\nabla_i u_n \rightharpoonup U_i \quad \text{in } L_p(G) \qquad \text{as} \quad n \to \infty.$$

It is a known fact that the extended difference quotients converge weakly to the generalized derivatives (cf. Temam (1970, M), p. 69). Hence,

$$U_i = D_i u$$

and therefore, $g = \omega(u)$.

Step 2: The operator A and the condition (C1).

By Proposition 26.10, there exists an operator $A: X \to X^*$ with

$$\langle Au, v \rangle = a(u, v) \qquad \text{for all} \quad u, v \in X.$$

The generalized problem (12*) is equivalent to

$$Au = b, \qquad u \in X. \tag{19}$$

Note that $f \in L_q(G)$ with $q^{-1} + p^{-1} = 1$ implies $b \in X^*$. Furthermore, by Proposition 26.10, equation (19) has exactly one solution $u \in X$. This is (C1).

In addition, it follows from Proposition 26.10 that

$$a(u, u - v) - a(v, u - v) \geq c\|u - v\|_{1,p,0}^p + s \int_G (u - v)^2 \, dx, \tag{20}$$

for all $u, v \in X$ and fixed $c > 0$.

Step 3: The operator A_n and the condition (H3).

Analogous to the proof (II) of Proposition 26.10, there results

$$a_{h_n}(u, u - v) - a_{h_n}(v, u - v)$$

$$\geq c|u - v|_{1,p,0}^p + s \int (u - v)^2 \, dx_{h_n} \qquad \text{for all} \quad u, v \in X_n. \tag{21}$$

In this connection, replace the derivative D_i by the difference quotient ∇_i and the integrals by discrete integrals. For $u \in X_n$, the mapping

$$v \mapsto a_{h_n}(u, v)$$

is a linear functional on the space X_n and, because $\dim X_n < \infty$, it is also continuous. Therefore, there exists an operator $A_n: X_n \to X_n^*$ with

$$\langle A_n u, v \rangle = a_{h_n}(u, v) \qquad \text{for all} \quad u, v \in X_n.$$

This implies

$$\|A_n u - A_n v\| \, \|u - v\| \geq |\langle A_n u - A_n v, u - v \rangle| \geq c\|u - v\|^p$$

and hence

$$\|A_n u - A_n v\| \geq c\|u - v\|^{p-1} \qquad \text{for all} \quad u, v \in X_n.$$

This is precisely the stability condition (H3).

The generalized discrete problem (13*) is equivalent to the operator equation

$$A_n u_n = b_n, \qquad u_n \in X_n, \tag{22}$$

where we write b_n for b_{h_n}.

Step 4: The weak consistency condition (H2).

We must show that

$$A_n R_n u \overset{d^*}{\to} Au \qquad \text{for all} \quad u \in X.$$

By definition, this is equivalent to

$$a_{h_n}(R_n u, v_n) \to a(u, v) \qquad \text{as} \quad n \to \infty, \tag{23}$$

for all sequences (v_n) with the property $v_n \in X_n$ for all $n \in \mathbb{N}$ as well as $\sup_n \|v_n\| < \infty$ and

$$E_n v_n \rightharpoonup \omega(v) \quad \text{in } F \qquad \text{as} \quad n \to \infty. \tag{24}$$

Explicitly, relation (23) means that, as $n \to \infty$,

$$\int_G \left(\sum_i |\nabla_i \bar{u}_n|^{p-2} \nabla_i \bar{u}_n \nabla_i v_n + s\bar{u}_n v_n \right) dx \to \int_G \left(\sum_i |D_i u|^{p-2} D_i u D_i v + suv \right) dx, \tag{25}$$

where $\bar{u}_n$ denotes the lattice function arising from the function $u \in X$, for all grid points $P \in \mathscr{G}_{h_n,1}$, by forming the mean values according to (14). More precisely, in relation (25), the symbols $\bar{u}_n, \nabla_i \bar{u}_n, v_n, \nabla_i v_n$ denote the corresponding extended lattice functions in the sense of Definition 35.11. Recall that $u_n = u_{h_n}$.

Let us prove (25). From

$$E_n v_n \rightharpoonup \omega(v) \qquad \text{as} \quad n \to \infty$$

it follows that, as $n \to \infty$,

$$v_n \rightharpoonup v \qquad \text{and} \qquad \nabla_i v_n \rightharpoonup D_i v \quad \text{in } L_p(G). \tag{25a}$$

The compatibility condition $E_n R_n u \to \omega(u)$ means that, as $n \to \infty$,

$$\bar{u}_n \to u \quad \text{and} \quad \nabla_i \bar{u}_n \to D_i u \quad \text{in } L_p(G). \tag{25b}$$

Since $q \le p$, the embedding $L_p(G) \subseteq L_q(G)$ is continuous. This implies that, as $n \to \infty$,

$$\bar{u}_n \to u \quad \text{and} \quad \nabla_i \bar{u}_n \to D_i u \quad \text{in } L_q(G). \tag{25c}$$

By Proposition 26.6, the Nemyckii operator $v \mapsto |v|^{p-2} v$ is continuous from $L_p(G)$ to $L_q(G)$, since $p/q = p - 1$. Thus, it follows from (25b) that, as $n \to \infty$,

$$|\nabla_i \bar{u}_n|^{p-2} \bar{u}_n \to |D_i u|^{p-2} D_i u \quad \text{in } L_q(G). \tag{25d}$$

From (25a, c, d) we obtain (25), by $A_2(35)$.

Step 5: Condition (H4).

We have to show that $b_n \overset{d^*}{\to} b$. Recall that

$$b(v) = \int_G fv \, dx \qquad \text{for all} \quad v \in X.$$

Since we write b_n for b_{h_n}, we have

$$b_n(v_n) = \int_G \bar{f}_{h_n} v_n \, dx \qquad \text{for all} \quad v_n \in X_n, \tag{26}$$

where $\bar{f}_h$ denotes the mean value in the sense of (14).

We need the following lemma.

Lemma 35.17. *If $f \in L_q(G)$, then $\bar{f}_{h_n} \to f$ in $L_q(G)$ as $n \to \infty$.*

We leave the proof to the reader as an exercise.

Now let (v_n) be a sequence with $v_n \in X_n$ for all $n \in \mathbb{N}$ such that $\sup_n \|v_n\| < \infty$ and

$$E_n v_n \rightharpoonup \omega(v) \quad \text{in } F \qquad \text{as} \quad n \to \infty.$$

This implies

$$v_n \rightharpoonup v \quad \text{in } L_p(G) \qquad \text{as} \quad n \to \infty,$$

and hence

$$b_n(v_n) \to b(v) \qquad \text{as} \quad n \to \infty,$$

by Lemma 35.17 and $A_2(35)$. This means that $b_n \overset{d^*}{\to} b$.

Step 6: Proof of Proposition 35.13.

By Theorem 35.A, the condition (C2) formulated there is identical to Proposition 35.13. In particular, the statements (15) and (16) follow from

$$E_n u_n \to \omega(u) \quad \text{in } F \qquad \text{as} \quad n \to \infty$$

and $u_n \overset{d}{\to} u$, i.e.,

$$\|u_n - R_n u\|_{X_n} \to 0 \qquad \text{as} \quad n \to \infty.$$

In fact, this implies (15) and $|u_n - \bar{u}|_{1,p,0} \to 0$ as $n \to \infty$, and hence we get (16).

In this connection, note that $\bar{u} \to u$ in $L_p(G)$ as $n \to \infty$ by Lemma 35.17, and hence $u_n \to \bar{u}$ in $L_p(G)$ as $n \to \infty$ by (15), i.e., $|u_n - \bar{u}|_p \to 0$ as $n \to \infty$.

The proof of Proposition 35.13 is complete. □

Problems

35.1. *Proof of Lemma* 35.5.

Solution: Ad(i). By (6) and (8a), it follows from $u_n \xrightarrow{d} u$ that, as $n \to \infty$,

$$\|E_n u_n - \omega(u)\| \le \|E_n u_n - E_n R_n u\| + \|E_n R_n u - \omega(u)\|$$
$$\le (\sup_n \|E_n\|)\|u_n - R_n u\| + \|E_n R_n u - \omega(u)\| \to 0.$$

Ad(ii). Let $u_n^* \xrightarrow{d*} u^*$. If $\sup_n \|u_n^*\| < \infty$ does not hold, then there is a subsequence, again denoted by (u_n^*), with

$$\|u_n^*\| > n \qquad \text{for all } n.$$

Because

$$\|u_n^*\| = \sup\{\langle u_n^*, u_n\rangle : \|u_n\| = 1\}, \tag{27}$$

there is a subsequence, again denoted by (u_n), such that

$$\|u_n\| = 1 \qquad \text{and} \qquad \langle u_n^*, u_n\rangle > n \qquad \text{for all } n. \tag{28}$$

Because $\sup_n \|E_n\| < \infty$, we have $\sup_n \|E_n u_n\| < \infty$. Since the B-space F is reflexive, perhaps after going over to a subsequence, we get

$$E_n u_n \rightharpoonup g \quad \text{in } F \qquad \text{as} \quad n \to \infty.$$

The synchronization condition (7) yields $g = \omega(u)$. Thus, $u_n^* \xrightarrow{d*} u^*$ leads to

$$\langle u_n^*, u_n\rangle \to \langle u^*, u\rangle \qquad \text{as} \quad n \to \infty.$$

This contradicts (28).

Ad(iii). Let $u_n^* \xrightarrow{d*} 0$. By (27), there exists a sequence (u_n) with $\|u_n\| = 1$ and

$$|\,\|u_n^*\| - \langle u_n^*, u_n\rangle| < 1/n \qquad \text{for all } n.$$

As in (ii), there exists a subsequence, again denoted by (u_n), such that

$$E_n u_n \rightharpoonup \omega(u) \quad \text{in } F \qquad \text{as} \quad n \to \infty.$$

Since $u_n^* \xrightarrow{d*} 0$, we obtain

$$\langle u_n^*, u_n\rangle \to 0 \qquad \text{as} \quad n \to \infty,$$

and hence $\|u_n^*\| \to 0$ as $n \to \infty$.

In fact, we have shown that every subsequence of $(\|u_n^*\|)$ possesses a subsequence which goes to zero. From this it follows that the entire sequence $(\|u_n^*\|)$ goes to zero according to the convergence principle (Proposition 10.13).

Ad(iv). Let $E_n R_n u \to \omega(u)$ as $n \to \infty$ for all $u \in U$, where U is dense in the B-space X. We want to show that, for all $v \in X$,

$$E_n R_n v \to \omega(v) \qquad \text{as} \quad n \to \infty.$$

To this end, we use an approximation argument. Let $v \in X$ and $\varepsilon > 0$ be fixed.

Then we obtain

$$\|E_nR_nv - \omega(v)\| \le \|E_nR_nv - E_nR_nu\| + \|E_nR_nu - \omega(u)\| + \|\omega(u) - \omega(v)\|$$
$$\le (\sup_n\|E_n\|\,\|R_n\| + \|\omega\|)\|v - u\| + \|E_nR_nu - \omega(u)\| < \varepsilon,$$

for all $n \ge n_0(\varepsilon)$, in the case where we choose $u \in U$ so that $\|u - v\|$ is sufficiently small.

35.2. *Proof of Proposition* 35.9.
Solution: Let $X_h = \mathring{W}_p^m(\mathscr{G}_h)$. Since X_h is a finite-dimensional space, and all norms are equivalent on such spaces, it suffices to verify that $|\cdot|_{m,p}$ and $|\cdot|_{m,p,0}$ are norms on X_h.
We note that $|u_h|_{m,p} = 0$ automatically implies $u_h \equiv 0$. Moreover, $|u_h|_{m,p,0} = 0$ yields that the mth order difference quotients vanish identically. However, since u_h vanishes at the lattice points of a sufficiently large boundary strip, all lower difference quotients also equal zero at the boundary lattice points. From this it follows that $u_h \equiv 0$.

35.3. *Proof of Proposition* 35.10.
Hint: Cf. Ladyženskaja (1985, M), p. 221.

35.4.* *Difference methods for quasi-linear elliptic differential equations of order* 2*m*.
Generalize Proposition 35.13 to this situation.
Hint: Cf. Schumann and Zeidler (1979). It is crucial that the weak consistency condition (H2) in Theorem 35.A can be verified easily. To this end, use the principle of majorized convergence for the Lebesgue integral and the convergence argument in Problem 26.4. The stability of the operator A_h results from the uniform monotonicity of the operator A which corresponds to the differential equation.

35.5. *Proof of Lemma* 35.16.
Hint: Cf. Schumann and Zeidler (1979).

35.6. *Proof of Lemma* 35.17.
Hint: Cf. Schumann and Zeidler (1979).

35.7. *Proof of Corollary* 35.15.
Solution: We use Proposition 26.8. To this end, we must verify:

$$\sum_P (g_P(u_h) - g_P(v_h))(u_h(P) - v_h(P)) \ge s \sum_P (u_h(P) - v_h(P))^2 \tag{29}$$

and

$$u_h \mapsto g_P(u_h) \quad \text{is locally Lipschitz continuous.} \tag{30}$$

Let $f(x) = |x|^{p-2}x$, $p \ge 2$. Then (30) follows from the local Lipschitz continuity of f. Moreover, (29) follows from the monotonicity of f, i.e.,

$$(f(x) - f(y))(x - y) \ge 0.$$

To be precise, after discrete integration by parts, (29) is identical to (21) in the case where $c = 0$.

References to the Literature

Discrete Sobolev spaces: Raviart (1967), Temam (1970, M), (1977, M), Ladyženskaja (1973, M).

Theorem 35.A and applications to quasi-linear elliptic differential equations: Schumann and Zeidler (1979).

Difference methods for nonlinear elliptic differential equations: Brézis and Sibony (1968), Aubin (1970), (1972, M), Sobolevskii and Tiunčik (1970), Müller (1974), Schumann and Zeidler (1979), Hackbusch and Trottenberg (1982, P), Hackbusch (1985, M).

Multigrid methods and fast solvers: Hackbusch and Trottenberg (1982, P), Birkhoff and Schoenstadt (1984, P), Hackbusch (1985, M) (standard work).

Finite elements: Ciarlet (1977, M), Temam (1977, M), Girault and Raviart (1986, M).

Quasi-linear elliptic equations and finite elements: Frehse (1977), (1982, S), Frehse and Rannacher (1978), Dobrowolski and Rannacher (1980), Kufner and Sändig (1987, M).

CHAPTER 36

Mapping Degree for A-Proper Operators

> The concept of degree of mapping, in all its different forms, is one of the most effective tools for studying the properties of existence and multiplicity of solutions of nonlinear equations.
>
> Felix E. Browder (1983)

In Part I we demonstrated the fundamental importance of the Leray–Schauder mapping degree for operator equations involving compact operators. In this chapter we will generalize the Leray–Schauder mapping degree $\deg(I - C, G, b)$ for compact operators $C: \bar{G} \subseteq X \to X$ to a mapping degree

$$\mathrm{DEG}(A, G, b)$$

for A-proper operators $A: \bar{G} \subseteq X \to Y$. In this connection, we use the same approximation process as in Chapter 12. However, in contrast to Chapter 12, now the mapping degree of the equations being approximated need *not* tend to a unique limit value; hence, $\mathrm{DEG}(A, G, b)$ is, in the general case, a *set of integers*.

Let X be a separable H-space over $\mathbb{K}$. Then, by Section 34.4, all the operators $A: X \to X$ of the form

$$A = \lambda(B - K - C) \tag{1}$$

are A-proper in the case where the following hold:

(i) $B: X \to X$ is continuous and strongly monotone (e.g., $B = I$), that is,

$$\mathrm{Re}(Bu - Bv|u - v) \geq c\|u - v\|^2 \qquad \text{for all} \quad u, v \in X \quad \text{and fixed} \quad c > 0.$$

(ii) $K: X \to X$ is Lipschitz continuous, that is, more precisely,

$$\|Ku - Kv\| \leq d\|u - v\| \qquad \text{for all} \quad u, v \in X$$

and fixed d with $0 \leq d < c$.

(iii) $C: X \to X$ is compact.
(iv) λ is a nonzero fixed number in $\mathbb{K}$.

Therefore, in particular, the general mapping degree $\mathrm{DEG}(A, G, b)$ includes the following classes of operators:

(a) For $\lambda = 1$, $B = I$, $K = 0$, we obtain in (1) the operators for which the Leray–Schauder mapping degree is defined.
(b) For $\lambda = 1$ and $B = I$, (1) contains the class $A = I - K - C$, where K is k-contractive and C is compact, i.e., $K + C$ is a k-set contraction.

The point of departure for our discussion is the approximation scheme in Section 34.1, i.e., we use:

$$\begin{array}{ccc} X & \xrightarrow{\;A\;} & Y \\ R_n \Big\downarrow\Big\uparrow E_n & & \Big\downarrow Q_n, \\ X_n & \xrightarrow{\;A_n\;} & Y_n \end{array} \qquad A_n = Q_n A E_n. \tag{2}$$

The basic idea of our approach is to approximate the operator equation

$$Au = b, \qquad u \in \bar{G},$$

by means of the equation

$$A_n u_n = Q_n b, \qquad u_n \in \bar{G}_n,$$

where $G_n = E_n^{-1}(G)$.

In order to simplify the exposition, we consider only operators A that are defined on the entire space X. However, the considerations can also be easily carried over to the case where A has the form $A: \bar{G} \subseteq X \to Y$. Then, in the definition of A-properness, parallel to Section 34.2, we have to consider only sequences from $E_n^{-1}(\bar{G})$.

In Problem 36.3 we study in detail a mapping degree for demicontinuous $(S)_+$-operators. This mapping degree is based on the convergence of the Galerkin method for $(S)_+$-operators, which we have proved in Proposition 27.4.

36.1. Definition of the Mapping Degree

We make the following assumptions, where the condition

$$Au \neq b \qquad \text{for all} \quad u \in \partial G \tag{3}$$

is crucial.

(H1) The operator $A: X \to Y$ is A-proper with respect to the admissible inner approximation schema (2) in the sense of Definition 34.2, and condition (3) holds.

(H2) G is a nonempty, open, and bounded subset of X. Let $G_n = E_n^{-1}(G)$. The sets G_n are *uniformly bounded*, i.e., there is an $R > 0$ such that $\|u_n\| \le R$ for all $u_n \in G_n$ and all n.
(H3) All the operators A_n are continuous on $\bar{G}_n$.
(H4) All the spaces X_n, Y_n are provided with a fixed orientation, i.e., we designate a fixed basis.

Proposition 36.1 *Under the assumptions* (H1) *through* (H4), *there exists a number n_0 such that the Brouwer mapping degree* $\deg(A_n, G_n, Q_n b)$ *exists for all* $n \ge n_0$.

We give the proof in Problem 36.1.
Proposition 36.1 enables us to make the following fundamental definition.

Definition 36.2. Let $a_n = \deg(A_n, G_n, Q_n b)$ for $n \ge n_0$. Under the assumptions (H1) through (H4), we set

$$\mathrm{DEG}(A, G, b) = \text{the set of cluster points of the sequence } (a_n).$$

If $G = \varnothing$, then we set $\mathrm{DEG}(A, G, b) = 0$.

Recall that the number a with $-\infty \le a \le \infty$ is called a cluster point of the sequence (a_n) iff some subsequence of (a_n) converges to a. Moreover, recall the fact that the integer $\deg(A_n, G_n, Q_n b)$ is a measure of the number of solutions u_n of the equation

$$A_n u_n = Q_n b, \qquad u_n \in \bar{G}_n,$$

where, roughly speaking, the solutions are counted according to their multiplicity (cf. Chapter 12).
The definition of $\mathrm{DEG}(A, G, b)$ goes back to Browder and Petryshyn (1969).
As the proof in Problem 36.1 shows, $\deg(A_n, G_n, Q_n b)$ depends only on the orientation of X_n and Y_n, but not on the basis chosen.

EXAMPLE 36.3. Let X be a real H-space with $\dim X = \infty$. Let G be a bounded region in X with $0 \in G$. Then

$$\mathrm{DEG}(-I, G, 0) = \{-1, 1\}$$

holds for $-I$, the negative of the identity, in the case where we choose the approximation scheme (34.10).

PROOF. By (34.10), $G_n = E_n^{-1}(G) = G \cap X_n$. It follows from formula (3) in Chapter 13 that

$$\deg(-I, G_n, 0) = (-1)^n, \qquad \text{where} \quad n = \dim X_n.$$

Now Definition 36.2 yields the assertion. □

36.2. Properties of the Mapping Degree

All the properties of the Brouwer mapping degree in Part I can easily be generalized to $\mathrm{DEG}(A, G, b)$. We give several examples of this.

Proposition 36.4 (Existence Principle). *If* $\mathrm{DEG}(A, G, b) \neq \{0\}$, *then the equation* $Au = b$ *has a solution* $u \in G$.

PROOF. Let $a_n = \deg(A_n, G_n, Q_n b)$. If $a \in \mathrm{DEG}(A, G, b)$ with $a \neq 0$, then, by definition, there exists a subsequence, again denoted by (a_n), such that $a_n \to a$ as $n \to \infty$. Consequently, $a_n \neq 0$ for all $n \geq n_0$. By the existence principle (D2) of the Brouwer mapping degree in Section 13.6, we obtain from $a_n \neq 0$ that the equation

$$A_n u_n = Q_n b, \qquad u_n \in \bar{G}_n,$$

has a solution u_n for all $n \geq n_0$. By assumption (H2) above, the sets G_n are uniformly bounded, i.e., $\sup_n \|u_n\| < \infty$. The A-properness of the operator A yields the existence of a subsequence, again denoted by (u_n), such that $E_n u_n \to u$ as $n \to \infty$ and $Au = b$. □

Proposition 36.5 (Homotopy Principle). *We have*

$$\mathrm{DEG}(A_0, G, b) = \mathrm{DEG}(A_1, G, b)$$

in the case where the following hold.

(i) *There exists a family* $\{A_t\}$ *of operators* $A_t \colon X \to Y$ *for all* $t \in [0, 1]$ *such that* A_t *satisfies the assumptions* (H1) *through* (H4) *in Section* 36.1 *for all* $t \in [0, 1]$.

(ii) *The mapping* $(u, t) \mapsto A_t(u)$ *is continuous on* $\bar{G} \times [0, 1]$.

(iii) *The mapping* $t \mapsto A_t(u)$ *is continuous on* $[0, 1]$ *uniformly with respect to* $u \in \bar{G}$, *i.e., for each* $\varepsilon > 0$ *and for each* $t \in [0, 1]$, *there exists a* $\delta(\varepsilon, t) > 0$ *so that* $\|A_t(u) - A_s(u)\| < \varepsilon$ *holds for* $|t - s| < \delta(\varepsilon, t)$ *and* $u \in \bar{G}$.

We give the proof in Problem 36.2.

36.3. The Antipodal Theorem for A-Proper Operators

Theorem 36.A (Antipodal Theorem). *Let* $b \in Y$. *The equation*

$$Au = b, \qquad u \in G, \tag{4}$$

has a solution u in the case where the following hold along with assumptions

(H1) *through* (H4) *in Section* 36.1:

(i) $Au \neq tb$ *for all* $u \in \partial G$, $t \in [0, 1]$.
(ii) $A(-u) = -A(u)$ *for all* $u \in \partial G$.

PROOF. We set $B_t = Au - tb$. The operator A is A-proper. By Corollary 34.6, this is also true for the compact perturbation B_t of A. Proposition 36.5 yields

$$\mathrm{DEG}(B_0, G, 0) = \mathrm{DEG}(B_1, G, 0).$$

We now consider the operator $A_n = Q_n A E_n$. Note that $\partial G_n \subseteq E_n^{-1}(\partial G)$ and that the operators E_n and Q_n are linear. Thus, it follows from (ii) that A_n is odd on ∂G_n. The classical antipodal theorem (Theorem 16.B) yields

$$\deg(A_n, G_n, 0) \neq 0.$$

By Definition 36.2, $\mathrm{DEG}(A, G, 0) \neq \{0\}$. Since $B_0 = A$, we obtain

$$\mathrm{DEG}(B_1, G, 0) \neq \{0\}.$$

Now, Proposition 36.4 yields the existence of a solution of the equation $B_1 u = 0$, $u \in G$. This equation is equivalent to the original equation (4). □

36.4. A General Existence Principle

As an another example we generalize the important fixed-point principle in Section 13.1. In this connection, the following condition on the omitted rays is crucial:

$$Au \neq -\alpha u \quad \text{for all} \quad u \in \partial G \quad \text{and all real} \quad \alpha \geq 0. \tag{5}$$

Theorem 36.B. *The equation*

$$Au = 0, \qquad u \in G, \tag{6}$$

has a solution in the case where the following assumptions are fulfilled:

(i) *X is a separable infinite-dimensional* H-*space, and G is an open bounded set in X with $0 \in G$.*
(ii) *The operator $A: X \to X$ is bounded, continuous, and condition* (5) *holds.*
(iii) *For all $\alpha \geq 0$, the operator $A + \alpha I$ is A-proper with respect to the approximation scheme* (10) *in Chapter* 34.

Corollary 36.6. *Condition* (5) *is satisfied in the case where*

$$\mathrm{Re}(Au|u) > 0 \qquad \textit{for all} \quad u \in \partial G.$$

Moreover, condition (iii) *is satisfied in the case where the operator A possesses the structure* (1).

PROOF. According to the approximation scheme (34.10), we obtain $Q_n = P_n$ and $Q_n I E_n = I$ on X_n. By property (D1) of the mapping degree in Section 13.6,

$$\deg(Q_n I E_n, G_n, 0) = \deg(I, G_n, 0) = 1.$$

This implies

$$\mathrm{DEG}(I, G, 0) = \{1\}.$$

We now consider the homotopy

$$A_t u = (1 - t)Au + tu.$$

For $t \neq 1$, we obtain

$$A_t = (1 - t)(A + t(1 - t)^{-1} I).$$

Thus, by (iii), the operator A_t is A-proper for all $t \in [0, 1]$. According to the key condition (5), we have

$$A_t u \neq 0 \qquad \text{for all} \quad (u, t) \in \partial G \times [0, 1].$$

The homotopy principle (Proposition 36.5) yields

$$\mathrm{DEG}(A_0, G, 0) = \mathrm{DEG}(A_1, G, 0).$$

Since $A_0 = A$ and $A_1 = I$, we get

$$\mathrm{DEG}(A, G, 0) = \{1\}.$$

Now, according to the existence principle (Proposition 36.4), equation (6) has a solution. □

PROBLEMS

36.1. *Proof of Proposition* 36.1.

Solution: By (H4) we can identify X_n and Y_n with $\mathbb{R}^n$. The mapping $E_n: X_n \to X$ is continuous. Therefore, the set $E_n^{-1}(G)$ is open and $E_n^{-1}(\bar{G})$ is closed. Consequently, the set G_n is open, bounded, and nonempty for all $n \geq n_1$ and $\partial G_n \subseteq E_n^{-1}(\partial G)$.

We show that there exist a $d > 0$ and an n_0 such that

$$\|Q_n A E_n u_n - Q_n b\| \geq d \tag{7}$$

for all $u_n \in \partial G_n$ and all $n \geq n_0$. Otherwise, there exists a sequence $(u_{n'})$ with $u_{n'} \in \partial G_{n'}$ for all n' and

$$\|A_{n'} u_{n'} - Q_{n'} b\| \to 0 \qquad \text{as} \quad n \to \infty.$$

Moreover, by (H2), $\sup_{n'} \|u_{n'}\| < \infty$. The operator A is A-proper. Consequently, there exists a subsequence denoted by (u_n) such that

$$E_n u_n \to u \quad \text{in } X \qquad \text{as} \quad n \to \infty$$

and $Au = b$. Because $E_n u_n \in \partial G$ for all n, we have $u \in \partial G$. This is a contradiction to $Au \neq b$ for all $u \in \partial G$.

From (7) we obtain

$$A_n u_n \neq Q_n b \qquad \text{for all} \quad u_n \in \partial G_n, \quad n \geq n_0. \tag{8}$$

The operator $A_n: X_n \to Y_n$ is continuous. Now, according to Chapter 12, the existence of the Brouwer mapping degree $\deg(A_n, G_n, Q_n b)$ follows from (8).

The proof of Proposition 36.1 is complete.

By Step 1 in the proof of Theorem 12.B, the mapping degree $\deg(A_n, G_n, Q_n b)$ does not change if one goes over to other bases in X_n and Y_n which have the same orientation as the bases fixed by (H4).

36.2. *Proof of Proposition* 36.5.

Solution: In (7) we replace the operator A by A_t and we show that the numbers d and n_0 can be chosen uniformly for all $t \in [0,1]$. By Problem 36.1 one first finds, for each $t \in [0,1]$, numbers $d(t)$ and $n_0(t)$. Because of the continuity of $t \mapsto A_t(u)$ on $[0,1]$, uniformly with respect to $u \in \bar{G}$, and $\sup_n \|Q_n\| < \infty$, one finds, for each $t \in [0,1]$, a neighborhood $U(t)$ such that d and n_0 can be chosen uniformly on $U(t)$. The set $[0,1]$ is compact. Therefore, finitely many $U(t_i)$ already cover the interval $[0,1]$. Now choose $d = \min_i\{d(t_i)\}$ and $n_0 = \max_i\{n_0(t_i)\}$.

We set

$$H_n(u,t) = Q_n A_t E_n u, \qquad H(u,t) = A_t(u).$$

From (7) with A_t instead of A we obtain

$$H_n(u,t) \neq Q_n b$$

for all $(u,t) \in \partial G_n \times [0,1]$ and $n \geq n_0$. The homotopy invariance (D4) of the Brouwer mapping degree in Section 13.6 yields

$$\deg(H_n(\cdot,0), G_n, Q_n b) = \deg(H_n(\cdot,1), G_n, Q_n b)$$

for all $n \geq n_0$. By Definition 36.2, it follows from this that

$$\mathrm{DEG}(A_0, G, b) = \mathrm{DEG}(A_1, G, b).$$

36.3. *Mapping degree for demicontinuous bounded operators with condition* $(S)_+$. Let X be a real separable reflexive B-space with $\dim X = \infty$, and let G be a nonempty bounded open set in X. Moreover, let $\mathscr{S}$ denote the set of all bounded demicontinuous operators

$$A: X \to X^*$$

which satisfy $(S)_+$. For example, the operator

$$Au = Mu + Cu + b$$

belongs to $\mathscr{S}$ in the case where the operator $M: X \to X^*$ is uniformly monotone, demicontinuous, and bounded, the operator $C: X \to X^*$ is compact, and b is a fixed element in X^*. Let $\{w_1, w_2, \ldots\}$ be a basis in X and let $X_n = \operatorname{span}\{w_1, \ldots, w_n\}$. We choose functionals $w_i^* \in X^*$ such that $\langle w_i^*, w_j \rangle = \delta_{ij}$ for all i, j.

We want to solve the equation

$$Au = 0, \qquad u \in G, \tag{9}$$

by means of the Galerkin equation

$$A_n u_n = 0, \qquad u_n \in G_n, \tag{10}$$

where $G_n = G \cap X_n$ and

$$A_n u = \sum_{j=1}^{n} \langle Au, w_j \rangle_X w_j^*.$$

36.3a. *Lemma.* Show that if $A \in \mathscr{S}$ and $Au \neq 0$ on ∂G, then there exists an n_0 such that

$$A_n u \neq 0 \quad \text{on } \partial G_n \qquad \text{for all} \quad n \geq n_0.$$

Solution: Otherwise, there exists a sequence $(u_{n'})$, briefly denoted by (u_n), such that

$$A_n u_n = 0, \qquad u_n \in \partial G_n \qquad \text{for all} \quad n \geq n_0.$$

Since (u_n) is bounded, there exists a subsequence, again denoted by (u_n), such that

$$u_n \rightharpoonup u \qquad \text{as} \quad n \to \infty.$$

We choose a sequence (v_n) with $v_n \in X_n$ for all n such that $v_n \to u$ as $n \to \infty$. Then

$$\langle Au_n, u_n - u \rangle = \langle Au_n, v_n - u \rangle \qquad \text{for all} \quad n \geq n_0,$$

since $\langle A_n u_n, w \rangle = \langle Au_n, w \rangle$ for all $w \in X_n$ and $A_n u_n = 0$. Hence

$$\langle Au_n, u_n - u \rangle \to 0 \qquad \text{as} \quad n \to \infty,$$

since (Au_n) is bounded and $v_n \to u$ as $n \to \infty$. Condition $(S)_+$ implies $u_n \to u$ as $n \to \infty$. Hence $u \in \partial G$ and $Au = 0$. This is a contradiction.

36.3b. *Definition.* Let $A \in \mathscr{S}$ and suppose that $Au \neq 0$ on ∂G. We define

$$\deg(A, G) = \deg(A_n, G_n) \qquad \text{for all} \quad n \geq n_0.$$

Show that this definition does not depend on the choice of n and the basis $\{w_1, w_2, \ldots\}$.

Solution: Use exactly the same argument as in the proof of Theorem 12.B concerning the Leray–Schauder degree.

36.3c. *Existence principle.* Show that if

$$\deg(A, G) \neq 0,$$

then the equation $Au = 0$, $u \in G$, has a solution.

Solution: By definition of the mapping degree, $\deg(A_n, G_n) \neq 0$ for all $n \geq n_0$.

According to the existence principle for the Brouwer mapping degree, the Galerkin equation (10) has a solution $u_n \in G_n$ for all $n \geq n_0$. By Proposition 27.4, there exists a subsequence $(u_{n'})$ with $u_{n'} \to u$ as $n \to \infty$ and $Au = 0$. Hence $u \in \bar{G}$. Since $Au \neq 0$ on ∂G, we get $u \in G$.

36.3d. *Additivity.* Let

$$\bar{G} = \bigcup_{i=1}^{n} \overline{M_i},$$

where $\{M_i\}$ is a family of pairwise disjoint open bounded sets. Suppose that

$A \in \mathscr{S}$ with

$$Au \neq 0 \quad \text{on } \partial G \qquad \text{and} \qquad Au \neq 0 \quad \text{on } \partial M_i \qquad \text{for all } i.$$

Show that

$$\deg(A, G) = \sum_{i=1}^{n} \deg(A, M_i).$$

Solution: This follows from the corresponding property of the Brouwer mapping degree.

36.3e. *Homotopy invariance.* Let $H: X \times [0, 1] \to X^*$ be a demicontinuous bounded operator with condition $(S)_+ \times [0, 1]$, i.e., from

$$u_n \rightharpoonup u \qquad \text{and} \qquad t_n \to t \qquad \text{as} \quad n \to \infty,$$

with $t_n \in [0, 1]$ for all n, and

$$\overline{\lim_{n\to\infty}} \langle H(u_n, t_n), u_n - u \rangle \leq 0$$

it follows that $u_n \to u$ as $n \to \infty$. Suppose that

$$H(u, t) \neq 0 \quad \text{on } \partial G \times [0, 1].$$

Show that

$$\deg(H(\cdot, 0), G) = \deg(H(\cdot, 1), G).$$

Solution: Use a similar argument as in Problem 36.2.
Note that $H(\cdot, 0)$ and $H(\cdot, 1)$ belong to $\mathscr{S}$.

36.3f. *Linear homotopy.* Let $A, B \in \mathscr{S}$. Set

$$H(u, t) = (1 - t)Au + tBu \qquad \text{for all} \quad (u, t) \in X \times [0, 1]$$

and show that H satisfies condition $(S)_+ \times [0, 1]$.

Solution: Let $u_n \rightharpoonup u$ and $t_n \to t$ as $n \to \infty$ with

$$\overline{\lim_{n\to\infty}} (1 - t_n)\langle Au_n, u_n - u \rangle + t_n \langle Bu_n, u_n - u \rangle \leq 0. \tag{11}$$

Since A is demicontinuous and satisfies $(S)_+$, we have

$$\underline{\lim_{n\to\infty}} \langle Au_n, u_n - u \rangle \geq 0.$$

Otherwise, there would exist a subsequence, again denoted by (u_n), such that

$$\lim_{n\to\infty} \langle Au_n, u_n - u \rangle < 0.$$

Condition $(S)_+$ implies $u_n \to u$ and hence $Au_n \rightharpoonup Au$; therefore, $\langle Au_n, u - u_n \rangle \to 0$ as $n \to \infty$. This is a contradiction.

Suppose, without loss of generality, that $t > 0$. From

$$\underline{\lim_{n\to\infty}} (1 - t_n)\langle Au_n, u_n - u \rangle \geq 0$$

and (11) we obtain

$$t \overline{\lim_{n\to\infty}} \langle Bu_n, u_n - u \rangle = \overline{\lim_{n\to\infty}} t_n \langle Bu_n, u_n - u \rangle \leq 0.$$

Condition $(S)_+$ implies $u_n \to u$ as $n \to \infty$.

36.3g. *Convexity of the class $\mathscr{S}$.* Show that if the operators $A, B: X \to X^*$ belong to the class $\mathscr{S}$ in the sense of the definition given above, then $A + B \in \mathscr{S}$ and $\lambda A \in \mathscr{S}$ for each $\lambda > 0$. In particular, $A, B \in \mathscr{S}$ implies $(1-t)A + tB \in \mathscr{S}$ for each $t \in [0, 1]$, i.e., $\mathscr{S}$ is convex. Moreover, if we set $C(u) = -A(-u)$ for all $u \in X$, then $A \in \mathscr{S}$ implies $C \in \mathscr{S}$.

Solution: Use the same argument as in Problem 36.3f and use the definition of $(S)_+$ in Section 27.1.

36.3h. *An existence theorem.* Let $A \in \mathscr{S}$ and suppose that

$$\langle Au, u\rangle > 0 \qquad \text{for all} \quad u \in \partial G. \tag{12}$$

Show that if $0 \in G$, then $\deg(A, G) = 1$, and the equation $Au = 0$, $u \in G$, has a solution.

Solution: Condition (12) implies that $A_n u \neq -\alpha u$ on ∂G_n for all $\alpha \geq 0$. By Theorem 13.A, $\deg(A_n, G_n) = 1$ for all $n \geq n_0$. Now use Problems 36.3b, c.

36.3i. *Antipodal theorem.* Let $G = \{u \in X: \|u\| < r\}$ and let $A \in \mathscr{S}$. Suppose that $Au \neq 0$ on ∂G and

$$\frac{Au}{\|Au\|} \neq \frac{A(-u)}{\|A(-u)\|} \qquad \text{for all} \quad u \in \partial G. \tag{13}$$

Then $\deg(A, G)$ is odd and the equation $Au = 0$, $u \in G$, has a solution.

Solution: We use the linear homotopy

$$H(u, t) = (1-t)Au + t(A(u) - A(-u)). \tag{14}$$

Hence

$$\deg(A, G) = \deg(H(\cdot, 1), G).$$

Note that $H(u, t) \neq 0$ on $\partial G \times [0, 1]$, by (13). Thus, (14) represents an admissible homotopy by Problems 36.f, g.

Since $H(-u, 1) = -H(u, 1)$, the classical antipodal theorem (Theorem 16.B) yields

$$\deg(H_n(\cdot, 1), G_n) = \text{odd} \qquad \text{for all} \quad n \geq n_0.$$

This completes the proof by Problems 36.3b, c.

Further details on this mapping degree can be found in Browder (1968/76, M) (e.g., applications to pseudomonotone operators) and in Skrypnik (1986, M). In particular, it is shown in Browder (1983) that, parallel to Chapter 12, this mapping degree is uniquely determined by a few natural axioms.

Applications of this mapping degree to quasi-linear elliptic differential equations are studied in detail in Skrypnik (1986, M).

36.4. *The nonlinear open mapping theorem.* We assume:

(i) Let G be an open set in the real separable reflexive B-space X.
(ii) The operator $A: G \to X^*$ is continuous and locally injective.
(iii) The operator A satisfies the condition $(S)_+$ on G (e.g., $A \in \mathscr{S}$). That is, it follows from $u_n \rightharpoonup u$ as $n \to \infty$ on G and

$$\overline{\lim_{n\to\infty}} \langle Au_n - Au, u_n - u\rangle \leq 0$$

that $u_n \to u$ as $n \to \infty$.

Show that the set $A(G)$ is open in X^*.

Hint: Use the same argument as in the proof of Theorem 16.C from Part I based on the *antipodal theorem*. Replace the Leray–Schauder degree by the mapping degree introduced above. Cf. Skrypnik (1986, M), p. 45.

36.5. *The invariance of domain theorem*. Assume (i) through (iii) from Problem 36.4. Show that if G is a region, then so is $A(G)$.

Solution: The set G is open and connected. By Problem 36.4, the set $A(G)$ is open. Since A is continuous, it sends connected sets onto connected sets (A_1(13c)). Hence $A(G)$ is connected, i.e., $A(G)$ is a region.

The fundamental importance of the open mapping theorem and the invariance of the domain theorem has been thoroughly discussed in Section 16.4 from Part I.

References to the Literature

Mapping degree for A-proper maps: Browder and Petryshyn (1969), Browder (1968/76, M), Petryshyn (1975, S, B, H), (1980).

Mapping degree for demicontinuous $(S)_+$-operators: Browder (1968/76, M), (1983), Skrypnik (1973, M), (1986, M).

Applications to quasi-linear elliptic differential equations: Skrypnik (1973, M), (1986, M).

Appendix

> In contrast to the classical Riemann integral, my new concept of the integral provides a better answer to the question of the connection between differentiation and integration (main theorem of calculus).
>
> Henri Leon Lebesgue (1902)

> Almost all concepts, which relate to the modern measure and integration theory, go back to the works of Lebesgue (1875–1941). The introduction of these concepts was the turning point in the transition from mathematics of the nineteenth century to mathematics of the twentieth century.
>
> Naum Jakovlevič Vilenkin (1975)

The purpose of this Appendix is to collect a number of important well-known analytical tools. This is done for the convenience of the reader. We consider the following topics:

(a) The Lebesgue measure on $\mathbb{R}^N$.
(b) Measurable functions.
(c) The Lebesgue integral for functions with values in B-spaces.
(d) Lebesgue spaces.
(e) Sobolev spaces.
(f) Galerkin schemes in B-spaces and the method of finite elements.
(g) Ordinary differential equations with measurable functions.
(h) Distributions and locally convex spaces.
(i) Construction of general measures and the main theorem of measure theory.
(j) General measure spaces.
(k) The integral with respect to general measures.
(l) Measures on topological spaces.
(m) The classical Stieltjes integral and Stieltjes operator integrals.

(n) Measures and the spectral theory for self-adjoint operators in H-spaces.
(o) The functional calculus for self-adjoint operators in H-spaces.
(p) Linear semigroups and fractional powers of operators.
(q) Interpolation theory.
(r) Hausdorff measure, Hausdorff dimension, and fractals.
(s) Functions of bounded variation.

These tools will be used frequently in Parts II–V. Further basic material can be found in the Appendix of Part I. In case the opposite is not stated explicitly, all B-spaces are assumed to be real or complex B-spaces.

Lebesgue Measure

(1) *Measurable sets.* The Lebesgue measure is a generalization of the volume in elementary geometry. The classical volume was only defined for a small class of sufficiently regular sets. The goal of Lebesgue's measure theory around 1900 was to find a sufficiently large class of sets S in $\mathbb{R}^N$ which can be measured by a number meas S, where

$$0 \leq \operatorname{meas} S \leq \infty.$$

Such sets are called Lebesgue measurable or briefly measurable. The precise definition of the Lebesgue measure will be given in (75h) below as a special case of a general construction (main theorem of general measure theory). The most important properties of measurable sets in $\mathbb{R}^N$ are the following:

(a) The measure of cuboids is equal to the classical volume.
(b) Open sets and closed sets are measurable.
(c) The union or the intersection of at most countably many measurable sets is again measurable.
(d) If $\{S_i\}$ is an at most countable family of *pairwise disjoint* measurable sets, then

$$\operatorname{meas}\left(\bigcup_i S_i\right) = \sum_i \operatorname{meas} S_i.$$

(e) The difference $S - T$ of two measurable sets S and T is again measurable.
(f) A subset S of $\mathbb{R}^N$ has *measure zero* iff, for each $\varepsilon > 0$, there exists a family of at most countably many cuboids C_i such that $S \subseteq \bigcup_i C_i$ and

$$\sum_i \operatorname{meas} C_i < \varepsilon.$$

Examples. A set of finitely many points or of countably many points in $\mathbb{R}^N$ has measure zero.

Sufficiently regular curves in $\mathbb{R}^2$ or $\mathbb{R}^3$ have measure zero.

Sufficiently regular surfaces in $\mathbb{R}^3$ have measure zero.

The set of all rational numbers in $\mathbb{R}^1$ is countable and hence it has measure zero.

The set of all points in $\mathbb{R}^N$ with rational coordinates has measure zero.

(2) "*Almost everywhere.*" By definition, a property P holds true almost everywhere iff P holds true for all points of $\mathbb{R}^N$ with the exception of a set of measure zero.

One also uses "almost all." For example, almost all real numbers are irrational.

The importance of the notion "almost everywhere" is shown by the following theorem.

(3) *Theorem of Lebesgue.* Let $f: J \subseteq \mathbb{R} \to \mathbb{R}$ be a monotone function on the finite or infinite interval J. Then, f is almost everywhere differentiable and therefore, almost everywhere continuous.

According to the famous Hungarian mathematician Fryges (Frederick) Riesz (1880–1956), this is one of the most surprising theorems of modern analysis.

Note that most sets in $\mathbb{R}^N$ are measurable. The construction of non-measurable sets is based on sophisticated arguments.

Measurable Functions

Let Y be a B-space.

(4) *Step functions.* A function $f: M \subseteq \mathbb{R}^N \to Y$ is called a step function iff f is piecewise constant. To be precise, we suppose that the set M is measurable and that there exist finitely many pairwise disjoint measurable subsets M_i of M such that $\operatorname{meas} M_i < \infty$ for all i and

$$f(x) = \begin{cases} a_i & \text{for } x \in M_i \text{ and all } i, \\ 0 & \text{otherwise.} \end{cases}$$

(5) The *integral* of a step function is defined to be

$$\int_M f\,dx = \sum_i (\operatorname{meas} M_i) a_i.$$

The general definition of the Lebesgue integral in (14) below will be based on this simple definition.

(6) *Example.* Let $f: [a, b] \to \mathbb{R}$ be a step function as in Figure 1. Then the integral in the sense of (5) is precisely the classical integral.

(7) *Definition of measurable functions.* The function

$$f: M \subseteq \mathbb{R}^N \to Y$$

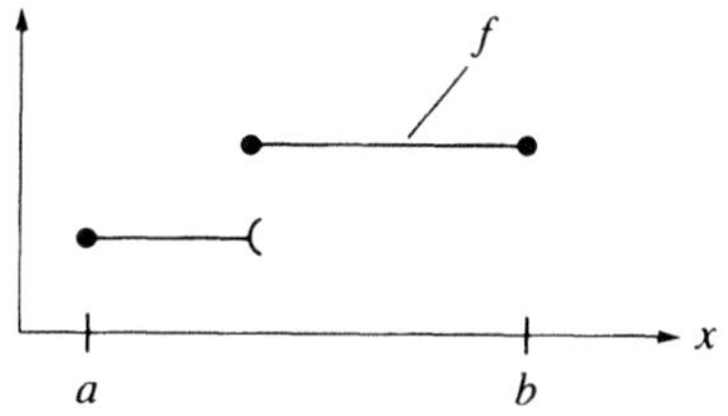

Figure 1

with values in the B-space Y is called measurable iff the following hold:

(i) The domain of definition M is measurable.
(ii) There exists a sequence (f_n) of step functions $f_n: M \to Y$ such that

$$\lim_{n\to\infty} f_n(x) = f(x) \qquad \text{for almost all} \quad x \in M.$$

(8) *Standard example.* The function $f: M \subseteq \mathbb{R}^N \to Y$ is measurable if the following hold:

(i) M is measurable and the B-space Y is separable.
(ii) f is continuous almost everywhere, i.e., there exists a set Z of measure zero in $\mathbb{R}^N$ such that $f: M - Z \to Y$ is continuous.

(9a) *Calculus.* Linear combinations, norm functions, and limits of measurable functions are again measurable.

More precisely, we set

$$F(x) = a(x)f(x) + b(x)g(x), \qquad G(x) = \|f(x)\|,$$

$$H(x) = \lim_{n\to\infty} f_n(x),$$

and we suppose that the functions

$$f, g, f_n: M \subseteq \mathbb{R}^N \to Y, \qquad a, b: M \to \mathbb{R}$$

are measurable for all n. Moreover, we assume that the limit $f_n(x) \to f(x)$ as $n \to \infty$ exists for almost all $x \in M$.

Then the functions F, G, and H are measurable.

(9b) *Modification of measurable functions.* If we change a measurable function at the points of a set of measure zero, then the function remains measurable.

In particular, the function H in (9a) may be defined arbitrarily at the points at which the limit does not exist.

(10) *Theorem of Pettis.* Let Y be a real *separable* B-space and let $f: M \subseteq \mathbb{R}^N \to Y$ be a measurable function. Then the following two statements are equivalent:

(i) The function f is measurable.

(ii) The real functions $x \mapsto \langle g, f(x) \rangle$ are measurable on M for all functionals $g \in Y^*$.

This important result reduces the measurability of functions with values in B-spaces to the measurability of real functions. We will frequently use this theorem.

(11a) *The theorem of Luzin and the characterization of measurable functions.* Let $f\colon M \subseteq \mathbb{R}^N \to \mathbb{R}$ be a function on the measurable set M with meas $M < \infty$. Then the following two statements are equivalent.

(i) f is measurable.
(ii) f is continuous up to small sets, i.e., for each $\delta > 0$, there exists a subset M_δ of M such that

$$f \text{ is continuous on the closed set } M - M_\delta \qquad \text{and} \qquad \operatorname{meas} M_\delta < \delta.$$

(11b) *The convergence theorem of Egorov.* Each convergent sequence of measurable functions on a set of finite measure is uniformly convergent up to small sets.

To be precise, let (f_n) be a sequence of measurable functions $f_n\colon M \subseteq \mathbb{R}^N \to Y$ with meas $M < \infty$ such that

$$f_n(x) \to f(x) \qquad \text{as} \quad n \to \infty \quad \text{for almost all} \quad x \in M.$$

Then the function $f\colon M \to Y$ is measurable and, for each $\delta > 0$, there exists a subset M_δ of M such that

$$f_n \rightrightarrows f \qquad \text{as} \quad n \to \infty \quad \text{on } M - M_\delta \qquad \text{and} \qquad \operatorname{meas} M_\delta < \delta.$$

(12) *Definition of measurable functions via substitution.* We set

$$F(x) = f(x, u(x)).$$

If the function $u\colon M \subseteq \mathbb{R}^N \to U$ is measurable, then the function $F\colon M \to Y$ is also measurable provided the following assumptions are satisfied:

(i) The set M is measurable and the B-spaces U and Y are real and separable.
(ii) The function $f\colon M \times U \to Y$ satisfies the Carathéodory condition, i.e.,

$$x \mapsto f(x, u) \text{ is measurable on } M \text{ for all } u \in U,$$

$$u \mapsto f(x, u) \text{ is continuous on } U \text{ for almost all } x \in M.$$

The Lebesgue Integral for Functions with Values in B-Spaces

> Basic definitions in mathematics must be simple.
>
> Folclore

Let Y be a B-space. We consider the function

$$f\colon M \subseteq \mathbb{R}^N \to Y$$

on the measurable set M. The definition of the integral is based on the very natural formula

$$\int_M f\,dx = \lim_{n\to\infty} \int_M f_n\,dx \tag{13}$$

together with the following two simple formulas:

$$f(x) = \lim_{n\to\infty} f_n(x) \qquad \text{for almost all} \quad x \in M, \tag{13a}$$

$$\int_M \|f_n(x) - f_m(x)\|\,dx < \varepsilon \qquad \text{for all} \quad n, m \geq n_0(\varepsilon). \tag{13b}$$

Here, the functions $f_n\colon M \to Y$ are step functions.

(14) *Definition of the integral.* The function $f\colon M \to Y$ is called *integrable* iff M is measurable and there exists a sequence (f_n) of step functions $f_n\colon M \to Y$ such that (13a) and (13b) hold.

To be precise, we demand that, for each $\varepsilon > 0$, there is an $n_0(\varepsilon)$ such that (13b) holds.

The integral for integrable functions is defined by (13) above.

(15) *Justification of the definition.* By (14), the integral is well defined because:

(i) the integrals over step functions have been defined in (5); and
(ii) the limit in (13) exists in Y and does not depend on the choice of the step functions f_n.

According to (13a), each integrable function is measurable.

Conditions (13a) and (13b) mean intuitively that the sequence (f_n) of step functions approximates f with a sufficiently high accuracy.

The value of the integral $\int_M f\,dx$ does not change in the case where the function f is changed on a set of measure zero.

(16) *Standard example.* The integral $\int_M f\,dx$ exists if the following two conditions are satisfied:

(i) The function $f\colon M \subseteq \mathbb{R}^N \to Y$ is measurable (e.g., Y is separable and f is continuous almost everywhere in the sense of (8)).
(ii) $\sup_{x\in M} \|f(x)\| < \infty$ and meas $M < \infty$.

Note that

$$\int_M f\,dx \in Y.$$

The classical Lebesgue integral corresponds to the special case $Y = \mathbb{R}$, i.e., the existence of $\int_M f\,dx$ implies

$$\left|\int_M f\,dx\right| < \infty \qquad \text{and} \qquad 0 \leq \text{meas}\, M \leq \infty.$$

Properties of the Integral

Let Y be a B-space.

(17) *Majorant criterion.* All the following integrals exist and we have the estimates

$$\left\| \int_M f\,dx \right\| \le \int_M \|f(x)\|\,dx \le \int_M g\,dx$$

provided the following two conditions are satisfied:

(i) $\|f(x)\| \le g(x)$ for almost all $x \in M$, and $\int_M g\,dx$ exists.
(ii) $f: M \subseteq \mathbb{R}^N \to Y$ is measurable.

(18) *Norm criterion.* Let $f: M \subseteq \mathbb{R}^N \to Y$ be measurable. Then

$$\int_M f(x)\,dx \text{ exists} \qquad \text{iff} \qquad \int_M \|f(x)\|\,dx \text{ exists.}$$

(19) *Majorized convergence.* We have

$$\lim_{n\to\infty} \int_M f_n\,dx = \int_M \lim_{n\to\infty} f_n(x)\,dx,$$

where all the integrals and limits exist, provided the following conditions are satisfied:

(i) $\|f_n(x)\| \le g(x)$ for almost all $x \in M$ and all $n \in \mathbb{N}$, and $\int_M g\,dx$ exists.
(ii) $\lim_{n\to\infty} f_n(x)$ exists for almost all $x \in M$, where $f_n: M \subseteq \mathbb{R}^N \to Y$ is measurable for all n.

This theorem of Lebesgue describes one of the most important properties of the integral and will be used frequently. The key condition is the majorant condition (i).

(19a) *Generalized majorized convergence.* Theorem (19) remains true if we replace the assumption (i) with the following more general assumption:

$$\|f_n(x)\| \le g_n(x) \qquad \text{for almost all} \quad x \in M \quad \text{and all} \quad n \in \mathbb{N}.$$

All the functions g_n, $g: M \to \mathbb{R}$ are integrable and we have the convergence $g_n \to g$ almost everywhere on M as $n \to \infty$ along with

$$\int_M g_n\,dx \to \int_M g\,dx \qquad \text{as} \quad n \to \infty.$$

Proof. Let $f(x) = \lim_{n\to\infty} f_n(x)$. Hence $\|f(x)\| \le g(x)$ and

$$g(x) + g_n(x) - \|f(x) - f_n(x)\| \ge 0.$$

In the following we write f instead of $f(x)$, etc.

By the majorant criterion (17) and the lemma of Fatou (19c) below, we obtain

$$\begin{aligned}\int 2g\,dx &\le \varliminf_{n\to\infty} \int (g + g_n - \|f - f_n\|)\,dx \\ &= \int 2g\,dx - \varlimsup_{n\to\infty} \int \|f - f_n\|\,dx.\end{aligned}$$

This implies $\varlimsup_{n\to\infty} \int \|f - f_n\|\,dx = 0$, i.e.,

$$\left\| \int (f - f_n)\,dx \right\| \le \int \|f - f_n\|\,dx \to 0 \quad \text{as} \quad n \to \infty. \qquad \square$$

(19b) *Monotone convergence.* We have

$$\lim_{n\to\infty} \int_M f_n\,dx = \int_M \lim_{n\to\infty} f_n(x)\,dx,$$

where all the integrals and limits exist,[1] provided the following two conditions are satisfied:

(i) $f_n\colon M \subseteq \mathbb{R}^N \to \mathbb{R}$ is integrable for all $n \in \mathbb{N}$.
(ii) The sequence (f_n) is *monotone* increasing (or monotone decreasing) and $\sup_n |\int_M f_n\,dx| < \infty$.

(19c) *Lemma of Fatou.* We have

$$\int_M \varliminf_{n\to\infty} f_n(x)\,dx \le \varliminf_{n\to\infty} \int_M f_n\,dx,$$

where all the integrals exist,[2] provided the following two conditions are satisfied:

(i) $f_n\colon M \subseteq \mathbb{R}^N \to \mathbb{R}$ is *nonnegative* and integrable for all $n \in \mathbb{N}$.
(ii) $\varliminf_{n\to\infty} \int_M f_n\,dx < \infty$.

This lemma generalizes the well-known relation

$$\varliminf_{n\to\infty} a_n + \varliminf_{n\to\infty} b_n \le \varliminf_{n\to\infty} (a_n + b_n)$$

for nonnegative real numbers a_n and b_n.

(20) *Absolute continuity of the integral.* Let $f\colon M \subseteq \mathbb{R}^N \to Y$ be integrable. Then, for each $\varepsilon > 0$, there exists a $\delta(\varepsilon) > 0$ such that

$$\left\| \int_H f\,dx \right\| \le \int_H \|f(x)\|\,dx < \varepsilon$$

holds for all subsets H of M with $\operatorname{meas} H < \delta(\varepsilon)$.

[1] In particular, it follows that $\lim_{n\to\infty} f_n(x) \ne \pm\infty$ for almost all $x \in M$.

[2] In particular, it follows that $0 \le \varliminf_{n\to\infty} f_n(x) < \infty$ for almost all $x \in M$.

(21) *Absolute equicontinuity and the convergence theorem of Vitali.* We have

$$\lim_{n\to\infty}\int_M f_n\,dx = \int_M \lim_{n\to\infty} f_n(x)\,dx,$$

where all the integrals and the left-hand limit exist, provided the following four conditions are satisfied:

(i) $f_n\colon M \subseteq \mathbb{R}^N \to \mathbb{R}$ is integrable for all $n \in \mathbb{N}$.
(ii) $\lim_{n\to\infty} f_n(x)$ exists for almost all $x \in M$ and is finite.
(iii) Equicontinuity. For each $\varepsilon > 0$, there exists a $\delta(\varepsilon) > 0$ such that

$$\sup_n \int_H |f_n(x)|\,dx < \varepsilon$$

holds for all subsets H of M with $\operatorname{meas} H < \delta(\varepsilon)$.
(iv) If $\operatorname{meas} M = \infty$, then, for each $\varepsilon > 0$, there exists a subset S of M such that $\operatorname{meas} S < \infty$ and

$$\sup_n \int_{M-S} |f_n(x)|\,dx < \varepsilon.$$

(22) *The convergence theorem of Vitali–Hahn–Saks.* Theorem (21) remains true if we replace assumption (iii) with the following weaker condition:

(iii*) $\lim_{n\to\infty} \int_H f_n\,dx$ exists and is finite for all measurable subsets H of M.

In addition, (iii*) implies (iii).

Iterated Integration

Our goal is the fundamental formula

$$\text{(I)}\qquad \begin{aligned}\int_M f(x,y)\,dx\,dy &= \int_{\mathbb{R}^N}\left(\int_{\mathbb{R}^L} f(x,y)\,dy\right)dx\\ &= \int_{\mathbb{R}^L}\left(\int_{\mathbb{R}^N} f(x,y)\,dx\right)dy.\end{aligned}$$

In this connection, we set $f(x, y) = 0$ outside M. Furthermore, let $x \in \mathbb{R}^N$ and $y \in \mathbb{R}^L$.

(23) *Theorem of Fubini.* Let $f\colon M \subseteq \mathbb{R}^{N+L} \to \mathbb{R}$ be integrable. Then formula (I) holds. To be precise, the inner integrals exist for almost all $x \in \mathbb{R}^N$ (resp. almost all $y \in \mathbb{R}^L$), and the outer integrals exist.

(24) *Theorem of Tonelli.* Let $f\colon M \subseteq \mathbb{R}^{N+L} \to \mathbb{R}$ be measurable. Then the following two conditions are equivalent:

(i) The function f is integrable.
(ii) There exists at least one of the iterated integrals in (I) if f is replaced by $|f|$, i.e., $\int(\int|f|\,dy)\,dx$ exists or $\int(\int|f|\,dx)\,dy$ exists.

If condition (ii) is satisfied, then all the assertions of the Theorem of Fubini (23) are valid.

Parameter Integrals

We consider the function

$$F(p) = \int_M f(x, p)\,dx$$

for all parameters $p \in P$. Here,

$$f: M \times P \to \mathbb{R}$$

is a function, where M is a measurable subset of $\mathbb{R}^N$ and P is a metric space (e.g., a subset of $\mathbb{R}^K$).

(25a) *Continuity*. The function $F: P \to \mathbb{R}$ is well defined and continuous if the following three conditions are satisfied:

(i) The function $x \mapsto f(x, p)$ is measurable on M for all $p \in P$.
(ii) There exists an integrable function $g: M \to \mathbb{R}$ such that

$$|f(x, p)| \leq g(x)$$

holds for all $p \in P$ and almost all $x \in M$.
(iii) The function $p \mapsto f(x, p)$ is continuous on P for almost all $x \in M$.

(25b) *Differentiability*. Let P be an open subset of $\mathbb{R}$. The function $F: P \to \mathbb{R}$ is differentiable and

$$F'(p) = \int_M f_p(x, p)\,dx \qquad \text{for all} \quad p \in P$$

provided the following two conditions are satisfied:

(i) The integral $\int_M f(x, p)\,dx$ exists for all $p \in P$.
(ii) There exists an integrable function $g: M \to \mathbb{R}$ such that

$$|f_p(x, p)| \leq g(x)$$

for all $p \in P$ and almost all $x \in M$. (This condition tacitly includes the existence of the derivative $f_p(x, p)$ for all $p \in P$ and almost all $x \in M$.)

Lebesgue's Main Theorem of Calculus

(25c) *Main theorem*. Let $F: [a, b] \to \mathbb{R}$ be a function, where $-\infty < a < b < \infty$. Then the following two conditions are equivalent:

(i) There exists an integrable function $f\colon [a, b] \to \mathbb{R}$ such that

$$F(x) = \int_a^x f(t)\,dt \qquad \text{for all} \quad x \in [a, b].$$

(ii) F is absolutely continuous.

Moreover, if (i) holds, then

$$F'(x) = f(x) \qquad \text{for almost all} \quad x \in [a, b].$$

(25d) Recall that $F\colon [a, b] \to \mathbb{R}$ is said to be *absolutely continuous* iff, for each $\varepsilon > 0$, there exists a $\delta > 0$ such that

$$\sum_k |F(b_k) - F(a_k)| < \varepsilon$$

holds for all finite systems of pairwise disjoint subintervals $[a_k, b_k]$ of $[a, b]$ with total length $< \delta$.

(25e) *Differentiation of integrals with respect to the domain.* Let $f\colon M \subseteq \mathbb{R}^N \to Y$ be an integrable function from the open set M to the B-space Y, and let C be a cube in $\mathbb{R}^N$ with $x \in C$ and measure $m(C)$. Then, for almost all $x \in M$,

$$\lim_{m(C) \to 0} \frac{1}{m(C)} \int_C f(t)\,dt = f(x),$$

$$\lim_{m(C) \to 0} \frac{1}{m(C)} \int_C \|f(t) - f(x)\|\,dt = 0.$$

In particular, these limit relations hold for all points x at which f is continuous.

Lebesgue Spaces

Let G be a nonempty bounded open set[1] in $\mathbb{R}^N$, $N \geq 1$.

(26) *Definition of the Lebesgue space $L_p(G)$.* Let $1 \leq p < \infty$. We set

$$\|u\|_p = \left(\int_G |u|^p\,dx\right)^{1/p}.$$

Let $L_p(G)$ denote the set of all measurable functions

$$u\colon G \to \mathbb{R} \qquad \text{with} \quad \|u\|_p < \infty.$$

Then $L_p(G)$ together with the norm $\|\cdot\|_p$ becomes a real B-space once we identify any two functions which differ only on a set of N-dimensional Lebesgue measure zero.

More precisely, the two measurable functions u, $v\colon G \to \mathbb{R}$ are called equivalent iff $u(x) = v(x)$ for almost all $x \in G$. Then, the elements of the space $L_p(G)$

[1] Most of the following statements remain true if G is an unbounded open set or, more generally, if G is a nonempty measurable set in $\mathbb{R}^N$, $N \geq 1$.

are precisely all the equivalence classes of measurable functions $u: G \to \mathbb{R}$ with respect to this equivalence relation.

The space $L_2(G)$ together with the scalar product

$$(u|v) = \int_G uv\,dx$$

becomes an H-space.

(27) $L_p(G)$ is separable and $C_0^\infty(G)$ is dense in $L_p(G)$.

(28) If $1 < p < \infty$, then the B-space $L_p(G)$ is uniformly convex and hence reflexive.

(29) *The Hölder inequality.* Let $1 < p_1, \ldots, p_n < \infty$ be given with $\sum_{i=1}^n p_i^{-1} = 1$ and let $u_i \in L_{p_i}(G)$ for all i. Then

$$\left| \int_G \prod_{i=1}^n u_i\,dx \right| \le \prod_{i=1}^n \left(\int_G |u_i|^{p_i}\,dx \right)^{1/p_i},$$

where all the integrals exist.

(30) *Important inequalities for real numbers.* Let $1 \le s < \infty$, $0 < r < \infty$, $1 < p < \infty$, $p^{-1} + q^{-1} = 1$. Then, for all nonnegative real numbers $\xi_1, \ldots, \xi_N$, ξ, η, we have the following inequalities:

(30a) $$a\left(\sum_{i=1}^N \xi_i^s\right)^{1/s} \le \sum_{i=1}^N \xi_i \le b\left(\sum_{i=1}^N \xi_i^s\right)^{1/s},$$

(30b) $$\left(\sum_{i=1}^N \xi_i\right)^r \le c \sum_{i=1}^N \xi_i^r,$$

(30c) $$\xi\eta \le \frac{\xi^p}{p} + \frac{\eta^q}{q},$$

$$\prod_{i=1}^N \xi_i \le \sum_{i=1}^N \frac{\xi_i^{p_i}}{p_i} \qquad \text{(Young's inequality)},$$

where p_i is given as in (29), and

(30d) $$\sum_{i=1}^N \xi_i\eta_i \le \left(\sum_{i=1}^N \xi_i^p\right)^{1/p} \left(\sum_{i=1}^N \eta_i^q\right)^{1/q} \qquad \text{(Hölder's inequality)}.$$

Here, the positive constants a, b, c depend only on N, s, r.

These inequalities are frequently used in analysis in order to obtain estimates for integrals.

(31a) *Example* 1. Let $1 \le q, p_i < \infty$ and

$$u \in L_q(G), \qquad v_i \in L_{p_i}(G) \qquad \text{for all } i.$$

From the estimate

$$|w(x)| \leq \text{const}\left(|u(x)| + \sum_{i=1}^{N} |v_i(x)|^{p_i/q}\right)$$

for all $x \in G$, it follows that $w \in L_q(G)$ and

$$\|w\|_q \leq \text{const}\left(\|u\|_q + \sum_{i=1}^{N} \|v_i\|_{p_i}^{p_i/q}\right).$$

Proof. By (30b),

$$|w(x)|^q \leq \text{const}\left(|u(x)|^q + \sum_i |v_i(x)|^{p_i}\right).$$

Integration over G yields

$$\|w\|_q^q \leq \text{const}\left(\|u\|_q^q + \sum_i \|v_i\|_{p_i}^{p_i}\right).$$

Now the assertion follows from (30b). □

(31b) *Example* 2. Let $1 < p < \infty$, $p^{-1} + q^{-1} = 1$ and

$$u_i \in L_p(G), \qquad v_i \in L_q(G) \qquad \text{for all } i.$$

Then

$$\sum_{i=1}^{N} \left|\int_G u_i v_i \, dx\right| \leq \sum_{i=1}^{N} \|u_i\|_p \|v_i\|_q$$

$$\leq \text{const}\left(\sum_{i=1}^{N} \|u_i\|_p^p\right)^{1/p} \left(\sum_{i=1}^{N} \|v_i\|_q^q\right)^{1/q}.$$

This follows from the Hölder inequalities (29) and (30d).

(32) *Mean continuity*. Let $1 \leq p < \infty$. Every function $u \in L_p(G)$ is p-mean continuous, i.e., for each $\varepsilon > 0$, there exists a $\delta(\varepsilon) > 0$ such that

$$\int_G |u(x+h) - u(x)|^p \, dx < \varepsilon$$

provided $|h| < \delta(\varepsilon)$. In this connection, we set $u(x) = 0$ outside G.

(33) *The compactness theorem of Riesz–Kolmogorov*. Let $1 \leq p < \infty$ and suppose that S is a bounded set in $L_p(G)$. Then the following two conditions are equivalent:

(i) S is relatively compact.
(ii) S is *p-mean equicontinuous*. i.e., for each $\varepsilon > 0$, there exists a $\delta(\varepsilon) > 0$ such that

$$\sup_{u \in S} \int_G |u(x+h) - u(x)|^p \, dx < \varepsilon$$

provided $|h| < \delta(\varepsilon)$. Here, we set $u(x) = 0$ outside G. Note that $\delta(\varepsilon)$ does not depend on u.

(34) *The dual space.* Let $1 < p, q < \infty$ and $p^{-1} + q^{-1} = 1$. Then

$$(L_p(G))^* = L_q(G).$$

This relation is to be understood in the following sense:

(i) Let $u \in L_q(G)$ and set

$$U(v) = \int_G uv\,dx \qquad \text{for all} \quad v \in L_p(G).$$

Then $U: L_p(G) \to \mathbb{R}$ is a linear continuous functional on $L_p(G)$, i.e., $U \in (L_p(G))^*$. Moreover, we have

$$\|U\| = \|u\|_p.$$

(ii) Conversely, each $U \in (L_p(G))^*$ can be obtained in this way, where $u \in L_q(G)$ is uniquely determined by U.

Consequently, there exists a normisomorphism $u \mapsto U$ from $L_q(G)$ onto $(L_p(G))^*$.

Identifying U with u, we may write

$$\langle u, v \rangle = \int_G uv\,dx \qquad \text{for all} \quad u \in L_q(G), \quad v \in L_p(G).$$

(34a) The convergence $u_n \to u$ in $L_p(G)$ as $n \to \infty$ means $\|u_n - u\|_p \to 0$, i.e.,

$$\int_G |u_n - u|^p\,dx \to 0 \qquad \text{as} \quad n \to \infty.$$

(34b) The weak convergence $u_n \rightharpoonup u$ in $L_p(G)$ as $n \to \infty$ means $\langle v, u_n \rangle \to \langle v, u \rangle$ for all $v \in L_q(G)$, i.e.,

$$\int_G u_n v\,dx \to \int_G uv\,dx \qquad \text{as} \quad n \to \infty \qquad \text{for all} \quad v \in L_q(G).$$

(35) *A convergence theorem.* Let $1 < p, q < \infty$ and $p^{-1} + q^{-1} = 1$. From

$$u_n \to u \quad \text{in } L_p(G) \qquad \text{as} \quad n \to \infty,$$
$$v_n \rightharpoonup v \quad \text{in } L_q(G) \qquad \text{as} \quad n \to \infty,$$

it follows that

$$\int_G u_n v_n\,dx \to \int_G uv\,dx \qquad \text{as} \quad n \to \infty.$$

(36) *Convergent subsequences.* Let $1 \le p < \infty$.

(36a) From

$$u_n \to u \quad \text{in } L_p(G) \qquad \text{as} \quad n \to \infty$$

it follows that there exists a subsequence $(u_{n'})$ such that

$$u_{n'}(x) \to u(x) \qquad \text{as} \quad n \to \infty \qquad \text{for almost all} \quad x \in G.$$

Moreover, there exists a function $v \in L_p(G)$ such that

$$|u_{n'}(x)| \le v(x) \qquad \text{for all } n' \quad \text{and almost all} \quad x \in G.$$

(36b) From the weak convergence

$$u_n \rightharpoonup u \quad \text{in } L_p(G) \qquad \text{as} \quad n \to \infty$$

and

$$u_n(x) \to v(x) \qquad \text{as} \quad n \to \infty \quad \text{for almost all} \quad x \in G,$$

it follows that $u(x) = v(x)$ for almost all $x \in G$.

(36c) Let $1 < p < \infty$. Then each bounded sequence (u_n) in $L_p(G)$ has a subsequence $(u_{n'})$ with

$$u_{n'} \rightharpoonup u \quad \text{in } L_p(G) \qquad \text{as} \quad n \to \infty.$$

(36d) Each bounded sequence (u_n) in $L_\infty(G)$ has a subsequence $(u_{n'})$ with

$$u_{n'} \overset{*}{\rightharpoonup} u \quad \text{in } L_\infty(G) \qquad \text{as} \quad n \to \infty,$$

i.e., for all $v \in L_1(G)$,

$$\int_G u_{n'} v \, dx \to \int_G uv \, dx \qquad \text{as} \quad n \to \infty.$$

The definition of the space $L_\infty(G)$ can be found in (39) below.

(37) *Surface measure and surface integral.* Let G be a bounded region in $\mathbb{R}^N$, $N \ge 2$, with a sufficiently smooth boundary, i.e., $\partial G \in C^{0,1}$ in the sense of Section 6.2.

For sufficiently regular subsets of ∂G one can introduce a classical surface area by using classical surface integrals. In terms of general measure theory this leads to a premeasure on ∂G. By the main theorem of general measure theory (75f) below, this premeasure can be uniquely extended to a measure which we call the surface measure on ∂G.

According to (78) below, to each measure μ there corresponds the notion of a Lebesgue integral $\int f \, d\mu$. Let $f: \partial G \to \mathbb{R}$ be an integrable function with respect to the surface measure on ∂G. Then the corresponding Lebesgue integral is denoted by

$$\int_{\partial G} f \, dO.$$

(38) *The Lebesgue space $L_p(\partial G)$.* Let $1 \le p < \infty$ and let G be given as in

(37). We set

$$\|u\|_{p,\partial G} = \left(\int_{\partial G} |u|^p \, dO \right)^{1/p}.$$

Let $L_p(\partial G)$ denote the set of all measurable functions

$$u: \partial G \to \mathbb{R} \qquad \text{with} \qquad \|u\|_{p,\partial G} < \infty.$$

Here, the measurability of u refers to the surface measure.

Then $L_p(\partial G)$ together with the norm $\|\cdot\|_{p,\partial G}$ becomes a real B-space once we identify any two functions which differ only on a set of surface measure zero on ∂G.

Let $N = 1$ and $G =]a, b[$ with $-\infty < a < b < \infty$. We set

$$\|u\|_{p,\partial G} = (|u(a)|^p + |u(b)|^p)^{1/p}.$$

Let $L_p(\partial G)$ denote the set of all functions $u: \partial G \to \mathbb{R}$. Then $L_p(\partial G)$ together with the norm $\|\cdot\|_{p,\partial G}$ is a real B-space.

(39) *The space* $L_\infty(G)$. Let G be a nonempty bounded open set in $\mathbb{R}^N$, $N \geq 1$. The number B is said to be an essential bound for the function $u: G \to \mathbb{R}$ iff

$$|u(x)| \leq B \qquad \text{for almost all} \quad x \in G.$$

We set

$$\|u\|_\infty = \inf\{B: B \text{ is an essential bound for } u\}.$$

Let $L_\infty(G)$ denote the set of all measurable functions

$$u: G \to \mathbb{R} \qquad \text{with} \qquad \|u\|_\infty < \infty.$$

Then $L_\infty(G)$ together with the norm $\|\cdot\|_\infty$ becomes a real B-space once we identify any two functions which differ only on a set of N-dimensional measure zero.

(39a) *Embedding theorem.* Let $1 \leq q \leq p \leq \infty$. Then the embedding

$$L_p(G) \subseteq L_q(G)$$

is continuous.

(40) *The dual space to* $L_1(G)$. We have

$$(L_1(G))^* = L_\infty(G).$$

This is to be understood in the sense of (34) with $p = 1$ and $q = \infty$. In particular,

$$\langle u, v \rangle = \int_G uv \, dx$$

for all $u \in L_\infty(G)$, $v \in L_1(G)$.

The spaces $L_1(G)$ and $L_\infty(G)$ are *not* reflexive. Hence the notion of weak* convergence on $L_\infty(G)$ becomes important.

(40a) *Weak convergence in* $L_1(G)$. The weak convergence

$$v_n \rightharpoonup v \quad \text{in } L_1(G) \qquad \text{as} \quad n \to \infty$$

means $\langle u, v_n \rangle \to \langle u, v \rangle$ for all $u \in L_\infty(G)$, i.e.,

$$\int_G uv_n\,dx \to \int_G uv\,dx \qquad \text{as} \quad n \to \infty \qquad \text{for all} \quad u \in L_\infty(G).$$

(40b) *Weak* convergence in* $L_\infty(G)$. The weak* convergence

$$u_n \overset{*}{\rightharpoonup} u \quad \text{in } L_\infty(G) \qquad \text{as} \quad n \to \infty$$

means $\langle u_n, v \rangle \to \langle u, v \rangle$ for all $v \in L_1(G)$, i.e.,

$$\int_G u_n v\,dx \to \int_G uv\,dx \qquad \text{as} \quad n \to \infty \qquad \text{for all} \quad v \in L_1(G).$$

(40c) *A convergence theorem.* From

$$u_n \overset{*}{\rightharpoonup} u \quad \text{in } L_\infty(G) \qquad \text{as} \quad n \to \infty,$$

$$v_n \to v \quad \text{in } L_1(G) \qquad \text{as} \quad n \to \infty,$$

it follows that $\int_G u_n v_n\,dx \to \int_G uv\,dx$ as $n \to \infty$.

Sobolev Spaces

Let G be a bounded region in $\mathbb{R}^N$ with $N \geq 1$. The Sobolev spaces

$$W_p^k(G) \qquad \text{and} \qquad \mathring{W}_p^k(G)$$

with $1 \leq p \leq \infty$ and $k = 0, 1, \ldots$ have been introduced in Section 21.2.

(41) $W_p^k(G)$ is separable for $1 \leq p < \infty$.

(42) $W_p^k(G)$ is uniformly convex and hence reflexive for $1 < p < \infty$.

(43) *Density.* Let $1 \leq p < \infty$. Then $C_0^\infty(G)$ is dense in $\mathring{W}_p^k(G)$.

Moreover, the set $C^\infty(G) \cap W_p^k(G)$ is dense in $W_p^k(G)$.

If the boundary ∂G is piecewise smooth,[1] i.e., $\partial G \in C^{0,1}$, then $C^\infty(\bar{G})$ is dense in $W_p^k(G)$.

The set $C_0^\infty(\mathbb{R}^N)$ is dense in $W_p^k(\mathbb{R}^N)$, and hence

$$W_p^k(\mathbb{R}^N) = \mathring{W}_p^k(\mathbb{R}^N).$$

(43a) *The universal extension theorem.* Let $1 \leq p \leq \infty$ and let $\partial G \in C^{0,1}$. Then there exists a linear continuous operator

$$E\colon W_p^k(G) \to W_p^k(\mathbb{R}^N)$$

[1] For $N = 1$, the condition "$\partial G \in C^{0,1}$" means that G is a bounded open interval.

with

$$(Eu)(x) = u(x) \qquad \text{for almost all} \quad x \in G.$$

In addition, the extension Eu of u is C^∞ outside $\bar{G}$.

The operator E is universal, i.e., E is independent of p and $k = 0, 1, \ldots$.

The same result holds if G is an open half-space in $\mathbb{R}^N$. In this case, the operator E can be constructed in such a way that the boundedness of supp u implies the boundedness of supp Eu.

(43b) Let $1 \leq p < \infty$ and $k = 0, 1, \ldots$. Let G be a nonempty open set in $\mathbb{R}^N$, $N \geq 1$. Then there exists a linear continuous operator

$$E_0: \mathring{W}_p^k(G) \to \mathring{W}_p^k(\mathbb{R}^N)$$

such that $E_0 u = u$ on G and $\|E_0 u\| = \|u\|$ for all $u \in \mathring{W}_p^k(G)$.

Further material on the extension of functions can be found in Problems 21.7 and 21.8.

(44a) The dual spaces $W_p^k(G)^*$ and $\mathring{W}_p^k(G)^*$, negative norms and distributions. See (71) below.

(44b) Sobolev spaces and the Fourier transform. See (74) below.

The Sobolev Embedding Theorems

Let $X \subseteq Y$. The embedding operator $E: X \to Y$ assigns to each $x \in X$ the corresponding element $x \in Y$. The embedding $X \subseteq Y$ is called continuous or compact iff the operator E is continuous or compact, respectively.

(45) *Main embedding theorem.* Let G be a bounded region in $\mathbb{R}^N$ with $N \geq 1$ and piecewise smooth boundary,[1] i.e., $\partial G \in C^{0,1}$. Let

$$0 \leq j < k, \qquad 1 \leq p, q < \infty, \qquad 0 < \alpha < 1.$$

We set

$$d = \frac{1}{p} - \frac{k-j}{N}.$$

Then the following fundamental assertions hold.

(45a) The embeddings

$$W_p^k(G) \subseteq W_q^j(G)$$

and

$$\mathring{W}_p^k(G) \subseteq \mathring{W}_q^j(G)$$

are continuous for $d \leq 1/q$ and *compact* for $d < 1/q$.

[1] For $N = 1$, the condition "$\partial G \in C^{0,1}$" means that G is a bounded open interval.

This statement for $\mathring{W}_p^k(G) \subseteq \mathring{W}_q^j(G)$ remains valid if the boundary of G is arbitrary.

(45b) The embeddings

$$W_p^k(G) \subseteq C^{j,\alpha}(\bar{G}) \subseteq C^j(\bar{G})$$

are *compact* for $d + \alpha/N < 0$. In particular, the embedding

$$W_p^k(G) \subseteq C^j(\bar{G})$$

is compact for $d < 0$, i.e., $k - j > N/p$.

More precisely, this means that a function $u \in W_p^k(G)$ also belongs to $C^{j,\alpha}(\bar{G})$ and $C^j(\bar{G})$ provided we change u on a suitable set of N-dimensional measure zero.

(45c) Let G be a bounded region in $\mathbb{R}^N$, $N \geq 1$. The embedding

$$W_p^k(G) \subseteq W_q^k(G)$$

is continuous for $1 \leq q \leq p \leq \infty$ and $k = 0, 1, 2, \ldots$.

(45d) The embedding

$$W_p^k(\mathbb{R}^N) \subseteq C^j(\mathbb{R}^N)$$

is continuous if $d < 0$, i.e., $k - j > N/p$ and $1 \leq p < \infty$. Here, we equip $C^j(\mathbb{R}^N)$ with the usual norm $\|u\|_{j,\infty}$ (cf. Section 21.2).

(45e) The embedding

$$W_p^k(\mathbb{R}^N) \subseteq W_q^j(\mathbb{R}^N)$$

is continuous if $0 \leq j < k$, $1 \leq p, q < \infty$ and $d \leq 1/q$.

(46a) *Standard example* 1. Let $1 \leq p < \infty$ and let G be a bounded region in $\mathbb{R}^N$, $N \geq 1$. Then the embeddings

$$L_p(G) \supseteq \mathring{W}_p^1(G) \supseteq \mathring{W}_p^2(G) \supseteq \cdots$$

are *compact*. If $\partial G \in C^{0,1}$, then the embeddings

$$L_p(G) \supseteq W_p^1(G) \supseteq W_p^2(G) \supseteq \cdots$$

are also compact.

(46b) *Standard example* 2. Let G be a bounded region in $\mathbb{R}^N$ with $\partial G \in C^{0,1}$. Then the embeddings

$$W_2^1(G) \subseteq L_q(G) \qquad \text{and} \qquad \mathring{W}_2^1(G) \subseteq L_q(G)$$

are compact if either

$$N = 1, 2 \qquad \text{and} \qquad 1 \leq q < \infty$$

or

$$N > 2 \qquad \text{and} \qquad 1 \leq q < 2N/(N-2).$$

This result remains valid for $\mathring{W}_2^1(G)$ if the boundary of G is arbitrary.

In Part IV we will use the compactness of the embedding

$$\mathring{W}_2^1(G) \subseteq L_4(G)$$

in $\mathbb{R}^3$ in order to give existence proofs for the Navier–Stokes equations.

(46c) *Standard example* 3. Let G be a bounded open interval in $\mathbb{R}^1$. Then the embeddings

$$W_2^{1+j}(G) \subseteq C^{j,\alpha}(\bar{G}) \subseteq C^j(\bar{G})$$

are compact for $0 < \alpha < 1/2$ and $j = 0, 1, \ldots$.

The following special cases of (45) show that this situation in $\mathbb{R}^1$ becomes worse for increasing dimension N.

(47) *Critical exponents and prototypes of the general Sobolev embedding theorems.* We define the following critical exponents:

$$q_{\text{crit}} = \begin{cases} \dfrac{pN}{N-p} & \text{for} \quad N > p, \\ \infty & \text{for} \quad N \leq p, \end{cases}$$

$$\alpha_{\text{crit}} = \begin{cases} 1 - \dfrac{N}{p} & \text{for} \quad N < p, \\ -\infty & \text{for} \quad N \geq p. \end{cases}$$

Obviously, $q_{\text{crit}} > p$. Let G be given as in (45).

Case 1: $N < p < \infty$. The embeddings

$$W_p^1(G) \subseteq C^\alpha(\bar{G}) \subseteq C(\bar{G}), \qquad 0 < \alpha < \alpha_{\text{crit}},$$

$$W_p^1(G) \subseteq L_q(G), \qquad 1 \leq q < \infty,$$

are compact. Here, $q_{\text{crit}} = \infty$. The embeddings

$$W_p^1(G) \subseteq C^\alpha(\bar{G}), \qquad \alpha = \alpha_{\text{crit}},$$

are continuous.

Case 2: $1 \leq p < N$. The embeddings

$$W_p^1(G) \subseteq L_q(G)$$

are compact for $1 \leq q < q_{\text{crit}}$ and continuous for $q = q_{\text{crit}}$.

Case 3: $p = N$. The embeddings

$$W_p^1(G) \subseteq L_q(G), \qquad 1 \leq q < \infty,$$

are compact. Here, $q_{\text{crit}} = \infty$.

Case 4: The embeddings

$$C^\alpha(\bar{G}) \subseteq C^\beta(\bar{G}) \subseteq C(\bar{G}), \qquad 0 < \beta < \alpha \leq 1,$$

$$C^{0,1}(\bar{G}) \subseteq C^\beta(\bar{G}) \subseteq C(\bar{G}), \qquad 0 < \beta < 1,$$

are compact.

Table 1

	$p = 2$		$p = 3$	
dimension N	q_{crit}	α_{crit}	q_{crit}	α_{crit}
1	∞	1/2	∞	2/3
2	∞	$-\infty$	∞	1/3
3	6	$-\infty$	∞	$-\infty$
4	4	$-\infty$	12	$-\infty$

Case 5: $N = 1$ and $1 \le p < \infty$. Let $G =]a, b[$ with $-\infty < a < b < \infty$. Then the space $W_p^1(G)$ consists of all absolutely continuous functions $u: [a, b] \to \mathbb{R}$ whose derivative u' (which exists almost everywhere) belongs to $L_p(G)$. More precisely, each $u \in W_p^1(G)$ has this property after changing the values on a set of measure zero. The embeddings

$$W_p^1(G) \subseteq C^{1-1/p}(\bar{G}) \subseteq C(\bar{G}), \qquad 1 \le p < \infty,$$

are continuous.

Table 1 contains the critical exponents for $p = 2, 3$. Note that $\alpha_{\text{crit}} = -\infty$ means that the favorable Case 1 is not at hand.

The general embedding theorems (45) follow from these prototypes.

(47a) *Example.* Let $N = 3$. By Case 2, the embeddings

$$W_2^2(G) \subseteq W_q^1(G), \qquad 1 \le q \le 6,$$

are continuous. By Case 1, the embeddings

$$W_6^1(G) \subseteq C^\alpha(\bar{G}) \subseteq C(\bar{G}), \qquad 0 < \alpha < 1/2,$$

are compact. Thus, the embeddings

$$W_2^2(G) \subseteq C^\alpha(\bar{G}) \subseteq C(\bar{G}), \qquad 0 < \alpha < 1/2,$$

are compact. The same result follows from (45b). Moreover, by Case 1, the embeddings

$$W_2^2(G) \subseteq W_6^1(G) \subseteq C^{1/2}(\bar{G})$$

are continuous.

(48) *Generalized boundary values.* Let G be a bounded region in $\mathbb{R}^N$ with $N \ge 1$ and $\partial G \in C^{0,1}$. Let $1 \le p < \infty$. Then, for each such p, there exists exactly one linear continuous operator

$$B_p: W_p^1(G) \to L_p(\partial G)$$

such that, for all $u \in C^1(\bar{G})$, the following holds:

$$B_p u = \textit{classical boundary function of } u \textit{ with respect to } \partial G.$$

If $u \in W_p^1(G)$, then we call $b = B_p u$ the generalized boundary function of u (or the trace of u).

The function b is uniquely determined as an element of the space $L_p(\partial G)$.

That is, for $N \geq 2$, b is uniquely determined up to changing the values on a set of surface measure zero.

In the case where $N = 1$, the embedding $W_p^1(G) \subseteq C(\bar{G})$ is continuous. That is, the function $u \in W_p^1(G)$ is continuous on $\bar{G} = [a, b]$ after changing the values on a set of measure zero on $\bar{G}$. Then $B_p u$ corresponds to the classical boundary values $u(a)$ and $u(b)$ of this uniquely determined continuous function.

(48a) For all $u \in W_p^1(G)$,

$$\|B_p u\|_{L_p(\partial G)} \leq \text{const}\, \|u\|_{W_p^1(G)}.$$

(48b) $u \in \mathring{W}_p^1(G)$ is equivalent to

$$u \in W_p^1(G) \qquad \text{and} \qquad u = 0 \quad \text{on } \partial G$$

in the sense of generalized boundary values.

(48c) If $u \in \mathring{W}_p^k(G)$, $k \geq 1$, then

$$D^\alpha u \in \mathring{W}_p^1(G) \qquad \text{for all} \quad \alpha\colon |\alpha| \leq k - 1.$$

This implies $D^\alpha u = 0$ on ∂G for all $\alpha\colon |\alpha| \leq k - 1$, in the sense of generalized boundary values.

(49) *Characterization of generalized boundary values.* Let G be a bounded region in $\mathbb{R}^N$ with $N \geq 2$ and $\partial G \in C^{0,1}$, and let $1 < p < \infty$. It is remarkable that there are functions $b \in L_p(\partial G)$ which are not the generalized boundary functions to a function $u \in W_p^1(G)$. Those boundary functions $b \in L_p(\partial G)$, which correspond to functions $u \in W_p^1(G)$, form the proper subset $W_p^{1-1/p}(\partial G)$ of $L_p(\partial G)$. This underlines the importance of the Sobolev spaces W_p^k of fractional order k which will be introduced in (50ff) below. More precisely, we have the following two results which are very important for the modern theory of boundary value problems.

(49a) For the boundary operator, the mapping

$$B_p\colon W_p^1(G) \to W_p^{1-1/p}(\partial G)$$

is linear, continuous, and *surjective*.

(49b) There exists a linear continuous operator

$$E_p\colon W_p^{1-1/p}(\partial G) \to W_p^1(G)$$

such that the following diagram is commutative:

$$\begin{array}{ccc} W_p^1(G) & \xrightarrow{B_p} & W_p^{1-1/p}(\partial G) \\ & {\scriptstyle id}\nwarrow \quad \swarrow{\scriptstyle E_p} & \\ & W_p^1(G). & \end{array}$$

Consequently, exactly the functions

$$b \in W_p^{1-1/p}(\partial G)$$

are generalized boundary functions of functions $u \in W_p^1(G)$, and the function $u = E_p b$ is an extension of the boundary function b to the region G.

Sobolev Spaces of Fractional Order

(50) *The space $W_p^k(G)$ for noninteger k.* Let G be a bounded region in $\mathbb{R}^N$ with $N \geq 1$ and $\partial G \in C^{0,1}$. Moreover, let $1 \leq p < \infty$ and

$$k = m + \mu, \qquad m = 0, 1, \ldots, \qquad 0 < \mu < 1.$$

We set

$$f(u) = \int_{G \times G} \frac{|u(x) - u(y)|^p}{|x - y|^{N + p\mu}} \, dx \, dy$$

and

$$\|u\|_{k,p} = \left(\|u\|_{m,p}^p + \sum_{|\alpha| \leq m} f(D^\alpha u) \right)^{1/p}.$$

By definition,

$$u \in W_p^k(G) \qquad \text{iff} \qquad u \in W_p^m(G) \qquad \text{and} \qquad \|u\|_{k,p} < \infty.$$

Then $W_p^k(G)$ together with the norm $\|\cdot\|_{k,p}$ becomes a real separable B-space which is reflexive for $1 < p < \infty$. The embedding

$$C^{m,\beta}(\bar{G}) \subseteq W_p^k(G)$$

is continuous for $0 < \mu < \beta \leq 1$.

According to the definition above, the space $W_p^k(G)$ can be regarded as an integral variant of the Hölder space $C^{m,\mu}(\bar{G})$.

(51a) *The boundary space $W_p^k(\partial G)$ for $0 < k < 1$.* Let G be a bounded region in $\mathbb{R}^N$ with $N \geq 2$ and $\partial G \in C^{0,1}$ and let $1 \leq p < \infty$. We set

$$f(b) = \int_{\partial G \times \partial G} \frac{|b(x) - b(y)|^p}{|x - y|^{N-1+pk}} \, dO \, dO$$

and

$$\|b\|_{W_p^k(\partial G)} = \left(\int_G |b|^p \, dO + f(b) \right)^{1/p}.$$

By definition,

$$b \in W_p^k(\partial G) \qquad \text{iff} \qquad b \in L_p(\partial G) \qquad \text{and} \qquad \|b\|_{W_p^k(\partial G)} < \infty.$$

Then $W_p^k(\partial G)$ together with the norm $\|\cdot\|_{W_p^k(\partial G)}$ becomes a real separable B-space which is reflexive for $1 < p < \infty$.

(51b) *The boundary space $W_p^k(G)$ for noninteger $k > 1$.* Let

$$k = m + \mu, \qquad m = 1, 2, \ldots, \qquad 0 < \mu < 1,$$

and let G be a bounded region in $\mathbb{R}^N$ with $N \geq 2$ and $\partial G \in C^{m,1}$. According to Section 6.2, the boundary ∂G can be covered by finitely many $C^{m,1}$-surfaces S_i. Furthermore, with respect to a suitable local coordinate system, the surface S_i is given by the equation

$$\zeta = \zeta(\xi), \qquad \xi \in C_i,$$

where $\zeta: C_i \to \mathbb{R}$ is a $C^{m,1}$-function on the open $(N-1)$-dimensional cube C_i. We set

$$\|b\|_{W_p^k(\partial G)} = \left(\sum_i \|b\|_{W_p^k(C_i)}^p\right)^{1/p}.$$

By definition,

$$b \in W_p^k(\partial G) \qquad \text{iff} \qquad b \in L_p(\partial G) \qquad \text{and} \qquad \|b\|_{W_p^k(\partial G)} < \infty.$$

This definition includes tacitly that $b \in W_p^k(C_i)$ for all i.

Then $W_p^k(\partial G)$ together with the norm $\|\cdot\|_{W_p^k(\partial G)}$ becomes a real separable B-space which is reflexive for $1 < p < \infty$.

Equivalent Norms on Sobolev Spaces

Let X be a B-space. Recall that the two norms $\|\cdot\|_1$ and $\|\cdot\|_2$ are equivalent on X iff there are positive constants c and d such that

$$c\|u\|_2 \leq \|u\|_1 \leq d\|u\|_2 \qquad \text{for all} \quad u \in X.$$

(52) *A general theorem.* Let G be a bounded region in $\mathbb{R}^N$ with $N \geq 1$ and $\partial G \in C^{0,1}$, and let $1 \leq p < \infty$, $k = 1, 2, \ldots$. We set

$$\|u\| = \left(\int_G \sum_{|\alpha|=k} |D^\alpha u|^p \, dx + \sum_{j=1}^J f_j(u)^p\right)^{1/p}.$$

Recall that

$$\|u\|_{k,p} = \left(\int_G \sum_{|\alpha|\leq k} |D^\alpha u|^p \, dx\right)^{1/p}.$$

Then the norms $\|\cdot\|$ and $\|\cdot\|_{k,p}$ are equivalent on the Sobolev space $W_p^k(G)$ provided the following two conditions are satisfied:

(i) The functions $f_j: W_p^k(G) \to \mathbb{R}$ are seminorms on $W_p^k(G)$ for all j with

$$0 \leq f_j(u) \leq \text{const}\,\|u\|_{k,p} \qquad \text{for all} \quad u \in W_p^k(G) \quad \text{and all } j.$$

(ii) If

$$f_j(P) = 0, \qquad j = 1, \ldots, J,$$

for a polynomial $P: \mathbb{R}^N \to \mathbb{R}$ of degree $\leq k-1$, then $P \equiv 0$.

(53a) *Standard example* 1. On the Sobolev space $W_p^1(G)$, each of the following three norms is equivalent to the original norm $\|u\|_{1,p}$:

$$\left(\int_G \sum_{i=1}^N |D_i u|^p\,dx + \left|\int_G u\,dx\right|^p\right)^{1/p},$$

$$\left(\int_G \sum_{i=1}^N |D_i u|^p\,dx + \left|\int_{\partial G} u\,dO\right|^p\right)^{1/p},$$

$$\left(\int_G \sum_{i=1}^N |D_i u|^p\,dx + \int_{\partial G} |u|^p\,dO\right)^{1/p}.$$

Note that in the case where $N = 1$ and $G =]a, b[$, $-\infty < a < b < \infty$, we set

$$\int_{\partial G} g\,dO = g(a) + g(b),$$

in contrast to our convention in Chapter 18.

Proof. This is a simple consequence of (52). Note that in this special case, condition (ii) ab*ove must be checked only for P =* constant. □

(53b) *Standard example* 2. On the Sobolev space $\mathring{W}_p^k(G)$, the norm

$$\|u\|_{k,p,0} = \left(\int_G \sum_{|\alpha|=k} |D^\alpha u|^p\,dx\right)^{1/p}$$

is equivalent to the original norm $\|\cdot\|_{k,p}$. This theorem remains valid if the assumption $\partial G \in C^{0,1}$ drops out.

(53c) *Standard example* 3. Let G be a bounded region in $\mathbb{R}^N$, $N \geq 2$, with $\partial G \in C^{0,1}$. Let $\partial_1 G$ and $\partial_2 G$ be disjoint open subsets of ∂G with

$$\partial G = \overline{\partial_1 G} \cup \overline{\partial_2 G}, \qquad \partial_1 G \neq \varnothing.$$

Then

$$\left(\int_G \sum_{i=1}^N |D_i u|^p\,dx + \int_{\partial_1 G} |u|^p\,dO\right)^{1/p}$$

is an equivalent norm on $W_p^1(G)$, $1 \leq p < \infty$.

Let $X = \{u \in W_p^1(G): u = 0 \text{ on } \partial_1 G\}$. Then

$$\left(\int_G \sum_{i=1}^N |D_i u|^p\,dx\right)^{1/p}$$

is an equivalent norm on X.

This fact is fundamental for the investigation of mixed boundary value problems.

The Gagliardo–Nirenberg Inequalities

The following results are fundamental for handling the nonlinearities of partial differential equations by means of Sobolev spaces.

(54a) Let $G = \mathbb{R}^N$, $N \geq 1$, or let G be a bounded region in $\mathbb{R}^N$ with $\partial G \in C^{0,1}$. Let $m = 1, 2, \ldots$ and $1 \leq p, q, r \leq \infty$. Then, for all

$$u \in W_p^m(G) \cap L_q(G),$$

we have

$$\|D^\beta u\|_r \leq \text{const}\, \|u\|_{m,p}^\theta \|u\|_q^{1-\theta}$$

provided that $0 \leq |\beta| \leq m - 1$, $\theta = |\beta|/m$, and

$$\frac{1}{r} = \frac{|\beta|}{N} + \theta\left(\frac{1}{p} - \frac{m}{N}\right) + (1 - \theta)\frac{1}{q}.$$

If $m - |\beta| - N/p$ is *not* a nonnegative integer, then the values $|\beta|/m \leq \theta \leq 1$ are allowed for θ.

(54b) If $G = \mathbb{R}^N$, then

$$\|D^\beta u\|_r \leq \text{const}\left(\sum_{|\alpha|=m} \|D^\alpha u\|_p\right)^\theta \|u\|_q^{1-\theta},$$

for all $u \in W_p^m(G) \cap L_q(G)$.

(54c) *Banach algebra.* We set $X = W_p^m(G)$, where $m = 1, 2, \ldots$ and $1 \leq p \leq \infty$. Let $G = \mathbb{R}^N$ or let G be a bounded region in $\mathbb{R}^N$ with $N \geq 1$ and $\partial G \in C^{0,1}$. Furthermore, let

$$mp > N.$$

Then the Sobolev space X forms a generalized Banach algebra, i.e., $u, v \in X$ implies $uv \in X$ and $\|uv\| \leq \text{const}\, \|u\|\, \|v\|$ for all $u, v \in X$.

This result remains true for $m = 0$ and $p = \infty$.

(55) *Abstract interpolation operator on Sobolev spaces.* Let Q be an open cuboid or an open simplex in $\mathbb{R}^N$, $N \geq 1$, where

$$h = \text{diameter of } Q;$$

$$\rho = \text{radius of the largest ball contained in } Q.$$

We set

$$(Au)(x) = u(Bx + b),$$

where $B\colon \mathbb{R}^N \to \mathbb{R}^N$ is an arbitrary linear bijective operator, and b is an arbitrary element in $\mathbb{R}^N$. Furthermore, let $1 \leq p \leq \infty$ and $0 \leq m \leq k + 1$, where m and k are integers.

Suppose that the linear continuous operator

$$\Pi\colon W_p^{k+1}(Q) \to W_p^m(Q)$$

has the following two properties:

(i) $\Pi u = u$ for all polynomials of degree $\leq k$.
(ii) Π is affinely invariant, i.e., we have $\Pi A = A\Pi$ for all the operators A introduced above.

Then, for all $u \in W_p^{k+1}(Q)$,

(55a) $$\|u - \Pi u\|_{m,p,0} \leq \text{const}\, h^{k+1} \rho^{-m} \|u\|_{k+1,p,0}.$$

The constant depends on N, k, m, and p, but *not* on Q.

This result will be used basically in (59) below in order to prove important approximation properties of finite elements. The proof proceeds analogously to the proof of Proposition 21.52 (cf. Ciarlet (1977, M), p. 121).

Galerkin Schemes in Sobolev Spaces via Polynomials

(56) *Polynomial basis in* $W_p^k(G)$. We set

$$w_{i_1 \dots i_N}(x) = \xi_1^{i_1} \xi_2^{i_2} \dots \xi_N^{i_N},$$

where $i_1, \dots, i_N = 0, 1, \dots, x = (\xi_1, \dots, \xi_N)$. Let G be a bounded region in $\mathbb{R}^N$ with $\partial G \in C^{0,1}$ and $N \geq 1$. Furthermore, let $1 \leq p < \infty$ and $k = 0, 1, 2, \dots$.

Then all the functions $w_{\dots}$ form a basis in $W_p^k(G)$. If we set

$$X_n = \text{span}\{w_{i_1 \dots i_N}\colon 0 \leq i_1 + \cdots + i_N \leq n\},$$

then X_n consists of all real polynomials of degree $\leq n$, and (X_n) forms a Galerkin scheme in $W_p^k(G)$.

(57) *Quasi-polynomial basis in* $\mathring{W}_p^k(G)$. We replace the functions $w_{\dots}$ above by

$$v_{i_1 \dots i_N}(x) = \varphi(x)^k w_{i_1 \dots i_N}(x),$$

where $i_1, \dots, i_N = 0, 1, \dots$. Let G and G_1 be bounded regions in $\mathbb{R}^N$ with $\bar{G} \subset G_1$ and $N \geq 1$. Furthermore, let $1 \leq p < \infty$ and $k = 0, 1, \dots$.

Let $\varphi\colon \bar{G}_1 \to \mathbb{R}$ be a fixed function with $\varphi \in C^{k+2}(G)$ and

$$\varphi(x) = 0 \qquad \text{iff} \qquad x \in \partial G.$$

Moreover, assume that

$$\sum_{i=1}^{N} \left(\frac{\partial \varphi(x)}{\partial \xi_i}\right)^2 \neq 0 \qquad \text{for all} \quad x \in \partial G.$$

We set $X = \mathring{W}_p^k(G)$ and

$$\mathring{X}_n = \text{span}\{v_{i_1 \dots i_N}\colon 0 \leq i_1, \dots, i_N \leq n\}.$$

For fixed $r = 0, 1, \dots, \gamma \in [0, 1]$ and $k = 1, 2, \dots$, let

$$Y = \{u \in C^{k+r,\gamma}(\bar{G})\colon D^\alpha u = 0 \text{ on } \partial G \text{ for all } \alpha\colon |\alpha| \leq k - 1\}.$$

Then:

(i) All the functions $v_{\dots}$ form a basis in X, and $(\mathring{X}_n)$ is a Galerkin scheme in X.

(ii) For all $u \in Y$,

$$\text{dist}_X(u, \mathring{X}_n) \leq \frac{\text{const}}{n^{r+\gamma}} \|u\|_Y, \qquad n = 1, 2, \dots.$$

The constant depends on k, r, γ, p, and G. Cf. Ciarlet, Schultz, and Varga (1969).

Galerkin Schemes in Sobolev Spaces and the Method of Finite Elements

(58) *Basic ideas.* Suppose we want to solve an elliptic boundary value problem via the Galerkin method. If we use a polynomial Galerkin scheme as described in (56) and (57) above, then, as a rule, the corresponding matrices are *not* sparse matrices, i.e., most of the entries are different from zero. This is a typical shortcoming of polynomial bases. In contrast to this unfavorable situation, both the difference method and the method of finite elements produce *sparse matrices.* Compared with the difference method, the method of finite elements is much more flexible. For example, we can vary the mesh size of the triangulation in different subregions depending on the expected subtle behavior of the solution (Figs. 2(a), (b)). Today, the method of finite elements is widely used in engineering and in the natural sciences. This method represents one of the important *achievements* of modern numerical mathematics.

(58a) *Prototype for finite elements in* $\mathbb{R}^1$. Let $\alpha < \beta < \gamma$. We start with

$$w(x) = \begin{cases} 0 & \text{if} \quad x = \alpha, \gamma, \\ 1 & \text{if} \quad x = \beta, \end{cases}$$

and extend w to a function $w: \mathbb{R} \to \mathbb{R}$ via linear interpolation. This way

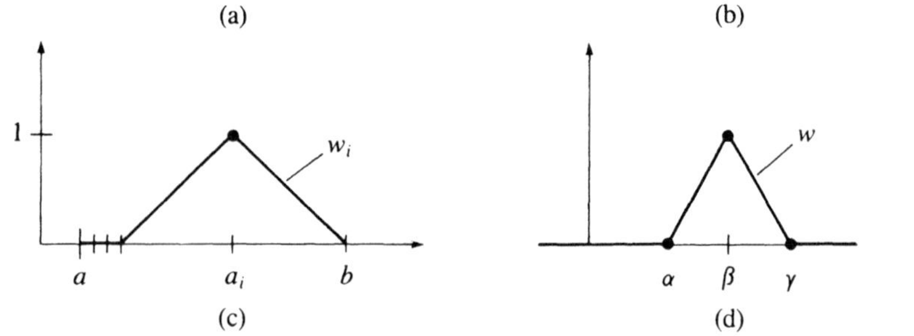

Figure 2

we obtain

$$w(x) = \begin{cases} 0 & \text{if} \quad x \notin [\alpha, \gamma], \\ (x - \alpha)/(\beta - \alpha) & \text{if} \quad x \in [\alpha, \beta], \\ (\gamma - x)/(\gamma - \beta) & \text{if} \quad x \in [\beta, \gamma], \end{cases}$$

(Fig. 2(d)). We now consider a triangulation $\mathscr{T}$ of the compact interval $[a, b]$, i.e., we choose nodes a_i such that $a = a_0 < a_1 < a_2 < \cdots < a_k = b$. The mesh size of $\mathscr{T}$ is defined through

$$h(\mathscr{T}) = \max_i (a_{i+1} - a_i).$$

We set

$$w_j(a_i) = \begin{cases} 1 & \text{if} \quad i = j, \\ 0 & \text{if} \quad i \neq j, \end{cases}$$

and extend w_j to a function $w_j\colon [a, b] \to \mathbb{R}$ via linear interpolation. The functions $w_0, w_1, \ldots, w_k$ are called *finite elements*. Furthermore, we set

$$X(\mathscr{T}) = \operatorname{span}\{w_0, \ldots, w_k\}$$

and

$$\mathring{X}(\mathscr{T}) = \operatorname{span}\{w_1, \ldots, w_{k-1}\}.$$

Then, $X(\mathscr{T})$ consists of all piecewise linear, continuous functions $w\colon [a, b] \to \mathbb{R}$ with arbitrarily prescribed values $w(a_0), \ldots, w(a_k)$ at the nodes (Fig. 2(b)). Moreover, we get

$$\mathring{X}(\mathscr{T}) = \{w \in X(\mathscr{T})\colon w(a) = w(b) = 0\}.$$

Obviously, we obtain the *generalized orthogonality* relations

$$\int_a^b w_i(x) w_j(x)\, dx = 0 \qquad \text{if} \quad |i - j| \geq 2.$$

These relations are responsible for the appearance of *sparse matrices* in connection with the Galerkin method.

Now, let $(\mathscr{T}_n)$ be a sequence of triangulations of the interval $[a, b]$, where the mesh size goes to zero as $n \to \infty$, i.e., $h(\mathscr{T}_n) \to 0$ as $n \to \infty$. Let $1 \leq p < \infty$ and set $h_n = h(\mathscr{T}_n)$ as well as $X_n = X(\mathscr{T}_n)$ and $\mathring{X}_n = \mathring{X}(\mathscr{T}_n)$. It follows from Section 21.14 that:

(a) (X_n) *forms a Galerkin scheme in* $X = W_p^1(a, b)$.

(b) *For all* $u \in W_p^2(a, b)$,

$$\operatorname{dist}_X(u, X_n) \leq \text{const}\, h_n \|u\|_{2,p},$$

i.e., the accuracy of approximation is proportional to the mesh size h_n.

(c) $(\mathring{X}_n)$ *forms a Galerkin scheme in* $\mathring{X} = \mathring{W}_p^1(a, b)$.

(d) *For all* $u \in W_p^2(a, b) \cap \mathring{X}$,

$$\operatorname{dist}_X(u, \mathring{X}_n) \leq \text{const}\, h_n \|u\|_{2,p}.$$

The constants in (b) and (d) depend on $b - a$ and p.

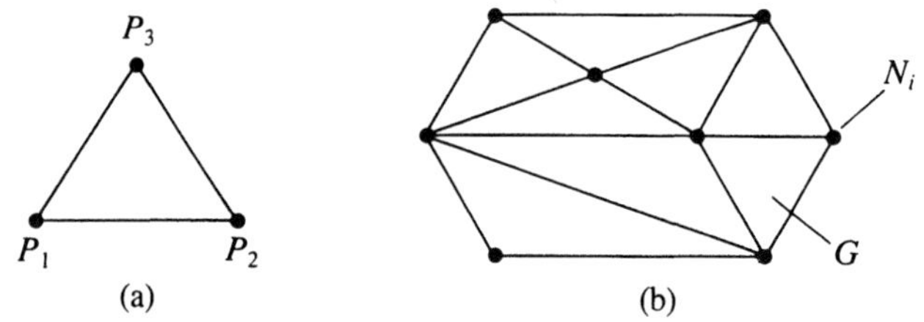

Figure 3

(59) *Piecewise linear finite elements in* $\mathbb{R}^2$. We want to generalize the previous results from $\mathbb{R}^1$ to $\mathbb{R}^2$. Let G be a bounded polygonal region in $\mathbb{R}^2$. We choose a triangulation $\mathscr{T}$ of G with nodes $N_1, \ldots, N_k$ (Fig. 3(b)). Moreover, we set

$$h(\mathscr{T}) = \text{maximal diameter of the triangles of } \mathscr{T},$$

$$\vartheta(\mathscr{T}) = \text{minimal angle of the triangles of } \mathscr{T}.$$

The following elementary result is decisive for the construction of finite elements.

(L) Let T be a triangle with vertices P_1, P_2, P_3. Then there exists a unique linear function $w(\xi, \eta) = a + b\xi + c\eta$ with prescribed values $w(P_1)$, $w(P_2)$, $w(P_3)$.

Construction (C). We now prescribe the values $w(N_1), \ldots, w(N_k)$ at the nodes and extend w uniquely to a function $w: \bar{G} \to \mathbb{R}$ via linear interpolation according to (L). Then w has the following properties.

(i) *w is continuous on $\bar{G}$.*
(ii) *w is piecewise differentiable, and $w \in W_p^1(G)$, $1 \leq p \leq \infty$.*

Proof. Ad(i). Let T_1 and T_2 be neighboring triangles with the joint side $\overline{P_1 P_2}$. Then w is uniquely determined on $\overline{P_1 P_2}$ by the values at the points P_1 and P_2.
Ad(ii). This follows from Example 21.6. □

The set of all the functions w constructed by (C) above is denoted by $X(\mathscr{T})$. In particular, if we set

$$w_j(N_i) = \begin{cases} 1 & \text{if } i = j, \\ 0 & \text{if } i \neq j, \end{cases}$$

and if we construct $w_j: \bar{G} \to \mathbb{R}$ by (C), then the so-called finite elements w_j form a basis of the space $X(\mathscr{T})$, i.e.,

$$X(\mathscr{T}) = \operatorname{span}\{w_1, \ldots, w_k\}.$$

Furthermore, we set

$$\mathring{X}(\mathscr{T}) = \{w \in X(\mathscr{T}): w = 0 \text{ on } \partial G\}.$$

Obviously, $\mathring{X}(\mathscr{T})$ consists of all $w \in X(\mathscr{T})$ with $w(N_i) = 0$ for all nodes $N_i \in \partial G$.

If the nodes N_i and N_j do not belong to neighboring triangles, then

$$(w_i|w_j)_{1,2} \equiv \int_G \left(w_i w_j + \sum_{r=1}^{2} D_r w_i D_r w_j \right) dx = 0.$$

This generalized orthogonality relation produces *sparse matrices* in connection with the Galerkin method.

Finally, we consider a sequence $(\mathscr{T}_n)$ of triangulations of G, where the mesh size goes to zero as $n \to \infty$, i.e.,

$$\lim_{n \to \infty} h(\mathscr{T}_n) = 0.$$

Moreover, we assume that $(\mathscr{T}_n)$ is *nondegenerated*, i.e., the minimal angles of the triangles are bounded below:

$$\inf_n \vartheta_n > 0.$$

We set $h_n = h(\mathscr{T}_n)$ as well as $X_n = X(\mathscr{T}_n)$ and $\mathring{X}_n = \mathring{X}(\mathscr{T}_n)$. Let $1 < p < \infty$. Then the following hold:

(a) (X_n) *is a Galerkin scheme in* $X = W_p^1(G)$.
(b) *For all* $u \in W_p^2(G)$,

$$\operatorname{dist}_X(u, X_n) \leq \text{const}\, h_n \|u\|_{2,p}.$$

(c) $(\mathring{X}_n)$ *is a Galerkin scheme in* $\mathring{X} = \mathring{W}_p^1(G)$.
(d) *For all* $u \in W_p^2(G) \cap \mathring{X}$,

$$\operatorname{dist}_X(u, \mathring{X}_n) \leq \text{const}\, h_n \|u\|_{2,p}.$$

The constants in (b) *and* (d) *depend only on* G, p, *and* $\inf_n \vartheta_n$.

Proof. We will use theorem (55) on abstract interpolation operators.

Ad(a), (b).

(I) Let T be a triangle of the fixed triangulation $\mathscr{T}_n$ of G. For $m = 0, 1$, we construct the operator

$$\Pi\colon W_p^2(G) \to W_p^m(G)$$

in the following way. Let $u \in W_p^2(G)$ be given. This implies $u \in C(\bar{G})$ by the Sobolev embedding theorems. Thus, it is meaningful to set

$$(\Pi u)(N_i) = u(N_i)$$

and to extend this to a function $\Pi u\colon \bar{G} \to \mathbb{R}$ via linear interpolation according to (L) above. Note the following:

(α) The construction of Πu is unique by (L), and we have $\Pi u \in W_p^m(G) \cap X_n$ by (ii) above.

(β) $\Pi u = u$ on the triangle T if u is a linear function on T.

(γ) Π is affinely invariant on T in the sense of (55).

Consequently, we can apply theorem (55) to the triangle T. For $m = 0$, 1, this yields

$$\|u - \Pi u\|_{W_p^1(T)}^p \leq C(h_n^2 + h_n^2/\rho_n)^p \|u\|_{W_p^2(T)}^p,$$

where C denotes a constant. Summing over all the triangles $T \in \mathscr{T}_n$, we find

$$\|u - \Pi u\|^p_{W^1_p(G)} \leq C(h_n^2 + h_n^2/\rho_n)^p \|u\|^p_{W^2_p(G)}.$$

(II) We now consider the sequence $(\mathscr{T}_n)$ of triangulations. We set $\Pi_n = \Pi$. From $\inf_n \vartheta_n > 0$ we obtain the key relation

$$\sup_n h_n/\rho_n < \infty.$$

Thus, we get

$$\|u - \Pi_n u\|_{W^1_p(G)} \leq \text{const}\, h_n \|u\|_{W^2_p(G)}.$$

Since $\Pi_n u \in X_n$, we obtain (b), i.e.,

$$\text{dist}_X(u, X_n) \leq \text{const}\, h_n \|u\|_{W^2_p(G)}.$$

(III) Density. Let $u \in W^1_p(G)$ be given. The set $C^\infty(\bar{G})$ is dense in $W^1_p(G)$. Thus, for each $\varepsilon > 0$, there exists a $v \in C^\infty(\bar{G})$ with $\|u - v\|_{1,p} < \varepsilon$. This implies

$$\begin{aligned}\|u - \Pi_n v\|_{1,p} &\leq \|u - v\|_{1,p} + \|v - \Pi_n v\|_{1,p} \\ &\leq \varepsilon + \text{const}\, h_n \|v\|_{2,p} \leq 2\varepsilon\end{aligned}$$

for all $n \geq n_0(\varepsilon)$. Since $\Pi_n v \in X_n$,

$$\lim_{n \to \infty} \text{dist}_X(u, X_n) = 0.$$

Therefore, (X_n) is a Galerkin scheme in $W^1_p(G)$.

Ad(c), (d). This follows similarly to (a), (b). □

(59a) *Generalization to piecewise quadratic finite elements.* The previous results can be easily extended to piecewise polynomial functions of degree $k \geq 2$. To explain this we consider the special case $k = 2$.

Let T be a triangle with vertices P_1, P_2, P_3. Denote by P_4, P_5, P_6 the midpoints of the sides of T (Fig. 4). The following elementary result will be crucial.

(Q) There exists a unique quadratic polynomial

$$w(\xi, \eta) = a + b\xi + c\eta + d\xi^2 + e\eta^2 + f\xi\eta$$

with prescribed values $w(P_1), \ldots, w(P_6)$.

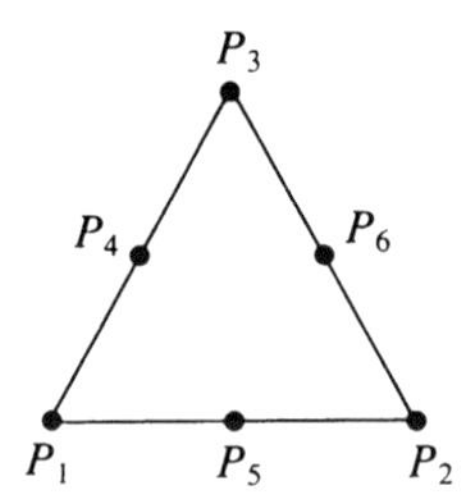

Figure 4

Let $\mathscr{T}$ be a triangulation of G (Fig. 3(b)). For each triangle $T \in \mathscr{T}$, we prescribe the values $w(P_1), \ldots, w(P_6)$. Using (Q), we extend these values to a function $w: \bar{G} \to \mathbb{R}$. Then w has the following properties:

(i) w is continuous on $\bar{G}$.
(ii) w is piecewise differentiable, and $w \in W_p^1(G)$, $1 \leq p \leq \infty$.

This follows from Example 21.6. Let $X^{(2)}(\mathscr{T})$ denote the set of all these piecewise quadratic polynomials. Moreover, we set

$$\mathring{X}^{(2)}(\mathscr{T}) = \{w \in X^{(2)}(T): w = 0 \text{ on } \partial G\}.$$

As in (59) above, we now consider a sequence $(\mathscr{T}_n)$ of triangulations with $h_n \to 0$ as $n \to \infty$ and $\inf_n \vartheta_n > 0$. Let $1 \leq p < \infty$. Set $Y_n = X^{(2)}(\mathscr{T}_n)$ and $\mathring{Y}_n = \mathring{X}^{(2)}(\mathscr{T}_n)$. Then the following hold:

(a) (Y_n) *is a Galerkin scheme in* $X = W_p^1(G)$.
(b) *For all* $u \in W_p^3(G)$,

$$\operatorname{dist}_X(u, Y_n) \leq \text{const}\, h_n^2 \|u\|_{3,p}.$$

(c) (Y_n) *is a Galerkin scheme in* $\mathring{X} = \mathring{W}_p^1(G)$.
(d) *For all* $u \in W_p^3(G) \cap \mathring{X}$,

$$\operatorname{dist}_X(u, \mathring{Y}_n) \leq \text{const}\, h_n^2 \|u\|_{3,p}.$$

The constants in (b) *and* (d) *depend only on* G, p, *and* $\inf_n \vartheta_n$.

Note that, in contrast to the piecewise linear finite elements in (59) above, we find here a higher-order accuracy of the approximation according to the appearance of h_n^2. Roughly speaking, it follows from our proof below that we have the following situation in the case where $0 \leq m \leq k + 1$:

In the Sobolev space $W_p^m(G)$, *the accuracy of approximation of* $W_p^{k+1}(G)$-*functions by piecewise polynomial functions of degree k is proportional to*

$$h^{k+1-m},$$

where h denotes the mesh size of the triangulation.

This result remains true for all bounded polyhedral regions in $\mathbb{R}^N$, $N \geq 1$, provided $(k + 1)p > N$ and $1 \leq p < \infty$.

Proof. We prove (a)–(d). Let $k = 2$ and $m = 0, 1$. We will use an analogous argument as in (59) above. The key to our proof is to construct the operator

$$\Pi: W_p^{k+1}(G) \to W_p^m(G)$$

by setting

$$(\Pi u)(P_i) = u(P_i), \qquad i = 1, \ldots, 6,$$

on each triangle $T \in \mathscr{T}_n$, and by extending these values to a quadratic polynomial on T via (Q) above. The point is that:

$$\Pi u = u \quad \text{on } T \quad \text{provided } u \text{ is a polynomial of degree} \leq k.$$

Hence, it follows from (55a) that

$$\|u - \Pi u\|_{W_p^1(T)} \le C(h_n^{k+1} + h_n^{k+1}/\rho_n)\|u\|_{k+1,p}.$$

This implies the assertions (a)–(d) as in the proof in (59) above. □

(60) *Piecewise polynomial functions on cubic grids.* Let G be a bounded region in $\mathbb{R}^N$ with $\partial G \in C^{0,1}$ and $N \ge 1$. Furthermore, let $m = 0, 1, \ldots$ and $1 \le p < \infty$. Our goal is to construct systematically finite elements in the Sobolev space $W_p^m(G)$.

(60a) *Special finite elements in* $\mathbb{R}^1$. We start with the function

$$\pi_1(x) = \begin{cases} 1 & \text{if } \ 0 < x < 1, \\ 0 & \text{otherwise,} \end{cases}$$

and define inductively

$$\pi_{m+1} = \pi_1 * \pi_m, \qquad m = 1, 2, \ldots,$$

where $f * g$ denotes the convolution, i.e.,

$$(f * g)(x) = \int_{-\infty}^{\infty} f(x - y)g(y)\,dy.$$

Figure 5 shows π_1, π_2, π_3. Explicitly,

$$\pi_2(x) = \begin{cases} x & \text{if } \ x \in [0, 1], \\ 2 - x & \text{if } \ x \in [1, 2], \\ 0 & \text{otherwise,} \end{cases}$$

and

$$\pi_3(x) = \begin{cases} x^2/2 & \text{if } \ x \in [0, 1], \\ -x^2 + 3x - 3/2 & \text{if } \ x \in [1, 2], \\ (x^2 - 6x + 9)/2 & \text{if } \ x \in [2, 3], \\ 0 & \text{otherwise.} \end{cases}$$

Generally, we have

$$\pi_{m+1}(x) = \sum_{j=0}^{m} \frac{(x - k)^j}{j!} \sum_{i=0}^{k} (-1)^i \binom{m+1}{i} \frac{(k - i)^{m-j}}{(m - j)!}$$

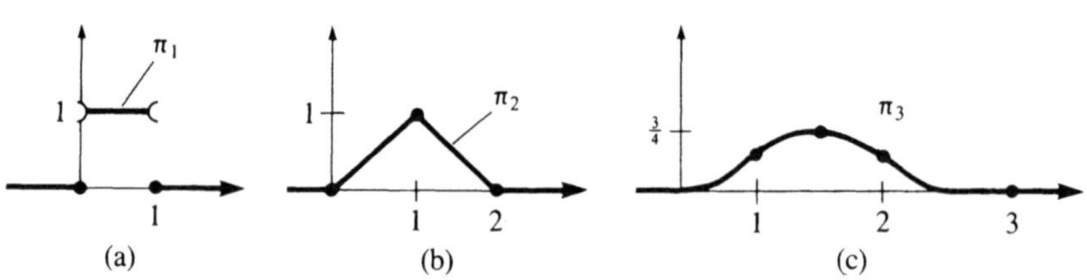

Figure 5

for all $x \in [k, k+1]$ with $k = 0, 1, \ldots, m$ and

$$\pi_{m+1}(x) = 0 \qquad \text{for all} \quad x \notin [0, m+1].$$

Here, we agree to set "$0^0 = 1$." For $m \geq 1$, these functions have the following properties:

(i) π_{m+1} has continuous $(m-1)$th derivatives on $\mathbb{R}$ and piecewise continuous mth derivatives on $\mathbb{R}$, i.e., $\pi_{m+1} \in C^{m-1}(\mathbb{R}) \cap W_p^m(\mathbb{R})$.
(ii) $\int_{-\infty}^{\infty} \pi_{m+1}(x)\,dx = 1$.

(60b) *Cubic grids in* $\mathbb{R}^N$. We consider a cubic grid in $\mathbb{R}^N$ of width h, i.e., the grid points are given by

$$y = (m_1 h, \ldots, m_N h),$$

where $m_1, \ldots, m_N$ are arbitrary integers. Set $x = (\xi_1, \ldots, \xi_N)$ and define

$$F_{m+1}(x) = \prod_{j=1}^{N} \pi_{m+1}\left(\frac{\xi_j}{h}\right).$$

Furthermore, we define the *finite elements* $F_{m+1}^y\colon \bar{G} \to \mathbb{R}$ by

$$F_{m+1}^y(x) = F_{m+1}(x - y),$$

where y denotes an arbitrary grid point. Finally, we set

$$X_h = \operatorname{span}\{F_{m+1}^y\colon y = \text{grid point}\}$$

and

$$\mathring{X}_h = \{F \in X_h\colon \operatorname{supp} F \subset G\}.$$

Then the following hold, where $\|\cdot\|_{k,2}$ refers to $W_2^k(G)$.

(i) For all $u \in W_2^{m+1}(G)$ and $0 \leq s \leq m$,

$$\inf_{v \in X_h} \|u - v\|_{s,2} \leq \text{const}\, h^{m+1-s} \|u\|_{m+1,2}.$$

(ii) For all $u \in W_2^{m+1}(G) \cap \mathring{W}_2^m(G)$ and $0 \leq s \leq m$,

$$\inf_{v \in \mathring{X}_h} \|u - v\|_{s,2} \leq \text{const}\, h^{m+1-s} \|u\|_{m+1,2}.$$

(iii) If (h_n) is a sequence with $h_n \to 0$ as $n \to \infty$, then

$$(X_{h_n}) \quad (\text{resp. } (\mathring{X}_{h_n}))$$

forms a *Galerkin scheme* in $W_p^m(G)$ (resp. $\mathring{W}_p^m(G)$).

The proof can be found in Aubin (1972, M).

Ordinary Differential Equations and Measurable Functions

For given $x_0 \in \mathbb{R}^N$ and $t_0 \in \mathbb{R}$, we consider the initial value problem

$$\text{(P)} \qquad \begin{aligned} x'(t) &= f(t, x(t)), \qquad t \in U, \\ x(t_0) &= x_0 \end{aligned}$$

along with the integral equation

$$(\mathrm{P}^*) \qquad x(t) = x_0 + \int_{t_0}^{t} f(s, x(s))\, ds, \qquad t \in U.$$

Let $U \subseteq J$ and set

$$J = \{t \in \mathbb{R}: |t - t_0| \le r_0\}, \qquad K = \{x \in \mathbb{R}^N: |x - x_0| \le r\},$$

where $r, r_0 > 0$ and $N \ge 1$. Letting $x = (\xi_1, \ldots, \xi_N)$ and $f = (f_1, \ldots, f_N)$, problem (P) can be written in the following form:

$$\xi_i'(t) = f_i(t, x(t)), \qquad i = 1, \ldots, N, \qquad t \in U,$$
$$\xi_i(t_0) = \xi_{i0}.$$

(61) *Theorem of Carathéodory.* We assume:

(i) The function $f: J \times K \to \mathbb{R}$ satisfies the *Carathéodory condition*, i.e., for all i,

$$t \mapsto f_i(t, x) \quad \text{is measurable on } J \text{ for each} \quad x \in K;$$
$$x \mapsto f_i(t, x) \quad \text{is continuous on } K \text{ for almost all} \quad t \in J.$$

(ii) *Majorant.* There exists an integrable function $M: J \to \mathbb{R}$ such that

$$|f_i(t, x)| \le M(t) \qquad \text{for all} \quad (t, x) \in J \times K \quad \text{and all } i.$$

Then:

(a) There exists an open neighborhood U of t_0 and a continuous function

$$x(\cdot): U \to \mathbb{R}^N,$$

which solves the integral equation (P*).

(b) For almost all $t \in U$, the derivative $x'(t)$ exists and (P) holds.

(c) The components $\xi_1(\cdot), \ldots, \xi_N(\cdot)$ of $x(\cdot)$ have generalized derivatives on U, and $x(\cdot)$ is a solution of (P) on U in the sense of generalized derivatives.

This result remains true if J is a one-sided neighborhood of t_0, i.e., $J = \{t \in \mathbb{R}: 0 \le t - t_0 \le r_0\}$. Then $U = \{t \in J: 0 \le t - t_0 \le r_1\}$, where $r_1 > 0$.

If the assumptions (i), (ii) are satisfied for $K = \mathbb{R}^N$, then $U = J$.

The proofs can be found in Coddington and Levinson (1955, M), p. 43, and in Kamke (1960, M), p. 193.

Distributions with Values in B-Spaces

Between 1930 and 1940, several mathematicians began to investigate systematically the concept of a "weak" solution of a linear partial differential equation, which appeared episodically (and without a name) in Poincaré's work....

It was one of the main contributions of Laurent Schwartz when he saw, in 1945, that the concept of distribution introduced by Sobolev (which he had rediscovered independently) could give a satisfactory generalization of the Fourier transform including all the preceding ones....

To the credit of Laurent Schwartz (born in 1915) must be added his persistent efforts to weld all the previous ideas into a unified and complete theory, which he enriched by many definitions and results (such as those concerning the tensor product and the convolution of distributions) in his now classical treatise (Schwartz, 1950). By his own research and those of his numerous students, he began to explore the potentialities of distributions, and gradually succeeded in convincing the world of analysts that this new concept should become central in all linear problems of analysis, due to the greater freedom and generality it allowed in the fundamental operations of calculus, doing away with a great many unnecessary restrictions and pathology.

The role of Schwartz in the theory of distributions is very similar to the one played by Newton and Leibniz in the history of calculus: contrary to popular belief, they of course did not invent it, for derivation and integration were practised by men such as Cavalieri, Fermat, and Roberval when Newton and Leibniz were mere schoolboys. But they were able to systematize the algorithms and notations of calculus in such a way that it became the versatile and powerful tool which we know, whereas before them it could only be handled via complicated arguments and diagrams.

Jean Dieudonné (1981)

Distributions generalize classical functions. In contrast to classical functions, distributions possess derivatives of arbitrary order. This crucial property of distributions is responsible for the fact that the modern theory of linear partial differential equations is based on the theory of distributions. The most important notions of the theory of distributions are:

derivative, tensor product, and convolution of distributions;
Fourier transform of distributions.

Our main applications to linear partial differential equations concern:

the fundamental solution,
the construction of solutions via convolution and fundamental solution, and
the generalized initial value problem.

In the following let G be a nonempty open set in $\mathbb{R}^N$ with $N \geq 1$, and let $\mathbb{K} = \mathbb{R}, \mathbb{C}$. For brevity, we set

$$\mathscr{D}(G) = C_0^\infty(G).$$

The *relation* between classical functions u and distributions U is based on the following simple formula:

$$U(\varphi) = \int_G u(x)\varphi(x)\,dx \qquad \text{for all} \quad \varphi \in \mathscr{D}(G). \tag{F}$$

However, note that there are distributions which cannot be represented by (F).

The *basic strategy* of the development of the theory of distributions consists in formulating properties of functions u in terms of distributions U via (F).

This strategy ensures that distributions can be regarded as *generalized functions*. For example, if the function u has the classical derivative $v = D^\alpha u$, then v corresponds to the distribution

$$V(\varphi) = \int_G D^\alpha u(x)\varphi(x)\,dx \qquad \text{for all} \quad \varphi \in \mathscr{D}(G).$$

Integration by parts yields

$$\int_G D^\alpha u(x)\varphi(x)\,dx = (-1)^{|\alpha|} \int_G u(x) D^\alpha \varphi(x)\,dx.$$

Hence

$$V(\varphi) = (-1)^{|\alpha|} U(D^\alpha \varphi).$$

This leads us in a natural way to the *definition* of the derivative $D^\alpha U$ of U by setting $D^\alpha U = V$, i.e.,

$$\text{(D)} \qquad (D^\alpha U)(\varphi) = (-1)^{|\alpha|} U(D^\alpha \varphi) \qquad \text{for all} \quad \varphi \in \mathscr{D}(G).$$

We shall show below that this definition is also meaningful for general distributions which do not correspond to a function via (F).

In the same way we will translate the following notions for functions to distributions: the tensor product, the convolution, and the Fourier transform.

(62) *Introductory example: the Dirac delta function.* Since the 1930's, physicists use the so-called "δ-function," which describes the density of a mass point with mass $m = 1$ at x_0. According to this physical interpretation, physicists set

$$\delta(x - x_0) = \begin{cases} 0 & \text{if} \quad x \neq x_0, \\ +\infty & \text{if} \quad x = x_0, \end{cases}$$

and

$$\text{(62a)} \qquad \int_{\mathbb{R}^N} \delta(x - x_0)\varphi(x)\,dx = \varphi(x_0) \qquad \text{for all} \quad \varphi \in \mathscr{D}(G).$$

In terms of mathematics, such a definition is *nonsense* because there is *no* function having these two properties. However, the intuition of physicists led them to an important new mathematical tool. The rigorous definition of δ will be given in (64a) below.

(63) *Convergence in the space $\mathscr{D}(G)$.* Denote by $\mathscr{D}(G, \mathbb{K})$ the set of all C^∞-functions

$$u\colon G \to \mathbb{K}$$

having compact support, i.e., there is a compact subset K of G such that $u = 0$ outside K. According to our earlier definition, $\mathscr{D}(G, \mathbb{R}) = \mathscr{D}(G) = C_0^\infty(G)$.

Let (φ_n) be a sequence in $\mathscr{D}(G, \mathbb{K})$ and let $\varphi \in \mathscr{D}(G, \mathbb{K})$. By definition, we write

$$\varphi_n \to \varphi \quad \text{in } \mathscr{D}(G, \mathbb{K}) \qquad \text{as} \quad n \to \infty$$

iff the following hold:

(i) For all α with $|\alpha| \geq 0$, we have the uniform convergence

$$D^\alpha \varphi_n \rightrightarrows D^\alpha \varphi \quad \text{on } G \qquad \text{as} \quad n \to \infty.$$

(ii) There exists a compact subset K of G such that $\varphi_n = \varphi$ outside K for all n.

(64) *Definition of distributions.* Let Y be a B-space over $\mathbb{K}$. By a distribution, we understand a linear, sequentially continuous map

$$U: \mathscr{D}(G, \mathbb{K}) \to Y, \tag{M}$$

i.e., U is linear and, as $n \to \infty$,

$$\varphi_n \to \varphi \quad \text{in } \mathscr{D}(G, \mathbb{K}) \qquad \text{implies} \qquad U(\varphi_n) \to U(\varphi) \quad \text{in } Y.$$

The set of these distributions is denoted by

$$\mathscr{D}'(G, Y).$$

In the case where $Y = \mathbb{R}$, we briefly write $\mathscr{D}'(G)$. In (72) below we will introduce a topology on $\mathscr{D}(G)$. In the sense of this topology, distributions are linear *continuous* maps of the form (M). Thus, $\mathscr{D}'(G)$ is the *dual space* to $\mathscr{D}(G)$.

(64a) *Standard example: the delta distribution.* For fixed $x_0 \in \mathbb{R}^N$, we set

$$\delta_{x_0}(\varphi) = \varphi(x_0) \qquad \text{for all} \quad \varphi \in \mathscr{D}(\mathbb{R}^N, \mathbb{K}).$$

This definition is based on the idea (62a) of physicists. Obviously, δ_{x_0} is a distribution, i.e., $\delta_{x_0} \in \mathscr{D}'(G, \mathbb{K})$. This distribution is called the Dirac *delta distribution* at x_0. We set $\delta = \delta_{x_0}$ if $x_0 = 0$.

(65) *Classical functions and regular distributions.* Let $u \in L_{1,\text{loc}}(G, Y)$, i.e., the function

$$u: G \to Y$$

is integrable on each compact subset of G. We set

$$U(\varphi) = \int_G u(x)\varphi(x)\,dx \qquad \text{for all} \quad \varphi \in \mathscr{D}(G, \mathbb{K}). \tag{R}$$

Then U is a distribution, i.e., $U \in \mathscr{D}'(G, Y)$. Furthermore, let $v \in L_{1,\text{loc}}(G, Y)$ and set

$$V(\varphi) = \int_G v(x)\varphi(x)\,dx \qquad \text{for all} \quad \varphi \in \mathscr{D}(G, \mathbb{K}).$$

Then

$$U = V \qquad \text{iff} \qquad u(x) = v(x) \qquad \text{for almost all} \quad x \in G. \tag{R*}$$

A distribution U is called *regular* iff it can be represented in the form (R), where $u \in L_{1,\text{loc}}(G, Y)$. In the sense of (R*), there exists a linear bijective map from $L_{1,\text{loc}}(G, Y)$ onto the set of regular distributions in $\mathscr{D}'(G, Y)$. In this sense, we can *identify* each function

$$u \in L_{1,\text{loc}}(G, Y)$$

with a regular distribution, i.e., $L_{1,\text{loc}}(G, Y) \subseteq \mathscr{D}'(G, Y)$.

This justifies the designation "*generalized functions*" for distributions. Note that the delta distribution is *not* regular.

(65a) *The support of a function.* Let $u: G \to Y$ be a function. Recall that the support of u is defined to be the set

$$\operatorname{supp} u = \text{closure of } \{x \in G: u(x) \neq 0\}.$$

(65b) *The support of a distribution.* Let $U \in \mathscr{D}'(G, Y)$ and let H be a nonempty open subset of G. By definition,

$$U = 0 \quad \text{on } H \qquad \text{iff} \qquad U(\varphi) = 0 \qquad \text{for all} \quad \varphi \in \mathscr{D}(H, \mathbb{K}).$$

The support of U is defined to be the set

$$\operatorname{supp} U = \{x \in \bar{G}: U \neq 0 \text{ on each } V(x)\},$$

where $V(x)$ denotes the intersection of G with an open neighborhood of x.

For example, if $u \in L_{1,\text{loc}}(G, Y)$ and if u is continuous, then

$$\operatorname{supp} u = \operatorname{supp} U,$$

where U denotes the regular distribution corresponding to the function u.

(65c) *Standard example.* For the delta distribution,

$$\operatorname{supp} \delta_{x_0} = \{x_0\}.$$

(65d) *Multiplication of a distribution by a C^∞-function.* Let $U \in \mathscr{D}'(G, Y)$ and let $a: G \to \mathbb{K}$ be a C^∞-function. We set

$$(aU)(\varphi) = U(a\varphi) \qquad \text{for all} \quad \varphi \in \mathscr{D}(G, \mathbb{K}).$$

Obviously, $aU \in \mathscr{D}'(G, Y)$.

(66) *Derivatives of distributions.* Let U be a distribution, i.e., $U \in \mathscr{D}'(G, Y)$. For each α with $|\alpha| \geq 0$, we define

$$(D^\alpha U)(\varphi) = (-1)^{|\alpha|} U(D^\alpha \varphi) \qquad \text{for all} \quad \varphi \in \mathscr{D}(G, \mathbb{K}).$$

Obviously, $D^\alpha U$ is again a distribution, i.e., $D^\alpha U \in \mathscr{D}'(G, Y)$. We call $D^\alpha U$ the D^α-derivative of U. Consequently, we obtain the following decisive result:

Distributions possess derivatives of arbitrary order.

(66a) *Standard example.* For fixed $x_0 \in \mathbb{R}$, consider the so-called Heaviside function

$$u(x) = \begin{cases} 1 & \text{if} \quad x_0 < x, \\ 0 & \text{if} \quad x \leq x_0. \end{cases}$$

Then, $u' = \delta_{x_0}$ in $\mathscr{D}'(\mathbb{R})$.

Proof. Set

$$U(\varphi) = \int_{\mathbb{R}} u(x)\varphi(x)\,dx \qquad \text{for all} \quad \varphi \in \mathscr{D}(\mathbb{R}).$$

By definition, $U'(\varphi) = -U(\varphi')$ for all $\varphi \in \mathscr{D}(\mathbb{R})$. Hence

$$U'(\varphi) = -\int_{x_0}^{\infty} \varphi'\,dx = \varphi(x_0),$$

i.e., $U'(\varphi) = \delta_{x_0}(\varphi)$. According to (65), we write u' instead of U'. □

(66b) *Example.* Let

$$u(x) = \begin{cases} 1 & \text{if} \quad x_0 < x, \\ -1 & \text{if} \quad x > x_0, \\ \text{arbitrary} & \text{if} \quad x = x_0. \end{cases}$$

Then, $u' = 2\delta_{x_0}$ in $\mathscr{D}'(\mathbb{R})$.

Proof. For all $\varphi \in \mathscr{D}(\mathbb{R})$, $U'(\varphi) = -U(\varphi')$, i.e.,

$$U'(\varphi) = \int_{-\infty}^{x_0} \varphi\,dx - \int_{x_0}^{\infty} \varphi\,dx = 2\varphi(x_0).$$ □

(66c) *Example.* For the delta distribution, we find

$$(D^\alpha \delta_{x_0})(\varphi) = (-1)^{|\alpha|}\delta_{x_0}(D^\alpha \varphi) = (-1)^{|\alpha|} D^\alpha \varphi(x_0),$$

for all $\varphi \in \mathscr{D}(\mathbb{R})$.

(66d) *Generalized plane waves.* We consider the wave equation

$$\text{(W)} \qquad \Box u \overset{\text{def}}{=} \frac{1}{c^2} u_{tt} - \Delta u = 0$$

for $u = u(x, t)$, where $x \in \mathbb{R}^3$ and $t \in \mathbb{R}$. Let n be a fixed unit vector and set $nx = \langle n|x \rangle$. Then:

(i) If $v \in C^2(\mathbb{R})$, then the function

$$\text{(W*)} \qquad u(x, t) = v(nx - ct)$$

is a classical solution of (W) on $\mathbb{R}^4$, which is called a plane wave.

(ii) If $v \in L_{\infty,\text{loc}}(\mathbb{R})$, then u in (W*) is a solution of (W) in the sense of distributions in $\mathscr{D}'(\mathbb{R}^4)$.

By (W*), the discontinuities of u propagate along the moving planes

$$nx - ct = \text{const}.$$

Recall that $v \in L_{\infty,\text{loc}}(\mathbb{R})$ iff the function $v: \mathbb{R} \to \mathbb{R}$ is measurable and bounded on each compact subset of $\mathbb{R}$, e.g., v is piecewise continuous.

This result shows that the theory of distributions is able to describe reasonable physical situations, which lack regularity.

Proof. Ad(i). This follows from a simple computation.

Ad(ii). For all $\varphi \in \mathscr{D}(\mathbb{R}^4)$, set

$$U(\varphi) = \int_{\mathbb{R}^4} u(x, t)\varphi(x, t)\, dx\, dt.$$

We have to show that $(1/c^2)U_{tt}(\varphi) - (\Delta U)(\varphi) = 0$, i.e.,

$$U\left(\frac{1}{c^2}\varphi_{tt} - \Delta\varphi\right) = 0 \qquad \text{for all} \quad \varphi \in \mathscr{D}(\mathbb{R}^4).$$

That means

$$J \overset{\text{def}}{=} \int_{\mathbb{R}^4} v(nx - ct)\Box\varphi(x, t)\, dx\, dt = 0 \qquad \text{for all} \quad \varphi \in \mathscr{D}(\mathbb{R}^4).$$

To show this fix $\varphi \in \mathscr{D}(\mathbb{R}^4)$. Since $\operatorname{supp}\varphi$ is compact, there exist an open ball B and an open bounded interval $\mathscr{B}$ such that $\operatorname{supp}\varphi \subset B$ and

$$nx - ct \in \mathscr{B} \qquad \text{for all} \quad (x, t) \in B.$$

Since $v \in L_1(\mathscr{B})$, there exists a sequence (v_k) in $\mathscr{D}(\mathbb{R})$ such that

$$v_k(\xi) \to v(\xi) \qquad \text{as} \quad k \to \infty \qquad \text{for almost all} \quad \xi \in \mathscr{B}.$$

Moreover, there is a $C > 0$ such that

$$|v_k(\xi)| \le C \qquad \text{for all} \quad \xi \in \mathscr{B} \quad \text{and all } k.$$

This follows from the properties of the smoothing operator in Section 18.14. Let

$$J_k \overset{\text{def}}{=} \int_B v_k(nx - ct)\Box\varphi(x, t)\, dx\, dt.$$

Integration by parts yields

$$J_k = \int_B \varphi(x, t)\Box v_k(nx - ct)\, dx\, dt = 0,$$

since $\Box v_k(nx - ct) = 0$ by (i). Using majorized convergence $A_2(19)$, we obtain

$$J_k \to J \qquad \text{as} \quad k \to \infty.$$

Hence $J = 0$. □

(67) *The tensor product of distributions.*

(67a) *Functions.* Let $u\colon \mathbb{R}^N \to \mathbb{K}$ and $v\colon \mathbb{R}^M \to \mathbb{K}$ be functions. We set

$$(u \otimes v)(x, y) \overset{\text{def}}{=} u(x)v(y) \qquad \text{for all} \quad (x, y) \in \mathbb{R}^N \times \mathbb{R}^M$$

and call the function $u \otimes v\colon \mathbb{R}^{N+M} \to \mathbb{K}$ the tensor product of u and v.

(67b) *Regular distributions.* Let $u \in L_{1,\mathrm{loc}}(\mathbb{R}^N, \mathbb{K})$ and $v \in L_{1,\mathrm{loc}}(\mathbb{R}^M, \mathbb{K})$. Denote by $U \otimes V$ the regular distribution corresponding to $u \otimes v$, i.e., for all $\varphi \in \mathscr{D}(\mathbb{R}^{N+M}, \mathbb{K})$,

$$\begin{aligned}(U \otimes V)(\varphi) &= \int_{\mathbb{R}^{N+M}} (u \otimes v)(x, y)\varphi(x, y)\,dx\,dy \\ &= \int_{\mathbb{R}^N} u(x)\left(\int_{\mathbb{R}^M} v(y)\varphi(x, y)\,dy\right)dx \\ &= U(V(\varphi(x, \cdot))).\end{aligned}$$

(67c) *General distributions.* Let U, V be given as in (67d) below. For all $\varphi \in \mathscr{D}(\mathbb{R}^{N+M}, \mathbb{K})$, we set

$$(U \otimes V)(\varphi) \overset{\mathrm{def}}{=} U(V(\varphi(x, \cdot))). \tag{T}$$

This is to be understood in the sense of

$$U(V(\varphi(x, \cdot))) = U(\psi), \qquad \text{where} \quad \psi(x) = V(\varphi(x, \cdot)).$$

This way we obtain a distribution, i.e., $U \otimes V \in \mathscr{D}'(\mathbb{R}^{N+M}, \mathbb{K})$. The distribution $U \otimes V$ is called the *tensor product* of U and V.

(67d) *Properties of the tensor product.* Let

$$U \in \mathscr{D}'(\mathbb{R}^N, \mathbb{K}), \qquad V \in \mathscr{D}'(\mathbb{R}^M, \mathbb{K}), \qquad W \in \mathscr{D}'(\mathbb{R}^L, \mathbb{K}).$$

(i) For all $\varphi \in \mathscr{D}(\mathbb{R}^{N+M}, \mathbb{K})$,

$$(U \otimes V)(\varphi) = U(V(\varphi(x, \cdot))) = V(U(\varphi(\cdot, y))).$$

This is briefly expressed by

$$U_x \otimes V_y = V_y \otimes U_x.$$

(ii) Let $\varphi(x, y) = a(x)b(y)$, where $a \in \mathscr{D}(\mathbb{R}^N, \mathbb{K})$ and $b \in \mathscr{D}(\mathbb{R}^M, \mathbb{K})$. Then,

$$(U \otimes V)(\varphi) = U(a)V(b).$$

(iii) $(U \otimes V) \otimes W = U \otimes (V \otimes W)$.
(iv) $D_x^\alpha(U \otimes V) = (D_x^\alpha U) \otimes V$ for all α: $|\alpha| \geq 0$.
(v) $D_y^\alpha(U \otimes V) = U \otimes (D_y^\alpha V)$ for all α: $|\alpha| \geq 0$.
(vi) $\operatorname{supp} U \otimes V = \operatorname{supp} U \times \operatorname{supp} V$.

(68) *Convolution of distributions.* Let U, $V \in \mathscr{D}(\mathbb{R}^N, \mathbb{K})$ and suppose that $\operatorname{supp} U$ is bounded. We set

$$(U * V)(\varphi) \overset{\mathrm{def}}{=} (U \otimes V)(\varphi_*) \qquad \text{for all} \quad \varphi \in \mathscr{D}(\mathbb{R}^N, \mathbb{K}),$$

where $\varphi_*(x, y) = \varphi(x + y)$. Then, $U * V$ is a distribution, i.e., $U * V \in \mathscr{D}'(\mathbb{R}^N, \mathbb{K})$. This distribution is called the convolution of U and V.

(68a) *Regular distributions.* Let $u, v \in L_{1,\mathrm{loc}}(\mathbb{R}^N, \mathbb{K})$ and suppose that $\operatorname{supp} u$

is bounded. For all $x \in \mathbb{R}^N$, we set

$$(u * v)(x) = \int_{\mathbb{R}^N} u(x - y)v(y)\,dy.$$

Then, $u * v \in L_{1,\text{loc}}(\mathbb{R}^N, \mathbb{K})$.

Let U and V be the regular distribution corresponding to u and v, respectively. Then, $U * V$ is the regular distribution corresponding to $u * v$.

(68b) *Properties of the convolution.* Let U, V, $W \in \mathscr{D}'(\mathbb{R}^N, \mathbb{K})$ and suppose that $\operatorname{supp} U$ is bounded. Let λ, $\mu \in \mathbb{K}$. Then:

(i) Linearity: $U * (\lambda V + \mu W) = \lambda(U * V) + \mu(U * W)$.
(ii) Commutativity: $U * V = V * U$.
(iii) Associativity: $(U * V) * W = U * (V * W)$ if $\operatorname{supp} U$ and $\operatorname{supp} V$ are bounded.
(iv) Derivative: $D^\alpha(U * V) = (D^\alpha U) * V = U * D^\alpha V$ for all α: $|\alpha| \geq 0$.
(v) $\delta * W = W * \delta = W$.

Distributions and Linear Partial Differential Equations

(69) *Fundamental solution.* We consider the linear differential equation with constant coefficients

$$\text{(E)} \qquad \sum_{|\alpha| \leq m} a_\alpha D^\alpha u = f \qquad \text{on } \mathbb{R}^N,$$

where $a_\alpha \in \mathbb{C}$ for all α. By a fundamental solution of (E), we understand a distribution $u \in \mathscr{D}'(\mathbb{R}^N, \mathbb{C})$ which solves (E) in the case where $f = \delta$.

If u_F is a special fundamental solution of (E), then all fundamental solutions of (E) are obtained through

$$u = u_F + v,$$

where $v \in \mathscr{D}'(\mathbb{R}^N, \mathbb{C})$ is an arbitrary solution of (E) with $f = 0$.

(69a) *The existence theorem of Ehrenpreis* (1954) *and Malgrange* (1955). For each equation (E), there exists a fundamental solution.

(69b) *The existence theorem for the inhomogeneous problem.* Let $f \in \mathscr{D}'(\mathbb{R}^N, \mathbb{C})$ be given, where $\operatorname{supp} f$ is bounded (e.g., f is a regular distribution corresponding to the function $\tilde{f} \in L_{1,\text{loc}}(\mathbb{R}^N, \mathbb{C})$ and the support of this function is bounded). Then, equation (E) has a solution $u \in \mathscr{D}'(\mathbb{R}^N, \mathbb{C})$ given by

$$u = u_F * f,$$

where u_F is an arbitrary fundamental solution of (E).

Proof. By (68),

$$\sum_\alpha a_\alpha D^\alpha(u_F * f) = \sum_\alpha (a_\alpha D^\alpha u_F) * f = \delta * f = f. \qquad \square$$

(69c) *Example* 1. The equation

$$u^{(n)} = \delta \quad \text{on } \mathbb{R}, \qquad n = 1, 2, \ldots,$$

has the solution

$$u(x) = \begin{cases} \dfrac{x^{n-1}}{(n-1)!} & \text{if} \quad x \geq 0, \\ 0 & \text{if} \quad x < 0, \end{cases}$$

in $\mathscr{D}'(\mathbb{R}, \mathbb{C})$. This is to be understood in the sense of regular distributions.

Proof. For all $\varphi \in \mathscr{D}(\mathbb{R}, \mathbb{C})$, set

$$U(\varphi) = \int_{\mathbb{R}} u(x)\varphi(x)\,dx.$$

Integration by parts yields

$$U(\varphi^{(n)}) = \int_0^\infty \frac{x^{n-1}}{(n-1)!}\varphi^{(n)}(x)\,dx = (-1)^n \varphi(0).$$

That means $U^{(n)}(\varphi) = (-1)^n U(\varphi^{(n)}) = \varphi(0) = \delta(\varphi)$, i.e., $U^{(n)} = \delta$. □

(69d) *Example* 2. Table 2 shows fundamental solutions for the Laplace equation and the heat equation (regular distributions), and for the wave equation (singular distribution).

(70) *The generalized initial value problem for the wave equation.* We consider the classical initial value problem

$$\text{(C)} \qquad \begin{aligned} &\Box u \overset{\text{def}}{=} u_{tt} - \Delta u = f, \qquad x \in \mathbb{R}^3, \quad t > 0, \\ &u(x, 0) = u_0(x), \qquad u_t(x, 0) = u_1(x, 0) \end{aligned}$$

Table 2

Differential equation	Fundamental solution ($x \in \mathbb{R}^3$)
$-\Delta u = \delta$ on $\mathbb{R}^3$	$u(x) = \dfrac{1}{4\pi\lvert x\rvert}$ (regular distribution)
$u_t - \Delta u = \delta$ on $\mathbb{R}^4$	$u(x,t) = \begin{cases} \dfrac{e^{-\lvert x\rvert^2/4t}}{8(\pi t)^{3/2}} & \text{if} \quad t > 0, \\ 0 & \text{if} \quad t \leq 0, \end{cases}$ (regular distribution)
$u_{tt} - \Delta u = \delta$ on $\mathbb{R}^4$	$u(\varphi) = \dfrac{1}{4\pi}\displaystyle\int_0^\infty \frac{1}{t}\left(\int_{\lvert x\rvert = t} \varphi(x,t)\,dO_x\right)dt$ for all $\varphi \in \mathscr{D}(\mathbb{R}^4, \mathbb{C})$

together with the so-called *generalized* initial value problem

$$\text{(C*)} \qquad \Box U = F + U_0 \otimes \delta' + U_1 \otimes \delta, \qquad \operatorname{supp} U \subseteq \tilde{\mathbb{R}}^4_+,$$

where $\tilde{\mathbb{R}}^4_+ = \{(x,t) \in \mathbb{R}^4 : t \geq 0\}$.

Problem (C*) is motivated by the following result. Let

$$f \in C(\tilde{\mathbb{R}}^4_+), \qquad u_0 \in C^1(\mathbb{R}^3), \qquad u_1 \in C(\mathbb{R}^3)$$

be given, and let $u \in C^2(\operatorname{int} \tilde{\mathbb{R}}^4_+) \cap C^1(\tilde{\mathbb{R}}^4_+)$ be a solution of (C). Set $u(x,t) = 0$ and $f(x,t) = 0$ if $(x,t) \notin \tilde{\mathbb{R}}^4_+$.

Then the corresponding regular distributions in $\mathscr{D}'(\mathbb{R}^4, \mathbb{C})$ satisfy (C*).

Proof. For all $\varphi \in \mathscr{D}(\mathbb{R}^4, \mathbb{C})$, integration by parts yields

$$\int_{\mathbb{R}^4} \varphi \Box u \, dx\, dt = \int_{\tilde{\mathbb{R}}^4_+} \varphi \Box u \, dx\, dt = -\int_{\mathbb{R}^3} \varphi(x,0) u_t(x,0)\, dx$$
$$+ \int_{\mathbb{R}^3} \varphi_t(x,0) u(x,0)\, dx + \int_{\tilde{\mathbb{R}}^4_+} u \Box \varphi \, dx\, dt.$$

Note that $\cos(n,t) = -1$ and $\cos(n,x) = 0$, where n denotes the outer unit normal to $\partial \tilde{\mathbb{R}}^4_+ = \mathbb{R}^3$. From (C) it follows that

$$\int_{\mathbb{R}^4} u \Box \varphi \, dx\, dt = \int_{\mathbb{R}^4} f\varphi \, dx - \int_{\mathbb{R}^3} \varphi_t(x,0) u_0(x)\, dx$$
$$+ \int_{\mathbb{R}^3} \varphi(x,0) u_1(x)\, dx.$$

This is (C*). In this connection, note that

$$(\Box U)(\varphi) = U(\Box \varphi) = \int_{\mathbb{R}^4} u \Box \varphi \, dx\, dt,$$

$$F(\varphi) = \int_{\mathbb{R}^4} f\varphi \, dx\, dt,$$

$$(U_1 \otimes \delta)(\varphi) = U_1(\delta(\varphi(x,t))) = U_1(\varphi(x,0))$$
$$= \int_{\mathbb{R}^3} u_1(x) \varphi(x,0)\, dx,$$

$$(U_0 \otimes \delta')(\varphi) = U_0(\delta'(\varphi(x,t))) = U_0(-\varphi_t(x,0))$$
$$= -\int_{\mathbb{R}^3} u_0(x) \varphi_t(x,0)\, dx. \qquad \Box$$

(70a) *The Poisson formula.* Our point of departure is the classical Poisson formula:

$$\text{(P)} \qquad u(x,t) = \frac{\theta(t)}{4\pi} \int_{|y-x| \leq t} \frac{f(y, t - |x-y|)}{|x-y|}\, dy$$
$$+ \frac{\theta(t)}{4\pi t} \int_{|y-x|=t} u_1(y)\, dO_y + \frac{1}{4\pi} \frac{\partial}{\partial t}\left(\frac{\theta(t)}{t} \int_{|y-x|=t} u_0(y)\, dO_y \right),$$

where $\theta(t) = 1$ if $t > 0$ and $\theta(t) = 0$ if $t \le 0$. Then:

(i) If $u_0 \in C^3(\mathbb{R}^3)$, $u_1 \in C^2(\mathbb{R}^3)$, and $f \in C^2(\tilde{\mathbb{R}}^4_+)$, then $u = u(x, t)$ in (P) is a classical solution of the initial value problem (C) above.

(ii) If $u_0, u_1 \in L_{1,\text{loc}}(\mathbb{R}^3)$, and $f \in L_{1,\text{loc}}(\mathbb{R}^4)$ with $\operatorname{supp} f \subseteq \tilde{\mathbb{R}}^4_+$, then the right-hand side of (P) represents a distribution $U \in \mathscr{D}'(\mathbb{R}^4, \mathbb{C})$, which is the unique solution of the generalized initial value problem (C*).

In (i), the initial condition is to be understood in the sense of $u(x, t) \to u_0(x)$ and $u_t(x, t) \to u_1(x)$ as $t \to +0$ for all $x \in \mathbb{R}^3$.

Distributions and Sobolev Spaces

(71a) *The space $W_p^m(G)$.* Let $m = 1, 2, \ldots$ and $1 \le p \le \infty$. In terms of distributions, the Sobolev space $W_p^m(G)$ can be characterized by

$$W_p^m(G) = \{u \in L_p(G) \colon D^\alpha U \in L_p(G) \text{ for all } \alpha \colon |\alpha| \le m\},$$

where U denotes the regular distribution corresponding to the function u, i.e.,

$$U(\varphi) = \int_G u(x)\varphi(x)\,dx \qquad \text{for all} \quad \varphi \in \mathscr{D}(G).$$

Moreover, $D^\alpha U \in L_p(G)$ means that $D^\alpha U$ is a regular distribution corresponding to an $L_p(G)$-function. If $u \in W_p^m(G)$, then, for all α with $|\alpha| \le m$, the distributional derivative $D^\alpha U$ corresponds to the generalized derivative $D^\alpha u$ in the sense of Section 21.1, i.e.,

$$(D^\alpha U)(\varphi) = \int_G (D^\alpha u)\varphi\,dx \qquad \text{for all} \quad \varphi \in \mathscr{D}(G).$$

(71b) *Negative norms and the space $W_q^{-m}(G)$.* Let $m = 1, 2, \ldots, 1 < p, q < \infty$, $p^{-1} + q^{-1} = 1$. For $U \in \mathscr{D}'(G)$ we define the so-called negative norm

$$\|U\|_{-m,q} = \sup_{\varphi \in \mathscr{D}(G)} \frac{|U(\varphi)|}{\|\varphi\|_{m,p}}.$$

By definition,

$$W_q^{-m}(G) = \{U \in \mathscr{D}'(G) \colon \|U\|_{-m,q} < \infty\}.$$

(71c) *The dual space $\mathring{W}_p^m(G)^*$.* Choose m, p, q as in (71b). We have

$$\mathring{W}_p^m(G)^* = W_q^{-m}(G)$$

in the following sense. Let $U \in \mathring{W}_p^m(G)^*$, i.e., U is a continuous linear functional on $\mathring{W}_p^m(G)$ with norm $\|U\|$. Since $\mathscr{D}(G) \subseteq \mathring{W}_p^m(G)$,

$$|U(\varphi)| \le \|U\|\,\|\varphi\|_{m,p} \qquad \text{for all} \quad \varphi \in \mathscr{D}(G).$$

This implies $U \in W_q^{-m}(G)$.

Conversely, let $U \in W_q^{-m}(G)$. Then,

$$|U(\varphi)| \leq \|U\|_{-m,q}\|\varphi\|_{m,p} \qquad \text{for all} \quad \varphi \in \mathscr{D}(G).$$

Since $\mathscr{D}(G)$ is dense in $\mathring{W}_p^m(G)$, U can be uniquely extended to a linear continuous functional on $\mathring{W}_p^m(G)$ with

$$\|U\| = \|U\|_{-m,q}.$$

(71d) *Representation theorem.* Choose m, p, q as in (71b). Let $u_\alpha \in L_q(G)$ for all α: $|\alpha| \leq m$. Let U_α be the regular distribution corresponding to u_α. We set

$$U = \sum_{|\alpha| \leq m} (-1)^{|\alpha|} D^\alpha U_\alpha. \tag{R}$$

Then $U \in W_q^{-m}(G)$. Conversely, each $U \in W_q^{-m}(G)$ can be represented in the form (R).

(71e) We have

$$\cdots \subseteq W_2^2(G) \subseteq W_2^1(G) \subseteq L_2(G) \subseteq W_2^{-1}(G) \subseteq W_2^{-2}(G) \subseteq \cdots.$$

This motivates the designation "W^{-m}."

Distributions and Locally Convex Spaces

In (64) we defined distributions without using a topology on the space $C_0^\infty(G)$. We only used the notion of an appropriate convergence. Now our goal is to introduce a topology on $C_0^\infty(G)$ in such a way that distributions are linear continuous functionals on $C_0^\infty(G)$. To this end, we need locally convex spaces and the strict inductive limit of F-spaces. In fact, the development of the theory of locally convex spaces around 1950 was mainly stimulated by applications to distributions.

The definition of locally convex spaces can be found in the Appendix to Part I.

(72) *Fréchet spaces.* A locally convex space is called a Fréchet space (F-space) iff it is a complete metric space, i.e., the topology is given by a complete metric.

(72a) *Standard example* 1. Each B-space is also an F-space.

(72b) *Standard example* 2—*the space* $C^\infty(\bar{G})$. Let G be a nonempty open set in $\mathbb{R}^N$. Then $C^\infty(\bar{G})$ is an F-space. The system of seminorms is given by

$$p_k(\varphi) = \sum_{|\alpha| \leq k} \sup_{x \in \bar{G}} |D^\alpha \varphi(x)|,$$

for $k = 0, 1, \ldots$, and the metric is given by

$$d(\varphi, \psi) = \sum_{k=0}^{\infty} \frac{p_k(\varphi - \psi)}{2^k(1 + p_k(\varphi - \psi))},$$

for all $\varphi, \psi \in C^\infty(\bar{G})$.

Let Y be a real B-space. Then the linear mapping

$$U: C^\infty(\bar{G}) \to Y$$

is *continuous* iff there exists a seminorm p_k and a constant C such that

$$\|U(\varphi)\| \leq Cp_k(\varphi) \qquad \text{for all} \quad \varphi \in C^\infty(\bar{G}).$$

(73) *Strict inductive limit of* F-*spaces.* Let X be a linear space over $\mathbb{K} = \mathbb{R}$, $\mathbb{C}$ with

$$X = \bigcup_{n=1}^{\infty} X_n \qquad \text{and} \qquad X_1 \subseteq X_2 \subseteq \cdots,$$

where all the spaces X_n are F-spaces over $\mathbb{K}$ and

$$X_n \text{ is a closed linear subspace of } X_{n+1} \text{ for all } n.$$

Then there exists a strongest locally convex topology τ on X such that all the embeddings

$$X_n \subseteq X, \qquad n = 1, 2, \ldots,$$

are continuous. We write

$$X = \liminf_{n \to \infty} X_n$$

in the case where X is equipped with the topology τ. Then X is called the strict inductive limit of (X_n) and the following hold:

(i) X is complete.
(ii) X_n is a linear closed subspace of X for all n.
(iii) Let Y be a locally convex space over $\mathbb{K}$. Then the linear mapping

$$U: X \to Y$$

is *continuous* iff the restrictions

$$U: X_n \to Y$$

are continuous for all n.

(73a) *Explicit construction of* τ. Let $S \subseteq X$. Then the set S is called circled iff, for $\lambda \in \mathbb{K}$,

$$x \in S \qquad \text{and} \qquad |\lambda| \leq 1 \qquad \text{implies} \qquad \lambda x \in S.$$

Moreover, S is called absorbing iff

$$\bigcup_{t>0} tS = X.$$

Let Σ be the system of all convex, circled, absorbing sets S in X with the additional property

$$S \cap X_n \text{ is open in } X_n \text{ for all } n.$$

Then a set M in X is *open* with respect to the inductive topology τ iff, for each

$x \in M$, there exists a set $S \in \Sigma$ such that

$$x + S \subseteq M.$$

(73b) *Strictly inductive topology on* $C_0^\infty(G)$. Let G be a nonempty open set in $\mathbb{R}^N$. We choose a sequence (G_n) of nonempty open sets G_n in $\mathbb{R}^N$ such that

$$G = \bigcup_{n=1}^{\infty} \bar{G}_n \qquad \text{and} \qquad \bar{G}_1 \subseteq \bar{G}_2 \subseteq \cdots \subset G.$$

We set $X = C_0^\infty(G)$ and

$$X_n = \{\varphi \in X : \operatorname{supp} \varphi \subseteq \bar{G}_n\}.$$

Then X_n is a closed linear subspace of the F-space $C^\infty(\bar{G}_n)$, i.e., X_n is an F-space. Thus, we have the same situation as in (73). We now equip $C_0^\infty(G)$ with the corresponding inductive topology τ, i.e., we set

$$C_0^\infty(G) = \liminf_{n \to \infty} C^\infty(\bar{G}_n).$$

Let Y be a real B-space and let

$$U: C_0^\infty(G) \to Y$$

be a linear mapping. Then the following three conditions are mutually equivalent:

(i) U is continuous.
(ii) U is a distribution in the sense of (64).
(iii) For each $n \in \mathbb{N}$, there is a constant C and a number $k = 0, 1, \ldots$ such that

$$\|U(\varphi)\| \leq C \sum_{|\alpha| \leq k} \sup_{x \in \bar{G}_n} |D^\alpha \varphi(x)| \qquad \text{for all} \quad \varphi \in C^\infty(\bar{G}_n).$$

An analogous result holds for distributions $U: \mathscr{D}(G, \mathbb{C}) \to Y$ with values in a complex B-space Y. In this connection, one has only to replace the real $C_0^\infty(G)$-functions by complex $C_0^\infty(G)$-functions, i.e., replace $\mathscr{D}(G, \mathbb{R})$ by $\mathscr{D}(G, \mathbb{C})$.

Fourier Transform and Sobolev Spaces

(74a) *The space* $\mathscr{S}$. We set

$$p_{k,m}(u) = \sup_{x \in \mathbb{R}^N} (|x|^k + 1) \sum_{|\alpha| \leq m} |D^\alpha u(x)|,$$

where $k, m = 0, 1, \ldots$ and $N = 1, 2, \ldots$. The space $\mathscr{S}(\mathbb{R}^N)$ consists of all the C^∞-functions $u: \mathbb{R}^N \to \mathbb{C}$ with

$$p_{k,m}(u) < \infty \qquad \text{for all } k, m,$$

i.e., the functions $u \in \mathscr{S}(\mathbb{R}^N)$ and their derivatives are rapidly going to zero as $|x| \to \infty$. In particular,

$$\mathscr{D}(\mathbb{R}^N, \mathbb{C}) \subseteq \mathscr{S}(\mathbb{R}^N).$$

The function $u(x) = e^{-|x|^2}$ belongs to $\mathscr{S}(\mathbb{R}^N)$, but not to $\mathscr{D}(\mathbb{R}^N, \mathbb{C})$.

$\mathscr{S}(\mathbb{R}^N)$ is an F-space equipped with the seminorms $\{p_{k,m}\}$. We have

$$u_n \to u \quad \text{on } \mathscr{S}(\mathbb{R}^N) \qquad \text{as} \quad n \to \infty$$

iff $p_{k,m}(u_n - u) \to 0$ as $n \to \infty$ for all k, m.

(74b) *The Fourier transform.* Let $u \in \mathscr{S}(\mathbb{R}^N)$. The classical Fourier transform $\hat{u} = Fu$ is given by

$$\hat{u}(y) = \frac{1}{(2\pi)^{N/2}} \int_{\mathbb{R}^N} e^{-i\langle y|x\rangle} u(x)\, dx.$$

This transformation, which represents one of the most important tools in analysis, has the following three fundamental properties:

(i) *Inverse transformation.* The operator

$$F\colon \mathscr{S}(\mathbb{R}^N) \to \mathscr{S}(\mathbb{R}^N)$$

is a linear homeomorphism. The inverse transformation $u = F^{-1}\hat{u}$ is given by

$$u(x) = \frac{1}{(2\pi)^{N/2}} \int_{\mathbb{R}^N} e^{i\langle y|x\rangle} \hat{u}(y)\, dy.$$

(ii) *Differentiation passes to multiplication.* From the last formula we obtain the key relation:

$$D^\alpha u(x) = \frac{1}{(2\pi)^{N/2}} \int_{\mathbb{R}^N} e^{i\langle y|x\rangle} i^{|\alpha|} y^\alpha \hat{u}(y)\, dy.$$

More precisely, for all multi-indices $\alpha = (\alpha_1, \ldots, \alpha_N)$ and all $u \in \mathscr{S}(\mathbb{R}^N)$, we obtain

$$F(D^\alpha u) = i^{|\alpha|} y^\alpha Fu.$$

Conversely,

$$D^\alpha(Fu) = (-i)^{|\alpha|} F(x^\alpha u),$$

where $y = (\eta_1, \ldots, \eta_N)$ and $y^\alpha = \eta_1^{\alpha_1}\eta_2^{\alpha_2}\ldots\eta_N^{\alpha_N}$.

(iii) *Invariance of the scalar product.* We have the so-called Parseval identity

$$(Fu|Fv) = (u|v) \qquad \text{for all} \quad u, v \in \mathscr{S}(\mathbb{R}^N),$$

where $(u|v) = \int_{\mathbb{R}^N} \bar{u}v\, dx$.

(74c) *Extension of the Fourier transform.* Since $\mathscr{S}(\mathbb{R}^N)$ is dense in $L_2^{\mathbb{C}}(\mathbb{R}^N)$, the Fourier transform $F\colon \mathscr{S}(\mathbb{R}^N) \to \mathscr{S}(\mathbb{R}^N)$ can be uniquely extended to a *unitary operator*

$$F\colon L_2^{\mathbb{C}}(\mathbb{R}^N) \to L_2^{\mathbb{C}}(\mathbb{R}^N)$$

such that $(Fu|Fv) = (u|v)$ for all $u, v \in L_2^{\mathbb{C}}(\mathbb{R}^N)$.

(74d) The *Sobolev space* $W_2^m(\mathbb{R}^N)$, $m = 1, 2, \ldots$, consists of exactly all the functions $u \in L_2(\mathbb{R}^N)$ with

$$\|u\|^*_{m,2} \overset{\text{def}}{=} \left(\int_{\mathbb{R}^N} (1 + |y|^2)^m |\hat{u}(y)|^2 \, dy\right)^{1/2} < \infty,$$

where $\hat{u} = Fu$. Furthermore, $\|\cdot\|^*_{m,2}$ is an equivalent norm on $W_2^m(\mathbb{R}^N)$.

If $u \in W_2^m(\mathbb{R}^N)$, then $D^\alpha u \in L_2(\mathbb{R}^N)$,

$$F(D^\alpha u) = i^{|\alpha|} y^\alpha Fu,$$

and $F(D^\alpha u) \in L_2^{\mathbb{C}}(\mathbb{R}^N)$ for all α: $|\alpha| \leq m$.

(74e) *The dual space* $\mathscr{S}'(\mathbb{R}^N)$. This space consists of all continuous linear functionals

$$T: \mathscr{S}(\mathbb{R}^N) \to \mathbb{C}.$$

Each $T \in \mathscr{S}'(\mathbb{R}^N)$ is called a *tempered distribution*. The Fourier transform

$$F: \mathscr{S}'(\mathbb{R}^N) \to \mathscr{S}'(\mathbb{R}^N)$$

is defined by

$$(FT)(\varphi) = T(F\varphi) \qquad \text{for all} \quad \varphi \in \mathscr{S}(\mathbb{R}^N).$$

A linear map $T: \mathscr{S}(\mathbb{R}^N) \to \mathbb{C}$ is a tempered distribution iff there exists a seminorm $p_{k,m}$ and a constant c such that

$$|T(\varphi)| \leq c p_{k,m}(\varphi) \qquad \text{for all} \quad \varphi \in \mathscr{S}(\mathbb{R}^N).$$

(74f) *Examples*. The following functions $u: \mathbb{R}^N \to \mathbb{C}$ belong to $\mathscr{S}'(\mathbb{R}^N)$:

(i) $u \in L_p^{\mathbb{C}}(\mathbb{R}^N)$, $1 \leq p \leq \infty$;
(ii) $u =$ polynomial.

The corresponding distributions $T \in \mathscr{S}'(\mathbb{R}^N)$ are given by

$$T(\varphi) = \int_{\mathbb{R}^N} u\varphi \, dx \qquad \text{for all} \quad \varphi \in \mathscr{S}(\mathbb{R}^N).$$

If there exists a measurable function $\hat{u}: \mathbb{R}^N \to \mathbb{C}$ with

$$\int_{\mathbb{R}^N} \hat{u}\varphi \, dx = \int_{\mathbb{R}^N} uF\varphi \, dx \qquad \text{for all} \quad \varphi \in \mathscr{S}(\mathbb{R}^N),$$

then the Fourier transform FT of T can be identified with $\hat{u}$ in the sense of

$$FT(\varphi) = \int_{\mathbb{R}^N} \hat{u}\varphi \, dx \qquad \text{for all} \quad \varphi \in \mathscr{S}(\mathbb{R}^N).$$

We briefly write $Fu = \hat{u}$.

(74g) *Further examples*. We have

$$\mathscr{S}'(\mathbb{R}^N) \subseteq \mathscr{D}'(\mathbb{R}^N, \mathbb{C}).$$

The following distributions from $\mathscr{D}'(\mathbb{R}^N, \mathbb{C})$ also belong to $\mathscr{S}'(\mathbb{R}^N)$:

(i) the delta distribution δ and all its derivatives $D^\alpha\delta$;
(ii) $D^\alpha u$ for $u \in L_p(\mathbb{R}^N)$, $1 \le p \le \infty$, and arbitrary α.

This is to be understood in the sense of

$$\delta(\varphi) = \varphi(0), \qquad (D^\alpha\delta)(\varphi) = (-1)^{|\alpha|}\delta(D^\alpha\varphi),$$

$$(D^\alpha u)(\varphi) = (-1)^{|\alpha|}u(D^\alpha\varphi) = (-1)^{|\alpha|}\int_{\mathbb{R}^N} uD^\alpha\varphi\,dx,$$

for all $\varphi \in \mathscr{S}(\mathbb{R}^N)$. Moreover,

$$F\delta = (2\pi)^{-N/2}.$$

This follows from

$$F\delta(\varphi) = \delta(F\varphi) = (F\varphi)(0) = \int_{\mathbb{R}^N} (2\pi)^{-N/2}\varphi\,dx,$$

for all $\varphi \in \mathscr{S}(\mathbb{R}^N)$.

The Sobolev Spaces W_p^m for Real m

We summarize the basic definitions for real and complex Sobolev spaces W_p^m of arbitrary real order m. For noninteger m, the spaces W_p^m are also called Sobolev–Slobodeckii spaces. Below we will consider the connection between the spaces W_2^m and the Fourier transform.

Let $1 \le p < \infty$ with $p^{-1} + q^{-1} = 1$, i.e., $p = 1$ implies $q = \infty$. Furthermore, let G be a nonempty open set in $\mathbb{R}^N$, $N \ge 1$. Finally, let $\mathbb{K} = \mathbb{R}, \mathbb{C}$. We set

$$f(u) = \int_{G\times G} \frac{|u(x) - u(y)|^p}{|x - y|^{N+p\mu}}\,dx\,dy,$$

$$\|u\|_{m,p} = \left(\sum_{|\alpha|\le m}\int_G |D^\alpha u|^p\,dx\right)^{1/p},$$

$$\|u\|_{k,p} = \left(\|u\|_{m,p}^p + \sum_{|\alpha|\le m} f(D^\alpha u)\right)^{1/p},$$

where $m = 0, 1, \ldots,$ and $k = m + \mu$, $0 < \mu < 1$.

(74h) Let $m = 0, 1, \ldots$. We set

$$W_p^m(G)_{\mathbb{K}} = \{u \in L_p^{\mathbb{K}}(G)\colon D^\alpha u \in L_p^{\mathbb{K}}(G) \text{ for all } \alpha\colon |\alpha| \le m\}.$$

This space together with the norm $\|\cdot\|_{m,p}$ becomes a B-space over $\mathbb{K}$.

(74i) Let $k = m + \mu$, $m = 0, 1, \ldots,$ and $0 < \mu < 1$. We set

$$W_p^k(G)_{\mathbb{K}} = \{u \in W_p^m(G)_{\mathbb{K}}\colon \|u\|_{k,p} < \infty\}.$$

This space together with the norm $\|\cdot\|_{k,p}$ becomes a B-space over $\mathbb{K}$.

(74j) For real $m \geq 0$, let

$$\mathring{W}_p^m(G)_{\mathbb{K}} = \text{closure of } \mathscr{D}(G, \mathbb{K}) \quad \text{in } W_p^m(G)_{\mathbb{K}}$$

and

$$W_q^{-m}(G)_{\mathbb{K}} = \mathring{W}_p^m(G)_{\mathbb{K}}^*.$$

As usual, the star denotes the dual space.

If $\mathbb{K} = \mathbb{R}$, then we set $W_p^m(G) = W_p^m(G)_{\mathbb{R}}$.

(74k) Let $-\infty < l \leq m < \infty$. Then the *embeddings*

$$\mathscr{D}(\mathbb{R}^N, \mathbb{C}) \subseteq \mathscr{S}(\mathbb{R}^N) \subseteq W_2^m(\mathbb{R}^N)_{\mathbb{C}} \subseteq W_2^l(\mathbb{R}^N)_{\mathbb{C}}$$
$$\subseteq \mathscr{S}'(\mathbb{R}^N) \subseteq \mathscr{D}'(\mathbb{R}^N, \mathbb{C})$$

are continuous provided the dual spaces $\mathscr{S}'(\mathbb{R}^N)$ and $\mathscr{D}'(\mathbb{R}^N, \mathbb{C})$ are equipped with the weak* topology (cf. $A_1(41)$).

(74l) Let $m > l + N/2$ with $l = 0, 1, \ldots$ and suppose that G is a bounded region in $\mathbb{R}^N$ with $\partial G \in C^{0,1}$. Then the *embedding*

$$W_2^m(G)_{\mathbb{K}} \subseteq C^l(\bar{G})_{\mathbb{K}}$$

is continuous. This also holds true if either $G = \mathbb{R}^N$ or G is an open half-space in $\mathbb{R}^N$. Here, the space $C^l(\bar{G})_{\mathbb{K}}$ consists of all continuous functions $u: \bar{G} \to \mathbb{K}$ with

$$\|u\|_l \stackrel{\text{def}}{=} \sum_{|\alpha| \leq l} \sup_{x \in \bar{G}} |D^\alpha u(x)| < \infty.$$

In this connection, we assume that all the derivatives $D^\alpha u: G \to \mathbb{K}$ are continuous up to the order l and that these derivatives can be continuously extended to the closure $\bar{G}$.

The Sobolev Spaces W_2^m for Real m and the Fourier Transform

Let $N \geq 1$, and let m be a real number. We set

$$H^m = W_2^m(\mathbb{R}^N), \qquad H_{\mathbb{C}}^m = W_2^m(\mathbb{R}^N)_{\mathbb{C}}.$$

(74m) Let

$$(u|v)_{m,2}^* = \int_{\mathbb{R}^N} \overline{\hat{u}(y)}\hat{v}(y)(1 + |y|^2)^m \, dy$$

and let $\|u\|_{m,2}^* = (u|u)_{m,2}^{*1/2}$, where $\hat{u}$ denotes the Fourier transform of u, i.e.,

$$\|u\|_{m,2}^* = \left(\int_{\mathbb{R}^N} |\hat{u}|^2 (1 + |y|^2)^m \, dy \right)^{1/2}.$$

Then

$$H_{\mathbb{C}}^m = \{u \in \mathscr{S}'(\mathbb{R}^N): \|u\|_{m,2}^* < \infty\}.$$

More precisely, the space $H_{\mathbb{C}}^m$ consists of all the distributions u in $\mathscr{S}'$ whose Fourier transform $\hat{u}$ is a function with $\|u\|_{m,2}^* < \infty$.

(74n) The space $H_{\mathbb{C}}^m$ together with the scalar product $(u|v)_{m,2}^*$ becomes a complex H-space. The set $\mathscr{D}(\mathbb{R}^N, \mathbb{C})$ is dense in $H_{\mathbb{C}}^m$.

(74o) The norm $\|\cdot\|_{m,2}^*$ is an equivalent norm on $W_2^m(\mathbb{R}^N)_{\mathbb{C}}$.

(74p) For all $m \geq 0$,

$$H^{-m} = (H^m)^*, \qquad H_{\mathbb{C}}^{-m} = (H_{\mathbb{C}}^m)^*.$$

(74q) *Interpolation.* Let $r, s \in \mathbb{R}$ with $s < r$ and let

$$m = (1 - \theta)r + \theta s, \qquad 0 < \theta < 1.$$

Then the embeddings

$$H_{\mathbb{C}}^r \subseteq H_{\mathbb{C}}^m \subseteq H_{\mathbb{C}}^s$$

are continuous, and for all $u \in H_{\mathbb{C}}^r$ and all $\varepsilon > 0$, we have

$$\|u\|_{m,2}^* \leq \|u\|_{r,2}^{*1-\theta} \|u\|_{s,2}^{*\theta}$$

and

$$\|u\|_{m,2}^* \leq \varepsilon \|u\|_{r,2}^* + \varepsilon^{1-1/\theta} \|u\|_{s,2}^*.$$

In the sense of (112) below, we obtain

$$[H_{\mathbb{C}}^r, H_{\mathbb{C}}^s]_{\theta,2} = H_{\mathbb{C}}^m,$$

where the corresponding norms are equivalent.

Construction of Measures and the Main Theorem of Measure Theory

The concept of measure is one of the most important tools of the mathematics of the twentieth century. For example, measures play an important role in the theory of probability, spectral theory, and representation theory of topological groups (together with applications in elementary particle physics). One distinguishes between

(i) measures on arbitrary sets, and
(ii) measures on topological spaces.

Primarily, the notion of measure has nothing to do with topology. Therefore, we use here an approach to measure theory which works on arbitrary sets without any topology. In particular, such an approach is important with respect to the general theory of probability. We follow the line:

$$\text{premeasure} \quad \Rightarrow \quad \text{measure} \quad \Rightarrow \quad \text{integral.}$$

The main theorem of measure theory, (75f) below, contains a construction for the unique extension of a premeasure to a measure. If we apply this construction to the volume of cuboids in $\mathbb{R}^N$, then we obtain the classical Lebesgue measure on $\mathbb{R}^N$. Similarly, one obtains measures on curves (arc length) or on surfaces (surface area).

The integral $\int f\,d\mu$ with respect to a measure μ is defined completely analogously to the Lebesgue integral $\int f\,dx$ in (14).

Measures on topological spaces will be considered in (83ff). In this connection, the representation theorem of Riesz–Markov and the compactness theorem of Prohorov are especially important. In particular, the Prohorov theorem is needed for the investigation of the Navier–Stokes equations by means of stochastic processes.

(75a) *Set algebras and σ-algebras.* Let S be a set. By definition, a *set algebra* of S is a nonempty system $\mathscr{A}$ of subsets of S which is closed with respect to the usual finite set operations, i.e., if the sets A, B, $A_1, \ldots, A_m$ belong to $\mathscr{A}$, then the sets

$$S - B, \qquad A - B, \qquad \bigcup_{n=1}^{m} A_n, \qquad \bigcap_{n=1}^{m} A_n,$$

also belong to $\mathscr{A}$. In particular, S and the empty set belong to $\mathscr{A}$.

By definition, a *σ-algebra* $\mathscr{A}$ is a set algebra which has the additional property that the sets

$$\bigcup_{n=1}^{\infty} A_n, \qquad \bigcap_{n=1}^{\infty} A_n$$

belong to $\mathscr{A}$ provided all the sets A_n belong to $\mathscr{A}$.

Let $\mathscr{A}$ be a set algebra of S. Then $\sigma(\mathscr{A})$ denotes the intersection of all the σ-algebras of S which contain $\mathscr{A}$. This intersection is nonempty and $\sigma(\mathscr{A})$ is the smallest σ-algebra of S with $\sigma(\mathscr{A}) \supseteq \mathscr{A}$.

Example. Let X be a topological space. By definition, the *Borel field* $\mathscr{B}(X)$ is the smallest σ-algebra of X which contains all the open sets in X.

(75b) *Measure.* Let $\mathscr{A}$ be a σ-algebra. By definition, a measure on $\mathscr{A}$ is a function

$$\mu\colon \mathscr{A} \to [0, \infty]$$

with the following two properties:

(i) *Additivity.* Let A and B be two arbitrary disjoint sets from $\mathscr{A}$. Then

$$\mu(A \cup B) = \mu(A) + \mu(B).$$

(ii) *σ-Additivity.* Let $\{A_n\}$ be an arbitrary countable family of pairwise disjoint sets A_n from $\mathscr{A}$. Then

$$\mu\left(\bigcup_{n=1}^{\infty} A_n\right) = \sum_{n=1}^{\infty} \mu(A_n).$$

The number $\mu(A)$ is called the measure of the set A. In terms of physics, measures may be interpreted as mass or charge. In the theory of probability, measures correspond to probability.

From the σ-additivity of a measure μ we obtain the following two continuity properties:

$$\lim_{n\to\infty}\left(\bigcup_{i=1}^{n} A_i\right) = \mu\left(\bigcup_{i=1}^{\infty} A_i\right),$$

$$\lim_{n\to\infty}\left(\bigcap_{i=1}^{n} A_i\right) = \mu\left(\bigcap_{i=1}^{\infty} A_i\right).$$

Here, it is assumed that all A_i belong to $\mathscr{A}$.

(75c) *Premeasure.* The notion of a premeasure is weaker than the notion of a measure. Let $\mathscr{A}$ be a set algebra. By definition, a premeasure is a function

$$\mu\colon \mathscr{A} \to [0, \infty]$$

with the following two properties:

(i) Additivity. Let A and B be two arbitrary disjoint sets from $\mathscr{A}$. Then

$$\mu(A \cup B) = \mu(A) + \mu(B).$$

(ii) Weak σ-additivity. Let $\{A_n\}$ be an arbitrary countable family of pairwise disjoint sets from $\mathscr{A}$ so that the union $\bigcup_{n=1}^{\infty} A_n$ also belongs to $\mathscr{A}$. Then

$$\mu\left(\bigcup_{n=1}^{\infty} A_n\right) = \sum_{n=1}^{\infty} \mu(A_n).$$

Special properties of measures. Let $\mathscr{A}$ be a σ-algebra of S. A measure μ on $\mathscr{A}$ is called *finite* iff $\mu(S) < \infty$.

A measure on $\mathscr{A}$ (or a premeasure on the set algebra $\mathscr{A}$) is called *σ-finite* iff there exists a decomposition of the form

$$S = \bigcup_{n=1}^{\infty} S_n$$

with $S_n \in \mathscr{A}$ and $\mu(S_n) < \infty$ for all n.

If $\mu(A) = 0$, then A is called a *zero set.* In particular, $\mu(\varnothing) = 0$.

(75d) *Standard example and motivation.* Intuitively, the prototype of a measure is given by a point x_0 in $\mathbb{R}^N$ with mass $m > 0$. To be precise, let $\mathscr{A}$ be the system of all sets in $\mathbb{R}^N$ and define

$$\mu_0(A) = \begin{cases} m & \text{for} \quad x_0 \in A, \\ 0 & \text{for} \quad x_0 \notin A. \end{cases}$$

Then μ_0 is a measure on the σ-algebra $\mathscr{A}$. Physically, $\mu_0(A)$ is the mass of the set A (Fig. 6 for $N = 1$).

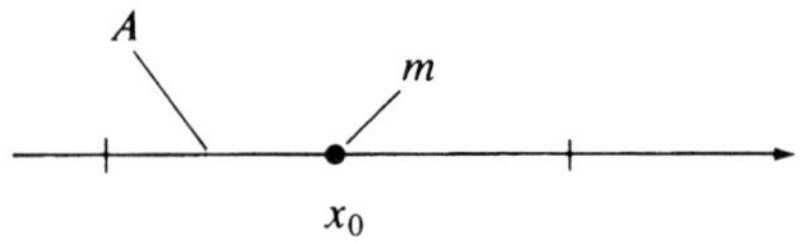

Figure 6

In order to obtain a continuous mass distribution on $\mathbb{R}^N$, we choose a continuous function $\rho: \mathbb{R}^N \to \mathbb{R}_+$ and define

$$\mu(A) = \int_A \rho \, dx,$$

where the integral is to be understood in the classical sense. Then we regard $\mu(A)$ as the mass of the set A, and we regard $\rho(x)$ as the mass density at the point x. The point is that the classical integral $\int_A \rho \, dx$ only exists for sufficiently regular sets A. Thus, we do not obtain a measure but merely a premeasure. However, by the following main theorem of measure theory (75f), this premeasure can be uniquely extended to a measure which is called the Stieltjes measure.

In physics, the concept of measure has the advantage that one can handle discrete and continuous mass distributions in a uniform way. In the physical literature one writes formally

$$\mu_0(A) = \int_A m\delta(x - x_0) \, dx,$$

where the "delta function" "$x \mapsto \delta(x - x_0)$" is called the mass density of a discrete mass $m = 1$ at the point $x = x_0$.

(75e) *Completion of a measure.* A measure μ on $\mathscr{A}$ is called *complete* iff every subset of a zero set is again a zero set, i.e., for all $A \in \mathscr{A}$,

$$\mu(A) = 0 \quad \text{and} \quad B \subseteq A \qquad \text{imply} \qquad B \in \mathscr{A} \quad \text{and} \quad \mu(B) = 0.$$

Let μ be a measure on the σ-algebra $\mathscr{A}$. Then there exists a smallest σ-algebra $\bar{\mathscr{A}}$ such that μ can be extended to a complete measure on $\bar{\mathscr{A}}$. This extension is unique. Here, $\bar{\mathscr{A}}$ is called the completion of $\mathscr{A}$ with respect to μ.

Explicitly, $\bar{\mathscr{A}}$ consists of all the sets

$$A \cup Z, \qquad A \in \mathscr{A},$$

where Z is an arbitrary subset of a zero set. Then the extension of μ to $\bar{\mathscr{A}}$ is given by $\mu(A \cup Z) = \mu(A)$.

(75f) *The main theorem of measure theory.* Let μ be a σ-finite premeasure on a set algebra $\mathscr{A}$. Then μ can be uniquely extended to a measure on $\sigma(\mathscr{A})$.

Moreover, this measure is σ-finite.

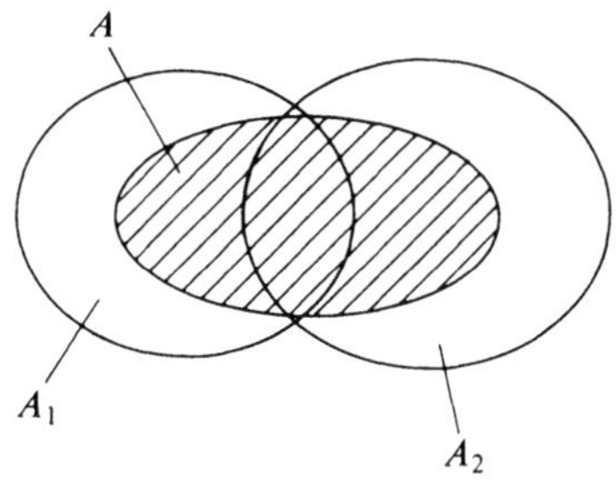

Figure 7

(75g) *Idea of proof*. The construction of the measure is based on the following steps.

Step 1: Outer measure μ^*.

Let $\mathscr{A}$ be a set algebra of the set S and let $A \subseteq S$. We choose an arbitrary and at most countable family $\{A_n\}$ of sets with

$$A \subseteq \bigcup_n A_n \qquad \text{and} \qquad A_n \in \mathscr{A} \qquad \text{for all } n.$$

Let

$$a = \sum_n \mu(A_n).$$

By definition,

$$\mu^*(A) = \inf a,$$

i.e., the outer measure $\mu^*(A)$ of A is equal to the infimum of all possible numbers a (Fig. 7).

Step 2: σ-algebra $\mu^*(\mathscr{A})$.

Let $A \subseteq S$. By definition, the set A belongs to $\mu^*(\mathscr{A})$ iff

$$\mu^*(M) = \mu^*(M \cap A) + \mu^*(M - A) \qquad \text{for all subsets } M \text{ of } S.$$

Step 3: Measure μ.

We set

$$\mu(A) \stackrel{\text{def}}{=} \mu^*(A) \qquad \text{for all} \quad A \in \mu^*(\mathscr{A}).$$

Then we have $\sigma(\mathscr{A}) \subseteq \mu^*(\mathscr{A})$ and the restriction of μ to $\sigma(\mathscr{A})$ is the desired measure.

Step 4: Completion of the measure μ on $\sigma(\mathscr{A})$.

One can show that this unique completion in the sense of (75e) coincides with μ on $\mu^*(\mathscr{A})$.

(75h) *Standard example* 1. Lebesgue measure on $\mathbb{R}^N$, $N \geq 1$. Let $-\infty \leq a_i < b_i \leq \infty$ and $x = (\xi_1, \ldots, \xi_N)$. Exactly all the sets

$$J = \{x \in \mathbb{R}^N \colon a_i \leq \xi_i < b_i \text{ for all } i\}$$

Figure 8

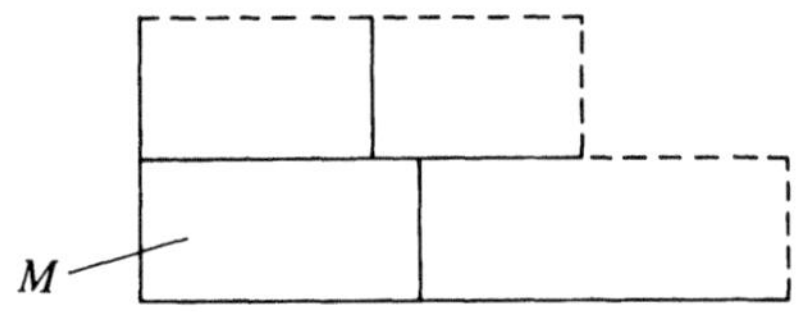

Figure 9

are called half-open intervals in $\mathbb{R}^N$ (Fig. 8). We set

$$\mu(J) = \text{volume of } J = \prod_{i=1}^{N} (b_i - a_i).$$

In order to extend μ to a measure, let $\mathscr{A}$ be the system of all the sets of the form

$$M = \bigcup_n J_n,$$

where $\{J_n\}$ is an arbitrary finite family of pairwise disjoint half-open intervals J_n in $\mathbb{R}^N$ (Fig. 9). We set

$$\mu(M) = \sum_n \mu(J_n)$$

and $\mu(\varnothing) = 0$. Then μ is a σ-finite premeasure on the set algebra $\mathscr{A}$.

We now use the main theorem of measure theory (75f), (75g). The extension of μ to the σ-algebra $\mu^*(\mathscr{A})$ is called the *Lebesgue measure* on $\mathbb{R}^N$.

This measure is σ-finite and complete. Moreover, the σ-algebra $\sigma(\mathscr{A})$ coincides with the Borel field $\mathscr{B}(\mathbb{R}^N)$. Note that $\sigma(\mathscr{A}) \subseteq \mu^*(\mathscr{A})$.

(75i) *Standard example* 2. Stieltjes measure on $\mathbb{R}$. Let $M: \mathbb{R} \to \mathbb{R}$ be a monotone increasing function which is continuous from the left (Fig. 10). Let $J = [a, b[$ be an half-open interval. We set

$$\mu(J) = M(b) - M(a).$$

In the case where $a = -\infty$ or $b = \infty$ this is to be understood in the sense of a corresponding limit.

We now use the same construction as in (75h). Then the corresponding extension of μ to a measure on $\mu^*(\mathscr{A})$ is called the *Stieltjes measure*. This

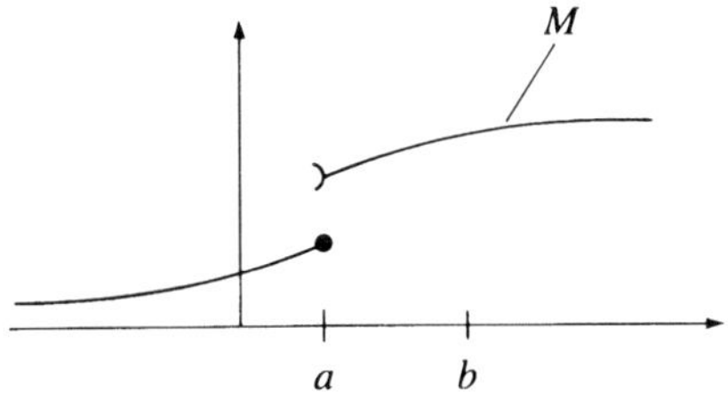

Figure 10

measure allows the following physical interpretation:

$$\mu(A) = \text{mass of the set } A.$$

In particular, a jump of the function M at the point a corresponds to a discrete point mass at a, i.e.,

$$\mu(\{a\}) = \lim_{b \to a+0} M(b) - M(a)$$

(cf. Fig. 10). In the special case where $M \equiv 1$ we obtain the Lebesgue measure.

(75j) *Measure space.* By definition, a measure space $(S, \mathscr{A}, \mu)$ is a triple consisting of a set S, a σ-algebra $\mathscr{A}$ of S, and a *σ-finite measure* μ on $\mathscr{A}$.

A measure space is called *complete* (resp. finite) iff the measure μ is complete (resp. finite, i.e., $\mu(S) < \infty$).

The sets A in $\mathscr{A}$ are called *measurable.*

In the following we shall see that many results for the classical Lebesgue measure can be generalized to measure spaces. This is very important for many branches of modern mathematics and mathematical physics.

Measurable Functions on Measure Spaces

Let $(S, \mathscr{A}, \mu)$ and $(T, \mathscr{B}, v)$ be measure spaces and let Y be a B-space. Note that there exist two different notions of measurable functions in the literature which will be introduced in (76a) and (76b) below. The relation between these two notions will be described in (76c) and (76d). Let $M \subseteq S$.

(76a) *Measurable functions.* Completely analogously to (7), a function

$$f: M \to Y$$

is called measurable iff M is measurable, i.e., $M \in \mathscr{A}$, and there exists a sequence (f_n) of step functions f_n: $M \to Y$ with

$$f(x) = \lim_{n \to \infty} f_n(x) \qquad \text{for almost all} \quad x \in M.$$

(76b) *p-Measurable functions.* The function

$$f\colon M \to T$$

is called p-measurable with respect to $\mathscr{A}$ and $\mathscr{B}$ iff

$$B \in \mathscr{B} \qquad \text{implies} \qquad f^{-1}(B) \in \mathscr{A}.$$

This notion is important for the theory of probability.

(76c) *Theorem.* Let the B-space Y be *separable* (e.g. $Y = \mathbb{R}$) and let the measure space $(S, \mathscr{A}, \mu)$ be complete.

We consider the function

$$f\colon M \to Y, \qquad M \in \mathscr{A}.$$

Then the following three conditions are mutually equivalent:

(i) f is measurable.
(ii) f is p-measurable with respect to $\mathscr{A}$ and the Borel field $\mathscr{B}(Y)$, i.e., $B \in \mathscr{B}(Y)$ implies $f^{-1}(B) \in \mathscr{A}$.
(iii) The preimages of open sets are measurable.

(76d) *Modified theorem.* Suppose that the B-space Y is not necessarily separable. Then we need the following assumption:

(A) There exists a zero set Z such that $f(M - Z)$ is separable, i.e., the set $f(M - Z)$ contains an at most countable dense subset.

Then theorem (76c) must be replaced by the following equivalences:

$$\text{(i)} \quad \Leftrightarrow \quad \text{(ii)} + \text{(A)} \quad \Leftrightarrow \quad \text{(iii)} + \text{(A)}.$$

(77) *Theorem of Egorov.* Let $(S, \mathscr{A}, \mu)$ be a complete measure space and let Y be a B-space. Suppose that the sequence (f_n) of measurable functions

$$f_n\colon M \to Y$$

converges almost everywhere on M to the function $f\colon M \to Y$, where $\mu(M) < \infty$. Then:

(i) f is measurable.
(ii) (f_n) is uniformly convergent up to sets of small measure, i.e., for each $\delta > 0$, there exists a subset M_δ of M such that $\mu(M_\delta) < \delta$ and

$$f_n \rightrightarrows f \quad \text{on } M - M_\delta \qquad \text{as} \quad n \to \infty.$$

Integrals with Respect to General Measures

Let $(S, \mathscr{A}, \mu)$ be a measure space and let Y be a B-space. We consider functions

$$f\colon M \subseteq S \to Y$$

with $M \in \mathscr{A}$. Completely analogously to (14), the integral $\int_M f\,d\mu$ is defined by the formula

$$\int_M f\,d\mu = \lim_{n\to\infty} \int_M f_n\,d\mu,$$

where (f_n) is a sequence of step functions $f_n\colon M \to Y$.

In the case where $M = \varnothing$ we set $\int_M f\,d\mu = 0$.

(78a) *Properties of the integral.* Completely analogously to the classical Lebesgue integral, the following theorems are valid for the integral $\int_M f\,d\mu$:

(i) Majorant criterion (17).
(ii) Norm criterion (18).
(iii) Majorized convergence (19).
(iv) Generalized majorized convergence, monotone convergence, and the lemma of Fatou (19a)–(19c).
(v) Absolute continuity of the integral (20).
(vi) Continuity and differentiability of parameter integrals (25a), (25b).

The theorem of Fubini–Tonelli on iterated integration and the theorem of Radon–Nikodym on absolutely continuous measures can be found in (81) and (82), respectively.

(78b) *σ-Additivity.* Suppose that $M = \bigcup_n M_n$, i.e., the set M is the union of an at most countable family of pairwise disjoint measurable sets M_n. Then

$$\int_M f\,d\mu = \sum_n \int_{M_n} f\,d\mu.$$

Here, the existence of the left-hand integral implies the existence of all the right-hand integrals and the convergence of the corresponding series.

(78c) *Convergence with respect to the domain.* Let $M_1 \subseteq M_2 \subseteq \cdots$ be a sequence of measurable sets with

$$M_n \to M \quad \text{as} \quad n \to \infty,$$

i.e., $M = \bigcup_n M_n$. Then

$$\int_M f\,d\mu = \lim_{n\to\infty} \int_{M_n} f\,d\mu.$$

Here, the existence of the left-hand integral implies the existence of all the right-hand integrals and the existence of the limit.

(79) *The Lebesgue space $L_p(M \to Y, \mu)$, $1 \le p < \infty$.* We set

$$\|f\|_p = \left(\int_M \|f(x)\|^p\,d\mu\right)^{1/p}.$$

Let $L_p(M \to Y, \mu)$ denote the set of all measurable functions $f\colon M \to Y$ with $\|f\|_p < \infty$.

Then $L_p(M \to Y, \mu)$ together with the norm $\|\cdot\|_p$ becomes a B-space once we identify any two functions which differ only on a zero set in M.

In the case where $Y = \mathbb{R}$ and $Y = \mathbb{C}$ we write briefly $L_p(M, \mu)$ and $L_p^{\mathbb{C}}(M, \mu)$, respectively. For $p = 2$, these two spaces are H-spaces with the scalar product

$$(f|g) = \int_M \bar{f} g \, d\mu.$$

More generally, if Y is an H-space, then $L_2(M \to Y, \mu)$ is also an H-space with the scalar product

$$(f|g) = \int_M (f(x)|g(x)) \, d\mu.$$

(79a) *Density*. The set of step functions $f: M \to Y$ is dense in $L_p(M \to Y, \mu)$, $1 \le p < \infty$.

(79b) *The Hölder inequality*. Let $1 < p$, $q < \infty$, $p^{-1} + q^{-1} = 1$ and let $f \in L_p(M \to Y, \mu)$, $g \in L_q(M \to Y, \mu)$. Then

$$\int_M \|f(x)\| \, \|g(x)\| \, d\mu \le \|f\|_p \|g\|_q.$$

(79c) *Convergence of subsequences*. Suppose that

$$f_n \to f \quad \text{in } L_p(M \to Y, \mu) \qquad \text{as} \quad n \to \infty$$

and $1 \le p < \infty$. Then there exists a subsequence with

$$f_{n'}(x) \to f(x) \qquad \text{as} \quad n \to \infty \quad \text{for almost all} \quad x \in M.$$

(79d) *Convergence in measure*. Let $f: M \to Y$ be a measurable function and let (f_n) be a sequence of measurable functions $f_n: M \to Y$. By definition, (f_n) converges to f in measure iff

$$\lim_{n \to \infty} \mu(\|f_n - f\| > \varepsilon) = 0 \qquad \text{for all} \quad \varepsilon > 0.$$

To be precise, this means $\mu(M_n) \to 0$ as $n \to \infty$ for all $\varepsilon > 0$, where $M_n = \{x \in M: \|f_n(x) - f(x)\| > \varepsilon\}$.

In the theory of probability one uses synonymously:

convergence almost everywhere = convergence almost certainly,

convergence in measure = stochastic convergence.

(79e) *Interrelations between different notions of convergence*. Let $(S, \mathscr{A}, \mu)$ be a measure space and let Y be a B-space. We consider an arbitrary sequence (f_n) of measurable functions

$$f_n: M \to Y.$$

Then all the statements of Figure 11 are valid. For example, convergence almost everywhere implies convergence in measure if $\mu(M) < \infty$, etc.

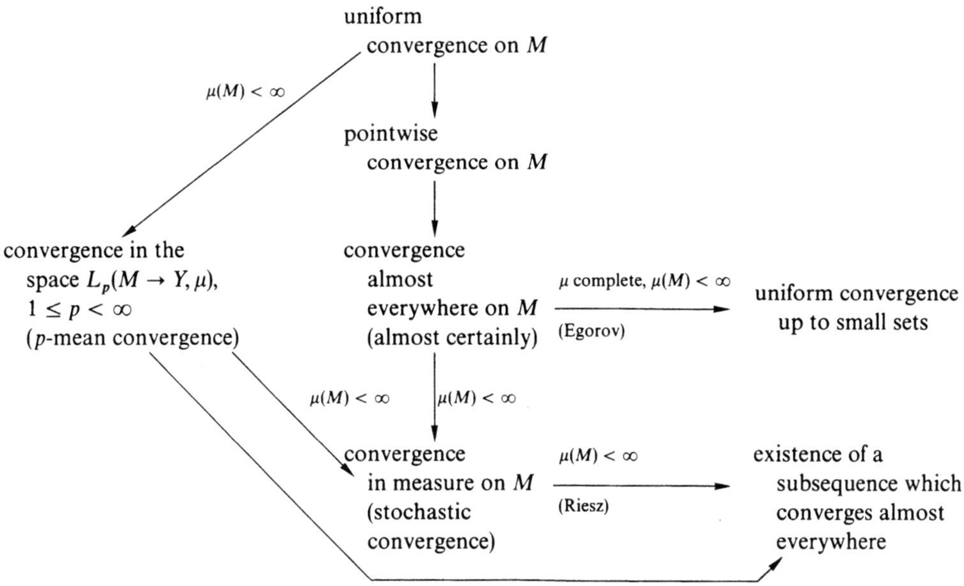

Figure 11

In particular, we may choose $S = \mathbb{R}^N$, $\mu =$ Lebesgue measure, and $M =$ bounded open set in $\mathbb{R}^N$. Then μ is complete and $\mu(M) < \infty$.

Product Measures

(80) *Definition.* Let $(S, \mathscr{A}, \mu)$ and $(T, \mathscr{B}, \nu)$ be measure spaces. We construct a new measure space (product space)

$$(S \times T, \mathscr{A} \times \mathscr{B}, \mu \times \nu)$$

in the following way:

(i) $\mathscr{A} \times \mathscr{B}$ denotes the smallest σ-algebra of $S \times T$ which contains all the product sets

$$A \times B \quad \text{with} \quad A \in \mathscr{A}, \quad B \in \mathscr{B}.$$

(ii) For all $A \in \mathscr{A}$, $B \in \mathscr{B}$, we set

$$(\mu \times \nu)(A \times B) = \mu(A)\nu(B).$$

Then there exists a unique extension of $\mu \times \nu$ to the σ-algebra $\mathscr{A} \times \mathscr{B}$. This measure is called the *product measure* $\mu \times \nu$.

(80a) *Explicit construction of the product measure.* Let $C \in \mathscr{A} \times \mathscr{B}$. We choose arbitrary sets $A_n \in \mathscr{A}$ and $B_n \in \mathscr{B}$ with

$$C \subseteq \bigcup_{n=1}^{\infty} A_n \times B_n$$

and set

$$a = \sum_{n=1}^{\infty} \mu(A_n)\nu(B_n).$$

Then

$$(\mu \times \nu)(C) = \inf a,$$

i.e., $(\mu \times \nu)(C)$ is equal to the infimum of all the possible numbers a.

(80b) *Example.* Let $S = T = \mathbb{R}$ and $\mu = \nu =$ Lebesgue measure on $\mathbb{R}$. Then $\mathscr{A} \times \mathscr{B}$ contains the Borel field $\mathscr{B}(\mathbb{R}^2)$ and the product measure $\mu \times \nu$ is equal to the Lebesgue measure on $\mathscr{A} \times \mathscr{B}$. The completion of this product measure yields the Lebesgue measure on $\mathbb{R}^2$.

An analogous result holds for $\mathbb{R}^n \times \mathbb{R}^m$.

(80c) *Product measure for arbitrarily many factors.* The following construction due to Kolmogorov is fundamental for the theory of stochastic processes.

By definition, a probability space is a measure space $(E, \mathscr{A}, \mu)$ with $\mu(E) = 1$. We are given a system of probability spaces

$$(E_j, \mathscr{A}_j, \mu_j), \qquad j \in J,$$

where J is an arbitrary finite or infinite nonempty index set. We want to construct a probability space $(E, \mathscr{A}, \mu)$ for the product set

$$E = \prod_j E_j.$$

(i) We construct $\mathscr{A} = \prod_j \mathscr{A}_j$. A set C in E is called a cylindrical set iff

$$C = \prod_j A_j$$

and $A_j \neq E_j$ holds for at most finitely many j. Let $\Pi_j \mathscr{A}_j$ denote the smallest σ-algebra of E which contains all the cylindrical sets of E.

(ii) We construct $\mu = \prod_j \mu_j$. For all cylindrical sets C, we define

$$\mu(C) = \prod_j \mu(A_j).$$

By the main theorem of measure theory (75f), the premeasure μ can be uniquely extended to a measure μ on $\mathscr{A}$.

(81) *The theorem of Fubini–Tonelli.* Let $(S, \mathscr{A}, \mu)$ and $(T, \mathscr{B}, \nu)$ be measure spaces and let the function

$$f: S \times T \to \mathbb{R}$$

be p-measurable with respect to $\mathscr{A} \times \mathscr{B}$ and the Borel field $\mathscr{B}(\mathbb{R})$, i.e., the preimages of open sets are measurable with respect to $\mathscr{A} \times \mathscr{B}$. Then the following two conditions are equivalent:

(i) $\int_{S \times T} f \, d(\mu \times \nu)$ exists.
(ii) $\int_S (\int_T |f(s,t)| \, d\nu(t)) \, d\mu(s)$ exists.

If (i) holds, then

$$\int_{S\times T} f\,d(\mu\times\nu) = \int_S\left(\int_T f\,d\nu\right)d\mu$$
$$= \int_T\left(\int_S f\,d\mu\right)d\nu.$$

Absolute Continuity of Measures

Let $(S, \mathscr{A}, \mu)$ and $(S, \mathscr{A}, \nu)$ be measure spaces.

(82a) *The theorem of Radon–Nikodym.* The following two conditions are equivalent:

(i) There exists an integrable function $f\colon S \to \mathbb{R}$ such that

$$\nu(A) = \int_A f\,d\mu \qquad \text{for all} \quad A \in \mathscr{A}.$$

(ii) The measure ν is *absolutely continuous* with respect to μ, i.e., by definition, $\mu(A) = 0$ implies $\nu(A) = 0$.

Moreover, if (ii) holds, then f in (i) is unique as an element of the space $L_1(S, \mu)$.

(82b) *The decomposition theorem of Lebesgue.* The measure ν allows a unique decomposition

$$\nu = \nu_{\text{ac}} + \nu_{\text{sing}}$$

with the following properties:

(i) The measure ν_{ac} on $\mathscr{A}$ is absolutely continuous with respect to the given measure μ on $\mathscr{A}$, i.e., $\mu(A) = 0$ implies $\nu_{\text{ac}}(A) = 0$.
(ii) The measure ν_{sing} on $\mathscr{A}$ is *singular* with respect to μ, i.e., there exists a set $A \in \mathscr{A}$ such that $\mu(A) = 0$ and $\nu_{\text{sing}}(S - A) = 0$.

Using the Radon–Nikodym theorem (82a), we may write

$$\nu(A) = \int_A f\,d\mu + \nu_{\text{sing}}(A) \qquad \text{for all} \quad A \in \mathscr{A},$$

where $f \in L_1(S, \mu)$ is uniquely determined by the measure ν.

(82c) *Special measures.* A measure μ on the σ-algebra $\mathscr{A}$ is called a *pure point* measure iff

$$\mu(A) = \sum_{x\in A} \mu(\{x\}) \qquad \text{for all} \quad A \in \mathscr{A}.$$

The measure μ is called *continuous* iff

$$\mu(A) = 0$$

for all one-point sets A.

A set $A \in \mathscr{A}$ is called an *atom* iff

$$\mu(A) > 0$$

and there is no subset $B \subseteq A$ with $0 < \mu(B) < \mu(A)$. The measure is called *nonatomic* iff there are no atoms. For example, the Lebesgue measure on $\mathbb{R}^N$ is nonatomic.

(82d) *The theorem of Ljapunov on vector measures.* Let $(S, \mathscr{A}, \mu_j), j = 1, \ldots, n$, be a family of finite measure spaces, i.e., $\mu_j(S) < \infty$ for all j. The family $(\mu_1, \ldots, \mu_n)$ is called a *vector measure* with values in $\mathbb{R}^n$ and the set

$$R = \{(\mu_1(A), \ldots, \mu_n(A)): A \in \mathscr{A}\}$$

is called the range of the vector measure.

Then the range R is a compact convex subset of $\mathbb{R}^n$ whenever each μ_j is nonatomic.

This theorem plays an important role in modern control theory (Bang–Bang control).

Measures on Topological Spaces

Measure theory on topological spaces culminates in the Riesz–Markov theorem (87) below. Roughly speaking, this theorem says that on compact spaces, continuous functions and measures are in *duality*. In the following let X denote a topological space.

(83) *Basic notions.* Figure 12 contains interrelations between important classes of topological spaces. The corresponding definitions may be found in the Appendix of Part I. In particular, a topological space is called *locally compact* iff every point has a compact neighborhood. A locally compact space is called *countable at infinity* iff it is the union of at most countably many compact subsets. The prototype of such a space is the $\mathbb{R}^N$.

Let $C(X)$ denote the set of continuous functions

$$f: X \to \mathbb{R}$$

and let $C_0(X)$ denote the set of all functions $f \in C(X)$ with compact support. We set

$$\|f\|_C = \max_{x \in X} |f(x)| \qquad \text{for all} \quad f \in C_0(X).$$

If X is compact, then $C(X) = C_0(X)$.

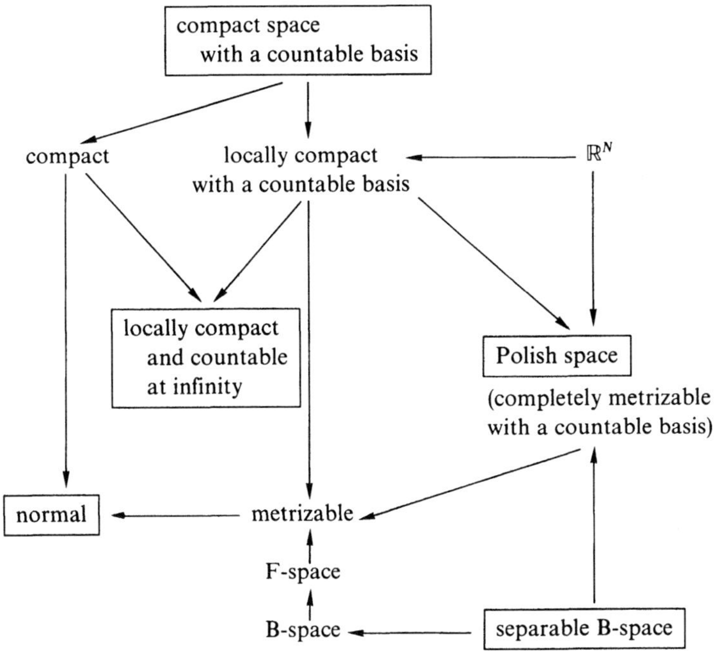

Figure 12

(84) *Borel sets and Baire sets.* By definition, the *Borel field* $\mathscr{B}(X)$ is the smallest σ-algebra of X which contains all open sets of X.

By definition, the *Baire field* $\mathscr{B}_0(X)$ is the smallest σ-algebra of X which contains all the preimages of open sets with respect to all the functions $f \in C(X)$. Obviously,

$$\mathscr{B}_0(X) \subseteq \mathscr{B}(X).$$

The sets in $\mathscr{B}_0(X)$ and $\mathscr{B}(X)$ are called *Baire sets* and *Borel sets*, respectively.

(i) $\mathscr{B}_0(X)$ is the smallest σ-algebra $\mathscr{A}$ of X with the property that all continuous functions $f: X \to \mathbb{R}$ are p-measurable with respect to $\mathscr{A}$ and $\mathscr{B}(\mathbb{R})$.

(ii) Let X be a normal space. Then the Baire field $\mathscr{B}_0(X)$ is the smallest σ-algebra of X which contains all the open sets O of the form

$$O = \bigcup_n C_n,$$

where $\{C_n\}$ is an at most countable family of closed sets C_n.

Note that, by Figure 12, both compact spaces and metrizable spaces are normal. In particular, the $\mathbb{R}^N$ and B-spaces are normal.

(iii) Let X be a metrizable space. Then $\mathscr{B}_0(X) = \mathscr{B}(X)$, i.e.,

$$\text{Borel set} = \text{Baire set}.$$

Note that the $\mathbb{R}^N$ and B-spaces are metrizable.

Let X and Y be topological spaces. Then a function $f: X \to Y$ is called a *Borel function* iff the preimages of Borel sets are again Borel sets.

(85) *Borel measure and Baire measure.* A measure μ on X is called a *Borel measure* (resp. Baire measure) iff it is defined on the Borel field (resp. Baire field) of X and

$$\mu(C) < \infty$$

holds for all compact sets C in X.

A Borel measure μ on X is called *regular* iff, for all Borel sets S, the measure μ has the following two important approximation properties:

$$\mu(S) = \inf\{\mu(O): S \subseteq O, O \text{ open}\},$$

$$\mu(S) = \sup\{\mu(C): C \subseteq S, C \text{ compact}\}.$$

Analogously, a Baire measure μ on X is called *regular* iff for all Baire sets S,

$$\mu(S) = \inf\{\mu(O): S \subseteq O, O \text{ open Baire set}\},$$

$$\mu(S) = \sup\{\mu(C): C \subseteq S, C \text{ compact Baire set}\}.$$

(i) Let X be a Polish space. Then:

$$\text{Borel set} = \text{Baire set},$$

$$\text{Borel measure} = \text{Baire measure} = \text{regular measure}.$$

This underlines the importance of Polish spaces for measure theory. Note that each separable B-space (e.g., the $\mathbb{R}^N$) is a Polish space.

(ii) Let X be a metrizable space. Then:

$$\text{Borel set} = \text{Baire set},$$

$$\text{Borel measure} = \text{Baire measure}.$$

(iii) Let X be a locally compact space which is countable at infinity. Then:

$$\text{Baire measure} = \text{regular and } \sigma\text{-finite Baire measure}.$$

Moreover, if μ is a Baire measure, then $C_0(X)$ is dense in $L_p(X, \mu)$ with $1 \le p < \infty$.

(iv) Let X be compact. Then each Baire measure is regular and can be uniquely extended to a regular Borel measure.

(85a) *Standard example.* The Lebesgue measure on $\mathbb{R}^N$ is a regular Borel measure and a regular Baire measure. Here, Borel sets and Baire sets are the same.

(86) By definition, a *positive linear functional* on $C_0(X)$ is a linear functional $L: C_0(X) \to \mathbb{R}$ such that

$$f \ge 0 \quad \text{on } X \qquad \text{implies} \qquad L(f) \ge 0.$$

If X is locally compact, then every positive linear functional on X is locally bounded, i.e., for each compact set K in X, we have

$$|L(f)| \leq \text{const} \|f\|_C,$$

for all $f \in C_0(X)$ with $\operatorname{supp} f \subseteq K$.

(87) *Theorem of Riesz–Markov.* Let X be a locally compact space which is countable at infinity.

Then there exists a bijective mapping $L \mapsto \mu$ between the positive linear functionals L on $C_0(X)$ and the Baire measures μ on X such that

$$L(f) = \int_X f\, d\mu \qquad \text{for all} \quad f \in C_0(X).$$

(87a) *The dual space* $C(X, \mathbb{K})^*$. Let X be a compact space and let $\mathbb{K} = \mathbb{R}$, $\mathbb{C}$. Moreover, let $C(X, \mathbb{K})$ denote the set of all continuous functions $f\colon X \to \mathbb{K}$. We set

$$\|f\| = \max_{x \in X} |f(x)|.$$

Then $C(X, \mathbb{K})$ together with the norm $\|\cdot\|$ becomes a B-space over $\mathbb{K}$. Exactly all the linear continuous functionals $F\colon C(X, \mathbb{K}) \to \mathbb{K}$ are given by the formula

$$F(f) = \int_X f\, d\mu \qquad \text{for all} \quad f \in C(X, \mathbb{K}).$$

In this connection, we have

$$\mu = \begin{cases} \mu_1 - \mu_2 & \text{for} \quad \mathbb{K} = \mathbb{R}, \\ \mu_1 - \mu_2 + i(\mu_3 - \mu_4) & \text{for} \quad \mathbb{K} = \mathbb{C}, \end{cases}$$

where all the μ_j denote Baire measures on X.

Moreover, all the μ_j are uniquely determined by F.

In this connection, we use the convention

$$\int_X f\, d\mu = \sum_j \alpha_j \int_X f\, d\mu_j$$

if $\mu = \sum_j \alpha_j \mu_j$, where the α_j are complex numbers and the μ_j are measures.

(88) *Vague and weak* convergence of Baire measures.* Let $\mathscr{M}_0(X)$ denote the set of all Baire measures on X and let $C_b(X)$ denote the set of all *bounded* continuous functions $f\colon X \to \mathbb{R}$. Moreover, let $\mu_n, \mu \in \mathscr{M}_0(X)$. By definition,

$$\mu_n \xrightarrow{\text{vague}} \mu \qquad \text{as} \quad n \to \infty$$

iff

$$\int_X f\, d\mu_n \to \int_X f\, d\mu \qquad \text{as} \quad n \to \infty \quad \text{for all} \quad f \in C_0(X).$$

Similarly,

$$\mu_n \xrightarrow{\text{weak*}} \mu \quad \text{as} \quad n \to \infty$$

iff

$$\int_X f\,d\mu_n \to \int_X f\,d\mu \quad \text{as} \quad n \to \infty \quad \text{for all } f \in C_b(X).$$

Since $C_0(X) \subseteq C_b(X)$,

$$\mu_n \xrightarrow{\text{weak*}} \mu \quad \text{implies} \quad \mu_n \xrightarrow{\text{vague}} \mu.$$

(88a) By definition, the *vague topology* on $\mathcal{M}_0(X)$ (resp. weak* topology) is the coarsest topology on $\mathcal{M}_0(X)$ with the property that for all $f \in C_0(X)$ (resp. $f \in C_b(X)$), the mapping

$$\mu \mapsto \int_X f\,d\mu$$

is continuous on $\mathcal{M}_0(X)$.

Explicitly, a set S in $\mathcal{M}_0(X)$ is *open* with respect to the vague topology (resp. weak* topology) iff, for each $\mu_0 \in S$ and for each $\varepsilon > 0$, there exists a finite number of functions $f_i \in C_0(X)$ (resp. $f_i \in C_b(X)$) such that all the measures $\mu \in \mathcal{M}_0(X)$ with

$$\sup_i \left| \int_X f_i\,d\mu - \int_X f_i\,d\mu_0 \right| < \varepsilon$$

belong to S.

(88b) *Vague compactness theorem.* Let X be a locally compact space which is countable at infinity and let S be a set in $\mathcal{M}_0(X)$. Then the following two conditions are equivalent:

(i) S is relatively compact with respect to the vague topology.
(ii) S is equibounded, i.e.,

$$\sup_{\mu \in S} \left| \int_X f\,d\mu \right| < \infty \quad \text{for all} \quad f \in C_0(X).$$

In addition, if X has a countable basis, then the vague topology on $\mathcal{M}_0(X)$ is metrizable, i.e., we have:

$$S \text{ is relatively compact} = S \text{ is relatively sequentially compact.}$$

In particular, we obtain the following result. Let X be a locally compact space with a countable basis and let (μ_n) be a sequence in $\mathcal{M}_0(X)$. Then the following two conditions are equivalent:

(a) There exists a subsequence with $\mu_{n'} \xrightarrow{\text{vague}} \mu$ as $n \to \infty$.
(b) $\sup_n |\int_X f\,d\mu_n| < \infty$ for all $f \in C_0(X)$.

(88c) *Weak* compactness theorem of Prohorov.* Let X be a complete separable metric space (e.g., a separable B-space) and let (μ_n) be a sequence in

$\mathscr{M}_0(X)$. Then the following two conditions are equivalent:

(i) There exists a subsequence with $\mu_{n'} \xrightarrow{\text{weak*}} \mu_n$ as $n \to \infty$.
(ii) $\sup_n \mu_n(X) < \infty$ and, for each $\varepsilon > 0$, there exists a compact set C in X with

$$\sup_n \mu_n(X - C) < \varepsilon.$$

In addition, if X is only a separable metric space, then (ii) implies (i).

The Stieltjes Integral

(89) *Physical motivation.* Let $M: \mathbb{R} \to \mathbb{R}$ be a monotone increasing function which is continuous from the left. We use the following physical interpretation:

$$M(b) = \text{mass on the interval }]-\infty, b[.$$

Hence if $b > a$, then

$$M(b) - M(a) = \text{mass on } [a, b[.$$

In particular, we obtain that

$$M(a + 0) - M(a) = \text{mass at the point } a.$$

Here, we set $M(a \pm 0) = \lim M(b)$ as $b \to a \pm 0$. For example, if there exists exactly one mass point on $\mathbb{R}$, say at a, then the mass distribution function M has the form of Figure 13, where m is the mass of the point.

(90) *Lebesgue–Stieltjes integral.* According to (75i), the function M generates a measure μ on $\mathbb{R}$ which is called a Stieltjes measure. The corresponding integral

$$\int f \, d\mu$$

is called a Lebesgue–Stieltjes integral.

(91) *The classical Stieltjes integral for piecewise continuous functions.* In order to prepare the definition of Stieltjes operator integrals which are important for the spectral theory of self-adjoint operators, we recall the definition of the classical Stieltjes integral which is a special case of the Lebesgue–

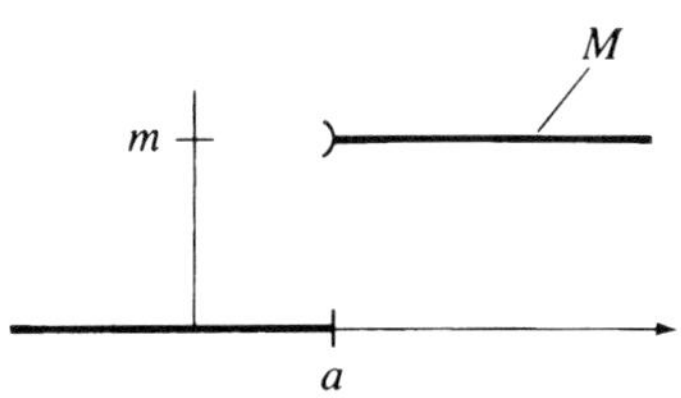

Figure 13

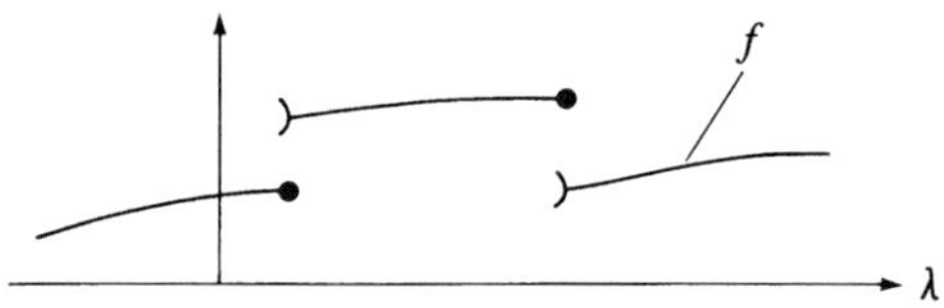

Figure 14

Stieltjes integral. In the following let the function

$$f: \mathbb{R} \to \mathbb{C}$$

be *piecewise continuous*, i.e., for all λ, the one-sided limits $f(\lambda \pm 0)$ exist and each bounded open interval contains at most a finite number of points λ with $f(\lambda + 0) \neq f(\lambda - 0)$ (Fig. 14). The basic idea of the following definition is the relation:

$$\int_J dM = \text{mass on } J$$

for all types of intervals J.

Definition.

Step 1: Let f be continuous on the bounded open interval $J =]a, b[$.
We define

$$\int_J f\,dM = \lim_{c \to a+0} \int_c^b f\,dM.$$

In this connection, the integral $\int_c^b f\,dM$ is defined as in Section 3.1 by the limit of partial sums

$$\int_c^b f\,dM = \lim_{n \to \infty} \sum_{k=1}^{n} f(\lambda_k)(M(\lambda_k) - M(\lambda_{k-1})),$$

where $\lambda_k = c + k(b - c)/n$. This limit exists according to a similar proof as in Section 3.1.

Step 2: Let $P = [a, a]$ be a point.
Then we define

$$\int_P f\,dM = f(a)(M(a + 0) - M(a)).$$

Step 3: Let J be an arbitrary bounded interval.
Then we use the disjoint decomposition

$$J = \bigcup_i J_i + \bigcup_j P_j,$$

where f is continuous on the open interval J_i, and P_j denotes a point. We define

$$\int_J f\,dM = \sum_i \int_{J_i} f\,dM + \sum_j \int_{P_j} f\,dM.$$

For example, we obtain

$$\int_{[a,b[} f\,dM = \int_{[a,a]} f\,dM + \int_{]a,b[} f\,dM.$$

Step 4: Finally, we define

$$\int_{-\infty}^{\infty} f\,dM = \lim \int_{]a,b[} f\,dM$$

as $a \to -\infty$ and $b \to +\infty$.

Spectral Theory for Self-Adjoint Operators

In the following let X denote an H-space over $\mathbb{K} = \mathbb{R}, \mathbb{C}$.

(92) *Basic ideas.* Let $A\colon \mathbb{R}^2 \to \mathbb{R}^2$ be a symmetric matrix. Then there exists a unitary matrix $U\colon \mathbb{R}^2 \to \mathbb{R}^2$ such that we obtain the diagonalization (first basic formula)

(92a) $$\begin{pmatrix} \lambda_1 & 0 \\ 0 & \lambda_2 \end{pmatrix} = UAU^{-1},$$

where λ_1 and λ_2 are the eigenvalues of A. Moreover, letting

$$P_1 = U^{-1}\begin{pmatrix} 1 & 0 \\ 0 & 0 \end{pmatrix}U, \qquad P_2 = U^{-1}\begin{pmatrix} 0 & 0 \\ 0 & 1 \end{pmatrix}U,$$

we obtain the second basic formula

(92b) $$A = \lambda_1 P_1 + \lambda_2 P_2.$$

The goal of spectral theory is to generalize this classical result to self-adjoint operators $A\colon D(A) \subseteq X \to X$ on H-spaces. The two main theorems are the following:

(i) Diagonalization theorem of von Neumann (102) (generalization of (92a)).
(ii) Decomposition theorem of Hilbert (95) (generalization of (92b)).

In this connection, formula (92b) is generalized to

$$A = \int_{-\infty}^{\infty} \lambda\,dE_\lambda.$$

This way it is possible to construct functions f of A by means of the formula

$$f(A) = \int_{-\infty}^{\infty} f(\lambda)\,dE_\lambda.$$

This basic formula of the functional calculus generalizes the classical formula

$$f(A) = U^{-1}\begin{pmatrix} f(\lambda_1) & 0 \\ 0 & f(\lambda_2) \end{pmatrix} U = f(\lambda_1)P_1 + f(\lambda_2)P_2,$$

where $A\colon \mathbb{R}^2 \to \mathbb{R}^2$ is a symmetric matrix.

(93) By definition, a *spectral family* $\{E_\lambda\}$ on X is a family of orthogonal projection operators $E_\lambda\colon X \to X$ for all $\lambda \in \mathbb{R}$ such that for all $\lambda, \mu \in \mathbb{R}$, $u \in X$, the following hold:

(i) $E_\lambda E_\mu = E_\mu E_\lambda = E_\mu$ for $\mu \le \lambda$.
(ii) $\lim_{\lambda \to -\infty} E_\lambda u = 0$ and $\lim_{\lambda \to +\infty} E_\lambda u = u$.
(iii) $\lim_{\lambda \to \mu - 0} E_\lambda u = E_\mu u$.

Consequently, if $\mu \le \lambda$, then

$$E_\mu(X) \subseteq E_\lambda(X)$$

and $E_\mu = E_\lambda$ on $E_\mu(X)$.

(94) *Stieltjes operator integrals.* For fixed $u \in X$, we set

$$M(\lambda) = \|E_\lambda u\|^2.$$

Then the function $M\colon \mathbb{R} \to \mathbb{R}$ is monotone increasing and continuous from the left. Let $f\colon \mathbb{R} \to \mathbb{K}$ be piecewise continuous. Then the integral

$$\int_{-\infty}^{\infty} f(\lambda)\, dE_\lambda u$$

is defined completely similarly to the classical Stieltjes integral in (91). This so-called operator integral exists iff

$$\int_{-\infty}^{\infty} |f(\lambda)|^2\, d\|E_\lambda u\|^2 < \infty$$

holds in the sense of (91). By definition, the integrals of the type

$$\int_{-\infty}^{\infty} f(\lambda)\, d(E_\lambda u|v)$$

are reduced in a natural way to Stieltjes integrals with respect to $M(\lambda) = \|E_\lambda w\|^2$ by means of the identity

$$(E_\lambda u|v) = \tfrac{1}{4}(\|E_\lambda(u+v)\|^2 - \|E_\lambda(u-v)\|^2 - i\|E_\lambda(u+iv)\|^2 + i\|E_\lambda(u-iv)\|^2).$$

In the case where X is real we set $i = 0$.

(95) *The first main theorem of spectral theory* (Hilbert (1906a), v. Neumann 1929)). Let $A\colon D(A) \subseteq X \to X$ be a self-adjoint operator on the H-space X over

$\mathbb{K}$. Then there exists exactly one spectral family $\{E_\lambda\}$ such that

$$Au = \int_{-\infty}^{\infty} \lambda \, dE_\lambda u \qquad \text{for all} \quad u \in D(A). \tag{95a}$$

In this connection, $u \in D(A)$ iff the integral in (95a) exists, i.e.,

$$\int_{-\infty}^{\infty} \lambda^2 \, d\|E_\lambda u\|^2 < \infty.$$

Conversely, each spectral family $\{E_\lambda\}$ on X generates a self-adjoint operator A on X by means of (95a).

Functional Calculus for Self-Adjoint Operators

(96a) *Definition.* Let $A: D(A) \subseteq X \to X$ be a self-adjoint operator on the H-space X over $\mathbb{K}$ and let $f, g: \mathbb{R} \to \mathbb{K}$ be piecewise continuous functions. We set

$$D(f(A)) = \left\{ u \in X: \int_{-\infty}^{\infty} |f(\lambda)|^2 \, d\|E_\lambda u\|^2 < \infty \right\}$$

and define the linear operator $f(A): D(f(A)) \subseteq X \to X$ by the formula

$$f(A)u = \int_{-\infty}^{\infty} f(\lambda) \, dE_\lambda u \qquad \text{for all} \quad u \in D(f(A)).$$

Note that the right-hand integral exists iff $u \in D(f(A))$. Hence our definition makes sense.

(96b) *Properties of the calculus.* In the following let B^* denote the adjoint operator to B.

(i) The operator $f(A)$ is graph closed and densely defined. If $f: \mathbb{R} \to \mathbb{K}$ is bounded, then the linear operator $f(A): X \to X$ is bounded with the operator norm

$$\|f(A)\| \leq \sup_{\lambda \in \mathbb{R}} |f(\lambda)|.$$

(ii) The adjoint operator $f(A)^*$ corresponds to the conjugate complex function $\overline{f}: \mathbb{R} \to \mathbb{K}$, i.e.,

$$f(A)^* u = \int_{-\infty}^{\infty} \overline{f(\lambda)} \, dE_\lambda u \qquad \text{for all} \quad u \in D(f(A)^*)$$

with

$$D(f(A)^*) = \left\{ u \in X: \int_{-\infty}^{\infty} |\overline{f(\lambda)}|^2 \, d\|E_\lambda u\|^2 < \infty \right\},$$

i.e., $D(f(A)^*) = D(f(A))$. In particular, if $f: \mathbb{R} \to \mathbb{R}$ is a real function, then the operator $f(A)$ is self-adjoint.

(iii) For all $u \in D(f(A))$ and $v \in X$,

$$(f(A)u|v) = \int_{-\infty}^{\infty} f(\lambda)\, d(E_\lambda u|v).$$

(iv) For $f(\lambda) = \lambda^n$, $n = 0, 1, \ldots$, we have

$$f(A) = A^n$$

with $A^0 = I$. In particular, for all $u \in X$,

$$u = \int_{-\infty}^{\infty} dE_\lambda u, \qquad \|u\|^2 = \int_{-\infty}^{\infty} d\|E_\lambda u\|^2.$$

(v) We have

$$\overline{f(A)g(A)} = \overline{g(A)f(A)} = (fg)(A).$$

Here, the bar denotes the closure of operators. If there exist positive constants c and d with

$$|g(\lambda)| \leq c|f(\lambda)g(\lambda)| + d \qquad \text{for all} \quad \lambda \in \mathbb{R},$$

then

$$f(A)g(A) = (fg)(A).$$

(vi) The dense set Y. Define

$$Y = \bigcup_{\lambda < \mu} (E_\mu - E_\lambda)(X),$$

where the union is taken over all indices $\lambda, \mu \in \mathbb{R}$ with $\lambda < \mu$. Moreover, let $X_{f(A)}$ denote the set $D(f(A))$ equipped with the graph scalar product

$$(u|v)_{f(A)} = (u|v) + (f(A)u|f(A)v).$$

Then $f(A)(Y) \subseteq Y$ and the set Y is dense in the H-spaces X and $X_{f(A)}$. For all $u \in Y$,

$$f(A)g(A)u = g(A)f(A)u = (fg)(A)u.$$

(97) *Majorant criterion.* Let J be an interval in $\mathbb{R}$. Suppose that the function $f: \mathbb{R} \times J \to \mathbb{K}$ has the property that

$$\lambda \mapsto f(\lambda, t)$$

is piecewise continuous on $\mathbb{R}$ for all $t \in J$. We set

$$f(A,t)u = \int_{-\infty}^{\infty} f(\lambda, t)\, dE_\lambda u \qquad \text{for all} \quad u \in D(f(A,t)).$$

By definition, $u \in D(f(A,t))$ iff

$$\int_{-\infty}^{\infty} |f(\lambda,t)|^2\, d\|E_\lambda u\|^2 < \infty.$$

Let $u \in \bigcap_{t \in J} D(f(A,t))$ and let

$$x(t) = f(A,t)u.$$

Then:

(i) The function $t \mapsto x(t)$ is continuous on J provided there exists a piecewise continuous function $g: \mathbb{R} \to \mathbb{R}$ satisfying the following majorant condition

$$|f(\lambda, t)| \leq g(\lambda) \qquad \text{for all} \quad (\lambda, t) \in \mathbb{R} \times J,$$

and $\int_{-\infty}^{\infty} g(\lambda)\, d\|E_\lambda u\|^2 < \infty$.

(ii) The function $t \mapsto x(t)$ is differentiable on J and we have

$$x'(t) = f_t(A, t)u \qquad \text{for all} \quad t \in J$$

provided the derivative $f_t(\lambda, t)$ exists for all $(\lambda, t) \in \mathbb{R} \times J$, the function $\lambda \mapsto f_t(\lambda, t)$ is piecewise continuous on $\mathbb{R}$ for all $t \in J$, and there exists a piecewise continuous function $h: \mathbb{R} \to \mathbb{R}$ satisfying the majorant condition

$$|f_t(\lambda, t)| \leq h(\lambda) \qquad \text{for all} \quad (\lambda, t) \in \mathbb{R} \times J,$$

and $\int_{-\infty}^{\infty} h(\lambda) d\|E_\lambda u\|^2 < \infty$. Here, we set

$$f_t(A, t)u = \int_{-\infty}^{\infty} f_t(\lambda, t)\, dE_\lambda u.$$

(98) *Generalization.* Let $A: D(A) \subseteq X \to X$ be a self-adjoint operator on the H-space X over $\mathbb{K}$ and let $f: \mathbb{R} \to \mathbb{K}$ be a bounded Borel function in the sense of (84). Then there exists exactly one linear continuous operator $B: X \to X$ with

$$(Bu|u) = \int_{-\infty}^{\infty} f(\lambda)\, d\|E_\lambda u\|^2 \qquad \text{for all} \quad u \in X.$$

This integral is to be understood in the sense of Lebesgue–Stieltjes (cf. (90)). We set $f(A) = B$. Then

$$f(A)^* = \bar{f}(A) \qquad \text{and} \qquad \|f(A)\| \leq \sup_{\lambda \in \mathbb{R}} |f(\lambda)|.$$

(99) *Example.* Let $A: \mathbb{R}^2 \to \mathbb{R}^2$ be a real symmetric matrix with the eigenvalues $\lambda_1 < \lambda_2$. Then there exists a unitary matrix $U: \mathbb{R}^2 \to \mathbb{R}^2$ such that

$$A = U^{-1}\begin{pmatrix} \lambda_1 & 0 \\ 0 & \lambda_2 \end{pmatrix} U.$$

We set

$$P_1 = U^{-1}\begin{pmatrix} 1 & 0 \\ 0 & 0 \end{pmatrix} U, \qquad P_2 = U^{-1}\begin{pmatrix} 0 & 0 \\ 0 & 1 \end{pmatrix} U.$$

Then the spectral family of A is given by

$$E_\lambda = \begin{cases} 0 & \text{for} \quad \lambda \leq \lambda_1, \\ P_1 & \text{for} \quad \lambda_1 < \lambda \leq \lambda_2, \\ P_1 + P_2 = I & \text{for} \quad \lambda_2 < \lambda. \end{cases}$$

We obtain

$$P_j = E_{\lambda_j+0} - E_{\lambda_j}, \qquad j = 1, 2.$$

Hence the formula $Au = \int \lambda\, dE_\lambda u$ corresponds to

$$A = \lambda_1 P_1 + \lambda_2 P_2,$$

and the formula $f(A)u = \int f(\lambda)\, dE_\lambda u$ corresponds to

$$f(A) = f(\lambda_1)P_1 + f(\lambda_2)P_2 = U^{-1}\begin{pmatrix} f(\lambda_1) & 0 \\ 0 & f(\lambda_2) \end{pmatrix} U.$$

(100) *Standard example.* Let $A: D(A) \subseteq X \to X$ be a self-adjoint operator on the separable H-space X over $\mathbb{K}$ and suppose that A has a complete orthonormal system $\{u_m\}$ of eigenvectors with $Au_m = \lambda_m u_m$, $m = 1, 2, \ldots$. Additionally, we assume that $\inf_m \lambda_m > -\infty$. These assumptions are satisfied if one of the following three conditions holds:

(i) The operator $A: X \to X$ is linear and symmetric, and $\dim X < \infty$.
(ii) The operator $A: X \to X$ is linear, symmetric, and compact.
(iii) The operator A is the Friedrichs extension of a linear symmetric and strongly monotone operator and the embedding $X_E \subseteq X$ is compact. Here, X_E denotes the energetic space of A (see Section 19.9).

The eigenvalues λ_m of A are counted according to their multiplicity. Let μ_1, $\mu_2, \ldots$ denote the different eigenvalues of A, i.e., $\mu_i \neq \mu_j$ iff $i \neq j$. Moreover, let $P_k: X \to X$ denote the orthogonal projection operator onto the eigenspace corresponding to the eigenvalue μ_k of A.

First suppose that $\mu_1 < \mu_2 < \cdots$ and that there exists a largest eigenvalue $\mu_{\max}$. Then the spectral family of A is given by

$$E_\lambda = \begin{cases} 0 & \text{for} \quad \lambda \leq \mu_1, \\ P_1 & \text{for} \quad \mu_1 < \lambda \leq \mu_2, \\ P_1 + \cdots + P_k & \text{for} \quad \mu_k < \lambda \leq \mu_{k+1}, \\ I & \mu_{\max} < \lambda, \end{cases}$$

where $k = 2, 3, \ldots$. In the general case, we define $\{E_\lambda\}$ such that

$$E_{\mu_k+0} - E_{\mu_k} = P_k \qquad \text{for all } k$$

and $E_{-\infty} = 0$, $E_{+\infty} = I$. Moreover, let $E_\lambda =$ constant on λ-intervals which do not contain eigenvalues of A. Finally, this definition has to be arranged in such a way that $E_\mu \to E_\lambda$ as $\mu \to \lambda - 0$ for all $\lambda \in \mathbb{R}$.

Then the formula $Au = \int_{-\infty}^{\infty} \lambda\, dE_\lambda u$ is identical with the well-known expansion

$$Au = \sum_m \lambda_m (u_m | u) u_m = \sum_k \mu_k P_k u \qquad \text{for all} \quad u \in D(A).$$

The formula $f(A)u = \int_{-\infty}^{\infty} f(\lambda)\, dE_\lambda u$ is identical to

$$f(A)u = \sum_m f(\lambda_m)(u_m|u)u_m = \sum_k f(\mu_k)P_k u \qquad \text{for all} \quad u \in D(f(A))$$

and

$$D(f(A)) = \left\{u \in X: \int_{-\infty}^{\infty} |f(\lambda)|^2\, d\|E_\lambda u\|^2 < \infty\right\}$$
$$= \left\{u \in X: \sum_m |f(\lambda_m)|^2 |(u_m|u)|^2 < \infty\right\}.$$

Note that $u \in D(f(A))$ iff the series $\sum_m f(\lambda_m)(u_m|u)u_m$ converges.

Diagonalization of Self-Adjoint Operators

(101) *Isometric and unitary operators.* Let X and Y be H-spaces over $\mathbb{K} = \mathbb{R}, \mathbb{C}$. A linear operator $U: X \to Y$ is called *isometric* iff

$$\|Uu\| = \|u\| \qquad \text{for all} \quad u \in X.$$

Such an operator is injective. Recall that $U: X \to Y$ is called *unitary* or an H-isomorphism iff U is linear, bijective, and

$$(Uu|Uv) = (u|v) \qquad \text{for all} \quad u, v \in X.$$

A linear operator $U: X \to X$ is unitary iff it is isometric and surjective.

(102) *The second main theorem of spectral theory* (v. Neumann (1949)). We need the basic formula

$$(\hat{A}\hat{u})(m) = \lambda(m)\hat{u}(m) \qquad \text{for almost all} \quad m \in M \tag{102a}$$

and the commutative diagram

$$\begin{array}{ccc} D(A) \subseteq X & \xrightarrow{A} & X \\ \uparrow{\scriptstyle U^{-1}} & & \downarrow{\scriptstyle U} \\ D(\hat{A}) \subseteq \hat{X} & \xrightarrow{\hat{A}} & \hat{X}. \end{array}$$

Let $A: D(A) \subseteq X \to X$ be a self-adjoint operator on the separable complex H-space X. Then:

(i) There exists a set M and a measure μ on M with $\mu(M) < \infty$. Let $\hat{X} = L_2^{\mathbb{C}}(M, \mu)$, i.e., $\hat{X}$ consists of all measurable functions $\hat{u}: M \to \mathbb{C}$ with

$$\int_M |\hat{u}|^2\, d\mu < \infty.$$

(ii) There exists a unitary operator $U: X \to \hat{X}$ and a function $\lambda: M \to \mathbb{R}$ which is finite almost everywhere on M.

(iii) The operator U generates the transformation

$$\hat{u} = Uu.$$

This way the operator A is transformed into the operator $\hat{A} = UAU^{-1}$.

(iv) The operator $\hat{A}$ has the *simple form* (102a) for all $\hat{u} \in D(\hat{A})$ where

$$D(\hat{A}) = \{\hat{u} \in \hat{X} : \lambda\hat{u} \in \hat{X}\}.$$

(v) $u \in D(A)$ iff $Uu \in D(\hat{A})$.

Remark. This famous theorem shows that the H-space X can be realized by the function space $\hat{X} = L_2^{\mathbb{C}}(M, \mu)$. In this connection, the self-adjoint operator A on X corresponds to the simple multiplication operator $\hat{A}$. The set M may be regarded as an index set for the generalized eigenvalues $\lambda(m)$ of A. By the classical Fourier transform, differentiation is transformed into multiplication. Hence the unitary operator U may be regarded as an abstract Fourier transform.

(103) *Example.* Let $A: \mathbb{C}^2 \to \mathbb{C}^2$ be a real symmetric or complex hermitian matrix, i.e., $\bar{a}_{ij} = a_{ji}$ for all i, j, with the eigenvalues $\lambda(1)$, $\lambda(2)$. We set

$$\hat{A} = \begin{pmatrix} \lambda(1) & 0 \\ 0 & \lambda(2) \end{pmatrix}.$$

Then there exists a unitary matrix $U: \mathbb{C}^2 \to \mathbb{C}^2$ such that

$$\hat{A} = UAU^{-1}. \tag{103a}$$

We set $X = \mathbb{C}^2$, $M = \{1, 2\}$ and

$$\begin{pmatrix} \hat{u}(1) \\ \hat{u}(2) \end{pmatrix} = U \begin{pmatrix} u(1) \\ u(2) \end{pmatrix}.$$

Then (103a) corresponds to

$$(\hat{A}\hat{u})(m) = \lambda(m)\hat{u}(m) \qquad \text{for all} \quad m \in M, \quad \hat{u} \in \mathbb{C}^2.$$

This is the basic formula (102a) above. In fact, if we define the measure μ on M by $\mu(m) = 1$ for all $m \in M$, then the space $\hat{X} = L_2^{\mathbb{C}}(M, \mu)$ is equal to $\mathbb{C}^2$.

In this example we have $\hat{X} = X$. Note that in the general case of theorem (102) above we have to leave the space X, i.e., $\hat{X} \neq X$.

Classification of the Spectrum of Self-Adjoint Operators

Let $A: \mathbb{R}^N \to \mathbb{R}^N$ be a symmetric matrix. Then all the eigenvalues of A are real and the set of all the eigenvalues of A forms the spectrum $\sigma(A)$ of A. In the case of general self-adjoint operators the structure of the spectrum can be much more complex. This fact reflects mathematically the possibly complex structure of the energy values of quantum systems (see (104d)).

(104) *Complex spaces.* Let $A: D(A) \subseteq X \to X$ be a self-adjoint operator on the complex H-space X. The case of real spaces X will be considered in (105). Let $X \neq \{0\}$.

By definition, the point $\lambda \in \mathbb{C}$ belongs to the *resolvent set* $\rho(A)$ of A iff the inverse $(A - \lambda I)^{-1}: X \to X$ exists as a linear continuous operator. The *spectrum* $\sigma(A)$ of A is defined to be the complement of $\rho(A)$, i.e.,

$$\sigma(A) = \mathbb{C} - \rho(A).$$

Here, $\sigma(A)$ is a closed subset of $\mathbb{R}$. We define the point spectrum of A by

$$\sigma_{\mathrm{p}}(A) = \text{set of all eigenvalues of } A.$$

Moreover, we define the *discrete spectrum* of A by

$$\sigma_{\mathrm{disc}}(A) = \text{set of all eigenvalues of } A \text{ of finite multiplicity.}$$

(104a) *First classification of the spectrum $\sigma(A)$ due to Hilbert.* Our goal is the not necessarily disjoint decomposition

$$\sigma(A) = \overline{\sigma_{\mathrm{p}}(A)} \cup \sigma_{\mathrm{c}}(A).$$

By definition, the operator A has a *pure point spectrum* iff the eigenvectors of A form a complete orthonormal system in X. In this case, we have $\sigma(A) = \overline{\sigma_{\mathrm{p}}(A)}$, and we set $\sigma_{\mathrm{c}}(A) = \varnothing$.

The operator A has a *purely continuous spectrum* in X iff A has no eigenvalues. In this case, we have $\sigma_{\mathrm{p}}(A) = \varnothing$, and we set $\sigma_{\mathrm{c}}(A) = \sigma(A)$.

We now consider the orthogonal decomposition

$$X = X_{\mathrm{pp}} \oplus X_{\mathrm{c}},$$

where X_{pp} denotes the closed linear hull of the set of all eigenvectors of A, and $X_{\mathrm{c}} = X_{\mathrm{pp}}^{\perp}$ denotes the orthogonal complement to X_{pp}. Then the operator A transforms the subspaces X_{pp} and X_{c} into itself, and A has a pure point spectrum on X_{pp} and a purely continuous spectrum of on X_{c}. We define the *continuous spectrum* of A by

$$\sigma_{\mathrm{c}}(A) = \text{spectrum of } A \text{ on } X_{\mathrm{c}}.$$

(104b) *Refined first classification of $\sigma(A)$.* Our goal is the disjoint decomposition

$$\sigma_{\mathrm{c}}(A) = \sigma_{\mathrm{ac}}(A) \cup \sigma_{\mathrm{sing}}(A)$$

along with the orthogonal decomposition

$$X_{\mathrm{c}} = X_{\mathrm{ac}} \oplus X_{\mathrm{sing}}.$$

To this end, for each fixed $u \in X$, we set

$$M(\lambda) = \|E_\lambda u\|^2.$$

By (90), the function $M: \mathbb{R} \to \mathbb{R}$ generates a Stieltjes measure μ_u on $\mathbb{R}$. We define:

(i) $u \in X_{\text{ac}}$ iff μ_u is absolutely continuous with respect to the Lebesgue measure on $\mathbb{R}$.
(ii) $u \in X_{\text{sing}}$ iff μ_u is a continuous measure which is singular with respect to the Lebesgue measure on $\mathbb{R}$.
Moreover, we have:

(iii) $u \in X_{\text{pp}}$ iff μ_u is a pure point measure.

The operator A transforms the subspaces X_{ac}, X_{sing}, and X_{pp} into itself. We define the absolutely continuous spectrum $\sigma_{\text{ac}}(A)$ and the singular spectrum $\sigma_{\text{sing}}(A)$ of A by

$$\sigma_{\text{ac}}(A) = \text{spectrum of } A \text{ on } X_{\text{ac}},$$

$$\sigma_{\text{sing}}(A) = \text{spectrum of } A \text{ on } X_{\text{sing}}.$$

The corresponding definitions for measures may be found in (82).

(104c) *Second classification of* $\sigma(A)$. We define the essential spectrum of A by

$$\sigma_{\text{ess}}(A) = \sigma(A) - \sigma_{\text{disc}}(A).$$

Thus, we have the disjoint decomposition

$$\sigma(A) = \sigma_{\text{disc}}(A) \cup \sigma_{\text{ess}}(A)$$

and $\sigma_{\text{disc}}(A) \subseteq \sigma_{\text{p}}(A)$ as well as $\sigma_{\text{c}}(A) \subseteq \sigma_{\text{ess}}(A)$. Let $\{E_\mu\}$ denote the spectral family of A. Then:

(i) The sets $\sigma(A)$ and $\sigma_{\text{ess}}(A)$ are closed subsets of $\mathbb{R}$.
(ii) $\lambda \in \sigma(A)$ iff $\{E_\mu\}$ is not constant on a neighborhood of λ.
(iii) $\lambda \in \sigma_{\text{disc}}(A)$ iff $0 < \dim(E_{\lambda+0} - E_\lambda)(X) < \infty$.
(iv) $\lambda \in \sigma_{\text{ess}}(A)$ iff $\dim(E_{\lambda+\varepsilon} - E_{\lambda-\varepsilon})(X) = \infty$ for all $\varepsilon > 0$.
(v) $\lambda \in \sigma_{\text{ess}}(A)$ iff there exists a Weyl sequence (u_n) for λ, i.e.,

$$Au_n - \lambda u_n \to 0 \qquad \text{as} \quad n \to \infty$$

and the bounded sequence (u_n) does not contain any convergent subsequence.

(104d) *Physical interpretation.* In quantum physics, each *state* of a quantum system is described by a vector $u \in X$ in a complex H-space X normalized by $\|u\| = 1$. The *energy* of the system is described by a self-adjoint operator $A: D(A) \subseteq X \to X$ which is called the Hamiltonian of the system. The states

$$u \in X_{\text{pp}} \qquad \text{and} \qquad u \in X_{\text{c}}$$

are called bound and free, respectively. In particular, if

$$Au = \lambda u, \qquad \|u\| = 1,$$

then u corresponds to a bound state of energy λ. Roughly speaking, if we compare the electron of the hydrogen atom with a body in celestial mechanics, then the bound and free states of the electron correspond to the orbits of planets and freely moving comets, respectively.

More generally, each *physical quantity* a of the system corresponds to a self-adjoint operator $A: D(A) \subseteq X \to X$ (e.g., momentum, angular momentum, etc.). Let $\{E_\lambda\}$ be the spectral family of A. Let the system be in the *state* u and suppose that we *measure* the quantity a. It is typical for quantum physics, that there exists a *probability* $p(B)$ for finding the measured value of a in the set B. This probability is given by the *fundamental formula*

$$p(B) = \int_B dM(\lambda),$$

where

$$M(\lambda) = \|E_\lambda u\|^2 \qquad \text{for all} \quad \lambda \in \mathbb{R},$$

i.e., M represents the *distribution function* for the measured values of a, and hence

$$M(\lambda) = \text{probability for finding } a \text{ in the interval }]-\infty, \lambda[.$$

In particular, if B is a subset of the resolvent set $\rho(A)$, then $p(B) = 0$, and hence, roughly speaking, only those values can be measured which lie in the spectrum $\sigma(A)$. According to the theory of probability, the expectation value $\bar{a}$ and the dispersion $(\Delta a)^2$ for the quantity a are given by

$$\bar{a} = \int_{-\infty}^{\infty} \lambda\, dM(\lambda), \qquad (\Delta a)^2 = \int_{-\infty}^{\infty} (\lambda - \bar{a})^2\, dM(\lambda),$$

and hence

$$\bar{a} = (Au|u), \qquad (\Delta a)^2 = ((A - \bar{a}I)^2 u|u).$$

This underlines the fundamental importance of the spectrum and the spectral family for quantum physics. We will study these questions in greater detail in Chapter 89 of Part V.

(105) *Real spaces.* Let $A: D(A) \subseteq X \to X$ be a self-adjoint operator on the real H-space X. By Problem 19.6, the complexification $A_{\mathbb{C}}$ is also a self-adjoint operator on the complexified H-space $X_{\mathbb{C}}$. By definition, all the spectral properties of A refer to $A_{\mathbb{C}}$. In particular, by definition, $\sigma(A) = \sigma(A_{\mathbb{C}})$, etc.

Note that the spectral family of $A_{\mathbb{C}}$ is equal to the complexification of the spectral family of A.

Linear Semigroups

In the following we summarize basic results on linear semigroups in B-spaces. This should help the reader to recognize important connections between the linear and nonlinear theory of semigroups. For brevity, semigroups of linear bounded operators on real or complex B-spaces are simply called linear semigroups. The basic definitions on semigroups are contained in Section 19.17a.

(106) *Importance of generators.* Two linear strongly continuous semigroups on a B-space are identical iff the generators are identical.

The generators of such semigroups are densely defined on the B-space and graph closed.

(107) *Main theorem on linear uniformly continuous semigroups.* Precisely all linear uniformly continuous semigroups $\mathscr{S} = \{S(t)\}$ on a B-space X are obtained through

$$S(t) = e^{tB} \qquad \text{for all} \quad t \geq 0,$$

where $B: X \to X$ is an arbitrary linear bounded operator. Moreover, B is the generator of $\mathscr{S}$.

If we consider e^{tB} for all $t \in \mathbb{R}$, then we obtain one-parameter groups with

$$\|S(t)\| \leq e^{|t|\,\|B\|} \qquad \text{for all} \quad t \in \mathbb{R}.$$

Hence it is *impossible* to describe typical *irreversible* processes in nature by linear uniformly continuous semigroups. The generators of such processes must be unbounded. This underlines the importance of *unbounded* operators for the mathematical description of nature.

(108) *Main theorem on linear strongly continuous semigroups* (Hille–Yosida theorem). If $\{S(t)\}$ is a linear strongly continuous semigroup on a B-space X, then there exist real numbers β and $M \geq 1$ such that

$$\text{(C)} \qquad \|S(t)\| \leq Me^{\beta t} \qquad \text{for all} \quad t \geq 0.$$

The operator $B: D(B) \subseteq X \to X$ is the generator of a linear strongly continuous semigroup with (C) iff the following hold:

(i) B is linear, densely defined on X, and graph closed.
(ii) All real numbers $\lambda > \beta$ belong to the resolvent set of B and

$$\|(\lambda I - B)^{-n}\| \leq M(\lambda - \beta)^{-n} \qquad \text{for all} \quad \lambda > \beta, \qquad n = 1, 2, \ldots.$$

The corresponding semigroup is obtained through

$$S(t)u = \lim_{\mu \to +0} e^{tB_\mu}u \qquad \text{for all} \quad t \geq 0 \quad \text{and all} \quad u \in X,$$

where $R_\mu = (I - \mu B)^{-1}$ and the operator

$$B_\mu = \mu^{-1}(R_\mu - I)$$

is called the *Yosida approximation* of B. We have $B_\mu = BR_\mu$.

The set of operators B with (i) and (ii) above is denoted by $G(X, M, \beta)$. With respect to (C), one has the following special cases:

$\beta = 0$: bounded semigroup;

$M = 1$: quasi-nonexpansive (or quasi-contractive) semigroup;

$M = 1, \beta = 0$: nonexpansive (or contractive) semigroup.

(108a) *The homogeneous Cauchy problem.* Let $B \in G(X, M, \beta)$ with the corresponding semigroup $\{S(t)\}$. Then $u(t) = S(t)u_0$ with $u_0 \in D(B)$ is the unique solution of the Cauchy problem

$$u'(t) = Bu(t), \qquad 0 \le t < \infty,$$
$$u(0) = u_0.$$

More precisely, we have the following statements:

(α) The map $t \mapsto S(t)u_0$ is continuous from $\mathbb{R}_+$ to X for all $u_0 \in X$. Recall that $\mathbb{R}_+ = [0, \infty[$.
(β) This map is differentiable at $t = 0$ (and equivalently, for all $t \ge 0$) iff $u_0 \in D(B)$. Moreover, if $u_0 \in D(B)$, then $S(t)u_0 \in D(B)$ for all $t \ge 0$ and

$$\frac{d}{dt}S(t)u_0 = S(t)Bu_0 = BS(t)u_0 \qquad \text{for all} \quad t \ge 0.$$

(γ) Let X_B denote the B-space $D(B)$ equipped with the graph norm of B. Then, for each $u \in X_B$, the mapping

$$t \mapsto S(t)u$$

is continuous from $\mathbb{R}_+$ to X_B, and the mapping

$$t \mapsto \frac{d}{dt}(S(t)u)$$

is continuous from $\mathbb{R}_+$ to X.

(108b) *The inhomogeneous Cauchy problem* (*strongly continuous semigroups*). Let $0 < T < \infty$. The Cauchy problem

$$\text{(P)} \qquad \begin{aligned} u'(t) &= Bu(t) + f(t), \qquad 0 < t < T, \\ u(0) &= u_0 \end{aligned}$$

has a unique solution if the following conditions are satisfied:

(i) B generates a linear strongly continuous semigroup $\{S(t)\}$ on the B-space X.
(ii) $u_0 \in D(B)$.
(iii) $f\colon [0, T] \to X$ is C^1.

This solution is given by

$$\text{(S)} \qquad u(t) = S(t)u_0 + \int_0^t S(t - s)f(s)\,ds,$$

where $u \in C([0, T[, X) \cap C^1([0, T[, X)$ and

$$u(t) \in D(B) \qquad \text{for all} \quad t \in [0, T[.$$

(108c) *The inhomogeneous Cauchy problem* (*analytic semigroups*). Let $0 < T < \infty$. The Cauchy problem (P) above has a unique solution if the following conditions are satisfied:

(i*) The operator $-B$ is *sectorial* on the complex B-space X.
(ii*) $u_0 \in X$.
(iii*) $f: [0, T] \to X$ is Hölder continuous.

This solution is given by (S) above. Here,

$$u \in C([0, T[, X) \cap C^1(]0, T[, X),$$

and $u(t) \in D(B)$ for all $t \in]0, T[$.

Condition (i*) is equivalent to the fact that B generates a linear analytic semigroup. Hence, (i*) is stronger than (i) in (108b). But observe that (ii*) and (iii*) is weaker than (ii) and (iii), respectively.

(109) *Characterization of linear nonexpansive semigroups.* Let $B: D(B) \subseteq X \to X$ be a densely defined linear operator on the B-space X. Then B is the generator of a linear nonexpansive semigroup iff one of the following two conditions is satisfied:

(i) $-B$ is *maximal accretive*, i.e., the inverse operators $(I - \mu B)^{-1}: X \to X$ exist for all $\mu > 0$ and are nonexpansive.
(ii) B is *maximal dissipative*, i.e., for every $u \in D(B)$ there exists an element $u^* \in J(u)$ with

$$\operatorname{Re}\langle u^*, Bu\rangle \leq 0$$

and $R(I - \mu B) = X$ for some $\mu > 0$, where $J: X \to 2^{X^*}$ is the duality map of X.

If X is an H-space, then condition (ii) is equivalent to

$$\operatorname{Re}(u|Bu) \leq 0 \qquad \text{for all} \quad u \in D(B)$$

and $R(I - \mu B) = X$ for some $\mu > 0$.

Sectorial Operators, Analytic Semigroups, and Parabolic Equations

(110) Let $A: D(A) \subseteq X \to X$ be a *sectorial operator* on the complex B-space X, i.e., the following conditions are satisfied:

(i) A is linear, graph closed, and densely defined on X.
(ii) There are numbers $c \in \mathbb{R}$, $M \geq 1$, and $\gamma \in]0, \pi/2[$ such that the open sector

$$\Sigma = \{\lambda \in \mathbb{C}: \gamma < |\arg(\lambda - c)| \leq \pi, \lambda \neq c\}$$

is a subset of the resolvent set of A and

$$\|(\lambda I - A)^{-1}\| \leq M|\lambda - c|^{-1} \qquad \text{for all} \quad \lambda \in \Sigma.$$

Then $-A$ is the generator of a linear analytic semigroup $\{e^{-tA}\}$, which is

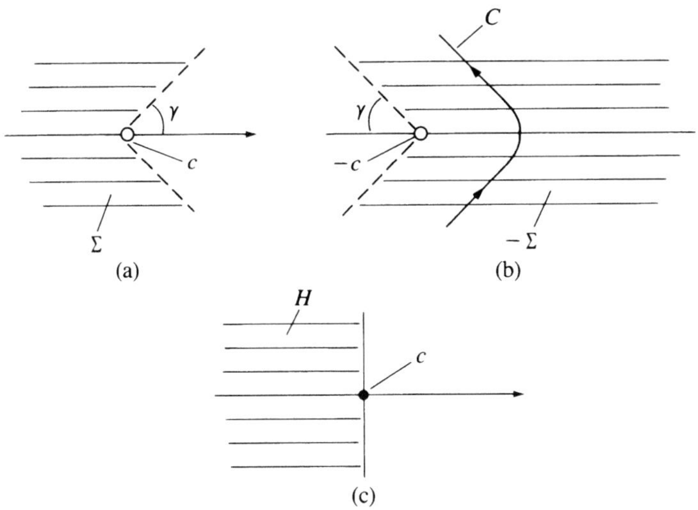

Figure 15

given by the formula

$$e^{-tA} = \frac{1}{2\pi i} \int_C (\lambda I + A)^{-1} e^{\lambda t} \, d\lambda.$$

Here, C is a smooth curve in $-\Sigma$ with $\arg \lambda \to \pm(\pi - \gamma)$ as $|\lambda| \to \infty$ (Figs. 15(a), (b)). Note that e^{-tA} is independent of the concrete shape of C.

A sectorial operator with $c = 0$ is called *strictly* sectorial.

(110a) *Standard example* 1. Let $A: D(A) \subseteq X \to X$ be a linear self-adjoint operator on the complex B-space X with

$$(Au|u) \geq c(u|u) \qquad \text{for all} \quad u \in D(A) \quad \text{and fixed} \quad c \in \mathbb{R}.$$

Then A is sectorial, and $-A$ generates a linear analytic semigroup $\{e^{-tA}\}$.

If, in addition, $c = 0$, i.e., A is monotone, then A is strictly sectorial, and $\{e^{-tA}\}$ is both a linear bounded analytic semigroup and a nonexpansive semigroup. More precisely, we have

$$\|e^{-tA}\| \leq 1 \qquad \text{for all} \quad t \in \mathbb{C} \quad \text{with} \quad \operatorname{Re} t \geq 0.$$

(110b) *Standard example* 2. Let $A: D(A) \subseteq X \to X$ be a linear, graph closed, and densely defined operator on the complex B-space X. Suppose that there is a real number c such that the closed half-space

$$H = \{\lambda \in \mathbb{C}: \operatorname{Re} \lambda \leq c\}$$

is a subset of the resolvent set of A (Fig. 15(c)) and

$$\|(A - \lambda I)^{-1}\| \leq \frac{\text{const}}{1 + |\lambda|} \qquad \text{for all} \quad \lambda \in H.$$

Then, the operator A is *sectorial* with respect to Σ in (110) for some $\gamma \in]0, \pi/2[$, and hence $-A$ generates a linear *analytic semigroup*. This follows from Problem 1.7.

(110c) *Standard example* 3 (*elliptic differential operators and parabolic equations*). We consider the parabolic differential equation of order $2m$:

$$\text{(I)}\qquad \begin{aligned} u_t + Au &= 0 && \text{on } G \times]0, \infty[, \\ B_\gamma u(x,t) &= 0 && \text{on } \partial G \times]0, \infty[\quad \text{for all} \quad \gamma\colon |\gamma| \leq m-1, \\ u(x,0) &= u_0(x) && \text{on } G, \end{aligned}$$

where $B_\gamma u = D^\gamma u$ and

$$Au = \sum_{|\alpha| \leq m} (-1)^{|\alpha|} D^\alpha (a_{\alpha\beta} D^\beta u).$$

We assume:

(i) G is a bounded region in $\mathbb{R}^N$ with $\partial G \in C^\infty$, where $N, m \geq 1$.
(ii) The differential operator A is *strongly elliptic*, where all the coefficient functions $a_{\alpha\beta}\colon \bar{G} \to \mathbb{R}$ are C^∞.

We set

$$X = L_p(G), \qquad 1 < p < \infty.$$

Let $D(A)$ be the closure of the set

$$\{u \in C^\infty(\bar{G})\colon B_\gamma u = 0 \text{ on } \partial G \text{ for all } \gamma\colon |\gamma| \leq m-1\}$$

in the Sobolev space $W_p^{2m}(G)$. Then the operator

$$A\colon D(A) \subseteq X \to X$$

is well defined in the sense of generalized derivatives. Moreover, it is of great importance that the operator A satisfies all the conditions of Standard example 2 in (110b) if we consider the complexification of A. This fact has the following fundamental consequence.

The original problem (I) can be written in the form:

$$\text{(II)}\qquad \begin{aligned} u'(t) + Au(t) &= 0, \qquad 0 < t < \infty, \\ u(0) &= u_0. \end{aligned}$$

Let $\{S(t)\}$ be the analytic semigroup generated by $-A$. Then, for each $u_0 \in D(A)$, the unique solution of (II) is given by

$$\text{(III)}\qquad u(t) = S(t)u_0.$$

If $u_0 \in X$, then we regard $u(\cdot)$ in (III) as a generalized solution of (II) and (I).

This important result can be extended to general classes of boundary conditions (cf. Friedman (1969, M), Theorem 19.4). This way it is possible to investigate general classes of parabolic equations (and parabolic systems) via the theory of semigroups.

(110d) *Characterization of linear bounded analytic semigroups.* Let X be a complex B-space. We define the open sector

$$\Sigma_\alpha = \{z \in \mathbb{C}: |\arg z| < \alpha, z \neq 0\}.$$

By a linear *bounded analytic* semigroup $\{S(t)\}$ we understand the following:

(a) $\{S(t)\}$ is a linear analytic semigroup on Σ_α for fixed $\alpha \in]0, \pi/2[$.

This means that the operator $S(t)$: $X \to X$ is linear and continuous for each $t \in \Sigma_\alpha$, and the mapping $t \mapsto S(t)$ is analytic from Σ_α into $L(X, X)$. Moreover, $S(0) = I$ and $S(t + s) = S(t)S(s)$ for all $t, s \in \Sigma_\alpha$. Finally, $S(t)u \to u$ as $t \to 0$ in Σ_α, for each $u \in X$.

(b) For each β with $0 < \beta < \alpha$,

$$\sup_{t \in \Sigma_\beta} \|S(t)\| < \infty.$$

If the operator $-A$ is the generator of such a linear bounded analytic semigroup, then the spectrum $\sigma(A)$ is contained in the closure of the sector $\Sigma_{(\pi/2)-\alpha}$.

More precisely, the following two statements are equivalent:

(i) The operator A: $D(A) \subseteq X \to X$ is *strictly* sectorial on the complex B-space X.
(ii) The operator $-A$ is the generator of a linear *bounded* analytic semigroup.

(110e) *Characterization of linear analytic semigroups.* The following two statements are equivalent:

(i) The operator A: $D(A) \subseteq X \to X$ is sectorial on the complex B-space X.
(ii) The operator $-A$ is the generator of a linear analytic semigroup.

Fractional Powers of Sectorial Operators

(111) Let the operator A: $D(A) \subseteq X \to X$ be sectorial as in (110), where X is a complex B-space with $X \neq \{0\}$. We set

$$B = A + aI$$

and choose the real number a in such a way that $\operatorname{Re} \sigma(B) > 0$, i.e., $\operatorname{Re} \sigma(A) + a > 0$, where $\sigma(A)$ denotes the spectrum of A.

For every $\alpha > 0$, we define

$$B^{-\alpha} = \frac{1}{\Gamma(\alpha)} \int_0^\infty t^{\alpha-1} e^{-tB}\, dt.$$

Then $B^{-\alpha}$: $X \to X$ is linear, bounded, and injective. Moreover, we set

$$B^\alpha = (B^{-\alpha})^{-1} \qquad \text{for all} \quad \alpha > 0,$$

and $B^0 = I$.

For all $\alpha, \beta \geq 0$, these fractional powers have the following properties:

(i) B^α is densely defined on X and graph closed.

(ii) If $0 \leq \alpha \leq \beta$, then $D(B^\beta) \subseteq D(B^\alpha)$, and

$$B^\alpha B^\beta = B^\beta B^\alpha = B^{\alpha+\beta} \quad \text{on } D(B^{\alpha+\beta})$$

as well as $B^{-\alpha}B^{-\beta} = B^{-\beta}B^{-\alpha} = B^{-\alpha-\beta}$ on X.

(iii) e^{-tA} maps X into $D(B^\alpha)$ for all $t > 0$.

(iv) For all $u \in D(B^\alpha)$ and $t > 0$,

$$B^\alpha e^{-tA}u = e^{-tA}B^\alpha u.$$

(v) For all $u \in D(B^\alpha)$ and $0 < \alpha \leq 1$,

$$\|B^\alpha u\| \geq \|B^{-\alpha}\|^{-1}\|u\|.$$

Therefore, the set $X_\alpha = D(B^\alpha)$, equipped with the graph norm

$$\|u\|_\alpha = \|B^\alpha u\|,$$

becomes a B-space (abstract Sobolev space).

If $0 \leq \alpha \leq \beta$, then the embedding

$$X_\beta \subseteq X_\alpha \tag{E}$$

is *continuous*, and X_β is dense in X_α. For $\alpha = 0$, we get $X_0 = X$.

If $0 \leq \alpha < \beta$ and if the operator A has compact resolvent, i.e., the operator $(\lambda I - A)^{-1}: X \to X$ is compact for each λ in the resolvent set $\rho(A)$, then the embedding (E) is *compact*.

(vi) For all $u \in D(B)$ and $0 \leq \alpha \leq 1$,

$$\|B^\alpha u\| \leq \text{const}\,\|Bu\|^\alpha \|u\|^{1-\alpha}.$$

In terms of the norm $\|\cdot\|_\alpha$ on X_α, this is the interpolation inequality

$$\|u\|_\alpha \leq \text{const}\,\|u\|_1^\alpha \|u\|_0^{1-\alpha} \qquad \text{for all} \quad u \in X_1.$$

(vii) Suppose that $\operatorname{Re}\sigma(A) > \delta > 0$. Then there are constants K and M depending on α such that

$$\|A^\alpha e^{-tA}\| \leq Kt^{-\alpha}e^{-\delta t}, \qquad 0 \leq \alpha < \infty,$$

$$\|(e^{-tA} - I)u\| \leq Mt^\alpha \|A^\alpha u\|, \qquad 0 \leq \alpha \leq 1,$$

for all $t > 0$ and $u \in D(A^\alpha)$.

Suppose that $\operatorname{Re}\sigma(A) > \delta$ with $\delta \in \mathbb{R}$. Then there is a constant K such that, for all $t > 0$,

$$\|Ae^{-tA}\| \leq Kt^{-1}e^{-\delta t}, \qquad \|e^{-tA}\| \leq Ke^{-\delta t}.$$

If we consider the special linear operator $A: X \to X$, where $X = \mathbb{C}$ and $A > 0$ is real, then the fractional power A^α in the sense above coincides with the classical notion for each real α.

Further important properties of fractional powers can be found in Henry (1981, L) and Pazy (1983, M).

Interpolation Theory in B-Spaces

The *basic idea* of interpolation theory is contained in the following diagram:

$$\begin{array}{ccc} V & \xrightarrow{T} & X \\ [V,H]_\theta & \xrightarrow{T} & [X,Y]_\theta \\ H & \xrightarrow{T} & Y, \qquad 0<\theta<1. \end{array}$$

For given linear continuous operators $T: V \to X$ and $T: H \to Y$ between B-spaces, we want to construct new B-spaces $[V,H]_\theta$ and $[X,Y]_\theta$ by "interpolation" such that the operator T *remains continuous.*

In this connection, we will describe the following important interpolation methods:

(i) the real method (K-method) for B-spaces;
(ii) the complex method for B-spaces; and
(iii) the interpolation between H-spaces by letting

$$[V,H]_\theta = D(B^{1-\theta}),$$

i.e., this method uses fractional powers of *monotone* self-adjoint operators B.

We shall show below that methods (i) and (ii) yield different results for the Sobolev spaces $W_p^m(G)$ if $p \neq 2$. Method (iii) is a special case of (i) and (ii).

The following two estimates are of fundamental importance for modern analysis. Let $\|\cdot\|_\theta$ denote the norm on the interpolation space $[V,H]_\theta$. Then we shall obtain estimates of the form

$$\|u\|_\theta \le \text{const}\, \|u\|_V^{1-\theta} \|u\|_H^\theta \qquad \text{for all} \quad u \in V \cap H$$

and

$$\|T\|_\theta \le \|T\|_0^{1-\theta} \|T\|_1^\theta,$$

where we set[1]

$$[V,H]_\theta = \begin{cases} V & \text{for} \quad \theta = 0 \\ H & \text{for} \quad \theta = 1. \end{cases}$$

Here, the operator norm $\|T\|_\theta$ refers to the corresponding interpolation spaces.

The classical prototype of interpolation theory is given by the famous Convexity Theorem of Marcel Riesz (1926) for $L_p(G)$-spaces which will be considered in (113) below. The general interpolation theory was created around 1960. The complex method was introduced independently by A. Calderón, S. Krein, and J. Lions. The K-method is due to J. Peetre.

There is the following philosophy behind interpolation theory. For example,

[1] This definition is very natural. However, note that there also exist different definitions in the literature which are, roughly speaking, based on continuity properties of $[V,H]_\theta$ as $\theta \to 0$ and $\theta \to 1$. Here, $[V,H]_\theta$ denotes an arbitrary interpolation method.

consider a differential equation briefly denoted by

$$Tu = f.$$

In order to study this concrete analytical problem via functional analysis, we regard T as an operator between certain function spaces. However, the point is that there are many different choices for the function spaces. Interpolation theory helps us to study *systematically* the properties of T in different spaces. In particular, interpolation theory may help us to find "natural" spaces A and B for T, i.e., the operator $T: A \to B$ is a homeomorphism. In fact, such natural results for linear elliptic partial differential equations may be found in Lions and Magenes (1968, M), and in Triebel (1978, M), (1983, M).

(112) *The* K-*method for* B-*spaces*. By definition, an *interpolation couple* $\{X, Y\}$ consists of two B-spaces X and Y over $\mathbb{K} = \mathbb{R}, \mathbb{C}$. Moreover, in order to have the sum $X + Y$ at hand, we assume that there is a topological vector space Z such that both the embeddings

$$X \subseteq Z \qquad \text{and} \qquad Y \subseteq Z$$

are continuous. For $z \in X + Y$ and $t \in \mathbb{R}$, we set

$$K(z,t) = \inf_{z=x+y} \{\|x\|_X + t\|y\|_Y\}.$$

Here, the infimum is taken over all possible decompositions $z = x + y$ with $x \in X$ and $y \in Y$. For $0 < \theta < 1$, we define

$$\|z\|_{\theta,r} = \begin{cases} \left(\displaystyle\int_0^\infty K(z,t)t^{-1-\theta r}\,dt\right)^{1/r} & \text{for} \quad 1 \le r < \infty, \\ \displaystyle\sup_{0<t<\infty} K(z,t)t^{-\theta} & \text{for} \quad r = \infty. \end{cases}$$

Now our *basic definition* reads as follows:

$$[X, Y]_{\theta,r} = \{z \in X + Y: \|z\|_{\theta,r} < \infty\}.$$

Here, $0 < \theta < 1$ and $1 \le r \le \infty$.

The following four theorems are important.

(a) B-*space*. The set $[X, Y]_{\theta,r}$ together with the norm $\|\cdot\|_{\theta,r}$ becomes a B-space over $\mathbb{K}$ with

$$X \cap Y \subseteq [X, Y]_{\theta,r} \subseteq X + Y$$

and

$$\|z\|_{\theta,r} \le c_{\theta,r}\|z\|_X^{1-\theta}\|z\|_Y^{\theta} \qquad \text{for all} \quad z \in X \cap Y,$$

where $c_{\theta,r}$ denotes a constant.

(b) *Density*. Let $0 < \theta < 1$ and $1 \le r < \infty$. Then the set $X \cap Y$ is dense in $[X, Y]_{\theta,r}$.

(c) *Linear continuous operators*. Let $0 < \theta < 1$, $1 \le r \le \infty$, and let $\{V, H\}$

and $\{X, Y\}$ be interpolation couples over $\mathbb{K}$. Moreover, let

$$T: V + H \to X + Y$$

be a linear operator and suppose that the two restrictions

$$T: V \to X,$$

$$T: H \to Y$$

are linear continuous operators with the corresponding norms $\|T\|_0$ and $\|T\|_1$. Then the restriction

$$T: [V, H]_{\theta,r} \to [X, Y]_{\theta,r}$$

is also a linear operator with the norm $\|T\|_{\theta,r}$ and

$$\|T\|_{\theta,r} \leq \|T\|_0^{1-\theta}\|T\|_1^{\theta}.$$

(d) *Duality*. Let $0 < \theta < 1$, $1 \leq p < \infty$, and $p^{-1} + q^{-1} = 1$. Suppose that $X \cap Y$ is dense both in X and Y. Then

$$([X, Y]_{\theta,p})^* = [X^*, Y^*]_{\theta,q}.$$

Let X, Y, and Z be B-spaces. By convention, a relation of the form

$$[X, Y]_{\ldots} = Z$$

below means that the set $[X, Y]_{\ldots}$ is equal to the set Z, and the corresponding norms are equivalent. The same convention will also be used for a relation of the form $X = Z$.

Standard Examples for the Interpolation of Function Spaces

(113) *Lebesgue spaces* (*Convexity Theorem of M. Riesz* (1926)). Let G be a nonempty measurable set in $\mathbb{R}^N$ with $N \geq 1$ (e.g., G is open or closed). Let $1 \leq p, q < \infty$, $0 < \theta < 1$, and

$$r^{-1} = (1 - \theta)p^{-1} + \theta q^{-1}.$$

Then

$$[L_p(G), L_q(G)]_{\theta,r} = L_r(G)$$

and from the Hölder inequality it follows that

$$\|u\|_r \leq \|u\|_p^{1-\theta}\|u\|_q^{\theta} \qquad \text{for all} \quad u \in L_p(G) \cap L_q(G).$$

Suppose that the operator

$$T: L_p(G) \to L_q(G) \tag{113a}$$

is linear for all $1 \leq p, q \leq \infty$. Let C be the set of all points (p^{-1}, q^{-1}) in $\mathbb{R}^2$ for which the operator (113a) is continuous and denote the corresponding operator norm by $\|T\|_{p,q}$. Then the set C is *convex* and the function

$$(p^{-1}, q^{-1}) \mapsto \ln \|T\|_{p,q}$$

is *convex* on C. This follows from (112c).

(114) *Hölder spaces.* Let G be a bounded region in $\mathbb{R}^N$ with $\partial G \in C^\infty$ and $N \geq 1$. For $r = 0, 1, \ldots$ and $0 < \alpha < 1$, we set

$$C^{r+\alpha}(\bar{G}) = C^{r,\alpha}(\bar{G}).$$

Then

$$[C^k(\bar{G}), C^m(\bar{G})]_{\theta,\infty} = C^j(\bar{G})$$

for real numbers $0 \leq k < j < m < \infty$ and

$$j = (1-\theta)k + \theta m, \qquad 0 < \theta < 1,$$

where j is *not* an integer. In the special case where $k = 0, m = 1$, and $0 < \theta < 1$, we obtain

$$[C(\bar{G}), C^1(\bar{G})]_{\theta,\infty} = C^\theta(\bar{G}).$$

(115) *Sobolev spaces (real method).* Let $1 < p < \infty$. Under the same assumptions as in (114), we obtain that

$$\text{(I)} \qquad [W_p^k(G), W_p^m(G)]_{\theta,p} = W_p^j(G).$$

This relation remains valid for $G = \mathbb{R}^N$.

If $p = 2$, then j is also allowed to be an integer.

(116) *Sobolev spaces (complex method).* Let $1 < p < \infty$ and let k, j, m be *integers.* Under the same assumptions as in (114), we obtain that

$$\text{(II)} \qquad [W_p^k(G), W_p^m(G)]_\theta = W_p^j(G).$$

This relation remains valid for $G = \mathbb{R}^N$.

Here, $[\cdot, \cdot]_\theta$ corresponds to the so-called complex method whose definition can be found in (116e) below. For many applications it is sufficient to know that the fundamental theorems (a)–(d) in (112) *remain valid for the complex method* if we use complex B-spaces and if we assume, in the duality theorem (d), that X or Y is reflexive.

In the special case where $k = 0, m = 2, 1 < p < \infty$, and $0 < \theta < 1$, we obtain that

$$[L_p(G), W_p^2(G)]_{\theta,p} = W_p^{2\theta}(G)$$

if $\theta \neq 1/2$ and

$$[L_p(G), W_p^2(G)]_{1/2} = W_p^1(G).$$

Furthermore, we have

$$[L_2(G), W_2^2(G)]_{\theta,2} = W_2^{2\theta}(G), \qquad 0 < \theta < 1.$$

(116a) *An important special case.* Let $p = 2$. If $G = \mathbb{R}^N$, $N \geq 1$, then the relations (I) and (II) above remain true for all real numbers k, m, θ with

$$-\infty < k < m < \infty \qquad \text{and} \qquad 0 < \theta < 1.$$

If G is a bounded region in $\mathbb{R}^N$ with smooth boundary, i.e., $\partial G \in C^\infty$, then (I) and (II) remain true if $0 \le k < m < \infty$ and $0 < \theta < 1$.

All these relations above remain valid if we replace the real Sobolev spaces by the corresponding complex Sobolev spaces.

(116b) *The fundamental role played by Besov spaces and Triebel–Lizorkin spaces.* Let the bounded or unbounded region G be given as in (116a) above. If $p \neq 2$, then the interpolation of Sobolev spaces may lead to *new* spaces. The prototype of this fact is given by the relation

$$[W_p^k(G)_{\mathbb{C}}, W_p^m(G)_{\mathbb{C}}]_{\theta,q} = B_{p,q}^j(G),$$

where $0 \le k < m < \infty$, $1 < p, q < \infty$, $0 < \theta < 1$, and $j = (1-\theta)k + \theta m$.

More precisely, for $-\infty < m < \infty$ and $1 \le p, q \le \infty$, it is possible to define the spaces

$$B_{p,q}^m(G) \qquad \text{and} \qquad F_{p,q}^m(G), \tag{S}$$

which are called *Besov spaces* and *Triebel–Lizorkin spaces*, respectively. The fairly simple definition via Fourier transform will be given in (116d) below. (In the case of the spaces $F_{p,q}^m$, the index $p = \infty$ is excluded).

All the spaces (S) are complex B-spaces whose elements are distributions. The point is that:

(α) important classical function spaces are special cases of (S), and
(β) the spaces (S) possess extremely elegant interpolation properties, in contrast to Sobolev spaces.

Moreover, the spaces (S) play a fundamental role in the theory of linear elliptic differential equations as discussed in Problem 22.10.

Standard example 1 (Sobolev spaces). Let $-\infty < m < \infty$, $1 < p < \infty$, and $G = \mathbb{R}^N$. Then

$$W_p^m(G)_{\mathbb{C}} = \begin{cases} F_{p,2}^m(G) & \text{if} \quad m = 0, \pm 1, \pm 2, \ldots, \\ B_{p,p}^m(G) & \text{otherwise.} \end{cases}$$

If G is bounded, then this relation remains true provided $m \neq -(n + p^{-1})$, $n = 0, 1, 2, \ldots$.

Standard example 2 (Hölder spaces). Let $m = 0, 1, 2, \ldots$ and $0 < \alpha < 1$. Then

$$C^{m,\alpha}(\bar{G})_{\mathbb{C}} = B_{\infty,\infty}^{m+\alpha}(G),$$

where the functions $f \in C^{m,\alpha}(\bar{G})_{\mathbb{C}}$ are complex-valued.

Example 3 (Bessel potentials). For $-\infty < m < \infty$ and $1 < p < \infty$, one also uses the notation

$$H_p^m(G)_{\mathbb{C}} = F_{p,2}^m(G).$$

The elements of these spaces are called Bessel potentials.

Example 4. For $-\infty < m < \infty$ and $1 < p < \infty$,

$$B^m_{p,p}(G) = F^m_{p,p}(G).$$

Moreover, if $-\infty < m < \infty$ and $G = \mathbb{R}^N$, then

$$W_2^m(G)_{\mathbb{C}} = H_2^m(G)_{\mathbb{C}} = B^m_{2,2}(G) = F^m_{2,2}(G).$$

This relation remains true if G is bounded provided $m \neq -\frac{1}{2}, -\frac{3}{2}, -\frac{5}{2}, \ldots$.

(116c) *The fundamental interpolation theorem.* Let the bounded or unbounded region G be given as in (116a). Let $-\infty < k < m < \infty$, $0 < \theta < 1$, and

$$j = (1 - \theta)k + \theta m$$

as well as $1 \le p_0, p_1, p, q_0, q_1, q \le \infty$ with

$$\frac{1}{p} = \frac{(1-\theta)}{p_0} + \frac{\theta}{p_1}, \qquad \frac{1}{q} = \frac{(1-\theta)}{q_0} + \frac{\theta}{q_1},$$

where we use the convention "$1/\infty = 0$". Then the K-method and the complex interpolation method yield the following fundamental relations:

$$[B^k_{p,q_0}(G), B^m_{p,q_1}(G)]_{\theta,q} = B^j_{p,q}(G), \tag{A}$$

$$[F^k_{p,q_0}(G), F^m_{p,q_1}(G)]_{\theta,q} = B^j_{p,q}(G), \tag{B}$$

$$[B^k_{p_0,q_0}(G), B^m_{p_1,q_1}(G)]_{\theta} = B^j_{p,q}(G), \tag{C}$$

$$[F^k_{p_0,q_0}(G), F^m_{p_1,q_1}(G)]_{\theta} = F^j_{p,q}(G). \tag{D}$$

Here, we exclude $p = \infty$ in (B) and $p = \infty$, $q = \infty$ in (D).

Relations (I) and (II) in (115) and (116) above are special cases of (A)–(D) because of (116b).

The proofs, along with further important properties of these spaces, can be found in Triebel (1978, M), (1983, M) (e.g., density, embedding, duality, and generalized boundary values). This theory is essentially based on the Fourier transform and on so-called Fourier multipliers.

(116d) *The basic definitions.*

Step 1: Let $G = \mathbb{R}^N$.

The *Besov space* $B^m_{p,q}(\mathbb{R}^N)$ consists precisely of all the tempered distributions $u \in \mathscr{S}'(\mathbb{R}^N)$ with

$$| \| 2^{mk} F^{-1} \varphi_k F u \|_p |_q < \infty. \tag{B}$$

Similarly, the *Triebel–Lizorkin* space $F^m_{p,q}(G)$ is obtained if we replace the condition (B) with

$$\| | 2^{mk} F^{-1} \varphi_k F u |_q \|_p < \infty. \tag{F}$$

The spaces $B^m_{p,q}(G)$ and $F^m_{p,q}(G)$ become complex B-spaces under the norms (B) and (F), respectively.

We now explain our notation. As usual, F denotes the Fourier transform. Moreover, $\{\varphi_k\}$ is a fixed system of functions

$$\varphi_k \in \mathscr{S}(\mathbb{R}^N), \qquad k = 0, 1, \ldots,$$

which forms a partition of unity, i.e.,

$$\sum_{k=0}^{\infty} \varphi_k(x) = 1 \qquad \text{for all} \quad x \in \mathbb{R}^N,$$

and $\varphi_0(x) \neq 0$ implies $|x| \leq 2$ as well as

$$\varphi_k(x) \neq 0 \qquad \text{implies} \qquad 2^{k-1} \leq |x| \leq 2^{k+1}$$

for $k = 1, 2, \ldots$. In addition, for each multi-index α,

$$\sup_{k,x} 2^{k|\alpha|} |D^\alpha \varphi_k(x)| < \infty.$$

Finally, $\|\cdot\|_p$ denotes the usual norm on the Lebesgue space $L_p(\mathbb{R}^N)$, and, for brevity of notation, we set

$$|a_k|_q = \begin{cases} \left(\sum_{k=0}^{\infty} |a_k|^q \right)^{1/q} & \text{if} \quad 1 \leq q < \infty, \\ \sup_k |a_k| & \text{if} \quad q = \infty. \end{cases}$$

Note that the functions $F\varphi_k F^{-1}u$ are analytic on $\mathbb{R}^N$ for $u \in \mathscr{S}'(\mathbb{R}^N)$. If we change the system $\{\varphi_k\}$, then the norms (B) and (F) above are replaced by equivalent norms.

Step 2: Let G be a bounded region in $\mathbb{R}^N$ with $\partial G \in C^\infty$.

By definition, $u \in B^m_{p,q}(G)$ iff $u \in \mathscr{D}'(G, \mathbb{C})$ and

(R) $$u = v \quad \text{on } G \qquad \text{for some} \quad v \in B^m_{p,q}(\mathbb{R}^N).$$

Then, $B^m_{p,q}(G)$ becomes a complex B-space under the norm

$$\|u\| = \inf_v \{\|v\|_{B^m_{p,q}(\mathbb{R}^N)}\},$$

where the infimum is taken over all $v \in B^m_{p,q}(\mathbb{R}^N)$ which have the restriction property (R).

Analogously, we obtain the space $F^m_{p,q}(G)$.

These definitions are due to Triebel (1973).

(116e) *Definition of the complex interpolation method.* We define the open strip

$$S = \{z \in \mathbb{C}: 0 < \operatorname{Re} z < 1\},$$

and, for $j = 0, 1$, we set

$$\partial_j S = \{z \in \mathbb{C}: \operatorname{Re} z = j\}.$$

Then $\partial S = \partial_0 S \cup \partial_1 S$. Let X and Y be two complex B-spaces which form an interpolation couple. Let $0 < \theta < 1$. By definition,

$$u \in [X, Y]_\theta \qquad \text{iff} \qquad u = f(\theta),$$

where $f: \bar{S} \to X + Y$ is some function which has the following properties:

(i) f is analytic on S;
(ii) the restrictions $f: \partial_0 S \to X$ and $f: \partial_1 S \to Y$ are continuous, and the sets $f(\partial_j S)$ are bounded for $j = 0, 1$.

In this connection, we equip $X + Y$ with the norm

$$\|w\| = \inf_{w=x+y} \{\|x\|_X + \|y\|_Y\},$$

for all $w \in X + Y$, where the infimum is taken over all the decompositions $w = x + y$ with $x \in X$ and $y \in Y$.

Then $[X, Y]_\theta$ becomes a complex B-space under the norm

$$\|u\|_\theta = \inf_{f(\theta)=u} \{\|f\|\},$$

where the infimum is taken over all the functions f which have the properties (i), (ii), and we set

$$\|f\| = \max\left\{ \sup_{z \in \partial_0 S} \|f(z)\|_X, \sup_{z \in \partial_1 S} \|f(z)\|_Y \right\}.$$

If X and Y are real B-spaces, then we define

$$[X, Y]_\theta = [X_{\mathbb{C}}, Y_{\mathbb{C}}]_\theta \cap (X + Y),$$

where $X_{\mathbb{C}}$ and $Y_{\mathbb{C}}$ denote the complexification of X and Y, respectively (cf. A_1(23h)).

(116f) *Monotone self-adjoint operators and interpolation theory.* Let

$$B: D(B) \subseteq X \to X$$

be a linear *monotone* self-adjoint operator on the H-space X over $\mathbb{K} = \mathbb{R}, \mathbb{C}$. Let $0 \le \alpha, \beta < \infty$ and

$$\gamma = (1 - \theta)\alpha + \theta\beta,$$

where $0 < \theta < 1$. We set

$$X_\alpha = D(B^\alpha).$$

Then, X_α becomes an H-space over $\mathbb{K}$ under the graph scalar product

$$(u|v)_\alpha = (u|v) + (B^\alpha u|B^\alpha v).$$

By the K-method and the complex method,

$$[X_\alpha, X_\beta]_{\theta,2} = [X_\alpha, X_\beta]_\theta = X_\gamma.$$

Interpolation Between H-Spaces, Fractional Powers, and the Friedrichs Extension

(117) *Definition.* Let

$$\text{“}V \subseteq H \subseteq V^*\text{”}$$

be an evolution triple where V and H are H-spaces. We construct a linear operator $A: D(A) \subseteq H \to H$ by the relation

$$(u|v)_V = (Au|v)_H \qquad \text{for all} \quad u \in D(A), \quad v \in H.$$

To be precise, let $D(A)$ be the set of all $u \in H$ such that the map $v \mapsto (u|v)_V$ is linear and continuous on H. Let $u \in D(A)$. By the Riesz theorem, there exists a $w \in H$ such that

$$(u|v)_V = (w|v)_H \qquad \text{for all} \quad v \in H.$$

Now let $Au = w$.

The operator A is self-adjoint and we have

$$(Au|u)_H = \|u\|_V^2 \geq c\|u\|_H^2 \qquad \text{for all} \quad u \in D(A) \quad \text{and fixed} \quad c > 0.$$

Define $B = A^{1/2}$ and the "graph scalar product"

$$(u|v)_\theta = (B^{1-\theta}u|B^{1-\theta}v)_H \qquad \text{for all} \quad u, v \in D(B^{1-\theta}).$$

Now our *basic definition* reads as follows:

$$[V, H]_\theta = D(B^{1-\theta}).$$

(118) *Basic properties.* The following four theorems are important.

(a) H-*space.* Let $0 \leq \theta \leq 1$. Then the set $[V, H]_\theta$, together with the scalar product $(\cdot|\cdot)_\theta$, becomes an H-space. For all $v \in V$,

$$\|v\|_\theta \leq \|v\|_V^{1-\theta}\|v\|_H^\theta.$$

Moreover, we obtain

$$V \subseteq [V, H]_\theta \subseteq H$$

and $[V, H]_0 = V$, $[V, H]_1 = H$.

(b) *Density.* The set V is dense in $[V, H]_\theta$ for all $0 \leq \theta \leq 1$.

(c) *Linear continuous operators.* Let "$V \subseteq H \subseteq V^*$" and "$X \subseteq Y \subseteq X^*$" be two evolution triples. Let

$$T: H \to Y$$

be a linear continuous operator and suppose that the restriction

$$T: V \to X$$

is continuous. Then all the restrictions

$$T: [V, H]_\theta \to [X, Y]_\theta, \qquad 0 \leq \theta \leq 1,$$

are continuous.

(d) $[V, V^*]_{1/2} = H$.

By our definition of an evolution triple "$V \subseteq H \subseteq V^*$" given in Section 23.4, the spaces V and H are real. However, the same method above also applies to the complex case, i.e., we may assume that V and H are separable H-spaces over $\mathbb{K} = \mathbb{R}, \mathbb{C}$ such that V is dense in H and the embedding $V \subseteq H$ is continuous. By (116f), the interpolation method above coincides with both the K-method $[\cdot, \cdot]_{2,\theta}$ and the complex method.

(119) *Standard example.* Let H be a real separable H-space and let $A: D(A) \subseteq H \to H$ be the Friedrichs extension of a linear, symmetric, strongly monotone operator. We set

$$V = H_E,$$

i.e., V is the energetic space of A. Let us assume that the embedding $H_E \subseteq H$ is compact. Then the operator A has a complete orthonormal system of eigenvectors (u_n) in H with $Au_n = \lambda_n u_n$. Moreover, we obtain

$$[V, H]_\theta = D(B^{1-\theta}),$$

where $B = A^{1/2}$, and $u \in [V, H]_\theta$ iff

$$\|u\|_\theta = \left(\sum_n \lambda_n^{1-\theta} |(u_n|u)|^2 \right)^{1/2} < \infty.$$

Special case. Let G be a bounded region in $\mathbb{R}^N$, $N \geq 1$. We set $H = L_2(G)$. Let A denote the Friedrichs extension of the negative Laplacian

$$-\Delta: C_0^\infty(G) \to H.$$

Then $H_E = \mathring{W}_2^1(G)$.

Application to Sobolev Spaces

(120) *Standard example.* Let $-\infty < k < m < \infty$ and $j = (1 - \theta)k + \theta m$ with $0 < \theta < 1$. Let $G = \mathbb{R}^N$, $N \geq 1$. Then

$$[W_2^k(G), W_2^m(G)]_\theta = W_2^j(G) \tag{R}$$

Now let G be a bounded region in $\mathbb{R}^N$ with smooth boundary, i.e., $\partial G \in C^\infty$. Let k, m, and j be given as above. If, in addition, k, m, and j are different from the critical values $-\frac{1}{2}, -\frac{3}{2}, -\frac{5}{2}, \ldots$, then relation (R) above remains true. Moreover, we get

$$[\mathring{W}_2^k(G), \mathring{W}_2^m(G)]_\theta = \mathring{W}_2^j(G)$$

provided $0 \leq k < m < \infty$ and k, m, and j are different from the critical values $\frac{1}{2}, \frac{3}{2}, \frac{5}{2}, \ldots$.

All these relations above remain valid if we replace the real Sobolev spaces by the corresponding complex Sobolev spaces.

(121) *The trace theorem.* Let "$V \subseteq H \subseteq V^*$" be an evolution triple. We set

$$\mathscr{W}_2^m = \{u \in L_2(0, T; V): u^{(m)} \in L_2(0, T; H)\},$$

where $m = 1, 2, \ldots$ and $0 < T < \infty$. This set becomes an H-space with the scalar product

$$(u|v) = (u|v)_{L_2(0,T;V)} + (u^{(m)}|v^{(m)})_{L_2(0,T;H)}.$$

Let $u \in \mathscr{W}_2^m$. Then, for $j = 0, 1, \ldots, m-1$, the derivatives have the following properties:

$$u^{(j)} \in L_2(0, T; [V, H]_{j/m}),$$

$$u^{(j)} \in C([0, T], [V, H]_{(j+1/2)/m}).$$

The so-called trace map

$$u \mapsto (u(0), u'(0), \ldots, u^{(m-1)}(0))$$

is *surjective* from $\mathscr{W}_2^m$ to $\prod_{j=0}^{m-1} [V, H]_{(j+1/2)/m}$.

Hausdorff Measure, Hausdorff Dimension, and Fractals

The following notions play an important role in the modern regularity theory for partial differential equations (partial regularity), and in the modern theory of evolution equations and dynamical systems (e.g., the Navier–Stokes equations). We recommend Federer (1969, M), Giaquinta (1983, M), Simon (1983, L), Giusti (1984, M), Temam (1983, M), (1988, M), and Mandelbrot (1982, M) (fractals).

(122) *Hausdorff measure.* Let X be a metric space (e.g., X is a subset of $\mathbb{R}^N$). Let $0 \leq m < \infty$ and $\delta > 0$. By definition, the m-dimensional Hausdorff measure of the nonempty set X is given by

$$H^m(X) = \lim_{\delta \to 0} H_\delta^m(X), \tag{122a}$$

where

$$H_\delta^m(X) = \inf \sum_i (2^{-1} \operatorname{diam} B_i)^m,$$

the infimum being taken over all at most countable coverings (B_i) of the set X by subsets B_i with

$$\operatorname{diam} B_i \leq \delta \qquad \text{for all } i.$$

If such a covering of X does not exist, then we set $H_\delta^m(X) = \infty$.

The number $H_\delta^m(X)$ is called the outer δ-Hausdorff measure of X. The limit in (122a) is well defined since $(H_\delta^m(X))$ is monotone increasing as $\delta \to 0$, and hence

$$H^m(X) = \sup_{\delta > 0} H_\delta^m(X).$$

Note that $0 \leq H^m(X) \leq \infty$. For $m = 0$, we use the convention "$0^m = 1$".

Consequently, if X consists of a single point, then $H^0(X) = 1$ and $H^m(X) = 0$ for all $m > 0$.

If the set X is empty, then we define $H^m(\varnothing) = H_\delta^m(\varnothing) = 0$ for all $m \geq 0$ and $\delta > 0$.

The number

$$\mathscr{H}^m(X) = V_m H^m(X)$$

is called the *normalized* Hausdorff measure of X, where

$$V_m = \Gamma(\tfrac{1}{2})^m/\Gamma\left(\frac{m}{2} + 1\right), \qquad \text{for all} \quad m \geq 0.$$

In particular, if $m = 1, 2, \ldots$, then V_m denotes the volume of the m-dimensional unit ball in $\mathbb{R}^m$.

Examples. Let $m = 1, 2, \ldots$. If X is an m-dimensional ball in $\mathbb{R}^m$, then the normalized m-dimensional Hausdorff measure $\mathscr{H}^m(X)$ is identical to the classical volume of X.

More generally, we have

$$\mathscr{H}^m(X) = \operatorname{meas} X$$

for all subsets X of $\mathbb{R}^m$ which are measurable with respect to the m-dimensional Lebesgue measure. This Lebesgue measure is denoted by $\operatorname{meas} X$. For arbitrary subsets X of $\mathbb{R}^m$, $\mathscr{H}^m(X)$ is identical to the outer m-dimensional Lebesgue measure, and $H^m(X) = H_\delta^m(X)$ for all $\delta > 0$.

If X denotes a sufficiently smooth m-dimensional surface in $\mathbb{R}^N$, $N > m$, then

$$\mathscr{H}^m(X) = \text{classical surface measure.}$$

More precisely, this relation holds if X is an m-dimensional C^1-submanifold of $\mathbb{R}^N$, $N > m$.

For $m = 0$, we get

$$H^0(X) = \mathscr{H}^0(X) = \begin{cases} \operatorname{card} X & \text{if} \quad X \text{ is a finite set,} \\ +\infty & \text{otherwise,} \end{cases}$$

where $\operatorname{card} X$ denotes the number of distinct points of X.

The proofs can be found in Simon (1983, L).

Note that there exist different definitions of the Hausdorff measure in the literature. For example, if we replace the arbitrary subsets B_i of X with closed *balls* B_i in X, then $H_\delta^m(X)$ and $H^m(X)$ in (122a) are replaced with $S_\delta^m(X)$ and $S^m(X)$, respectively. Here, $S^m(X)$ is called the m-dimensional *spherical* Hausdorff measure of X. Similarly, $\mathscr{S}^m(X) = V_m S^m(X)$ is called, the normalized m-dimensional spherical Hausdorff measure. Obviously,

$$H^m(X) \leq S^m(X) \qquad \text{and} \qquad \mathscr{H}^m(X) \leq \mathscr{S}^m(X) \qquad \text{for all} \quad m \geq 0.$$

However, note that there exist pathological subsets X of $\mathbb{R}^N$ such that $H^m(X) \neq S^m(X)$ for some m, and hence $\mathscr{H}^m(X) \neq \mathscr{S}^m(X)$.

(123) *Hausdorff dimension.* The Hausdorff dimension of the metric space X in (122) above is defined to be

$$\dim_H X = \inf\{m \geq 0: H^m(X) = 0\}.$$

Roughly speaking, sets with a *low* Hausdorff dimension are "small". If $H^m(X) = 0$ for some m, then $\dim_H X \leq m$.

In particular, $H^m(\varnothing) = 0$ for all $m \geq 0$ and hence $\dim_H \varnothing = 0$.

One can show that

$$H^m(X) = \begin{cases} 0 & \text{if} \quad m > \dim_H X, \\ +\infty & \text{if} \quad m < \dim_H X, \end{cases}$$

i.e., the Hausdorff dimension $\dim_H X$ corresponds to a "phase transition" of the Hausdorff measure $H^m(X)$ of X.

If $\dim_H X < \infty$, then X is homeomorphic to a subset of some $\mathbb{R}^N$.

(123a) *Example.* Let X be an at most countable subset of a metric space, and let n be the number of distinct points in X, i.e., $0 \leq n \leq \infty$. Then

$$H^m(X) = \begin{cases} 0 & \text{if} \quad m > 0, \\ n & \text{if} \quad m = 0. \end{cases}$$

Hence $\dim_H X = 0$.

(124) *Fractals.* A metric space is called a fractal iff its Hausdorff dimension strictly exceeds its topological dimension.

The notion of the topological dimension can be found in Definition 13.15.

(124a) *Example.* Let X be the Cantor ternary set, i.e., X is the collection of all real numbers represented by the formula

$$z = \frac{c_1}{3^1} + \frac{c_2}{3^2} + \frac{c_3}{3^3} + \cdots,$$

where the coefficients c_n arbitrarily assume one of the two values 0 or 2. Then

$$\dim_H X = \log 2/\log 3 = 0.6309\ldots,$$

whereas the topological dimension of X is equal to zero, i.e., X is a fractal. Furthermore, the set X has the one-dimensional Lebesgue measure zero.

This example shows that the Lebesgue measure is not always the appropriate tool for measuring the size of sets.

Functions of Bounded Variation

Functions of bounded variation play a fundamental role in the modern calculus of variations. In what follows, the two *key* results are given by

(i) the general integration by parts formula; and
(ii) the general existence theorem for parametric minimal surfaces.

(125) *Basic definition.* Let G be a nonempty open set in $\mathbb{R}^N$, $N \geq 1$. The function $u: G \to \mathbb{R}$ is called a function of *bounded variation* iff $u \in L_1(G)$ and there exists a real constant $K \geq 0$ such that

$$\left|\sum_{i=1}^{N} \int_G u D_i \varphi_i \, dx\right| \leq K \max_{x \in G} |\varphi(x)| \tag{125a}$$

for all $\varphi_1, \ldots, \varphi_N \in C_0^\infty(G)$, where $|\varphi(x)|$ denotes the Euclidean norm of $\varphi(x) = (\varphi_1(x), \ldots, \varphi_N(x))$. The smallest possible constant $K \geq 0$ in (125a) is denoted by $\int_G |Du|$, i.e., we set

$$\int_G |Du| = \inf K.$$

Let $BV(G)$ denote the space of all the functions $u: G \to \mathbb{R}$ of bounded variation. Then, $BV(G)$ becomes a real B-space under the norm

$$\|u\|_{BV(G)} = \|u\|_{L_1(G)} + \int_G |Du|.$$

More generally, the definition of $\int_G |Du|$ also makes sense if $u \in L_{1,\mathrm{loc}}(G)$. Finally, we set $\int_G |Du| = \infty$ iff there is *no* real $K \geq 0$ such that inequality (125a) holds true. Summarizing, we have

$$\int_G |Du| = \sup \left| \int_G u \operatorname{div} \varphi \, dx \right|,$$

where the supremum is taken over all the functions $\varphi \in C_0^\infty(G, \mathbb{R}^N)$ with $|\varphi(x)| \leq 1$ on G. As usual, we set $\operatorname{div} \varphi = \sum_{i=1}^N D_i \varphi_i$.

(126) *Standard example* 1. Let $u \in C^1(\bar{G})$, where G is a nonempty bounded open set in $\mathbb{R}^N$, $N \geq 1$. Then, $u \in BV(G)$ and

$$\int_G |Du| = \int_G |\operatorname{grad} u| \, dx, \tag{126a}$$

where $|\operatorname{grad} u| = (\sum_{i=1}^N (D_i u)^2)^{1/2}$.

Proof. For all $\varphi_1, \ldots, \varphi_N \in C_0^\infty(G)$, integration by parts yields

$$\begin{aligned}
\left|\int_G \sum_{i=1}^N u D_i \varphi_i \, dx\right| &= \left|\int_G \sum_{i=1}^N \varphi_i D_i u \, dx\right| \\
&\leq \int_G |\varphi(x)| |\operatorname{grad} u(x)| \, dx \\
&\leq \left(\int_G |\operatorname{grad} u| \, dx\right) \max_{x \in G} |\varphi(x)|.
\end{aligned}$$

Hence $\int_G |Du| \le \int_G |\text{grad}\, u|\, dx$. An additional argument shows that $\int_G |Du| < \int_G |\text{grad}\, u|\, dx$ is impossible. □

More generally, if G is a nonempty bounded open set in $\mathbb{R}^N$, $N \ge 1$, then

$$W_1^1(G) \subseteq BV(G),$$

and relation (126a) holds true for all $u \in W_p^1(G)$, $1 \le p \le \infty$.

(127) *Standard example* 2. Let E be a bounded region in $\mathbb{R}^N$, $N \ge 1$, with smooth boundary, i.e., $\partial E \in C^2$. Let χ_E be the characteristic function of E, i.e.,

$$\chi_E(x) = \begin{cases} 1 & \text{if } x \in E, \\ 0 & \text{if } x \in \mathbb{R}^N - E. \end{cases}$$

Then we get the *fundamental formula*

$$\int_{\mathbb{R}^N} |D\chi_E| = \textit{surface measure of the boundary } \partial E. \tag{127a}$$

More generally, if G is a nonempty open set in $\mathbb{R}^N$, then

$$\int_G |D\chi_E| = \textit{surface measure of the set } \partial E \cap G.$$

In the case where the boundary ∂E is not sufficiently smooth, we call $\int_{\mathbb{R}^N} |D\chi_E|$ the *generalized* $(N-1)$-dimensional surface measure of ∂E.

(128) *Generalized derivatives*. Let $u \in BV(G)$. It follows from (125a) via the Riesz theorem that there exist a measure μ on the set G and μ-measurable functions $\alpha_i \colon G \to \mathbb{R}$, $i = 1, \ldots, N$, such that

$$\int_G u D_i\varphi\, dx = -\int_G \varphi\alpha_i\, d\mu \qquad \text{for all} \quad \varphi \in C_0^\infty(G), \tag{128a}$$

and $i = 1, \ldots, N$. In particular, if G is a bounded region in $\mathbb{R}^N$ and $u \in C^1(\bar{G})$, then $u \in BV(G)$ and

$$\int_G u D_i\varphi\, dx = -\int_G \varphi D_i u\, dx \qquad \text{for all} \quad \varphi \in C_0^\infty(G).$$

Hence formula (128a) holds with

$$\alpha_i = D_i u \qquad \text{and} \qquad \mu = \text{Lebesgue measure}.$$

In terms of the theory of distributions, relation (128a) tells us the following:

The partial derivative $D_i u$ of the function $u \in BV(G)$ corresponds to the measure $\alpha_i\, d\mu$.

In fact, let U denote the distribution corresponding to the function $u \in BV(G)$, i.e.,

$$U(\varphi) = \int_G u\varphi \, dx \qquad \text{for all} \quad \varphi \in C_0^\infty(G).$$

Then, $D_i U(\varphi) = -U(D_i\varphi)$ for all $\varphi \in C_0^\infty(G)$. By (128a),

$$D_i U(\varphi) = \int_G \varphi \alpha_i \, d\mu \qquad \text{for all} \quad \varphi \in C_0^\infty(G).$$

(129) *The general integration by parts formula.* Let G be a bounded region in $\mathbb{R}^N$, $N \geq 1$, with $\partial G \in C^{0,1}$. Then, for each function $u \in BV(G)$, there exists a function $u_0 \in L_1(\partial G)$ such that

$$\int_G u D_i \varphi \, dx = -\int_G \varphi \alpha_i \, d\mu + \int_{\partial G} u_0 \varphi n_i \, dO \tag{129a}$$

for all $\varphi \in C_0^\infty(\mathbb{R}^N)$ and $i = 1, \ldots, N$. Here, $n = (n_1, \ldots, n_N)$ denotes the outer unit normal to the boundary ∂G.

Obviously, this is a generalization of the classical integration by parts formula

$$\int_G u D_i \varphi \, dx = -\int_G \varphi D_i u \, dx + \int_{\partial G} u_0 \varphi n_i \, dO,$$

where $u_0 = u$ on ∂G.

(130) *Compactness.* Let G be a nonempty bounded open set in $\mathbb{R}^N$, $N \geq 1$. Then the embedding

$$BV(G) \subseteq L_1(G)$$

is compact.

(131) *Semicontinuity.* Let G be a nonempty bounded open set in $\mathbb{R}^N$, $N \geq 1$, and let (u_n) be a sequence in $BV(G)$ such that

$$u_n \to u \quad \text{in } L_1(G) \qquad \text{as} \quad n \to \infty.$$

Then

$$\int_G |Du| \leq \varliminf_{n \to \infty} \int_G |Du_n|.$$

(132) *Caccioppoli sets.* A subset E of $\mathbb{R}^N$ is called a Caccioppoli set iff E is a Borel set and $\chi_E \in BV(G)$, i.e.,

$$\int_G |D\chi_E| < \infty,$$

for every nonempty *bounded* open set G in $\mathbb{R}^N$.

(133) *The general existence theorem for parametric minimal surfaces.* We

consider the following minimum problem:

$$\int_{\mathbb{R}^N} |D\chi_E| = \min!, \qquad E \text{ is a measurable set in } \mathbb{R}^N,$$
$$E = C \quad \text{outside } G. \tag{133a}$$

Let G be a given bounded open set in $\mathbb{R}^N$, $N \geq 1$, and let C be a given Caccioppoli set in $\mathbb{R}^N$.

Then the minimum problem (133a) *has a solution* E_0.

Interpretation. Recall that $\int_{\mathbb{R}^N} |D\chi_E|$ denotes the generalized $(N-1)$-dimensional surface measure of the boundary ∂E of the set E. Thus, roughly speaking, the boundary ∂E_0 of the solution E_0 of problem (133a) minimizes the *area* among all surfaces with boundary $\partial C \cap \partial G$.

The simple proof of this theorem via compactness (130) and semicontinuity (131) can be found in Giusti (1984, M). This monograph also contains sophisticated results about the *regularity* of the set ∂E_0 as well as applications of the space $BV(G)$ to the classical nonparametric minimal surface problem:

$$\int_G \sqrt{1 + u_\xi^2 + u_\eta^2}\, dx = \min!,$$
$$u = g \quad \text{on } \partial G,$$

where $x = (\xi, \eta)$.

References to the Literature

Function spaces: Kufner, John, and Fučik (1977, M) (introduction), Triebel (1978, M), (1983, M).

Sobolev spaces: Nečas (1967, M) and Kufner, John, and Fučik (1977, M) (introduction), Adams (1975, M) (standard work), Lions and Magenes (1968, M), Friedman (1969, M) (inequalities of Gagliardo–Nirenberg), Triebel (1972, M), (1978, M), (1983, M) (Sobolev spaces and interpolation theory), Nikolskii (1975, M), Mazja (1985, M) (sharp embedding theorems), Wloka (1982, M) (Fourier transform and Sobolev spaces).

Weighted Sobolev spaces and partial differential equations: Kufner (1985, M), Kufner and Sändig (1987, M).

Sobolev spaces of infinite order and differential equations: Dubinskii (1984, L).

Sobolev spaces on manifolds: Aubin (1982, M), Eichhorn (1988) (cf. also the References to the Literature for Chapter 85).

Measure theory and integration: Lebesgue (1902) (classical work), Halmos (1954, M), Kamke (1960, M), Hewitt and Stromberg (1965, M), Bauer (1968, M) (applications to the theory of probability, measures on topological spaces), Dieudonné (1968, M), Vol. 2, Günther (1972, M), Vol. 3, Bögel and Tasche (1974, M), Mukherjea and Pothoven (1978, M), Royden (1988, M).

History of the Lebesgue integral: Hawkins (1970, M).

Geometric measure theory: Federer (1969, M), Simon (1983, L).

Functions of bounded variation and minimal surfaces: Simon (1983, L), Giusti (1984, M).

Hausdorff measure, Hausdorff dimension, and partial regularity of the solutions of partial differential equations: Giaquinta (1983, M), Simon, (1983, L), Giusti (1984, M), Temam (1983, M), (1988, M).

Fractals and their applications: Mandelbrot (1982, M).

Radon measures on topological spaces: Schwartz (1973, L).

Vector measures: Dinculeanu (1966, M), Holmes (1975, M).

Integration of functions with values in B-spaces: Dunford and Schwartz (1958, M), Yosida (1965, M), Bögel and Tasche (1974, M).

Distributions: Kanwal (1983, M) (elementary introduction with many applications), Schwartz (1950, M) (classical standard work), Gelfand and Šilov (1964, M), Vols. 1–5, Yosida (1965, M), Horváth (1966, M), Reed and Simon (1971, M), Vol. 1 (introduction), Vladimirov (1971, M).

Distributions and partial differential equations: Hörmander (1964, M), (1983, M), Vols. 1–4.

Spectral theory of unbounded operators: Riesz and Nagy (1956, M), Dunford and Schwartz (1958, M), Vols. 1–3, Kato (1966, M), Maurin (1967, M), Triebel (1972, M) (functional calculus).

Spectral theory and applications in quantum theory: Reed and Simon (1971, M), Vols. 1–4 (standard work), Triebel (1972, M).

Linear semigroups: Davies (1980, M), Pazy (1983, M).

Sectorial operators and analytic semigroups: Henry (1981, L).

Interpolation theory: Reed and Simon (1971, M), Vol. 2 (introduction), Lions and Magenes (1968, M), Vol. 1, Bergh and Löfström (1976, M), Triebel (1978, M) (comprehensive representation), Krein, Petunin, and Semenov (1982, M).

Interpolation theory and semigroups: Butzer and Berens (1967, M).

Interpolation theory and partial differential equations: Lions and Magenes (1968, M), Triebel (1978, M), (1983, M).

Galerkin schemes: Ciarlet, Schultz and Varga (1969), Michlin (1969, M), (1976, M), Aubin (1972, M), Meis and Marcowitz (1978, L).

Finite elements: Ciarlet (1977, M) (standard work), Zlámal (1968), Strang and Fix (1973, M), Bathe and Wilson (1976, M), Akin (1982, M), Babuška and Szabo (1988, M), Ciarlet and Lions (1988, M), Vol. 2 (handbook of numerical analysis).

Software for finite elements: Sewell (1985, M).

Nonlinear differential equations and finite elements: Temam (1977, M) and Girault and Raviart (1986, M) (Navier–Stokes equations), Frehse (1977), (1982, S), Frehse and Rannacher (1978), Dobrowolski (1980), Dobrowolski and Rannacher (1980), Thomée (1984, M), Kufner and Sändig (1987, M) (weighted Sobolev spaces), Feistauer and Ženišek (1988).

References

Ablowitz, M. and Segur, H. (1981): *Solitons and the Inverse Scattering Transform.* SIAM, Philadelphia, PA.

Ablowitz, M., Beals, R., and Tenenblat, K. (1986): *On the solution of the generalized wave and generalized sine–Gordon equations.* Stud. Appl. Math. **74**, 177–203.

Adams, R. (1975): *Sobolev Spaces.* Academic Press, New York.

Agmon, S., Douglis, A., and Nirenberg, L. (1959): *Estimates near the boundary for solutions of elliptic partial differential equations*, I. II. Comm. Pure Appl. Math. **12** (1959), 623–727; **17** (1964), 35–92.

Agmon, S. (1965): *Lectures on Elliptic Boundary Value Problems.* Van Nostrand, Princeton, NJ.

Akin, J. (1982): *Application and Implementation of Finite Element Methods.* Academic Press, New York.

Albeverio, S. *et al.* (1986): *Nonstandard Methods in Stochastic Analysis and Mathematical Physics.* Academic Press, New York.

Albrecht, P. (1979): *Die numerische Behandlung gewöhnlicher Differentialgleichungen.* Akademie-Verlag, Berlin.

Aleksandrov, P. [ed.] (1971): *Die Hilbertschen Probleme.* Geest & Portig, Leipzig.

Alt, H. and Luckhaus, S. (1983): *Quasilinear elliptic–parabolic equations.* Math. Z. **183**, 311–341.

Alt, H. (1985): *Lineare Funktionalanalysis.* Springer-Verlag, Berlin.

Alt, H. and Struwe, M. (1989): *Nonlinear Functional Analysis and Variational Problems* (monograph to appear).

Altman, M. (1986): *A Unified Theory of Nonlinear Operator and Evolution Equations with Applications.* Dekker, New York.

Amann, H. (1969): *Ein Existenz- und Eindeutigkeitssatz für die Hammersteinsche Gleichung in Banachräumen.* Math. Z. **111**, 175–190.

Amann, H. (1969a): *Zum Galerkin–Verfahren für die Hammersteinsche Gleichung.* Arch. Rational Mech. Anal. **35**, 114–121.

Amann, H. (1972): *Existence theorems for equations of Hammerstein type.* Applicable Anal. **1**, 385–397.

Amann, H. (1972a): *Über die näherungsweise Lösung nichtlinearer Integralgleichungen.* Numer. Math. **19**, 19–45.

Amann, H. (1976): *Fixed-point problems and nonlinear eigenvalue problems in ordered Banach spaces.* SIAM Rev. **18**, 620–709.

Amann, H. (1984): *Existence and regularity for semilinear parabolic evolution equations.* Ann. Scuola Norm. Sup. Pisa, Cl. Sci. Serie IV, **11**, 593–676.

Amann, H. (1985): *Global existence for semilinear parabolic systems.* J. Reine. Angew. Math. **360**, 47–83.

Amann, H. (1986): *Quasilinear evolution equations and parabolic systems.* Trans. Amer. Math. Soc. **293**, 191–227.

Amann, H. (1986a): *Quasilinear parabolic systems under nonlinear boundary conditions.* Arch. Rational Mech. Anal. **92**, 153–192.

Amann, H. (1986b): *Semigroups and nonlinear evolution equations.* Linear Algebra Appl. **84**, 3–32.

Amann, H. (1986c): *Parabolic evolution equations with nonlinear boundary conditions.* In: Browder, F. [ed.] (1986), Part 1, pp. 17–27.

Amann, H. (1987): *On abstract parabolic fundamental solutions.* J. Math. Soc. Japan **39**, 93–116.

Amann, H. (1988): *Remarks on quasilinear parabolic systems* (to appear).

Amann, H. (1988a): *Dynamic theory of quasilinear parabolic equations*, I, II (to appear).

Amann, H. (1988b): *Parabolic evolution equations in interpolation and extrapolation spaces.* J. Funct. Anal. **78**, 233–277.

Amann, H. (1988c): *Parabolic equations and nonlinear boundary conditions.* J. Differential Equations **72**, 201–269.

Ambrosetti, A. and Prodi, G. (1972): *On the inversion of some differentiable mappings with singularities between Banach spaces.* Ann. Mat. Pura Appl. **93**, 231–246.

Ambrosetti, A. and Rabinowitz, P. (1973): *Dual variational methods in critical point theory and applications.* J. Funct. Anal. **14**, 349–381.

Ambrosetti, A. (1983): *Existence and multiplicity for some classes of nonlinear problems.* In: Warsaw (1983), Vol. 2, pp. 1125–1132.

Ames, W. (1965): *Nonlinear Partial Differential Equations in Engineering.* Vols. 1, 2. Academic Press, New York.

Ames, W. (1977): *Numerical Methods for Partial Differential Equations.* Academic Press, New York.

Anselone, P. (1971): *Collectively Compact Operator Approximation Theory.* Prentice-Hall, Englewood Cliffs, NJ.

Ansorge, R. (1978): *Differenzenapproximation partieller Anfangswertaufgaben.* Teubner, Stuttgart.

Appell, J. (1987): *The superposition operator (Nemyckii operator) in function spaces: A survey.* Mathematical Institute, Report No. 141, University of Augsburg.

Arms, J., Marsden, J., and Moncrief, V. (1982): *The structure of the space of solutions of Einstein's equations. II: Several Killing fields and the Einstein–Yang–Mills equations.* Ann. Physics **144**, 81–106.

Atiyah, M. and Hitchin, N. (1988): *The Geometry and Dynamics of Magnetic Monopoles.* University Press, Princeton, NJ.

Attouch, H. (1984): *Variational Convergence for Functionals and Operators.* Pitman, London.

Aubin, J. (1967): *Behavior of the error of the approximate solution of boundary value problems for linear elliptic operators.* Ann. Scuola Norm. Sup. Pisa Cl. Sci. **21**, 599–637.

Aubin, J. (1970): *Approximation des problèmes aux limites non homogènes pour des operateurs non linéaires.* J. Math. Anal. Appl. **30**, 510–521.

Aubin, J. (1972): *Approximation of Elliptic Boundary Value Problems.* Wiley, New York.

Aubin, T. (1982): *Nonlinear Analysis on Manifolds: Monge–Ampère Equations.* Springer-Verlag, New York.

Babič, V. *et al.* (1967): *Lineare Differentialgleichungen der Mathematischen Physik.* Akademie-Verlag, Berlin.

Babin, A. and Višik, M. (1983): *Regular attractors of semigroups and evolution equations.* J. Math. Pures Appl. **62**, 441–491.

Babin, A. and Višik, M. (1983a): *Attractors of evolution equations and estimates for their dimensions.* Uspekhi Mat. Nauk **38** (4), 133–187 (Russian).

Babin, A. and Višik, M. (1986): *Unstable invariant sets of semigroups of nonlinear operators and their perturbations.* Uspekhi Mat. Nauk **41** (4), 3–34 (Russian).

Babuška, I. and Szábo, B. (1988): *Finite Element Methods: Concepts and Applications.* Springer-Verlag, New York.

Bahri, A. and Coron, J. (1988): *On a nonlinear elliptic equation: the effect of the topology on the domain.* Comm. Pure Appl. Math. **41**, 253–294.

Baire, R. (1899): *Sur les fonctions des variables réelles.* Ann. di Mat. **3**, 1–123.

Ballman, J. and Jeltsch, R. [eds.] (1988): *Hyperbolic Problems.* Lecture Notes in Mathematics, Springer-Verlag, Berlin (Proceedings to appear).

Banach, S. and Steinhaus, H. (1927): *Sur le principe de condensation des singularités.* Fund. Math. **9**, 50–61.

Banach, S. (1929): *Sur les fonctionelles linéaires.* Studia Math. **1**, 211–216, 223–229.

Banach, S. (1932): *Théorie des opérations linéaires.* Warszawa.

Bannerjee, P. and Butterfield, R. (1981): *Boundary Element Methods in Engineering Sciences.* McGraw-Hill, London.

Barbu, V. (1976): *Nonlinear Semigroups and Differential Equations in Banach Spaces.* Noordhoff, Leyden.

Bardos, C. (1982): *Introduction aux problèmes hyperbolique non linéaires.* Lecture Notes in Mathematics, Vol. 1047. Springer-Verlag, Berlin, pp. 1–76.

Bardos, C., Degond, P., and Golse, F. (1986): *A priori estimates and existence results for the Vlasov and Boltzmann equations.* Lectures in Applied Mathematics, Vol. 23, Amer. Math. Soc., Providence, RI, pp. 189–207.

Bathe, K. and Wilson, E. (1976): *Numerical Methods in Finite Element Analysis.* Prentice-Hall, Englewood Cliffs, NJ.

Bauer, H. (1968): *Wahrscheinlichkeitstheorie und Grundzüge der Maßtheorie.* De Gruyter, Berlin.

Bauer, F., Garabedian, P., and Korn, D. (1972/77): *A Theory of Supercritical Wing Sections, with Computer Examples,* Vols. 1–3. Springer-Verlag, New York.

Bauer, F., Betancourt, O., Garabedian, P. (1978): *A Computational Method in Plasma Physics.* Springer-Verlag, New York.

Bauer, F., Betancourt, O., and Garabedian, P. (1984): *Magnetohydrodynamic Equilibrium and Stability of Stellarators.* Springer-Verlag, New York.

Bauer, F., Betancourt, O., and Garabedian, P. (1986): *Monte Carlo calculation of transport for three-dimensional magnetohydrodynamic equilibria.* Comm. Pure Appl. Math. **39**, 595–621.

Bazley, N. and Weinacht, R. (1983): *A class of operator equations in nonlinear mechanics.* Applicable Anal. **15**, 7–14.

Bazley, N. and Miletta, P. (1986): *Forced Liénard equations with nonlocal nonlinearities.* Applicable Anal. **21**, 339–346.

Bebernes, J. and Eberly, D. (1988): *A description of self-similar blow-up for dimension $n \geq 3$.* Ann. Inst. Henri Poincaré, Analyse Non Linéaire **5**, 1–21.

Beckert, H. (1976): *Variationsrechnung und Stabilitätstheorie.* Math. Nachr. **75**, 47–59.

Beckert, H. (1984): *Nichtlineare Elastizitätstheorie.* Ber. Sächs. Akad. Wiss. Leipzig. Math.-nat. Kl. **117**, 2. Akademie-Verlag, Berlin.

Belleni-Morante, A. (1979): *Applied Semigroups and Evolution Equations.* Clarendon, Oxford.

Benilan, P., Crandall, M., and Pazy, A. (1989): *Nonlinear Evolution Governed by Accretive Operators* (monograph to appear).

Benkert, F. (1985): *Über den globalen Lösungszusammenhang einiger Scharen von Variationsproblemen.* Dissertation, Karl-Marx-Universität, Leipzig.

Benkert, F. (1989): *On the number of stable local minima of some functionals.* Z. Anal. Anwendungen (to appear).

Bensoussan, A., Lions, J., and Papanicolaou, G. (1978): *Asymptotic Methods in Periodic Structures.* North-Holland, Amsterdam.

Berezin, F. and Shubin, M. (1983): *The Schrödinger Equation.* University Press, Moscow (Russian).

Berger, M. and Podolak, E. (1974): *On nonlinear Fredholm operator equations.* Bull. Amer. Math. Soc. **80**, 861–864.

Berger, M. and Podolak, E. (1975): *On the solutions of a nonlinear Dirichlet problem.* Indiana Univ. Math. J. **24**, 837–846.

Berger, M. (1977): *Nonlinearity and Functional Analysis.* Academic Press, New York.

Berger, M. and Schechter, M. (1977): *On the solvability of semilinear gradient operator equations.* Adv. in Math. **25**, 97–132.

Berger, M. (1979): *Nonlinear problems with exactly three solutions.* Indiana Univ. Math. J. **28**, 689–698.

Berger, M. and Church, P. (1979): *Complete integrability and perturbation of a nonlinear Dirichlet problem.* Indiana Univ. Math. J. **28**, 935–952; **29**, 715–735; **30**, 799.

Berger, M., Church, P., and Timourian, J. (1985): *Folds and cusps in Banach spaces with applications to nonlinear differential equations.* Indiana Univ. Math. J. **34**, 1–19.

Berger, M. and Kohn, R. (1988): *A rescaling algorithm for the numerical calculation of blowing-up solutions.* Comm. Pure Appl. Math. **41**, 848–863.

Bergh, J. and Löfström, J. (1976): *Interpolation Spaces.* Springer-Verlag, New York.

Berkeley (1983): *Nonlinear Functional Analysis.* Cf. Browder, F. [ed.] (1986).

Berkeley (1986): *Proceedings of the International Conference of Mathematicians in Berkeley.* Vols. 1, 2. Amer. Math. Soc., Providence, RI.

Berndt, W. (1982): *Solution of a degenerated elliptic equation of second order in an unbounded domain.* Z. Anal. Anwendungen **1** (3), 53–68.

Bers, L., John, F., and Schechter, M. (1964): *Partial Differential Equations.* Wiley, New York.

Besov, O., Iljin, V., and Nikolskii, S. (1975): *Integral Representations of Functions and Embedding Theorems.* Nauka, Moscow (Russian).

Betancourt, O. (1988): *Betas, a spectral code for three-dimensional magnetohydro-*

dynamic equilibrium and nonlinear stability calculations. Comm. Pure Appl. Math. **41**, 551–568.

Birkhoff, G. [ed.] (1973): *A Source Book in Classical Analysis.* Harvard University Press, Cambridge, MA.

Birkhoff, G. and Lynch, R. (1984): *Numerical Solution of Elliptic Problems.* SIAM, Philadelphia, PA.

Birkhoff, G. and Schoenstadt, A. [eds.] (1984): *Elliptic Problem Solvers.* Academic Press, New York.

Birkhoff, G. (1989): *The software systems ELLPACK and ITPACK* (to appear).

Blanchard, P. and Brüning, E. (1982): *Direkte Methoden der Variationsrechnung.* Springer-Verlag, Wien.

Blaschke, W. (1942): *Nicht-Euklidische Geometrie und Mechanik.* Teubner, Leipzig.

Blaschke, W. (1957): *Reden und Reisen eines Geometers.* Verl. der Wiss., Berlin.

Blumenthal, O. (1932): *Lebensgeschichte von Hilbert.* In: Hilbert, D. (1932), Vol. 3, pp. 388–429.

Bobenko, A. and Bordag, L. (1987): *The qualitative analysis of multi-cnoidal waves of the Korteweg–de Vries equation via automorphic functions.* Zapiski Naučnych Seminarov LOMI, Vol. 165, pp. 31–41. Nauka, Leningrad (Russian). (English translation in J. Sov. Math. (to appear).)

Bobenko, A., Its, A., and Matveev, V. (1991): *Analytic Theory of Solitons.* Nauka, Moscow (Russian) (monograph to appear).

Boffi, V. and Neunzert, H. [eds.] (1984): *Applications of Mathematics in Technology.* Teubner, Stuttgart.

Bögel, K. and Tasche, M. (1974): *Analysis in normierten Räumen.* Akademie-Verlag, Berlin.

Böhm, M. and Showalter, R. (1985): *A nonlinear pseudoparabolic diffusion equation.* SIAM J. Math. Anal. **16**, 980–999.

Böhme, R. (1972): *Die Lösung der Verzweigungsgleichungen für nichtlineare Eigenwertprobleme.* Math. Z. **127**, 105–126.

Böhmer, K. and Stetter, H. [eds.] (1984): *Defect Correction Methods: Theory and Applications.* Springer-Verlag, New York.

Bona, J. and Smith, R. (1975): *The initial-value problem for the Korteweg–de Vries equation.* Philos. Trans. Roy. Soc. London **A278**, 555–601.

Bona, J. and Scott, R. (1976): *Solution of the Korteweg–de Vries equation in fractional order Sobolev spaces.* Duke Math. J. **43**, 87–99.

Bonsall, F. and Duncan, J. (1971): *Numerical Range of Operators of Normed Spaces and of Elements of Normed Algebras.* University Press, Cambridge, England.

Bony, J. (1981): *Calcul symbolique et propagation des singularités pour les équations aux dérivées partielles non linéaires.* Ann. Sci. École Norm. Sup. **14**, 209–246.

Bony, J. (1984): *Propagation et interaction des singularités pour les solutions des équations aux dérivees partielles non linéaires.* In: Warsaw (1983), Vol. 2, pp. 1133–1147.

Booss, B. and Bleecker, D. (1985): *Topology and Analysis.* Springer-Verlag, New York.

Borisovič, Ju. *et al.* (1977): *Nonlinear Fredholm maps and the Leray–Schauder theory.* Uspekhi Mat. Nauk **32** (4), 3–54 (Russian).

Botha, J. and Pindler, G. (1983): *Fundamental Concepts in the Numerical Solutions of Differential Equations.* Wiley, New York.

Bottazzini, U. (1986): *The "Higher Calculus". A History of Real and Complex Analysis from Euler to Weierstrass.* Springer-Verlag, New York.

Bourbaki, N. (1940): *Topologie générale*. Hermann, Paris.

Bourbaki, N. (1952): *Intégration*. Hermann, Paris.

Braess, D. (1986): *Nonlinear Approximation Theory*. Springer-Verlag, New York.

Bramble, J. and Osborne, J. (1973): *Rate of convergence estimates for non-self-adjoint eigenvalue approximations*. Math. Comp. **27**, 525–549.

Brandt, A. (1982): *Guide to multigrid development*. In: Hackbusch, W. and Trottenberg, U. [eds.] (1982), pp. 220–312.

Bratley, P., Bennett, L., and Schrage, L. (1987): *A Guide to Simulation*. Springer-Verlag, New York.

Brebbia, C. *et al.* (1984): *Boundary Element Techniques: Theory and Applications in Engineering*. Springer-Verlag, New York.

Brézis, H. (1968): *Équations et inéquations non linéaires dans les espaces vectoriels en dualité*. Ann. Inst. Fourier **18**, 115–175.

Brézis, H. and Sibony, M. (1968): *Méthodes d'approximation et d'itération pour les opérateurs monotones*. Arch. Rational Mech. Anal. **28**, 59–82.

Brézis, H. (1970): *On some degenerate parabolic equations*. In: Browder, F. [ed.] (1970c), pp. 28–38.

Brézis, H., Crandall, M., and Pazy, A. (1970): *Perturbations of maximal monotone sets*. Comm. Pure Appl. Math. **23**, 123–144.

Brézis, H. and Pazy, A. (1970): *Accretive sets and differential equations in Banach spaces*. Israel J. Math. **8**, 367–383.

Brézis, H. (1973): *Opérateurs maximaux monotones*. North-Holland, Amsterdam.

Brézis, H. and Browder, F. (1975): *Nonlinear integral equations and systems of Hammerstein type*. Adv. in Math. **18**, 115–144.

Brézis, H. and Haraux, A. (1976): *Image d'une somme d'opérateurs monotones et applications*. Israel J. Math. **23**, 165–186.

Brézis, H. and Browder, F. (1978): *Linear maximal monotone operators and singular nonlinear integral equations of Hammerstein type*. In: Cesari, L. [ed.] (1978), pp. 31–42.

Brézis, H. and Nirenberg, L. (1978): *Characterization of the ranges of some nonlinear operators and applications to boundary value problems*. Ann. Scuola Norm. Sup. Pisa Cl. Sci., Ser. IV, **5**, 225–326.

Brézis, H. and Nirenberg, L. (1978a): *Forced vibrations for a nonlinear wave equation*. Comm. Pure Appl. Math. **31**, 11–30.

Brézis, H. and Lions, J. (1979): *Nonlinear PDEs and Their Applications*. Collège de France Seminar. Vols. 1–8 (to be continued).

Brézis, H., Coron, J., and Nirenberg, L. (1980): *Free vibrations for a nonlinear wave equation and a theorem of P. Rabinowitz*. Comm. Pure Appl. Math. **33**, 667–689.

Brézis, H. (1983): *Analyse fonctionelle*. Masson, Paris.

Brézis, H. and Nirenberg, L. (1983): *Positive solutions of nonlinear elliptic equations involving critical Sobolev exponents*. Comm. Pure Appl. Math. **36**, 437–477.

Brézis, H. and Coron, J. (1984): *Multiple solutions of H-systems and Rellich's conjecture*. Comm. Pure Appl. Math. **37**, 149–187.

Brézis, H. (1986): *Elliptic equations with limiting Sobolev exponents: the impact of topology*. Comm. Pure Appl. Math. **39**, Supplement S17–S39.

Brézis, H., Crandall, M., and Kappel, F. [eds.] (1986): *Semigroups and Applications*, Vols. 1, 2. Wiley, New York.

Brézis, H. (1987): *Liquid crystals and energy estimates for S^2-valued maps*. In: Ericksen, J. and Kinderlehrer, D. [eds.] (1987), pp. 31–52.

Brezzi, F. (1986): *New applications of mixed finite element methods.* In: Berkeley (1986), Vol. 2, pp. 1335–1347.

Brouwer, L. (1912): *Über Abbildungen von Mannigfaltigkeiten.* Math. Ann. **71**, 97–115.

Browder, F. (1954): *Strongly elliptic systems of differential equations.* In: Bers, L. [ed.], *Contributions to the Theory of Partial Differential Equations*, pp. 14–52. University Press, Princeton, NJ.

Browder, F. (1961): *On the spectral theory of elliptic differential operators.* Math. Ann. **142**, 22–130.

Browder, F. (1962): *Functional analysis and differential equations.* Math. Ann. **138**, 55–79; **145**, 81–226.

Browder, F. (1963): *Nonlinear elliptic boundary value problems.* Bull. Amer. Math. Soc. **69**, 862–874.

Browder, F. (1963a): *Variational boundary value problems for quasilinear elliptic differential equations of arbitrary order*, I–III. Proc. Nat. Acad. Sci. U.S.A. **50**, 31–37; 592–598; 794–798.

Browder, F. (1963b): *The solvability of nonlinear functional equations.* Duke Math. J. **30**, 557–566.

Browder, F. (1965): *Existence and uniqueness theorems for solutions of nonlinear boundary-value problems.* Proc. Sympos. Appl. Math. **17**, 24–49. Amer. Math. Soc., Providence, RI.

Browder, F. (1966): *On the unification of the calculus of variations and the theory of monotone nonlinear operators in Banach spaces.* Proc. Nat. Acad. Sci. U.S.A. **56**, 419–425.

Browder, F. (1966a): *Problèmes non linéaires.* Les Presses de l'Université de Montréal.

Browder, F. (1967): *Approximation-solvability of nonlinear functional equations in normed linear spaces.* Arch. Rational Mech. Anal. **26**, 33–42.

Browder, F. and Petryshyn, W. (1967): *Construction of fixed points of nonlinear mappings in Hilbert space.* J. Math. Anal. Appl. **20**, 197–228.

Browder, F. (1968): Cf. Browder, F. (1968/76).

Browder, F. (1968a): *Nonlinear maximal monotone mappings in Banach spaces.* Math. Ann. **175**, 81–113.

Browder, F. (1968/76): *Nonlinear Operators and Nonlinear Equations of Evolution in Banach Spaces.* Proc. Sympos. Pure Math., Vol. 18, Part 2 (the entire issue). Amer. Math. Soc., Providence, RI, 1976 (preprint version 1968).

Browder, F. (1969): *Remarks on nonlinear interpolation in Banach spaces.* J. Funct. Anal. **4**, 390–403.

Browder, F. (1969a): *Nonlinear variational inequalities and maximal monotone mappings in Banach spaces.* Math. Ann. **183**, 213–231.

Browder, F. and Gupta, C. (1969): *Monotone operators and nonlinear integral equations of Hammerstein type.* Bull. Amer. Math. Soc. **75**, 1347–1353.

Browder, F. and Petryshyn, W. (1969): *Approximation methods and the generalized degree for nonlinear mappings in Banach spaces.* J. Funct. Anal. **3**, 217–245.

Browder, F. (1970): *Existence theorems for nonlinear partial differential equations.* Proc. Sympos. Pure Math., Vol. 16, pp. 1–62. Amer. Math. Soc., Providence, RI.

Browder, F. (1970a): *Pseudomonotone operators and the calculus of variations.* Arch. Rational Mech. Anal. **38**, 268–277.

Browder, F. [ed.] (1970b): *Functional Analysis and Related Fields.* Springer-Verlag, New York.

Browder, F. [ed.] (1970c): *Functional Analysis.* Proc. Sympos. Pure Math., Vol. 18, Part 1. Amer. Math. Soc., Providence, RI.

Browder, F. (1971): *Nonlinear functional analysis and nonlinear integral equations of Hammerstein and Urysohn type.* In: Zarantonello, E. [ed.] (1971), pp. 425–500.

Browder, F. (1971a): *Normal solvability and the Fredholm alternatives for mappings into infinite-dimensional manifolds.* J. Funct. Anal. **8**, 250–274.

Browder, F. and Hess, P. (1972): *Nonlinear mappings of monotone type.* J. Funct. Anal. **11**, 251–294.

Browder, F. (1975): *The relation of functional analysis to concrete analysis in 20th century mathematics.* Historia Math. **2**, 577–590.

Browder, F. [ed.] (1976): *Mathematical Developments Arising from Hilbert's Problems.* Amer. Math. Soc., New York.

Browder, F. (1977): *On a principle of H. Brézis and its applications.* J. Funct. Anal. **25**, 356–365.

Browder, F. (1978): *Image d'un opérateur maximal monotone et le principle de Landesman–Lazer.* C. R. Acad. Sci. Paris, Sér. A **287**, 715–718.

Browder, F. (1983): *Fixed-point theory and nonlinear problems.* In: Browder, F. [ed.] (1983a), Part 2, pp. 49–87.

Browder, F. [ed.] (1983a): *The Mathematical Heritage of Henri Poincaré.* Proc. Sympos. Pure Math., Vol. 39, Parts 1, 2. Amer. Math. Soc., Providence, RI.

Browder, F. (1984): *Coincidence theorems, minimax theorems, and variational inequalities.* Contemp. Math. **26**, 67–80.

Browder, F. [ed.] (1986): *Nonlinear Functional Analysis and Its Applications.* Proc. Sympos. Pure Math., Vol. 45, Parts 1, 2. Amer. Math. Soc., Providence, RI.

Bruck, R. (1983): *Asymptotic behavior of nonexpansive mappings.* Contemp. Math. **18**, 1–47.

Bullough, R. and Caudrey, P. [eds.] (1980): *Solitons.* Springer-Verlag, New York.

Burenkov, V. (1976): *On a method of extending differentiable functions.* Trudy Mat. Inst. Steklova, Akad. Nauk SSSR **140**, 27–67 (Russian).

Burenkov, V. (1985): *The extension of functions under preservation of the Sobolev seminorm.* Trudy Mat. Inst. Steklova, Akad. Nauk SSSR **172**, 71–84 (Russian).

Burkhardt, H. and Meyer, F. (1900): *Potentialtheorie.* In: *Encyklopädie der Mathematischen Wissenschaften* (1899ff), Vol. 2, Part 1.1, pp. 464–503.

Butcher, J. (1987): *The Numerical Analysis of Ordinary Differential Equations.* Wiley, New York.

Butzer, P. and Berens, H. (1967): *Semigroups of Operators and Approximations.* Springer-Verlag, Berlin.

Caffarelli, L., Nirenberg, L., and Spruck, J. (1984/88): *The Dirichlet problem for nonlinear second-order elliptic equations,* I-V. Comm. Pure Appl. Math. **37**, 369–402; **41**, 47–70.

Calderon, A. and Zygmund, A. (1956): *On singular integrals.* Amer. J. Math. **78**, 289–309.

Calogero, F. and Degasperis, A. (1982): *Spectral Transform and Solitons.* North-Holland, Amsterdam.

Calude, C. (1984): *Theories of Computational Complexity.* North-Holland, Amsterdam.

Cannon, J., Ford, W., and Lair, A. (1976): *Quasilinear parabolic systems.* J. Differential Equations **20**, 441–472.

Carasso, C. and Raviart, P. [eds.] (1987): *Nonlinear Hyperbolic Problems.* Lecture Notes in Mathematics, Vol. 1270. Springer-Verlag, Berlin.

Carleman, T. (1923): *Sur les équations intégrales singulières à noyau réel et symétrique.* Uppsala.

Caroll, R. (1969): *Abstract Methods in Partial Differential Equations.* Harper and Row, New York.

Caroll, R. and Showalter, R. (1976): *Singular and Degenerate Cauchy Problems.* Academic Press, New York.

Carrier, G. and Pearson, C. (1976): *Partial Differential Equations.* Academic Press, New York.

Céa, J. (1964): *Approximation variationelle des problèmes aux limites.* Ann. Inst. Fourier **14**, 345–444.

Céa, J. (1971): *Optimisation: théorie et algorithmes.* Dunod, Paris.

Cesari, L. (1964): *Functional analysis and Galerkin's method.* Michigan Math. J. **11**, 385–414.

Cesari, L. (1976): *Functional analysis, nonlinear differential equations, and the alternative method.* In: Cesari, L. [ed.], *Nonlinear Functional Analysis*, pp. 1–197. Lecture Notes in Pure and Applied Mathematics, Vol. 19, Dekker, New York.

Cesari, L. *et al.* [eds.] (1978): *Nonlinear Analysis: A Collection of Papers in Honor of Erich Rothe.* Academic Press, New York.

Cesari, L. [ed.] (1987): *Contributions to the Modern Calculus of Variations.* Pitman, London.

Chazarain, J. and Pirion, A. (1982): *Introduction to the Theory of Linear Partial Differential Equations.* North-Holland, Amsterdam.

Cheney, E. (1967): *Introduction to Approximation Theory.* McGraw-Hill, New York.

Chern, S. [ed.] (1984): *Seminar on Nonlinear Partial Differential Equations.* Springer-Verlag, New York.

Choquet-Bruhat, Y. and Yorke, J. (1980): *The Cauchy problem for the Einstein equations.* In: Held, A. [ed.], *General Relativity and Gravitation*, Vol. 1, pp. 99–136. Plenum, New York.

Choquet-Bruhat, Y. and Christodoulou, D. (1981): *Existence of global solutions of the Yang–Mills, Higgs, and spinor field equations in (3 + 1)-dimensions.* Ann. Sci. École Norm. Sup. IV, **14**, 481–500.

Choquet-Bruhat, Y. and Christodoulou, D. (1982): *Existence of global solutions of the Yang–Mills–Higgs fields in 4-dimensional Minkowski space*, I, II. Comm. Math. Phys. **83**, 171–191; 193–212.

Choquet-Bruhat, Y., DeWitt-Morette, C., and Dillard-Bleick, M. (1982a): *Analysis, Manifolds, and Physics.* North-Holland, Amsterdam.

Chorin, A. *et al.* (1978): *Product formulas and numerical algorithms.* Comm. Pure Appl. Math. **31**, 205–256.

Chow, S. and Hale, J. (1982): *Methods of Bifurcation Theory.* Springer-Verlag, New York.

Chow, S. and Lauterbach, R. (1988): *A bifurcation theorem for critical points of variational problems.* Nonlinear Anal. **12**, 51–61.

Christodoulou, D. (1986): *Global solutions of nonlinear hyperbolic equations for small initial data.* Comm. Pure Appl. Math. **39**, 267–282.

Christodoulou, D. and Klainerman, S. (1990): *Nonlinear Hyperbolic Equations* (monograph to appear).

Ciarlet, P., Schultz, M., and Varga, R. (1969): *Numerical methods of high order accuracy for nonlinear boundary value problems.* Numer. Math. **13**, 51–77.

Ciarlet, P. (1977): *Numerical Analysis of the Finite Element Method for Elliptic Boundary Value Problems.* North-Holland, Amsterdam.

Ciarlet, P. (1988): *Mathematical Elasticity*, Vols. 1, 2. North-Holland, Amsterdam.

Ciarlet, P. and Lions, J. [eds.] (1988): *Handbook of Numerical Analysis*. Vol. 1: *Finite Element Method*. Vol 2: *Finite Difference Method* (Vols. 3ff to appear). North-Holland, Amsterdam.

Cioră nescu, I. (1974): *Aplicatii de dualitate in analiza functională neliniară*. Editura Academiei, Bucuresti.

Clarke, F. (1983): *Optimization and Non-Smooth Analysis*. Wiley, New York.

Clarkson, J. (1936): *Uniformly convex spaces*. Trans. Amer. Math. Soc. **40**, 396–414.

Clément, P. *et al.* (1987): *One-Parameter Semigroups*. North-Holland, Amsterdam.

Coddington, E. and Levinson, N. (1955): *Theory of Ordinary Differential Equations*. McGraw-Hill, New York.

Cohen, P. and Feigenbaum, E. (1981): *Handbook of Artificial Intelligence*, Vols. 1–3. Addison-Wesley, New York.

Coleman, T. (1984): *Large Sparse Numerical Optimization*. Lecture Notes in Computer Science, Vol. 165. Springer-Verlag, New York.

Colombeau, J. (1984): *New Generalized Functions and Multiplication of Distributions*. North-Holland, Amsterdam.

Colombeau, J. (1985): *Elementary Introduction to New Generalized Functions*. North-Holland, Amsterdam.

Collatz, L. (1960): *The Numerical Treatment of Ordinary Differential Equations*. Springer-Verlag, Berlin.

Collatz, L. (1963): *Eigenwertaufgaben mit technischen Anwendungen*. Geest & Portig, Leipzig.

Collatz, L. (1964): *Funktionalanalysis und numerische Mathematik*. Springer-Verlag, Berlin.

Connes, A. (1977): *Sur la théorie non-commutative de l'intégration*. Lecture Notes in Mathematics, Vol. 725. Springer-Verlag, Berlin.

Courant, R. (1920): *Über die Eigenwerte bei den Differentialgleichungen der mathematischen Physik*. Math. Z. **7**, 1–57.

Courant, R., Friedrichs, K., and Lewy, H. (1928): *Über die partiellen Differentialgleichungen der mathematischen Physik*. Math. Ann. **100**, 32–74.

Courant, R. and Hilbert, D. (1937): *Methoden der mathematischen Physik*, Vol. 2. Springer-Verlag, Berlin.

Courant, R. (1943): *Variational methods for the solution of problems of equilibrium and vibrations*. Bull. Amer. Math. Soc. **49**, 1–23.

Courant, R. (1950): *Dirichlet's Principle, Conformal Mappings, and Minimal Surfaces*. Interscience, New York.

Courant, R. and Hilbert, D. (1953): *Methods of Mathematical Physics*, Vols. 1, 2. Interscience, New York.

Crandall, M. and Pazy, A. (1969): *Semigroups of nonlinear contractions and dissipative sets*. Israel J. Math. **21**, 261–278.

Crandall, M. and Liggett, T. (1971): *Generation of semigroups of nonlinear transformations on general Banach spaces*. Amer. J. Math. **93**, 265–298.

Crandall, M. and Evans, L. (1975): *On the relation of the operator* $\partial/\partial s + \partial/\partial t$ *to evolution governed by accretive operators*. Israel J. Math. **21**, 261–278.

Crandall, M. [ed.] (1978): *Nonlinear Evolution Equations*. Academic Press, New York.

Crandall, M. and Lions, P. (1983): *Viscosity solutions of Hamilton–Jacobi equations*. Trans. Amer. Math. Soc. **277**, 1–42.

Crandall, M. and Lions, P. (1985): *Hamilton–Jacobi equations in infinite dimensions*, I–III. J. Funct. Anal. **63**, 379–396; **65**, 368–405.

Crandall, M. (1986): *Nonlinear semigroups and evolution governed by accretive operators*. In: Browder, F. [ed.] (1986), Part 1, pp. 305–338.

Crandall, M. and Sougandis, P. (1986): *Convergence of difference approximations of quasilinear evolution equations*. Nonlinear Anal. **10**, 425–445.

Crandall, M., Rabinowitz, P., and Turner, R. [eds.] (1987): *Directions in Partial Differential Equations*. Academic Press, New York.

Cronin, J. (1973): *Equations with bounded nonlinearities*. J. Differential Equations **14**, 581–596.

Crouch, S. and Storfield, A. (1983): *Boundary Element Methods in Solid Mechanics*. Allen, London.

Cryer, C. (1982): *Numerical Functional Analysis*. Clarendon, Oxford.

Dacarogna, B. (1982): *Weak Continuity and Weak Lower Semicontinuity of Nonlinear Functionals*. Lecture Notes in Mathematics, Vol. 922. Springer-Verlag, Berlin.

Dafermos, C. and Hrusa, W. (1985): *Energy methods for quasilinear hyperbolic initial-value problems: Applications to elastodynamics*. Arch. Rational Mech. Anal. **87**, 267–292.

Dafermos, C. (1986): *Quasilinear systems with involutions*. Arch. Rational Mech. Anal. **94**, 373–389.

Dancer, E. and Hess, P. (1988): *On stable solutions of quasilinear periodic–parabolic problems* (to appear).

Dautray, D. and Lions, J. (1984): *Analyse mathématique et calcul numérique pour les sciences et les techniques*, Vols. 1–3. Masson, Paris.

Davies, E. (1980): *One-Parameter Semigroups*. Academic Press, New York.

Davis, P. and Hersh, R. (1981): *The Mathematical Experience*. Birkhäuser, Boston, MA.

Davis, P. and Hersh, R. (1987): *Descartes' Dream: The World According to Mathematics*. Houghton Miflin Company, Boston, MA.

Davydov, A. (1984): *Solitons in Molecular Systems*. Reidel, Boston, MA.

Debrunner, H. and Flor, P. (1964): *Ein Erweiterungssatz für monotone Mengen*. Arch. Math. **15**, 445–447.

De Figueiredo, D. (1971): *An existence theorem for pseudomonotone operator equations*. J. Math. Anal. Appl. **34**, 151–156.

De Figueiredo, D. and Karlovitz, L. (1972): *The extension of contractions and the intersection of balls in Banach spaces*. J. Funct. Anal. **11**, 168–178.

De Figueiredo, D., Lions, P., and Nussbaum, R. (1982): *A priori estimates and existence of positive solutions of semilinear elliptic equations*. J. Math. Pures Appl. **61**, 41–63.

De Figueiredo, D. and Solimini, S. (1984): *A variational approach to superlinear elliptic problems*. Comm. Partial Differential Equations **9**, 699–717.

De Figueiredo, D. (1988): *Lectures on the Ekeland Variational Principle with Applications and Detours*. International Centre for Theoretical Physics, Trieste, Italy.

De Giorgi, E. (1957): *Sulla differenziabilità e l'analiticità delle estremali degli integrali multipli regolari*. Mem. Accad. Sci. Torino **3**, 25–43.

De Giorgi, E. (1968): *Un esempio di estremali discontinue per un problema variazionale di tipo ellittico*. Boll. Un. Mat. Ital. (4) **1**, 135–137.

De Giorgi, E. (1983): *G-operators and Γ-convergence*. In: Warsaw (1983), Vol. 2, pp. 1175–1191.

Deimling, K. (1985): *Nonlinear Functional Analysis.* Springer-Verlag, New York.

Dekker, K. and Verwer, J. (1984): *Stability of Runge–Kutta Methods for Stiff Nonlinear Differential Equations.* North-Holland, Amsterdam.

Delves, L. and Walsh, J. (1974): *Numerical Solution of Integral Equations.* Clarendon, Oxford.

Deuel, J. and Hess, P. (1975): *A criterion for the existence of solutions of non-linear elliptic boundary value problems.* Proc. Roy. Soc. Edinburgh Sect. A **74**, 49–54.

Deuel, J. (1978): *Non-linear parabolic boundary-value problems with sub- and super-solutions.* Israel J. Math. **29**, 92–104.

Dieudonné, J. (1968): *Foundations of Modern Analysis*, Vols. 1–9. Academic Press, New York.

Dieudonné, J. (1978): *Abrégé d'histoire des mathématiques, 1700–1900*, Vols. 1, 2. Hermann, Paris.

Dieudonné, J. (1981): *History of Functional Analysis.* North-Holland, Amsterdam.

Dinculeanu, N. (1966): *Vector Measures.* Akademie-Verlag, Berlin.

DiPerna, R. (1983): *Convergence of the viscosity method for isentropic gas dynamics.* Comm. Math. Phys. **91**, 1–30.

Ditzian, Z. and Totik, V. (1987): *Moduli of Smoothness.* Springer-Verlag, New York.

Dobrowolski, M. (1980): *L_∞-convergence of linear finite element approximations to nonlinear parabolic equations.* SIAM J. Numer. Anal. **17**, 663–674.

Dobrowolski, M. and Rannacher, R. (1980): *Finite element methods for nonlinear elliptic systems of second order.* Math Nachr. **94**, 155–172.

Dodd, R. *et al.* (1982): *Solitons and Nonlinear Waves.* Academic Press, New York.

Dolezal, V. (1979): *Monotone Operators and Applications in Control and Network Theory.* Elsevier, New York.

Douglas, J. and Dupont, T. (1970): *Galerkin methods for parabolic equations.* SIAM J. Numer. Anal. **7**, 575–626.

Dubinskii, Ju. (1968): *Quasilinear elliptic and parabolic equations of arbitrary order.* Uspekhi Mat. Nauk **23** (139), 45–90 (Russian).

Dubinskii, Ju. (1976): *Nonlinear elliptic and parabolic equations.* Itogi Nauki i Tekhniki, Sovremennye problemy matematikii **9**, 1–130 (Russian).

Dubinskii, Ju. (1984): *Sobolev Spaces of Infinite Order and Differential Equations.* Teubner, Leipzig.

Dubrovin, B., Matveev, V., and Novikov, S. (1976): *On nonlinear equations of the Korteweg–de Vries type.* Uspekhi Mat. Nauk **31** (1), 55–136 (Russian).

Duff, I., Erisman, A., and Reid, J. (1986): *Direct Methods for Sparse Matrices.* Clarendon, Oxford.

Dunford, N. and Schwartz, J. (1958): *Linear Operators*, Vols. 1–3. Interscience, New York.

Eardley, D. and Moncrieff, V. (1982): *The global existence of Yang–Mills–Higgs fields in 4-dimensional Minkowski space.* Comm. Math. Phys. **83**, 171–191; 193–212.

Eberlein, W. (1947): *Weak compactness in Banach spaces.* Proc. Nat. Acad. Sci. U.S.A. **33**, 51–53.

Edwards, R. (1965): *Functional Analysis.* Holt, Rinehart, and Winston, New York.

Ehrenpreis, L. (1954): *Solutions of some problems of division.* Amer. J. Math. **76**, 883–903.

Ehrling, G. (1954): *On a type of eigenvalue problems for certain differential operators.* Math. Scand. **2**, 267–287.

Eichhorn, J. (1988): *Elliptic operators on non-compact manifolds.* In: Schulze, B. and Triebel, H. [eds.], *Seminar Analysis* 1986/87, pp. 4–170. Teubner, Leipzig, 1988.

Ekeland, I. and Temam, R. (1974): *Analyse convexe et problèmes variationnels.* Dunod, Paris. (English edition: North-Holland, New York, 1976.)

Encyklopädie der Mathematischen Wissenschaften (1899ff): Edited by A. Burkhardt, W. Wirtinger, and R. Fricke. Teubner, Leipzig.

Ericksen, J. *et al.* [eds.] (1986): *Homogenization and Effective Moduli of Materials.* Springer-Verlag, New York.

Ericksen, J. and Kinderlehrer, D. [eds.] (1987): *Theory and Applications of Liquid Crystals.* Springer-Verlag, New York.

Escobar, J. (1987): *Positive solutions for some semilinear elliptic equations with critical Sobolev exponents.* Comm. Pure Appl. Math. **40**, 623–657.

Euler, L. (1744): *Methodus inveniendi lineas curvas maximi minimive proprietate gaudentes, sive solution problematis isoperimetrici latissimo sensu accepti.* Lausanne. (German Translation: See Stäckel, P. (1894)).

Euler, L. (1911): *Opera omnia (Collected Works)*, Vols. 1–72. Leipzig–Berlin; later Basel–Zürich.

Evans, L. (1982): *Classical solutions of fully nonlinear, convex, second-order elliptic equations.* Comm. Pure Appl. Math. **35**, 333–363.

Evans. L. (1983): *Classical solutions of the Hamilton–Jacobi–Bellman equations for uniformly elliptic operators.* Trans. Amer. Math. Soc. **121**, 211–232.

Faddeev, L. and Takhtadjan, L. (1987): *The Hamiltonian Approach to the Theory of Solitons.* Springer-Verlag, New York.

Fattorini, H. (1983): *The Cauchy Problem.* Addison-Wesley, London.

Federer, H. (1969): *Geometric Measure Theory.* Springer-Verlag, New York.

Feistauer, M., Mandel, J., and Nečas, J. (1985): *Entropy regularization of the transonic flow problem.* Comment Math. Univ. Carolin. **25** (3), 431–443.

Feistauer, M. (1987): *On the finite element approximation of a cascade flow problem.* Numer. Math. **50**, 655–684.

Feistauer, M. and Ženišek, A. (1988): *Compactness method in the finite element theory of nonlinear elliptic problems.* Numer. Math. **52**, 147–163.

Feistauer, M. and Nečas, J. (1988): *Viscosity method in transonic flow.* Comm. Partial Differential Equations (to appear).

Feng Kang (1983): *Finite element method and natural boundary reduction.* In: Warsaw (1983), Vol. 2, pp. 1439–1453.

Fenyö, S. and Stolle, H. (1982): *Theorie und Praxis der linearen Integralgleichungen*, Vols. 1–4. Verl. der Wiss., Berlin.

Fichera, G. (1965): *Linear Elliptic Differential Systems and Eigenvalue Problems.* Lecture Notes in Mathematics, Vol. 8. Springer-Verlag, Berlin.

Fichera, G. (1978): *Numerical and Quantitative Analysis.* Pitman, London.

Fichtenholz, G. (1972): *Differential- und Integralrechnung*, Vols. 1–3. Verl. der Wiss., Berlin. (Modified English edition: Cf. Fikhtengol'ts, G. (1965).)

Fife, P. (1979): *Mathematical Aspects of Reacting and Diffusing Systems.* Lecture Notes in Biomathematics, Vol. 28. Springer-Verlag, Berlin.

Fikhtengol'ts, G. (1965): *The Fundamentals of Mathematical Analysis*, Vols. 1, 2. Pergamon, Oxford.

Finn, R. (1985): *Equilibrium Capillary Surfaces.* Springer-Verlag, New York.

Fischer, E. (1905): *Über quadratische Formen mit reellen Koeffizienten.* Monatsh. Math. Phys. **16**, 234–249.

Fischer, A. and Marsden, J. (1972): *The Einstein evolution equations as a first-order quasilinear symmetric hyperbolic system.* Comm. Math. Phys. **28**, 1–38.

Fitzpatrick, P., Hess, P., and Kato, T. (1972): *Local boundedness of monotone-type operators.* Proc. Japan Acad. **48**, 275–277.

Fitzpatrick, P. (1973): *Surjectivity results from a Banach space to its dual.* Math. Ann. **204**, 177–188.

Fitzpatrick, P. and Petryshyn, W. (1975): *Fixed-point theorems and the fixed-point index for multivalued mappings in cones.* J. London Math. Soc. **12**, 75–85.

Fitzpatrick, P. (1977): *Continuity of nonlinear monotone operators.* Proc. Amer. Math. Soc. **62**, 111–116.

Fitzpatrick, P. (1978): *Existence results for equations involving noncompact perturbations of Fredholm mappings.* J. Math. Anal. Appl. **66**, 151–177.

Fitzpatrick, P. and Petryshyn, W. (1978): *On the nonlinear eigenvalue problem $Tu = \lambda Cu$ involving abstract and differential operators.* Boll. Un. Mat. Ital. **15B**, 80–107.

Fitzpatrick, P. and Petryshyn, W. (1979): *Some applications of A-proper mappings.* Nonlinear Anal. **3**, 525–532.

Fitzpatrick, P. and Petryshyn, W. (1982): *On connected components of solutions of nonlinear eigenvalue problems.* Houston J. Math. **8**, 477–500.

Fleming, W. (1986): *Controlled Markov Processes and Viscosity Solutions of Nonlinear Evolution Equations.* Lezione Fermiane, Pisa.

Fletcher, C. (1984): *Computational Galerkin Methods.* Springer-Verlag, New York.

Florian, H., Perktold, K., and Gruber, K. (1984): *Simulation der Partikelbahnen im pulsierenden Blutstrom einer Carotis-Verzweigung und finite Elemente.* Sitzungsber. Österr. Akad. Wiss. Math.-Naturwiss. Kl., Vol. 193, pp. 105–113.

Forsythe, G. and Wasow, W. (1960): *Finite Difference Methods for Partial Differential Equations.* Wiley, New York.

Fortin, M. and Glowinski, R. [eds.] (1983): *Augmented Lagrangian Methods: Application to the Numerical Solution of Boundary-Value Problems.* North-Holland, Amsterdam.

Fourier, J. (1822): *La théorie analytique de la chaleur.* Didot, Paris.

Frank, P. and von Mises, R. (1961): *Die Differential- und Integralgleichungen der Mechanik und Physik.* Vieweg, Braunschweig.

Fredholm, I. (1900): *Sur une nouvelle méthode pour la résolution du problème de Dirichlet.* Kong. Vetenskaps-Akademiens Förh. Stockholm, pp. 39–46.

Fredholm, I. (1903): *Sur une class d'équations fonctionelles.* Acta Math. **27**, 365–390.

Freed, D. and Uhlenbeck, D. (1984): *Instantons and Four-Manifolds.* Springer-Verlag, New York.

Frehse, J. (1977): *Optimale gleichmäßige Konvergenz der Methode der finiten Elemente bei quasilinearen n-dimensionalen Randwertproblemen.* Z. Angew. Math. Mech. **57**, 229–231.

Frehse, J. (1977a): *On Signorini's problem and variational problems with thin obstacles.* Ann. Scuola Norm. Sup. Pisa Cl. Sci., Ser. IV, **4**, 343–362.

Frehse, J. and Rannacher, R. (1978): *Asymptotic L_∞-error estimates for linear finite element approximations of quasilinear boundary-value problems.* SIAM J. Numer. Anal. **15**, 418–431.

Frehse, J. (1982): *Zur asymptotischen Konvergenz der Methode der finiten Elemente.* Mitt. Math. Ges. DDR, 1982, Heft 3/4, pp. 8–30.

Frehse, J. (1982a): *Capacity methods in the theory of partial differential equations.* Jahresber. Deutsch. Math.-Verein. **84**, 1–44.

Frehse, J. (1983): *Uniform asymptotic error estimates for finite element approximations.* In: Proceedings of the China–France Symposium on Finite Elements. Science Press, Beijing, China, pp. 253–264.

Frehse, J. (1983a): *On the smoothness of solutions of variational inequalities with obstacles.* Banach Center Publications, Vol. 10, PWN, Warsaw, pp. 87–128.

Friedlander, F. (1976): *The Wave Equation on a Curved Space–Time.* University Press, Cambridge, England.

Friedman, A. (1964): *Partial Differential Equations of Parabolic Type.* Prentice-Hall, Englewood Cliffs, NJ.

Friedman, A. (1969): *Partial Differential Equations.* Holt, Rinehart, and Winston, New York. (Second edition 1976.)

Friedman, A. (1982): *Variational Principles and Free-Boundary-Value Problems.* Wiley, New York.

Friedrichs, K. (1929): *Ein Verfahren der Variationsrechnung das Minimum eines Integrals als das Maximum eines anderen Ausdrucks darzustellen.* Nachr. Ges. Wiss. Göttingen, Math.-Phys. Kl., pp. 13–20.

Friedrichs, K. (1934): *Spektraltheorie halbbeschränkter Operatoren,* I. II. Math. Ann. **109**, 465–487; 685–713.

Friedrichs, K. (1939): *On differential operators in Hilbert spaces.* Amer. J. Math. **61**, 523–544.

Friedrichs, K. (1944): *The identity of weak and strong extensions of differential operators.* Trans. Amer. Math. Soc. **55**, 132–153.

Friedrichs, K. (1953): *On the differentiability of solutions of linear elliptic differential equations.* Comm. Pure Appl. Math. **6**, 299–326.

Friedrichs, K. (1954): *Symmetric hyperbolic linear differential equations.* Comm. Pure Appl. Math. **7**, 345–392.

Friedrichs, K. (1980): *Von Neumann's Hilbert space theory and partial differential equations.* SIAM Rev. **22**, 486–493.

Friedrichs, K. (1986): *Selecta,* Vols. 1, 2 (*Selected Works*). Birkhäuser, Boston.

Fučik, S., Nečas, J., Souček, J. & V. (1973): *Spectral Analysis of Nonlinear Operators.* Lecture Notes in Mathematics, Vol. 346. Springer-Verlag, Berlin.

Fučik, S., Kratochvil, A., and Nečas, J. (1973): *The Kačanov–Galerkin method.* Comment. Math. Univ. Carolin. **14**, 651–659.

Fučik, S. (1977): *Ranges of Nonlinear Operators.* Lecture Notes, Prague University. See Fučik, S. (1981).

Fučik, S. and Kufner, A. (1980): *Nonlinear Differential Equations.* Elsevier, New York.

Fučik, S. (1981): *Solvability of Nonlinear Equations and Boundary Value Problems.* Reidel, Boston.

Fujita, H. and Kato, T. (1964): *On the Navier–Stokes initial-value problem.* Arch. Rational Mech. Anal. **16**, 269–315.

Funk, P. (1962): *Variationsrechnung und ihre Anwendung in Physik und Technik.* Springer-Verlag, Berlin.

Gagliardo, E. (1959): *Ulteriori proprietà di alcune classi di funzioni in piu variabli.* Ricerche Mat. **8**, 102–137.

Gaines, R. and Mawhin, J. (1977): *Coincidence Degree and Nonlinear Differential Equations.* Lecture Notes in Mathematics, Vol. 568. Springer-Verlag, Berlin.

Gajewski, H. (1970): *Iterations–Projektions- und Projektions–Iterationsverfahren zur Berechnung visco-plastischer Strömungen.* Z. Angew. Math. Mech. **50**, 485–490.

Gajewski, H. (1970a): *Über einige Fehlerabschätzungen bei Gleichungen mit monotonen Potentialoperatoren in Banachräumen.* Monatsber. Dt. Akad. Wiss. Berlin **12**, 571–579.

Gajewski, H. and Kluge, R. (1970): *Projektions–Iterationsverfahren und nichtlineare Probleme mit monotonen Operatoren.* Monatsber. Dt. Akad. Wiss. Berlin **12**, 90–115.

Gajewski, H. (1972): *Zur funktionalanalytischen Formulierung stationärer Kriechprozesse.* Z. Angew. Math. Mech. **52**, 485–490.

Gajewski, H., Gröger, K., and Zacharias, K. (1974): *Nichtlineare Operatorgleichungen und Operatordifferentialgleichungen.* Akademie-Verlag, Berlin.

Gajewski, H. and Gröger, K. (1986): *On the basic equations for carrier transport in semiconductors.* J. Math. Anal. Appl. **113**, 12–35.

Galerkin, B. (1915): *Rods and plates.* Vestnik Inženerov **19** (Russian).

Garabedian, P. (1964): *Partial Differential Equations.* Wiley, New York.

Garabedian, P. (1972/77), (1978), (1984) (1986): Cf. Bauer, F. *et al.* (1972/77), (1978), (1984), (1986).

Gårding, L. (1953): *Dirichlet's problem for linear elliptic differential equations.* Math. Scand. **1**, 55–72.

Gardner, C., Green, J., Kruskal, M., and Miura, R. (1967): *Method for solving the Korteweg–de Vries equation.* Phys. Lett. **19**, 1095–1097.

Gauss, C. (1839): *Allgemeine Lehrsätze über die im verkehrten Verhältnisse des Quadrats der Entfernung wirkenden Anziehungs- und Abstoßungskräfte.* In: Gauss, C. (1863), Vol. 5, pp. 195–242.

Gauss, C. (1863): *Werke (Collected Works)*, Vols. 1–12. Göttingen, 1863–1929.

Gear, C. (1971): *Numerical Initial-Value Problems in Ordinary Differential Equations.* Prentice-Hall, Englewood Cliffs, NJ.

Gekeler, E. (1984): *Discretization Methods for Stable Initial-Value Problems.* Lecture Notes in Mathematics, Vol. 1044. Springer-Verlag, Berlin.

Gelfand, I. and Šilov, G. (1964): *Generalized Functions*, Vols. 1–5. Academic Press, New York.

Germain, P. and Nayroles, B. [eds.] (1976): *Applications of Methods of Functional Analysis to Problems in Mechanics.* Lecture Notes in Mathematics, Vol. 503. Springer-Verlag, Berlin.

Giaquinta, M. (1983): *Multiple Integrals in the Calculus of Variations and Nonlinear Systems.* University Press, Princeton, NJ.

Giaquinta, M. and Hildebrandt, S. (1989): *The Calculus of Variations*, Vols. 1–3 (monograph to appear).

Gidas, B., Ni, W., and Nirenberg, L. (1979): *Symmetry and related properties via the maximum principle.* Comm. Math. Phys. **68**, 209–243.

Gilbarg, D. and Trudinger, N. (1983): *Elliptic Partial Differential Equations of Second Order.* Springer-Verlag, New York (second enlarged edition).

Ginibre, J. and Velo, G. (1983): *On the global Cauchy problem for some nonlinear Schrödinger equations.* Ann. Inst. H. Poincaré. Anal. Non Linéaire **1**, 309–324.

Ginibre, J. and Velo, G. (1986): *The Cauchy problem in the energy space for the nonlinear Schrödinger and Klein–Gordon equations.* In: Brézis, H., Crandall, M., and Kappel, F. [eds.] (1986), Vol. 1, pp. 121–133.

Girault, V. and Raviart, P. (1986): *Finite Element Approximation of the Navier–Stokes Equations.* Springer-Verlag, New York.

Gittel, H. (1987): *Studies on transonic flow problems by nonlinear variational inequalities.* Z. Anal. Anwendungen **6**, 449–458.

Giusti, E. and Miranda, C. (1968): *Sulla regularità delle soluzioni deboli di una classe di sistemi ellittici quasilineari.* Arch. Rational Mech. Anal. **31**, 173–184.

Giusti, E. (1984): *Minimal Surfaces and Functions of Bounded Variation.* Birkhäuser, Boston.

Giusti, E. [ed.] (1985): *Harmonic Mappings and Minimal Immersions.* Lecture Notes in Mathematics, Vol. 1161, Springer-Verlag, Berlin.

Glashoff, K. and Werner, B. (1979): *Inverse monotonicity of monotone L-operators with applications to quasilinear and free boundary value problems.* J. Math. Anal. Appl. **72**, 89–105.

Glassey, R. and Strauss, W. (1987): *Absence of shocks in an initially dilute collisionless plasma.* Comm. Math. Phys. **113**, 191–208.

Glimm, J. (1965): *Solutions in the large for nonlinear hyperbolic systems of equations.* Comm. Pure Appl. Math. **18**, 697–715.

Glimm, J. and Jaffe, A. (1981): *Quantum Physics.* Springer-Verlag, New York.

Glimm, J. (1988): *The interaction of nonlinear hyperbolic waves: a survey.* Comm. Pure Appl. Math. **41**, 569–589.

Glowinski, R. and Lions, J. [eds.] (1974): *Computing Methods in Applied Sciences and Engineering,* I–VII. Springer-Verlag, New York, 1974, 1976, 1977, 1979; North-Holland, Amsterdam, 1980, 1986.

Glowinski, R., Lions, J., and Trémolières, R. (1976): *Analyse numérique des inéquations variationelles,* Vols. 1, 2. Gauthier-Villars, Paris. (English edition: North-Holland, Amsterdam, 1981.)

Glowinski, R. (1984): *Numerical Methods for Nonlinear Variational Problems.* Springer-Verlag, New York.

Godunov, S. and Ryabenkii, V. (1987): *Difference Schemes.* North-Holland, Amsterdam.

Golomb, M. (1935): *Zur Theorie der nichtlinearen Integralgleichungen.* Math. Z. **39**, 45–75.

Gould, S. (1966): *Variational Methods for Eigenvalue Problems.* University Press, Toronto.

Greenberg, W. *et al.* (1986): *Boundary Value Problems in Abstract Kinetic Theory.* Birkhäuser, Boston.

Grigorieff, R. (1973): *Zur Theorie approximations–regulärer Operatoren,* I, II. Math. Nachr. **55**, 233–249; 251–263.

Grillakis, M. (1988): *Linearized instability for nonlinear Schrödinger and Klein–Gordon equations.* Comm. Pure Appl. Math. **41**, 747–774.

Grisvard, P. (1985): *Elliptic Problems in Nonsmooth Domains.* Pitman, London.

Gröger, K. (1986): *On the existence of steady states of certain reaction–diffusion systems.* Arch. Rational Mech. Anal. **92**, 297–306.

Gröger, K. (1989): *W_p^1-Estimates for solutions to mixed boundary problems for second-order elliptic differential equations.* Math. Ann. (to appear).

Günter, N. (1957): *Die Potentialtheorie und ihre Anwendung auf Grundaufgaben der mathematischen Physik.* Teubner, Leipzig.

Günther, P. *et al.* (1972): *Grundkurs Analysis,* Vols. 1–4. Teubner, Leipzig.

Günther, P. (1988): *Huygens' Principle for Linear Partial Differential Operators of Second Order.* Academic Press, New York.

Günther, M. (1987): *On nonlinear interpolation theory and global solutions of the Korteweg–de Vries equation* (unpublished paper).

Günther, M. (1988): *Beiträge zur lokalen Lösbarkeit partieller Differentialgleichungen auf Mannigfaltigkeiten.* Dissertation B, Karl-Marx-Universität, Leipzig.

Günther, M. (1988a): *Über eine spezielle Klasse nichtlinearer, lokal stetig invertierbarer Operatoren.* Math Nachr. **136**, 167–184.

Gupta, C. (1983): *Some theorems on the sum of nonlinear mappings of monotone type in Banach spaces.* J. Math. Anal. Appl. **97**, 38–45.

Guseinov, A. and Myhtarov, H. (1980): *Introduction to the Theory of Nonlinear Integral Equations.* Nauka, Moscow (Russian).

Gustafson, K. and Sather, D. (1972): *Large nonlinearities and monotonicity.* Arch. Rational Mech. Anal. **48**, 109–122.

Hackbusch, W. (1979): *On the fast solution of nonlinear elliptic equations.* Numer. Math. **32**, 83–95.

Hackbusch, W. and Trottenberg, U. [eds.] (1982): *Multigrid Methods.* Lecture Notes in Mathematics, Vol. 960. Springer-Verlag, Berlin.

Hackbusch, W. (1985): *Multigrid Methods and Applications.* Springer-Verlag, New York.

Hackbusch, W. (1986): *Theorie und Numerik elliptischer Differentialgleichungen.* Teubner, Stuttgart.

Hackbusch, W. and Trottenberg, U. [eds.] (1986): *Multigrid Methods II.* Lecture Notes in Mathematics, Vol. 1228. Springer-Verlag, Berlin.

Hageman, L. and Young, D. (1981): *Applied Iterative Methods.* Academic Press, New York.

Hahn, H. (1927): *Über lineare Gleichungssysteme in linearen Räumen.* J. Reine Angew. Math. **157**, 214–229.

Hale, J. (1988): *Asymptotic Behavior of Dissipative Systems.* Amer. Math. Soc., Providence, RI.

Halmos, P. (1954): *Measure Theory.* Van Nostrand, Toronto.

Hamilton, R. (1982): *The inverse function theorem of Nash and Moser.* Bull. Amer. Math. Soc. (N.S.) **7**, 65–222.

Hamilton, R. (1982a): *Three-manifolds with positive Ricci curvature.* J. Differential Geometry **17**, 255–306.

Hammerstein, A. (1930): *Nichtlineare Integralgleichungen nebst Anwendungen.* Acta Math. **54**, 117–176.

Haraux, A. (1981): *Nonlinear Evolution Equations: Global Behavior of Solutions.* Lecture Notes in Mathematics, Vol. 841. Springer-Verlag, Berlin.

Hartman, P. and Stampacchia, G. (1966): *On some nonlinear elliptic differential equations.* Acta Math. **115**, 271–310.

Hawking, S. and Ellis, G. (1973): *The Large Scale Structure of Space–Time.* University Press, Cambridge, England.

Hawkins, T. (1970): *History of Lebesgue's Theory of Integration.* Chelsea, New York.

Heinrich, B. (1987): *Finite Difference Methods in Irregular Networks.* Akademie-Verlag, Berlin.

Heinz, E. (1986): *Über die Regularität schwacher Lösungen nichtlinearer elliptischer Systeme.* Nachr. Akad. Wiss. Göttingen, Math.-Phys. Kl., pp. 1–15.

Helly, E. (1921): *Über Systeme linearer Gleichungen mit unendlich vielen Unbekannten.* Monatsh. Math. Phys. **31**, 60–91.

Henry, D. (1981): *Geometric Theory of Semilinear Parabolic Equations.* Lecture Notes in Mathematics, Vol. 840. Springer-Verlag, Berlin.

Hess, P. (1971): *On nonlinear equations of Hammerstein type in Banach spaces.* Proc. Amer. Math. Soc. **30**, 308–312.

Hess, P. (1972): *On the Fredholm alternative for nonlinear functional equations in Banach spaces.* Proc. Amer. Math. Soc. **33**, 55–61.

Hess, P. (1973): *On nonlinear problems of monotone type with respect to two Banach spaces.* J. Math. Pures Appl. **52**, 13–26.

Hess, P. (1974): *On a theorem by Landesman and Lazer.* Indiana Univ. Math. J. **23**, 827–829.

Hess, P. (1974a): *On semicoercive nonlinear problems.* Indiana Univ. Math. J. **23**, 645–654.

Hess, P. (1978): *Nonlinear elliptic problems in unbounded domains.* In: Kluge, R. [ed.] (1978), *Theory of Nonlinear Operators, Constructive Aspects*, pp. 105–110. Akademie-Verlag, Berlin.

Hess, P. (1981): *On multiple solutions of nonlinear elliptic eigenvalue problems.* Comm. Partial Differential Equations **6**, 951–961.

Hestenes, M. (1951): *Applications of the theory of quadratic forms in Hilbert space to the calculus of variations.* Pacific J. Math. **1**, 525–581.

Hewitt, E. and Stromberg, K. (1965): *Real and Abstract Analysis.* Springer-Verlag, Berlin.

Hilbert, D. (1897): *Die Theorie der algebraischen Zahlkörper.* In: Hilbert, D. (1932), Vol. 1, pp. 63–482.

Hilbert, D. (1900): *Mathematische Probleme* (Paris Lecture). German version: Cf. Aleksandrov, P. [ed.] (1971), pp. 22–80. (English translation: Cf. Hilbert, D. (1902).)

Hilbert, D. (1900a): *Über das Dirichletsche Prinzip.* Jahresber. Deutsch. Math.-Verein. **8**, 184–188.

Hilbert, D. (1901): *Lecture on the Dirichlet principle.* Published in Hilbert, D. (1904).

Hilbert, D. (1902). *Mathematical Problems.* Lecture delivered before the First International Congress of Mathematicians at Paris 1900. Bull. Amer. Math. Soc. **8**, 437–479. (Reprinted in Browder, F. [ed.] (1976), Vol. 1, pp. 1–34.)

Hilbert, D. (1904): *Über das Dirichletsche Prinzip.* Math. Ann. **59**, 161–186.

Hilbert, D. (1904a): *Grundzüge einer allgemeinen Theorie der linearen Integralgleichungen. Erste Mitteilung (first note).* Göttinger Nachr., 1904, pp. 49–91. (Reprinted in Hilbert, D. (1912).)

Hilbert, D. (1906): *Zur Variationsrechnung.* Math. Ann. **62**, 351–370.

Hilbert, D. (1906a): Cf. Hilbert, D. (1912), *Vierte und fünfte Mitteilung (fourth and fifth note).*

Hilbert, D. (1910): Cf. Hilbert, D. (1912), *Sechste Mitteilung (sixth note).*

Hilbert, D. (1912): *Grundzüge einer allgemeinen Theorie der linearen Integralgleichungen.* Teubner, Leipzig.

Hilbert, D. (1932): *Gesammelte Abhandlungen (Collected Works)*, Vols. 1–3. Springer-Verlag, Berlin. (Second edition 1970.)

Hildebrandt, S. (1969): *Boundary behavior of minimal surfaces.* Arch. Rational Mech. Anal. **35**, 47–82.

Hildebrandt, S. and Widman, K. (1977): *On the Hölder continuity of weak solutions of quasilinear elliptic systems of second order.* Ann. Scoula Norm. Sup. Pisa Cl. Sci., Ser. IV, **4**, 144–178.

Hildebrandt, S. (1982): *Liouville theorems for harmonic mappings and an approach to the Bernstein theorem.* Annals of Mathematical Studies, Vol. 102, pp. 107–132, Princeton, NJ.

Hildebrandt, S. (1984): *Minimal surfaces with free boundaries and related problems.* Astérisque **118**, 69–88.

Hildebrandt, S. (1984a): *Calculus of variations today, reflected in the Oberwolfach meetings.* In: Jäger, W. *et al.* (1984), pp. 321–336.

Hildebrandt, S. (1985): *Lectures on harmonic mappings of Riemannian manifolds.* In: Giusti, E. [ed.] (1985), pp. 1–117.

Hildebrandt, S. and Tromba, A. (1985): *Mathematics and Optimal Form.* Scientific American Library, Freeman, New York.

Hildebrandt, S. (1986): *Free boundary value problems for minimal surfaces and related questions.* Comm. Pure Appl. Math. **39**, Supplement S111–S138.

Hildebrandt, S. and Giaquinta, M. (1989): *The Calculus of Variations*, Vols. 1–3 (monograph to appear).

Hildebrandt, S. and Leis, R. [eds.] (1989): *Partial Differential Equations and Calculus of Variations.* Springer-Verlag, New York (to appear).

Hille, E. (1948): *Functional Analysis and Semigroups.* Amer. Math. Soc., New York.

Hille, E. and Phillips, R. (1957): *Functional Analysis and Semigroups.* Amer. Math. Soc., Providence, RI.

Hirzebruch, F. and Scharlau, W. (1971): *Einführung in die Funktionalanalysis.* Bibliograph. Inst., Mannheim.

Hirzebruch, F. (1974): Cf. Otte, M. [ed.] (1974).

Hofer, H. (1982): *Variational and topological methods in partially ordered Hilbert spaces.* Math. Ann. **261**, 493–514.

Hofer, H. (1984): *A note on the topological degree at a critical point of mountain-pass type.* Proc. Amer. Math. Soc. **90**, 309–325.

Holmes, R. (1975): *Geometric Functional Analysis and Its Applications.* Springer-Verlag, New York.

Hörmander, L. (1955): *On the theory of general partial differential operators.* Acta Math. **94**, 162–248.

Hörmander, L. (1964): *Linear Partial Differential Operators.* Springer-Verlag, Berlin.

Hörmander, L. (1983): *The Analysis of Linear Partial Differential Operators*, Vols. 1–4. Springer-Verlag, New York.

Horváth, J. (1966): *Topological Vector Spaces and Distributions.* Addison-Wesley, New York.

Hughes, T., Kato, T., and Marsden, J. (1977): *Well-posed quasilinear second-order hyperbolic systems with applications to nonlinear elastodynamics and general relativity.* Arch. Rational Mech. Anal. **63**, 273–294.

Hünlich, R. (1970): *Über einige Näherungsverfahren zur Lösung nichtlinearer Funktionalgleichungen mit Nebenbedingungen.* Math. Nachr. **47**, 171–182.

IMSL (1987): *Software System for General Classes of Mathematical Problems.* Houston, Texas.

Isaacson, E. and Keller, H. (1966): *Analysis of Numerical Methods.* Wiley, New York.

Its, A. and Matveev, V. (1975): *Schrödinger operators with finite-gap spectrum and N-soliton solutions of the Korteweg–de Vries equation.* Theoret. Math. Phys. **23** (1), 51–67 (Russian).

Ivanov, A. (1982): *Quasilinear degenerate and non-uniformly elliptic and parabolic second-order differential equations.* Trudy Mat. Inst. Steklov., Leningrad, Vol. 160 (Russian).

Jackson, D. (1912): *On approximation by trigonometric sums and polynomials.* Trans. Amer. Math. Soc. **13**, 491–515.

Jacobi, C. (1837): Cf. Stäckel, P. [ed.] (1894).

Jäger, W., Moser, J., and Remmert, R. [eds.] (1984): *Perspectives in Mathematics.* Birkhäuser, Basel.

Jameson, A. (1988): *Computational Transonics.* Comm. Pure Appl. Math. **41**, 507–550.

Janenko, N. (1969): *Die Zwischenschrittmethode zur Lösung mehrdimensionaler Probleme der mathematischen Physik.* Lecture Notes in Mathematics, Vol. 91. Springer-Verlag, Berlin.

Jeffrey, A. and Taniuti, T. (1964): *Nonlinear Wave Propagation with Applications to Physics and Magnetohydrodynamics.* Academic Press, New York.

Jeffrey, A. (1976): *Quasilinear Hyperbolic Systems and Waves.* Pitman, London.

John, F. (1955): *Plane Waves and Spherical Means Applied to Partial Differential Equations.* Interscience, New York.

John, F. (1974): *Formation of singularities in one-dimensional nonlinear wave propagation.* Comm. Pure Appl. Math. **27**, 377–405.

John, F. (1976): *Delayed singularity formation in solutions of nonlinear wave equations in higher dimensions.* Comm. Pure Appl. Math. **29**, 649–681.

John, F. (1982): *Partial Differential Equations.* Springer-Verlag, New York.

John, F. (1983): *Lower bounds for the life span of solutions of nonlinear wave equations in three dimensions.* Comm. Pure Appl. Math. **36**, 1–35.

John, F. (1984): *A walk through partial differential equations.* In: Chern, S. [ed.] (1984), pp. 73–84.

John, F. and Klainerman, S. (1984): *Almost global existence to nonlinear wave equations.* Comm. Pure Appl. Math. **37**, 443–455.

John, F. (1985): *Collected Papers*, Vols. 1, 2. Birkhäuser, Boston.

John, F. (1986): *Long time effects of nonlinearity in second-order hyperbolic equations.* Comm. Pure Appl. Math. **39**, Supplement S139–S148.

John, F. (1987): *Existence for large times of strict solutions of nonlinear wave equations in three dimensions for small initial data.* Comm. Pure Appl. Math. **40**, 79–109.

John, F. (1988): *Almost global existence of elastic waves of finite amplitude arising from small initial disturbances.* Comm. Pure Appl. Math. **41**, 615–666.

Jost, J. (1984): *Harmonic Maps between Surfaces.* Lecture Notes in Mathematics, Vol. 1062. Springer-Verlag, Berlin.

Jost, J. (1985): *Lectures on Harmonic Maps with Applications to Conformal Mappings and Minimal Surfaces.* Lecture Notes in Mathematics, Vol. 1161. Springer-Verlag, Berlin.

Jost, J. (1988): *Nonlinear Methods in Riemannian and Kählerian Geometry.* Birkhäuser, Boston, MA.

Jost, J. (1988a): *Das Existenzproblem für Minimalflächen.* Jahresber. Deutsch. Math.-Verein. (to appear).

Kačur, J. (1985): *Method of Rothe in Evolution Equations.* Teubner, Leipzig.

Kačurovskii, R. (1960): *On monotone operators and convex functionals.* Uspekhi Mat. Nauk **15** (4), 213–215 (Russian).

Kačurovskii, R. (1970): *On the Fredholm theory for nonlinear operators.* Dokl. Akad. Nauk SSSR **192**, 969–972 (Russian).

Kadec, M. (1959): *Spaces normisomorphic to a locally uniformly convex space.* Izv. Vysš. Učebn. Zaved. Mat. **6** (13), 51–57 (Russian).

Kamke, E. (1960): *Das Lebesgue–Stieltjes Integral.* Teubner, Leipzig.

Kantorovič, L. (1948): *Functional analysis and applied mathematics.* Uspekhi Mat. Nauk **3** (6), 89–185 (Russian).

Kantorovič, L. and Akilov, G. (1964): *Funktionalanalysis in normierten Räumen.* Akademie-Verlag, Berlin. (English edition: Pergamon Press, Oxford, 1964.)

Kanwal, R. (1983): *Generalized Functions.* Academic Press, New York.

Karpman, V. (1977): *Nichtlineare Wellen.* Akademie-Verlag, Berlin.

Kato, T. (1966): *Perturbation Theory for Linear Operators.* Springer-Verlag, Berlin.

Kato, T. (1975): *Quasilinear equations of evolution with applications to partial differential equations.* Lecture Notes in Mathematics, Vol. 448, pp. 25–70. Springer-Verlag, New York.

Kato, T. (1975a): *The Cauchy problem for quasilinear symmetric hyperbolic systems.* Arch. Rational Mech. Anal. **58**, 181–205.

Kato, T. (1976): *Linear and quasilinear equations of evolution of hyperbolic type.* In: *Hyperbolicity*, Centro Internationale Matematico Estivo II Ciclo 1976, Italy, pp. 125–191.

Kato, T. (1983): *On the Cauchy problem for the generalized Korteweg–de Vries equations.* In: Guillemin, V. [ed.] (1983), *Studies in Applied Mathematics*, Vol. 8, pp .93–128. Academic Press, New York.

Kato, T. (1984): *Locally coercive nonlinear equations with applications to some periodic solutions.* Duke Math. J. **51**, 923–936.

Kato, T. and Lai, C. (1984): *Nonlinear evolution equations and the Euler flow.* J. Funct. Anal. **56**, 15–28.

Kato, T. (1985): *Abstract Differential Equations and Nonlinear Mixed Problems.* Lezione Fermiane, Pisa.

Kato, T. (1986): *Nonlinear equations of evolution in* B-*spaces.* In: Browder, F. [ed.] (1986), Part 2, pp. 9–24.

Kato, T. and Masuda, K. (1986): *Nonlinear evolution equations and analyticity I.* Ann. Inst. H. Poincaré. Anal. Non Linéaire **3**, 455–467.

Kawohl, B. (1985): *Rearrangements and Convexity of Level Sets in Partial Differential Equations.* Lecture Notes in Mathematics, Vol. 1150. Springer-Verlag, Berlin.

Kazdan, J. and Warner, F. (1975): *Remarks on some quasilinear elliptic equations.* Comm. Pure Appl. Math. **28**, 567–597.

Kazdan, J. (1985): *Prescribing the Curvature of a Riemannian Manifold.* Amer. Math. Soc., Providence, RI.

Kemnochi, N. (1974): *Nonlinear operators of monotone type in reflexive Banach spaces and nonlinear perturbations.* Hiroshima Math. J. **4**, 229–263.

Kenderov, P. (1976): *The set-valued monotone mappings are almost everywhere single-valued.* Studia Math. **56**, 199–203.

Kichenassamy, S. (1987): *Quasi-linear problems with singularities.* Manuscripta Math. **57**, 281–313.

Kinderlehrer, D. and Stampacchia, G. (1980): *An Introduction to Variational Inequalities and Their Application.* Academic Press, New York.

Kirszbraun, M. (1934): *Über die kontraktiven und Lipschitz-stetigen Transformationen.* Fund. Math. **22**, 77–108.

Klainerman, S. and Majda, A. (1980): *Formation of singularities for wave equations including the nonlinear vibrating string.* Comm. Pure Appl. Math. **33**, 241–263.

Klainerman, S. and Majda, A. (1981): *Singular limits of quasilinear hyperbolic systems with large parameters and the incompressible limit of compressible fluids.* Comm. Pure Appl. Math. **34**, 481–524.

Klainerman, S. (1983): *Long time behavior of solutions to nonlinear wave equations.* In: Warsaw (1983), Vol. 2, pp. 1209–1215.

Klainerman, S. and Ponce, G. (1983): *Global small amplitude solutions to nonlinear evolution equations.* Comm. Pure Appl. Math. **36**, 133–141.

Klainerman, S. (1985): *Uniform decay estimates and the Lorentz invariance of the classical wave equation.* Comm. Pure Appl. Math. **38**, 321–332.

Klainerman, S. (1985a): *Global existence of small amplitude solutions to nonlinear Klein–Gordon equations in four space–time dimensions.* Comm. Pure Appl. Math. **38**, 631–641.

Klainerman, S. (1987): *Einstein geometry and hyperbolic equations.* In: Crandall, M., Rabinowitz, P., and Turner, R. [eds.] (1987), pp. 113–144.

Klein, F. (1926): *Vorlesungen über die Geschichte der Mathematik des 19. Jahrhunderts.* Springer-Verlag, Berlin.

Klötzler, R. (1971): *Mehrdimensionale Variationsrechnung.* Verl. der Wiss., Berlin.

Kluge, R. (1979): *Nichtlineare Variationsungleichungen und Extremalaufgaben.* Verl. der Wiss., Berlin.

Kluge, R. (1985): *Zur Parameterbestimmung in nichtlinearen Problemen.* Teubner, Leipzig.

Knauer, B. (1971): *Untere Schranken für die Einschließung selbstadjungierter positiv-definiter Operatoren.* Numer. Math. **17**, 166–171.

Kneser, H. (1950): *Eine direkte Ableitung des Zornschen Lemmas aus dem Auswahlaxiom.* Math. Z. **53**, 110–113.

Knörrer, H. (1986): *Integrable Hamiltonsche Systeme und algebraische Geometrie.* Jahresber. Deutsch. Math.-Verein. **88**, 82–103.

Knuth, D. (1968): *The Art of Computer Programming*, Vols. 1–3. Addison-Wesley, New York.

Knuth, D. and Greene, E. (1982): *Mathematics for the Analysis of Algorithms.* Birkhäuser, Boston, MA.

Knuth, D., Graham, R., and Patashnic, O. (1989): *Concrete Mathematics: A Foundation for Computer Science.* Addison-Wesley, New York.

Kohn, R. (1984): *The method of partial regularity as applied to the Navier–Stokes equations.* In: Chern, S. [ed.] (1984), pp. 117–128.

Kolomý, J. (1976): *Some properties of monotone (potential) operators.* Beiträge Anal. **8**, 77–84.

Kolomý, J. (1978): *Some methods for finding eigenvalues and eigenvectors of linear and nonlinear operators.* In: Kluge, R. [ed.], *Theory of Nonlinear Operators*, Akademie-Verlag, Berlin.

Komura, Y. (1967): *Nonlinear semigroups in Hilbert space.* J. Math. Soc. Japan **19**, 493–507.

Kondratjev, V. (1967): *Boundary value problems for elliptic equations on domains with conical or angular points.* Trudy Moskov. Mat. Obščestva **16**, 209–292 (Russian).

Kondratjev, V. and Oleinik, O. (1983): *Boundary value problems for partial differential equations in nonsmooth domains.* Uspekhi Mat. Nauk **38**(2), 3–76 (Russian).

König, H. (1985): *Eigenvalue Distributions of Compact Operators.* Birkhäuser, Boston.

Konopelchenko, B. (1987): *Nonlinear Integrable Equations.* Springer-Verlag, New York.

Koshelev, A. and Chelkak, S. (1985): *Regularity of Solutions of Quasilinear Elliptic Systems*. Teubner, Leipzig.

Koshelev, A. (1986): *Regularity of Solutions of Quasilinear Elliptic Systems*. Nauka, Moscow (Russian).

Köthe, G. (1969): *Topological Vector Spaces*, Vols. 1, 2. Springer-Verlag, New York.

Krasnoselskii, M. (1956): *Topological Methods in the Theory of Nonlinear Integral Equations*. Gostehizdat, Moscow (Russian). (English edition: McMillan, New York, 1964).

Krasnoselskii, M. *et al.* (1966): *Integral Operators in Spaces of Summable Functions*. Nauka, Moscow (Russian). (English edition: Noordhoff, Leyden, 1976.)

Krasnoselskii, M. *et al.* (1973): *Näherungsverfahren zur Lösung von Operatorgleichungen*. Akademie-Verlag, Berlin.

Krasnoselskii, M. and Zabreiko, P. (1975): *Geometrical Methods in Nonlinear Analysis*. Nauka, Moscow (Russian). (English edition: Springer-Verlag, New York, 1984.)

Krauss, E. (1985): *A representation of maximal monotone operators by saddle functions*. Rev. Roumaine Math. Pures Appl. **30**, 823–837.

Krauss, E. (1985a): *On the range of maximal monotone operators in nonreflexive Banach spaces*. Math. Nachr. **120**, 195–201.

Krauss, E. (1985b): *A representation of arbitrary maximal monotone operators via subgradients of skew-symmetric saddle functions*. Nonlinear Anal. **9**, 1381–1399.

Krauss, E. (1986): *Maximal monotone operators and saddle functions*. Z. Anal. Anwendungen **5**, 333–346.

Krauss, E. (1988): *On the maximality of the sum of monotone operators* (to appear).

Krein, S. (1967): *Linear Differential Equations in Banach Spaces*. Nauka, Moscow (Russian).

Krein, S. (1982): *Linear Equations in Banach spaces*. Birkhäuser, Basel.

Krein, S., Petunin, E., and Semenov, E. (1982): *Interpolation of Linear Operators*. Amer. Math. Soc., Providence, RI.

Kružkov, S. and Faminskii, A. (1985): *Generalized solutions of the Cauchy problem for the Korteweg–de Vries equation*. Math. USSR-Sbornik **48**, 391–421.

Krylov, N. (1985): *Nonlinear Elliptic and Parabolic Equations of Second Order*. Nauka, Moscow (Russian). (English edition: Reidel, Boston, MA, 1987.)

Kufner, A., John, O., and Fučik, S. (1977): *Function Spaces*. Academia, Prague.

Kufner, A. (1985): *Weighted Sobolev Spaces*. Wiley, New York.

Kufner, A. and Sändig, A. (1987): *Some Applications of Weighted Sobolev Spaces*. Teubner, Leipzig.

Kurpel, N. (1976): *Projection–Iterative Method for Solution of Operator Equations*. Amer. Math. Soc., Providence, RI.

Kuttler, J. and Sigilito, V. (1987): *Estimating Eigenvalues with A Posteriori/A Priori Inequalities*. Pitman, London.

Ladde, G., Lakshmikantham, V., and Vabala, A. (1985): *Monotone Iterative Techniques for Nonlinear Differential Equations*. Pitman, London.

Ladyženskaja, O. (1953): *On the Mixed Problem for Hyperbolic Differential Equations*. Gostehizdat, Moscow.

Ladyženskaja, O. and Višik, M. (1956): *Boundary-value problems for partial differential equations and a class of operator equations.* Uspekhi Mat. Nauk **11** (6), 41–97 (Russian).

Ladyženskaja, O. and Uralceva, N. (1964): *Linear and Quasilinear Elliptic Equations.* Nauka, Moscow. (English edition: Academic Press, New York, 1968; Second enlarged Russian edition, 1973.)

Ladyženskaja, O., Solonnikov, V., and Uralceva, N. (1967): *Linear and Quasilinear Parabolic Equations.* Nauka, Moscow (Russian).

Ladyženskaja, O. (1973): *The Boundary-Value Problems of Mathematical Physics.* Nauka, Moscow (Russian). (Revised English edition: Springer-Verlag, 1985.)

Ladyženskaja, O. and Uralceva, N. (1973): Second enlarged edition of Ladyženskaja O. and Uralceva, N. (1964).

Ladyženskaja, O. (1985): Revised English edition of Ladyženskaja, O. (1973). Springer-Verlag, New York.

Ladyženskaja, O. and Uralceva, N. (1986): *A survey on the solvability of uniformly elliptic and parabolic quasilinear equations of second order having singularities.* Uspekhi Mat. Nauk **41** (5), 59–83.

Ladyženskaja, O. (1986): *On some directions of the research in mathematical physics at the Steklov Institute in Leningrad.* In: Vladimirov, V. [ed.] (1986), pp. 217–245 (Russian).

Ladyženskaja, O. (1987): *On the construction of minimal global attractors for the Navier–Stokes equations and other partial differential equations.* Uspekhi Mat. Nauk **42** (6), 25–60.

Lagrange, J. (1762): *Essai d'une nouvelle méthode pour déterminer les maxima et minima des formules intégrales indéfinies.* In: Lagrange, J. (1867), Vol. 1, pp. 335–362.

Lagrange, J. (1867): *Oeuvres (Collected Works)*, Vols. 1–14. Paris, 1867–1892.

Landau, L. and Lifšic, E. (1962): *Course of Theoretical Physics*, Vols. 1–10. Pergamon, Oxford. (German edition, Akademie-Verlag, Berlin, 1962.)

Landes, R. (1985): *Quasilinear hyperbolic variational inequalities.* Arch. Rational Mech. Anal. **91**, 267–282.

Landesman, E. and Lazer, A. (1969): *Nonlinear perturbations of linear elliptic boundary value problems at resonance.* J. Math. Mech. **19**, 609–623.

Langenbach, A. (1959): *Variationsmethoden in der nichtlinearen Elastizitäts- und Plastizitätstheorie.* Wiss. Z. Humboldt-Univ. Berlin Math.-Nat. Reihe **9**, 145–164.

Langenbach, A. (1965): *Verallgemeinerte und exakte Lösungen des Problems der elastisch-plastischen Torsion von Stäben.* Math. Nachr. **28**, 219–234.

Langenbach, A. (1966): *Über Gleichungen mit Potentialoperatoren und Minimalfolgen nichtquadratischer Funktionale.* Math. Nachr. **32**, 9–24.

Langenbach, A. (1976): *Monotone Potentialoperatoren in Theorie und Anwendungen.* Verl. der Wiss., Berlin.

Langenbach, A. (1984): *Vorlesungen zur Höheren Analysis.* Verl. der Wiss., Berlin.

Laplace, P. (1782): *Théorie des attractions des sphéroides et de la figure des planètes.* In: Laplace, P. (1878), Vol. 10, pp. 341–419.

Laplace, P. (1878): *Oeuvres (Collected Works)*, Vols. 1–14. Paris, 1878–1912.

Lasota, A. and Yorke, J. (1973): *The generic property of existence of solutions of differential equations on Banach spaces.* J. Differential Equations **13**, 1–12.

Laurent, P. (1972): *Approximation et optimisation.* Hermann, Paris.

Lax, P. (1953): *On the equivalence between stability and convergence of difference methods.* Lecture held in a seminar at New York University. Cf. Richtmyer, R. and Morton, K. (1967), Chapter 3.

Lax, P. and Milgram, N. (1954): *Parabolic Equations.* Annals of Mathematics Studies, Vol. 33, pp. 167–190, University Press, Princeton, NJ.

Lax, P. and Richtmyer, R. (1956): *Survey of the stability of linear finite difference equations.* Comm. Pure Appl. Math. **19**, 437–492.

Lax, P. (1964): *Development of singularities of solutions of nonlinear hyperbolic equations.* J. Math. Phys. **5**, 611–613.

Lax, P. and Wendroff, B. (1964): *Difference schemes for hyperbolic systems with high order accuracy.* Comm. Pure Appl. Math. **17**, 381–395.

Lax, P. and Nirenberg, L. (1966): *On the stability of difference schemes: a sharp form of Gårding's inequality.* Comm. Pure Math. **19**, 473–492.

Lax, P. (1968): *Integrals of nonlinear equations of evolution and solitary waves.* Comm. Pure Appl. Math. **21**, 467–490.

Lax, P. (1973): *Hyperbolic Systems of Conservation Laws and the Mathematical Theory of Shock Waves.* SIAM, Philadelphia, PA.

Lax, P. (1983): *Problems solved and unsolved concerning linear and nonlinear partial differential equations.* In: Warsaw (1983), Vol. 1, pp. 119–138.

Lax, P. (1984): *Shock waves, increase of entropy, and loss of information.* In: Chern, S. [ed.] (1984), pp. 129–171.

Lebeau, G. (1986): *Interaction des singularités pour les équations aux dérivées partielles non linéaires d'après J. Bony.* Astérisque **133/34**, 209–222.

Lebesgue, H. (1902): *Intégrale, longeure, aire.* Annali Mat. Pura Appl. **7**, 231–359.

Lee, T. (1981): *Particle Physics and Introduction to Field Theory.* Harwood, New York.

Lee, J. and Parker, T. (1987): *The Yamabe problem.* Bull. Amer. Math. Soc. (N.S.) **17**, 37–91.

Legendi, T. and Szentivanyi, T. (1983): *Leben und Werk des John von Neumann.* Bibliograph. Inst., Mannheim.

Legendre, A. (1786): Cf. Stäckel, P. [ed.] (1894).

Leis, R. (1986): *Initial-Boundary Value Problems in Mathematical Physics.* Wiley, New York.

Leray, J. (1934): *Essai sur les mouvements plans d'une liquide visceux que limitent des paraois.* J. Math. Pures Appl. **13**, 331–418.

Leray, J. (1952): *Hyperbolic Differential Equations.* Institute for Advanced Studies, Princeton, NJ.

Leray, J. and Lions, J. (1965): *Quelques résultats de Višik sur les problèmes elliptiques non linéaires par les méthodes de Minty–Browder.* Bull. Soc. Math. France **93**, 97–107.

Leray, J. and Ohya, Y. (1967): *Equations et systèmes non linéaires hyperbolic non-strict.* Math. Ann. **170**, 167–205.

Levitan, B. (1984): *The Inverse Sturm–Liouville Problem.* Nauka, Moscow (Russian).

Lichnerowicz, A. (1967): *Relativistic Hydrodynamics and Magnetohydrodynamics.* Benjamin, New York.

Lichnewsky, A. and Saguez, C. [eds.] (1987): *Supercomputing: the State of the Art.* North-Holland, Amsterdam.

Lichtenstein, L. (1921): *Neuere Entwicklung der Potentialtheorie.* In: *Encyklopädie der Mathematischen Wissenschaften* (1899ff), Vol. 2, Part 3.1, pp. 181–377.

Lieberman, G. and Trudinger, S. (1986): *Nonlinear oblique boundary value problems for nonlinear elliptic equations.* Trans. Amer. Math. Soc. **295**, 509–546.

Linz, P. (1988): *A critique of numerical analysis.* Bull. Amer. Math. Soc. (N.S.) **19**, 407–416.

Lions, J. (1955): *Problèmes aux limites en théorie des distributions.* Acta Math. **94**, 13–153.

Lions, J. (1959): *Théorèmes de traces et d'interpolation*, I–V; I–II, Ann. Scuola Norm. Sup. Pisa Cl. Sci. **13** (1959), 389–403, **14** (1960), 317–331; III, J. Math. Pures Appl. **42** (1963), 196–203; IV, Math. Ann. **151** (1963), 42–56; V, Anais de Acad. Brasileira de Ciecias **35** (1963), 1–110.

Lions, J. and Magenes, E. (1960): *Problèmes aux limites non homogènes*, I–VII; I, III, IV, V, Ann. Scuola Norm. Sup. Pisa Cl. Sci. **14** (1960), 259–308; **15** (1961), 39–101; 311–326; **16** (1962), 1–44; II, Ann. Inst. Fourier **11** (1961), 137–178; VI, J. Analyse Math. **11** (1963), 165–188; VII, Annali Mat. Pura Appl. **4**, LXIII (1963), 201–224.

Lions, J. (1961): *Équations différentielles opérationelles et problèmes aux limites.* Springer-Verlag, Berlin.

Lions, J. and Stampacchia, G. (1967): *Variational inequalities.* Comm. Pure Appl. Math. **20**, 493–519.

Lions, J. (1968): *Controle optimal des systèmes gouvernés par des équations aux dérivées partielles.* Dunod, Paris. (English edition: Springer-Verlag, Berlin, 1971.)

Lions, J. and Magenes, E. (1968): *Problèmes aux limites non homogènes et applications*, Vols. 1–3. Dunod, Paris. (English edition: Springer-Verlag, Berlin, 1972.)

Lions, J. (1969): *Quelques méthodes de résolution des problèmes aux limites non linéaires.* Dunod, Paris.

Lions, J. and Dautray, R. (1984): *Analyse mathématique et calcul numérique pour les sciences et les techniques*, Vols. 1–3. Masson, Paris.

Lions, J. and Ciarlet, P. [eds.] (1988): *Handbook of Numerical Analysis*, Vols. 1, 2. Vol 1: *Finite Element Method.* Vol. 2: *Finite Difference Method* (Vols. 3ff to appear). North-Holland, Amsterdam.

Lions, P. (1982): *On the existence of positive solutions of semilinear elliptic equations.* SIAM Rev. **24**, 441–467.

Lions, P. (1983): *Generalized Solutions of Hamilton–Jacobi Equations.* Pitman, London.

Lions, P. (1983a): *Hamilton–Jacobi–Bellman equations and the optimal control of stochastic systems.* In: Warsaw (1983), Vol. 2, 1403–1417.

Lions, P. (1984): *The concentration–compactness principle in the calculus of variations.* Ann. Inst. H. Poincaré. Anal. Non Linéaire **1**, 109–145.

Lions, P. (1985): *Sur les équations de Monge–Ampère.* Arch. Rational Mech. Anal. **89**, 93–122.

Ljapunov, A. (1906): *Sur les figures d'equilibre.* In: Ljapunov, A. (1959), Vol. 4, 225 pp.

Ljapunov, A. (1959): *Collected Works*, Vols. 1–5. Academy of Sciences, Moscow (Russian).

Majda, A. (1984): *Compressible Fluid Flow and Systems of Conservation Laws in Several Space Variables.* Springer-Verlag, New York.

Malgrange, B. (1955): *Existence et approximation des solutions des équations aux dérivées partielles et des équations de convolution.* Ann. Inst. Fourier **6**, 271–355.

Mandelbrot, B. (1982): *The Fractal Geometry of Nature.* Freeman, San Francisco, CA.

Marchenko, V. (1988): *Nonlinear Equations and Operator Algebras.* Reidel, Boston, MA.

Marchuk, G. and Shaidurov, V. (1983): *Difference Methods and Their Extrapolations.* Springer-Verlag, New York.

Marsden, J. and Hughes, T. (1983): *Mathematical Foundations of Elasticity.* Prentice-Hall, Englewood Cliffs, NJ.

Martin, R. (1976): *Nonlinear Operators and Differential Equations in Banach Spaces.* Wiley, New York.

Martin, J. [ed.] (1988): *Performance Evaluation of Supercomputers.* North-Holland, Amsterdam.

Maslennikova, V. (1988): *L_p-theory for elliptic boundary value problems with nonsmooth and noncompact boundary* (to appear).

Maso, G. and Mosco, U. (1986): *Wiener criteria and energy decay for relaxed Dirichlet problems.* Arch. Rational Mech. Anal. **95**, 345–387.

Masuda, K. and Suzuki, T. (1988): *Recent Topics in Nonlinear PDEs III.* North-Holland, Amsterdam.

Mathematics: The Unifying Thread in Science (1986): Notices of the American Mathematical Society, Vol. 33, pp. 716–733.

Matveev, V. *et al.* (1986): *Algebraic–geometrical principles of the superposition of finite-gap solutions of integrable nonlinear equations.* Uspekhi Mat. Nauk **41** (2), 3–42 (Russian).

Maurin, K. (1967): *Methods of Hilbert Spaces.* PWN, Warsaw.

Maurin, K. (1982): *Plato's cave parable and the development of modern physics.* Rend. Sem. Mat. Univ. Politec. Torino **40**, 1–31.

Mawhin, J. (1972): *Equivalence theorems for nonlinear operator equations and coincidence degree for some mappings in locally convex spaces.* J. Differential Equations **12**, 610–636.

Mawhin, J. (1979): *Topological Degree Methods in Nonlinear Boundary Value Problems.* Amer. Math. Soc., Providence, RI.

Mawhin, J. (1981): Cf. the Preface to Fučik, S. (1981).

Mawhin, J. (1986): *Variational methods and semilinear elliptic equations.* Arch. Rational Mech. Anal. **95**, 269–277.

Mawhin, J. (1986a): *Necessary and sufficient conditions for the solvability of nonlinear equations through the dual least action principle.* In: International Workshop on Applied Differential Equations, Beijing, pp. 91–108. World Scientific, Singapore.

Mazja, V. (1985): *Sobolev Spaces.* Springer-Verlag, New York.

Mazur, S. (1933): *Über konvexe Mengen in linearen normierten Räumen.* Studia Math. **5**, 70–84.

Meis, T. and Marcowitz, U. (1978): *Numerische Behandlung partieller Differentialgleichungen.* Springer-Verlag, Berlin. (English edition: *Numerical Solutions of Partial Differential Equations.* Springer-Verlag, New York, 1981.)

Michlin, S. (1962): *Variationsmethoden der mathematischen Physik.* Akademie-Verlag, Berlin.

Michlin, S. (1969): *Numerische Realisierung von Variationsmethoden.* Akademie-Verlag, Berlin. (English edition: Cf. Mikhlin, S. (1971)).

Michlin, S. and Smolickii, C. (1969): *Näherungsmethoden zur Lösung von Differential- und Integralgleichungen.* Teubner, Leipzig. (English edition: Mikhlin, S. and Smolitskiy, K., *Approximate Methods for Solution of Differential and Integral Equations.* American Elsevier, New York, 1967.)

Michlin, S. (1975): *Lehrgang der mathematischen Physik.* Akademie-Verlag, Berlin.

Michlin, S. (1976): *Approximation auf dem kubischen Gitter.* Akademie-Verlag, Berlin.

Michlin, S. and Prößdorf, S. (1980): *Singuläre Integralgleichungen.* Akademie-Verlag, Berlin. (English edition: *Singular Integral Equations*, Berlin, 1982.)

Michlin, S. (1981): *Konstanten in einigen Ungleichungen der Analysis.* Teubner, Leipzig. (English edition: Mikhlin, S., *Constants in Some Inequalities of Analysis*, Wiley, New York, 1986.)

Michlin, S. (1985): *Fehler in numerischen Prozessen.* Akademie-Verlag, Berlin.

Miersemann, E. (1982): *Zur Regularität verallgemeinerter Lösungen von quasilinearen elliptischen Differentialgleichungen zweiter Ordnung in Gebieten mit Ecken.* Z. Anal. Anwendungen **1** (4), 59–71.

Miersemann, E. (1988): *Asymptotic expansions of solutions of the Dirichlet problem for quasi-linear elliptic equations of second order near a conical point.* Math. Nachr. **135**, 239–274.

Miersemann, E. (1988a): *Asymptotic expansions at a corner for the capillary problem.* Pacific J. Math. **134**, 299–311.

Mihailov, A., Šabat, A., and Jamilov, R. (1987): *A symmetric approach to the classification of nonlinear equations: A full classification of integrable systems.* Uspekhi Mat. Nauk **42** (4), 3–53 (Russian).

Mikhlin, S. (1971): *The Numerical Performance of Variational Methods.* Noordhoff, Groningen.

Minty, G. (1962): *Monotone operators in Hilbert spaces.* Duke Math. J. **29**, 341–346.

Minty, G. (1963): *On a monotonicity method for the solution of non-linear equations in Banach spaces.* Proc. Nat. Acad. Sci. U.S.A. **50**, 1038–1041.

Minty, G. (1978): *On variational inequalities for monotone operators.* Adv. in Math. **30**, 1–7.

Miranda, C. (1970): *Partial Differential Equations of Elliptic Type.* Springer-Verlag, Berlin.

Miranker, W. (1981): *Numerical Methods for Stiff Equations and Singular Perturbation Problems.* Reidel, Boston.

Morawetz, C. [ed.] (1986): *Frontiers of the Mathematical Sciences.* Comm. Pure Appl. Math. **39**, Supplement S. The entire issue.

Morrey, C. (1940): *Existence and differentiability theorems for the solutions of variational problems for multiple integrals.* Bull. Amer. Math. Soc. **46**, 439–458.

Morrey, C. (1966): *Multiple Integrals in the Calculus of Variations.* Springer-Verlag, Berlin.

Morse, P. and Feshbach, H. (1953): *Methods of Theoretical Physics*, Vols. 1, 2. McGraw-Hill, New York.

Mosco, U. (1973): *An Introduction to the Approximate Solution of Variational Inequalities.* Ed. Cremonese, Roma.

Moser, J. (1961): *On Harnack's theorem for elliptic differential equations.* Comm. Pure Appl. Math. **14**, 577–591.

Moser, J. (1966): *A rapidly convergent iteration method and nonlinear partial differential equations.* Ann. Scuola Norm. Sup. Pisa Cl. Sci. **20**, 265–315; 499–535.

Moser, J. (1981): *Integrable Hamiltonian Systems and Spectral Theory.* Lezione Fermiane, Pisa.

Moser, J. (1986): *Minimal solutions of variational problems on a torus.* Ann. Inst. H. Poincaré. Anal. Non Linéaire **3**, 229–272.

Mukherjea, A. and Pothoven, K. (1978): *Real and Functional Analysis.* Plenum, New York.

Müller, M. (1974): *Das Differenzenverfahren für quasilineare elliptische Differentialgleichungen höherer Ordnung.* Math. Nachr. **63**, 255–275.

Mumford, D. (1983): *Tata Lectures on Theta.* Birkhäuser, Boston, MA.

Murat, F. (1987): *A survey of compensated compactness.* In: Cesari, L. [ed.] (1987), pp. 145–183.

Murman, E. [ed.] (1985): *Progress and Supercomputing in Computational Fluid Dynamics.* Birkhäuser, Boston.

Myint, U. and Debnath, L. (1987): *Partial Differential Equations for Scientists and Engineers.* North-Holland, New York.

Nagel, R. *et al.* (1986): *One-Parameter Linear Semigroups of Positive Operators.* Springer Lecture Notes in Mathematics, Vol. 1184. Springer-Verlag, Berlin.

Nash, J. (1958): *Continuity of solutions of parabolic and elliptic equations.* Amer. J. Math. **80**, 931–954.

Natanson, I. (1955): *Konstruktive Funktionentheorie.* Akademie-Verlag, Berlin.

Nečas, J. (1967): *Les méthodes directes en théorie des equations elliptiques.* Academia, Prague.

Nečas, J. (1973): *On the ranges of nonlinear operators with linear asymptotes which are not invertible.* Comment. Math. Univ. Carolin. **14**, 63–72.

Nečas, J. (1977): *Example of an irregular solution to a nonlinear elliptic system with analytic coefficients and conditions for regularity.* Abh. Akad. Wiss. DDR, Mathematik **1**, 197–206.

Nečas, J. and Hlaváček, I. (1981): *Mathematical Theory of Elastic and Elasto-Plastic Bodies.* Elsevier, Amsterdam.

Nečas, J. (1982): *The solution of the* 19th *Hilbert problem.* In: Kurke, H. *et al.* [eds.] (1982), *Recent Trends in Mathematics*, pp. 214–223. Teubner, Leipzig.

Nečas, J. (1983): *Introduction to the Theory of Nonlinear Elliptic Equations.* Teubner, Leipzig.

Neumann, J. v. (1929): *Allgemeine Eigenwerttheorie hermitescher Funktionaloperatoren.* Math. Ann. **102**, 49–131.

Neumann, J. v. (1929a): *Zur Theorie der unbeschränkten Matrizen.* J. Reine Angew. Math. **161**, 208–236.

Neumann, J. v. (1932): *Mathematische Grundlagen der Quantenmechanik.* Springer-Verlag, Berlin. (English edition: *Mathematical Foundations of Quantum Mechanics*, University Press, Princeton, NJ, 1955.)

Neumann, J. v. and Richtmyer, R. (1947): *On the numerical solution of partial differential equations of parabolic type.* In: Neumann, J. v. (1961), Vol. 5, pp. 652–663.

Neumann, J. v. (1949): *On rings of operators.* Ann. Math. **50**, 401–485.

Neumann, J. v. (1961): *Collected Works*, Vols. 1–6. Pergamon, New York.

Newell, A. *et al.* (1987): *Soliton Mathematics.* Les Presses de l'Université de Montréal.

Ni, W. and Serrin, J. (1985): *Nonexistence theorems for partial differential equations.* Rend. Circ. Mat. Palermo (2), Suppl. **8**, 171–185.

Nicolaenko, B. *et al.* [ed.] (1986): *Nonlinear Systems of PDEs in Applied Mathematics*, Parts I, II. Amer. Math. Soc., Providence, RI.

Nikolskii, S. (1975): *Approximation of Functions of Several Variables and Embedding Theorems.* Springer-Verlag, New York.

Nirenberg, L. (1955): *Remarks on strongly elliptic differential equations.* Comm. Pure Appl. Math. **8**, 649–675.

Nirenberg, L. (1959): *On elliptic partial differential equations*. Ann. Scuola Norm. Sup. Pisa Cl. Sci. **13**, 1–48.

Nirenberg, L. (1966): *An extended interpolation inequality*. Ann Scuola Norm. Sup. Pisa Cl. Sci. **20**, 733–737.

Nirenberg, L. (1970): *An application of generalized degree to a class of nonlinear problems*. Troisième Colloq. Analyse, Liège, Math. Vander, pp. 57–74.

Nirenberg, L. (1972): *An abstract form of the nonlinear Cauchy–Kovalevskaja theorem*. J. Differential Geom. **6**, 561–576.

Nirenberg, L. (1974): *Topics in Nonlinear Functional Analysis*. Courant Institute, New York University.

Nirenberg, L. (1979): Cf. Gidas, B. *et al.* (1979).

Nirenberg, L. (1981): *Variational and topological methods in nonlinear problems*. Bull. Amer. Math. Soc. (N.S.) **4**, 267–302.

Nirenberg, L. (1984/88): Cf. Caffarelli, L. *et al.* (1984/88).

Nishida, T. (1977): *On a theorem of Nirenberg*. J. Differential Geom. **12**, 629–633.

Nitsche, J. (1970): *Lineare Spline-Funktionen und die Methode von Ritz für elliptische Randwertprobleme*. Arch. Rational Mech. Anal. **36**, 348–355.

Nohel, J., Hrusa, W., and Renardy, M. (1987): *Mathematical Problems in Viscoelasticity*. Wiley, New York.

Nohel, J., Rogers, R., and Tzavaras, A. (1988): *Weak solutions for a nonlinear system in viscoelasticity*. Comm. Partial Differential Equations **13**, 97–127.

Novikov, S. *et al.* (1980): *Theory of Solitons*. Nauka, Moscow (Russian). (English edition: Plenum, New York, 1984.)

Oberguggenberger, M. (1986): *Propagation of singularities for semilinear hyperbolic initial–boundary value problems in one space dimension*. J. Differential Equations **61**, 1–39.

Oberguggenberger, M. (1987): *Generalized solutions to semilinear hyperbolic systems in two variables*. Monatsh. Math. **103**, 133–144.

Oberguggenberger, M. (1988): *Weak limits of solutions to semilinear hyperbolic systems*. Math. Ann. (to appear).

Oleinik, O. (1957): *Discontinuous solutions of nonlinear differential equations*. Uspekhi Mat. Nauk **12** (3), 3–73 (Russian).

Oleinik, O. and Kružkov, S. (1961): *Quasilinear parabolic equations of second order*. Uspekhi Mat. Nauk **16** (5), 115–155 (Russian).

Oleinik, O. *et al.* (1979): *Homogenization and G-convergence of differential operators*. Uspekhi Mat. Nauk **34** (5), 65–133 (Russian).

Oleinik, O. and Kondratjev, V. (1983): *Boundary value problems for partial differential equations in nonsmooth domains*. Uspekhi Mat. Nauk **38** (2), 3–76 (Russian).

Osborn, J. (1975): *Spectral approximation for compact operators*. Math. Comp. **29**, 712–725.

Osserman, R. (1986): *A Survey of Minimal Surfaces*. Dover, New York.

Ostrowski, A. (1951): *Vorlesungen über Differential- und Integralrechnung*, Vols. 1–3. Birkhäuser, Basel.

Otte, M. [ed.] (1974): *Mathematiker über Mathematik*. Springer-Verlag, Berlin.

Ovsjannikov, L. (1976): *Cauchy problem in a scale of Banach spaces and its application to the shallow water justification*. In: Germain, P. and Nayroles, B. [eds.] (1976), pp. 426–437.

Pao, C. (1978): *Asymptotic behavior and nonexistence of global solutions for a class of nonlinear boundary value problems of parabolic type.* J. Math. Anal. Appl. **65**, 616–637.

Pao, C. (1979): *Bifurcation analysis on a nonlinear diffusion system in reactor dynamics.* Applicable Anal. **9**, 107–119.

Pao, C. (1980): *Nonexistence of global solutions for an integro-differential system in reactor dynamics.* SIAM J. Math. Anal. **11**, 559–564.

Pascali, D. and Sburlan, S. (1978): *Nonlinear Mappings of Monotone Type.* Academiei, Bucuresti.

Pavel, N. (1987): *Nonlinear Evolution Operators and Semigroups.* Springer-Verlag, New York.

Payne, L. (1976): *Improperly Posed Problems in Partial Differential Equations.* SIAM, Philadelphia, PA.

Pazy, A. (1983): *Semigroups of Linear Operators and Applications to Partial Differential Equations.* Springer-Verlag, New York.

Peitgen, H., Saupe, D., and Schmitt, K. (1981): *Nonlinear elliptic boundary value problems versus their finite difference approximations: numerically irrelevant solutions.* J. Reine Angew. Math. **322**, 74–117.

Petryshyn, W. (1967): *Iterative construction of fixed points of contractive type mappings in Banach spaces.* In: *Numerical Analysis of Partial Differential Equations*, pp. 307–339. C.I.M.E., Edizione Cremonese, Roma.

Petryshyn, W. (1968): *On the approximation-solvability of nonlinear equations.* Math. Ann. **177**, 156–164.

Petryshyn, W. (1970): *Nonlinear equations involving noncompact operators.* In: Browder, F. [ed.] (1970c), pp. 206–233.

Petryshyn, W. (1970a): *A characterization of strict convexity of Banach spaces and other uses of duality mappings.* J. Funct. Anal. **6**, 282–291.

Petryshyn, W. (1975): *On the approximation-solvability of equations involving A-proper and pseudo-A-proper mappings.* Bull. Amer. Math. Soc. **81**, 223–312.

Petryshyn, W. (1976): *On the relationship of A-properness to mappings of monotone type with applications to elliptic equations.* In: Swaminatha, S. [ed.] (1976): *Fixed-Point Theory and Its Applications*, pp. 149–174. Academic Press, New York.

Petryshyn, W. (1980): *Using degree theory for densely defined A-proper maps in the solvability of semilinear equations with bounded and noninvertible linear part.* Nonlinear Anal. **4**, 259–281.

Petryshyn, W. (1980a): *Solvability of linear and quasilinear elliptic boundary value problems via the A-proper mapping theory.* Numer. Funct. Anal. Optim. **2**, 591–635.

Petryshyn, W. (1981): *Variational solvability of quasilinear elliptic boundary value problems at resonance.* Nonlinear Anal. **5**, 1095–1108.

Petryshyn, W. and Yu, Z. (1985): *On the solvability of an equation describing the periodic motions of a satellite in its elliptic orbit.* Nonlinear Anal. **9**, 969–975.

Petryshyn, W. (1986): *Approximation-solvability of periodic boundary value problems via the A-proper mapping theory.* Proc. Sympos. Pure Math. Vol. 45, pp. 261–282, Amer. Math. Soc., Providence, RI.

Petryshyn, W. (1986a): *Solvability of various boundary value problems for the equation* $x'' = f(t, x, x', x'') - y$. Pacific J. Math. **122**, 169–195.

Petryshyn, W. (1987): *Existence of positive eigenvectors and fixed points for A-proper type maps in cones.* In: Liu, B. [ed.] (1987), Proceedings of the International Conference in Honor of Ky Fan. Dekker, New York, pp. 59–63.

Petryshyn, W. (1987a): *Multiple positive fixed points of multivalued condensing mappings with some applications.* J. Math. Anal. Appl. **124**, 237–253.

Piessens, R. *et al.* (1983): *QUADPACK: A Subroutine Package for Automatic Integration.* Springer-Verlag, New York.

Pietsch, A. (1987): *Eigenvalues and s-Numbers.* Geest & Portig, Leipzig.

Piran, T. and Weinberg, S. (1988): *Strings and Superstrings.* World Scientific, Singapore.

Pissanetzky, S. (1986): *Sparse Matrix Technology.* Academic Press, New York.

Pogorzelski, W. (1966): *Integral Equations and Their Applications.* PWN, Warsaw.

Pohožaev, S. (1965): *On eigenfunctions of the equation* $-\Delta u = \lambda f(u)$. Doklady Akad. Nauk SSSR **165** (1), 36–39 (Russian).

Pohožaev, S. (1967): *On the solvability of nonlinear equations involving odd operators.* Funktional. Anal. i Priložen. **1** (3), 66–73 (Russian).

Pohožaev, S. (1969): *On nonlinear operators with weakly closed range and quasi-linear elliptic equations.* Mat. Sbornik **78**, 237–259 (Russian).

Poincaré, H. (1890): *Sur les équations aux dérivées partielles de la physique mathématique.* Amer. J. Math. **12**, 211–294.

Poincaré, H. (1916): *Oeuvres (Collected Works)*, Vols. 1–11. Gauthier-Villars, Paris.

Pöschel, J. and Trubowitz, E. (1987): *Inverse Spectral Theory.* Academic Press, New York.

Press, H. *et al.* (1986): *Numerical Recipes and Computer Programs: The Art of Scientific Computing.* University Press, Cambridge, England.

Prößdorf, S. and Silbermann, B. (1977): *Projektionsverfahren und die näherungsweise Lösung singulärer Gleichungen.* Teubner, Leipzig.

Rabier, P. (1986): *A general study of nonlinear problems with three solutions in Hilbert spaces.* Arch. Rational Mech. Anal. **95**, 123–154.

Rabinowitz, P. (1974): *Variational methods for nonlinear eigenvalue problems.* In: Prodi, G. [ed.], *Eigenvalues of Nonlinear Problems*, pp. 141–195, Ed. Cremonese, Roma.

Rabinowitz, P. (1977): *A bifurcation theorem for potential operators.* J. Funct. Anal. **26**, 48–67.

Rabinowitz, P. (1978): *Free vibrations for a semilinear wave equation.* Comm. Pure Appl. Math. **31**, 31–68.

Rabinowitz, P. (1984): *Minimax methods and their applications to partial differential equations.* In: Chern, S. [ed.] (1984), pp. 307–320.

Rabinowitz, P. (1984a): *Large amplitude time-periodic solutions of a semilinear wave equation.* Comm. Pure Appl. Math. **37**, 189–206.

Rabinowitz, P. (1986): *Methods in Critical Point Theory with Applications to Differential Equations.* Amer. Math. Soc., Providence, RI.

Rajaraman, R. (1982): *Solitons and Instantons.* North-Holland, Amsterdam.

Rannacher, R. (1980): *On finite element approximation of general boundary value problems in nonlinear elasticity.* Estratto da Calcolo **17**, 174–193.

Rauch, J. and Reed, M. (1980): *Propagation of singularities for semilinear hyperbolic equations in one space variable.* Ann. Math. **111**, 531–552.

Rauch, J. and Reed, M. (1981): *Jump discontinuities of semilinear, strictly hyperbolic systems in two variables.* Comm. Math. Phys. **81**, 203–227.

Rauch, J. and Reed, M. (1982): *Nonlinear microlocal analysis of semilinear hyperbolic systems in one space dimension.* Duke Math. J. **49**, 397–475.

Rauch, J. and Reed, M. (1988): *Nonlinear superposition and absorbtion of delta waves in one space dimension.* J. Funct. Anal. (to appear).

Raviart, P. (1967): *Sur l'approximation de certaines equations d'evolution linéaires.* J. Math. Pures Appl. **46**, 11–183.

Rebbi, C. and Soliani, G. [eds.] (1984): *Solitons and Particles.* World Scientific, Singapore.

Reed, M. and Simon, B. (1971): *Methods of Modern Mathematical Physics*, Vols. 1–4. Academic Press, 1971.

Reed, M. (1976): *Abstract Nonlinear Wave Equations.* Lecture Notes in Mathematics, Vol. 507. Springer-Verlag, Berlin.

Reich, S. (1984): *New results concerning accretive operators and nonlinear semigroups.* J. Math. Phys. Sci. **18**, 91–97.

Reid, C. (1970): *Hilbert.* Springer-Verlag, New York.

Reid, C. (1976): *Courant in Göttingen and New York.* Springer-Verlag, New York.

Reinhardt, H. (1985): *Analysis of Approximation Methods for Differential- and Integral Equations.* Springer-Verlag, New York.

Rektorys, K. (1977): *Variational Methods in Mathematics, Science and Engineering.* Reidel, Boston, MA.

Rektorys, K. (1982): *The Method of Discretization in Time and Partial Differential Equations.* Reidel, Boston, MA.

Rellich, F. (1930): *Ein Satz über mittlere Konvergenz.* Nachr. Akad. Wiss. Göttingen, pp. 30–35.

Rempel, S. and Schulze, B. (1982): *Index Theory of Elliptic Boundary Value Problems.* Akademie-Verlag, Berlin.

Renardy, M., Hrusa, W., and Nohel, J. (1987): *Mathematical Problems in Viscoelasticity.* Wiley, New York.

Rhodius, A. (1977): *Der numerische Wertebereich und die Lösbarkeit linearer und nichtlinearer Operatorgleichungen.* Math. Nachr. **79**, 343–360.

Rice, J. (1983): *Numerical Methods, Software, and Analysis.* McGraw-Hill, New York.

Rice, J. and Boisvert, R. (1984): *Solving Elliptic Problems Using ELLPACK.* Springer-Verlag, New York.

Richtmyer, R. and Morton, K. (1967): *Difference Methods for Initial-Value Problems.* Interscience, New York.

Riemann, B. (1851): *Grundlagen für eine allgemeine Theorie der Funktionen einer veränderlichen komplexen Größe.* Dissertation. In: Riemann, B. (1892), pp. 3–43.

Riemann, B. (1857): *Theorie der Abelschen Funktionen.* In: Riemann, B. (1892), pp. 88–142.

Riemann, B. (1892): *Gesammelte mathematische Werke* (*Collected Works*), Vols. 1, 2. Leipzig, 1892. Nachträge (Supplements), Leipzig, 1902.

Riesz, F. (1918): *Über lineare Funktionalgleichungen.* Acta Math. **41**, 71–98.

Riesz, F. (1934): *Zur Theorie des Hilbertschen Raumes.* Acta Sci. Math. (Szeged) **7**, 34–38.

Riesz, F. and Nagy, B. (1956): *Vorlesungen über Funktionalanalysis*, Verl. der Wiss., Berlin. (English edition: *Functional Analysis*, Ungar, New York, 1955.)

Riesz, F. (1960): *Oeuvres complètes* (*Collected Works*), Vols. 1–4. Hungarian Academy of Sciences, Budapest.

Riesz, M. (1926): *Sur les maxima des formes bilinéaires et sur les fonctionnelles linéaires.* Acta Math. **49**, 465–497.

Ritz, W. (1909): *Über eine neue Methode zur Lösung gewisser Variationsprobleme der mathematischen Physik.* J. Reine Angew. Math. **135**, 1–61.

Ritz, W. (1909a): *Theorie der Transversalschwingungen einer quadratischen Platte mit freien Rändern.* Ann. Physik **28**, 737–786.

Rjabenki, V. and Filippov, A. (1960): *Über die Stabilität von Differenzengleichungen.* Verl. der Wiss., Berlin.

Rockafellar, R. (1966): *Characterization of the subdifferential of convex functions.* Pacific J. Math. **17**, 497–510. (Cf. also Rockafellar, R. (1970b).)

Rockafellar, R. (1968): *Convex functions, montone operators, and variational inequalities.* In: Proceedings of the NATO Advanced Study Institute, pp. 35–65. Ed. Oderisi, Gubbio, Italy.

Rockafellar, R. (1969): *Local boundedness of nonlinear monotone operators.* Michigan Math. J. **16**, 397–407.

Rockafellar, R. (1970): *On the maximal monotonicity of sums of nonlinear monotone operators.* Trans. Amer. Math. Soc. **149**, 75–88.

Rockafellar, R. (1970a): *Convex Analysis.* University Press, Princeton, NJ.

Rockafellar, R. (1970b): *On the maximal monotonicity of subdifferential mappings.* Pacific J. Math. **33**, 209–216.

Rockafellar, R. (1970c): *Monotone operators associated with saddle-functions and minimax problems.* In: Browder, F. [ed.] (1970c), pp. 241–250.

Rockafellar, R. (1970d): *On the virtual convexity of the range of a nonlinear maximal operator.* Math. Ann. **185**, 81–90.

Rockafellar, R. (1985): *Maximal monotone relations and the second derivatives of nonsmooth functions.* Ann. Inst. H. Poincaré. Anal. Non Linéaire **2**, 167–184.

Rosinger, E. (1987): *Generalized Solutions of Nonlinear Partial Differential Equations.* North-Holland, Amsterdam.

Rothe, F. (1984): *Global Solutions of Reaction–Diffusion Systems.* Springer-Verlag, New York.

Rothe, E. (1986): *Introduction to Various Aspects of Degree Theory in Banach Spaces.* Amer. Math. Soc., Providence, RI.

Royden, H. (1988): *Real Analysis.* Macmillan, New York.

Roždestvenskii, B. and Janenko, N. (1978): *Systems of Quasilinear Equations and Their Applications to Gas Dynamics.* Nauka Moscow. (English edition: Amer. Math. Soc., Providence, RI, 1983.)

Ruf, B. (1986): *Multiplicity results for nonlinear elliptic equations.* In Krbec, M. *et al.* [eds.], *Nonlinear Analysis*, pp. 109–138, Teubner, Leipzig.

Runst, T. (1985): *Paradifferential operators in spaces of Triebel–Lizorkin and Besov type.* Z. Anal. Anwendungen **4**, 537–573.

Sacks, J. and Uhlenbeck, K. (1981): *The existence of minimal immersions of 2-spheres.* Ann. Math. **113**, 1–24.

Sakamoto, R. (1982): *Hyperbolic Boundary Value Problems.* University Press, Cambridge, England.

Samarskii, A. (1984): *Theorie der Differenzenverfahren.* Geest & Portig, Leipzig.

Sard, A. (1942): *The measure of the critical points of differentiable maps.* Bull. Amer. Math. Soc. **48**, 883–890.

Sauer, R. (1958): *Anfangswertprobleme bei partiellen Differentialgleichungen.* Springer-Verlag, Berlin.

Sauer, R. and Szabó, I. (1967): *Mathematische Hilfsmittel des Ingenieurs*, Vols. 1–4. Springer-Verlag, Berlin.

Schauder, J. (1930): *Über lineare vollstetige Funktionaloperationen.* Studia Math. **2**, 103–196.

Schauder, J. (1935): *Das Anfangswertproblem einer quasilinearen hyperbolischen Differentialgleichung zweiter Ordnung mit beliebig vielen unabhängigen Variablen.* Fund. Math. **24**, 213–246.

Schechter, M. (1971): *Spectra of Partial Differential Operators.* North-Holland, Amsterdam.

Schechter, M. (1973): *A nonlinear elliptic boundary value problem.* Ann. Scuola Norm. Sup. Pisa Cl. Sci. **27**, 1–10.

Schechter, M. (1977): *Modern Methods in Partial Differential Equations.* McGraw-Hill, New York.

Schechter, M. and Snow, M. (1978): *Solution of the nonlinear problem $Au = Nu$ in a Banach space.* Trans. Amer. Math. Soc. **241**, 69–78.

Schechter, M. (1981): *Operator Methods in Quantum Mechanics.* North-Holland, Amsterdam.

Schechter, M. (1986): *Derivatives of mappings with applications to nonlinear differential equations.* Trans. Amer. Math. Soc. **293**, 53–69.

Scheidt, J. v. and Purkert, W. (1983): *Random Eigenvalue Problems.* Akademie-Verlag, Berlin.

Scheurle, J. (1977): *Ein selektives Projektions–Iterationsverfahren und Anwendung auf Verzweigungsprobleme.* Numer. Math. **29**, 11–35.

Schmidt, E. (1907a): *Entwicklung willkürlicher Funktionen nach Systemen vorgeschriebener Funktionen.* Math. Ann. **63**, 433–476.

Schmidt, E. (1907b): *Auflösung der allgemeinen linearen Integralgleichung.* Math. Ann. **64**, 161–174.

Schmidt, E. (1908): *Zur Theorie der linearen und nichtlinearen Integralgleichungen.* Math. Ann. **65**, 370–399.

Schoen, R. and Yau, S. (1979): *On the proof of the positive mass conjecture in general relativity.* Comm. Math. Phys. **65**, 45–76; **79**, 231–260.

Schoen, R. (1984): *Conformal deformation of a Riemannian metric to a constant scalar curvature.* J. Differential Geom. **20**, 479–495.

Schoen, R. (1986): *Recent progress in geometric partial differential equations.* In: Berkeley (1986), pp. 121–135.

Schoen, R. (1988): *The existence of weak solutions with prescribed singular behavior for a conformally invariant scalar equation.* Comm. Pure Appl. Math. **41**, 317–392.

Schumann, R. and Zeidler, E. (1979): *The finite difference method for quasilinear elliptic equations of order 2m.* Numer. Funct. Anal. Optimiz. **1**, 161–194.

Schumann, R. (1987): *The convergence of Rothe's method for parabolic differential equations.* Z. Anal. Anwendungen **6**, 559–574.

Schumann, R. (1988): *Eine neue Methode zur Gewinnung starker Regularitätsaussagen für das Signoriniproblem in der linearen Elastizitätstheorie.* Dissertation B, Karl-Marx-Universität, Leipzig.

Schumann, R. (1989): *Regularity for Signorini's problem in elasticity.* Manuscripta Math. *63*, 255–291; 455–468.

Schwartz, L. (1950): *Théorie des distributions*, Vols. 1, 2. Hermann, Paris.

Schwartz, L. (1973): *Radon Measures on Arbitrary Topological Spaces and Cylindrical Measures.* University Press, Oxford.

Serrin, J. (1964): *Local behavior of solutions of quasi-linear equations.* Acta Math. **111**, 247–302.

Serrin, J. (1965): *Isolated singularities of solutions of quasi-linear equations.* Acta Math. **113**, 219–240.

Sewell, G. (1985): *Analysis of a Finite Element Method: PDE/PROTRAN.* Springer-Verlag, New York.

Shampine, L. and Gordon, M. (1975): *Computer Solutions of Ordinary Differential Equations.* Freeman, San Francisco, CA.

Shampine, L. and Gear, C. (1979): *A user's view of solving stiff ordinary differential equations.* SIAM Rev. **21**, 1–17.

Shatah, J. (1982): *Global existence of small solutions to nonlinear evolution equations.* J. Differential Equations **46**, 409–425.

Shibata, Y. and Tsumutsi, Y. (1986): *On a global existence theorem of small amplitude solutions for nonlinear wave equations in an exterior domain.* Math. Z. **191**, 165–199.

Shokin, Ju. (1983): *Method of Differential Approximation.* Springer-Verlag, New York.

Showalter, R. (1977): *Hilbert Space Methods for Partial Differential Equations.* Pitman, London.

Showalter, W. (1978): *Mechanics of Non-Newtonian Fluids.* Pergamon, Oxford.

Shubin, M. (1986): *Pseudodifferential Operators and Spectral Theory.* Springer-Verlag, New York.

Simader, C. (1972): *On Dirichlet's Boundary Value Problem.* Lecture Notes in Mathematics, Vol. 268, Springer-Verlag, Berlin.

Simader, C. (1976): *Über schwache Lösungen des Dirichletproblems für streng nichtlineare elliptische Differentialgleichungen.* Math. Z. **150**, 1–26.

Simon, L. (1983): *Lectures on Geometric Measure Theory.* University of Canberra, Australia.

Skrypnik, I. (1973): *Nonlinear Elliptic Equations of Higher Order.* Naukova Dumka, Kiev (Russian).

Skrypnik, I. (1986): *Nonlinear Elliptic Boundary Value Problems.* Teubner, Leipzig.

Smale, S. (1965): *An infinite-dimensional version of Sard's theorem.* Amer. J. Math. **87**, 861–866.

Smale, S. (1985): *Efficiency of algorithms of analysis.* Bull. Amer. Math. Soc. (N.S.) **13**, 94–118.

Smirnov, W. (1956): *Lehrgang der höheren Mathematik*, Vols. 1–5. Verl. der Wiss., Berlin. (English edition: *A Course in Higher Mathematics.* Addison-Wesley, Reading, MA.)

Smith, G. (1965): *Numerical Solutions of Partial Differential Equations.* University Press, Oxford.

Smoller, J. (1983): *Shock Waves and Reaction–Diffusion Equations.* Springer-Verlag, New York.

Šmuljan, V. (1940): *Über lineare topologische Räume.* Mat. Sbornik **7** (49), 425–448.

Sobolev, S. (1936): *Méthode nouvelle a resoudre le problème de Cauchy pour les équations linéaires hyperboliques normales.* Mat. Sbornik **1**, 39–72.

Sobolev, S. (1937): *On a boundary value problem for the polyharmonic equation.* Mat. Sbornik **2**, 467–500 (Russian).

Sobolev, S. (1938): *On a theorem in functional analysis.* Mat. Sbornik **4**, 471–497 (Russian).

Sobolev, S. (1950): *Applications of Functional Analysis to Mathematical Physics.* University Press, Leningrad (Russian). (English edition: Amer. Math. Soc., Providence, RI, 1963.)

Sobolevskii, P. and Tiunčik, M. (1970): *The finite difference method for the approximate solution of quasilinear elliptic and parabolic equations.* Trudy Mat. Fak. Voronež. Univ. **2**, 82–106 (Russian).

Sommerfeld, A. (1900): *Randwertaufgaben in der Theorie der partiellen Differentialgleichungen.* In: *Encyklopädie der Mathematischen Wissenschaften* (1899ff); Vol. 2, Part 1.1, pp. 504–570.

Stäckel, P. [ed.] (1894): *Abhandlungen über Variationsrechnung von Johann und Jacob Bernoulli, Euler, Lagrange, Legendre, und Jacobi.* In: Ostwald's *Klassiker der exakten Wissenschaften,* Vols. 46/47. Engelmann, Leipzig.

Stein, E. (1970): *Singular Integrals and Differentiability Properties of Functions.* University Press, Princeton, NJ.

Stetter, H. (1973): *Analysis of Discretization Methods for Ordinary Differential Equations.* Springer-Verlag, Berlin.

Stoer, J. and Bulirsch, R. (1976). *Einführung in die numerische Mathematik,* Vols. 1, 2. Springer-Verlag, Berlin. (English edition: *Introduction to Numerical Analysis.* Springer-Verlag, New York, 1980.)

Stone, M. (1932): *On one-parameter unitary groups in Hilbert space.* Ann. of Math. **33**, 643–648.

Stone, M. (1932a): *Linear Transformations in Hilbert Space.* Amer. Math. Soc., Providence, RI.

Stone, M. (1970): *Remarks of Professor Stone.* In: Browder, F. [ed.] (1970b), pp. 236–241.

Strang, G. and Fix, G. (1973): *An Analysis of the Finite Element Method.* Prentice-Hall, Englewood Cliffs, NJ.

Strauss, W. (1969): *The Energy Method in Nonlinear Partial Differential Equations.* IMPA, Rio de Janeiro.

Strocchi, F. *et al.* [eds.] (1981): *Topics in Functional Analysis.* Scuola Norm. Sup., Pisa.

Strocchi, F. (1981a): *Classification of solutions of nonlinear hyperbolic equations and nonlinear elliptic equations.* In: Strocchi, F. [ed.] (1981), pp. 1–36.

Struwe, M. (1985): *On the evolution of harmonic mappings of Riemannian surfaces.* Comm. Math. Helv. **60**, 558–581.

Struwe, M. and Vivaldi, M. (1985a): *On the Hölder continuity of bounded weak solutions of quasilinear parabolic inequalities.* Ann. Mat. Pura Appl. **139**, 175–189.

Struwe, M. (1986): *Nonuniqueness in the plateau problem for surfaces of constant mean curvature.* Arch. Rational Mech. Anal. **93**, 135–157.

Stüben, K. and Trottenberg, U. (1982): *Multigrid methods.* In: Hackbusch, W. and Trottenberg, U. [eds.] (1982), pp. 1–176.

Stummel, F. (1970): *Diskrete Konvergenz linearer Operatoren,* I. II. Math. Ann **190**, 45–92; Math. Z. **120**, 231–264.

Švarc, A. (1964): *The homotopic topology of Banach spaces.* Dokl. Akad. Nauk SSSR **154**, 61–63 (Russian).

Synge, J. (1957): *The Hypercircle in Mathematical Physics.* University Press, Cambridge, England.

Tanabe, H. (1979): *Equations of Evolution.* Pitman, London.

Tanaka, M. (1986): *Boundary Elements.* North-Holland, Amsterdam.

Taniuti, T. and Nishihara, K. (1983): *Nonlinear Waves.* Pitman, London.

Tartar, L. (1972): *Interpolation non linéaire et régularité.* J. Funct. Anal. **9**, 469–489.

Tartar, L. (1983): *Compacité par compensation: résultats et perspectives.* Res. Notes Math. **84**, 350–369.

Tartar, L. (1986): *Ocsillations in Nonlinear PDEs: Compensated Compactness and Homogenization.* Lectures in Applied Mathematics, Vol. 23, Amer. Math. Soc., Providence, RI, pp. 243–266.

Taubes, C. (1982): *The existence of a non-minimal solution to the* SU(2) *Yang–Mills–Higgs equations on* $\mathbb{R}^3$. Comm. Math. Phys. **86**, 257–320.

Taubes, C. (1982a): *Self-dual Yang–Mills connections on non-self-dual* 4*-manifolds.* J. Differential Geom. **17**, 139–170.

Taubes, C. (1984): *Self-dual connections on* 4*-manifolds with indefinite intersection matrix. J. Differential Geom.* **19**, 517–560.

Taubes, C. (1986): *Physical and mathematical applications of gauge theories.* Notices of the American Mathematical Society **33**, 707–715.

Taylor, M. (1981): *Pseudodifferential Operators.* University Press, Princeton, NJ.

Taylor, C. *et al.* [eds.] (1987): *Numerical Methods for Nonlinear Problems.* Pineridge Press, Swansea, England.

Teller, E. [ed.] (1981): *Fusion,* Vols. 1, 2. Academic, New York.

Temam, R. (1968): *Sur la stabilité et la convergence de la méthod des pas fractionnaires.* Ann. Mat. Pura Appl. **79**, 191–379.

Temam, R. (1970): *Analyse numérique.* Université de France, Paris. (English edition: Reidel, Boston, MA, 1973.)

Temam, R. (1977): *Navier–Stokes Equations: Theory and Numerical Analysis.* North-Holland, Amsterdam. (Third revised edition, 1984.)

Temam, R. (1983): *Navier–Stokes Equations and Nonlinear Functional Analysis.* SIAM, Philadelphia, PA.

Temam, R. (1986): *Infinite-dimensional dynamical systems in fluid dynamics.* In: Browder, F. [ed.] (1986), Part 2, pp. 431–445.

Temam, R. (1988): *Infinite-Dimensional Dynamical Systems in Mechanics and Physics.* Springer-Verlag, New York.

Thirring, W. (1983): *A Course in Mathematical Physics,* Vols. 1–4. Springer-Verlag, Wien.

Thomée, V. (1984): *Galerkin Finite Element Methods for Parabolic Problems.* Lecture Notes in Mathematics, Vol. 1054. Springer-Verlag, Berlin.

Thompson, D. (1917): *On Growth and Form.* University Press, Cambridge, England.

Thorin, G. (1939): *An extension of a convexity theorem due to M. Riesz.* Kungl. Comm. Sém. Lund **4**, 1–5.

Toda, M. (1981): *Theory of Nonlinear Lattices.* Springer-Verlag, New York.

Tomi, F. and Tromba, A. (1988): *Existence Theorems for Minimal Surfaces of Non-Zero Genus Spanning a Contour.* Mem. Amer. Math. Soc., Providence, RI.

Tonelli, L. (1921): *Fondamenti di calcolo delle variazioni,* Vols. 1, 2. Bologna.

Törnig, W., Lipser, M, and Kaspar, B. (1985): *Numerische Lösung partieller Differentialgleichungen.* Teubner, Stuttgart.

Traub, J. and Wozniakowski, H. (1980): *A General Theory of Optimal Algorithms.* Academic, New York.

Trefftz, E. (1927): *Ein Gegenstück zum Ritzschen Verfahren.* Verhandlungen des II. Kongresses für technische Mechanik, Zürich, pp. 131–138.

Trefftz, E. (1928): *Konvergenz und Fehlerabschätzung beim Ritzschen Verfahren*. Math. Ann. **100**, 503–521.

Treves, F. (1975): *Basic Linear Partial Differential Operators*. Academic Press, New York.

Treves, F. (1982): *Introduction to Pseudodifferential and Fourier Integral Operators*. Plenum, New York.

Triebel, H. (1972): *Höhere Analysis*. Verl. der Wiss., Berlin.

Triebel, H. (1973): *Spaces of distributions of Besov type on Euclidean n-space. Duality, interpolation*. Arkiv Mat. **11**, 13–64.

Triebel, H. (1978): *Interpolation Theory, Function Spaces, Differential Operators*. Verl. der Wiss., Berlin.

Triebel, H. (1981): *Analysis und mathematische Physik*. Teubner, Leipzig. (English edition: Teubner, Leipzig, 1983.)

Triebel, H. (1983): *Theory of Function Spaces*. Geest & Portig, Leipzig.

Tromba, A. [ed.] (1987): *Seminar on New Results in Nonlinear Partial Differential Equations*. Vieweg, Braunschweig.

Troyanski, S. (1971): *On locally uniformly convex and differentiable norms in certain non-separable spaces*. Studia Math. **37**, 173–180.

Turner, R. (1974): *Positive solutions of nonlinear eigenvalue problems*. In: Prodi, G. [ed.], *Eigenvalues of Nonlinear Problems*, pp. 213–239, Ed. Cremonese, Roma.

Tychonov, A. and Samarskii, A. (1959): *Differentialgleichungen der Mathematischen Physik*. Verl. der Wiss., Berlin. (English edition: *Partial Differential Equations of Mathematical Physics*, Holden Day, New York, 1964.)

Uhlenbeck, K. (1982): *Connections with L_p-bounds on curvature*. Comm. Math. Phys. **83**, 31–42.

Ulam, S. (1976): *Adventures of a Mathematician*. Scribners, New York.

Uralceva, N. (1987): *On the regularity of the solutions of variational inequalities*. Uspekhi Mat. Nauk **42** (6), 151–174.

Vahanija, N. *et al.* (1985): *Random Distributions in Banach Spaces*. Nauka, Moscow (Russian).

Vainberg, M. (1956): *Variational Methods for the Study of Nonlinear Operators*. Gostehizdat, Moscow (Russian). (English edition: Holden Day, San Francisco, CA, 1964.)

Vainberg, M. (1972): *The Variational Method and the Method of Monotone Operators*. Nauka, Moscow (Russian). (English edition: *Monotone Operators*, Wiley, New York, 1973.)

Vainikko, G. (1976): *Funktionalanalysis der Diskretisierungsverfahren*. Teubner, Leipzig.

Vainikko, G. (1977): *Über die Konvergenz und Divergenz von Näherungsmethoden bei Eigenwertproblemen*. Math. Nachr. **78**, 145–164.

Vainikko, G. (1978): *Approximative methods for nonlinear equations*. Nonlinear Anal. **2**, 647–687.

Vainikko, G. (1979): *Regular convergence of operators and the approximate solution of equations*. Itogi nauki i tehniki, Mat. Anal. **16**, 5–53 (Russian).

Valent, T. (1988): *Boundary Value Problems of Finite Elasticity*. Springer-Verlag, New York.

Valentine, F. (1945): *A Lipschitz condition preserving extension for a vector function.* Amer. J. Math. **67**, 83–93.

Varga, R. (1962): *Matrix Iterative Analysis.* Prentice-Hall, Englewood Cliffs, NJ.

Varga, R. (1971): *Functional Analysis and Approximation Theory in Numerical Analysis.* SIAM, Philadelphia, PA.

Varga, R. (1984): *A survey of recent results on iterative methods for solving large sparse linear systems.* In: Birkhoff, G. and Schoenstadt, A. [eds.] (1984), pp. 197–217.

Vejvoda, O. *et al.* (1981): *Partial Differential Equations.* Noordhoff, Groningen.

Velte, W. (1976): *Direkte Methoden der Variationsrechnung.* Teubner, Stuttgart.

Velte, W. (1984): *Bounds for critical values and eigenfrequencies of mechanical systems.* In: Boffi, V. and Neunzert, H. [eds.] (1984), pp. 469–484.

Vilenkin, N. (1975): Cf. Virčenko, N. [ed.] (1983), p. 228.

Vinogradov, A., Krasilčik, I., and Lyčagin, V. (1986): *Introduction to the Geometry of Nonlinear Partial Differential Equations.* Nauka, Moscow (Russian).

Virčenko, N. [ed.] (1983): *Mathematics: A Collection of Quotations.* Vyšča Skola, Kiev (Russian).

Višik, M. (1949): *The method of orthogonal projection and direct decomposition in the theory of differential equations.* Mat. Sbornik **25**, 189–234 (Russian).

Višik, M. (1951): *On strongly elliptic differential equations.* Mat. Sbornik **29**, 615–676 (Russian).

Višik, M. (1952): *On general boundary value problems for elliptic differential equations.* Trudy Moskov. Mat. Obščestva **1**, 187–246 (Russian).

Višik, M. (1961): *Boundary value problems for quasilinear strongly elliptic equations.* Dokl. Akad. Nauk SSSR **138**, 518–521 (Russian).

Višik, M. (1962): *On initial–boundary value problems for quasilinear parabolic equations of higher order.* Mat. Sbornik **59**, 289–325 (Russian).

Višik, M. (1963): *Quasilinear strongly elliptic systems of differential equations in divergence form.* Trudy Moskov. Mat. Obščestva **12**, 125–184 (Russian).

Višik, M. and Fursikov, A. (1981): *Mathematical Problems in Statistical Hydromechanics.* Nauka, Moscow (Russian). (German edition: *Mathematische Probleme der statistischen Hydromechanik.* Teubner, Leipzig, 1986.)

Višik, M. and Kuksin, S. (1986): *Perturbation of quasilinear elliptic equations and Fredholm manifolds.* Mat. Sbornik **130**, 222–242 (Russian).

Vladimirov, V. (1971): *Equations of Mathematical Physics.* Dekker, New York.

Vladimirov, V. [ed.] (1986): *Collection of Survey Articles on the Occasion of the* 50th *Anniversary of the Steklov Institute.* Trudy Mat. Inst. Steklov. **175** (Russian).

Vladimirov, V. [ed.] (1986a): *A Collection of Problems on the Equations of Mathematical Physics.* Mir. Moscow (English edition).

Wahl, W. v. (1978): *Über die stationären Gleichungen von Navier–Stokes, semilineare elliptische und parabolische Gleichungen.* Jahresber. Deutsch. Math.-Verein. **80**, 129–149.

Wahl, W. v. (1978a): *Die stationären Gleichungen von Navier–Stokes und semilineare elliptische Systeme.* Amer. J. Math. **100**, 1173–1184.

Wahl, W. v. (1982): *Nichtlineare Evolutionsgleichungen.* In: Kurke, H. *et al.* [eds.], *Recent Trends in Mathematics*, pp. 294–302. Teubner, Leipzig.

Wahl, W. v. (1982a): *The equation* $u' + A(t)u = f$ *in Hilbert space and* L_p*-estimates for parabolic equations.* J. London Math. Soc. (2), **25**, 483–497.

Wahl, W. v. (1985): *The Equations of Navier–Stokes and Abstract Parabolic Equations*. Vieweg, Braunschweig.

Walter, W. (1964): *Differential- und Integralungleichungen*. Springer-Verlag, Berlin.

Walter, W. (1985): *Functional differential equations of the Cauchy–Kovalevskaja type*. Aequationes Math. **28**, 102–103.

Warsaw (1983): *Proceedings of the International Congress of Mathematicians in Warsaw*, Vols. 1, 2. PWN, Warsaw.

Wegert, E. (1987): *Topological methods for strongly nonlinear Riemann–Hilbert problems for holomorphic functions*. Math. Nachr. **134**, 201–230.

Weierstrass, K. (1870): *Über das sogenannte Dirichletsche Prinzip*. In: Weierstrass, K. (1894), Vol. 2, pp. 49–54.

Weierstrass, K. (1894): *Mathematische Werke* (*Collected Works*), Vols. 1–7. Berlin, 1894–1927.

Weinberg, S. (1986): *Cf. Mathematics: The Unifying Thread in Science* (1986).

Weinberger, H. (1974): *Variational Methods for Eigenvalue Approximation*. SIAM, Philadelphia, PA.

Weinstein, A. and Stenger, W. (1972): *Methods of Intermediate Problems for Eigenvalues*. Academic Press, New York.

Wells, R. [ed.] (1988): *The Mathematical Heritage of Hermann Weyl*. Amer. Math. Soc., Providence, RI.

Wendland, W. (1979): *Elliptic Systems in the Plane*. Pitman, London.

Wendland, W. (1984): *On some mathematical aspects of boundary element methods for elliptic problems*. In: Whiteman, J. [ed.], *The Mathematics of Finite Elements and Applications*, pp. 193–227. Academic Press, New York.

Wente, H. (1986): *Counterexample to a conjecture of Heinz Hopf*. Pacific J. Math. **121**, 193–243.

Weyer, J. (1982): *Maximal monotonicity of operators with sufficiently large domain and applications to the Hartree problem*. Manuscripta Math. **38**, 163–174.

Weyl, H. (1911): *Über die asymptotische Verteilung der Eigenwerte*. Göttinger Nachr. pp. 110–117.

Weyl, H. (1940): *The method of orthogonal projection in potential theory*. Duke Math. J. **7**, 411–444.

Weyl, H. (1944): *David Hilbert and his mathematical work*. Bull. Amer. Math. Soc. **50**, 612–654.

Weyl, H. (1968): *Collected Works*. Vols. 1–4. Springer-Verlag, New York.

Whitham, G. (1974): *Linear and Nonlinear Waves*. Wiley, New York.

Widder, D. (1975): *The Heat Equation*. Academic Press, New York.

Wille, F. (1972): *Galerkins Lösungsannäherungen bei monotonen Abbildungen*. Math. Z. **127**, 10–16 (*A poem*).

Wilkinson, J. (1965): *The Algebraic Eigenvalue Problem*. Clarendon, Oxford.

Willoughby, R. [ed.] (1974): *Stiff Differential Systems*. Plenum, New York.

Wilson, K. (1982): *The renormalization group and critical phenomena*. In: *Nobel Prize Lectures*, Vol. 1982, pp. 57–87, Stockholm.

Witten, E. (1986): *Physics and Geometry*. In: Berkeley (1986), pp. 267–306.

Wloka, J. (1982): *Partielle Differentialgleichungen*. Teubner, Stuttgart. (English edition: *Partial Differential Equations*, University Press, Cambridge, England, 1987.)

Wolfersdorf, L. v. (1982): *Monotonicity methods for two classes of nonlinear boundary value problems with semilinear first-order systems in the plane*. Math. Nachr. **109**, 215–238.

Wolfersdorf, L. v. (1983): *Monotonicity methods for nonlinear singular integro-differential equations.* Z. Angew. Math. Mech. **63**, 249–259.

Wolfersdorf, L. v. (1987): *A class of nonlinear Riemann–Hilbert problems with monotone nonlinearities.* Math. Nachr. **130**, 111–119.

Wollman, S. (1984): *An existence and uniqueness theorem for the Vlasov–Maxwell system.* Comm. Pure Appl. Math. **37**, 457–477.

Wussing, H. and Arnold, W. (1975): *Biographien bedeutender Mathematiker.* Volk und Wissen, Berlin.

Yau, S. (1978): *On the Ricci curvature of a compact Kähler manifold and the complex Monge–Ampère equation.* Comm. Pure Appl. Math. **31**, 339–411.

Yau, S. (1986): *Nonlinear Analysis in Geometry.* L'Enseignement Mathématique, Geneva.

Yosida, K. (1948): *On the differentiability and the representation of one-parameter semigroups of linear operators.* J. Math. Soc. Japan **1**, 15–21.

Yosida, K. (1965): *Functional Analysis.* Springer-Verlag, Berlin.

Zabreiko, P. and Krasnoselskii, M. (1975): *Integral Equations.* Noordhoff, Leyden.

Zarantonello, E. (1960): *Solving functional equations by contractive averaging.* Mathematics Research Center Rep. # 160, Madison, WI.

Zarantonello, E. (1964): *The closure of the numerical range contains the spectrum.* Bull. Amer. Math. Soc. **70**, 781–787.

Zarantonello, E. [ed.] (1971): *Contributions to Nonlinear Analysis.* Academic Press, New York.

Zarantonello, E. (1973): *Dense single-valuedness of monotone operators.* Israel J. Math. **15**, 158–166.

Zarantonello, E. (1981): *Conical spectral theory.* In: Strocchi, F. [ed.] (1981), pp. 37–116.

Zeidler, E. (1980): *The Ljusternik–Schirelman theory for indefinite and not necessarily odd nonlinear operators and its applications.* Nonlinear Anal. **4**, 451–489.

Zeidler, E. (1986): *Ljusternik–Schnirelman theory on general level sets.* Math. Nachr. **129**, 235–259.

Zlámal, M. (1968): *On the finite element method.* Numer. Math. **12**, 394–409.

Zurmühl, R. (1965): *Praktische Mathematik.* Springer-Verlag, Berlin (Fifth edition 1984).

List of Symbols

We use the following abbreviations:

B-space	Banach space
H-space	Hilbert space
M–S sequence	Moore–Smith sequence
F-derivative	Fréchet derivative
G-derivative	Gâteaux derivative

$A_i(10)$ means (10) in the Appendix to Part i

General Notation

$\mathscr{A} \Rightarrow \mathscr{B}$	$\mathscr{A}$ implies $\mathscr{B}$
iff	if and only if
$\mathscr{A} \Leftrightarrow \mathscr{B}$	$\mathscr{A}$ iff $\mathscr{B}$
$f(x) \stackrel{\text{def}}{=} 2x$	$f(x) = 2x$ by definition
$x \in S$	x is an element of the set S
$x \notin S$	x is not an element of S
$\{x: \ldots\}$	set of all x with the property ...
$S \subseteq T$	the set S is contained in the set T
$S \subset T$	S is properly contained in T
$S \subset\subset T$	identical to $\overline{S} \subset T$, where $\overline{S}$ denotes the closure of S
$\cap, \cup, -$	intersection, union, difference
$\emptyset$	empty set
2^S	set of all subsets of the set S, the power set of S

$X \times Y$	product set, $X \times Y = \{(x, y): x \in X, y \in Y\}$
$\mathbb{N}$	set of the natural numbers $1, 2, \ldots$
$\mathbb{R}, \mathbb{C}, \mathbb{Q}, \mathbb{Z}$	set of real, complex, rational, integer numbers
$\mathbb{K}$	$\mathbb{R}$ or $\mathbb{C}$
$\mathbb{R}_+$	set of nonnegative real numbers $\xi \geq 0$
$\mathbb{R}_>$	set of positive real numbers $\xi > 0$
$\mathbb{R}^N$	set of all real N-tuples $x = (\xi_1, \ldots, \xi_N)$
$\mathbb{R}^N_+$	set of all $x \in \mathbb{R}^N$ with $\xi_i \geq 0$ for all i
$\operatorname{Re} z$, $\operatorname{Im} z$	real part of the complex number z, imaginary part of z
$[a, b]$, $]a, b[$, $[a, b[$	closed, open, half-open real interval
I, id	identity mapping
$f: S \subseteq X \to Y$	mapping from the set S into the set Y with $S \subseteq X$
$D(f)$	domain of f, $D(f) = S$
$R(f)$	range of f, $R(f) = f(S)$
$N(f)$	null space of f, $N(f) = \{x: f(x) = 0\}$
$G(f)$	graph of f, $G(f) = \{(x, f(x)): x \in S\}$
$\operatorname{dom} f$, $\operatorname{im} f$	identical to $D(f)$, $R(f)$
$\ker f$	identical to $N(f)$
f surjective	mapping onto Y, i.e., $f(S) = Y$
f injective	one-to-one mapping
f bijective	one-to-one mapping onto Y, i.e., f is surjective and injective
$f(A)$	image of the set A, $f(A) = \{f(x): x \in A\}$
$f^{-1}(B)$	preimage of the set B, $f^{-1}(B) = \{x: f(x) \in B\}$
$f\|_A$	restriction of the map f to the set A
$f \circ g$	f applied to g, $(f \circ g)(x) = f(g(x))$
$f: S \to 2^T$	multivalued mapping, $f(x)$ is a subset of T
$R(f)$	range of the multivalued mapping f, $R(f) = \bigcup_{x \in S} f(x)$
$G(f)$	graph of the multivalued mapping f, $G(f) = \{(x, y): x \in S, y \in f(x)\}$
$\operatorname{dom} f$, $D(f)$	the effective domain of the multivalued mapping f, $\operatorname{dom} f = \{x: f(x) \neq \emptyset\}$
δ_{ij}	Kronecker symbol, $\delta_{ij} = 1$ if $i = j$ and $\delta_{ij} = 0$ if $i \neq j$
$g(x) = o(f(x))$, $x \to a$	$g(x)/f(x) \to 0$ as $x \to a$
$g(x) = O(f(x))$, $x \to a$	$\|g(x)\| \leq \text{const}\,\|f(x)\|$ for all x in a neighborhood of the point a
B^N	closed unit ball in $\mathbb{R}^N$, $B^N = \{x \in \mathbb{R}^N: \|x\| \leq 1\}$
S^N	N-dimensional unit sphere, $S^N = \{x \in \mathbb{R}^N: \|x\| = 1\}$
$\operatorname{sgn} a$	signum of the real number a

Special Notation

In persuing the following symbols, the reader should pay attention to the possible danger of confusion.
The page numbers I10 and 10 refer to page 10 of Part I and the present Part II, respectively.

		Page
X^*	dual space to X	I774
A^*	dual operator to A in a B-space, transposed matrix	I775
$A^{*'}$	adjoint operator to A in an H-space, adjoint matrix (transposed and conjugate complex)	I786, 124

Observe that if the continuous linear operator $A: X \to X$ is defined on the H-space X, then the operators $A^*: X^* \to X^*$ and $A^{*'}: X \to X$ are defined on *different* spaces.

F'	F-derivative or G-derivative of the operator F (the text always refers precisely to the momentary meaning)	I135
F_x	partial F-derivative or G-derivative of F with respect to the variable x (the text always refers precisely to the momentary meaning)	I140
$(x\|y)$	inner (scalar) product on an H-space	7
(x, y)	ordered pair, an element from the product set $X \times Y$	
$\langle x\|y\rangle$	inner (scalar) product in $\mathbb{R}^N$ or $\mathbb{C}^N$	I771
$\langle f, x\rangle$	value of the linear functional f at the point x, $f(x)$	
$x_n \to x$	convergence in norm	I769
$x_n \rightharpoonup x$	weak convergence	255
$f_n \overset{*}{\rightharpoonup} f$	weak* convergence of functionals	260
$f_n \rightrightarrows f$	uniform convergence of functions	I801
$x \mapsto f(x)$	another notation for the mapping f	
$f(\cdot)$	another notation for the mapping f	
$(x_{n'})$	subsequence of (x_n)	
$\\|x\\|$	norm of x in a B-space	I768
$\|x\|$	Euclidean norm of x, $\|x\| = \langle x\|x\rangle^{1/2}$	
$\\|x\\|_p$, $\\|f\\|_p$	p-norm of the point x in $\mathbb{R}^N$ or norm of the function f in the Lebesgue space $L_p(G)$ (the text always refers precisely to the momentary meaning)	I771, 35, 36
$dF(u; h)$	F-differential (Fréchet differential) of the mapping F at the point u in direction h	I135
$d^nF(u; h_1, \ldots, h_n)$	nth F-differential of F at the point u in the directions $h_1, \ldots, h_n$	I143
$d^nF(u; h)$	identical to $d^nF(u; h, \ldots, h)$	
$\delta F(u; h)$	first variation of the mapping F at the point u in direction h	I143, 689

$\delta^n F(u; h_1, \ldots, h_n)$	nth variation of F at the point u in the directions $h_1, \ldots, h_n$	I143
$\delta^n F(u; h)$	nth variation of F at the point u in direction h (identical to $\delta^n F(u; h, \ldots, h)$)	689
$F'(u)h$	$F'(u)$ applied to h	I135
$F''(u)hk$	identical to $(F''(u)h)k$	I136, I144
$F''(u)h^2$	identical to $F''(u)hh$	

Note the following. The nth F-derivative $F^{(n)}(u)$ exists at the point u iff the nth F-differential $d^n F(u; \ldots)$ exists at the point u. In this case, we have

$$F^{(n)}(u)h_1 \ldots h_n = d^n F(u; h_1, \ldots, h_n) = \delta^n F(u; h_1, \ldots, h_n)$$

for all $h_1, \ldots, h_n$. In particular, we get

$$F^{(n)}(u)h^n = d^n F(u; h) = \delta^n F(u; h)$$

for all h (see page 144 of Part I).

General Notation Introduced in Part I

∂S	boundary of the set S	I751
$\bar{S}$	closure of S	I751
int S	interior of S	I751
$U(x)$	neighborhood of the point x	I751
$U(x, R)$	open ball with center x and radius R, $U(x, R) = \{y: \|y - x\| < R\}$	
$\bar{U}(x, R)$	closed ball, $\bar{U}(x, R) = \{y: \|y - x\| \leq R\}$	
supp f	support of the function (distribution) f	I756, 1048
$\underline{\lim}$, $\overline{\lim}$	lower, upper limit	I761
diam S, $d(S)$	diameter of the set S	I762
dist(x, S), $d(x, S)$	distance of the point x from the set S	I762
$\text{dist}_X(x, S)$	distance of x from S in the space X	
dist(S, T), $d(S, T)$	distance between the sets S and T	I762
$S + T$	sum of the sets S and T	I764
λS	product of the set S by the number λ	I764
span S	linear hull of S	I764
co S	convex hull of S	I764
$\overline{\text{co}}\, S$	closed convex hull of S	I764
dim L	dimension of the linear subspace L	I765
codim L	codimension of the linear subspace L	I765
X/Y	factor space	I765, I770
$X \oplus Y$	direct sum or topological direct sum (the text always refers precisely to the momentary meaning)	I765, I766

$X^{\perp}$	complement of X with respect to a direct sum or orthogonal complement to X (the text always refers precisely to the momentary meaning)	I766, 65
$X \times Y$	product space	I755, I770
$\prod_\alpha X_\alpha$	general product space	I755
$X_{\mathbb{C}}$	complexification of the real B-space X	I770
$A_{\mathbb{C}}$	complexification of the linear operator A	I770
rank A	rank of the linear operator A, rank $A = \dim R(A)$	
ind A	index of the linear operator A, ind $A = \dim N(A) - \operatorname{codim} R(A)$	
ind F	index of the nonlinear Fredholm operator F	667
det A	determinant of the linear operator A	I179
$\rho(A)$	resolvent set of the linear operator A	I795
$\sigma(A)$	spectrum of the linear operator A	I795
$r(A)$	spectral radius of the linear operator A	I795

In real B-spaces, $\rho(A)$, $\sigma(A)$, and $r(A)$ always refer to the complexification $A_{\mathbb{C}}$, i.e., $\rho(A) = \rho(A_{\mathbb{C}})$, etc. I770

Function Spaces Introduced in Part I

In the following, G denotes a nonempty bounded open set in $\mathbb{R}^N$, $N \geq 1$. Let $-\infty < a < b < \infty$. Moreover, let $k = 0, 1, \ldots$ and $0 < \alpha \leq 1$.

$\partial G \in C^{k,\alpha}$	boundary property of the set G (If $k \geq 1$, then the boundary ∂G of G is smooth.)	I232
$\partial G \in C^{0,1}$	piecewise smooth boundary	I233, 18
$\partial G = \overline{\partial_1 G} \cup \overline{\partial_2 G}$	special decomposition of ∂G	522
$L(X, Y)$	space of linear continuous operators from X into Y	I135
$C(X, Y)$	space of continuous operators from X into Y	I148
$C^k(X, Y)$	space of k-times continuously F-differentiable functions from X into Y	I148
$C^0(X, Y)$	identical to $C(X, Y)$	
$H_\alpha(u)$	Hölder constant of u, i.e., $H_\alpha(u)$ is the smallest constant L with $\lvert u(x) - u(y)\rvert \leq L\lvert x - y\rvert^\alpha$ for all $x, y \in \bar{G}$.	
$\lVert u\rVert_{C[a,b]}$	identical to $\sup_{a \leq x \leq b} \lvert u(x)\rvert$	
$\lVert u\rVert_{C^k[a,b]}$	identical to $\sum_{j=0}^k \sup_{a \leq x \leq b} \lvert u^{(j)}(x)\rvert$	
$\lVert u\rVert_{C^{k,\alpha}[a,b]}$	identical to $\lVert u\rVert_{C^k[a,b]} + H_\alpha(u^{(k)})$, where the Hölder constant H_α refers to $\bar{G} = [a, b]$	

$\|u\|_{C(\bar{G})}$	identical to $\sup_{x \in \bar{G}} \|u(x)\|$
$\|u\|_{C^k(\bar{G})}$	identical to $\sum_{\|\beta\| \le k} \sup_{x \in \bar{G}} \|D^\beta u(x)\|$
$\|u\|_{C^{k,\alpha}(\bar{G})}$	identical to

$$\|u\|_{C^k(\bar{G})} + \sum_{|\beta|=k} H_\alpha(D^\beta u).$$

Observe that $\|u\|_{C(\bar{G})} = \|u\|_\infty$ for continuous functions $u: \bar{G} \to \mathbb{R}$ (see page 36).

$C[a, b]$	real B-space of all continuous functions $u: [a, b] \to \mathbb{R}$ with the norm $\|u\|_{C[a,b]}$
$C^k[a, b]$	real B-space of all k-times continuously differentiable functions $u: [a, b] \to \mathbb{R}$ with the norm $\|u\|_{C^k[a,b]}$
$C^{k,\alpha}[a, b]$	real B-space of all functions $u \in C^k[a, b]$ with $\|u\|_{C^{k,\alpha}[a,b]} < \infty$. The norm on $C^{k,\alpha}[a, b]$ is given by $\|u\|_{C^{k,\alpha}[a,b]}$.
$C(\bar{G})$	real B-space of all continuous functions $u: \bar{G} \to \mathbb{R}$ with the norm $\|u\|_{C(\bar{G})}$
$C^k(\bar{G})$	real B-space of all k-times continuously differentiable functions $u: \bar{G} \to \mathbb{R}$ with the norm $\|u\|_{C^k(\bar{G})}$. (More precisely, $C^k(\bar{G})$ consists of all continuous functions $u: \bar{G} \to \mathbb{R}$ which are k-times continuously differentiable on G and all of whose partial derivatives up to kth order can be continuously extented to the closure $\bar{G}$ of G.)
$C^{k,\alpha}(\bar{G})$	real B-space of all functions $u \in C^k(\bar{G})$ with $\|u\|_{C^{k,\alpha}(\bar{G})} < \infty$. The norm on $C^{k,\alpha}(\bar{G})$ is given by $\|u\|_{C^{k,\alpha}(\bar{G})}$.
$C^0(\bar{G})$	identical to $C(\bar{G})$
$C^{0,\alpha}(\bar{G})$	identical to $C^\alpha(\bar{G})$ for $0 < \alpha < 1$.

Observe that $C^{0,1}(\bar{G})$ denotes a space of Lipschitz continuous functions, whereas $C^1(\bar{G})$ denotes a space of continuously differentiable functions.

$C_b(\mathbb{R}^N)$	real B-space of all *bounded* continuous functions $u: \mathbb{R}^N \to \mathbb{R}$ with the norm $\|u\|_{C(\mathbb{R}^N)}$
$C_b^k(\mathbb{R}^N)$	real B-space of all k-times continuously differentiable functions $u: \mathbb{R}^N \to \mathbb{R}$ with $\|u\|_{C^k(\mathbb{R}^N)} < \infty$. The norm on $C_b^k(\mathbb{R}^N)$ is given by $\|u\|_{C^k(\mathbb{R}^N)}$.

If we speak of the B-space $C(\bar{G})$ with $G = \mathbb{R}^N$, then $C(\bar{G})$ is understood to be $C_b(\mathbb{R}^N)$.

X^n	product space of X, i.e., X^n consists of all $u = (u_1, \ldots, u_n)$, where $u_i \in X$ for all i	

If X is a B-space over $\mathbb{K} = \mathbb{R}, \mathbb{C}$, then X^n is also a B-space over $\mathbb{K}$ with the norm

$$\|u\| = \sum_{i=1}^{n} \|u_i\|.$$

This way, we obtain the real B-spaces $C(\bar{G})^n$, $L_p(G)^n$, $W_p^m(G)^n$, etc., and the spaces $C^\infty(\bar{G})^n$, $C_0^\infty(G)^n$, etc.

Function Spaces Introduced in the Present Part II

Observe that the notations $C_0^\infty(G)$, $C^\infty(\bar{G})$, $L_p(G)$, $W_p^m(G)$, and $\mathring{W}_p^m(G)$ are of fundamental importance for the present Part II.

$C^\infty(G)$	space of infinitely continuously differentiable functions $u: G \to \mathbb{R}$	17
$C^\infty(\bar{G})$	$u \in C^\infty(\bar{G})$ iff $u \in C^k(\bar{G})$ for all $k = 1, 2, \ldots$	
$C_0^\infty(G)$	space of all functions $u \in C^\infty(G)$ with compact support in G	18
$\|u\|_p$	identical to $\left(\int_G \lvert u\rvert^p\,dx\right)^{1/p}$, where $1 \le p < \infty$	
$\|u\|_\infty$	identical to $\operatorname{ess\,sup}_{x \in G} \lvert u(x)\rvert$	36
$\|u\|_{p,\partial G}$	identical to $(\int_{\partial G} \lvert u\rvert^p\,dx)^{1/p}$, where $1 \le p < \infty$	
$\|u\|_{\infty,\partial G}$	identical to $\operatorname{ess\,sup}_{x \in \partial G} \lvert u(x)\rvert$	
$\|u\|_{m,p}$	identical to $\left(\sum_{\lvert\alpha\rvert \le m} \int_G \lvert D^\alpha u\rvert^p\,dx\right)^{1/p}$; norm on the Sobolev space $W_p^m(G)$, where $1 \le p < \infty$ and $m = 0, 1, 2, \ldots$	
$\|u\|_{m,p,0}$	identical to $\left(\sum_{\lvert\alpha\rvert = m} \int_G \lvert D^\alpha u\rvert^p\,dx\right)^{1/p}$, where $1 \le p < \infty$ and $m = 0, 1, \ldots$	
$\|u\|_{m,\infty}$	identical to $\sum_{\lvert\alpha\rvert \le m} \|D^\alpha u\|_\infty$; norm on the Sobolev space $W_\infty^m(G)$, where $m = 0, 1, \ldots$	
$\|u\|_{m,\infty,0}$	identical to $\sum_{\lvert\alpha\rvert = m} \|D^\alpha u\|_\infty$	
$(u \mid v)_2$	identical to $\int_G \bar{u}v\,dx$;	

	scalar product on $L_2(G)$ and $L_2^{\mathbb{C}}(G)$. (Note that $\bar{u} = u$ on $L_2(G)$, where the bar denotes the conjugate complex number.)	
$(u\|v)_{m,2}$	identical to $$\sum_{\|\alpha\| \leq m} \int_G D^\alpha \bar{u} D^\alpha v \, dx;$$ scalar product on $W_2^m(G)$ and $W_2^m(G)_{\mathbb{C}}$. (Note that $\bar{u} = u$ on $W_2^m(G)$.)	

Observe that $\|\cdot\|_{m,p}$ and $\|\cdot\|_{m,p,0}$ are equivalent norms on the Sobolev space $\mathring{W}_p^m(G)$, where $1 \leq p < \infty$ and $m = 0, 1, 2, \ldots$

$L_p(G)$	*Lebesgue space* of all measurable functions $u: G \to \mathbb{R}$ with $$\|u\|_p < \infty,$$ where $1 \leq p \leq \infty$	35, 36
$L_p(\partial G)$	Lebesgue space of all measurable functions $u: \partial G \to \mathbb{R}$ with $\|u\|_{p,\partial G} < \infty$, where $1 \leq p \leq \infty$. (Note that the measurability of u refers to the surface measure on ∂G.)	
$L_p^{\mathbb{K}}(G)$	Lebesgue space of all measurable functions $u: G \to \mathbb{K}$ with $\|u\|_p < \infty$, where $\mathbb{K} = \mathbb{R}$, $\mathbb{C}$ and $1 \leq p \leq \infty$. (Note that $L_p^{\mathbb{R}}(G) = L_p(G)$.)	
$W_p^m(G)$	*Sobolev space* of functions $u: G \to \mathbb{R}$, where $m = 0, 1, \ldots$ and $1 \leq p \leq \infty$	236, 237
$\mathring{W}_p^m(G)$	closure of $C_0^\infty(G)$ in $W_p^m(G)$	237
$\mathscr{W}_p^m(G)$	closure of $C^\infty(\bar{G})$ in $W_p^m(G)$	242
$W_p^m(\partial G)$	Sobolev space of functions $u: \partial G \to \mathbb{R}$ for *real* $m \geq 0$	1031
$W_p^{1/2}(\partial_j G)$	special Sobolev space for functions $u: \partial G \to \mathbb{R}$	530
$W_p^m(G)_{\mathbb{K}}$	Sobolev space of functions $u: G \to \mathbb{K}$ for *real* m, where $\mathbb{K} = \mathbb{R}, \mathbb{C}$	1061
$\mathring{W}_p^m(G)_{\mathbb{K}}$	closure of $C_0^\infty(G)_{\mathbb{K}}$ in $W_p^m(G)_{\mathbb{K}}$ for $m \geq 0$	

Observe that $W_p^m(G) = W_p^m(G)_{\mathbb{R}}$ and that

$$W_q^{-m}(G) = \mathring{W}_p^m(G)^* \tag{E}$$

for all real $m \geq 0$, $1 < p < \infty$, and $p^{-1} + q^{-1} = 1$. Here, the asterisk denotes the dual space. Equation (E) remains true if we replace $W_q^{-m}(G)$ and $\mathring{W}_p^m(G)$ with the corresponding complex Sobolev spaces $W_q^{-m}(G)_{\mathbb{C}}$ and $\mathring{W}_p^m(G)_{\mathbb{C}}$, respectively.

H^m	identical to $W_2^m(\mathbb{R}^N)$
$H_{\mathbb{C}}^m$	identical to $W_2^m(\mathbb{R}^N)_{\mathbb{C}}$.

Observe that

$$H^m \subseteq H^k \quad \text{and} \quad H_{\mathbb{C}}^m \subseteq H_{\mathbb{C}}^k \quad \text{for all real} \quad m \geq k.$$

In particular, we obtain that

$$\ldots H^2 \subseteq H^1 \subseteq H^0 \subseteq H^{-1} \subseteq H^{-2} \ldots,$$

where $H^0 = L_2(\mathbb{R}^N)$.

$W_p^0(G)$	identical to $L_p(G)$ for $1 \le p \le \infty$	
$W_p^0(G)_{\mathbb{C}}$	identical to $L_p^{\mathbb{C}}(G)$ for $1 \le p \le \infty$	
$L_{p,\text{loc}}(G)$	space of all functions $u: G \to \mathbb{R}$ with $u \in L_p(G)$ for all compact subsets C of G	
"$V \subseteq H \subseteq V^*$"	evolution triple	416
$L_p(0, T; X)$	Lebesgue space of functions $u:]0, T[\to X$	417
$W_p^1(0, T; V, H)$	special Sobolev space of functions $u:]0, T[\to V$ with respect to the evolution triple "$V \subseteq H \subseteq V^*$"	422
$C_w([0, \infty[, H)$	space of weakly continuous functions $u: [0, \infty[\to H$	784
$B_{p,q}^j$	Besov space	1106
$F_{p,q}^j$	Triebel–Lizorkin space	1106
$L_{p,\text{loc}}(G, Y)$	space of all functions $u: G \to Y$ with $\int_C \|u(x)\|_Y^p\,dx < \infty$ for all compact subsets C of G	1047
$L_p(M \to Y, \mu)$	Lebesgue space of functions $u: M \to Y$ with respect to the measure μ	1071
$BV(G)$	space of all functions $u: G \to \mathbb{R}$ of bounded variation	1114
$\mathscr{G}_h, \mathscr{G}_{h,m}, \partial\mathscr{G}_h$	special sets of grid points	986
$\mathscr{L}_p(\mathscr{G}_h)$	*discrete* Lebesgue space corresponding to $L_p(G)$	987
$\mathring{\mathscr{W}}_p^m(\mathscr{G}_h)$	*discrete* Sobolev space corresponding to $\mathring{W}_p^m(G)$	987
$\lvert u\rvert_{m,p}, \lvert u\rvert_{p,m,0}$	*discrete* Sobolev norm corresponding to $\|u\|_{m,p}$, $\|u\|_{m,p,0}$, respectively	987
$\mathscr{D}(G, \mathbb{K})$	space of infinitely continuously differentiable functions $u: G \to \mathbb{K}$ with compact support in G	1046
$\mathscr{D}(G)$	identical to $C_0^\infty(G)$	18, 1045
$\mathscr{D}'(G, Y)$	space of distributions with values in the B-space Y	1047
$\mathscr{D}'(G)$	identical to $\mathscr{D}'(G, \mathbb{R})$	
$\mathscr{S}(\mathbb{R}^N)$	space of infinitely continuously differentiable functions which go rapidly to zero as $\lvert x\rvert \to \infty$	1058
$\mathscr{S}'(\mathbb{R}^N)$	space of tempered distributions	1060
$H^m(A)$	m-dimensional Hausdorff measure of the set A	1111
$\mathscr{H}^m(A)$	m-dimensional normalized Hausdorff measure of the set A	1112

Notation Introduced in the Present Part II

$X = X^*$	identification of the *real* H-space X with the dual space X^*	254
$\bar{X}^*$	antidual space to the *complex* H-space X	67
$D_i = \partial/\xi_i$	partial derivative with respect to ξ_i (in the classical or generalized sense)	61, 232

$\partial G\colon F_0 \cong F_1 \pmod{y}$	homotopy mod y (see page 569 of Part I)	
$\partial G\colon f_0 \cong f_1$	homotopy (see page 527 of Part I)	
J	duality map	67, 860
$U * V$	convolution of distributions	1051
$U \otimes V$	tensor product of distributions	1051
δ_x	Dirac's delta distribution at the point x	1047
δ	identical to δ_x for $x = 0$	
$u_n \xrightarrow{+} u$	weak$^+$ convergence	784
$u_n \xrightarrow{d} u$	discrete convergence	982
$u_n^* \xrightarrow{d^*} u^*$	discrete* convergence	982
$S \simeq T$	the set S is almost equal to the set T, i.e., $\operatorname{int} S = \operatorname{int} T$ and $\bar{S} = \bar{T}$	906
$S'_{\varphi,c}(x)$	viscosity derivative	950
$[X, Y]_{\theta,p}$	interpolation space (K-method)	1102
$[X, Y]_\theta$	interpolation space (complex method)	1108

List of Theorems

Mathematics is endangered by a loss of unity and interaction.
Richard Courant (1888–1972)

List of the Most Important Definitions

(Cf. also the List of the Most Important Definitions on page 862 of Part I.)

List of Schematic Overviews

List of Important Principles

Monotonicity

Partial Differential Equations

Approximation of Solutions

Consider the operator equation

$$Au = b, \qquad u \in X. \tag{E}$$

If the approximation method is *consistent* and *stable*, then the following three conditions are mutually equivalent:

(i) Equation (E) has a solution.
(ii) Equation (E) is *uniquely approximation-solvable*.
(iii) The operator A is A-proper 966, 983

Index

The reader should also consult the detailed Contents of this volume and the index material (List of Theorems, etc.). Moreover, the reader may also consult the Index of Part I. If several page numbers belong to the same catch word, then the primary reference is *italized*. The page numbers I10 and 10 refer to page 10 of Part I and the present Part II, respectively.

Important properties of operators and sets are summarized under the catch word "*operator*" and "*set*", respectively. See also the catch words "*functional*" and "*multi-valued mapping*".

CPSIA information can be obtained at www.ICGtesting.com
Printed in the USA
LVOW100423220512

282728LV00006B/36/A